4th International Conference on Civil and Environmental Engineering

IOP Conference Series: Earth and Environmental Science
Volume 1117

Kuta, Bali, Indonesia
3 – 5 August 2022

ISBN: 978-1-7138-6524-7
ISSN: 1755-1307

Printed from e-media with permission by:

Curran Associates, Inc.
57 Morehouse Lane
Red Hook, NY 12571

Some format issues inherent in the e-media version may also appear in this print version.

Institute of Physics
Dirac House, Temple Back
Bristol BS1 6BE UK

Phone: 44 1 17 929 7481
Fax: 44 1 17 920 0979

techtracking@iop.org

Additional copies of this publication are available from:

Curran Associates, Inc.
57 Morehouse Lane
Red Hook, NY 12571 USA
Phone: 845-758-0400
Fax: 845-758-2633
Email: curran@proceedings.com
Web: www.proceedings.com

TABLE OF CONTENTS

Preface by the Chairman of Organizing Committee of 4TH ICCEE 2022

The organizing committee would like to extend a warm welcome to everyone who is attending and contributing to the "4th International Conferences on Civil and Environmental Engineering (ICCEE) 2022". The theme of the conference is Toward Smart and Green Infrastructure for Achievement of Sustainability and Development Goals. The conference, held in Bali, Indonesia, aims to promote scientific and technological advancement, knowledge exchange, and cooperation among engineers and researchers in infrastructure development, civil and environmental studies, and other engineering-related fields.

ICCEE 2022 is the fourth international conference held by the Departments of Civil Engineering and Environmental Engineering, following the successful 2010 conference attended by 260 participants, 284 participants in 2015, and 292 participants in 2019, all of whom were local and foreign participants. ICCEE 2022 is a continuation of the first, second, and third conferences, and it will provide opportunities for researchers to present and publish their research, as well as serve as a venue for strengthening national, regional, and international research collaboration networks for the Departments of Civil Engineering and Environmental Engineering. We will have one honorable keynote speaker and four invited speakers present during the day of plenary sessions. The parallel sessions will take place over the course of a single day, and 136 papers will be presented.

We would like to thank the scientific committees, reviewers, all participants, keynote and invited speakers, presenters, and moderators for their contributions to this conference. Our thanks go to our sponsors: Centre of Technology Hasanuddin University, PT. Bosowa Beton Indonesia, PT. Amythas, PT. Adhi Karya Persero, PT. WIKA Beton, PT. Bumi Karsa, PT. Nindya Karya, PT Yodya Karya, and our co-organizers: Kyushu University, Universitas Udayana Bali, Politeknik Negeri Bali, Politeknik Transportasi Darat Bali, and Universitas Pendidikan Nasional.

We sincerely hope that the 4th ICCEE 2022, themed "Toward Smart and Green Infrastructure for Achievement of Sustainability and Development Goals" will provide us with an invaluable opportunity to share knowledge and experiences.

Dr. Eng. Muh. Akbar Caronge, S.T., M.Eng.
Chairman of Organizing Committee

Committee

Advisory Committee

Rector of Universitas Hasanuddin

<u>Advisory Committee (AC)</u>
Prof. Dr. Ir. Jamaluddin Jompa, M.Sc (Universitas Hasanuddin, Indonesia)
Prof. Dr. Ir. Muhammad Saleh Pallu, M.Eng. (Universitas Hasanuddin, Indonesia)
Prof. Dr. Eng. Muhammad Isran Ramli, ST., MT (Universitas Hasanuddin, Indonesia)
Prof. Dr. Ir. Muhammad Arsyad Thaha, MT. (Universitas Hasanuddin, Indonesia)
Prof. Dr.-Ing. Ir. Herman Parung, M.Eng. (Universitas Hasanuddin, Indonesia)
Prof. Sakti Adji Adisasmitha, M.Si., M.Eng.Sc, Ph.D (Universitas Hasanuddin, Indonesia)
Prof. Muhammad Zaly Zhah (Universiti Teknologi Malaysia, Malaysia)
Prof. Yusak Octavius Susilo D.Eng., MT., ST (University of Natural Resources and Life Sciences Vienna, Austria)
Assoc. Prof. Dr. Eng. Yusihiro Arai (Tokyo Metropolitan University, Japan)
Dr. Dimas B. E. Dharmowijoyo (Universiti Teknologi PETRONAS)
Prof. Danang Parikesit (University Network on Indonesia Infrastructure Development)
Prof. Leksmono Suryo Putranto (Intern-University on Transportation Study Forum)
Prof. Dr. Eng. Rudy Djamaluddin, ST., M.Eng. (Universitas Hasanuddin, Indonesia)
Dr. Efendi Prih Raharjo, ST., S.SiT., MT (Politeknik Transportasi Darat, Indonesia)

<u>Organizing Committee (OC)</u>

Chairman	:	Dr. Eng. Muhammad Akbar Caronge, ST., M.Eng.
Secretary	:	Zarah Arwieny Hanami, ST., M.T.
Treasurer	:	Hajriyanti Yatmar, ST., M.Eng
Members	:	Dr. Eng. Fakhruddin, ST., M.Eng
		Dr. A.Ilda Dwi Puspita, ST., MT
		Muhammad Akbar Walenna, ST., M.Sc., Ph.D
		M. Farid Maricar, B.Eng., M.Eng
		Rasdiana Zakaria, ST., MT.
		Annisa Dwi Damayanti ST. MT
		Hardianti Alimuddin ST., M.Eng
		Nur An-nisa Putry Mangarengi, ST., M.Sc.
		Nurul Masyiah Rani ST., M.Eng
		Nurjannah Oktorina Abdullah, ST., MT.
		Dr. Miswar Tumpu ST., MT

International Scientific Committee (ISC)

Chairman:

Prof. Dr. Eng. Wihardi Tjaronge, ST., M.Eng. (Universitas Hasanuddin, Indonesia)
Vice Chairman: Dr. Eng. Muralia Hustim, ST., MT (Universitas Hasanuddin, Indonesia)
Secretary: Dr. M. Asad Abdurrahman, ST., M.Eng. (Universitas Hasanuddin, Indonesia)
Vice Secretary: Dr. Eng. Bambang Bakri, ST., MT (Universitas Hasanuddin, Indonesia)

Members:

Prof. Dr. Ir. Sumarni, Hamid Aly, MT (Universitas Hasanuddin, Indonesia)
Prof. Dr. Ir. Sarwoko Mangkoedihardjo, MScES (Institut Teknologi Sepuluh November)
Prof. Dr. Ir Lilik Sudiajeng, M.Erg. (Politeknik Negeri Bali, Indonesia)
Prof. Ir. I Nyoman Arya Thanaya, ME., Ph.D (Universitas Udayana, Indonesia)
Professor Dr. Mohd Suffian Yusoff (Universitas Sains Malaysia)
Dimas, PhD. (Universiti of Teknologi Petronas, Malaysia)
Dr. Muhammad Zahly Shah (Universiti of Teknologi Malaysia, Malaysia)
Dr. Yasuhiro Arai (Tokyo Metropolitan University, Japan)
Dr. Yasutaka Sagawa (Kyushu University, Japan)
Dr. Eng. Nurazuwa Md. Noor (Universiti Tun Hussein Onn, Malaysia)
Rahmita Sari Rafdinal, ST., M.Eng (PS. Mitsubishi Area Indonesia)
Dr. Ryohei Yanagida (Tokyo Institute of Technology, Japan)
I Gusti Ngurah Putu Dharmayasa, ST., MT., Ph.D (Universitas Pendidikan Nasional, Indonesia)
Dr. Ir. Putu Hermawati, MT (Politeknik Negeri Bali, Indonesia)
Dr. I Ketut Sutapa, S.ST., MT (Politeknik Negeri Bali, Indonesia)
Gede Pringgana, ST., MT., Ph.D (Universitas Udayana, Indonesia)
Dr. Eng Januarti Jaya Eka Putri, ST., MT (Institut Teknologi Sepuluh November, Indonesia)
Angga Fajar Setiawan, ST., M.Eng., Ph.D (Universitas Gadjah Mada, Indonesia)
Dr.Eng. Heriansyah Putra, ST., MT (Institut Pertanian Bogor, Indonesia)
Dr.Eng. Pinta Astuti, ST., MT (Universitas Muhammadiyah Yogyakarta, Indonesia)
Pertiwi Andarani, ST., M.Eng. M.Sc., Ph.D (Universitas Diponegoro)
Dr. Febriane Paulina Makalew, ST, MUDD (Politeknik Negeri Manado, Indonesia)
Ir. Agus Putu Abiyasa, B.Eng, Ph.D, IPM (Universitas Pendidikan Nasional, Indonesia)
Ir. Eddy Setiadi Soedjono, PhD (Institut Teknologi Sepuluh November, Indonesia)
Dr. Eng. Eka Juliafad, ST, M.Eng (Universitas Negeri Padang)
Dr. Pornpen Limpaninlachat (Mahidol University, Thailand)
Inamullah Inam, M. Eng (Laghman University, Afghanistan)
Kazuyo Hirose (Japan Space System)
Tomomi Takeda (Japan Space System)
Christian Gerald Daniel, S.T., M.Sc (Universitas Pelita Harapan, Indonesia)
Dr. Eng. Dahlia Patah, S.T., M.Eng (Universitas Sulawesi Barat)
Dr. Amry Dasar, M.Eng (Universitas Sulawesi Barat)
Dr. Ir. James Thoengsal, M.T., IPM (Sulawesi University of Technology)
Rafid Mahful, S.T, M.Eng, IAP (Universitas Terbuka, Indonesia)
Dr. Richard Frans, ST., MT (Universitas Atmajaya Makassar, Indonesia)
Suharman Hamzah, Ph.D. (Universitas Hasanuddin, Indonesia)

Dr. Eng. Ir. Rita Tahir Lopa, MT. (Universitas Hasanuddin, Indonesia)
Dr. Eng. Tri Harianto, ST., MT. (Universitas Hasanuddin, Indonesia)
Dr. Eng. Rita Irmawaty, ST., MT. (Universitas Hasanuddin, Indonesia)
Dr. Eng Ardy Arsyad, M.Eng.Sc (Universitas Hasanuddin, Indonesia)
Dr. Eng. Achmad Yasir Baeda, ST., MT. (Universitas Hasanuddin, Indonesia)
Dr. Hasdinar Umar, ST., MT (Universitas Hasanuddin, Indonesia)
Dr. A. Ildha Dwi Puspita, ST. MT (Universitas Hasanuddin, Indonesia)
Dr.Eng. Asiyanthi Tabran Lando, ST., MT (Universitas Hasanuddin, Indonesia)

Peer Review Statement

All papers published in this volume have been reviewed through processes administered by the Editors. Reviews were conducted by expert referees to the professional and scientific standards expected of a proceedings journal published by IOP Publishing Publishing.

- **Type of peer review:** Single Anonymous
- **Conference submission management system:** Morressier
- **Number of submissions received:** 168
- **Number of submissions sent for review:** 153
- **Number of submissions accepted:** 126
- **Acceptance Rate (Submissions Accepted / Submissions Received × 100):** 75
- **Average number of reviews per paper:** 2
- **Total number of reviewers involved:** 39
- **Contact person for queries:**
  Name: M Asad Abdurrahman
  Email: asad@unhas.ac.id
  Affiliation: Civil Engineering Department, Universitas Hasanuddin

IOP Conf. Series: Earth and Environmental Science 1117 (2022) 012001 doi:10.1088/1755-1315/1117/1/012001

Setting time of coconut fiber ash-based geopolymer binder

I N S Widnyana[1], I M A K Salain[2], I N Sutarja[2] and I B R Widiarsa[2]

[1]Doctoral Program of Engineering Sciences, Universitas Udayana, Denpasar, Indonesia
[2]Department of Civil Engineering, Universitas Udayana, Badung, Indonesia

Email: gussuta@gmail.com

Abstract: This watchfulness goal is to decide the beginning and ending binding of geopolymer binders using coconut fiber ash (CFA). Experiments were carried out by comparing the effects of ratio precursors and base activators on geopolymer binders. The precursor material is derived from CFA, which consists of 8.24% silicon dioxide, 2.3% diphosphate pentaoxide, 70.6% potassium oxide, 2.25% iron dioxide, and 14.1% chlorine. Using two types of alkaline activators, namely NaOH (SH) with a molar concentration of 14M and Na_2SiO_3 (SS), which is used as a solution in the manufacturing process. A comparison of the precursor and alkaline activator mixtures with three variations, namely 70%:30%, 75%:25%, and 80%:20%, with SS/SH 1:1, 1.5, and 2:1 ratio by weight. The specimens are cast in a vicat ring with the top and bottom diameters ∅60mm and ∅70mm, as well as a height of 40mm. The setting time test standard used is SNI 2049:2015. The results showed that the beginning and ending binding times of the geopolymer binder got shorter as the P/A and SS/SH got higher, with times 5%–20% decreased on average.

1. Introduction

The increasing demand for Portland cement has increased its use as a concrete binder. During production, enormous energy is required due to the heating of the material to form ordinary cement, until it reaches a temperature of 1400°C–1500°C [1]–[3]. The main ingredients of cement binders that are processed by heating include clay, limestone, iron ore, and silica sand. Meanwhile, high heat is also applied to limestone with fossil fuels, resulting in the release of carbon dioxide (CO_2) gas, which is a major producer of greenhouse gas formation [4]–[6]. It was discovered that 1 ton of cement clinker during the process is approximately 1 ton of CO_2 detached to space due to calcination of limestone and emissions from fossil fuels. A previous study stated that some cement-producing countries in the world can produce around 4.1 billion tons per year, potentially producing the same number of billions of tons of CO_2 in the atmosphere [1]. This indicates that CO_2 emissions are proportional to Portland cement production. Therefore, efforts must be made to resolve this issue by decreasing the use of common cement in the concrete industry.

In recent decades, the exertion of regular cement (Portland cement) has begun to be minimized to reach agreements with construction services in the manufacture of geopolymer binders. The word "geopolymer" was first researched and coined by J. Davidovits in 1978. In his research, he found that the presence of alkaline activator polymerization bonds with RHA (rice husk ash) and FA (fly ash) produces a binder that can be common as a surrogate for ordinary adhesive. They are also materials affluent in Si (silica) and Al (alumina) that occur naturally or as a by-product of industrial waste. When these materials are mixed with alkaline activators at certain concentrations, they form polymerizing bonds in the binder manufacturing process.

Inorganic alumina-silica polymeric substances are synthesized from materials with high amounts of silica and alumina derived from nature or by-products of industrial waste. According to previous studies, they are often referred to as mineral and glass polymers because of their mechanical properties [7]. Although the chemical content of the geopolymer material is almost similar to that of zeolite, the difference lies in the microstructure of the reaction [8],[9]. In the process of synthesis, silica and alumina form chemical bonds with a structure identical to that of normal stone. Chemical reactions also occur between alkali activators and silica and alumina-containing minerals during bond polymerization to create three-dimensional polymer chain bonds and stable Si-O-Al-O structural bonds [10]. acceleration of chemical reactions in geopolymers using alkaline activators, which include KOH, Na_2SiO_3, and NaOH [11]. The alkaline activator commonly used for optimal results is a mixture of Na_2SiO_3 and NaOH at a molar concentration of 8M to 14M with a ratio of 0.4 to 2.5 [5],[12].

The basic materials that are derived from industrial waste contain several silica and alumina elements, which contribute to the manufacturing of geopolymer binders. However, binder-forming material differs in origin, which will affect the final resistance [13]. Several factors that influence macro and micro properties in the manufacture of geopolymer binders are raw materials, particle size, activator solution, calcination temperature, and processing techniques [15],[16]. The molar concentration of NaOH and the ratio of the activator modulus in the appropriate composition also affect the product characteristics [5],[20]–[25].

In addition to the mechanical properties of the geopolymer binder, physical tests are also required which are prerequisites for beginning and ending bonding. The previous study [22] stated that the binding time on a mixture of 40% BFS (blast furnace slag) and 60% FA (fly ash) had beginning and ending times of 5–58 minutes and 65–105 minutes, respectively. This is because the variation of precursors to activators greatly affects the binding time of the geopolymer binder. Furthermore, SS is an activator that can shorten the beginning and last setting times, while the addition of water to the mixture reduces the geopolymer binder's binding time. The use of materials containing Si/Al elements can shorten the binding time [23]. In this study, the application of CFA precursors and 2 types of alkaline activators, namely SS and SH, at a molar concentration of 14M was tested against the beginning and ending setting times.

2. Material and Method

2.1. Preparation of materials

Using coconut fiber ash (CFA) precursor, which passed filter no. 180 (80 m), as shown in Figure 1, and the X-RF test results are shown in Table 1. Two types of alkaline activators used were SS and SH at a molar concentration of 14M. The manufacture of geopolymer binders was carried out in 3 groups of mixed variants of precursor and alkaline activator, namely 70%:30, 75%:25%, and 80%:20%. Furthermore, the ratio of SS to SH solution was 1:1, 1.5:1, and 2:1 as shown in Table 2. The samples met the reference standard for measuring the time it takes to set [24].

Table 1. XRF test result from coconut fiber ash

Compound	Percentage (%)
SiO_2	8.24
P_2O_5	2.3
SO_3	1.1
Cl	1.1
K_2O	70.6
TiO_2	0.19
MnO	0.05
Fe_2O_3	2.25
CuO	0.071
ZnO	0.04
Br	0.042
Rb_2O	0.25
SrO	0.20
BaO	0.08
Eu_2O_3	0.45
Re_2O_7	0.09

Figure 1. Coconut fiber ash (CFA)

2.2. Testing and mixing

The process of making geopolymer binders is carried out by mixing SS and SH activators with a concentration of 14M in a predetermined ratio, as shown in Table 2. The CFA precursor is added to a 10-liter mixing bowl, the machine is turned on for 15 seconds at slow speed, and these 2 types of alkali are added slowly. The engine is started again for 30 seconds at a slightly faster speed and then turned off until mixing is complete. Next, after the precursor and alkali activator were thoroughly mixed, the binder was prepared and collected by hand to form balls, which were tossed six times at a distance of 15 cm from one hand to the other. The ball is placed in a Vicat mold, pressed, and leveled with a leveling knife. The dimensions of the vivat ring diameter are 60 mm above, 70 mm below, and 40 mm high. The binder in the Vicat molded ring is placed on a glass plate with a diameter of 70 mm, and excess binding from the small diameter hole is removed with only one end of the leveling knife. The mold, which has been filled with binder, is immediately put into the oven at 80°C as shown in Figure 2. The setting times test was carried out every 15 min till the beginning times were met, with a decrease in the Vicat needle to a profound of 25 mm and 0 mm for ending binding times as shown in Figure 3. For each group of variants, 3 samples of test objects were taken.

Table 2. A geopolymer binder mixture ratio

Group	Code	P/A	SS/SH
I	Y701	70%:30%	1:1
	Y702	70%:30%	1.5:1
	Y703	70%:30%	2:1
II	Y751	75%:25%	1:1
	Y752	75%:25%	1.5:1
	Y753	75%:25%	2:1
III	Y801	80%:20%	1:1
	Y802	80%:20%	1.5:1
	Y803	80%:20%	2:1

Figure 2. The specimen to be tested is placed in an oven at 80°C.

Figure 3. Geopolymer binder setting time test

3. Results and Discussion

Geopolymer binders were prepared using CFA precursors with different ratios of P/A and SS/SH to test the beginning and ending setting times.

3.1. Ratio of Precursor to Activator

Results of the beginning and ending time test for each group of P/A and SS/SH variations are shown in Figures 4, 5, and 6 and Table 3. The P/A composition (80%:20%) resulted in faster geopolymer bonding times. However, the beginning and ending binding times for the Y701 model (P/A = 70%:30%) was 210 minutes and 450 minutes, respectively. By increasing the amount of precursor in the specimen, it is possible to accelerate the appearance of macromolecular bonds, which causes the sample to harden more rapidly. The binding time will also be shorter as the P/A ratio continues to increase and the SS/SH ratio remains constant. Based on the analysis, the shortest binding time occurred in specimen Y803, which indicates that the geopolymer binder performed better at the higher amount of precursor. During production, it was found that the CFA precursors contained sufficient silica and alumina, which could accelerate the polymerization [25].

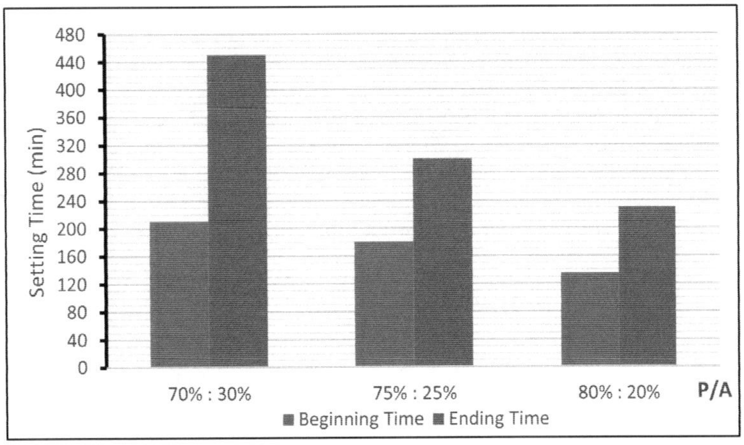

Figure 4. Geopolymer binder with a ratio of SS/SH (1:1)

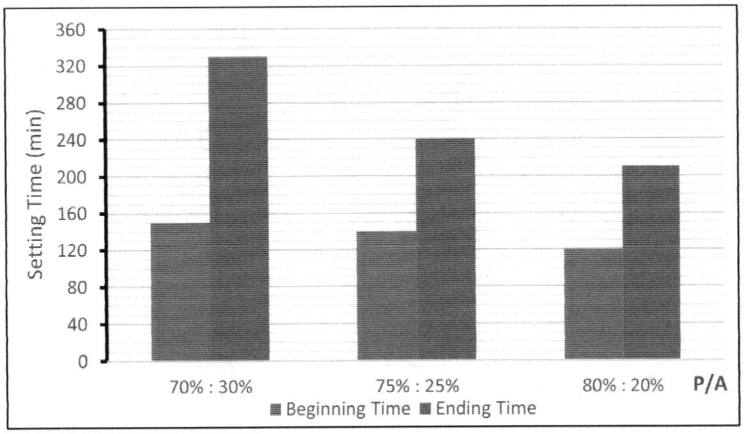

Figure 5. Geopolymer binder with a ratio of SS/SH (1.5:1)

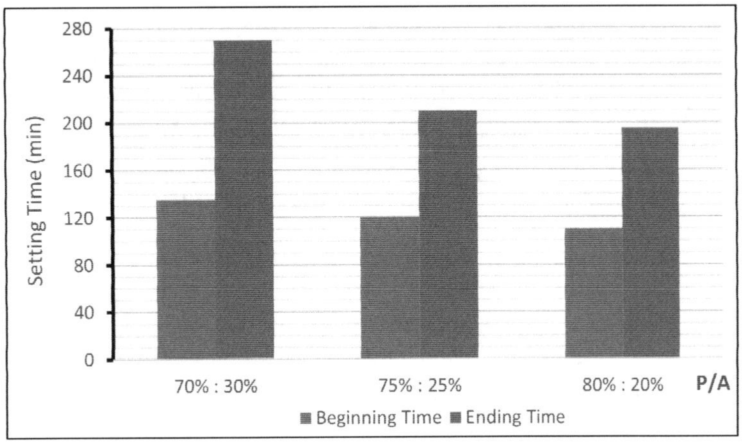

Figure 6 Geopolymer binder with a ratio of SS/SH (2:1)

Table 3. Result setting time in SS/SH variation group.

Group	Code	P/A	SS/SH	Beginning Time (min)	Ending Time (min)
I	Y701	70%:30%	1:1	210	450
	Y751	75%:25%	1:1	180	300
	Y801	80%:20%	1:1	135	230
II	Y702	70%:30%	1,5:1	150	330
	Y752	75%:25%	1,5:1	140	240
	Y802	80%:20%	1,5:1	120	210
III	Y703	70%:30%	2:1	135	270
	Y753	75%:25%	2:1	120	210
	Y803	80%:20%	2:1	110	195

3.2. SS/SH Ratio

At figures 7, 8, and 9 and Table 4 illustrate the impact of the SS/SH ratio on geopolymer binder production. The results show that beginning and ending times are faster as the SS/SH ratio increases. The mean for reductions in groups 1, 2, and 3 was 17% (sample Y701-Y703), 13% (sample Y751-Y753), and 7% (sample Y801-Y803), respectively. Ending binding time was also reduced in groups 1, 2, and 3 by 15%, 12%, and 4%, respectively. This shows that the highest SS and SH content will form the fastest polymerization bond. However, fewer reports of SS/SH will lead to longer binding times. The increase in SS can also contribute to the formation of geopolymer chain monomers, namely Si-O monomer bonds [26]. A steady rise in SS speeds up the formation of polymerization bonds [2], while a steady rise in SS/SH decreases the time it takes for the geopolymer binder to stick together [27].

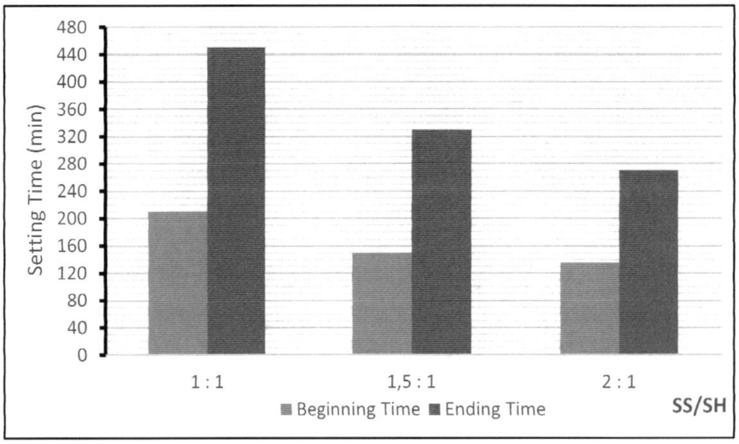

Figure 7. Geopolymer binder with a ratio of P/A (70%:30%)

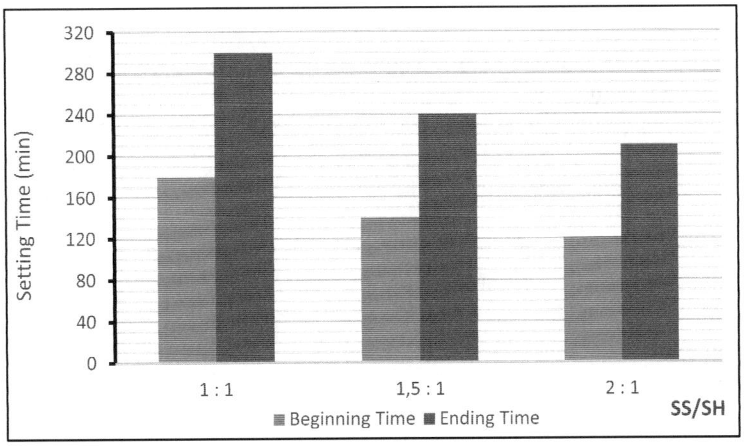

Figure 8. Geopolymer binder with a ratio of P/A (75%:25%)

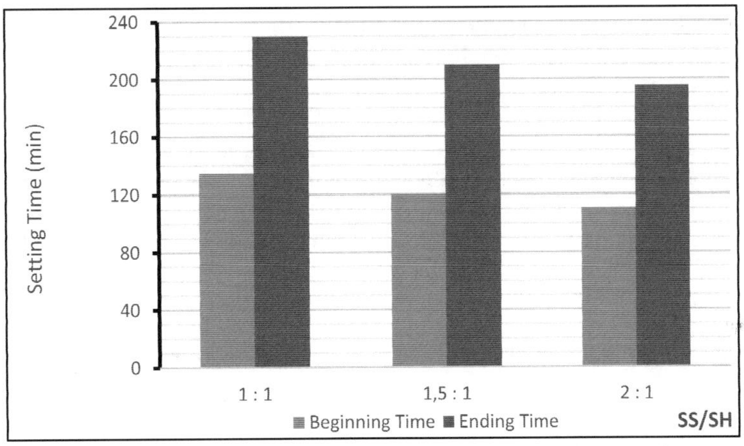

Figure 9. Geopolymer binder with a ratio of P/A (80%:20%)

Table 4. Results of the setting time test at the group ratio of P/A

Group	Code	P/A	SS/SH	Beginning Time (min)	Ending Time (min)
I	Y701	70%:30%	1:1	210	450
	Y702	70%:30%	1.5:1	150	330
	Y703	70%:30%	2:1	135	270
II	Y751	75%:25%	1:1	180	300
	Y752	75%:25%	1.5:1	140	240
	Y753	75%:25%	2:1	120	210
III	Y801	80%:20%	1:1	135	230
	Y802	80%:20%	1.5:1	120	210
	Y803	80%:20%	2:1	110	195

4. Conclusion

The amount of silica and alumina in coconut fiber ash (CFA) is sufficient to cause rapid hardening. Beginning and ending bonding times showed that CFA-based binders with a high proportion of mixed P/A and SS/SH resulted in faster bonding. At ratios of SS/SH (2:1) and P/A (80%:20%), the beginning and ending binding times were 110 min and 195 min, respectively.

Acknowledgments

Authors are grateful to the Dean of the Faculty of Engineering and the coordinator of the Engineering Doctoral Program at Udayana University for supporting this study. The authors are so grateful to Prof. Dr. Ir. I Made Alit Karyawan Salain, DEA, as the promoter and Dr. Ir. I Nyoman Sutarja, MS, and Ida Bagus Rai Widiarsa, ST., MA, Sc., Ph.D, as co-promoters who provided input, suggestions, and tips in the search process.

References

[1] L. D. Ellis, A. F. Badel, M. L. Chiang, R. J. Y. Park, and Y. M. Chiang, "Toward Electrochemical Synthesis Of Cement—An Electrolyzer-Based Process For Decarbonating Caco3 While Producing Useful Gas Streams," *Proc. Natl. Acad. Sci. U. S. A.*, vol. 117, no. 23, pp. 12584–12591, 2019.

[2] Z. Kubba *et al.*, "Impact of curing temperatures and alkaline activators on compressive strength and porosity of ternary blended geopolymer mortars," *Case Stud. Constr. Mater.*, vol. 9, p. e00205, 2018.

[3] R. R. Irawan, S. Hardono, Y. I. Budiman, O. Soeherman, D. D. Triani, and G. Gunawan, *Beton dengan Sedikit Semen Portland dan Tanpa Semen Portland dengan Memanfaatkan Abu Terbang PLTU Batubara*. 2015.

[4] N. Mahasenan, S. Smith, and K. Humphreys, "The Cement Industry And Global Climate Change : Current And Potential Future Cement Industry CO2 Emissions," *Greenh. Gas Control Technol.*, vol. II, no. 1, pp. 995–1000, 2003.

[5] D. Hardjito, S. E. Wallah, M. J. Sumajouw, and B. V. Rangan, "On the development of fly ash-based geopolymer concrete," *ACI Mater. J.*, vol. 101, no. 6, pp. 467–472, 2004.

[6] M. A. M. Ariffin, M. W. Hussin, and M. A. R. Bhutta, "Mix Design And Compressive Strength Of Geopolymer Concrete Containing Blended Ash From Agro-Industrial Wastes," *Adv. Mater. Res.*, vol. 339, no. 1, pp. 452–457, 2011.

[7] U. Sidik, "Sintesis Metakaolin Dan Abu Terbang Sebagai Prekursor Geopolimer," *Fmipa Ui*, pp. 1–95, 2012.

[8] J. Davidovits, "Geopolymers - Inorganic polymeric new materials," *J. Therm. Anal.*, vol. 37, no. 8, pp. 1633–1656, 1991.

[9] P. Rozek, M. Krol, and W. Mozgawa, "Geopolymer-zeolite composites: A review," *J. Clean. Prod.*, vol. 230, pp. 557–579, 2019.

[10] J. Davidovits, "Properties of Geopolymer Cements," *First Int. Conf. Alkaline Cem. Concr.*, pp. 131–149, 1994.

[11] N. A. Lloyd and B. V. Rangan, "Geopolymer concrete with fly ash," *2nd Int. Conf. Sustain. Constr. Mater. Technol.*, vol. 7, pp. 1493–1504, 2010.

[12] W. A. H. Putri, M. Olivia, and L. Darmayanti, "Karateristik Mortar Geopolimer Abu Sawit dengan Variasi Modulus Aktivator," *Tek. Sipil, Univ. Riau*, no. 1994, 2009.

[13] M. Olivia, "Durability Related Properties of Low Calcium Fly Ash Based Geopolymer Concrete," no. May, 2011.

[14] M. Amin and Suharto, "Pembuatan semen geopolimer ramah lingkungan berbahan baku mineral basal guna menuju lampung sejahtera," *J. Inov. Pembang.*, vol. 05 No. 01, pp. 30–45, 2017.

[15] Y. Mustakim, Nurlina, and I. Syahbanu, "Sintesis dan Karakterisasi Geopolimer Berbahan Dasar Kaolin Capkala dengan Variasi Rasio Mol SiO2/Al2O3," *Indones. J. Pure Appl. Chem.*, vol. 2, no. 2, pp. 84–93, 2019.

[16] P. Venkata, A. Sailalasa, and M. S. Reddy, "A Study on Granite Saw Dust & Fly Ash Blended Geopolymer Concrete Behavior with Various NaOH Molarities," *Int. J. Sci. Res. ISSN (Online Index Copernicus Value Impact Factor*, vol. 14, no. 9, pp. 2319–7064, 2013.

[17] P. Zhang, Y. Zheng, K. Wang, and J. Zhang, "A review on properties of fresh and hardened geopolymer mortar," *Compos. Part B Eng.*, vol. 152, pp. 79–95, 2018.

[18] J. . Ekaputri, Triwulan, and O. Damayanti, "Sifat Mekanik Beton Geopolimer Berbahan Dasar Fly Ash Jawa Power Paiton Sebagai Material Alternatif," vol. 13, no. Desember, 2007.

[19] A. . De Vargas, D. C. C. Dal Molin, A. C. F. Vilela, F. J. . Silva, B. Pavão, and H. Veit, "The effects of Na2O/SiO2 molar ratio, curing temperature and age on compressive strength, morphology and microstructure of alkali-activated fly ash-based geopolymers," *Cem. Concr. Compos.*, vol. 33, no. 6, pp. 653–660, 2011.

[20] S. J. Choi, S. S. Jun, J. E. Oh, and P. J. M. Monteiro, "Properties of alkali-activated systems with stone powder sludge," *J. Mater. Cycles Waste Manag.*, vol. 12, no. 4, pp. 275–282, 2010.

[21] A. Arulrajah, T. A. Kua, S. Horpibulsuk, C. Phetchuay, C. Suksiripattanapong, and Y. j. Du, "Strength and microstructure evaluation of recycled glass-fly ash geopolymer as low-carbon masonry units," *Constr. Build. Mater.*, vol. 114, pp. 400–406, 2016.

[22] Sahana R, "Setting Time, Compressive Strength and Microstructure of Geopolymer Paste," *Int. J. Innov. Res. Sci.*, vol. 2, no. 1, pp. 311–316, 2013.

[23] A. A. Siyal, K. A. Azizli, Z. Man, and H. Ullah, "Effects of Parameters on the Setting Time of Fly Ash Based Geopolymers Using Taguchi Method," *Procedia Eng.*, vol. 148, pp. 302–307, 2016.

[24] SNI 2049, "Semen portland," pp. 1–147, 2015.

[25] M. B. Karakoc, F. Kantarci, I. Turkmen, R. Demirboga, M. M. Maras, and M. U. Toprak, "Mechanical properties and setting time of geopolymer paste and mortar produced from ferrochrome slag," *Proc. 2013 Int. Conf. Renew. Energy Res. Appl. ICRERA 2013*, no. October, pp. 52–57, 2013.

[26] J. Davidovits, *Geopolymer Chemistry and Applications*. 1999.

[27] S. Pangdaeng, V. Sata, and P. Chindaprasirt, "Effect of sodium hydroxide concentration and sodium silicate to sodium hydroxide ratio on properties of calcined kaolin-white portlandcement geopolymer," *Int. J. GEOMATE*, vol. 14, no. 46, pp. 121–128, 2018.

IOP Conf. Series: Earth and Environmental Science 1117 (2022) 012002 doi:10.1088/1755-1315/1117/1/012002

The Evaluation of Physical and Mechanical Properties of Synthetic Polymer Modified Hot and Warm Mix Asphalt

C.G. Daniel[1*], J. Widjajakusuma[1], I. Otto[1], C. Saputan[1]

[1]Department of Civil Engineering, Pelita Harapan University, Tangerang, 15811, Indonesia

*Email: christian.daniel@uph.edu

Abstract. The application of polymer in an asphalt concrete (AC) mixture has reached wide popularity lately to deal with the increasing demand for higher quality roads with heavier traffic loading. Ethylene-vinyl acetate (EVA), classified as a thermoplastic, is currently one of the most popular polymers employed. This research attempts to evaluate the effect of adding EVA by means of the dry method, namely by pouring the additive directly into the mixture, which produces both hot and warm Polymer-Modified Asphalt Concrete (PMAC) mix. FTIR result indicates the polymer as a thermoplastic material. Moreover, the volumetric and mechanical properties of the mixture were examined through the density and semi-circular bending (SCB) tests. The outcomes show that the inclusion of the polymer could slightly modify the density and VMA of the AC mix. Meanwhile, the tensile strength, fracture energy, crack resistance index (CRI), and flexibility index (FI) results describe substantial enhancement brought by the polymer, with a generally increasing trend up 200% of the control mix. Conclusively, the application of polymer could enhance the physical and mechanical properties of hot and warm AC mixes with the optimum dosages of 5 and 6%. Additionally, the properties of warm PMA are somewhat comparable to the regular hot AC mix.

1. Introduction

The popularity of polymer-modified asphalt (PMA) mix has risen exponentially throughout the decades, in line with the increased demand for high-quality roads to support economic growth and the mobility of society. This is supported by the fact that a polymer product can be easily discovered either in nature (natural polymer) or obtained through a particular manufacturing process (synthetic polymer). Moreover, polymeric products can be divided into three categories, namely elastomeric, thermoplastic and thermoset. Elastomeric is a long-chain, rubber-like polymer with a high degree of flexibility and excellent elastic recovery properties, meaning that it can be easily deformed under external load and recover rapidly without any residual stress [1]. This is similar to the behaviour of thermoplastic polymers since they could also be processed repeatedly by melting them. However, repetitive processes could introduce a deterioration to the material quality and shape with irrecoverable strain or residual stress [2]. Several examples of thermoplastic are polyethylene, polyester, polypropylene, and PET. Meanwhile, there has been an attempt to combine the strength and stiffness of thermoplastic with the flexibility of the elastomeric product by the cross-linking process to form thermoplastic elastomeric (TPE). This results in an interesting material whose behaviour is reaching a thermoplastic state below its glass transition temperature, making it stiff enough to become brittle while becoming more elastic above its melting temperature [3]. The development of ethylene-vinyl acetate (EVA) is an example of a TPE product, where its tendency towards thermoplastic or elastomeric behaviour is predominantly determined by the vinyl acetate (VA) content, with the latter characteristic is reached by increasing the

VA fraction in the mix [4]. Finally, thermoset generally has higher strength and stiffness than thermoplastic products due to their strong covalent bond. However, its deformation when subjected to loading or heat is completely irreversible [5]. One example of a thermoset is epoxy.

Polymers can be added to an asphalt concrete (AC) mixture using two methods, namely the wet and dry mixing methods. The scheme of both approaches is shown in **Figure 1**.

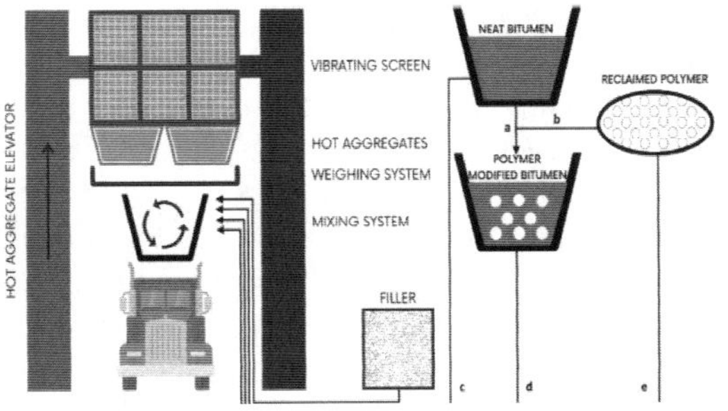

Figure 1. Scheme of dry and wet mixing method [6]

Firstly, a polymer can be mixed with heated, liquid bitumen to form a polymer-modified bitumen (PMB). This method is called the wet mixing method. From **Figure 1**, the process of making an AC mix with PMB requires a dedicated chamber to blend the virgin binder and the polymer (points a and b) first prior to being delivered to the mixing chamber with aggregates and filler. Various types of polymers have been utilised using this approach, each with different effects. For example, the use of rubber to modify has become a widely acknowledged approach, with much research accounting for the increase in fatigue life and stabilized dynamic stiffness at various temperatures, as well as improved resistance to rutting and cracking caused by temperature and ageing effect. Moreover, crumb rubber has been widely utilised to alter the rheological behaviour of bitumen, reported to have a superior fatigue life compared to the neat binder, as well as a dramatic increase in complex modulus and decreased flow activation energy that indicates lower thermal susceptibility. All these effects are enabled by the cross-linking polymer network inside the bitumen, which accounts for the enhanced stiffness. In addition, the crumb rubber-modified bitumen has been reported to have an increased softening point with a decreased penetration value, confirming the effect of the enhanced stiffness. Finally, the molecular weight of the bitumen is also added by the presence of the crumb rubber [7]–[9]. The crumb-rubber modified binder has been reported to yield bituminous mixtures with lower air void and higher density, a higher Marshall stability ratio, tensile strength, and resistance to permanent deformation [10][11]. Another example is polyethylene (PE), whose usage has spun for a long time and is being recorded by numerous researchers. The application of PE to modify a bitumen has been known to generate high complex modulus as well as low phase angle and ductility, resulting in an improved rutting characteristic [12][13]. However, this modified binder is prone to phase separation due to the inability of polyethylene to form a strong bond with the bitumen caused by different molecular weights and chain structures; thus, the maximum polymer fraction of 5% is suggested to avoid phase inversion and altering the bonding by introducing cross-linking copolymer by adding substances such as polyphosphoric acid, sulphur, Trans-polyoctenamer, nano-silicate, and maleic anhydride [14]–[20]. Another research also attempted to examine the impact of long-chain wax additive on the rheological properties of the polyethylene-modified bitumen, resulting in an increased rutting resistance, as well as a better ageing behaviour [21]. Lastly, PE-modified bitumen (PEMB) has been found to generate increased Marshall stability and decreased flow values [22]. Meanwhile, ethylene-vinyl acetate (EVA) has been widely utilised to create a polymer-modified bitumen (PMB) product by enhancing its stiffness and reducing its penetration value, as well as a better creep stiffness at a lower temperature. Another finding was that the amount of VA fraction in EVA contributed to the difference in the behaviour of a modified binder. Several reports have indicated the shift from a more thermoplastic to elastomeric behaviour due to the increase in VA content, which alters binder stiffness and softening point, as well as improves its elongation capacity [23]–[25]. All these effects can be explained by looking at the reaction mechanism inside the modified binder, where the polymer will absorb thc lighter maltene fraction from the bitumen to form a polymer-rich phase. However, due to the difference in molecular weight and solubility, the asphaltene-rich phase becomes separated, and as such, it was discovered that the application of EVA by larger than 5% would cause the polymer phase to swell, and phase separation could easily occur; thus, a constant stirring at hot temperature also becomes critical to ensure its stability [26][27]. Consequently, employing a cross-linking agent to the copolymer becomes compulsory to enhance the compatibility of the modified bitumen product. Several products have been discovered to bring varying degrees of impact to the properties of the bitumen, such as maleic anhydride, thermoplastic vulcanizates, and polyoctenamer. The use of these agents has been observed to increase elastic properties, storage stability, resistance to permanent deformation and low-temperature cracking, thermal stability, and lower viscosity [28]–[30].

On the other hand, the polymer can be directly poured into the mixing chamber and blended with the heated bitumen and hot aggregates from the bin; this is called the dry mixing method. In contrast to the wet method, the initial stage of the dry method is described in **Figure 1** from (c), where neat bitumen is directly transferred to the mixing chamber with filler and the aggregate, while the polymer is sent simultaneously (e) to the mixing process with other components. Fibre-shaped polymer is typically applied in this method to modify the characteristics of the AC mix. Numerous types of polymers have been employed throughout the decades, such as polyethylene, polypropylene, polyethylene terephthalate (PET), aramid, and a combination of those. There are several key factors that make fibre an excellent additive for an AC mix, namely strength, stiffness, density, and the network matrix formed in the body of a specimen through its length/diameter ratio [31]. Aramid is well-known for its high thermal and chemical stability in addition to its stiffness and tenacity [32][33], enabling it to form a highly durable fibrous network inside an AC mix that will lead to the improvement in fatigue life and rutting resistance [34]–[36]. Moreover, its combination with PE fibre has been discovered to enhance strength, fatigue life, resistance to permanent deformation, and thermal cracking resistance of both dense and Stone Mastic (SMA) AC mixtures under various loading and temperature modes, where the bonding mechanism between the fibre and asphaltic matrix can vary in response to different loading modes and speeds [37]–[42].

This research attempts to employ the granulated-shaped EVA+ LDPE polymer manufactured by Iterchemica by means of the dry method. The use of this polymer has been firstly found to improve the dynamic modulus of AC mix specimen, indicating the enhancement regarding mixture stiffness and resistance to permanent deformation [43]. The product is shown in **Figure *2***, coming in a granulated shape.

Figure 2. Superplast EVA polymer

2. Specimen preparations

This research was divided into three stages. In the first phase, the physical and volumetric properties of aggregate, bitumen, and additives were examined. Two parameters, namely density and absorption of the aggregate, were taken from the density test result. Meanwhile, several tests, such as penetration, density, flash point, and softening point, were conducted on the bitumen. All test results must comply with the boundary values taken from the national code called Bina Marga 2018, shown in **Table *1***.

Table 1. Bina Marga 2018 standards for the asphaltic mixture components

Materials	Properties	Unit	Limit values
Bitumen	Penetration	PEN	60 - 70
	Density	kg/m^3	> 1000
	Softening Point	°C	> 48
	Flash Point	°C	-
Coarse Aggregate	Density	kg/m^3	> 2500
	Absorption	%	< 3.0
Fine Aggregate	Density	kg/m^3	> 2500
	Absorption	%	< 3.0
Cement Filler	Density	kg/m^3	-

Fourier Transform Infrared (FTIR) tests were conducted to examine the chemical characteristic of the polymer prior to being used in the further stage. The result is the series of IR band-wave number curves that could be used to analyse the chemical composition of the product based on the intensity of each band. The polymer is manufactured by Iterchemica under the brand name Superplast, and the properties are described in **Table** *2*.

Table 2. Properties of Superplast by Iterchemica [44][45]

Physical properties	Value
Softening point	150°C
Fusion point	160°C
Composition	LDPE + EVA (Unknown proportion)

The mix design of the asphalt concrete (AC) specimens was developed in accordance with the national norm for a dense asphaltic mixture shown in **Figure** *3*. The bitumen content of 5.5% w/t was selected for this research to produce the Semi-Circular Bending (SCB) samples.

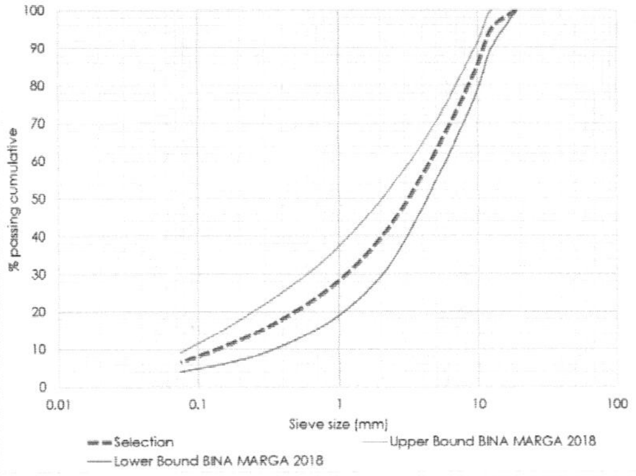

Figure 3. Mix design of Asphalt Concrete (AC) mix for Semi-Circular Bending (SCB) test specimens

Finally, the Semi-Circular Bending (SCB) tests were performed at room temperature with the adjusted loading speed of 10 mm/min following the requirement set in NEN 12697:44. For this purpose, half-cylindrical specimens were prepared with a diameter of 150mm and a thickness of 60mm. Notches with a depth of 10mm were fabricated at the midspan of the samples. The load was subjected to a specimen until reaching its breaking point, with a crack propagated from the notch location (**Figure *4***).

Figure 4. Setup of Semi-Circular Bending (SCB) test

The examinations were carried out by the Universal Testing Machine (UTM) connected to a PC, from which the data can be acquired for further analysis. The outcome is displayed as a series of force-displacement data, which can be converted to several parameters. The first one is tensile strength, defined by Equation (1). Furthermore, fracture energy is defined as the sum of the area under the force-displacement curve and is described in Equation (2). Additionally, the crack resistance index (CRI) is a parameter that can be used to describe the extent of specimen integrity to carry load before failure occurs, as displayed in Equation (3). Finally, flexibility index (FI) is a formula developed by the Illinois Flexibility Index Test (I-FIT) to illustrate the extent of deformability of the AC sample after the crack initiates, which is expected to start from the centre of the specimen, where the notch is formed (Equation (4)) [46]. However, the result needs to be adjusted to the standard expected air void of 7% to reduce the effect of air void on the result [47], as shown by Equation (5).

$$\sigma_{max} = \frac{4{,}263 P_{max}}{t \times d} \ N/mm^2 \tag{1}$$

$$G_f = 10^3 \times \frac{W_f}{A_{lig}} = 10^3 \times \frac{W_f}{t \times L} \ J/m^2 \tag{2}$$

$$CRI = \frac{G_f}{P_{max}} \tag{3}$$

$$FI = \frac{G_f}{|m|} \times 0.01 \tag{4}$$

$$FI_{corrected} = \frac{7}{Air\ Void\ (\%)} \times FI \tag{5}$$

Where:
P_{max} = maximum load
W_f = work of fracture (area under the force-displacement curve)
A_{lig} = area of ligament
L = ligament length = specimen radius – notch depth
m = slope of post-peak force-displacement curve

3. Results and Discussions

3.1. Examination of Asphalt Concrete (AC) Mix Components

The preliminary stage of this research is examining the physical properties of the mixture component. The results are shown in **Table 3**.

Table 3. Preliminary test (physical properties) result

Materials	Properties	Unit	Value	Boundaries (Bina Marga 2018)
Bitumen	Penetration	PEN	65.5	60 - 70
	Density	kg/m^3	1061	> 1000
	Softening Point	°C	52	> 48
	Flash Point	°C	267	-
Coarse Aggregate	Density	kg/m^3	2503	> 2500
	Absorption	%	2.705	< 3.0
Fine Aggregate	Density	kg/m^3	2530	> 2500
	Absorption	%	2.923	< 3.0
Cement Filler	Density	kg/m^3	2919	-

Thus, all the materials used in this project have satisfied the criteria and become eligible to use in further stages.

3.2. FTIR test for the polymer

Since the polymer product used in this research is EVA with an unknown fraction of vinyl acetate (VA), Fourier Transform Infrared (FTIR) test was conducted to examine the possible quantity of VA content in this product. The result will be compared to the current database of EVA with known properties and can be depicted in **Figure 5**.

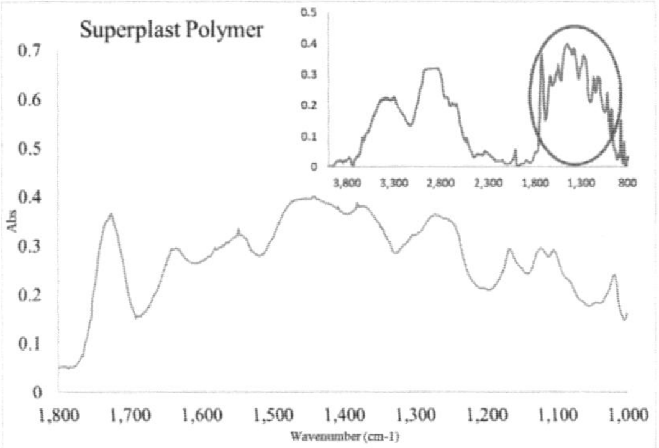

Figure 5. FTIR result of EVA- Superplast

It is known that the variation of VA content mainly alters the IR band of the carbonyl region (1700-1850 cm^{-1}), where the aldehyde compound becomes the dominant factor herein. Another critical point is in the fingerprint area of the C-H band (1460), indicating a medium-intensity alkane compound. Lastly, the 1000 and 1200 bands and the carbonyl chain depict the ester fractions of EVA. Evidently, by comparing the result to other research outcomes [48][49], it can be concluded that the VA fraction in this polymeric product is around 5%, judging by the comparison of intensity in both bands. Moreover, the high intensity in the alcohol (O-H) region of band 3500 and alkane (C-H) of band 2800-3000 indicate the ethylene groups in EVA. Therefore, the polymer used herein can be classified as the thermoplastic-dominant group, where the VA only plays a minor part in the blend.

3.3. Volumetric properties of polymer-modified asphalt

The results of volumetric tests of SCB specimens are the mixture density and void in mineral aggregate (VMA), shown in **Figure 6**.

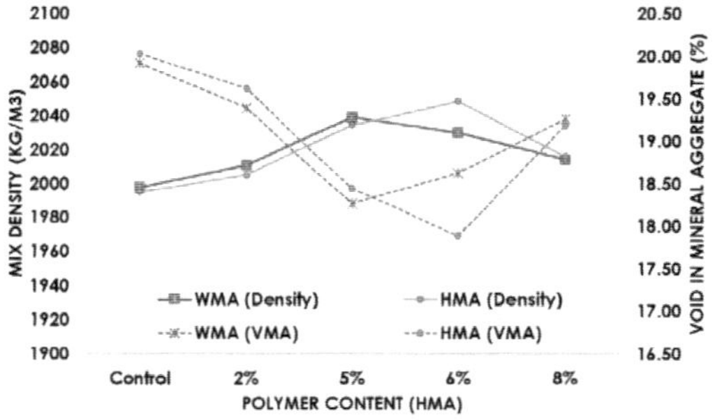

Figure 6. Density and Air Void of Polymer-Modified AC mix

Generally, it is apparent that polymer inclusion could improve the mix density and decrease the VMA value. Such a phenomenon could occur due to the existence of the polymer with considerable molecular weight to fill the gap inside the mixture while also contributing to forming a bond with bitumen and creating a complex chain of polymeric network, thereby reducing the void in the mixture. However, since the difference in the values is relatively insignificant, it can be concluded that the polymer does not affect the volumetric properties of the bituminous mix to a large extent. Further effort can be made to produce a higher density value, such as better compaction.

3.4. Semi-Circular Bending test result

The research outcome comes in the form of several parameters, such as tensile strength, fracture energy, crack resistance index (CRI), and flexibility index (FI). All the results are presented in **Figure** *7* to **Figure 10**.

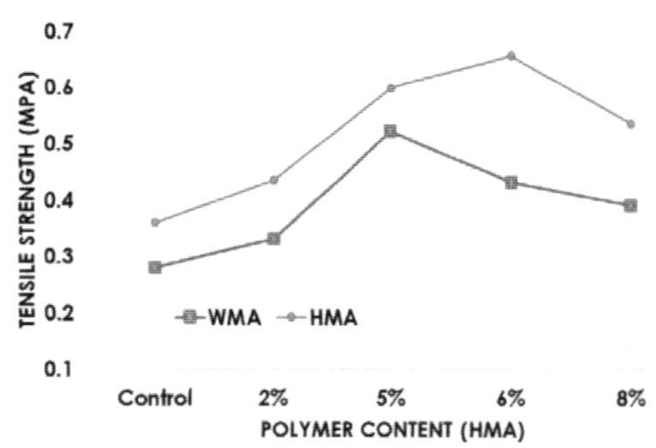

Figure 7. Tensile strength of Polymer-Modified AC mix (HMA and WMA)

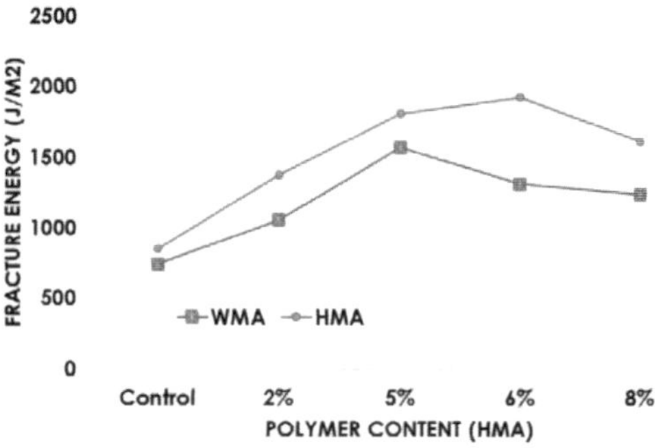

Figure 8. Fracture energy of Polymer-Modified AC mix (HMA and WMA)

As depicted in **Figure** *7*, the addition of polymer into the hot and warm PMA specimens has increased their tensile strength by almost 50% due to the modified stiffness of the mixture brought by the polymer. It is also apparent that the strength of polymer-modified WMA can surpass the strength of the control HMA mix, indicating the benefit of using the polymer. Nevertheless, there is an inflection point at the value of 5% for the WMA and 6% for the HMA, upon which the strength value declines. This could occur due to the different degrees of compaction for the specimens due to the increasing polymer dosage since it has been discovered that the enhanced specimen stiffness due to the polymer addition could introduce trouble in the compaction effort [12]. Moreover, the fracture energy (**Figure 8**), which indicates both the factor of strength and the deformation capacity, shows a similar trend of 1.5 to 2 times higher, with the polymer dosage of 5-6% being found to generate the highest values in both parameters in hot warm and hot bituminous mix, respectively. Additionally, the analogous trend occurring in the fracture energy indicates that the deformation behaviour is affected in a similar manner to its strength.

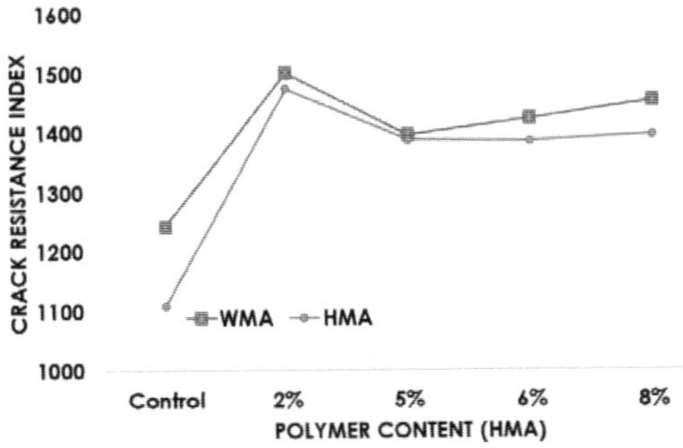

Figure 9. Crack Resistance Index of Polymer-Modified AC mix (HMA and WMA)

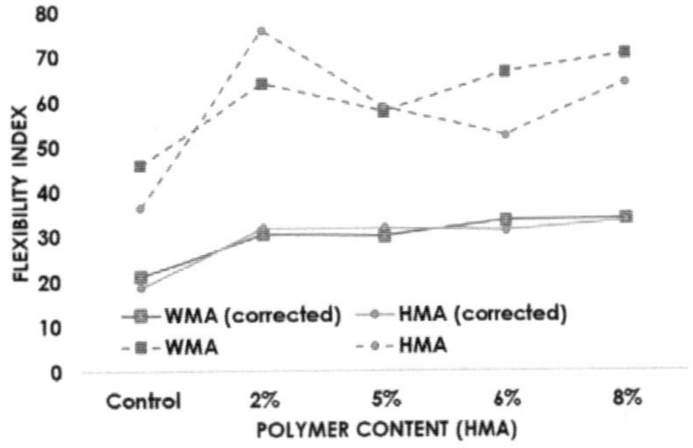

Figure 10. Flexibility Index of Polymer-Modified AC mix (HMA and WMA)

Both the crack resistance index (CRI) and flexibility index (FI) can be used to display the deformation capacity of the SCB specimens to support the result of fracture energy in a slightly different manner. The CRI results in **Figure 9** describe the deformation capacity of a specimen from the initial stage to the failure and conclude that the increase of polymer dosage can enhance the deformation of the specimens, both hot and warm mix asphalt, in general. Nonetheless, a reduction in the dosage of 5% compared to 2% might occur due to the heterogeneous distribution of the polymer inside the specimen, which can introduce an anomaly in the outcome. Compared to the research in EVA-modified binder, which found a vast amount of improvement in elongation capacity in a direct proportion to the increase in dosage, such behaviour is different in the PMA product due to the production method itself, wherein the polymer is not given a dedicated moment to develop a modified binder; instead, the reaction only occurs during the mixing process, which might limit the extent of binder modification. Moreover, the close difference between the CRI of WMA and HMA indicates that the polymer greatly affects the deformation capacity of WMA.

On the other hand, the flexibility index (FI) (**Figure 10**), highlighting the extent of deformation occurring in the post-peak stage, can describe the improvement brought by the polymer by enhancing the bonding with the aggregates and bitumen while increasing the stiffness of the mix. Nonetheless, the change in dosage has no apparent impact on the result. While this can be linked to the heterogeneous distribution of the polymer inside the specimen, this could also be attributed to the missing load carrying capacity due to the shape of the polymer itself, in contrast to the shape of the fibre that enables such a mechanism especially after the crack propagates. Moreover, the variation of air void in the mixture influences the result, as apparent from the curves of the non-corrected results. Conclusively, the behaviour of specimens beyond the crack initiation phase is highly dependent on their air void, such that an adjustment needs to be performed to decrease the variation. Finally, based on the classification by the research of Ozer et al. [46], the polymer-modified AC products can be categorised as a high-performance, stiff and flexible mix with low rut depth (less than 10mm).

Finally, it is evident that the polymer could elevate the mechanical properties of WMA specimens to match the HMA control mixture. This has shown a window of opportunity for WMA mixtures for future application, with more observation needing to be conducted to verify its potential. Moreover, the dosage of 5-6% can be recommended for both HMA and WMA mixtures, given the results. Further research needs to examine other properties of polymer-modified bituminous mix, such as the chemical, rheological and mechanical properties under cyclic loading.

4. Conclusion

This research focuses on evaluating the impact of incorporating thermoplastic ethylene-vinyl acetate (EVA) polymer to the blended bitumen-aggregate to form a polymer-modified asphalt (PMA) mix. Firstly, the comparison of the polymer chemical composition between the test results and other research outcomes for EVA with different vinyl acetate (VA) content concludes that the VA fraction in Superplast products is as high as 5%, thereby putting it into the thermoplastic category. Furthermore, the mechanical properties, namely tensile strength, fracture energy, crack resistance index, and flexibility index of the PMA mix with polymer dosages of 2%, 5%, 6%, and 8% of bitumen weight show improved values to various extent in both hot and warm mix asphalt, resulting in the AC mixes classified into the stiff and flexible against rut group. Overall, the increase of polymer dosage has been found to enhance the mechanical properties of the mixture aligned with the increasing physical properties. Nonetheless, the dosages of 5% and 6% have shown to bring the most considerable improvement to the behaviour of the hot and warm bituminous mixes, respectively. The properties of the warm mix asphalt modified with polymer are evidently comparable to that of the control hot mix asphalt. Conclusively, the inclusion of EVA can enhance the physical and mechanical properties of both hot and warm mix asphalt in relatively direct proportion to the increase of the dosage, with the optimum content of approximately 5 – 6% of the bitumen weight. Moreover, the modification to the warm bituminous mixture can actually elevate the behaviour to be comparable with the standard hot mix asphalt, which opens the opportunity for warm mix application in the future. Future research should examine the chemical and rheological properties of the polymer-modified asphalt, as well as other mechanical properties under various situations.

5. Acknowledgments

The authors wish to acknowledge the support from Pelita Harapan University for this project under the project code of 414/LPPM-UPH/XII/2021 and P-08-FaST/XII/2021, as well as the support from PT. Enceha Pacific with the Superplast polymer material for this research.

6. References

[1] N. Dishovsky and M. Mihaylov, *ELASTOMER-BASED COMPOSITE MATERIALS Mechanical, Dynamic, and Microwave Properties and Engineering Applications*. 2018.

[2] P. K. Mallick, "Thermoplastics and thermoplastic-matrix composites for lightweight automotive structures," 2010. doi: 10.1016/B978-1-84569-463-0.50004-3.

[3] J. G. Drobny, *Handbook of Thermoplastic Elastomers*. William Andrew, 2014.

[4] L. Wang, L. Zhang, Y. Shi, and Z. Wang, "Thermoplastic Elastomers Based on Ethylene-vinyl Acetate Copolymer and Waste Nitrile Butadiene Rubber Powder Blends Compatibilized by Chlorinated Polyethylene," *Journal of Macromolecular Science, Part B: Physics*, vol. 57, no. 4, pp. 305–316, Apr. 2018, doi: 10.1080/00222348.2018.1460975.

[5] Hanna Dodiuk, *Handbook of Thermoset Plastics*. William Andrew, 2014.

[6] S. Nizamuddin, Y. J. Boom, and F. Giustozzi, "Sustainable polymers from recycled waste plastics and their virgin counterparts as bitumen modifiers: A comprehensive review," *Polymers*, vol. 13, no. 19. MDPI, Oct. 01, 2021. doi: 10.3390/polym13193242.

[7] H. Wang, X. Liu, M. van de Ven, G. Lu, S. Erkens, and A. Skarpas, "Fatigue performance of long-term aged crumb rubber modified bitumen containing warm-mix additives," *Constr Build Mater*, vol. 239, Apr. 2020, doi: 10.1016/j.conbuildmat.2019.117824.

[8] H. Wang, X. Liu, P. Apostolidis, and T. Scarpas, "Rheological behavior and its chemical interpretation of crumb rubber modified asphalt containing warm-mix additives," *Transp Res Rec*, vol. 2672, no. 28, pp. 337–348, Jan. 2018, doi: 10.1177/0361198118781376.

[9] N. Lushinga, L. Cao, Z. Dong, C. Yang, and C. O. Assogba, "Performance Evaluation of Crumb Rubber Asphalt Modified with Silicone-Based Warm Mix Additives," *Advances in Civil Engineering*, vol. 2020, 2020, doi: 10.1155/2020/4840825.

[10] S. M. Soleimani, A. Faheiman, and Z. Mowaze, "The Effects of Using Crumb Rubber Modified Binder in an Asphalt Pavement," *American Journal of Engineering and Applied Sciences*, vol. 13, no. 2, pp. 237–253, Feb. 2020, doi: 10.3844/ajeassp.2020.237.253.

[11] M. A. Bilema *et al.*, "Moisture Sensitivity of Crumb Rubber Modified Modifier Warm Mix Asphalt Additive for Two Different Compaction Temperatures," in *IOP Conference Series: Earth and Environmental Science*, Apr. 2018, vol. 140, no. 1. doi: 10.1088/1755-1315/140/1/012072.

[12] K. A. Ghuzlan, G. G. Al-Khateeb, and Y. Qasem, "Rheological Properties of Polyethylene-Modified Asphalt Binder."

[13] S. Sharma, S. Sharma, and N. Upadhyay, "Composition Based Physicochemical Analysis of Modified Bitumen by HDPE/LDPE," *Oriental Journal of Chemistry*, vol. 35, no. 3, pp. 1167–1173, Jun. 2019, doi: 10.13005/ojc/350336.

[14] F. M. Nejad, K. Naderi, and R. Zarroodi, "Effect of cross-linkers on the performance of polyethylene-modified asphalt binders," *Proceedings of Institution of Civil Engineers: Construction Materials*, vol. 170, no. 4, pp. 186–193, Aug. 2017, doi: 10.1680/jcoma.16.00021.

[15] C. Fuentes-Audén *et al.*, "Evaluation of thermal and mechanical properties of recycled polyethylene modified bitumen," *Polym Test*, vol. 27, no. 8, pp. 1005–1012, 2008, doi: 10.1016/j.polymertesting.2008.09.006.

[16] E. S. Okhotnikova, I. N. Frolov, Y. M. Ganeeva, A. A. Firsin, and T. N. Yusupova, "Rheological behavior of recycled polyethylene modified bitumens," *Pet Sci Technol*, vol. 37, no. 10, pp. 1136–1142, May 2019, doi: 10.1080/10916466.2019.1578796.

[17] Z. Xie and J. Shen, "Effect of cross-linking agent on the properties of asphalt rubber," *Constr Build Mater*, vol. 67, no. PART B, pp. 234–238, Sep. 2014, doi: 10.1016/j.conbuildmat.2014.03.039.

[18] T. Mandal, R. Sylla, H. U. Bahia, and S. Barmand, "Effect of cross-linking agents on the rheological properties of polymer-modified bitumen," *Road Materials and Pavement Design*, vol. 16, pp. 349–361, May 2015, doi: 10.1080/14680629.2015.1029683.

[19] P. Ahmedzade *et al.*, "Use of maleic anhydride grafted recycled polyethylene treated by irradiation in bitumen modification," Jan. 2017. doi: 10.14311/ee.2016.155.

[20] M. Porto, P. Caputo, V. Loise, S. Eskandarsefat, B. Teltayev, and C. O. Rossi, "Bitumen and bitumen modification: A review on latest advances," *Applied Sciences (Switzerland)*, vol. 9, no. 4. MDPI AG, Feb. 20, 2019. doi: 10.3390/app9040742.

[21] L. Desidery, M. Lanotte, and S. Devasahayam, "Effect of Waste Polyethylene and Wax-Based Additives on Bitumen Performance," 2021, doi: 10.3390/polym.

[22] D. B. Eme and C. Nwaobakata, "Effect of low density polyethylene as bitumen modifier on some properties of hot mix asphalt," *Nigerian Journal of Technology*, vol. 38, no. 1, p. 1, Jan. 2019, doi: 10.4314/njt.v38i1.1.

[23] V. v. Yadykina, S. N. Navolokina, and A. M. Gridchin, "The dependence of the modified bitumen properties on the amount of vinyl acetate in the sevilen composition," in *Materials Science Forum*, 2020, vol. 974 MSF, pp. 175–180. doi: 10.4028/www.scientific.net/MSF.974.175.

[24] I. B. Joohari and F. Giustozzi, "Effect of different Vinyl-Acetate contents in hybrid SBS-EVA modified bitumen 1."

[25] A. Yuliestyan, A. A. Cuadri, M. García-Morales, and P. Partal, "Selection of ethylene-vinyl-acetate properties for modified bitumen with enhanced end-performance," *Rheol Acta*, vol. 57, no. 1, pp. 71–82, Jan. 2018, doi: 10.1007/s00397-017-1057-5.

[26] N. Saboo and P. Kumar, "Optimum Blending Requirements for EVA Modified Binder," in *Transportation Research Procedia*, 2016, vol. 17, pp. 98–106. doi: 10.1016/j.trpro.2016.11.065.

[27] Tony. McNally and Petra. Pötschke, *Polymer modified bitumen: Properties and Characterisation*. Woodhead Pub, 2011.

[28] N. Kunanusont, B. Sangpetngam, and A. Somwangthanaroj, "Asphalt incorporation with ethylene vinyl acetate (Eva) copolymer and natural rubber (nr) thermoplastic vulcanizates (tpvs): Effects of tpv gel content on physical and rheological properties," *Polymers (Basel)*, vol. 13, no. 9, May 2021, doi: 10.3390/polym13091397.

[29] R. K. Padhan, A. A. Gupta, and A. Sreeram, "Effect of cross-linking agent on ethylene vinyl acetate/polyoctenamer modified bitumen," *Road Materials and Pavement Design*, vol. 20, no. 7, pp. 1615–1623, 2019, doi: 10.1080/14680629.2018.1467335.

[30] W. Q. Luo and J. C. Chen, "Preparation and properties of bitumen modified by EVA graft copolymer," *Constr Build Mater*, vol. 25, no. 4, pp. 1830–1835, Apr. 2011, doi: 10.1016/j.conbuildmat.2010.11.079.

[31] N. Mashaan, M. Karim, F. Khodary, N. Saboo, and A. Milad, "Bituminous Pavement Reinforcement with Fiber: A Review," *CivilEng*, vol. 2, no. 3, pp. 599–611, Jul. 2021, doi: 10.3390/civileng2030033.

[32] M. Ertekin, "Aramid fibers," in *Fiber Technology for Fiber-Reinforced Composites*, Elsevier, 2017, pp. 153–167. doi: 10.1016/B978-0-08-101871-2.00007-2.

[33] R. H. Gong and X. Chen, "Technical Yarns," in *Handbook of Technical Textiles: Second Edition*, vol. 1, Elsevier Inc., 2016, pp. 43–62. doi: 10.1016/B978-1-78242-458-1.00003-0.

[34] M. D. Nazzal, P. Ahmad Al-Hosainat, S.-S. Kim, A. R. Abbas, and A. Hudaib, "Analysis of Aramid Synthetic Fibers in Asphalt Mixes on Local Roads Final Report," 2021.

[35] S. Badeli, A. Carter, and S. Saliani, "Effect of Short Aramid Fibers on Asphalt Performance," 2017. [Online]. Available: http://substanceen.etsmtl.ca/effectshortaramidfibersasphalt/

[36] A. Alnadish and Y. Aman, "Evaluation of aramid fibre-reinforced asphalt mixtures," in *Lecture Notes in Civil Engineering*, vol. 9, Springer, 2019, pp. 1377–1388. doi: 10.1007/978-981-10-8016-6_99.

[37] S. Majer and B. Budziński, "Influence of Aramid-Polyolefin Fibers on the Properties of Bituminous Mixtures," *IOP Conf Ser Mater Sci Eng*, vol. 1203, no. 3, p. 032093, Nov. 2021, doi: 10.1088/1757-899x/1203/3/032093.

[38] H. Noorvand, S. C. Brockman, M. Mamlouk, and K. Kaloush, "Effect of Aramid Fibers on Balanced Mix Design of Asphalt Concrete," *CivilEng*, vol. 3, no. 1, pp. 21–34, Dec. 2021, doi: 10.3390/civileng3010002.

[39] D. Wiśniewski, M. Słowik, J. Kempa, A. Lewandowska, and J. Malinowska, "Assessment of impact of aramid fibre addition on the mechanical properties of selected asphalt mixtures," *Materials*, vol. 13, no. 15, Aug. 2020, doi: 10.3390/ma13153302.

[40] P. Apostolidis, X. Liu, C. G. Daniel, S. Erkens, and T. Scarpas, "Effect of synthetic fibres on fracture performance of asphalt mortar," *Road Materials and Pavement Design*, vol. 21, no. 7, pp. 1918–1931, Oct. 2020, doi: 10.1080/14680629.2019.1574235.

[41] C. G. Daniel, X. Liu, P. Apostolidis, S. Erkens, and A. Scarpas, "Impact of synthetic fibres on asphalt concrete mix," *Bituminous Mixtures and Pavements VII*, no. September, pp. 0–0, 2019, doi: 10.1201/9781351063265-96.

[42] C. G. Daniel, X. Liu, P. Apostolidis, S. M. J. G. Erkens, and A. Scarpas, "Low-temperature fracture behaviour of synthetic polymer-fibre reinforced warm mix asphalt," in *Green and Intelligent Technologies for Sustainable and Smart Asphalt Pavements*, 1st ed., vol. 1, no. 3, Taylor & Francis, 2021, pp. 358–362. doi: https://doi.org/10.1201/9781003251125.

[43] Eng. F. Montanelli and I. srl, "Fiber/Polymeric Compound for High Modulus Polymer Modified Asphalt (PMA)," *Procedia Soc Behav Sci*, vol. 104, pp. 39–48, Dec. 2013, doi: 10.1016/j.sbspro.2013.11.096.

[44] "SUPERPLAST POLYMER COMPOUND HIGH PERFORMANCE POLYMER MODIFIED ASPHALT (PMA)."

[45] "SUPERPLAST.1."

[46] H. Ozer, I. L. Al-Qadi, P. Singhvi, T. Khan, J. Rivera-Perez, and A. El-Khatib, "Fracture characterization of asphalt mixtures with high recycled content using Illinois semicircular bending test method and flexibility index," *Transp Res Rec*, vol. 2575, pp. 130–137, 2016, doi: 10.3141/2575-14.

[47] D. Batioja-Alvarez, J. Lee, and J. E. Haddock, "Understanding the Illinois Flexibility Index Test (I-FIT) using Indiana Asphalt Mixtures," *Transp Res Rec*, vol. 2673, no. 6, pp. 337–346, Jun. 2019, doi: 10.1177/0361198119841282.

[48] M. Giurginca, L. Popa, and T. Zaharescu, "Thermo-oxidative degradation and radio-processing of ethylene vinyl acetate elastomers," *Polym Degrad Stab*, vol. 82, no. 3, pp. 463–466, 2003, doi: 10.1016/S0141-3910(03)00200-3.
[49] Shimadzu, "Application Notc Quantitation of vinyl acctatc in EVA-FTIR." [Onlinc]. Availablc: www.shimadzu.de

Reinforced Concrete Beams Strengthened with Externally Bonded Hybrid FRP Sheets

J S Limbongan[1], **R Djamaluddin**[1], **R Irmawaty**[1] **and Jasman**[1]

[1] Civil Engineering Department, Engineering Faculty, Hasanuddin University, Makassar, Indonesia

joeylimbongan@gmail.com

Abstract. Fiber Reinforced Polymer (FRP) is commonly used in structural repair and strengthening due to its high tensile strength, non-corrosive and lightweight. On the other side FRP is relatively brittle compared to steel. This study used hybrid FRP consisting of GFRP and CFRP, by combining two type of FRP with different characteristic it is to be expected that the more brittle FRP will fail first followed by the more elastic one, therefore creating a sequential failure. However CFRP have a higher tensile strength in comparison to GFRP, therefore it's necessary to reduce its tensile strength by reducing its volume. This study is divided into two topics, hybrid FRP tensile strength test and beam flexural test. For the hybrid FRP tensile strength test, a total of three variation of hybrid FRP will be compared (GC30, GC40, and GC50). The variation that shown a sequential failure behaviour, will then be applied to a reinforced concrete beam and tested under monotonic point load device. The test result shown that sequential failure occur on GC30 and GC40. However GC40 had a higher ultimate load at 24.86 kN. For the beam flexural test, there will be two variation, i.e. control beam (CB) and beam strengthened with hybrid FRP (BGC). When compared, BGC had a 40.0% increase in yield strength and 55.43% increase for ultimate load.

Keywords: Reinforced Concrete Beam, Strengthening, Hybrid FRP

1. Introduction

Damage to reinforced concrete structures may occur due to the age of the structure, changes in loading or natural disasters. Most damaged structures will be dismantled immediately without considering the possibility of repair or retrofitting. Nowadays, the development of science and technology is growing rapidly, including developments in the field of civil engineering. One method of structural improvement that is currently being developed is to use a fiber layer reinforced with a polymer matrix or commonly known as Fiber Reinforced Polymer (FRP). Utilization of FRP material in building structures is useful for increasing the ability and restoring the strength of reinforced concrete structures. The use of an externally bonded FRP system is an effective reinforcement to strengthen damaged structural elements. Repair using FRP is also a more economical alternative when compared to building reconstruction.

In its application, FRP sheet is a material that is resistant to corrosion, has a relatively high tensile strength and good workability despite its light weight. But on the other hand FRP has brittle properties so that if a structure fails, the structure will fail suddenly. So it is necessary to create a solution that can be a factor of safety.

One solution that can be done is to combine two layer of FRP which have different characters, namely, Glass Fiber Reinforced Polymer (GFRP) and Carbon Fiber Reinforced Polymer (CFRP). GFRP has more elastic properties than CFRP. By combining the two it is hoped that if the load carried is excessive, the more brittle CFRP will fail first so that sudden collapse can be prevented.

The next problem is that CFRP has a better tensile strength than GFRP, so adjustments need to be made to reduce the tensile strength of CFRP. The method used is to reduce the width of CFRP. With this reduction, it is expected that CFRP will fail before GFRP.

2. Literature Review

Mufti Amir Sultan [1] in their research on the flexural capacity and ductility of reinforced concrete beam strengthened with GFRP-S. The beams tested with a static loading system with two-point load. The experimental results show that the installation of GFRP-S on the beam can increase the flexibility and ductility of the beam, respectively 37.96% and 25%.

Saddam Husein [2] present in their writings the analysis of flexural capacity and failure patterns of concrete beams without a blanket using GFRP bar reinforcement material. The research design is an experimental laboratory with a recapitulation of 6 samples consisting of 2 concrete blocks using steel reinforcement with concrete covers, 2 concrete blocks using GFRP bars with concrete covers, 2 concrete blocks using GFRP bars without concrete covers. The results of this study indicate that the flexural capacity of the beam with GFRP bar reinforcement is greater than that of the steel reinforcing beam and is able to increase the flexural capacity of the beam to withstand the load by 39.76%.

Koosha Khorramian [3] conducted a study using longitudinal CFRP strips to increase the flexural stiffness of columns for strengthening slender columns and eccentrically loaded columns where additional flexural stiffness is required for buckling control. The test results show that the use of a longitudinal CFRP unlaminated sheath is more effective than the proposed hybrid system for reinforcement of small-scale concrete columns subjected to axial loads.

Szmigiera et al. [4] explored the combination of Basalt Fiber Reinforced Polymer (BFRP) with CFRP with the aim of obtaining Hybrid Fiber Reinforced Polymer (HFRP) with better mechanical properties than BFRP and more economical than CFRP. The variations of the test objects used are, (C:B): 1:1, 1:2, 1:3, 1:4, and 1:9. Based on the test, the results of the 1:3 combination showed an increase in the modulus of elasticity by 68% and tensile strength by 15.8% when compared to BFRP.

3. Experimental Programme

3.1. Test Specimen

This study is divided into two topics, hybrid FRP tensile strength test and beam flexural test. In the hybrid FRP tensile strength test, GFRP is used as a base to determine the width of the CFRP. Trough out this test the width of GFRP is 40 mm. While for CFRP it is varied to 30% (GC30), 40% (GC40), and 50% (GC50) of GFRP width, which is equal to 16 mm, 18 mm, and 20 mm. Sample measurement based on ASTM D3039 [5] as shown in Figure 1.

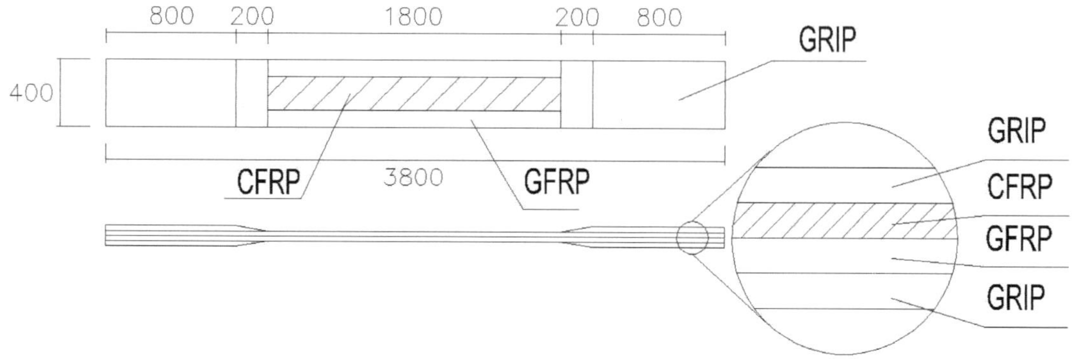

Figure 1. Hybrid FRP sample design

For the beam flexural test, there will be two variation, i.e. control beam (CB) and beam strengthened with hybrid FRP (BGC). The detail of the beam dimensions are 150x200x3000 mm. Sample beams are reinforced with steel reinforcement 3 D13 for the tension area, 2 Ø8 for the compression area, and Ø8-80 for shear reinforcement. For more details can be seen in Figure 2.

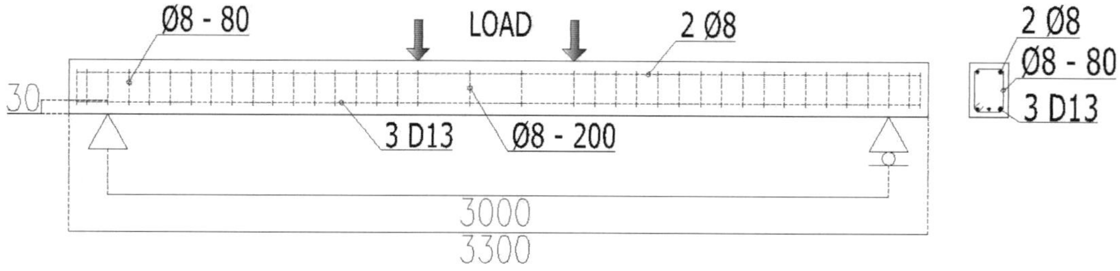

Figure 2. Test Beam Details

3.2. Material Properties

3.2.1. FRP composite

Mechanical properties of FRP composite used are as shown in Table 1.

Table 1. Mechanical Properties of FRP Composite. [6,7]

Properties	CFRP	GFRP	Epoxy
Tensile strength (MPa)	986	575	72.4
Modulus of Elasticity (GPa)	95.8	26.1	1.45
Thickness (mm)	1.0	1.3	
Elongation (%)	1.0	2.2	5

[a] Manufacturer's data

3.2.2. Concrete and Steel

Based on the compression test on 6 concrete cylinder samples, the average compression strength of concrete used in this experiment is 20 MPa. For tensile, shear, and compression reinforcements, yield strength of the rebar used is 315 MPa.

3.3. Beams Strengthening and Instrumentation

After curing process for 28 days, the test beam will be strengthened with hybrid FRP and allowed to dry for 3 days. Installation of hybrid FRP is done by applying a layer of epoxy resin to the concrete surface. In a separate place, GFRP is smeared with epoxy resin and then affixed to the surface of the concrete that has been applied with resin. The same steps are carried out for CFRP, then applied to GFRP. Detailed measurement can be seen in Figure 3 and finished strengthened beam in Figure 4.

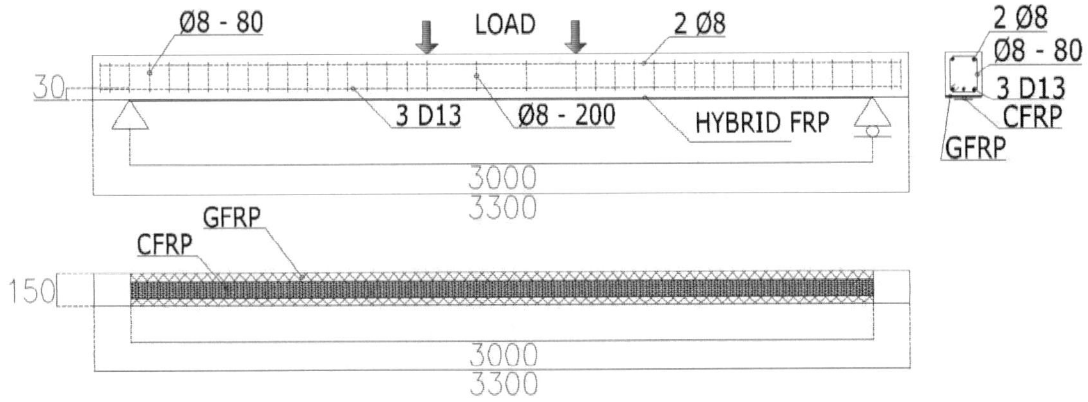

Figure 3. Hybrid FRP strengthened beam design

Figure 4. Hybrid FRP strengthened beam

All beams are subjected to static monotonic point load using displacement control with increments of 0.05mm/s. Two-point load each applied 300 mm from mid-span. Deflection was measured using an LVDT below the mid-span and two more under the point load. Details can be seen in Figure 5.

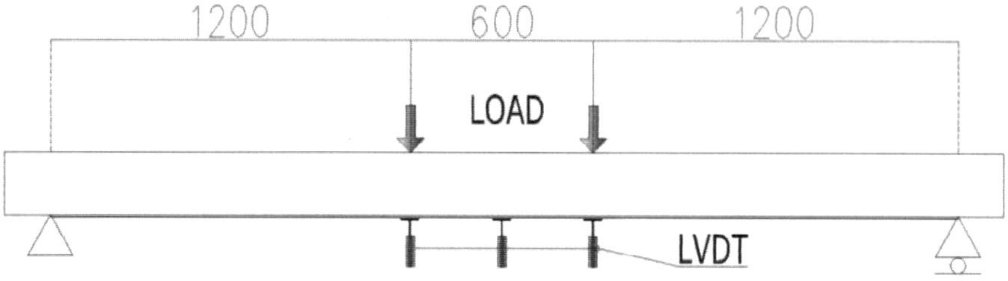

Figure 5. Test beam setting

4. Result and Discussion

4.1. Behavior of Hybrid FRP

The code for every variation used in this test is as mentioned in 3.1. Based on the test result, GC50 has the highest ultimate load at 27.83 kN, followed by GC40 with 24.86 kN, and the last GC30 with 21.91 kN. While it is the opposite in strain, with GC50 being the lowest at 0.00452197, GC40 with 0.00627847, and GC30 with 0.00656938. However for GC50 while having the highest ultimate load doesn't show the sequential failure behavior, both GFRP and CFRP fail at the same time. For GC30 and GC40, CFRP fails first before GFRP creating the sequential failure behavior. This behavior can be observed in Figure 6, where GC30 and GC40 had a decrease in load nearing its ultimate load. It is caused by the CFRP already failing beforehand, leaving GFRP to carry the load.

Therefore GC40 will be used for beam strengthening due to it is higher in ultimate load than GC30 while still showing the sequential behavior.

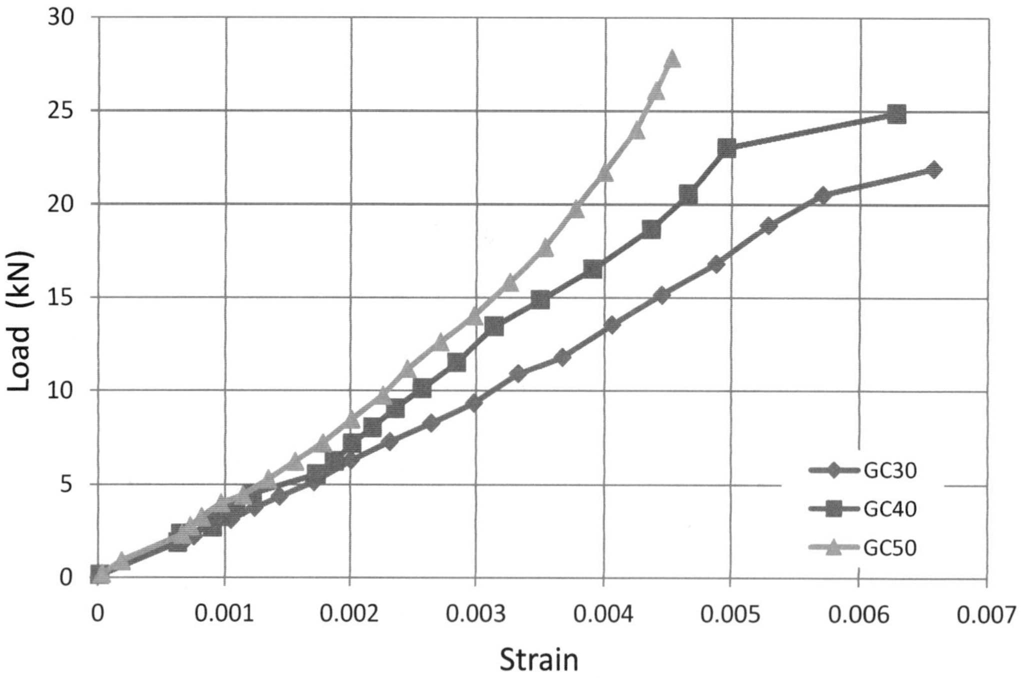

Figure 6. Hybrid FRP Tensile Strength Test Result

4.2. Beam Flexural Test

The control beam (CB) is observed to have a yield load of 25.32 kN and an ultimate load of 29.14 kN. The steel elastic phase ended at a displacement of 16.40 mm. The hybrid FRP strengthened beam (BGC) had a yield load of 35.45 kN and an ultimate load of 43.60 kN. When compared, the test beam had a 40.0% increase in yield strength and a 55.43% ultimate load increase. Beam behavior could be observed in Figure 7. The load on the specimen BGC drops from 42.51 to 34.98 kN due to delamination caused by a lack of bonding strength between the hybrid FRP and the beam surface when the displacement of the BGC at 38.7 mm leaving the rebar to carry the load. The test stopped soon after.

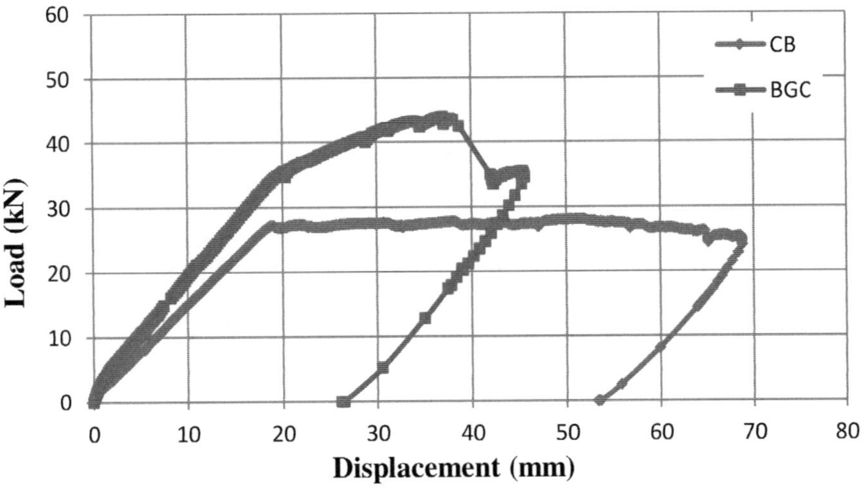

Figure 7. Beams Flexural Test Result

4.2.1. Failure mode

Though the BGC shows an increase in capacity compared to CB, the sequential failure behavior of the Hybrid FRP didn't occur due to delamination caused by a lack of bonding strength between the hybrid FRP and the beam surface. Delamination started from a crack under the point load.

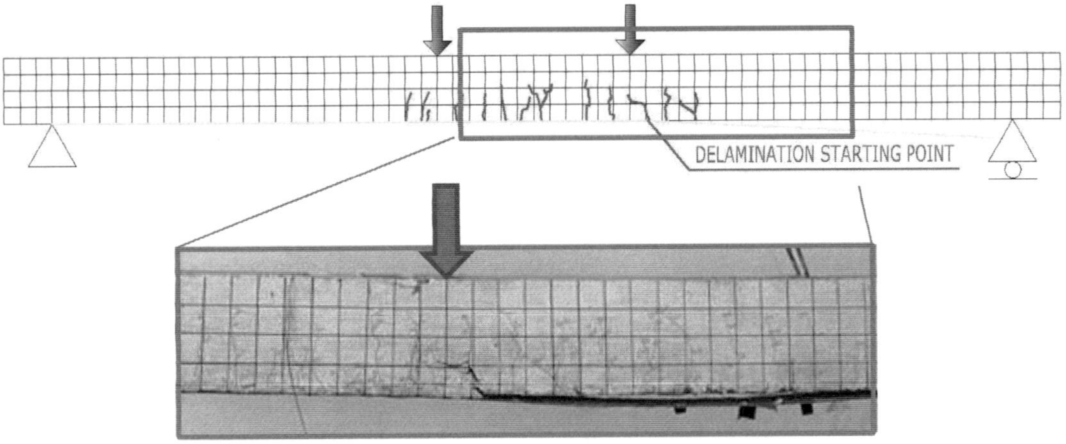

Figure 8. BGC post flexural test

Through observation in Figure 8, the delamination is located directly below the point load to the end of the beam.

5. Conclusion

From this experiment, it can be concluded as below:

- From 3 variation of hybrid FRP, only GC30 and GC40 shown the sequential failure behaviour, where GC40 had a higher load capacity compared to GC30
- By strengthening a beam using hybrid FRP (GFRP 100%, CFRP 40%), the yield strength increased by 40.0% and ultimate strength by 55.43%
- Failure mode of BGC (hybrid FRP strengthened beam) is delamination, caused by a lack of bonding strength between the hybrid FRP and the beam surface

References

[1] Sultan, M. A. (2020). Ductility and Flexural Capacity of GFRP-S Strengthened Reinforced Concrete
[2] Husein, S. (2020). Failure Mode Analysis of Concrete Beams using GFRP Rebar without Concrete Cover
[3] Khorramian K. (2021). Hybrid System of Longitudinal CFRP Laminates and GFRP Wraps for Strengthening of Existing Circular Concrete Columns
[4] Szmigiera E. D. (2019). Mechanical Properties of Hybrid FRP Bars and Nano-Hybrid FRP Bars.
[5] ASTM D3039, Standard Test Method for Tensile Properties of Polymer Matrix Composite Materials 2017.
[6] Fyfe. Co. LLC (2012), Characteristic CFRP (Carbon Fiber Reinforced Polymer)
[7] Fyfe. Co. LLC (2012), Characteristic GFRP (Glass Fiber Reinforced Polymer) type SHE-51A

Study of Mineralogy and Engineering Properties of Volcanic Rock at the Base of the Pamukkulu DAM Foundation, South Sulawesi, Indonesia

Hendra Pachri[1], Ratna Husain[2], Adi Tonggiroh[3], Muhammad Ichwanto[4] and Regina Glori Samanta[5]

[1,2,3,5] Department of Geological Engineering, Faculty of Engineering, Hasanuddin University, Gowa, Indonesia, 92171
[4] PT. Wijaya Karya (Persero), Jakarta 13340

E-mail: hendrapachri@unhas ac.id

Abstract. At the location of the Pamukkulu dam, it has engineering geological characteristics that are less supportive of the construction above it, especially for the foundation of the dam. By taking data directly in the field and by laboratory testing, information is needed regarding the characterization of mineralogy and the engineering properties of rocks. Mineralogical analysis was performed with petrography and X-Ray Diffraction (XRD) while the engineering properties of the rock were performed with Plate Bearing Tests (PBT), Standard Penetration Tests (SPT), and Rock Mass Rating (RMR). Based on mineralogical data, rocks have undergone weathering and converted the chemical-physical properties of minerals into porous, fine-grained minerals, forming mineral oxides as well as clay minerals. The carrying capacity of surface soils in the dam area based on the plate bearing test is relatively poor. As for subsurface rocks, the weight of the rock mass rating belongs to the class (II) Good to (III) Medium. Meanwhile, from the density value of hard soils, there are different depths from a depth of 1 meter to 6 meters. Weathering conditions of rocks affect the engineering value of soils and rocks where due to weathering minerals formed in the form of porous clay minerals, they are not well consolidated.

1. Introduction

In order to improve the welfare of the people in Indonesia, especially in South Sulawesi, the government has made various efforts to provide various kinds of infrastructure. Currently, the acceleration of agricultural irrigation infrastructure work that is considered strategic and has a high urgency is carried out to be realized in a short period of time. One of the infrastructures presented to the people of South Sulawesi is the Pamukkulu Dam, located in Takalar Regency, South Sulawesi Province.

Dam is structure built across a stream, a river, or an estuary to retain water. As function, dams are built to provide water for human consumption, for irrigating arid and semiarid lands, or for use in industrial processes. With age, the dam will experience a decrease in quality both in terms of physique and function. For this reason, the design and manufacturing processes in planning must be taken into account properly to avoid the risk of failure or sub optimal function of the dam after the construction is completed.

The role of geological science for engineering purposes is very important to ensure the influence of geological factors on the location, design, construction, implementation of development (operation), and maintenance of the results of engineering work or engineering works. In engineering construction, geological analysis is required for construction planning [1].

The geological engineering investigations carried out in the construction of dams are to determine the geological state, including the distribution of soil and rocks, geological structures, natural processes that occur, the arrangement of rock layers (stratigraphy), geological structures, and physical properties of rocks for the purposes of civil engineering planning [3].

The foundation of a dam in the form of compact and massive rocks can be considered a good foundation (ideal), but these conditions are rarely encountered [3]. In fact, these foundation rock layers often contain a number of fracturings, cracks, joint, weathered faults and discontinuities with other rocks, so geological investigations are needed to study the physical properties, techniques, and passage of water in the Baturappe-Cindako Volcanic Formation (Tpbv), composed of lava lithologies and volcanic breccia and basalt rocks [4].

The purpose of this study is to know the characteristics of rock mineralogy and the engineering properties of rocks for the base of dam foundations. Therefore, this research focuses on geological and geotechnical investigations of the main dam area at several points and laboratory testing of samples taken in the field. Furthermore, the benefit of this research in general for the wider community is that it serves as a reference for research on the study of engineering geology and gets a clearer picture of the geological conditions of subsurface engineering based on the main dam structure planned to be built as well as various geological engineering methods carried out in dam construction activities.

2. Geology Setting

Morphologically, the research area has two eroded cones which are more narrowly distributed in the western and northern parts of Mount Lompobatang. To the west is Mount Baturape, reaching a height of 1,124 m and to the north is Mount Cindako, reaching a height of 1500 m. These two eroded cone shapes are composed of Pliocene volcanic bottoms [4]. The area west of Mount Cindako and north of Mount Baturape is a rugged hilly area in the east and smooth in the west. The eastern part reaches a height of about 500 m, while the western part is less than 50 m above sea level and is almost a plain. This morphological form is composed of volcanic clastic rocks of Miocene age. The elongated hills scattered in this area lead to Mount Cindako and Mount Baturape in the form of basalt cracks.

The stratigraphy of the study area is composed of Baturape-Cindako Volcanic Rocks consisting of lava and volcanic breccias, with a few inserts of tuff and conglomerate [4]. Basal lava shows characteristics in the form of porphyry with pyroxene phenocrysts measuring more than 1 cm, dark gray green to black in color; the characteristics of the lava show partly joint and layered. Volcanic breccia rocks show fragments and matrices measuring 15 cm to 60 cm, mainly basalt and slightly andesite, with coarse-grained tuff cement to lapilli, rich in pyroxene minerals [4].

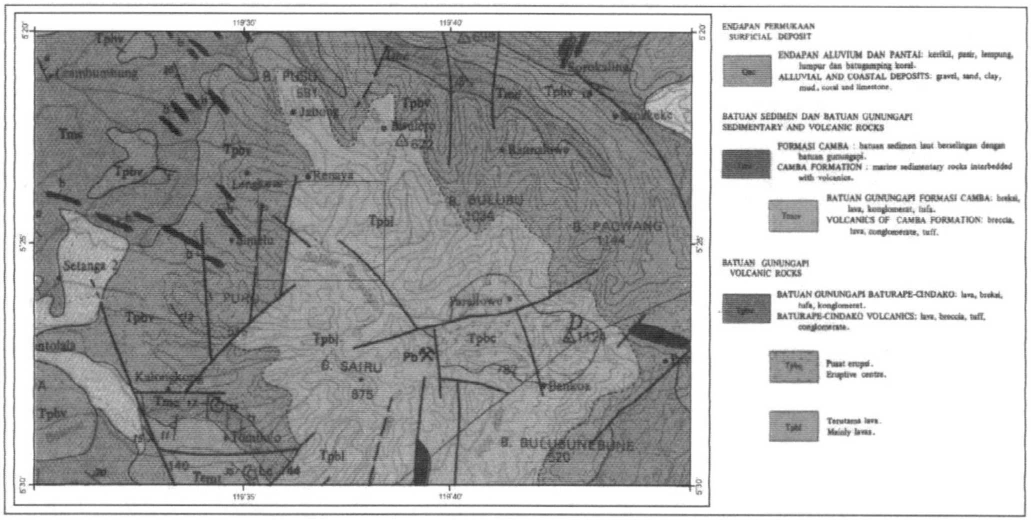

Figure 1 Map shows the Regional Geological of the Dam Area (Sukamto, R & Supriatna, S.1982)

Diorite complexes in the form of stocks and hacks at Baturape and Cindako are thought to have been former eruption centers (Tpbc); the surrounding rock is changed to be strong, amygdaloidal with secondary minerals zeolite and calcite. The galena mineral in Baturape is probably related to this diorite, and around the Baturape and Cindako areas the rocks are dominated by lava (Tpbl). This rock unit has a thickness of more than 1,250 m and is estimated to be of Late Pliocene age [4].

3. Data and Methods

The research method carried out to determine the mineralogy and engineering properties of rocks consists of taking surface and subsurface data. The method carried out is aimed at obtaining surface and subsurface data. Surface data is taken to determine the mineralogy and carrying capacity of surface rocks that are still in the form of weathered soil for the dam foundation. The subsurface data is known to find out the characteristics of the rocks for the base of dam foundation.

3.1 Surface Data

This study used surface data directly in the field, in the form of:

3.1.1. Geological Mapping. This geological mapping is carried out to obtain surface geological information on the dam area. In geological mapping, the data taken is in the form of geomorphological information, petrology, and geological structures. The result of this mapping is in the form of a geological map of the dam area. This data is data that is observed directly in the field.

3.1.2. Plate Bearing Test (PBT). PBT is a test method used to obtain soil carrying capacity values in the basic or subgrade soil layer simply by applying pressure to the soil [5].

The test was carried out using a circular bearing plate, which was then placed on a bearing plate that was pressed at a certain interval of time, and then the plate bearing test continued until the most maximum loading or the most maximum decrease.

The sampling on the surface is carried out to test the mineralogy of rocks by the method of:

3.1.2.1 Petrography. Petrography is carried out to optimize geological interpretation in the form of knowing the texture and structure of rock minerals, rock mineral composition, types, and names of rocks. This analysis was carried out by observations of thin incisions of rocks under a polarizing microscope to observe differences in the optical properties of minerals. Then it is determined the naming of rocks is based on the classification of rocks.

3.1.2.2 *X-Ray Diffraction (XRD).* XRD analysis is a method that can provide information about the types of minerals contained in a material. From the results of XRD, the mineral content of clay in the soil material in the dam area will be analyzed to know the weathering processes.

3.2. Subsurface Data

The test method of determining the boundaries between the soil layer and the rock layer using rock coring data, lithological determination, and Rock Mass Rating (RMR) and soil carrying capacity is used to collect subsurface data. Technical methods are as follows:

3.2.1. Core Drilling. Core drilling testing aims at further identification, including the determination of rock classes according to CRIEPI (1992) [6], mineral composition identification, and determination of layer thickness.

3.2.2. Standard Penetration Test (SPT). This aim SPT tool is to determine the dynamic resistance of the soil to the growth technique. This test is a dynamic experiment carried out in a borehole by inserting a sample tube using a pusher (hammer) period weighing 63.5 kg, which falls freely from a height of 760 mm. The number of blows of the hammer to insert the sample tube 305 deep is expressed as the value of N. The implementation is carried out in three stages, of which the first stage is a temporary stand. The number of strokes for entering the second and third stages is summed up to produce the value of the blow N, or the resistance of the tax return expressed in strokes per 30 cm.

3.2.3. Rock Mass Rating (RMR). RMR classification is a geomechanical classification that uses an empirical method to determine the weighting of rock masses to evaluate the resistance of rock masses by focusing on six parameters: rock compressive strength, Rock Quality Designation (RQD), joint space, joint conditions, groundwater conditions, and discontinuity orientation. The main reason for using RMR is its ease and flexibility for a variety of practical purposes in engineering [7].

4. Analysis and Discussion

4.1. Geology of the Surface

The grouping of geomorphological units in the study area is based on morphological and morphogenesic approaches. The morphographic approach, which is an approach based on the shape of the earth's surface found in the field, namely in the form of hilly topography, and this aspect needs to pay attention to the parameters of each topography, such as elevations, slope shapes, and valley shapes encountered in the field. Furthermore, geomorphology by morphogenesic approach are classified by the processes of formation and those affecting the development of landforms.

Based on the basis for naming the type of landscape unit, the study area is owned by one landscape unit, namely the Denudasional Hills landscape unit (Figure 2).

Figure 2. Morphological condition shows the denudational hill geomorphological unit of the Dam Area

The result of geological mapping is the describes of outcrops with physical characteristics with a fresh color of gray-black and a brownish-gray weathered color; the texture of hypocrystalline crystallinity; aphanitic granularity; subhedral-anhedral shape; inequigranular relations; and massive structure. Based on its physical characteristics, the name of this rock is *Basalt* (Travis, 1955).

Figure 3. Basalt Outcrops at As Dam area

Furthermore, it is found that the volcanic breccia outcrop megascopically has physical characteristics with a fresh gray color and a brownish-gray weathered color, which indicates a massif structure. It has basal fragments measuring 8–256 mm, with a matrix measuring 4–64 mm, and cement measuring 1/8–1/256 mm.

Figure 4. Volcanic Breccia Outcrop at As Dam area

4.2 Rock Mineralogy

Based on their physical appearance, the rocks of the dam area have undergone weathering. Weathering affects the carrying capacity of the rock for the foundation of the dam later. The minerals contained in these rocks can be seen microscopically using petrography analysis and also using XRD analysis.

There were four petrography samples tested from the rocks constituting the dam in the surface data. The four rocks based on petrography analysis are alkaline volcanic igneous rocks, namely Porphyry Basalt (ST-250 Basalt, ST-260 Basalt, and ST-260 Volcanic Breccia) and Amygdaloidal Basalt (ST-250 Breccia). Porphyritic textured rocks are composed of the minerals clinopyroxene (28-40%), plagioclase (24-24%), hornblende (10%), orthoclase (5-7%), opaque minerals (1-3%), and the base mass of volcanic glass (10–29%). Based on the formation of secondary minerals and individual physico-chemical changes of mineral crystals, rocks have undergone further processes due to deformation in the form of geological structures (faults) and interactions with the groundwater layer. Deformation processes that can form fracture or burial structures, identified by secondary mineral types, veins and fractures in crystals filled or unfilled by secondary minerals (ST-250 Basalt, ST-260 Basalt).

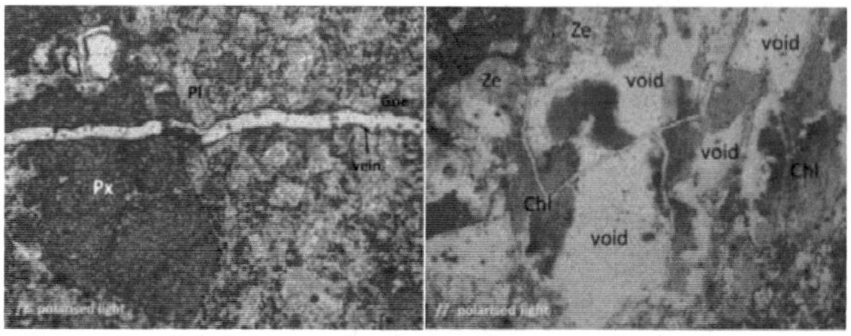

Figure 5. Photomicrographs showing secondary mineral types, vein and fractures on ST-250 and ST-260 Basalt samples

The secondary mineral zeolite alters the volcanic glass of the analcime type (ST.250 Basalt) and, at greater depths, forms the type of wairakite along with the mineral chlorite (ST-260 Basalt). The formation of these minerals is interpreted to be due to the presence of geological structures that cause low-level metamorphism with a temperature of 100–200^{0}C and a pressure of 0.2–0.6GPa. This zeolite mineral is a group of aluminum silicate whose mineral structure is similar to that of mineral clay, namely the porous structure, where in the layer the structure can accommodate H_2O [8]. The content of the zeolite mineral group in rocks between 10.1 and 20.9% can be at risk of landslides.

In the ST-260 Breccia sample, there were indications of rocks being intruded by younger rocks. Rocks close to intrusion are dark in color due to the backing effect. In addition, a metal mineral (oxide) is formed in the form of a subhedral measuring up to 0.15 mm.

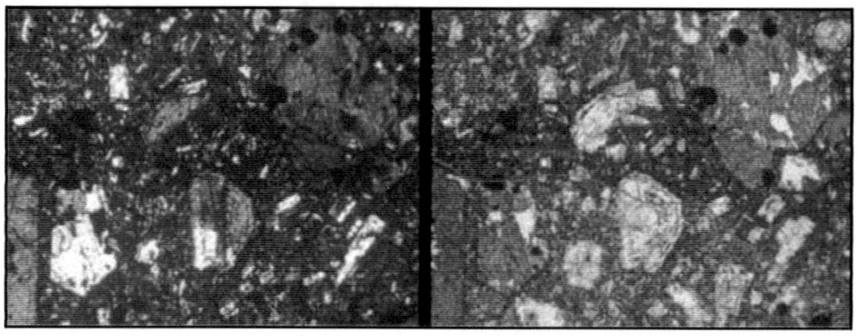

Figure 6. Photomicrograph showing rock close to intrusion are dark in color due to the backing effect on a ST-260 Breccia sample.

X-Ray diffraction analysis (XRD) has been conducted by a nondestructive technique that provides detailed information about the crystallographic structure, chemical composition, and physical properties of Basalt dan Volcanic Breccia rock. It is based on the constructive interference of monochromatic X-rays and a crystalline sample. Therefore, X-rays are shorter wavelength electromagnetic radiation that are generated when electrically charged particles with sufficient energy are decelerated.

The results of the XRD test were found in ST-250 which Basalt samples found clinopyroxene 68.4%, analcime 20.9%, cristobalite 9.8%, and montmorillonite 0.9%, which are clay mineral groups (Figure 7).

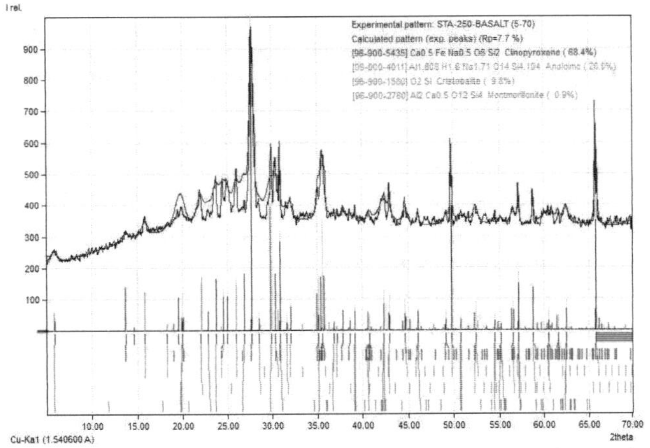

Figure 7. Graph showing XRD test of STA-250 Basalt Sample

The results of the XRD test on the STA-250 Breccia sample found feldspar 58.8%, caolinite 22%, cristobalite 9.8%, and montmotillonite 0.9%, which are clay mineral groups (Figure 8).

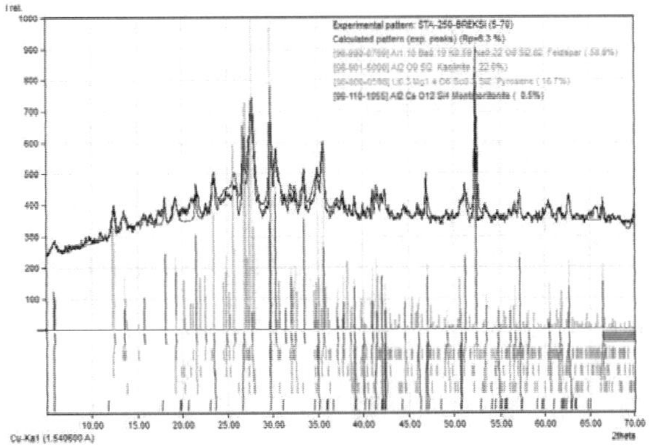

Figure 8. Graph showing XRD test of STA-250 Breccia

The results of the XRD test on the STA-260 Basalt sample found diopside 36.7%, feldspar 35.6%, chlorite 17.6%, and wairakite 10.1%, which are clay mineral groups (Figure 9).

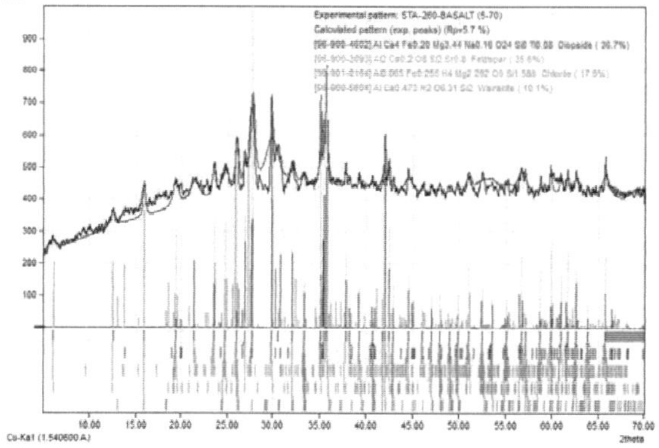

Figure 9. Graph showing XRD test of STA-260 Basalt Sample

The results of the XRD test on the STA-250 Breccia sample found Feldspar 58.8%, Caolinite 22%, Cristobalite 9.8%, and Montmorillonite 0.9%, which are clay mineral groups (Figure 10).

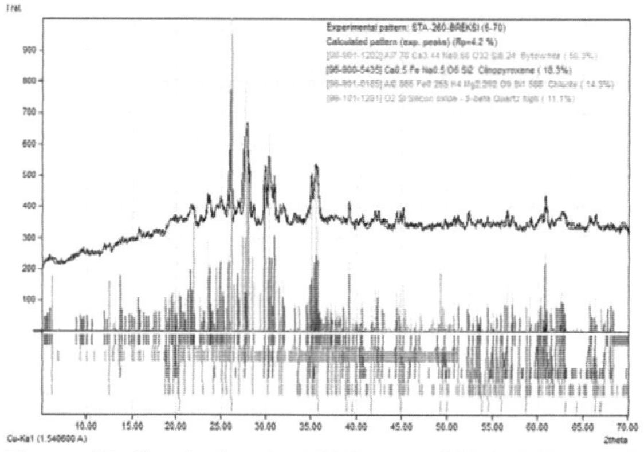

Figure 10. Graph showing XRD test of STA-260 Breccia

4.3. Carrying Capacity of the Soil

The carrying capacity of the soil in the dam area is obtained from the results of the Plate Bearing Test (PBT). The Plate Bearing Test was carried out in the field at as many as seven points in the main dam area (Table 1.). The value of the decrease and the value of the ultimate bearing capacity are known from these seven points. Ultimate bearing capacity is the minimum pressure that causes shear failure of the soil rapidly downwards.

The results of the plate bearing test showed that the value of the ultimate bearing capacity and its decrease at each point of investigation varied relatively. The results of the plate bearing test are presented in the table as follows.

Table 1. PBT test results

No	Test Pit	Settlement (cm)	Ultimate Bearing Capacity (kg/cm^2)	Description
1	PB.4	0.10	3.7	Poor
2	PB.6	1.30	11.3	Poor
3	PB.7	0.40	15.0	Poor
4	PB.8	0.20	11.4	Poor
5	PB.3	1.00	15.0	Poor
6	PB.10	1.35	12.2	Poor
7	PB.2	1.50	7.9	Poor

The results of the plate bearing test at the seven points where the soil has a limit carrying capacity are achieved when the decrease ranges from 0.1 to 1.5 cm with pressures ranging from 3.7 kg / cm2 to 15 kg / cm2.

From the results of the plate bearing test (Table 1), it shows that the carrying capacity of the soil at all seven points has a value of 40 and is classified as poor based on the classification of bearing capacity (Yavuz and Atilla 2013) [9]. Poor soil carrying capacity is very much the foundation of the dam. If the soil carrying capacity is **poor**, then an excessive decrease or collapse of the soil can occur when pressured from the dam foundation.

4.4 Relationship of Mineralogy and Soil Carrying Capacity

Based on the results of soil carrying capacity tests that show poor carrying capacity, the condition of the rocks in the dam area has undergone strong weathering. All rock samples based on rock mineralogy have undergone surface weathering of up to 40%, which changes the physico-chemical properties of minerals to porous, fine-grained, formed mineral oxides and clay minerals. The openings or voids contained in almost all samples are caused by the release of mafic minerals due to an imbalance in their chemical content. The openings are partly filled with clay minerals and some are empty spaces that are very vulnerable to being filled by rainwater or water in the soil layer. Clay minerals formed in the form of montmorillonite and kaolinite are minerals that are not well consolidated and can accommodate water in their crystal structure, so the presence of these minerals can potentially cause a potential instability in the dam foundation.

4.5 Core Drilling

Based on the surface data of the dam area in the form of rock distribution, rock mineralogy, and soil carrying capacity, it is felt necessary to know subsurface conditions. Surface data indicates less favorable conditions for the foundation of the dam. Therefore, subsurface investigations are necessary to find out the carrying capacity of subsurface rocks against the foundation of the dam.

The method of retrieving subsurface data uses core drilling. This core drilling was carried out at eight points, including BW-10, BW-11, PL-21, PL-25, PL-26, PL-46, PL-52, and PL-60. The spread of these drill points can be seen on the map of Figure 2. Core drilling is carried out in order to obtain information on the characteristics of rocks, types of lithologies, rock classes, and depth of subsurface rocks.

The results of the core drilling can be seen in the drill log in **figure 11**. It is known that the subsurface rock layers of the dam area are predominantly composed of volcanic breccia. There is also basalt that appears locally. Rock classes range from very soft rocks (*D Class*) to fairly dense rocks (*CH Class*). D Class rock ranges from the surface to a depth of between 1.7-8 meters, while the CH class rock dominates the subsurface area of the axle dam to the limit of drilling depth.

The characteristics of each class of rock found from the results of core drilling include:

- A fairly dense rock with fresh to lightly weathered characteristics and a high hardness (CH Class).
- A soft rock with light to medium weathered characteristics and a medium hardness (CM Class).
- A soft rock with the characteristics of a rock that has been subjected to severe weathering (CL Class).
- A very soft rock with very weathered rock characteristics and no visible original rock texture (D Class).

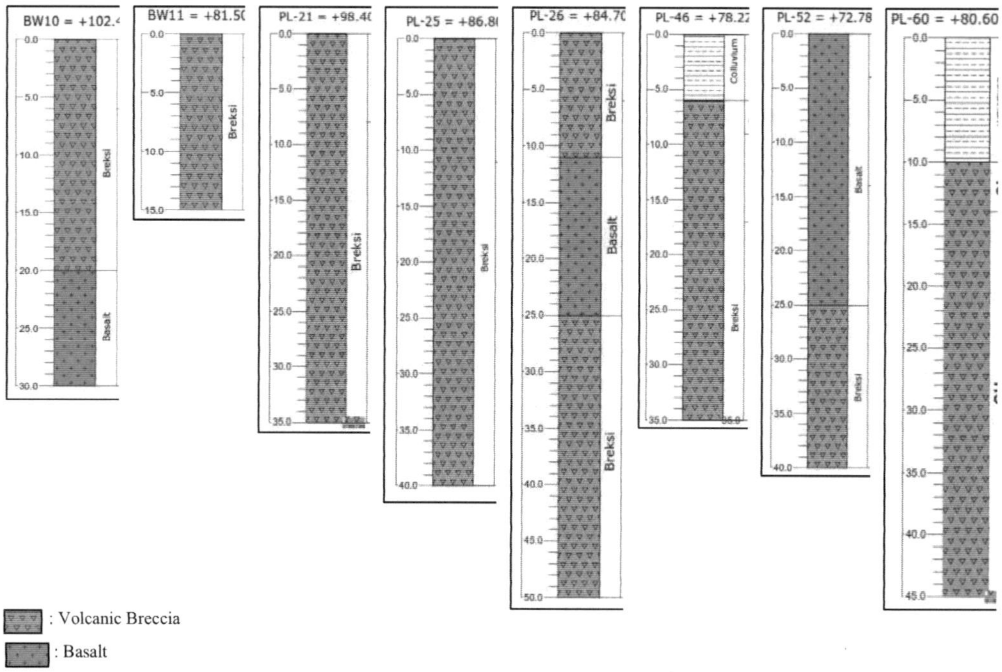

Figure 11. Core Drilling Log on Main DAM area

4.6 Rock Mass Rating (RMR)

The results of the core drilling work are also carried out and quality determination is based on the RMR classification. The weight of the RMR assessment is based on five parameters, namely whole rock strength, RQD, discontinuity spacing, discontinuity conditions, and groundwater discharge (Bieniawski, 1989). The weighting results of each RMR parameter are presented in the following graph.

On the weighting of the strength value of the rock, it is based on the value of the uniaxial compressive strength of the rock. The compressive strength value of rocks in basalt lithology is about 246 MPa, while the compressive strength value of volcanic breccia rocks is about 260 MPa. Based on this data, at all eight points the drill has a rock strength value of more than 250 and is given a weight of 15.

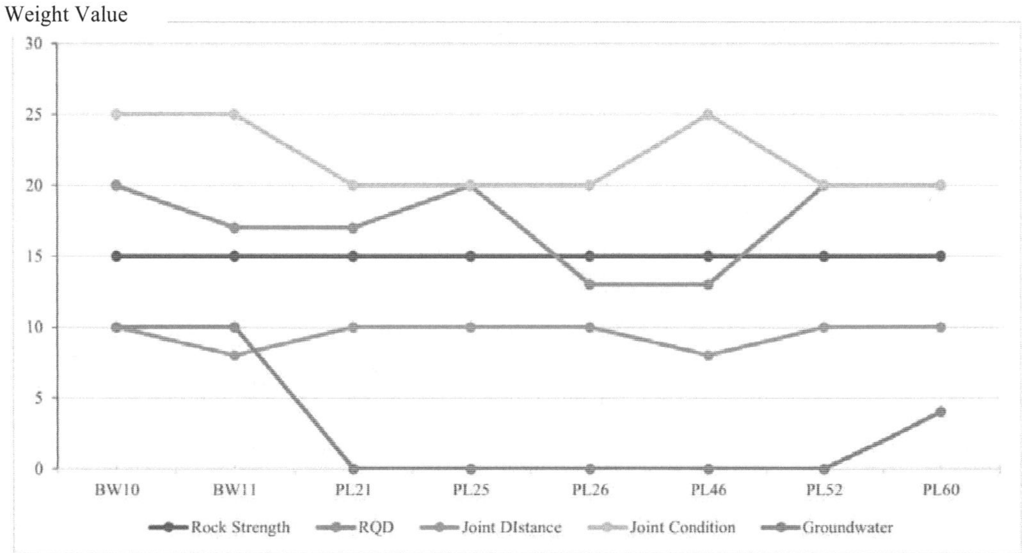

Figure 12. The Graph show the result of weighting of RMR parameters at each drill point.

In RQD weighting, it is obtained directly from measurements on core drilling samples. The average RQD value at each drill point is about 59 to 95. Based on such data, at all eight points the drill has a weight varying from 13 to 20. In the weighting of the joint distance, the value of the distance between the stocks is obtained from the perpendicular distance between two consecutive discontinuities along the measurement line. The average subsurface rock spacing distance at each drill point is about 10 cm to 60 cm. Based on such data, at all eight points the drill has a weight from 8 to 10.

In joint condition weighting, the value of the joint condition is obtained from the joint persistence, joint openings, joint roughness, stuffing and weathering rate. Each of these parameters is then added up and a general joint condition weight is obtained. Based on observations of joint conditions, at all eight points the drill has a weight from 8 to 10. In weighting the condition of groundwater, the value of the condition of groundwater is obtained by paying attention to the general condition with the category of dry, humid, wet, dripping, and flowing. The condition of the groundwater found at each drill point includes flowing, dripping, and moist. Based on observations of groundwater conditions, at all eight points the drill has a weight from 0 to 10.

Based on the weighting of RMR parameters, RMR weights were obtained at the bobor points of BW-10, BW-11, PL-21, PL-25, PL-26, PL-46, PL-52, and PL-60, respectively, namely 80, 75, 62, 65, 58, 61, 65, and 69. The rock masa class based on RMR values in subsurface rocks belongs to class (II) Good except at drill points PL-26, including class (III) Medium.

4.7 Soil Density

One of the parameters of the carrying capacity of a soil is the density of the soil. Soil density is obtained from the Standard Penetration Test (SPT) in the field. The results of the SPT test were subsequently correlated to the relative density of clay where the soil boundary was hard or had a tax return of more than 30 [10]. To get more detailed data with a wide spread, SPT was carried out in the main dam area with a total of 20 test points. Based on the results of the SPT test, each drill point gets a SPT value of more than 30 or hard soil at different depths, from a depth of 1 meter to 6 meters. These results become a reference for design changes and subsequent excavation work.

Table 2. SPT test results

No.	Nama	Kedalaman (m)	Nilai SPT (N)	No.	Nama	Kedalaman (m)	Nilai SPT (N)
1	SPT-1	1	30	12	SPT-12	1	33
		2	53			2	19
		2.5	70			3	54
2	SPT-2	1	30			3,5	69
		1.5	87	13	SPT-13	1	35
3	SPT-3	1	13			2	32
		1,5	14			3	40
		2	18			3,5	56
		2,5	17	14	SPT-14	1	29
		3	14			2	35
		3,5	10			3	61
		4	4			3,5	77
		4,5	7	15	SPT-15	1	32
		5,5	22			2	50
		6	18			2,5	69
		6,5	48	16	SPT-16	1	30
		7	69			2	65
4	SPT-4	1	14			2,5	80
		2	26	17	SPT-17	1	35
		3	40			2	55
		3,5	68			2,5	69
5	SPT-5	30	1	18	SPT-18	1	28
		34	2			2	35
		66	3			3	40
		76	3,5			4,5	70
6	SPT-6	59	1			5	79
		86	1,5	19	SPT-19	1	27
7	SPT-7	1	33			2	28
		2	53			3	34
		2,5	48			4	36
8	SPT-8	1	48			4,5	54
		2	61			4.5	67
		2,5	82	20	SPT-20	1	37
9	SPT-9	1	15			2	30
		2	68			3	25
		2,5	87			4	22
10	SPT-10	1	69			5	28
11	SPT-11	1	34			6	39
		2	66			6.5	69

4.8 Rock Hardness

The investigation of the quality of the rocks was carried out on the basis of RMR and SPT. Based on the SPT value, it can be seen that at some point, the depth of soft soil (SPT< 30) ranges from 1 to 3 meters. This condition is the result of weathering of the rocks below.

From the results of the tax return that shows the quality of the soil, the results of soft weathering can be attributed to the RMR of the rock. The RMR value of rocks generally has a good value, but based on weighting, the weight of groundwater is low because it ranges from flowing to wet. These conditions are based on field conditions where the groundwater level is shallow so that it accelerates weathering at the bottom.

Weathering conditions that affect the value of SPT and RMR are caused by the mineralogy of rocks that have changed as a result of weathering. Weathered minerals form in the form of porous clay minerals. The mineral is not well consolidated, so it can form less stable areas.

5. Conclusion

The rocks in the dam area are porphyritic textured volcanic rocks, composed of the minerals clinopyroxene, plagioclase, hornblende, orthoclase, and opaque minerals, and the base mass of glass volcanic. From mineralogical analysis, rocks have undergone weathering and transformed the chemical-physical properties of minerals into porous, fine-grained minerals, forming mineral oxides as well as clay minerals.

The surface soil carrying capacity in the dam area based on the plate bearing test at seven points has a value of 40 and is classified as poor. As for the engineering properties of rocks in subsurface, the weight of the rock mass rating belongs to the class (II) Good to (III) Medium. While the density value of hard soils is different at different depths, from a depth of 1 meter to 6 meters. Weathering conditions of rocks affect the engineering value of rocks where the result of weathering minerals is formed in the form of clay minerals that are porous and not well consolidated.

Based on the results of core drilling, SPT, and PBT, weak areas were still found in the dam area caused by fairly high weathering in the rocks. This condition is feared to be a weak point that affects the stability of the dam. More detailed analysis is needed to find out the desired depth limit of hard rock.

References

[1] Attewell, P. & Farmer, I.. (1976). Principles of Engineering Geology. 10.1007/978-94-009-5707-7.

[2] Sani, Asrul. 2008. Analisis Kapasitas Waduk dengan Metode Ripple dan Behaviour (Studi Kasus Pada Waduk Mamak Sumbawa). Yogyakarta: Universitas Muhammadiyah Yogyakarta.

[3] Direktorat Jenderal Sumber Daya Air. 2011. Pedoman Desain dan Konstruksi Bendungan Urugan Batu Membran Beton, Kementrian Pekerjaan Umum.. Jakarta

[4] Sukamto, R & Supriatna, S.1982. Geologi Lembar Ujung Pandang, Benteng dan Sinjai, Pusat Penelitian dan Pengembangan Geologi, Direktorat Geologi dan Sumber Daya Mineral, Departemen Pertambangan dan Energi. Bandung

[5] Kwandy, Reyner, Gregorius Sandjaja Sentosa and Ali Iskandar. "Prediksi Penurunan Gedung dan Daya Dukung Menggunakan Hasil Uji Plate Bearing di Labuan Bajo, NTT." Jurnal Mitra Teknik Sipil 2.2 (2019): 75-84.

[6] JSEG. 1992. Rock Mass Classification of the Central Research Insitute of Electric Power Industry (CRIEPI), In Japan Society of Engineering Geology. Chiba, Japan

[7] Bieniawski, Z.T. 1976. Rock Mass Classification in Rock Engineering. In Bieniawski, Z.T., Ed., Symposium Proceedings of Exploration for Rock Engineering, 1, 97-106. Pennsylvania State University.

[8] Lestari, D,Y., 2010, Kajian Modifikasi dan Karakterisasi Zeolit Alam dari Berbagai Negara, Prosiding, UNY

[9] Gül, Yavuz & Ceylanoğlu, Atilla. (2013). Evaluation of plate loading tests on some rock formations for assessing the ground bearing capacity. Bulletin of Engineering Geology and the Environment. 72. 10.1007/s10064-013-0462-y.

[10] Terzaghi, K., dan Peck, R., 1943. Theoritical Soil Mechanic. John Willey & Sons, New York

IOP Conf. Series: Earth and Environmental Science 1117 (2022) 012005 doi:10.1088/1755-1315/1117/1/012005

Marshall Tests For Asphalt Concrete Wearing Course Asb Lawele Containing Capsule Calsium Alginate as Self-Healing Additive

M.S. Prasetia[1 2], Ludfi Djakfar[2], Wisnumurti [2] and Akhmad Sabarudin[3]

[1] Department of Civil Engineering, Faculty of Engineering, Halu Oleo University, Kendari, Indonesia
[2] Department of Civil Engineering, Faculty of Engineering, Brawijaya University, Malang Indonesia
[3] Department of Chemistry, Faculty of Science, Brawijaya University, Malang, Indonesia

Email: prasetya_syarif@uho.ac.id

Abstract. Self-healing on pavement is solution to reduce the handling of road damage without doing heavy construction. Calcium Alginate capsules are mixed in AC-WC (Asphalt Concrete Waring Coarse) Asb Lawele as an added ingredient to healing from cracks. Marshall Test specification is reference to assess the pavement can use pavement Asphalt Concrete. Optimum Asphalt Content of 6.2% is used to make AC-WC Modification Mix Asb Lawele with Calcium Alginate. AC-WC Asb Lawele contains 2% Asphalt from Buton Island and 4.2% Asphalt Pen.60/70. The percentage of Additive Addition of Calcium Alginate Capsules ranging from 1% to 2% of the combined weight with a different oil/water composition of each calcium alginate capsule. There is similarity of IR spectrum between asphalt binder and Calcium Alginate capsules, allowing strong bonding if reviewed from Nitrogen group. However, with the addition of calcium alginate capsules showed a decrease in Marshall Stability but still within the limits of fairness, it was proven that the higher the percentage of additive capsules the lower the marshall stability, but didn't have a significant effect on flow, while the VMA and VIM continued to increase as the percentage of Additives increased so that VIM out of the specifications on AC-WC Asb Lawele.

1. Introduction

Asphalt Concrete (AC) has become a pavement that is widely used as pavement because of good stability then has the advantage of high durability and more resistant to water so as to provide comfort to road users [1][2]. There has been an increase in the use of asphalt in Indonesia, especially as a binder for Asphalt Concrete. So that another binder is needed that is able to guarantee the need for the binder.

Natural Asphalt located on Buton Island is the second largest reserve of natural fasteners in the world after Trinidad Tobago, Binders originating from the island of Buton are believed to be able to increase the roughness of asphalt concrete mixture [3] deposit of ±680 million tons [4] so as to be able to meet the needs of national asphalt by substituting some of the use of asphalt oil [5]. Modified paved mixtures that use natural asphalt from Buton Island can improve road service life when compared to mixtures that do not use buton asphalt [6]. Asphalt mixture technology to heal itself from cracks in asphalt so as to prevent the occurrence of holes is still not widely explored, to protect the asphalt from moisture to penetration and to prevent quality degradation then used liquid fasteners (additives) for road repair[7].

The self-healing capabilities of asphalt materials have been known for many years, and have been widely developed in various countries [8]. The materials used for self-healing also have many types, ranging from natural materials to materials that require special treatment to obtain materials suitable for self-healing asphalt materials. Road pavement performance with asphalt mixture contains suitable capsules can meet the specifications then this mixture gives a positive influence to improve self-healing performance in pavement construction [9]. So it is expected that AC-WC Asb Lawele which uses added ingredients Calcium Alginate Capsules will be able to provide a comfortable service and have a longer life because it is able to do self-healing independently if experiencing cracks. This study aims to find out the changes in marshall parameters due to the addition of calcium alginate capsules as a self-healing ingredient in the pavement ac-WC Asb Lawele.

2. Material and Methods

2.1. Aggregate

The Aggregate Gradations used in this study use a gradation of the specifications qualifies by the Ministry of Public Works of the Republic of Indonesia [10]. The gradation used is the middle gradation with 94% aggregate and 6% aggregate in the form of granular asphalt buton. Table 1 is the gradation composition that will be used in this study, where the mean value of the specification is used for the manufacture of the sampel.

Table 1. Gradation Aggregate

Sieve		Spec. (%)		Value (%)
Inchi	mm	Min	Max	
1"	25.4	100	100	100
3/4"	19.1	100	100	100
1/2"	12.7	90	100	95
3/8"	9.7	77	90	83.5
No.4	4.76	53	69	61
No.8	2.38	33	53	43
No.16	1.18	21	40	30.5
No.30	0.595	14	30	22
No.50	0.29	9	22	15.5
No.100	0.15	6	15	10.5
No.200	0.074	4	9	6.5

2.2. Asphalt

Asphalt Pen. 60/70 Ex.Pertamina is a binding material that is then combined with Aspal Buton B 50/30 which is a natural asphalt from Buton Island containing Bitumen and Aggregate in accordance with the specific requirements by the Ministry of Public Works of the Republic of Indonesia[10].

Table 2. Characteristics of asphalt and asphalt pen.60/70

Properties	Value
Asbuton	
Asbuton Bitumen Content, %	25.42
Penetration at 25°C (0.1 mm)	45,20
Softening Point, °C	50.5
Ductility at 25°C, cm	116
Specific gravity	1.063
Asphalt Pen 60/70	
Penetration at 25°C (0.1 mm)	54.3
Softening Point, °C	334
Ductility at 25°C, cm	150
Flash Point, °C	320
Specific gravity	1.041

2.3. Procedure Fabrication Capsule Calsium Algiante

Produce Capsule Calcium Alginate Capsule:

1. Put sunflower oil in pure water according to the type of capsule mixture;
2. Stir the oil and water mixture for 2 minutes then enter the sodium alginate then stir at a turning speed of 1350 rpm to 10 minutes;
3. use a syringe pump to pump the sodium alginate emulsion into a solution of Calcium Chloride in order to become granules of calcium alginate capsules;
4. Eliminate moisture content contained in calcium alginate capsule granules;

Table 3. Type Capsul Calsium Alginate

Type Capsule	Sodium Alginate Emulsion			o/w ratio
	Sodium Alginate (g)	Sun Flower Oil (ml)	Water (ml)	
CA-1	15	100	500	1 : 5
CA-2	15	150	450	1 : 3
CA-3	15	200	400	1 : 2

2.4. Optimum Bitumen Content

Fig.1 Shows the asphalt content value of 6.0% to 6.5% according to specifications, then in Fig.2 the VFB value of asphalt content that meets specifications is at 6.0% to 7% asphalt content. Fig.3 shows the VMA value decreases as the asphalt content increases, while Fig.4, which is the flow parameter, shows that the variation in the percentage of asphalt content used only gives a Flow value of 3.8 mm to 4.2 mm. The stability value produced by varying the asphalt content of 4.0% - 7.0% indicates the stability value tends to increase with the increase in the percentage of asphalt content.

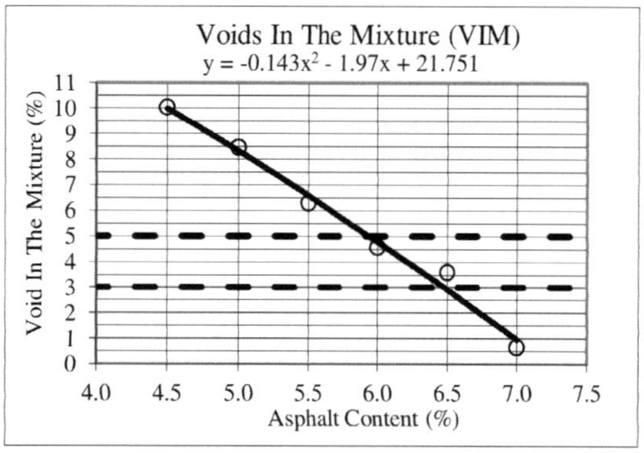

Figure 1. Void In Mixture (VIM)

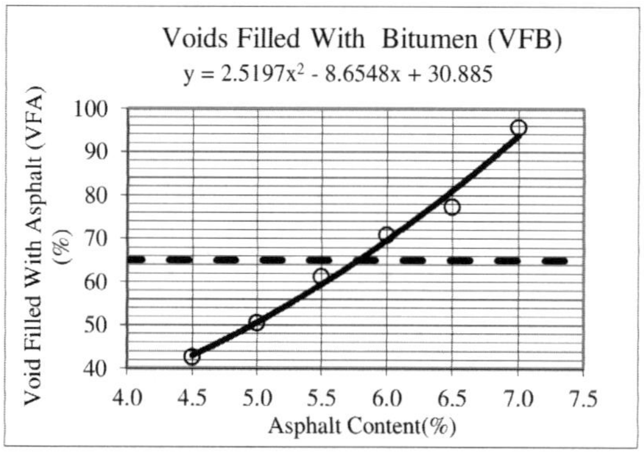

Figure 2. Void Filled With Bitumen(VFB)

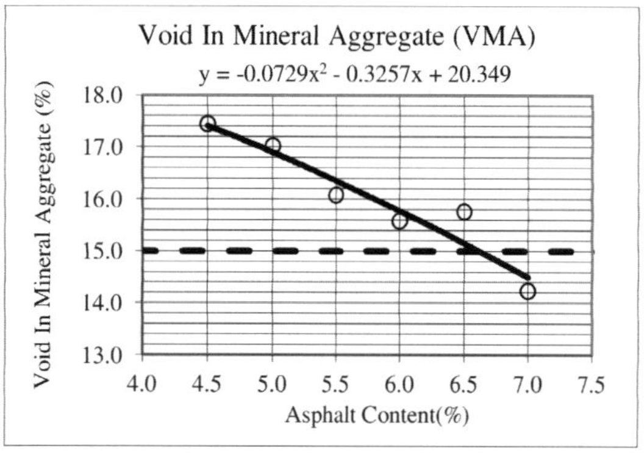

Figure 3. Void In Mineral Aggregate (VMA)

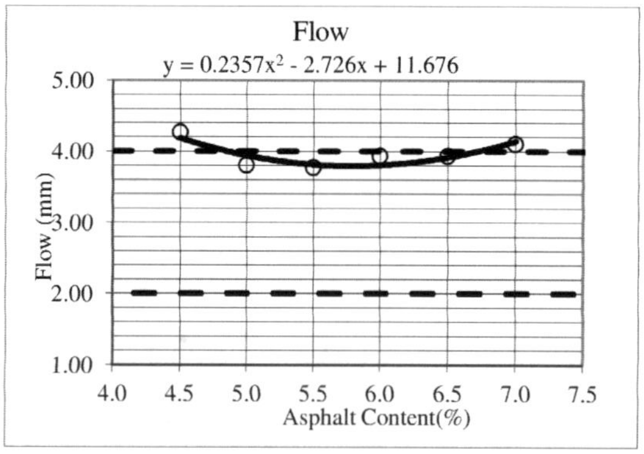

Figure 4. Flow

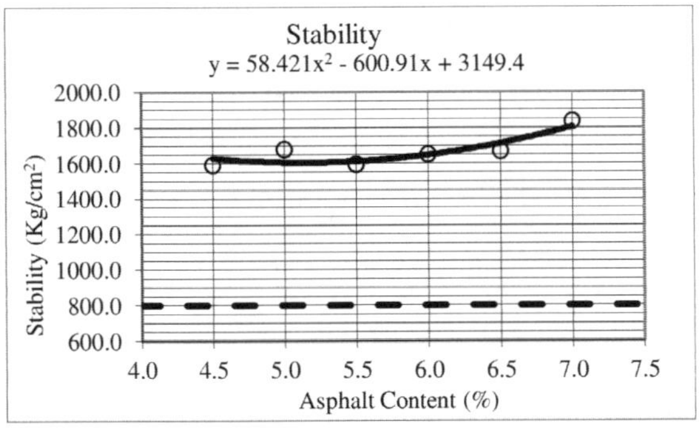

Figure 5. Stability

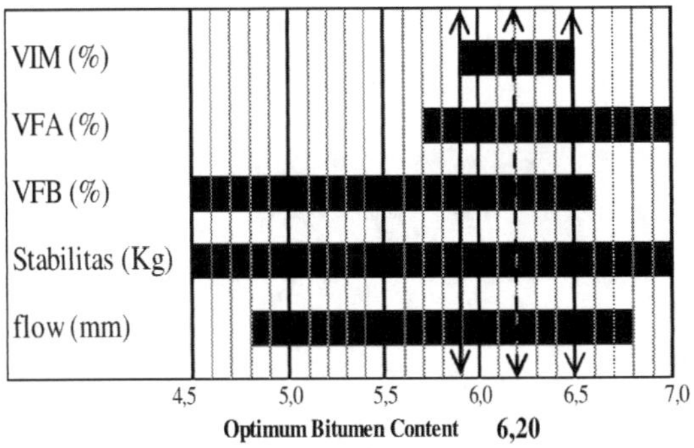

Figure 6. Optimum Bitumen Content

Based on calculations using marshall charts on Fig.6 obtained optimum asphalt content that meets the marshall parameters between 5.9% - 6.5%. Optimum Asphalt Content is used 6.2% of the mixed weight divided between 2% Asphalt Buton and 4.2% Asphalt Pen 60/70, asphalt buton substitutes 32% use of asphalt Pen. 60/70 as a binder.

3. Result And Discussion

3.1. FTIR

Based on FTIR spectra, it is seen that there are similarities between the Asbuton B50/30 FTIR spectra, Asphalt Pen 60/70, and Calcium Alginate Capsules synthesis results. Here is a comparison table of the material spectra.

Table 4. Binder Wave Number

Wave Number (cm^{-1})			Appropriate Functional Group
Asbuton B 50/30	Asphalt pen. 60/70	Capsule Calcium Alginate	
3405,79	3448,58	3461,42	NH *(stretching vibration)*
2923,73	2923,73	2926,59	C-H *streching asymetric*
2855,28	2852,42	2855,28	C-H streching *symmetric* SH *(stretching vibration)*
1615,90	1603,06	1633,01	C=C *aromatic*
1433,34	1451,88	1460,44	C-C *aromatic*

The presence of nitrogen groups in capsules is detected by peaks with a wave number of 3461.42 cm^{-1}. Contained nitrogen group has an important function in binding aggregate in paved mixture [11], so that the capsules synthesized as a plus for self-healing can bind together in the bonds of the pavement mixture because it has a similar function group.

3.2. Marshall Paramter

Before testing the marshall specimen gets collisions 75 times for each side [12] specimen is 101,6 mm in diameter with a height between 64 - 72 mm, the specimen is then soaked into water with a temperature of 60°C for 30 minutes before marshall stability and flow measurements are taken.

Figure 3 shows marshall stability chart decreasing as the percentage of calcium alginate capsules increases, CA-1 capsule type decreases by 28%, then CA-2 and CA-3 decrease by 31% and 32%, respectively, but the decrease is still within Asb Lawele AC-WC specificity threshold of 800 Kg. Figure 9 and figure 10 show the Void In Mineral Aggregate (VMA) and Flow parameters meet the specification of flow parameters between 2.96 mm – 3.13 mm, but the increase in the percentage of calcium alginate capsules is inversely proportional to the flow value in the pavement mixture. The value of VMA increases if the addition of calcium alginate capsules also increases, as well as the higher the oil content in the calcium alginate capsules eats the VMA value also increases, seen in the value of VMA paved mixture with type CA-3.

IOP Conf. Series: Earth and Environmental Science 1117 (2022) 012005 doi:10.1088/1755-1315/1117/1/012005

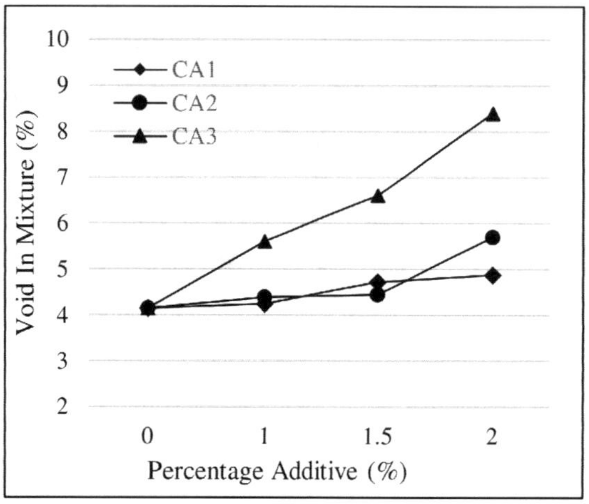

Figure 7. Relationship between within in Capsule Calcium Alginate and VIM

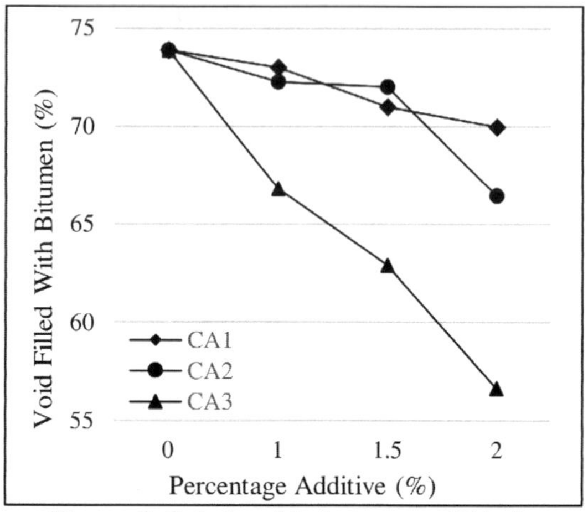

Figure 8. Relationship between within in Capsule Calcium Alginate and VFB

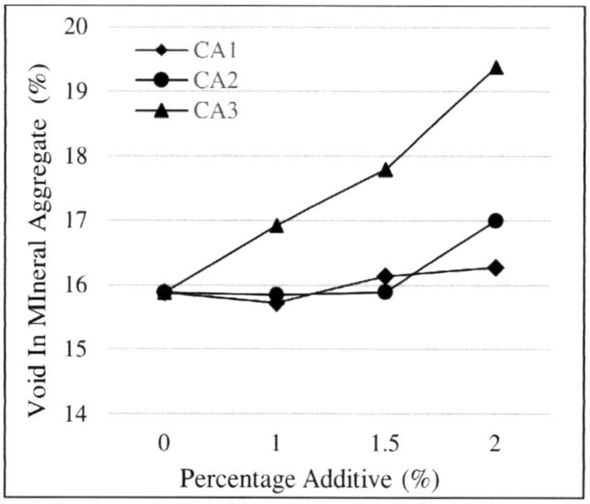

Figure 9. Relationship between within in Capsule Calcium Alginate and VMA

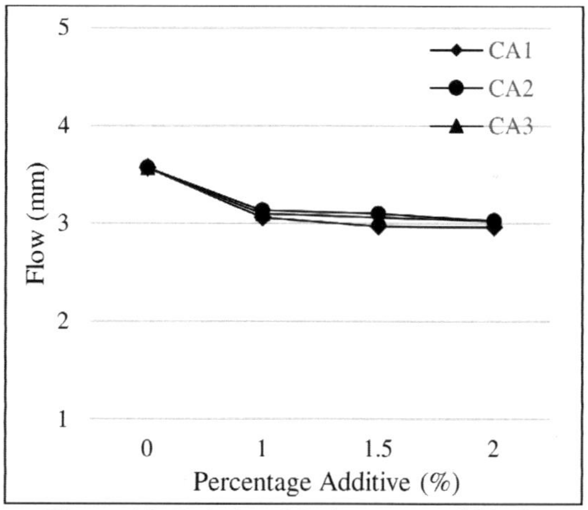

Figure 10. Relationship between within in Capsule Calcium Alginate and Flow

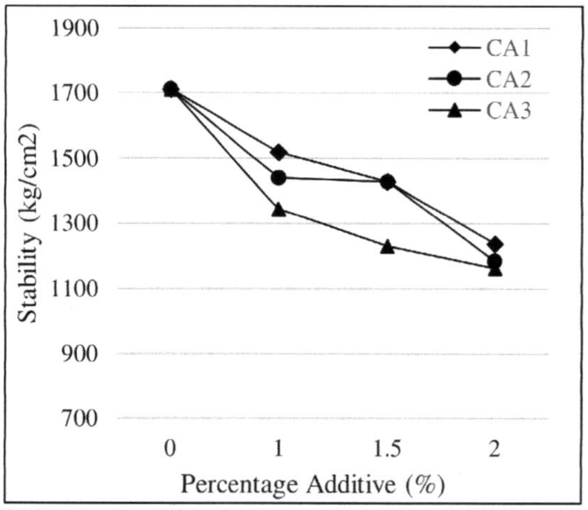

Figure 11. Relationship between within in Capsule Calcium Alginate and Marshall Stability

The higher the percentage of addition of Calcium Alginate Capsules causes void in mixture (VIM) values to come out of the required specifications, the higher the VIM value will lead to a permanent decrease in deformation resistance to paved mixtures[13], while VIM values below 3% will also cause decreased friction between aggregates [14]. Then the Void Filled In Bitumen (VFB) parameter continues to decrease with the increase of calcium alginate capsules contained in the paved mixture.

4. Conclusion
From the result of the study the following can be concluded :
1. There are similarities between the function groups of asphalt pen 60/70, Asbuton B 50/30 and calcium alginate capsules.
2. The addition of capsules of calcium alginate is not more than 1.5% of the combined weight, as well as the optimum comparison of o/w calcium alginate is 1 : 5 and 1 : 3.
3. The addition of additional additives of calcium alginate capsules will decrease marshall stability, but still within the limits of fairness but will make vim and VFB parameters do not meet the requirements.

5. References
[1] S. Uzun and S. Terzi, "Evaluation of andesite waste as mineral filler in asphaltic concrete mixture," *Construction and Building Materials*, vol. 31. pp. 284–288, 2012, doi: 10.1016/j.conbuildmat.2011.12.093.
[2] M. Yilmaz, B. V. Kök, and N. Kulolu, "Effects of using asphaltite as filler on mechanical properties of hot mix asphalt," *Constr. Build. Mater.*, vol. 25, no. 11, pp. 4279–4286, 2011, doi: 10.1016/j.conbuildmat.2011.04.072.
[3] S. P. Hadiwardoyo, E. S. Sinaga, and H. Fikri, "The influence of Buton asphalt additive on skid resistance based on penetration index and temperature," *Constr. Build. Mater.*, vol. 42, pp. 5–10, 2013, doi: 10.1016/j.conbuildmat.2012.12.018.
[4] F. Affandi, "The Performance Of Bituminous Mixes Using Indonesia Natural Asphalt," in *25th ARRB Confrence*, 2012, pp. 1–12.
[5] F. Affandi, "Characteristics of Bitumen Asbuton Granules For Hot Asphalt Mixtures," *J. Jalan-Jembatan*, vol. 25, no. 3, pp. 350–368, 2008.
[6] M. Karami and H. Nikraz, "Using advanced materials of granular BRA modifier binder to improve the flexural fatigue performance of asphalt mixtures," *Procedia Eng.*, vol. 125, pp. 452–460, 2015, doi: 10.1016/j.proeng.2015.11.120.
[7] K. Chung, S. Lee, W. Cho, J. Seo, and Y. Hong, "Rheological analysis of self-healing property of microcapsule-containing asphalt," *J. Ind. Eng. Chem.*, vol. 64, pp. 284–291, 2018, doi: 10.1016/j.jiec.2018.03.026.
[8] Q. Lv, W. Huang, and F. Xiao, "Laboratory evaluation of self-healing properties of various modified asphalt," *Constr. Build. Mater.*, vol. 136, pp. 192–201, 2017, doi: 10.1016/j.conbuildmat.2017.01.045.
[9] Y. Li, P. Hao, and M. Zhang, "Fabrication , characterization and assessment of the capsules containing rejuvenator for improving the self-healing performance of asphalt materials : A review," *J. Clean. Prod.*, p. 125079, 2020, doi: 10.1016/j.jclepro.2020.125079.
[10] Minister For Public Works and Human Settlements; Directorate General of Highways, "General Specifications of Bina Marga 2018 for Road and Bridge Works," 2018.
[11] Pertamina, *Asphalt Mixed Structural Engineering*. Bandung, Indonesia, 2005.
[12] Budiamin, M. W. Tjaronge, S. H. Aly, and R. Djamaluddin, "Mechanical characteristics of hotmix cold laid containing buton granular asphat (BGA) and flux oil as wearing course," *ARPN J. Eng. Appl. Sci.*, vol. 10, no. 12, pp. 5200–5205, 2015.
[13] R. Collins, H. Shami, and J. S. Lai, "Use of Georgia loaded wheel tester to evaluate rutting of asphalt samples prepared by superpave gyratory compactor," *Transp. Res. Rec.*, no. 1545, pp. 161–168, 1996, doi: 10.3141/1545-21.
[14] G. C. Page, J. A. Musselman, and D. C. Romano, "Effects of aggregate degradation on air voids of structural asphalt mixture in Florida," *Transp. Res. Rec.*, no. 1583, pp. 19–27, 1997, doi: 10.3141/1583-03.

IOP Conf. Series: Earth and Environmental Science 1117 (2022) 012006 doi:10.1088/1755-1315/1117/1/012006

CHARACTERISTICS MOTORCYCLE TAXI AND THE EFFECT ON FARMERS

Hakzah[1], Syarifuddin Yusuf[2], A Irmayani Pawelloi[3], Kurniawan[4]

[1,4] Department of Civil Engineering, Universitas Muhammadiyah Parepare, Indonesia;
[2] Department of Economics, Universitas Muhammadiyah Parepare, Indonesia;
[3] Department of Electrical Engineering, Universitas Muhammadiyah Parepare, Indonesia;

hakzahs@gmail.com

Abstract. Agricultural products in Sidenreng Rappang, South Sulawesi, Indonesia were abundant, the road infrastructure was not good and the farmers used existing modes of transportation. The purpose of the study is to determine the characteristics of farmers, operators and motorcycle taxis as well as farmer satisfaction. The survey research uses primary data based on interview surveys in May – June 2021. The analysis was a simple regression using the SPSS application. The results showed farmers: 93% male, main occupation 87% farmer, ownership of land area 63% one to above 5 Ha, education level 31% was the diploma, 63% completed junior high school and school upper middle. Operator: 100% male, 99% side job, 100% own vehicle. Income during harvest season, 77% IDR 200,000 – IDR 300,000/day. The results of the analysis of the service quality variable for motorcycle taxis provide a significant level of satisfaction for farmers.

1. Introduce

Motor taxis have been widely used by people in Indonesia, both in urban and rural areas. Motor taxi services are used primarily for passenger transport, but they can also be used for the transport of goods, as well as to provide courier and delivery services [1]. Motor taxis provide a solution to existing problems by offering timely transportation of people and goods, without waiting time and brought directly to the required destination at an agreed cost [2][3] Other capabilities of motorcycle taxis include being used to carry agricultural produce to local markets or to shelters [4], motorcycle taxis often reach remote villages and agricultural fields that are connected to feeder roads only by footpath. Motor taxis are created according to the needs and conditions of their operational areas [5].

In South Sulawesi, especially Sidenreng Rappang Regency, it is known as a city of rice or rice granaries with extensive agricultural areas, but road infrastructure is still not good and there are still many road networks that have not been connected. crops are damaged and the price is cheap. Research has ever been done, that plants remain unharvested or become damaged after harvesting due to the unavailability of vehicles during harvest [6]. Mobility in rural areas can be hampered by the lack of transportation facilities and the unavailability of good roads [7]. The poor condition of rural roads in the region has made it difficult for motorized vehicles to operate effectively [8]. The results also suggest that if road quality improves, farmers have lower marketing costs and gain access to a wider market [9].

Other phenomena are low levels of education and skills, lack of employment in both urban and rural areas in developing countries, which causes unemployment and poverty [10]. Becoming a motorcycle taxi operator is an opportunity for the community to be used as a livelihood even though it is still an informal transportation sub-sector to earn additional income [11]. The author conducted a study to determine and analyze the characteristics of farmers, operators motorcycle taxi transportation of grain and determine their effect on farmer satisfaction.

2. Literature review

2.1. Service quality

According to statement [12] that the quality of service in the service will be realized, if the consumer expectations are in accordance with the performance of the service provider provided and high service quality results in high customer satisfaction as well. If the service received or perceived meets or even exceeds customer expectations, then the service is considered of high quality and satisfies the customer [13]. The basic meaning of service quality is all forms of activities carried out to meet consumer needs. While service is defined as a service provided by the owner in the form of convenience, speed, relationship, ability and friendliness shown by the attitude and nature in providing services for customer satisfaction.

2.2. Customer satisfaction

According to [14], customer satisfaction is a person's feeling of pleasure caused by the perceived performance or results of the service product, compared to his expectations. In other words, customer satisfaction is the customer's perception of expectations that have been met or exceeded. Meanwhile, according to [12], customer satisfaction can be achieved if a service meets or exceeds customer expectations, and feels satisfied. The closer the customer's expectations to the expected service, the more likely it is to achieve satisfaction. Customer satisfaction as a response to the perceived fit between expectations and actual service performance.

3. Methodology and survey

3.1. Study object and location

The objects of study are farmers, operators and motorcycle taxis operating in the area of Sidenreng Rappang Regency, South Sulawesi, Indonesia. Coordinates: 3°43' – 4°09' South Latitude, 119°41' – 120°10 East Longitude. The area is 1,883.25 km2, and the population is 264,955 people, while the agricultural area is around 44,958 m2, with an area of agricultural land harvested from an area of 80,331.78 Ha, rice production reaching 524,214 tons. According to the 2018 BPS data [15].

Figure 1. Interview and motorcycle taxi activities

3.2. Scope and objectives of the study

Interview survey of farmers, motorcycle taxi operators operating in the Sidenreng Rappang district as well as the activities of transporting agricultural products carried out by operators are shown (Figure 1). A total of 403 farmer respondents and 202 operators have been surveyed. It is carried out by visiting their place of residence or to a gathering place with their community who are then interviewed according to the question items in the questionnaire in May – June 2021. The survey includes details related to the characteristics of farmers and operators such as: gender, age, education, marital status, land ownership, motorcycle ownership, land area, income per day, job jatisfaction and others. This paper discusses the results of data analysis and draws a number of conclusions about the existence of motorcycle taxi transportation, additional employment, aspects of job satisfaction and aspects of operator income per day during the harvest season. The simple regression analysis uses the SPSS v24 applications to determine the effect of the independent variable on the quality of motorcycle taxi service (X) on farmer satisfaction (Y).

4. Questionnaire analysis

The questionnaire was analyzed according to the characteristics of farmers, operators and tested the causal relationship between the independent variable of motorcycle taxi service quality and the dependent variable of farmer satisfaction.

4.1. Characteristics of farmers

Table 1. Results of analysis characteristics of farmers

Characteristics		Frequency (%) 403 Farmers
Gender	Man	93
	Woman	7
Education	Primary school	6
	Junior high school	20
	Senior High School	43
	Diploma	31
Age	<20 years	7
	20 – 35 years	23
	35 – 50 years	36
	>50 years	34
Status	Married	92
	Unmarried	8
Motorcycle Ownership	Ownership	28
	No	72
Work	Main	87
	Side work	13
Land Area	<1 Ha	37
	1 – 3 Ha	45
	3 – 5 Ha	14
	>5 Ha	4

Source: Survey data results 2021

Table 1 shows that farmers are 93% male and 7% female. 43% of education completed at high school and 31% at the diploma level, the rest finished elementary school and junior high school. The large percentage of the level of education that is still relatively low indicates that the community still lacks awareness or attention to higher schools in order to increase the expertise of each individual and prefer to work as a farmer who can be seen in the percentage of work. The age of farmers is 36% between 35 – 50 years old, 34% over 50 years old, 23% 20 – 35 years old and the rest are under 20 years old. As for the marital status of farmers, 92% are married and 8% are unmarried. Working as a farmer 87% only 13% make farmers a side job from government employees. The land area owned is 45% between 1 – 3 Ha, 14% between 3 – 5 Ha, 4% over 5 Ha and 37% under 1 Ha. The extent of land owned by individuals indicates that they prefer to work as farmers according to the education level of each farmer.

4.2. *Characteristics of operators*

Table 2. Results of analysis characteristics of operator motorcycle taxi

Characteristics		Frequency (%) 202 Operators
Gender	Man	100
	Woman	0
Education	Primary school	36
	Junior high school	32
	Senior High School	28
	Diploma	4
Age	<20 years	8
	20 – 35 years	45
	35 – 50 years	34
	>50 years	13
Status	Married	91
	Unmarried	9
Motorcycle Ownership	Ownership	100
	No	0
Income per day	< IDR 200.000	11
	IDR 200.000 – 250.000	42
	IDR 250.000 – 300.000	35
	> IDR 300.000	12
Work	Main	1
	Side work	99
Job Satisfaction	Satisfied	71
	No	29

Source: Survey data results 2021

Table 2 shows that the 100% taxi motorcycle operators are male and there are no women working as operators. The education level is 36% completed at the elementary school level, 32% at the junior secondary level, 28% at the high school level and 4% at the diploma level. The low level of education makes the operation of motorcycle taxis an alternative choice for them. This indicates that formal employment opportunities are still lacking for people with low education. Operating a motorcycle taxi is considered as a temporary solution for unemployment among elementary, junior high and high school graduates. Operator age 45% between 20 – 35 years, 34% 35 – 50 years, 13% over 50 years

IOP Conf. Series: Earth and Environmental Science 1117 (2022) 012006 doi:10.1088/1755-1315/1117/1/012006

and 8% under 20 years. This indicates that the operator is still relatively young and has a strong physique, 91% are married, 9% are unmarried. Not a few farming communities who have two-wheeled vehicles have been converted to taxi motorcycle and show that the ownership of motorcycle taxis is 100% their own. As many as 99% of operators consider it as an additional job because in general, they have a main job, namely as a farmer. The daily income is quite large every harvest season, 42% between IDR 200,000 – IDR 250,000, 35% IDR 250,000 – IDR 300,000, 12% above IDR 300,000, 11% below IDR 200,000, giving satisfaction to motorcycle taxi operators by 71% and only 29% show that they are not satisfied with the work done as an operator transporting agricultural products.

4.3. Validation test results

Validation tests are carried out to determine the accuracy or not of a question and to measure the accuracy of the questions processed in the SPSS application, the results of data processing that have been carried out for the quality of motorcycle taxi services such as aspects of service, cost, speed, ability and income are tabulated as follows.

Table 3. Validation test results

Variable		Question	Correlation result (rxy)	r table (5%)	Information
Service quality (X)	X1	Motorcycle taxi (McT) is a good mode of transportation for farmers' agricultural products	0,399	0,092	Valid
	X2	McT quickly distribute agricultural products	0,499	0,092	Valid
	X3	McT have a good service time	0,666	0,092	Valid
	X4	McT operators serve consumers well	0,577	0,092	Valid
	X5	McT operators provide transportation fares according to consumer desires	0,669	0,092	Valid
	X6	McT available when needed	0,231	0,092	Valid
Farmer satisfaction (Y)	Y1	McT transports people's agricultural produce with ease.	0,577	0,092	Valid
	Y2	McT operators provide a fast response to requests for transportation of agricultural products.	0,663	0,092	Valid
	Y3	Agricultural products are transported by McT according to the time that consumers expect.	0,671	0,092	Valid
	Y4	McT operators serve consumers well	0,359	0,092	Valid
	Y5	The McT rate is affordable.	0,215	0,092	Valid
	Y6	Easy to get McT transportation services.	0,135	0,092	Valid

Source: SPSS test results

Table 3 above shows that each of the question item variables has a correlation result of r count greater than r table (r count > r table 0.092) and is positive. Thus, the question items are stated to be well correlated and valid.

4.4. Reliability test results

The reliability test in table 4 below is to determine the extent to which the measurement results remain consistent, if two or more measurements are made of the same symptoms using the same measuring instrument. The results of the reliability test are as follows:

Table 4. Reliability statistics test results

Cronbach's Alpha	N of Items
,654	12

Source: SPSS test results

The results of the reliability test obtained that Cronbach's Alpha value of 0.654 was greater than 0.6, so the results concluded that the variable was reliable or consistent. The results of the reliability test based on the research variable instrument showed that the results of the independent variable test of the quality of the motorcycle taxi service quality (X) as well as the dependent variable of farmer satisfaction (Y) showed that the data obtained were reliable.

4.5. Data normality test

Table 5. One-Sample Kolmogorov-Smirnov test

N		405
Normal Parameters[a,b]	Mean	,0000000
	Std. Deviation	1,52397750
Most Extreme Differences	Absolute	,041
	Positive	,041
	Negative	-,035
Test Statistic		,041
Asymp. Sig. (2-tailed)		,095[c]

Source: SPSS test results

Based on the output of the Kolmogorof-Smirnov One-Sample normality test in table 5, it was found that the significance value of 0.095 is greater than 0.05 (0.095 > 0.05), the results mean that the tested data is normally distributed and the regression model has met the assumption of normality.

4.6. Hypothesis testing

4.6.1. Simple linear regression analysis.

Simple linear regression analysis to predict how much positive relationship data quality motorcycle taxi service (X) to farmer satisfaction (Y). The results of the simple linear regression analysis test are shown in table 6.

Table 6. Simple linear regression analysis and T-test results

Model	B	Beta	t	Sig.
(Constant)	9,428		20,295	,000
Service quality	,593	,711	13,332	,000

Source: SPSS test results

Table 6 shows that the constant value obtained from simple linear regression analysis is 9.428 and the variable value of the service quality variable for motorcycle taxis (X) with a coefficient value of 0.593. These results can be entered and form a simple regression equation as follows:

$$Y= 9,428 + 0,593X$$

The results of the above equation can be translated that the constant is 9.428 which contains the satisfaction variable value of 9.428 and the value of the X regression coefficient is 0.593 which states that the addition of 1% of the value of motorcycle taxi service quality will increase farmer satisfaction by 0.593 and vice versa. The regression coefficient is positive, so it can be said that the direction of the influence of motorcycle taxi service quality on farmer satisfaction is positive. The results with a significance value obtained from table 6 of 0.000 < 0.05 means that the level of trust given is 100% greater than the required level of confidence of 99% with a value of $\alpha = 5\%$, so it can be concluded that the variable motorcycle service quality taxi (X) has a significant effect on the variable of farmer satisfaction (Y).

The results of the partial hypothesis test or t-test in table 6 show that the t-count value is 13.332 with a significance value of 0.000. It can be concluded that the quality of motorcycle taxi services has a positive and significant effect on farmer satisfaction because the significant value of 0.000 is less than 0.05 or $\alpha = 5\%$. It shows that the existing data on motorcycle taxi service quality (X) has a positive and significant influence on farmer satisfaction (Y) in Sidenreng Rappang Regency, South Sulawesi Province.

4.6.2. F-test result

Table 7. F-test results

Model	df	F	Sig.
Regression	1	411,875	,000[b]
Residual	403		
Total	404		

Source: SPSS test results

The results of the F test in table 7 are known that F count = 411,875 with a significant level of 0.000 less than 0.05 or $\alpha = 5\%$, the regression model can be used to predict the satisfaction variable or in other words that there is an influence of the motorcycle taxi service quality variable (X) on the variable of farmer satisfaction (Y).

4.7. The result of the coefficient of determination

Determinant coefficients to find out how much influence the quality of motorcycle taxi service quality (X) on farmer satisfaction (Y), statistical calculations were carried out using the coefficient of determination (KD) as obtained in table 8.

Table 8. Results of the coefficient of determination

Model	R	R Square	Adjusted R Square
1	,711[a]	,505	,504

Source: SPSS test results

The value of the coefficient of determination R Square is 0.505 or 50.1% in table 8. This figure is obtained by squaring the correlation coefficient (0.711 x 0.711) and then multiplied by 100%. Based on this, it can be concluded that the variable quality of motorcycle taxi service quality (X) contributes to the variable of farmer satisfaction (Y) by 51%. While the remaining 49% is influenced by other factors that are not or have not been studied. The calculation of the data above shows that there is a positive and significant relationship between the quality of motorcycle taxi services (X) and farmer satisfaction (Y).

Conclusion

Characteristics of farmers are 93% male and 7% female. Educational level of 43% completed at high school education, 31% at diploma level, the rest at elementary school and junior high school. Age 36% between 35 – 50 years, 34% over 50 years, 23% between 20 – 35 years, the rest under 20 years. The main occupation as a farmer is 87%, 13% is a farmer as a sideline as a government employee. Own 63% of land between 1 Ha to more than 5 Ha. The characteristics of operators are 100% male, 99% used as additional work, 36% education completed at the elementary school level, 32% junior high school, 28% high school and 4% completed diploma level. The age of the operators is 45% between 20 – 35 years, 34% 35 – 50 years, 13% over 50 years and 8% under 20 years, indicating that the operators are still relatively young. 100% own motorcycle ownership, 99% make work as an additional job. Daily income earned during harvest season 42% between IDR 200,000 – IDR 250,000, 35% IDR 250,000 – IDR 300,000, 12% above IDR 300,000, 11% below IDR 200,000. The results of the test using simple linear regression analysis, namely the variable quality of service quality for motorcycle taxis and farmer satisfaction indicate that the variable service quality for motorcycle taxis has a positive and significant effect on farmer satisfaction in Sidenreng Rappang Regency, South Sulawesi Province.

Acknowledgments

This paper is an attempt to publish some parts of the existing field research. The authors would like to thank to The Ministry of Research and Technology/National Research and Innovation Agency - Indonesia and Universitas Muhammadiyah Parepare for their support in financing excellent applied research for higher education for the 2021 fiscal year. As well as useful comments and input from team members and parties involved participated in this study (Research contract number: 070/SP2H/LT/DRPM/2021).

References

[1] Tuffour, Y.A., Appiagyei, D.K.N., 2014. Motorcycle taxis in public transportation services within the Accra metropolis. Am. J. Civil Eng. 2 (4), 117–122.doi:10.11648/j.ajce.20140204.12

[2] Dinata, J., Wahab, Z., Widiyanti, M., Shihab, M.S., 2019. The Effect of Quality of Services and Prices on The Gojek. http://doi.org/10.35409/IJBMER.2019.2406

[3] Afukaar, F., Derry, J.D., Peters, K., Starkey, P., 2019. Rural Transport Services Indicators: Using a new mixed-methods methodology to inform policy in Ghana. Published by Elsevier Ltd. an open access article under the CC BY license http://dx.doi.org/10.1016/j.trip.2019.100074Another reference.

[4] Ehebrechta, D., Heinrichs, D., Lenz, B., 2018. Motorcycle-taxis in sub-Saharan Africa: Current knowledge, implications for the debate on informal transport and research needs. Journal of Transport Geography 69 (2018) 242–256: https://doi.org/10.1016/j.jtrangeo.2018.05.006

[5] Musso, A., Vuchic, V,R., Bruun, E., Corazza, M.V., 2010. A Research Agenda For Public Policy Towards Motorized Two wheelers In Urban Transport TRB. Annual Meeting https://www.researchgate.net/publication/264436062

[6] Ahmed, R., Rustagi, N.,1987. Marketing and Price Incentives in African and Asian Countries: A Comparison. Reprinted from Agricultural Marketing Strategy and Pricing, Iolicy (ed. Dieter Elz) Washington, D.C: International Bank for Reconstruction and Development. [https://invenio.unidep.org/invenio/record/4124/files/RP-107.pdf]

[7] Damme, V.P., 2007. Rural-urban Marketing Linkages. An Infrastructure Identification and Survey. Guide. ISBN: 92-5-105387-1, ISSN 1010-1365. Economic Botany, 61(1), 108–108. doi:10.1663/0013-0001 (2007) 61 [108b:rmlaii] 2.0.co;2

[8] Karema, F.M., Irandu, E.M., 2017. Role Of Commercial Motorcycles In Promoting Agricultural Production In Kenya: A Case Study Of Laikipia East Sub–County. International Journal of Education and Research ISSN: 2411-5681. 01.pdf (ijern.com)

[9] Rabirou, K., Ayanwale, A., Idowu, E.O., Williams, S.B., 2012. Effect of rural transportation system on agricultural productivity in Oyo State, Nigeria. Journal of Agriculture and Rural Development in the Tropics and Subtropics Vol. 113 No. 1 (2012) 13–19 ISSN: 1612-9830. Effect_of_rural_transportation_system_on_agricultu.pdf

[10] Ogunrinola, I.O., 2011. Informal Self-Employment and Poverty Alleviation: Empirical Evidence. International Journal of Economics and Finance. ISSN 1916-971X E-ISSN 1916-9728, doi:10.5539/ijef.v3n2p176 from Motorcycle Taxi Riders in Nigeria

[11] Al-Hasan A,Z., Momoh, S., Eboreime, L., 2015. Urban poverty and informal motorcycle transport services in a Nigerian intermediate settlement: a synthesis of operative motives and satisfaction. ISSN: (Print) 2165-0020 (Online):Urban Planning and Transport Research, http://dx.doi.org/10.1080/21650020.2014.978950

[12] Zeithaml VA, et al. 2010. Service Marketing Strategy Journal Zeithaml Services_Marketing_Strategy.pdf DOI: 10.1002/9781444316568.wiem01055

[13] Parasuraman A, et al. 2006. Perceived service quality as a customer-based performance measure: An empirical examination of organizational barriers using an extended service quality model. (PDF) Perceived service quality as a customer-based performance measure: An empirical examination of organizational barriers using an extended service quality model (researchgate.net). DOI:10.1002/hrm.3930300304

[14] Kotler, P., Keller K.L., 2006. Marketing Management. https://www.researchgate.net/publication/225084026_Marketing_Management.

[15] Badan Pusat statistik/BPS 2018. (Central Bureau of Statistics - South Sulawesi, Indonesia). [https://sulsel.bps.go.id/subject/153/geografi.html#subjekViewTab3]

Mechanical properties of dredged soil reinforced with natural pozzolan and shredded tires

K A Utama[1], T Harianto[2], A B Muhiddin[2] and A Arsyad[2]

[1]Doctoral Student of Civil Engineering, Hasanuddin University, Gowa, South Sulawesi, Indonesia, 92171

[2]Departement of Civil Engineering, Hasanuddin University, Gowa, South Sulawesi, Indonesia, 92171

E-mail: utamaka19d@student.unhas.ac.id

Abstract. The utilization of waste materials is still an interesting subject to investigate. The dredged soil (DS) was classified as a waste material that requires more effort to improve its mechanical properties. Adding natural pozzolan (NP) material and shredded tires (ST) can be an exciting field to investigate and quite friendly to the environment. This research objective investigates the mechanical properties of the dredged soil stabilized with natural pozzolan material and shredded tires as a new way of green engineering. This research involved trass as a natural pozzolan and shredded tires as car-used tires waste retreading factory. Basic and mechanical properties tests were investigated, including standard proctor, unconfined compressive strength (UCS), and direct shear of the stabilized dredged soil. The proportions of involved trass were 3%, 6%, 9%, and 12%, while the proportions of shredded tires were 2% and 3% to the maximum dry weight of the material sample. These mechanical properties tests were conducted in a curing period of 7, 14, and 28 days. This recent research shows an increase in the maximum dry weight of the stabilizing soil to the untreated soil by 0.5% - 1.7%. While for the compressive strength value, an increase in the UCS value by 175% - 324% was revealed. Furthermore, there was an increased shear strength value, and it was found that the increase in cohesion value (c) was 6% - 83%, and the increase in internal friction angle value (ϕ) was 31% - 98%.

Keywords: dredged soil, shredded tires, natural pozzolan.

1. Introduction

As a result of the landslide at the caldera of Mount Bawakaraeng in 2004, it increased the sedimentation rate in the Bilibili Dam. It is necessary to dredge sediment soil to provide an efficient river channel, construct an effective dam's storage, and maintain the downstream area of the river reach [1]. Dredged soil refers to any soil excavated or dredged from the water regime, pond, or dam's bottom. This dredged soil offers a vast and sustainable resource for various practical applications, including fill material, sub-grade construction, reclamation, landscaping, etc. However, many researchers report that the dredged soil cannot be used directly due to its poor geotechnical characteristics. [2] [3] [4] [5] [6]. Some of these properties are bearing capacity (qu) and shear strength (ϕ). Furthermore, it is necessary to investigate and find the solutions to the dredged soil problem.

IOP Conf. Series: Earth and Environmental Science 1117 (2022) 012007 doi:10.1088/1755-1315/1117/1/012007

There are many ways to stabilize the dredged soil. One of them is by adding cement. However, cement has stepped forward as a stabilizing agent with environmental consequences. CO2 emissions from cement production come from two sources: combustion (38%) and calcination (62%). CO2 emissions from combustion are related to fuel use, whereas emissions from calcination are produced when raw materials (mainly limestone and clay) are heated and CO2 is liberated from the decomposed limestone. Therefore, it is necessary to propose other stabilization efforts using natural pozzolan minerals. Adding a stabilizing agent like this is considered to be included in the stabilization effort that is friendly to the environment [7].

On the other hand, the growth of rubber waste as an impact of the waste tires automobile industry is increasing annually across the country. It poses the problem of disposal and adds to environmental contamination and health risks, utilization of such refuse, and industrial wastes [8]. Therefore, it takes an innovative effort to reuse this kind of waste. A relatively progressive thought will be involved in this research by combining the used shredded tires into dredged soil to produce a new alternative geo-material of the embankment that can be utilized. This research deals with the dredged soil from dredging activities at the Bilibili Dam and deals with waste materials from cars used tires factory activities.

Blending natural pozzolan minerals with shredded tires has been shown to provide benefits, specifically in terms of improving soil mechanical properties on the one hand and reducing environmental damage on the other. These properties can be enhanced by controlled compaction using mechanical equipment or by adding appropriate admixtures such as cement, fly ash, lime, or by reinforcing the soil with shredded tires, crumb rubber, plastic waste, etc [9] [10].

2. Materials

This research provides dredged soil, trass as a natural pozzolan, and shredded tires as the materials research. The following is an explanation of these three materials:

2.1. Dredged soil (DS)

The dredged soil used was obtained through dredging activities at the Bilibili Dam. This dredged soil has a specific gravity (Gs) of 2.66 and is mostly silt in grain size (85.3%). It has a plasticity index (PI) of 7% and a light cream color. According to the USCS soil classification system, this soil is classified as ML or silt with low plasticity. The basic properties of the dredged soil used are listed in the table below:

Table 1. Soil characteristic

Basic properties	Value	Unit
Gs	2.66	-
Sieve analysis		
Sand	11.8	%
Silt	85.3	%
Clay	2.9	%
Atterberg's limits		
PI	7	%
LL	39	%
PL	32	%
SL	30	%
USCS Classification	ML	-

2.2. Natural pozzolan (NP)
The natural pozzolan used in this study is trass. The trass was taken from a quarry in the Gorontalo Province district of Bone Bolango. The used trass has a specific gravity (Gs) of 2.69 and a grain size that passes through the #200 sieve. The dominant mineralogy content in this trass is silicon dioxide (SiO_2) at 61%, aluminium trioxide (Al_2O_3) at 32%, calcium oxide (CaO) at 3%, and iron trioxide (Fe_2O_3) at 2%, and the remaining minerals at 2%. According to ASTM C-618, this material is classified as a pozzolan material class-N if the percentage of SiO2 + Al2O3 + Fe2O3 is greater than 70%. The characteristics of the materials used in this study are presented below:

Table 2. Mineralogy of materials used

Mineral	SiO_2 %	Al_2O_3 %	Fe_2O_3 %	CaO %	Others %	G_s -	Grain size pass #200 %
Dredged soil	50	36	5	4	5	2.66	88
Natural Pozzolon (Trass)	61	32	2	3	2	2.69	100

2.3. Shredded tires (ST)
The shredded tires used in this study are from the shred of used tires in the vehicle tire retreading industry. It is black and contains only 0.80% moisture. The shredded tires retained at the #10 sieve are 29.72% of the volume of this waste material. The length of these shred materials is 20-30 mm, the thickness is 1-2 mm, and the width is 0.5-1 mm. The percentage of shredded tires used in this study was 2% and 3% by weight of the sample's dry weight.

Figure 1. Material used, (a) Dreddeg soil; (b) Trass; and (c) Shredded tires

3. Experimental Methods and Test
A comprehensive series of laboratory tests on dredged soil was conducted to investigate the potential uses of treated dredged soil as fill material in transportation-related projects. The dredged soil was mixed with trass, a natural pozzolan material (3%, 6%, 9%, and 12% by dry weight), as well as shredded tires (2% and 3%). Index properties, compaction characteristics, unconfined compressive strength, and shear stress were tested on trass-rubber treated dredged soil.

The mechanical properties of the dredged soil mixed with trass and shredded tires will be determined. The maximum dry density and optimum moisture content of soil stabilized with trass and shredded tires were investigated using a standard proctor test (ASTM: D698-07). Unconfined compressive strength tests were also performed to determine the effect of this stabilizing effort (ASTM: D2166-06). A testing machine with a 2 mm/minute compression speed is used to evaluate the unconfined compressive strength sample. The element test samples were made by remoulding and compacted in three layers, with compaction energy is 600 kN/m^3.

Meanwhile, a simple direct shear strength test (ASTM: D3080-04) was performed by remoulding the sample at the maximum density of each predetermined composition to determine the shear strength parameter. A sample size of 6 cm in diameter and 2 cm in height was created. The sample was tested with three axial load variations with a shear load speed of 1.25 mm/minute. All element samples were cured for 7, 14, and 28 days, respectively.

4. Result and Discussion

4.1. Compaction with a standard proctor apparatus

The compaction value results from the standard proctor experiment of trass stabilization soil and shredded tires revealed that the value of maximum dry density of the soil increased by 0.5% -1.7%, and the optimum moisture content increased by 3.8% - 4.7% as well. The following table shows the value of the density and moisture content of stabilized dredged soil:

Table 3. Maximum dry density (MDD) dan Optimum moisture content (OMC) of the stabilized soil

The proportion of Dredged Soil (DS) + Natural Pozzolan (NP) + Shredded Tires (ST)	Max. Dry Density (MDD) (γ)	Opt. Moisture Content (OMC) (ω)	Enhancement Percentage (%)	
	kN/m^3	%	γ	ω
0%	14.06	25.04	-	-
DS + 3% NP + 2% ST	14.15	26.22	0.6	4.7
DS + 6% NP + 2% ST	14.16	26.20	0.7	4.6
DS + 9% NP + 2% ST	14.25	26.10	1.3	4.2
DS + 12% NP + 2% ST	14.30	25.98	1.7	3.8
DS + 3% NP + 3% ST	14.12	26.17	0.4	4.5
DS + 6% NP + 3% ST	14.13	26.15	0.5	4.5
DS + 9% NP + 3% ST	14.22	26.05	1.1	4.1
DS + 12% NP + 3% ST	14.27	25.93	1.5	3.6

According to Table 3 above, the maximum dry density of all sediment dredged soil stabilized with natural pozzolan and shredded tires is greater than that of the original soil. The increase in the maximum dry density value can be an initial indicator of improved soil geotechnical properties due to mechanical effort.

4.2. Unconfined Compression Strength

The stabilized soil's compressive strength showed an increase in the value of the UCS. Adding 2% and 3% shredded tires increased 175% to 324% for acquiring 3%, 6%, 9%, and 12% natural pozzolans, respectively. The increase is depicted in the graphic below:

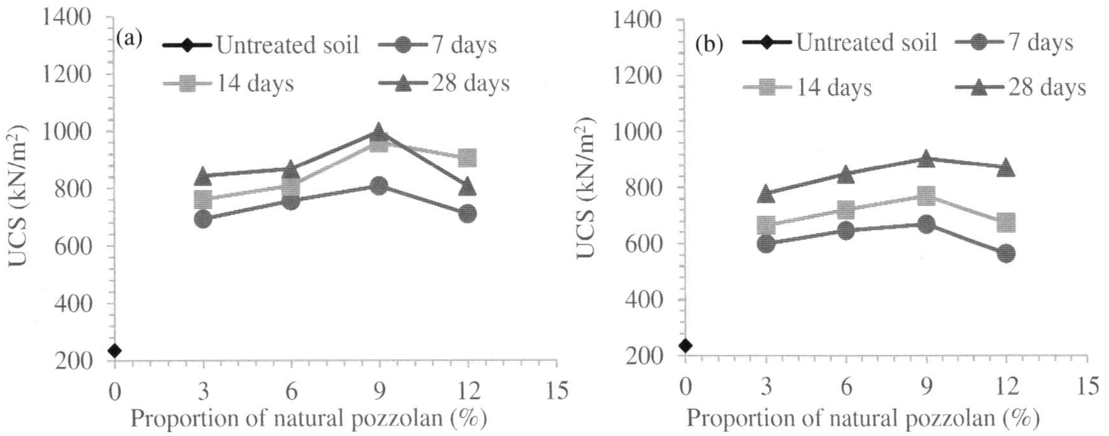

Figure 2. Increased UCS value, (a) soil with 2% shredded tires, (b) soil with 3% shredded tires

According to Figure 2, there is an increase in the UCS value in the stabilized soil. This increase occurred gradually and reached its peak with the addition of 9% natural pozzolan, both for 2% and 3% shredded tires. This condition indicates that the maximum amount of natural pozzolan added to the mixture of dredged soil and shredded rubber is 9%. The following table summarizes the normalized value of compressive strength test results on sediment dredged soil samples stabilized with natural pozzolan and shredded tires:

Table 4. The normalized value of the UCS increasing

Proportion	Normalized value (q_u/q_{uo}) of UCS					
	Adding 2% shredded tires			Adding 3% shredded tires		
	7 days	14 days	28 days	7 days	14 days	28 days
3%	2.9	3.2	3.6	2.5	2.8	3.3
6%	3.2	3.4	3.7	2.8	3.1	3.6
9%	3.4	4.1	4.2	2.8	3.3	3.8
12%	3.0	3.8	3.4	2.4	2.9	3.7

According to Table 4, all the compositions of shredded tires and natural pozzolan increased 2.5 to 4.2 times at various curing times of 7, 14, and 28 days. This value demonstrates that the value of the unconfined compressive strength, which is related to the bearing capacity of this material, is of high quality and should be considered for use as embankment material. The increase in the value of the unconfined compressive strength of the stabilizing soil indicates that the incorporation of natural pozzolan and shredded tires significantly influences the improvement of its geotechnical properties, particularly for determining the bearing capacity of the embankment soil. It can be used to justify the decision to use this material as an alternative to a new soil embankment type.

4.3. Shear strength parameter

The direct shear strength test results indicated that the value of cohesion (c) increased by 6 % to 83%, and the value of the internal shear angle (ϕ) increased by 30% to 98%. This increase is substantial in enhancing the soil's shear strength. The increased parameters of the shear strength of the stabilized material also indicate an improvement in the mechanical properties of the stabilized soil. The following figures are the results of the direct shear strength test of the dredged soil stabilized with shredded tires and natural pozzolan:

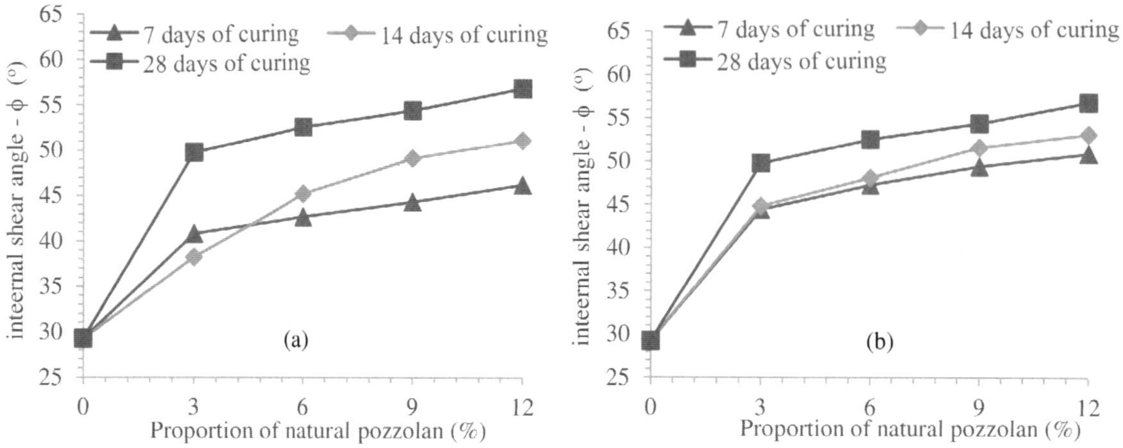

Figure 3. Increased of internal shear angle value, (a) soil with 2% shredded tires, (b) soil with 3% shredded tires

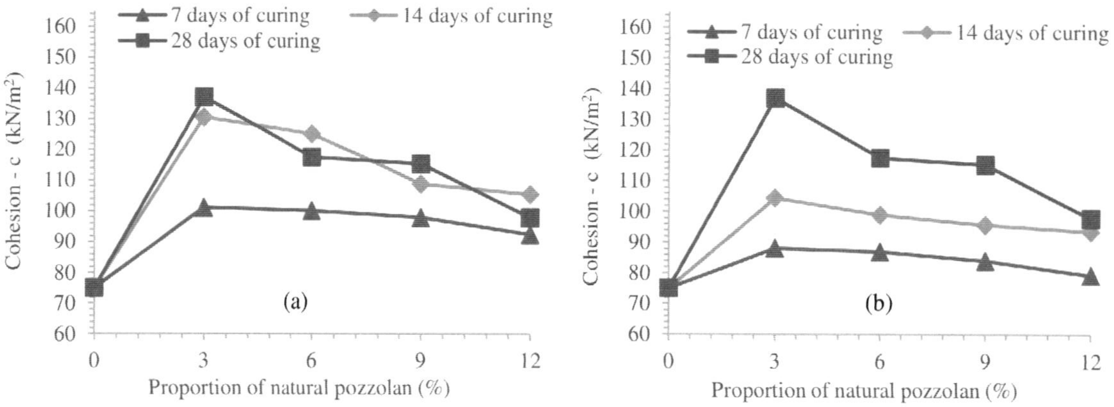

Figure 4. Cohesion value, (a) soil with 2% shredded tires, (b) soil with 3% shredded tires

Figure 3 demonstrates that the internal shear angle of the stabilized dredged soil, regardless of the proportion of natural pozzolan and shredded tires added, increased significantly compared to the untreated dredged soil. This increase is linearly proportional to the length of the curing period. It indicates that the stabilized dredged soil is improving over time. The increasing value of the internal shear angle demonstrates that the geotechnical conditions are improving, making it more suitable for use as a steep-sloped embankment.

Meanwhile, as shown in Figure 4, the cohesion value of the stabilized dredged soil increased when compared to the cohesion value of the untreated dredged soil. This higher cohesion value than the previous one contributed to the geotechnical properties of the dredged soil.

5. Conclusion

This study investigated the basic and mechanical properties of dredged soil stabilized with natural pozzolan and shredded tires using compaction, unconfined compression strength, and direct shear tests. Based on the results of these tests, it found that dredged soil of the Bilibili dam in the Gowa District, South Sulawesi, is typical of silt soil with a low plasticity index. Furthermore, there was an increase in the maximum dry density of the stabilized soil to the untreated soil by 0.5% - 1.7%, and the optimum moisture content increased by 3.8% - 4.7%. It also revealed that the compressive strength test of the stabilized soil had shown an increase of UCS value by 175% - 324%. Finally, there was an increased shear strength value, and it was found that the increase in cohesion value (c) was 6% - 83%, and the increase in internal friction angle value (ϕ) was 31% - 98%.

6. References

[1] Smith T Braden, Howard L Isaac and Vahedifard F 2018 Lightly cemented dredged sediments for sustainable reuse *Journal of Environmental Geotechnics* **5** 324

[2] Beeghly J and Schrock M 2010 Dredge material stabilisation using the pozzolanic or sulfo pozzolanic reaction of lime by-product to make an engineered structural fill *Internasional Journal of Soil, Sediment and Water* **3**, 1

[3] Yu H, Yin J, Soleimanbeigi A and Likos J William 2017 Effects of curing time and fly ash content on properties of stabilized dredged material *J. Mater. Civ. Eng.* **29** 1

[4] Nguyen M Thuy, Rabbanifar S, Brak N, Qian Q, Kibodeaux K and Croche H 2018 Stabilization of silty clayey dredged material *Journal of Materials in Civil Engineering* **30** 1

[5] Jan O and Mir B 2019 Proceedings of the 2nd GeoMEast International Congress and Exhibition on Sustainable Civil Infrastructures ed H. Shehata and H. Pulllos (Cairo: Egypt) p 132

[6] Harianto T and Sitepu F 2019 Strength improvement of cement stabilized soil by binder mineral additive *Lowland Technology International* **21** 90

[7] Cicek B, Martins N, Brumaud C, Habert G and Plotze M 2020 *IOP Conf. Ser.: Earth Environ. Sci.* **588**

[8] Agawane N, Pachpute S, Jadhav T, Ingle J and Motharkar S 2019 Soil stabilization using waste rubber *International Journal of Scientific & Engineering Research* **10** 71

[9] Kumar S, Singh J and Mishra M 2019 Stability investigation of clayey soil using shredded tyre and lime *International Journal of Advanced Scientific Research and Management* **4** 31

[10] Harianto T, Marfu'ah N, Rauf I and Leatemia T B C 2020 *IOP Conf. Ser.: Materials Science and Engineering* **875**

Acknowledgements

The authors would like to express their deepest gratitude for the financial support from the Education Fund Management Institute (LPDP), Ministry of Finance of the Republic of Indonesia.

The Effect of Bamboo Water Content on The Tensile Strength of Bamboo

A A Amiruddin[1)] H Parung[1)], N Sibela[2)], M N Fajar[2)] H Arifin[2)]

[1)] Civil Engineering Department, Hasanuddin University, Indonesia
[2)] Magister Student, Civil Engineering Department, Hasanuddin University, Indonesia

Coresponding Author : arwin_amiruddin@unhas.ac.id

Abstract. In today's world of construction, many use steel as reinforcement for reinforced concrete, where steel is one of the most widely produced materials to meet infrastructure needs. The use of steel is so rapid but has several drawbacks such as a less economical price and not environmentally friendly. In addition, the characteristics of bamboo plant materials that can be an alternative, where bamboo has a fiber that is good enough to withstand the same tensile strength as steel reinforcement that works in concrete construction. However, some things that must be considered are the air content which is quite influential on the tensile strength. This research focuses on the effect of water content in bamboo on the tensile strength of bamboo itself. The bamboo material used is Buluhpring bamboo from the Gowa Region, the bamboo is dried and tested for its characteristics. From the results of the water content test, the average water content was 34.44% on the bottom of bamboo, 25.89% on the middle bamboo and 14.89% on the upper bamboo where the Upper segment of bamboo have met ISO 22157 standards. -01-2004. The bamboo material was also tested for tensile strength of 277.54 Mpa at the bottom, 308.51 Mpa at the middle and 357.10 Mpa at the top. From the test results, it can be seen that the relationship between the water content of bamboo and the tensile strength of bamboo is that the smaller the water content contained in the bamboo, the higher the tensile strength of the bamboo material accepted by bamboo.

Keywords : Bamboo Buluhpring, Water Content, Tensile Strength

1. Introduction

Because of its many benefits, including its high compressive strength, resistance to fire and water, and ease of processing, concrete is becoming more and more popular in the construction industry. However, concrete cannot be used alone in its application, so steel reinforcement is used to cover the mechanical properties of the concrete. weak under stress. The issue arises from the fact that iron ore, which is the primary component of steel, is a non-renewable natural resource [1] and that, as a result of repeated consumption, its supply in nature is becoming depleted. Consequently, the cost of steel reinforcement is fluctuating [2].

Steel reinforcement corrosion is another issue that requires attention, particularly in buildings located in areas that are susceptible to corrosion, such as coastal areas. Steel reinforcement for

concrete can be replaced by non-metallic reinforcement, both synthetic and natural. In reinforced concrete structures, bamboo can be used as reinforcement in place of natural elements like steel. Morisco has looked into the viability of employing bamboo as reinforcement (1999) [3]
Morisco observed that the tensile strength of ori bamboo bark was quite high, being close to 5000kg/cm2 or around double the yield stress of steel, based on his research in 1994–1999 comparing the tensile strength of ori and petung bamboo with structural steel having a yield stress of 2400kg/cm2. The average tensile strength for specimens made of petung bamboo is higher than the yield stress of steel, with only one specimen having a lower tensile strength [4].
In addition to its great degree of versatility, bamboo has a number of drawbacks, including: difficult to work with due to its propensity to break or fracture; often attacked by insects that destroy wood; dimensional differences; and non-uniform segment length. Analyzing the physical characteristics of bamboo is crucial to overcome its flaws. The moisture level of bamboo is one of the physical characteristics that influences its strength; for bamboo to be utilized as a construction material, it must have a low water content to prevent significant shrinking. The percentage of water in a piece of bamboo is indicated by the term "water content.".

The goal of the problem formulation is to ascertain how water content affects tensile strength. The goal of this study was to determine whether bamboo pering type reeds could be utilized as a substitute for steel reinforcement and how much the water content value of bamboo's tensile strength was affected by it.

2. Material

2.1. Bamboo

The Buluhpring type of bamboo, which grows in the Gowa Regency area and is collected straight from the field by choosing bamboo for at least two years or more, is used as a substitute for steel reinforcement. The average trunk length of bamboo is between 800mm - 900mm. The bamboo is then separated into three sections, the bottom, middle, and top, each measuring around 300mm in length. Additionally, the three components of the bamboo will undergo mechanical and air testing as well as a parallel bamboo strength test to assess their properties. Example of the Buluhpring type of bamboo can be seen in figure 1:

Figure 1. Buluhpring Bamboo

3. Testing Methods

3.1. Bamboo water content

For the bamboo characteristic test, the water content of bamboo was examined by selecting three samples from each part of bamboo measuring 20 mm x 30 mm. According to ISO 22157-01-2004 [5], bamboo can have a maximum moisture content of 20% when used as building material. The following formula can be used to determine how much water is in bamboo:

$$Bamboo\ Water\ Content\ (\%) = \frac{initial\ weight - oven\ dry\ weight}{oven\ dry\ weight} \times 100\% \quad (1)$$

3.2. Bamboo tensile strength

Using a universal testing machine, a tensile test of bamboo will be performed to establish its mechanical characteristics in the form of tensile strength parallel to the fiber (UTM). Whereas in this test, a test object with dimensions of 300 millimeters in length, 10 to 20 millimeters in breadth, and 5 to 10 millimeters in thickness of bamboo was used.:

Figure 2. Bamboo Tensile Strength Test

In order to determine the tensile strength of bamboo, data from tensile strength tests conducted with the Universal Testing Machine (UTM) will be used to calculate the Maximum Tensile Load, which is then divided by the cross-sectional area of the test object.:

$$Tensile\ Strength\ (MPa) = \frac{Maximum\ Load\ (N)}{Bamboo\ Cross\ Sectonal\ Area\ (mm)} \quad (2)$$

4. Result and Discussion

4.1. Bamboo water content test result

Table 1 shows the results of the bamboo moisture content test that was performed on the three parts of the bamboo, namely the bottom, middle, and Upper parts:

Table 1. Bamboo Water Content Test Result

Segment	No. Sampel	Wet Weight (gr)	Dry Weight (gr)	Moisture Content (%)
Bottom	B1	15.86	11.80	34.44
Middle	B2	10.45	8.30	25.89
Upper	B3	9.19	8.00	14.89

Table 1 shows that the lower and middle bamboo have relatively high moisture contents of 34.44% at the bottom and 25.89% at the middle due to the water content. this is because the middle bamboo has a large yield since the middle has a lot of nodia, and the lower bamboo sticks are close to the ground level. Additionally, the results of testing bamboo's upper moisture content with ISO 22157-01-2004 standards, which is 14.89% where the result is less than 20%, can be observed in figure 3. Figure 3 compares the results of testing bamboo's upper moisture content with ISO 22157-01-2004 standards.

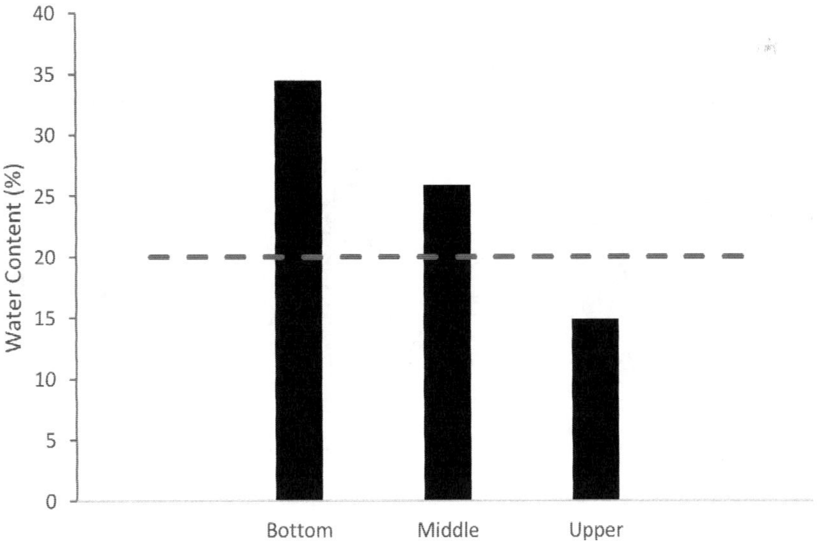

Figure 3. Water Content Diagram

4.2. Bamboo tensile strength test

Table 2 and figure 4 shows the results of the bamboo tensile strength test on the bottom, middle, and upper parts:

Table 2. B*amboo Tensile Strength Test Result*

Segment	No. Uji	Weight (gr)	Wide (mm)	Thick (mm)	Max. Load (N)	Tensile Strenght (Mpa)	Average (Mpa)
	B1	24.51	5.00	8.64	15480.00	358.33	
Bottom	B2	32.11	5.00	8.36	10000.00	239.23	277.54
	B3	35.75	5.00	9.24	10860.00	235.06	
	T1	21.18	5.00	5.50	8340.00	303.27	
Middle	T2	21.80	5.00	5.48	8720.00	318.25	308.51
	T3	21.92	5.00	5.50	8360.00	304.00	
	A1	16.92	5.00	4.30	9700.00	451.16	
Upper	A2	17.35	5.00	4.38	8360.00	381.74	357.10
	A3	17.91	5.00	5.00	5960.00	238.40	

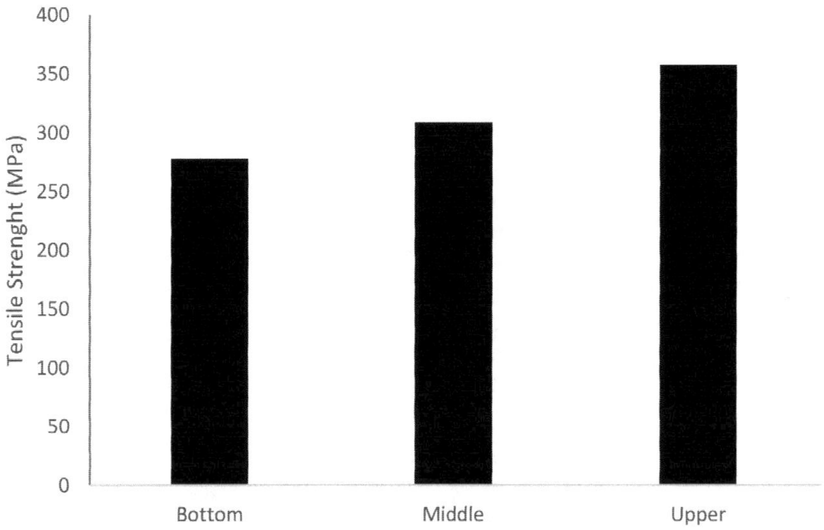

Figure 4. Tensile Strength Diagram

the lower bamboo has an average tensile strength of 107.81MPa, the middle bamboo has a tensile strength of 397.08MPa, and the upper bamboo has a tensile strength of 318.52MPa. The middle of the three bamboo pieces examined had the highest tensile strength findings, according to the test.

4.3. Correlation of water content and tensile strength

The test findings are shown in the graph, which shows that the allowable tensile strength of bamboo decreases with increasing water content and increases with decreasing water content, as shown by the test results.

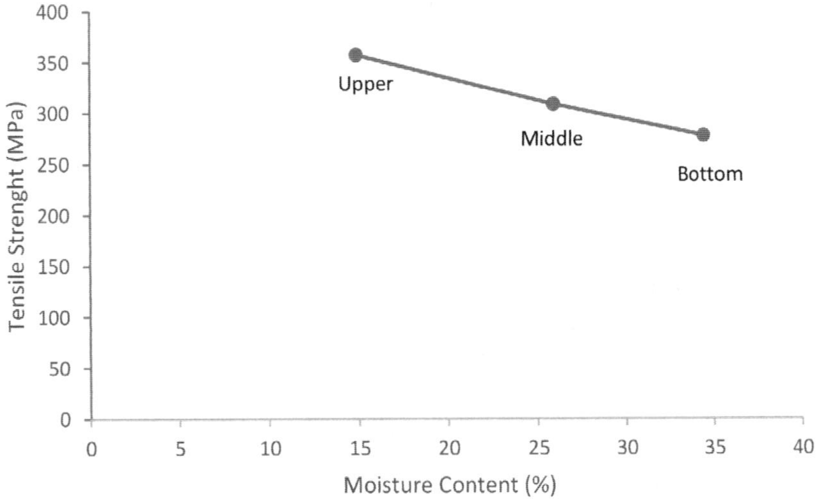

Figure 5. Correlation of Water Content and Tensile Strength

5. Conclusion

According to the findings of testing the water content of bamboo, the middle and top bamboo both passed the requirements of ISO 22157-01-2004, with an average water content of 24.27% on the bottom bamboo, 12.87% on the middle bamboo, and 19.96% on the upper bamboo. Tensile strength tests were also performed on the bamboo material, yielding results of 107.81MPa at the bottom, 397.08MPa in the middle, and 318.52MPa at the top. According to the test results, there is a direct correlation between bamboo's water content and its tensile strength: the less water that is present in bamboo, the more tensile strength the bamboo material can withstand.

6. References

[1] Budi, Setiya, Agus, Dkk, 2013, "*Model Balok Beton Bertulangan Bambu Sebagai Pengganti Tulangan Baja*". Prosiding Konferensi Nasional Teknik Sipil 7. Surakarta: UNS. Hal: S-245 – S-252.

[2] Surya, S dan Farlin, R. (2019) "*Variasi rajutan bambu petung sebagai pengganti tulangan baja dari kuat lentur fs 4,7 MPa*", Jurnal forum mekanika: Universitas Bina Darma-Palembang.

[3] I Ketut, S. Dharma, P. dan I Gusti APWP. "*Kuat Lekat Tulangan Bambu Petung dan Bambu Tali dalam Beton Normal*", Program Studi Teknik Sipil, Fakultas Teknik, Universitas Udayana.

[4] Alif, SA. Agus, SB. dan Sunarmasto. "*Kuat Lekat Tulangan Bambu Petung Bertakikan Tipe "U" Jarak 15 Cm*", Teknik Sipil, Universitas Sebelas Maret: Surakarta

[5] International Organization for Standardization (ISO) 2004 . *Bamboo_Determination of Physicial and Mechanical Properties.* ISO 22157-01-2004. International Organization for Standardization.

Optimization of Self-Healing Concrete Parameters using Experimental Design Taguchi Method

D Abdurrahman[1]**, and H Putra**[2]

[1,2] Department of Civil and Environmental Engineering, IPB University, Dramaga, Bogor, 16680, Indonesia

Correspondence author's : heriansyahptr@apps.ipb.ac.id

Abstract. The development of self-healing concrete (SHC) is intensively implemented to improve stability and reduce maintenance costs. However, the experiments require many parameter variations, increasing the cost. Taguchi methods are used in experimental design to obtain variations of the optimal parameters and reduce the cost of the experiment. The stages of this study are secondary data collection of the increase of compressive strength, testing Taguchi methods, determining contribution parameters using ANOVA, and developing the estimated model. The parameters analyzed consist of the type, the concentration, the application of bacteria, and the curing time conducted as experimental analysis of 2, 3, and 4 parameters. The results of the parameter analysis using the Taguchi have optimal variation results in a row *Bacillus subtilis,* 10^5 cells /ml, encapsulation methods, and 28 days. Bacterial concentrations have a very significant value with contributions of 82.94%, 78.90%, and 74.36% in the analysis of 2, 3, and 4 parameters, respectively. The slightest error value was obtained in the analysis of the 3-parameter experiment with a value of 2.99%. SHC modelling obtained two equations consisting of direct mixing and encapsulation, which have a coefficient of determination of 0.89 and 0.82.

1. Introduction

Cracking is one of the main problems affecting a concrete structure's strength and durability, resulting in water and various corrosive chemicals entering the concrete [1]. Untreated cracks in concrete will become an influx chamber for particles that can cause damage over time. Cracks that occur in concrete are progressive. Therefore, the potential cracks in concrete must be prevented early, which can affect the strength and carrying capacity of concrete. If cracks occur on an ongoing basis, the strength of the concrete can be reduced so that the cost of repairing structures using concrete can increase [2]. In the Netherlands, a third of the annual budget for civil engineering work is spent on inspection, monitoring, maintenance, upgrading and repair. The UK government spends 45% of the infrastructure budget on concrete maintenance costs [3].

The development of self-healing concrete (SHC) is intensively carried out to repair cracks in concrete and reduce maintenance costs. Among the many approaches, microbial methods are considered the most promising commercialisation in the coming years. As an environmentally friendly method, the most significant advantage of microbial self-healing technology is that it can repair cracks through the metabolic activity of bacteria in places that are difficult to access [1]. According to Khaliq and Ehsan [4], using microorganisms has been shown to increase the compressive strength of concrete by closing the pores with lime so that the concrete is more potent and denser. In addition, this application is expected to prevent premature cracking of concrete by producing minerals by microbial activity in concrete. The type of bacteria used for SHC is *Bacillus* sp. the genus. Such as *Bacillus subtilis*, *Bacillus cereus*, and *Bacilus sphaericus*. Genus *Bacillus sp.* can withstand high alkaline conditions such as found in cement used for concrete. Significantly, it can reduce the cost of crack detection and repair difficulty and improve the service time of concrete structures.

According to Srinivasa *et al.* [5], The cost of producing SHC is 2.3 to 3.9 times that of low-quality ordinary concrete. However, the trials need several parameter modifications, which raises the expense. This is one of the primary reasons why bacteria-infused concrete is not mass-produced and is not extensive employ in the building sector. Therefore, an experimental design is needed before researching SHC to obtain an optimal composition based on statistical analysis such as the Taguchi method. The Taguchi technique is one well-known experimental design. The Taguchi method's experimental design is used to enhance the quality of a product and process while reducing costs and resources to a minimum. Furthermore, the Taguchi method is low-cost and time-efficient since it reduces the quantity of mixture to asses while considering each level of parameters. The Taguchi method approach also gives a more detailed study of the impact of each parameter level on the resulting attribute examined [6]. The Taguchi method experimental design has been applied based on research on soil strengthening parameters using MICP (microbially induced carbonate precipitation) conducted Sadeghi *et al.* [7]. The study evaluated the parameters using bacterial concentration, urea concentration, inoculum size, curing time, and soil type using the Taguchi method design to obtain optimal conditions. The result of the confidence level obtained was 95%, so no confirmation test was needed in the study. Based on this, it is necessary to analyze and evaluate SHC parameters using the Taguchi method's experimental design.

2. Method

2.1. Parameter Identification and Analysis

This research begins with identifying and analyzing the parameters that affect self-healing concrete (SHC). These parameters were obtained by conducting a literature study conducted from 2011-2022. The parameters specified in this study are presented in Table 1. These parameters are identified and analyzed to determine the parameters that affect SHC.

Table 1. Selected parameters along with a level

No	Parameter	Level			
		1	2	3	4
1	Type of bacteria	*Bacillus subtilis*	*Bacillus sphaericus*	-	-
2	Bacterial concentration (cells/ml)	10^3	10^5	10^6	10^7
3	Bacterial application	Encapsulation	Direct mixing	-	-
4	Curing time (day)	7	28	-	-

The parameters and levels presented in Table 1 are adopted from several studies. SHC research conducted by Jena *et al.* [8] used *Bacillus subtilis* bacteria with concentrations of 10^2-10^6 cells/ml and conducted compressive strength testing on 7, 14, and 28 days. The bacteria are applied to the concrete encapsulated using gelatin. Kashyap and Radhakrishna [9] used *Bacillus sphaericus* bacteria with concentrations of 10^5-10^8 cells/ml with compressive strength testing at 7 and 28 days. *Bacillus sphaericus* is encapsulated using agar into the concrete mixture. Reddy and Revathi [10] used *Bacillus sphaericus* bacteria mixed directly into a concrete mixture at concentrations of 10^3-10^7 cells/ml with compressive strength testing at 3, 7, 28, 56, and 90 days. Rao *et al.* [11] used *Bacillus subtilis* bacteria mixed directly into a concrete mixture with concentrations of 10^4-10^7 cells/ml. Nimafar *et al.* [12] used the bacteria *Bacillus sphaericus* and *Sporosarcina pasteurii* mixed into concrete encapsulated using agar with a concentration of 10^7 cells/ml. Compressive strength testing is carried out when the concrete is 28 days old. The plan compressive strength value of the four studies ranged from 20-40 Mpa.

Based on these studies, this study used four parameters: the type of bacteria, the concentration of bacteria, the application of bacteria, and the time of curing. The parameters of the type and application of bacteria are determined as control variables consisting of *Bacillus subtilis* and *Bacillus sphaericus* by mixing into the concrete through encapsulation and direct mixing. In addition, the parameters of bacterial concentration along with the curing time become free variables. The bacterial concentration was 10^3, 10^5, 10^6, and 10^7 cells/ml with a curing time of 7 and 28 days. These parameters are the criteria for SHC testing observed using the Taguchi method. The bound variables that respond to SHC testing are the increase in compressive strength. Data on the increase in compressive strength based on such literature are presented in Table 2.

Table 2. Compressive strength increase data from literature study

Type of Bacteria	Source	Bacterial Concentration (cells/ml)	Compressive Strength Increase of 7 Day (%)	Compressive Strength Increase of 28 Day (%)	Bacterial Application
Bacillus subtilis	Jena *et al.* (2020)	0	0	0	Encapsulation
		10^3	14.034	18.15	
		10^5	27.099	31.82	
		10^6	15.848	21.62	
	Rao *et al.* (2017)	0	0	0	Direct mixing
		10^5	20.63	19.26	
		10^6	15.46	10.42	
		10^7	7.48	5.50	
Bacillus sphaericus	Reddy and Revathi (2019)	0	0	0	Direct mixing
		10^3	9.06	11.52	
		10^5	16.70	20.74	
		10^7	13.43	16.59	
	Kashyap and Radhakrishna (2013)	0	0	0	Encapsulation
		10^6	3.78	17.93	
	Nimafar *et al.* (2021)	0	0	0	
		10^7	6.40	6.40	

2.2. Testing the Taguchi Method

The parameters specified in the previous stage are processed and analyzed using the Taguchi method using Minitab 20 software to determine the number of tests to be carried out. However, the response from the test used secondary data in the form of an increase in compressive strength. In this study, these parameters were analyzed and designed in three experimental analyses presented in Table 3.

Table 3. Experimental design based on Taguchi Method

Parameter	Trial Analysis		
	2 Parameters	3 Parameters	4 Parameters
Type of Bacteria	√	√	√
Bacterial Concentration	√	√	√
Bacterial Application	-	√	√
Curing Time	-	-	√

√ : Analyzed parameters
- : Unanalyzed parameters

Based on the table, experiments 1. 2, and 3 have different numbers of parameters to analyze. In this study, the test combination of each experiment was designed by Minitab statistical software using an orthogonal matrix or orthogonal array. According to Sutoni [13], orthogonal arrays are designed matrices, where the design of the matrix is determined based on the number of degrees of freedom. In addition, orthogonal arrays are used to determine the number of combinations of experiments that can inform as much as possible of all the factors affecting the response. This matrix is used to determine the number of experiments to a minimum [14]. The orthogonal arrays selected in the analysis of experiments 1, 2, and 3, respectively, were L8 (4^1 x 2^1), L8 (4^1 x 22), and L8 (4^1 x 2^3). Tables 4-6 show the number of tests with an array orthogonal test matrix. The number of tests carried out was eight times.

Table 4. *Orthogonal array* L8 (4^1 x 2^1) of 2 parameters experimental analysis

Trial Number	Controlling Factors and Levels	
	A	B
1	1	1
2	2	1
3	1	2
4	2	2
5	1	3
6	2	3
7	1	4
8	2	4

Table 5. Orthogonal array L8 (4^1 x 2^2) of 3 parameters experimental analysis

Trial Number	Controlling Factors and Levels		
	A	B	C
1	1	1	1
2	2	1	2
3	1	2	1
4	2	2	2
5	1	3	2
6	2	3	1
7	1	4	2
8	2	4	1

Table 6. Orthogonal array L8 (4^1 x 2^3) of 4 parameters experimental analysis

Trial Number	Controlling Factors and Levels			
	A	B	C	D
1	1	1	1	1
2	2	1	2	2
3	1	2	1	2
4	2	2	2	1
5	1	3	2	1
6	2	3	1	2
7	1	4	2	2
8	2	4	1	1

The following process is carried out the signal-to-noise ratio (SNR) analysis. SNR analysis was performed to determine the factors affecting response variations in the Taguchi method. The SNR characteristic used in this study is SNR the more significant, the better. The SNR, the larger better it is used whenever the desired quality characteristics increase the value [15]. The equation SNR larger the better can be seen in Equation (1).

$$SNR_{ltb} = -10 \log\left(\frac{1}{n}\sum_{i=1}^{n}\frac{1}{y_i^2}\right) \quad (1)$$

2.3. Contribution of Each Parameter and Modelling

The contribution of each parameter is obtained using the Analysis of Variance method. Testing using the ANOVA method was carried out using Minitab 20 software. This test was carried out to determine the per cent value of the contribution, and the error value can be obtained in the analysis of experiments 2, 3, and 4 parameters. The significance rate used was 5% (0.05). The value can be specified as the error limit at this stage. Furthermore, the estimation model in this study was determined from the degree of significance of each parameter obtained from the per cent contribution and considered the results of the Taguchi test.

3. Results and Discussion

3.1. Taguchi Method Analysis

The type, concentration, application of bacteria, and curing time can affect the increase of compressive strength in self-healing concrete (SHC). It can be analyzed using the Taguchi Method to determine the parameters and the optimal level. These parameters are analyzed using the Taguchi Method to determine the optimal value of the variations present in each parameter. The Taguchi experimental design experiment was carried out in three parts: the analysis of experiments of 2, 3, and 4 parameters. The orthogonal array analysis of the experiment, the value of the increase of compressive strength and SNR larger the better, can be seen in Tables 7-9. The increase of compressive strength is secondary data, while the value of the SNR larger the better is calculated using Equation (1).

The SNR is used to identify factors influencing response variations in the Taguchi method. Based on analysis using Minitab 20, the results obtained are in the form of a graph of the relationship between each level in the parameter to the SNR provision larger is a better analysis of experiments 2, 3, and 4 parameters. The SNR is used to measure the variance of a design experiment. The SNR ratio is the ratio of the mean (signal) to standard deviation (noise) [16]. Graph response of the effect values analysis of experiments 2,3, and 4 parameters can be seen in Figures. 1-3, respectively.

Table 7. Test result response of 2 parameters

Bacterial Concentration (cells/ml)	Type of Bacteria	The Increases of Compressive Strength (%)	SNR Larger the Better
10^3	*Bacillus subtilis*	18.15	25.18
10^3	*Bacillus sphaericus*	11.52	21.23
10^5	*Bacillus subtilis*	31.82	30.05
10^5	*Bacillus sphaericus*	20.74	26.33
10^6	*Bacillus subtilis*	21.62	26.70
10^6	*Bacillus sphaericus*	17.93	25.07
10^7	*Bacillus subtilis*	5.50	14.81
10^7	*Bacillus sphaericus*	6.40	16.12

Table 8. Test result response of 3 parameters

Bacterial Concentration (cells/ml)	Type of Bacteria	Bacterial Application	The Increases of Compressive Strength (%)	SNR Larger the Better
10^3	*Bacillus subtilis*	*Encapsulation*	18.15	25.18
10^3	*Bacillus sphaericus*	*Direct mixing*	11.52	21.23
10^5	*Bacillus subtilis*	*Encapsulation*	31.82	30.05
10^5	*Bacillus sphaericus*	*Direct mixing*	20.74	26.33
10^6	*Bacillus subtilis*	*Direct mixing*	10.42	20.36
10^6	*Bacillus sphaericus*	*Encapsulation*	17.93	25.07
10^7	*Bacillus subtilis*	*Direct mixing*	5.50	14.81
10^7	*Bacillus sphaericus*	*Encapsulation*	6.40	16.12

Table 9. Test result response of 4 parameters

Bacterial Concentration (cells/ml)	Type of Bacteria	Bacterial Application	Curing Time (day)	The Increase of Compressive Strength (%)	SNR Larger the Better
10^3	*Bacillus subtilis*	*Encapsulation*	7	14.03	22.94
10^3	*Bacillus sphaericus*	*Direct mixing*	28	11.52	21.23
10^5	*Bacillus subtilis*	*Encapsulation*	28	31.82	30.05
10^5	*Bacillus sphaericus*	*Direct mixing*	7	16.70	24.45
10^6	*Bacillus subtilis*	*Direct mixing*	7	15.46	23.78
10^6	*Bacillus sphaericus*	*Encapsulation*	28	17.93	25.07
10^7	*Bacillus subtilis*	*Direct mixing*	28	5.50	14.81
10^7	*Bacillus sphaericus*	*Encapsulation*	7	6.40	16.13

The analysis showed that each parameter had a different response value to SNR larger, the better. As shown in figures 1-3, there are some similarities in the results. The bacterial concentrations of the three experimental analyses had the same optimal value of 10^5 cells/ml. Bacterial type parameters have optimal results in *Bacillus subtilis* with the application of bacteria by encapsulation and optimal curing time for 28 days. The SNR value in each optimal parameter is the average based on the orthogonal array in each experiment analysis. *Bacillus subtilis* had SNR values of 24.18, 22.60, and 22.90, while *Bacillus sphaericus* was 22.19, 22.19, and 21.72 from the analysis of experiments 2, 3, and 4 parameters. Concentrations of 10^5 cells/ml had the most significant SNR values compared to other concentrations, namely 25.88, 22.72, and 24.43, from the analysis of experiments 2, 3, and 4 parameters. The optimal method of applying bacteria, namely encapsulation, has SNR values of 24.11 and 23.55 in the analysis of experiments 3 and 4 parameters. The optimal curing time parameter was found at 28 days with an SNR value of 22.79 in the 4-parameter experimental analysis. These parameters become optimal because they adjust the response provisions, namely larger, the better. A high SNR number indicates that the signal is substantially more robust than the random noise effect. Noise variables investigate potential sources of performance fluctuation and why the product failed. Noise factors are anything that causes a measured product's or process's characteristics to differ from its target value [17]. Therefore, the more excellent SNR value is the best. The type and application of bacteria are one of the parameters that became the control variable in this study.

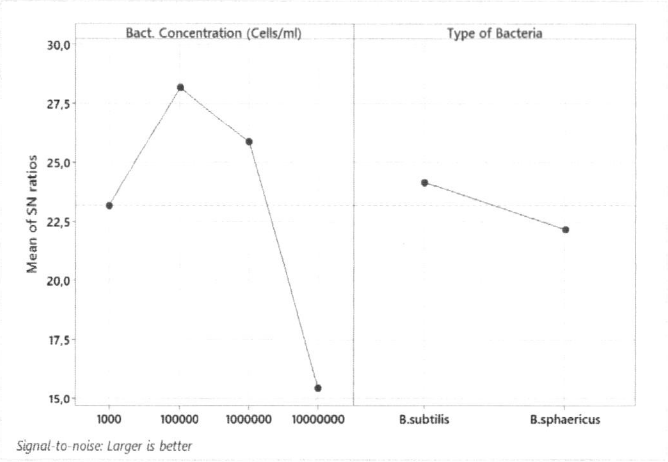

Figure 1. The response graph of the 2 parameter experimental analysis effect value

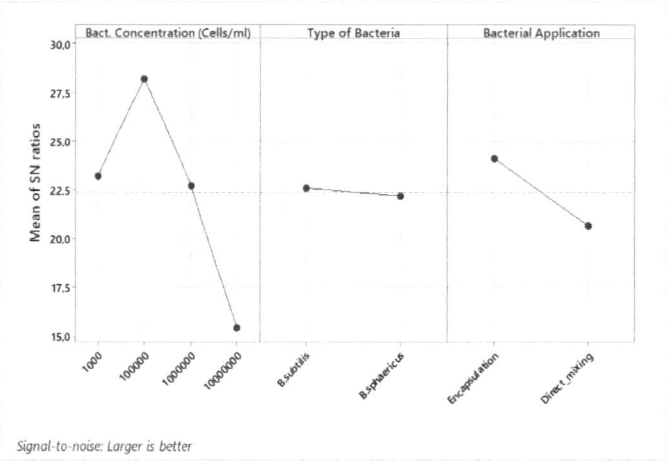

Figure 2. The response graph of the 3 parameter experimental analysis effect value

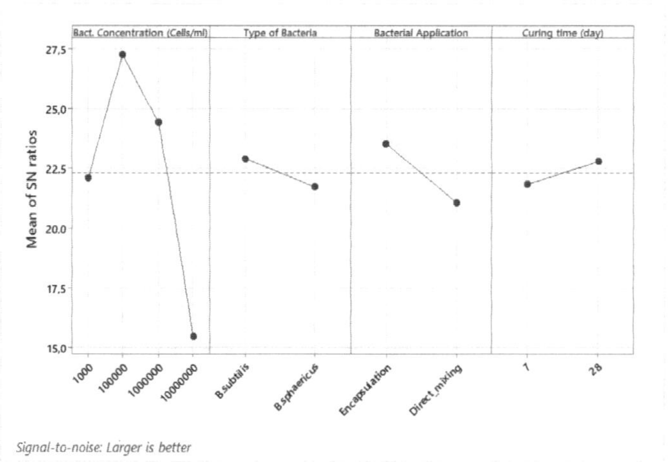

Figure 3. The response graph of the 4 parameter experimental analysis effect value

3.2. The Influence of The Type and Application of Bacteria
Based on the analysis of the Taguchi method, the type and application of bacteria that are optimal or have the greatest increase in compressive strength are found in the type of *Bacillus subtilis* by the encapsulation method. According to Danish *et al.* [18], encapsulation to produce SHC is a better choice because it shows good reaction results between unhydrated cement particles and healing agents. In addition, the difference in direct mixing and encapsulation methods can be seen in the life resistance of bacterial cells in concrete. The lifespan resistance in concrete is affected by a high alkaline pH that can affect bacteria [19]. In the direct mixing method of bacteria into concrete, microbial metabolic activity is influenced by high pH and alkali and concrete conditions that tend to be dry, causing bacteria to be prone to die [20]. As a result, encapsulation becomes necessary protection for this technique because it can maintain the life of bacteria inside the concrete matrix in the long term [21].

Elmahdy *et al.* [22] conducted SHC testing using several types of bacteria of the genus *Bacillus*. One of the two is *Bacillus subtilis* and *Bacillus sphaericus*. Based on the results obtained, the increase in compressive strength in 28-day-old concrete using *Bacillus subtilis* and *Bacillus sphaericus* by 22.2% and 8.7% compared to normal concrete aged 28 days. As a result, an increase in compressive strength employing *Bacillus subtilis* yields more than *Bacillus sphaericus*. However, compared to standard samples, most *Bacillus* bacteria had a favourable influence on the compressive strength and bending strength of concrete. *Bacillus sp.* is a bacterium that can thrive in alkaline environments using metabolic activities such as sulfate reduction, photosynthesis, and urea hydrolysis. *Bacillus sp.* bacteria are not harmful to humans in any way. [23]. Thus, the two types of bacteria are the types of bacteria that are most often used in SHC research.

3.3. Optimization for Bacterial Concentration
The optimal bacterial concentration obtained 10^5 cells/ml results in all three experimental analyses. In addition, optimal concentration results occur in the types of bacteria *Bacillus subtilis* and *Bacillus sphaericus,* as well as direct mixing and encapsulation applications. The increase in compressive strength using the concentration of these bacteria produces the most significant value compared to the concentration of other bacteria. The compressive strength value in SHC is higher when the concentration of bacteria used is more excellent. However, the concentration of these bacteria has optimal results at 10^5 cells/ml; if the concentration of bacteria is increased again, it will cause a decrease in the compressive strength value. According to Zawad *et al.* [24], using bacterial culture helps fill the pores between cement and microstructures in concrete to obtain greater compressive strength and become denser. Manjunath *et al.* [25] tested the mechanical properties of concrete, chloride permeability and water absorption and fly ash replacing cement at 10% and 20% with bacterial solutions of 10^3, 10^5, and 10^7 cells/ml using *Bacillus sphaericus* at the age of 28 days. In general, the mechanical properties of concrete are enhanced by bacteria. Better results were obtained in a bacterial solution of 10^5 cells/ml. In addition, Akindahunsi *et al.* [26] performed the same analyzes with *Bacillus subtilis* at concentrations of 10^0, 10^5, 10^7, and 10^9 cells/ml. The highest compressive strength was 29.9 Mpa at 10^5 cells/ml concentration. Therefore, the optimal concentration of the two tests was 10^5 cells/ml, according to the results obtained in this study.

Research conducted by [27] shows that the compressive strength of SHC is maximum when the bacterial concentration is 10^5 cells/ml. The precipitation of microbial-induced calcium carbonate precipitation reduces water absorption in bacterial concrete at such concentrations. After 28 days of treatment, the percentage of water absorption decreased by 50% in bacterial concrete with an optimal concentration compared to bacteria-free concrete. In contrast, water absorption decreased by 16-34 per cent in bacterial concrete with a concentration of bacteria other than 10^5 cells/ml. Adding water to fresh concrete aggregates can improve the adherence of cement paste to the aggregate, causing the concrete to harden. This can result in increased concrete strength and durability. [28]. In the study Achal *et al.* [29], the decrease in the permeability of mortar specimens treated with bacteria can be seen from water absorption experiments. The deposition of a layer of calcium carbonate crystals on the surface decreases permeation properties. Fresh concrete that produces $CaCO_3$ has more cohesive properties than OPC concrete because the pleasing filler effect of $CaCO_3$ particles ensures optimal water absorption of cement particles to carry out the hydration process and can improve the compactness and uniformity of fresh concrete. The fineness of the $CaCO_3$ powder causes this effect. The fine $CaCO_3$ powder physically disperses the cement floc, allowing more paste to use to lubricate the aggregate and increase working ability. The filler replaces the sand with fine surface particles because the filler articles enclose the cavity, reducing friction between the particles and boosting the slump. Because of the superior packing of particles in cement and $CaCO_3$ and the low degree of water absorption, the presence of suitable $CaCO_3$ particles lowers the pores inside the concrete. Lower water absorption in $CaCO_3$ concrete can be due to enhanced aggregate matrix bonding generating less porous transition zones [30]. Therefore, SHC's water absorption is smaller than normal concrete.

3.4. *Optimization of Curing Time*

Concrete curing aims to maintain sufficient humidity and temperature to encourage cement hydration on concrete for a sufficient period. Proper curing of concrete is essential to obtain maximum design strength and durability, especially for concrete exposed to extreme environmental conditions early on [31]. Curing is a process of controlling the rate and water loss from concrete during cement hydration. Overall, the strength of concrete, its durability and other physical properties are influenced by cure [32]. In general, concrete curing time consists of several days, namely, 3, 7, 14, 21, 28, 56, or 90 days. However, in this study, the curing time was compared for 7 and 28 days.

The Taguchi method test results obtained an optimal curing time of 28 days. The compressive strength value has increased from 7 to 28 days during the curing period. The curing time of 28 days produces the most robust compressive strength in each specified control variable, namely the type and application of bacteria. According to the American Society for Testing and Materials [33], the ratio of the increase in compressive strength in normal concrete to the curing time has resulted in the most significant increase in compressive strength at 28 days, which is 99%, while in 7 days it has the result of an increase in compressive strength by 65%.

3.5. *Contribute Each Parameters*

The contribution value of each parameter is obtained using the variance analysis method. The contribution value of each parameter in each experiment has a different result. In addition, the error value of each test analysis is also obtained. Per cent graph of each parameter's contribution to the experiment's analysis can be seen in Figure. 4.

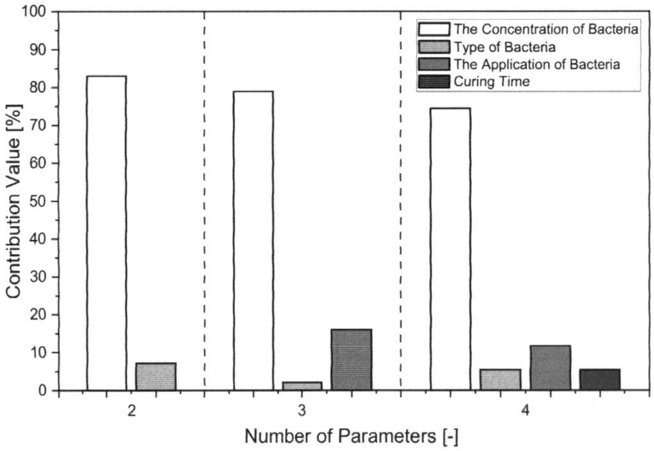

Figure 4. Percent contribution of parameters of each experimental analysis

Based on Figure. 4, the per cent contribution of each parameter has a different result. From each analysis of the experiment, the per cent of the most significant parameter contribution was the concentration of bacteria. The percentage of the contribution of bacterial concentration obtained in the analysis of experiments 2, 3, and 4 parameters was 82.94%, 78.90%, and 74.36%, respectively. This bacterial cell covers the tiny fractures and holes in the concrete matrix and functions as a micro-filter. Additionally, by manufacturing the enzyme urease, bacterial cells play a crucial part in the precipitation of calcium carbonate. More bacterial cells will result in the production of more urease enzymes. The urease enzyme hydrolyzes the urea into carbonate, which combines with calcium to create $CaCO_3$ on the cell surface. The urolithic bacteria will be found and encapsulated by this calcium carbonate, eventually producing sizable bacterial groups. [34]. This causes the concentration of bacteria to be a significant parameter affecting the compressive strength, resulting in SHC.

The parameters of application of bacteria become the parameters with the most significant contribution after the concentration of bacteria. The application of bacteria by direct mixing or encapsulation is used to increase the resistance or age of bacteria in concrete from the high alkaline properties of concrete and concrete conditions that tend to be dry [19]. However, the bacterial type parameter has a minor contribution to the analysis of experiments 2 and 3 but not in the analysis of experiment 4 parameters. According to Abdullah *et al.* [35], any bacterium can create autogenous healing concrete if it has the metabolic capacity to convert calcium sources into calcium carbonate. Additionally, the species of bacteria identified in this study are included in the category of microorganisms frequently employed in SHC research. Curing time has a minor contribution to the analysis of the four parameters experiment because the relationship between the curing time and the increase in compressive strength is linear.

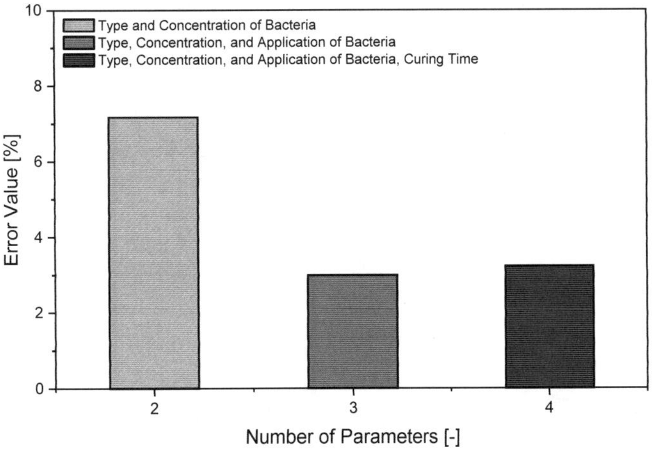

Figure 5. Comparison graph of the error value of each trial

A comparison of the error values on each experiment is presented in Figure. 5. The significance level used by 5% is the threshold limit for error values in each test analysis. The degree of significance indicates the probability that the event could occur by chance. If the rate is low enough, the probability of occurrence by chance is relatively small, so the event can be said to be significant [36]. The slightest error value can be seen in the analysis of the three parameters experiment by 2.99% compared to the analysis of experiments 2 and 4 parameters by 7.17% and 3.23%. Compared with the degree of significance, the error values obtained in experiments 3 and 4 are already below the limit but not in the analysis of experiment 2 parameters. In the 2-parameter analysis experiment, the error value was above 5% because there was a significant difference in contribution to the parameters analyzed. The contribution values of the parameters of the concentration and type of bacteria in the experiment were 82.94% and 9.90%. Thus, the type of bacteria does not affect SHC.

In addition, the error value in the analysis of the 4-parameter experiment was 3.23% greater than the 3-parameter experiment because one parameter did not have a significant effect, namely the curing time. The curing time contributed 5.33%, a minor contribution to the experiment. In addition, the curing time affects the continuous hydration process, increasing the concrete's strength. The hydration of cement is significantly impacted by curing, which has a significant impact on the character of concrete. By maintaining a warm, humid atmosphere conducive to the hydration process, proper concrete curing minimizes porosity in hydrated cement pastes and raises the density of microstructures in concrete [37]. Therefore, the relationship between the curing time and the compressive strength of concrete is positively correlated.

3.6. Self-healing Concrete Modelling

At this stage, modelling is carried out based on previously analyzed parameters. The modelling is in the form of a polynomial equation that is analyzed using the regression method through Minitab Statistical Software 20. In this modelling, the increase in compressive strength in SHC is influenced by free variables in the form of bacterial concentration and curing time, as well as control variables in the form of bacterial type and application. However, in this estimation model, the curing time is determined for 28 days because, in the previous stage, it was produced that the curing time has the most negligible contribution value. The relationship between curing time and the increase in compressive strength is linear, so it does not affect the increase in compressive strength. In addition, the type of bacteria used in the time estimation model is only *Bacillus subtilis*, where the type selection is based on analysis in the previous stage. In the previous stage, the type of bacteria had a small contribution value in the analysis of experiments of 2, 3, and 4 parameters, namely 9.90%, 2.04%, and 5.39%. *Bacillus subtilis* bacteria have the most significant SNR compared to *Bacillus sphaericus*, which indicates that *Bacillus subtilis* is the optimal type of bacteria. *Bacillus subtilis* had SNR values of 24.18, 22.60, and 22.90, while *Bacillus sphaericus* was 22.19, 22.19, and 21.72 from the analysis of experiments 2, 3, and 4 parameters.

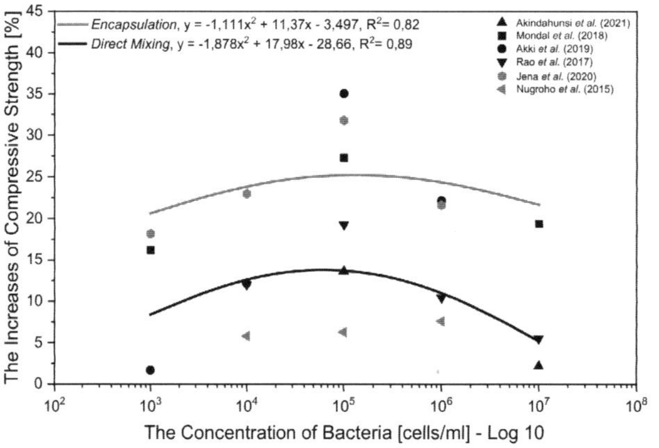

Figure 6. The curve of the relationship between the value of the increase in compressive strength with the concentration of bacteria

The modelling obtained can be seen in Figure. 6. Based on the figure, the equation obtained is divided into two parts: the effect of increasing compressive strength on the concentration of bacteria in encapsulation and direct mixing. Both equations have a coefficient of determination of 0.82 in encapsulation and 0.89 in direct mixing. The coefficient of determination is used to assess the fit of variables in linear regression models [38]. The value of the coefficient of determination indicates how accurately the variables are free to predict bound variables. If the value of the coefficient is close to 1 (one) and away from 0 (zero), the specified free variables can explain or have a significant influence on the bound variable [39]. According to Chin [40], the value of the coefficient of determination classifies as being strong if it is more significant than 0.67, moderate if it is greater than 0.33 but less than 0.67, and weak if it is more significant than 0.19 but less than 0.33. As a result, the value of the coefficient of determination found in this investigation is encouraging. In addition, the trendline formed in Figure. 7 is compared with the results of SHC research from several existing pieces of literature. SHC research conducted by Rao *et al.* [11], Mondal and Ghosh [41], Akki *et al.* [42], and Akindahunsi *et al.* [26] shows the same trend in the form of the best concentration for SHC of 10^5 cells/ml in the direct mixing method. The encapsulation method research by Jena *et al.* [8] has the same trend: the optimal bacterial concentration is 10^5 cells/ml. However, the research conducted by Nugroho *et al.* [43] produced an optimal concentration of bacteria at 10^6 cells/ml. The difference is because there are still error values in these two equations, namely 11.82% (direct mixing) and 21.39% (encapsulation), so data validation is needed.

4. Conclusions

Based on this research, an analysis has been carried out to produce the following conclusions:

- The type, concentration, and application of bacteria as self-healing parameters affect concrete to increase compressive strength. The results of parameter analysis using the Taguchi method in each experimental analysis have optimal variation results, namely a bacterial concentration of 10^5 cells/ml, the type of bacteria used by *Bacillus subtilis*, the application of bacteria with the encapsulation method, and the optimal curing time of 28 days.
- The per cent contribution of each parameter is determined using the analysis of variance method. The concentration of bacteria has a very significant value with a per cent contribution of 82.94%, 78.90%, and 74.36%. Successively on the analysis experiments of 2, 3, and 4 parameters. In addition, the type of bacteria produced a minor contribution to the analysis of experiments of 2 and 3 parameters. In comparison, the curing time became the parameter with a minor contribution in the analysis of experiments of 4 parameters. Therefore, the type of bacteria and the curing time have little effect on increasing the compressive strength of SHC. The effectiveness of each experimental analysis of the SHC parameters is shown by comparing error values in each experiment. The slightest error value was obtained in the analysis of the experiment of 3 parameters with a value of 2.99%.
- SHC modelling is determined using the regression method to obtain two equations consisting of direct mixing and encapsulation. The values of the coefficients of determination of the two equations are 0.89 and 0.82. Based on these values, the specified free variable affects the increase in the compressive strength of SHC. In addition, the value of the coefficient of determination obtained is relatively good. However, both equations have an error value of more than 5%, so direct testing is required as validation.

Acknowledgments

The research described in this paper was financially supported by the Minister of Education, Culture, Research, and Technology of the Republic of Indonesia, grant number 082/E5/PG.02.00.PT/2022

References

[1] Zhang X, Jin Z, Li M, Qian C. Effects of carrier on the performance of bacteria-based self-healing concrete. Constr Build Mater 2021;305:1–17. https://doi.org/10.1016/j.conbuildmat.2021.124771.
[2] Tziviloglou E, Wiktor V, Jonkers HM, Schlangen E. Bacteria-based self-healing concrete to increase liquid tightness of cracks. Constr Build Mater 2016;122:118–25. https://doi.org/10.1016/j.conbuildmat.2016.06.080.
[3] Rooij M, van Tittelboom K, Belie N, Schlangen E. Self-Healing Phenomena in Cement-Based Materials: State-of-the-Art Report of RILEM Technical Committee. 2013.
[4] Khaliq W, Ehsan MB. Crack healing in concrete using various bio influenced self-healing techniques. Constr Build Mater 2016;102:349–57. https://doi.org/10.1016/j.conbuildmat.2015.11.006.
[5] Srinivasa R V., Rao MVS, Sushma S. Feasibility Study on Bacterial Concrete as an Innovative Self Innovative Self-Crack Healing System. Int J Mod Trends Eng Res 2015;2:642–7.
[6] Gao L, Adesina A, Das S. Properties of eco-friendly basalt fibre reinforced concrete designed by Taguchi method. Constr Build Mater 2021;302:1–16.

https://doi.org/10.1016/j.conbuildmat.2021.124161.
[7] Sadeghi MM, Modarresnia AR, Shafiei F. Parameters Effects Evaluation of Microbial Strengthening of Sandy Soils in Mixing Experiments Using Taguchi Methodology. Geomicrobiol J 2015;32:453–65. https://doi.org/10.1080/01490451.2014.958206.
[8] Jena S, Basa B, Panda KC, Sahoo NK. Impact of *Bacillus subtilis* bacterium on the properties of concrete. Mater Today Proc 2020;32:651–6. https://doi.org/10.1016/j.matpr.2020.03.129.
[9] Kashyap VN, Radhakrishna. a Study on Effect of Bacteria on Cement Composites. Int J Res Eng Technol 2013;2:356–60. https://doi.org/10.15623/ijret.2013.0213067.
[10] Reddy BMR, Revathi D. An experimental study on effect of *Bacillus sphaericus* bacteria in crack filling and strength enhancement of concrete. Mater Today Proc 2019;19:803–9. https://doi.org/10.1016/j.matpr.2019.08.135.
[11] Rao MVS, Reddy VS, Sasikala C. Performance of Microbial Concrete Developed Using *Bacillus Subtilus* JC3. J Inst Eng Ser A 2017;98:501–10. https://doi.org/10.1007/s40030-017-0227-x.
[12] Nimafar M, Samali B, Hosseini SJ, Akhlaghi A. Use of bacteria externally for repairing cracks and improving properties of concrete exposed to high temperatures. Crystals 2021;11:1–18. https://doi.org/10.3390/cryst11121503.
[13] Sutoni A. Uji Kuat Tekan dan Daya Serap pada Batako dengan Menggunakan Metode Taguchi. J Sist Dan Manaj Ind 2018;2:93–100. https://doi.org/10.30656/jsmi.v2i2.711.
[14] Karabulut Ş. Optimization of surface roughness and cutting force during AA7039/Al2O3 metal matrix composites milling using neural networks and Taguchi method. Meas J Int Meas Confed 2015;66:139–49. https://doi.org/10.1016/j.measurement.2015.01.027.
[15] Anggraini D, Dewi SK, Saputro TE. Aplikasi Metode Taguchi Untuk Menurunkan Tingkat Kecacatan Pada Produk Paving. J Tek Ind 2015;16:1–9. https://doi.org/10.22219/jtiumm.vol16.no1.1-9.
[16] Ginting E, Tambunan MM. Selection of Optimal Factor Level from Process Parameters in Palm Oil Industry. IOP Conf Ser Mater Sci Eng 2018;288. https://doi.org/10.1088/1757-899X/288/1/012056.
[17] Hadiwidodo YS, Mohd SB. Taguchi Experiment Design for Investigation of Freshened Properties of Self-Compacting Concrete. Am J Eng Appl Sci 2010;3:300–6. https://doi.org/10.3844/ajeassp.2010.300.306.
[18] Danish A, Mosaberpanah MA, Salim MU. Past and present techniques of self-healing in cementitious materials: A critical review on efficiency of implemented treatments. J Mater Res Technol 2020;9:6883–99. https://doi.org/10.1016/j.jmrt.2020.04.053.
[19] Pacheco F, Loeff A, Muller V, Ehrenbring HZ, Christ R, Modolo RCE, et al. Evaluation of concrete self-healing with different insertion techniques of chemical and bacterial solutions. Rev La Asoc Latinoam Control Calidad, Patol y Recuper La Const 2022;12:32–46. https://doi.org/https://doi.org/10.21041/ra.v12i1.559.
[20] Wang JY, De Belie N, Verstraete W. Diatomaceous earth as a protective vehicle for bacteria applied for self-healing concrete. J Ind Microbiol Biotechnol 2012;39:567–77. https://doi.org/10.1007/s10295-011-1037-1.
[21] Algaifi HA, Alqarni AS, Alyousef R, Bakar SA, Ibrahim MHW, Shahidan S, et al. Mathematical prediction of the compressive strength of bacterial concrete using gene expression programming. Ain Shams Eng J 2021;12:3629–39. https://doi.org/10.1016/j.asej.2021.04.008.
[22] Elmahdy MAR, Elshami AA, Yousry ESM, Ahmad SSE. Self-healing mortar using different types, content, and concentrations of bacteria to repair cracks. Frat Ed Integrita Strutt 2021;16:486–513. https://doi.org/10.3221/IGF-ESIS.59.32.
[23] Stanaszek-Tomal E. Bacterial concrete as a sustainable building material? Sustain 2020;12:1–13. https://doi.org/10.3390/su12020696.
[24] Zawad MFS, Rahman MA, Priyom SN. Bio-Engineered Concrete: A Critical Review on The Next Generation of Durable Concrete. J Civ Eng Forum 2021;7:335–58. https://doi.org/10.22146/jcef.65317.

[25] Manjunath PM, Kadapure PSA, Kalaje AA. An Experimental Investigation on the Strength and Durability Aspects of Bacterial Concrete with Fly Ash. Civ Environ Res 2014;6:61–9.
[26] Akindahunsi AA, Adeyemo SM, Adeoye A. The use of bacteria (*Bacillus subtilis*) in improving the mechanical properties of concrete. J Build Pathol Rehabil 2021;6:1–8. https://doi.org/10.1007/s41024-021-00112-7.
[27] Dhanya S, Soman K, Smitha M, Shanthi DM, D DS, Rajesh AR. Strength and Water Absorption of Bacterial Concrete Prepared with M.sand – A Study:-. Int J Eng Res 2016;5:354–6. https://doi.org/10.17577/ijertv5is090346.
[28] Domagała L. The effect of lightweight aggregate water absorption on the reduction of water-cement ratio in fresh concrete. Procedia Eng 2015;108:206–13. https://doi.org/10.1016/j.proeng.2015.06.139.
[29] Achal V, Mukherjee A, Reddy MS. Microbial concrete: A way to enhance durability of building structures. 2nd Int Conf Sustain Constr Mater Technol 2010;23:23–8. https://doi.org/10.1061/(asce)mt.1943-5533.0000159.
[30] Ali M, Abdullah MS, Saad SA. Effect of Calcium Carbonate Replacement on Workability and Mechanical Strength of Portland Cement Concrete. Adv Mater Res 2015;1115:137–41. https://doi.org/10.4028/www.scientific.net/amr.1115.137.
[31] Wang K, Cable JK, Zhi G. Investigation Into Improved Pavement Curing Materials and Techniques: Part I (Phases I and II) 2002;1:52.
[32] James T, Malachi A, Gadzama EW, V A. The Effect of Curing Methods on Compressive Strength of Concrete. Niger J Technol 2011;30:14–20. https://doi.org/10.1088/1755-1315/921/1/012007.
[33] [ASTM] American Society for Testing and Materials. Standard specification for portland cement. Pennsylvania (US): American Society for Testing and Materials; 2019. https://doi.org/10.1520/C0150.
[34] Cuthbert MO, Riley MS, Handley-Sidhu S, Renshaw JC, Tobler DJ, Phoenix VR, et al. Controls on the rate of ureolysis and the morphology of carbonate precipitated by S. Pasteurii biofilms and limits due to bacterial encapsulation. Ecol Eng 2012;41:32–40. https://doi.org/10.1016/j.ecoleng.2012.01.008.
[35] Hizami Abdullah MA, Harmiza Abdullah NA, Tompang MF. Development and Performance of Bacterial Self-healing Concrete - A Review. IOP Conf Ser Mater Sci Eng 2018;431:1–9. https://doi.org/10.1088/1757-899X/431/6/062003.
[36] Riffenburgh RH. Managing Results of Analysis. Academic Press; 2012. https://doi.org/10.1016/b978-0-12-384864-2.00015-9.
[37] Uddin MA, Jameel M, Sobuz HR, Hasan NMS, Islam MS, Amanat KM. The effect of curing time on compressive strength of composite cement concrete. Appl Mech Mater 2012;204–208:4105–9. https://doi.org/10.4028/www.scientific.net/AMM.204-208.4105.
[38] Cheng CL, Shalabh, Garg G. Coefficient of determination for multiple measurement error models. J Multivar Anal 2014;126:137–52. https://doi.org/10.1016/j.jmva.2014.01.006.
[39] Ghozali I. Ghozali, Imam. (2016). Aplikasi Analisis Multivariate dengan Program IBM SPSS 23. Semarang: BPFE Universitas Diponegoro. IOSR J Econ Financ 2016;3:98.
[40] Chin WW. The partial least squares approach for structural equation modeling. Houston: Lawrence Erlbaum Associates; 1998.
[41] Mondal S, Ghosh A (Dey). Investigation into the optimal bacterial concentration for compressive strength enhancement of microbial concrete. Constr Build Mater 2018;183:202–14. https://doi.org/10.1016/j.conbuildmat.2018.06.176.
[42] Akki R V., Sunil SK, Jitendra S, Dhananjay M. Compressive strength of bacterial concrete by varying concentrations of e.Coli and JC3 bacteria for self-healing concrete. Int J Innov Technol Explor Eng 2019;9:3659–61. https://doi.org/10.35940/ijitee.A4749.119119.
[43] Nugroho A, Satyarno I, Subyakto. Bacteria as self-healing agent in mortar cracks. J Eng Technol Sci 2015;47:279–95. https://doi.org/10.5614/j.eng.technol.sci.2015.47.3.4.

Bonding Strength of Buluhpring Bamboo Reinforcement Parallel and Non-Parallel Notches U-Type in Concrete

H Arifin[1), A A Afriansya[2), and A A Amiruddin[3)

[1) Magister Student, Civil Engineering Department, Hasanuddin University, Indonesia
[2) Undergraduate Student, Civil Engineering Department, Hasanuddin University, Indonesia
[3) Civil Engineering Department, Hasanuddin University, Indonesia

Abstract. Bamboo plants are one of the natural materials that are often seen growing and quite fast; Bamboo has been extensively researched for use as an alternative to steel reinforcement in concrete construction. Bamboo is not only more accessible, it is also more cost effective and beneficial for the environment. This study examines the bonding strength of bamboo reinforcement in concrete. Buluhpring bamboo from the Gowa area which has been dried and checked for quality is the bamboo material used. A simple cylindrical concrete specimen measuring 150 mm x 300 mm was used for testing the compressive strength of concrete after 28 days and obtained a result of 25.35 MPa. The Pull-Out test was carried out on 28-day cylindrical concrete measuring 150 mm x 300 mm which was planted with parallel and non-parallel U type bamboo reinforcement in it to determine the bond strength of the test object. The result of concrete with parallel bamboo reinforcement type U is 0.79 MPa and concrete with non-parallel bamboo reinforcement type U is 0.69 MPa. Based on the test, concrete with U-type parallel notch bamboo reinforcement is stronger than U-type non-parallel bamboo notch reinforcement.

Keywords : *Buluhpring Bamboo, Bonding Strength Bamboo, Bamboo Characteristics*

1. Introduction

In the current environment, the number of infrastructure and housing projects constructed with traditional materials, such as steel and concrete, has risen during the past 20 years or more.[1] Concrete has historically had a substantial negative impact on the environment [2]. As a manufacturing material, concrete releases dangerous gases like CO_2 into the atmosphere, which is a disadvantage of utilizing these materials. Concrete and steel are the two materials that the building industry today uses most frequently. Consumption has an impact, for example 4.3 billion metric tons of cement were consumed globally in 2014 [3] and 840 million metric tons of steel used for reinforcing [4]. This uses a lot of resources and harms the environment.

Additionally, because concrete has a low tensile strength, it needs tensile reinforcing for added strength. The most popular kind of reinforcement is steel reinforcement, which has been utilized for many years in concrete building applications. However, one significant disadvantage of steel is its vulnerability to corrosion, according to several studies [5–7], corrosion is frequently started by either the carbonation of the concrete or the exposure of the concrete components to chloride ions. In developing nations, the increased expenditure needed for tourism has created several problems. For instance, relatively few poor nations have the resources necessary to build steel infrastructure. Only two of the 54 African nations, according to a report released in 2015 by the World Steel Association, really manufacture steel in significant numbers [8]. According to an estimate by the United Nations Population Division, the world's population will rise to 11 billion by 2100, with the majority of this growth taking place in emerging nations, particularly those in Africa and Asia [9]. As a result, countries will confront challenging obstacles in trying to fulfill the demand for steel reinforcing in residential and commercial building. Although the majority of the world's construction materials are consumed by high-rise skyscrapers and other dominating structures in industrialized nations, such as India and other South Asian nations, consumption in low-rise homes cannot be disregarded. The majority of these low-rise buildings are made of reinforced concrete [10].

In the past, individuals became interested in finding alternate applications for locally accessible materials to boost the demand for concrete constructions due to high costs and limitations in the manufacture of steel reinforcement on a global scale [11]. Since the 1960s, research on substitute materials has been ongoing. Several naturally occurring fibers are combined with concrete. These components are used to enhance the mechanical qualities of ordinary concrete. Bamboo is one of the naturally occurring materials. Bamboo is regarded as a rapidly growing, sustainable, and eco-friendly material. Bamboo really absorbs 1 ton of CO_2 from the atmosphere for each bamboo stem as it grows. In the tropics, where the majority of poor nations are found, bamboo is one of the materials that is most widely accessible [12-14]. It has been determined via earlier study that bamboo material may be utilized in place of steel reinforcement, especially in locations where steel is hard to come by [15]. Making notches on the left and right sides is how the bamboo reinforcement model method is carried out. It is anticipated that this would strengthen the link between bamboo and concrete. Due to the interlocking components between the surface of the reinforcement and the concrete, notched bamboo reinforcement can lessen the effect or expansion caused by air content [16].

In light of the aforementioned issues, there is a great desire to do research on concrete building using one form of bamboo (Buluhpring) as reinforcement in place of the often used steel reinforcement. In some straightforward building construction tasks, such those for residential homes, the feature of bamboo that has sufficient tensile strength will be an option. This study uses innovations and variations of parallel and non-parallel U notch types as a comparison in bamboo reinforcement in an effort to better understand how the physical and mechanical properties of bamboo and the adhesive strength that occur in the use of bamboo as concrete reinforcement due to the slippery nature of bamboo and the frequent occurrence of slippage in concrete.

2. Experiment

2.1 Material
The material used is concrete material and bamboo material as concrete reinforcement.

2.1.1. Concrete
A combination of fine aggregate, coarse aggregate, cement, and water is used to create concrete. With a cylinder sample of 150 mm x 300 mm and parallel and non-parallel notch bamboo reinforcement type U implanted in the centre of the cylinder with a distribution length of 200 mm, concrete with a compressive strength of 25 MPa was employed. On the 28th day following the printing of the sample, the concrete sample will be evaluated.

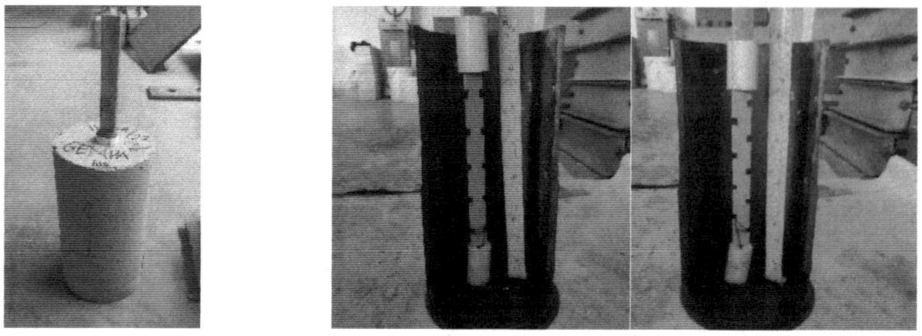

(a) (b)

Figure 1. (a) Sample after cast, (b) Bamboo reinforcement position's

2.1.2. Bamboo

The Buluhpring variety of bamboo, which grows in the Gowa Region and is obtained straight from the field by selecting bamboo with a minimum age of at least two years, is used as a substitute for steel reinforcement. To improve the binding strength of the concrete, the bamboo that passed the characteristic test is next given parallel and non-parallel U-type notches.

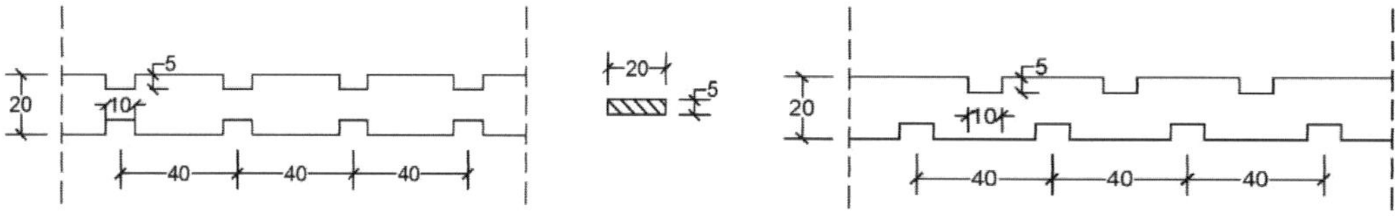

(a) (b)

Figure 2. (a) U type parallel notch details, (b) U type non-parallel notch details

2.2 Testing Methods

2.2.1. Sample

Each specimen is a reinforced concrete cylinder with parallel and non-parallel U-shaped notches made of bamboo. The bamboo material is dried before being assessed for its quality. If the bamboo passes the inspection, then the bamboo can be processed into reinforcement with variations of parallel and non-parallel U-shaped notches according to the model size criteria. When the concrete is 28 days old, the compressive strength and adhesive strength tests will be carried out on the concrete cylinder.

2.2.2. Bamboo water content

The amount of water present in bamboo material is used to reflect its water content. According to the ISO 22157-01-2004 [17] standard, which was analyzed in this study to determine the water content proportion limit, the maximum moisture content of bamboo under dry air conditions was 20%. The formula for calculating the water content of bamboo is:

$$Water\ Content\ (\%) = \frac{Initial\ Weight - Oven\ Dry\ Weight}{Oven\ Dry\ Weight} \times 100\% \quad (1)$$

As testing equipment, an oven and a scale were used to determine the moisture content of dried bamboo samples in the laboratory. The water content of the sample was determined by weighing it before and after 24 hours in an oven heated to 300°.

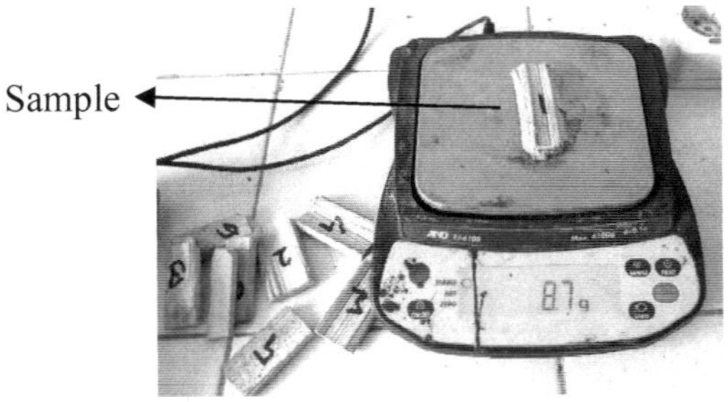

(a)

(b)

Figure 3. (a) Scale, (b) Oven

2.2.3. Tensile Stength

On the mechanical qualities of bamboo, particularly evaluating the strength of bamboo. The tensile strength of bamboo is the capacity of bamboo to endure tensile stresses, as well as steel reinforcement. The bamboo in this study will undergo a tensile force such that the tensile strength of the bamboo may be measured by the formula:

$$\sigma_{tr} = \frac{P}{A} \quad (2)$$

In this study, bamboo reinforcement material was evaluated for its tensile strength using a *Universal Testing Machine* (UTM) in the laboratory, and the value of the material's tensile strength will be determined when the bamboo breaks.

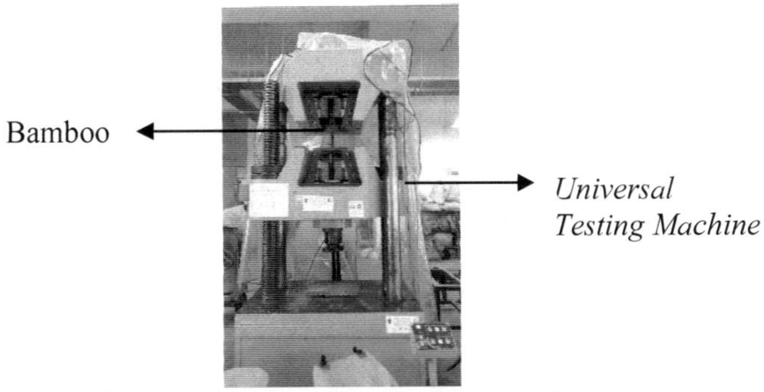

Figure 4. Bamboo Tensile Strength Test

2.2.4. Compressive Strength of Concrete

After 28 days, the compressive strength of the test item in the shape of a cylinder of concrete will be evaluated. The projected compressive strength at the outset of the job mix design will be evaluated for achievement. The compressive strength of concrete is the resistance of the concrete to a particular compressive load's compressive strength. The formula may be used to calculate the compressive strength of concrete. :

$$f'c = \frac{F}{A} \quad (3)$$

Normal concrete cylindrical specimens are evaluated for their compressive strength in a laboratory using a Compression Testing Machine (CTM) by first weighing the test object, then inserting it in the tool to be tested for pressure, and then stopping the test after the test item has been crushed.

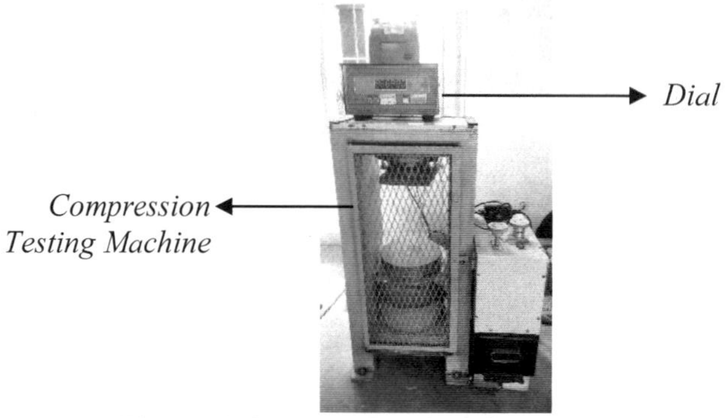

Figure 5. Concrete Compressive Strength Machine

2.2.5. Bonding Strength
After 28 days, the adhesive strength of both parallel and non-parallel notch bamboo reinforced concrete cylinder specimens of type U was evaluated using the Pull-Out test. Using the Universal Testing Machine (UTM) in the laboratory, this test is designed to determine the adhesive strength of bamboo reinforcement on concrete. Bonding strength calculation formula :

$$\mu = \frac{P}{\pi \emptyset \, Ld} \tag{4}$$

The test is conducted by placing the cylinder on the machine with the bamboo reinforcement extending upwards, and then applying a weight. The value of loading and strain will be saved on the computer, and loading will cease when the maximum load is reached, as shown by the machine pointer returning to its starting position (zero) and the other pointer ceasing to point to the maximum load.

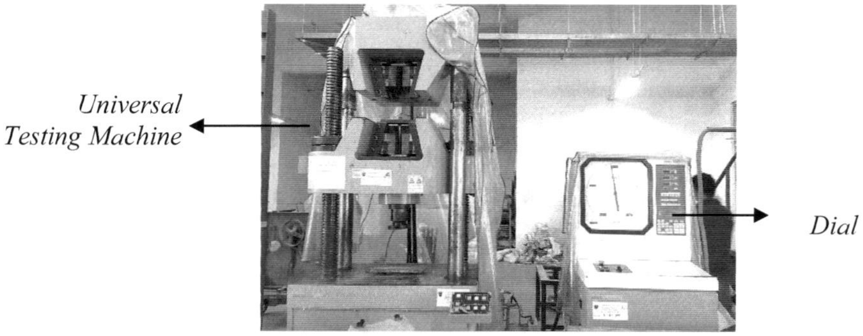

Figure 6. Bonding Strength Machine

3. Results and Discussion

3.1. Water content
The bamboo test consists of the bottom, middle, and top portions of bamboo. Based on the results of testing the moisture content of the top bamboo material, the results obtained meet ISO 22157-01-2004 [16] where the water content value of bamboo in maximum water content conditions is 20%, whereas the bottom and middle of bamboo does not meet because it is believed to be the most active part of the bamboo in receiving water from the soil.

Table 1. Water Content Test Result

Segment	No. Sampel	Wet Weight (gr)	Dry Weight (gr)	Water Content (%)
Bottom	B1	15.86	11.80	34.44
Middle	B2	10.45	8.30	25.89
Top	B3	9.19	8.00	14.89

3.2. Tensile strength

From the results of tensile strength testing on bamboo, the results in the table show that the highest tensile strength is at the top of the bamboo which has the lowest water content.

Table 2. Tensile Strength Test Result

Segment	No. Uji	Weight (gr)	Wide (mm)	Thick (mm)	Max. Load (N)	Tensile Strenght (Mpa)	Average (Mpa)
	B1	24.51	5.00	8.64	15480.00	358.33	
Bottom	B2	32.11	5.00	8.36	10000.00	239.23	277.54
	B3	35.75	5.00	9.24	10860.00	235.06	
	T1	21.18	5.00	5.50	8340.00	303.27	
Middle	T2	21.80	5.00	5.48	8720.00	318.25	308.51
	T3	21.92	5.00	5.50	8360.00	304.00	
	A1	16.92	5.00	4.30	9700.00	451.16	
Top	A2	17.35	5.00	4.38	8360.00	381.74	357.10
	A3	17.91	5.00	5.00	5960.00	238.40	

3.3. Concrete compressive strength result

The concrete compressive test yielded an average concrete compressive strength of 25.35 MPa when the concrete cylinder specimen reached the age of 28 days.

Table 3. Compressive Strength Result

Sample	Concrate Age (Day)	Area (mm^2)	Load (kN)	Compressive Strength (MPa)	Average (MPa)
C 1		7853.98	197.00	25.10	
C 2	28	7853.98	225.00	28.66	25.35
C 3		7853.98	175.00	22.29	

IOP Conf. Series: Earth and Environmental Science 1117 (2022) 012010 doi:10.1088/1755-1315/1117/1/012010

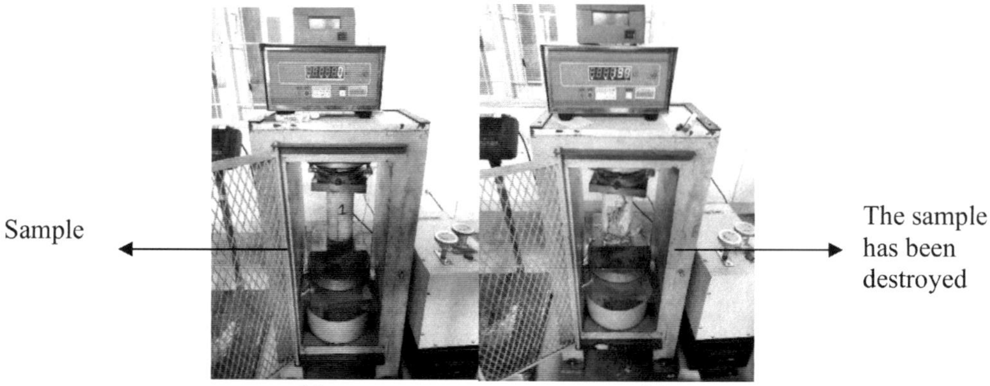

(a) (b)

Figure 7. (a) Before Testing, (b) After Testing

3.4. Bonding strength result

To assess the compressive strength of the test sample, a *Pull-Out* test was performed on the specimen after 28 days using the *Universal Testing Machine* (UTM) and the *Linear Variable Differential Transformer* (LVDT) instrument to measure the displacement.

Figure 8. Bonding Strength Test

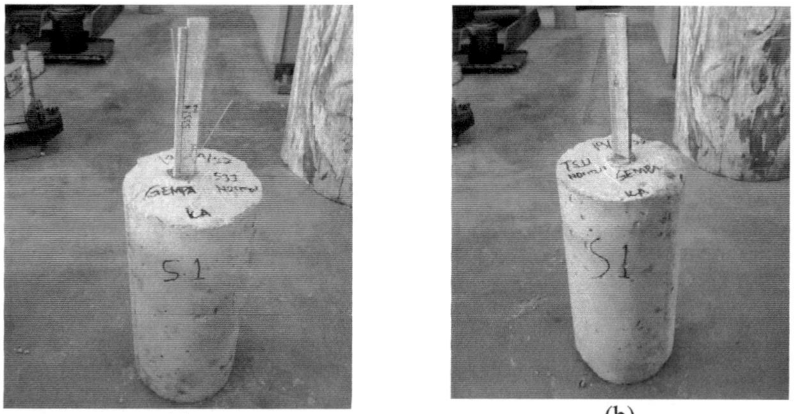

(a) (b)

Figure 9. (a) test result sample of notch parallel type U, (b) test result sample of notch non-parallel type U

The results obtained from the average bond strength of U type parallel notch bamboo specimens of 0.79 Mpa and the results of the average adhesive strength of U type non-parallel notch bamboo specimens of 0.69 Mpa. where it is found that the bonding strength of bamboo reinforced concrete with parallel notches of type U is higher than that of concrete with bamboo reinforcement not parallel to type U.

Table 4. Bonding Strength Result

Sample	Max Load (kN)	Bond Strength (MPa)	Avarage (MPa)
BP1	7.86	0.78	
BP2	8.56	0.85	0.79
BP3	7.23	0.72	
BNP1	6.91	0.69	
BNP2	7.68	0.76	0.69
BNP3	6.06	0.60	

BP : u-type parallel notch bamboo
BNP : u-type non-parallel notch bamboo

The graph demonstrates that the U type parallel notch bamboo specimen has a greater load and a lower displacement than the U type non parallel notch bamboo specimen, which has a smaller load and a greater displacement.

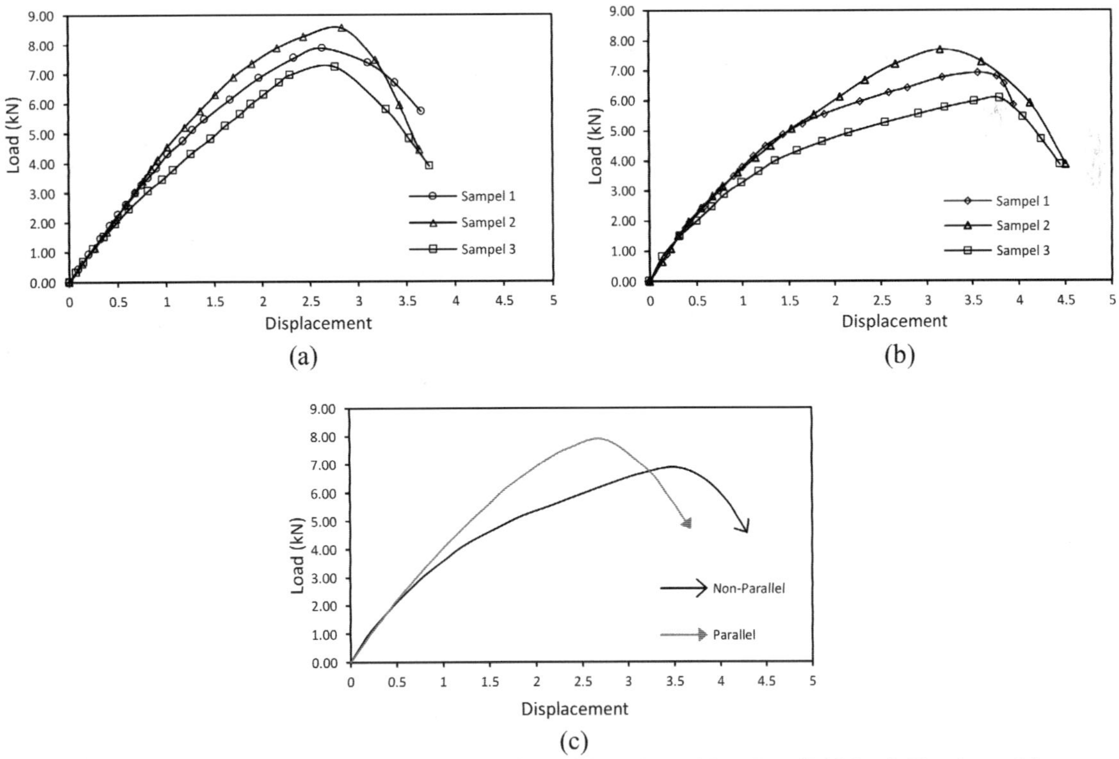

Figure 10. (a) U-Type Parallel Notch Bamboo, (b) U-Type Non-Parallel Notch Bamboo, (c) U-Type Parallel and Non-Parallel Notch Bamboo

The results of comparing two test samples with different types of notches indicate that type U parallel notch bamboo reinforcement has superior bond strength and a smaller displacement, making it safer than type U non-parallel notch bamboo reinforcement.

4. Conclusion

After the *Pull-Out* test, it was determined that the bonding strength of concrete with U-type parallel notch Buluhpring bamboo reinforcement was 0.79 Mpa, while the bonding strength of concrete with U-type non-parallel notch Buluhpring bamboo reinforcement was 0.69 Mpa. Therefore, it can be concluded that concrete with bamboo reinforcement with parallel notches of type U has a higher bond strength than concrete with bamboo reinforcement with non-parallel notches of type U.

5. References

[1] P R. Mali, D Datta, "*Experimental Evaluation of Bamboo Reinforced 1 Concrete Beams*", Journal of Building Engineering (2019).

[2] Alireza Javadian , Ian F.C. Smith and Dirk E. Hebel, "*Application of Sustainable Bamboo-Based Composite Reinforcement in Structural-Concrete Beams: Design and Evaluation*", Multidisiplin Digital Publishing Institute (2020).

[3] PCA Report, "*Global Cement Consumption On The Rise*", 2015. http://www.cement.org/news/2015/06/03/global-cement-consumption-on-the-rise.

[4] "*Tata Steel To Focus On Improving Productivity*", 2016. http://steelworld.com/Newsletter/2016/Jan16/NewsIndian0116-7.pdf

[5] Bentur, A.; Berke, N.; Diamond, S. "*Steel Corrosion in Concrete: Fundamentals and Civil Engineering Practice*"; Taylor & Francis: Derby, UK, 1997.

[6] Slater, J.E. "*Corrosion of Metals in Association with Concrete: A Manual Sponsored by ASTM Subcommittee G01. 14 on Corrosion of Reinforcing Steel, and Metal Properties Counci"l*; ASTM International: West Conshohocken, PA, United States, 1983.

[7] Broomfield, J.P. "*Corrosion Of Steel In Concrete: Understanding, Investigation And Repair*"; CRC Press: Boca Raton FL, USA, 2002.

[8] Yoon, S.; Wang, K.; Weiss, W.J.; Shah, S.P. "Interaction Between Loading, Corrosion, And Serviceability Of Reinforced Concrete". *Mater. J.* 2000, *97*, 637–644.

[9] Tan, Z.; Hansson, C. "*Effect Of Surface Condition On The Initial Corrosion Of Galvanized Reinforcing Steel Embedded In Concrete*". *Corros. Sci.* 2008, *50*, 2512–2522.

[10] S Anil and SUnnikrishnan, "*Investigation On Elastic Properties Of Bamboo And Behavior Of Bamboo Reinforced Concrete Beams*", International of Journal Earth Sciences Engineering ISSN 0974-5904, Volume 10, No. 02 (2017).

[11] Atul Agarwal and Damodar Maity "*Experimental Investigation on Behavior of Bamboo Reinforced Concrete Members*", 16th International Conference on Composite Structures, ICCS vol 16

[12] A. Anandamurthy, V. Guna, M. Ilangovan and N. Reddy, ''*A review of fibrous reinforcements of concrete*''. *Journal of Reinforced Plastics and Composites* 36(7):519-552, 2017.

[13] Y. Xiao, R.Z. Yang, B. Shan, "*Production, environmental impact and mechanical properties of glubam*". *Construction and Building Materials*. 1(44):765-773, 2013.

[14] B. Sharma, A. Gatoo, M. Bock, H. Mulligan, M. Ramage, "*Engineered Bamboo: State Of The Art*". *Proceedings of the Institution of Civil Engineers-Construction Materials.*

[15] M. Yamaguchi, K Murakami, K. Takeda, "*Flexural Performance of Bamboo-Reinforced concrete beams using bamboo as main rebars and stirrups*," 2013.

[16] A Azadeh, K Gavami, H Savastono Jr, RDT Filho, "*Static Flexural Behavior Bamboo As A Functionally Graded Material And The Effect Of Heat On Dynamic Flexural Modulus*" Journal of Building Engineering 34(2020) : 101949

[17] International Organization for Standardization (ISO) 2004 . *"Bamboo_Determination of Physicial and Mechanical Properties"*. ISO 22157-01-2004. International Organization for Standardization.

Sustainable Construction Framework Model on Development of Likupang Special Economic Zone

Daisy Pangemanan[1], R. U. Latief[2], and S. Hamzah[3]

[1]Doctoral program student of Civil Engineering Department, Faculty of Engineering, Hasanuddin University, South Sulawesi, Indonesia
[2,3]Associate Professor of Civil Engineering Department, Faculty of Engineering, Hasanuddin University, South Sulawesi, Indonesia

daisydebora88@gmail.com

Abstract. Infrastructure development must pay attention to basic aspects such as technical and social infrastructure. A special approach is needed to achieve sustainable development that adopts three main pillars in infrastructure development such as economic, environmental and social disparities. The absence of a holistic sustainable construction model requires a comprehensive solution by conducting an in-depth study to formulate the right model in implementing sustainable construction. The aim of this study is to build a holistic sustainable construction proposal model, which is expected to be a standard model so that it can be a guidance for stakeholders in implementing Sustainable Construction in the Likupang special economic zone (SEZ). This study used a quantitative method by structural equation modeling (SEM), as a multivariate analysis technique to examine the relation between complex variables and to obtain a comprehensive picture of the entire model. The sample used was 80 respondents. The results of this study indicate that the model built 56.8% has been fit and ready to be implemented. The results also show that economic and government factors have a positive and significant impact on sustainable construction. This study concludes that the model built is good and can be implemented in the Likupang SEZ.

Keywords: Construction, SEZ, Sustainable construction framework model

1. Introduction

Infrastructure development has developed rapidly, so that environmental issues due to global warming have become a priority aspect in the development process. The development of the Likupang SEZ is one of the central government's priority projects in the tourism sector, especially in the Eastern Indonesia region. Therefore, a model that can be implemented easily and in accordance with the technical, environmental, social and cultural context of the community is needed. The absence of this model has caused serious problems, so that the government through the Ministry of Public Works and Public Housing (PUPR) issued Ministerial Regulation Number 9 of 2021 about Guidelines for the Implementation of Sustainable Construction which is implemeted in sustainable construction.

Unfortunately, there are still constraints in the context of the implementation of sustainable construction that is diverse in society. Modeling of sustainable construction has been tried by considering some crucial factors, such as government, cost, knowledge and information, workforce, client and market factors. This model has been implemented in Malaysia [1]. The factor that needs to be more considered is the cultural factor as a factor that is trying to be included as an important factor in environmental sustainability by taking into account the factors of attitudes, company culture and social responsibility [2]. The solution to the absence of a holistic sustainable construction model has not been able to solve the problem comprehensively, especially local wisdom that is not paid attention to in the context of implementing sustainable construction. Hence, it is necessary to conduct an in-depth study by including cultural factors/local wisdom as a new factor forming the success of sustainable construction. The purpose of this study is to build a holistic proposal model, so that it can be a guidance for stakeholders in implementing sustainable construction in the Likupang SEZ.

2. Literature Review

2.1.The general concept of sustainable construction

Sustainable construction is a concept developed to explain the responsibility of the construction industry in realizing sustainable development [3]. The emphasis of the concept of sustainable construction focuses on three main pillars, namely environmental friendliness, social life, and economic welfare [4]. Sustainable construction can be understood as responsible development and maintenance of environmental health that is built based on ecological principles and efficient use of resources [2]. The development of sustainable construction trends tends to focus on the relationship of construction stakeholders, human development and environmental aspects.

2.2 Management concept of Sustainable Project

The concept of sustainable construction has 3 (three) main focus areas, namely the environment, social life, and economic welfare [5]. Each factor has its own variables according to the needs and conditions of the local community. Research related to the determinants of the success of construction project work is generally analyzed separately to determine the existence of a positive relationship between the object of influence and the affected [6]. Several studies have identified constraints on sustainable construction which include variables of government support and investment [1]. Human resource factors are included as a determining factor in the success of sustainable construction management [7].

3. Research Methodology

3.1 Flowchart

The flowchart of this research begins by identifying the problems to be solved in this research. Problems are obtained by conducting initial observations and the results of literature studies as well as regulatory/policy studies. After the problem has been defined, the research is continued by conducting interviews with stakeholders. The interview technique uses a semi-structured interview technique to obtain research data. The data were analyzed using the Miles and Huberman Model, to confirm and verify the factors that influence the implementation of sustainable construction in the development of the Special Economic Zone (SEZ), Likupang.

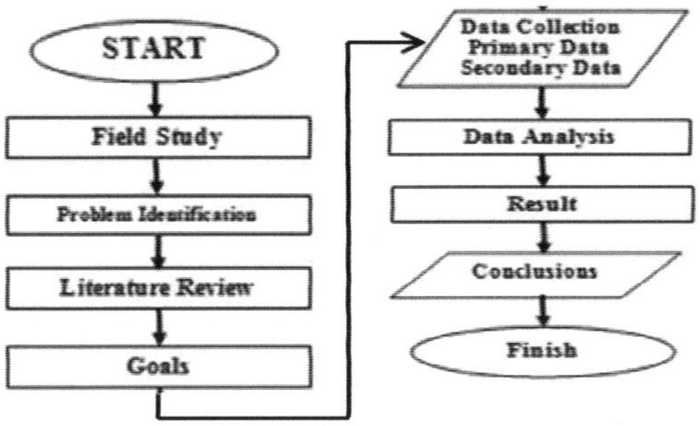

Figure 1. Research Flowchart

Data analysis activities consist of data reduction, data display, and conclusion drawing/verification. Data presentation is an effort to compile a set of information into a matrix or configuration that is easy to understand. An understandable data is the main key to analyze valid data.

IOP Conf. Series: Earth and Environmental Science 1117 (2022) 012011 doi:10.1088/1755-1315/1117/1/012011

3.2 Data collection and Quantitative data analysis

At this stage, using an instrument in the form of a questionnaire, which involves respondents from service providers, namely Contractors and Consultants who have handled sustainable construction project work, the government (Central, North Sulawesi Province, Regency Government, and local government), Investors, Academics, and communities in the Likupang SEZ development environment.

The results of field data collection are then processed and analyzed. In data analysis, SEM (Structural Equotion Modeling) will be used. The results of data processing and analysis are used as recommendations for the design of models and strategies for implementing sustainable construction in the construction of the Likupang SEZ. Descriptive statistical analysis was conducted to obtain the results of the study of the factors that influence the implementation of sustainable construction in the construction of the Likupang SEZ. Furthermore, the analysis was carried out using the statistical technique of one way analysis of variance to examine these factors in the application of sustainable construction. The data obtained through questionnaires have been analyzed, the information obtained from the FGDs is then discussed in a comprehensive manner and the modeling is designed which will be used as a guide by the North Sulawesi provincial government in the implementation of SEZ development with a sustainable construction approach.

3.3 Research Design

This study focuses on the development of the Likupang SEZ which is designed with a sustainable construction approach by starting with several factors, namely social factors, economic factors, environmental factors, government support factors, business investment factors, human resource factors and cultural factors.

Table 1. Research Variables and References

Variables	Definition	References
X1	Economic Factor	Abd Jamil Fathi, 2016, Ismail et al; 2017, Teng, 2018, Ervianto, 2018, Shurrab; 2019,
X2	Environmental Factor	Ismail et al; 2017, Ervianto,2018 Shurrab et al 2019, Li. X. et al, 2019, Ajibike et al, 2021.
X3	Social Factor	Ismail et al; 2017, Teng, 2018, Shurrab; 2019,
X4	Governmental Factor	Durdyev et al, 2018, Willar, et.al 2019
X5	Investment Factor	Durdyev et al, 2018, Sfakianaki; 2015
X6	Human Resource Factor	Sahadi. S dan Wibowo, A, 2015, Karunasena et al; 2016
X7	Cultural Factor	Fauzi; 2004, Froner; 2017
Y	Sustainable Construction (SC) Factor	Araujo, et al 2020, Lima, et.al 2021, Murtagh, et al. 2020, Ajibike, et al 2021

The factors that become variables in this research model are determined in connection with the implementation of construction project development with a sustainable construction approach table 1. In this study, we will use the Structural Equation Modeling (SEM) equation, where the SEM model is a multivariate analysis technique used to examine the relationship between complex variables to obtain a comprehensive picture of the entire model. For this modeling using PLS-SEM (Partial Least Squares Structural Equation Modeling) program. Latent variables formed in PLS-SEM, the indicators are reflective and formative. Figure 2 shows a research model with reflective and formative indicators in relation to latent variables.

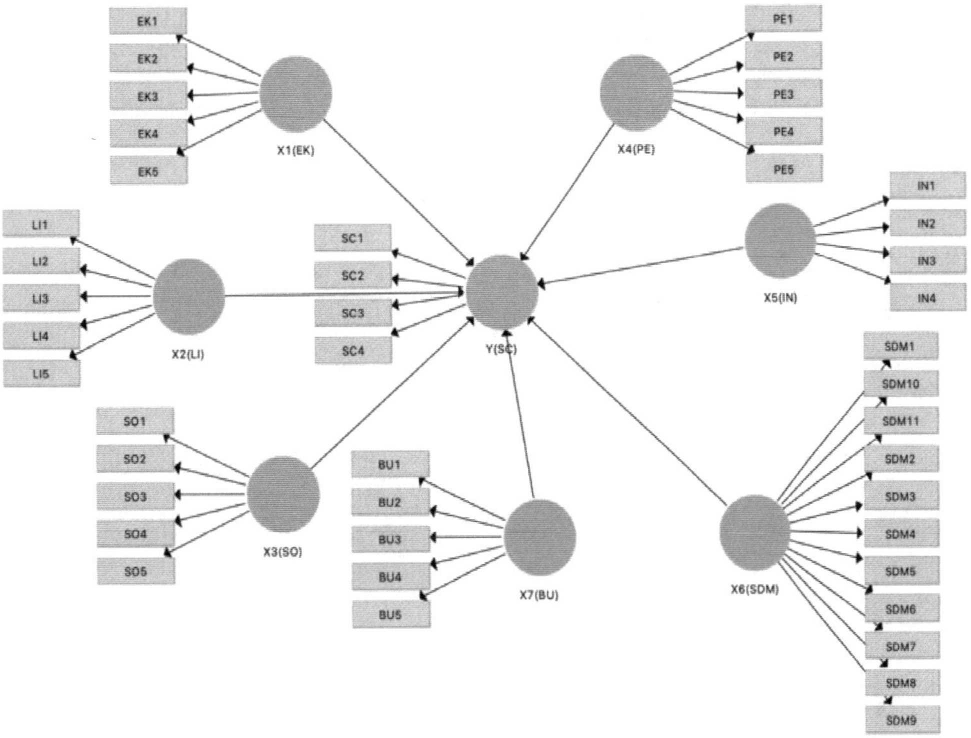

Figure 2. Research Model Design

3.4 Population and Sample

The population in this study is each element of the construction project Stakeholder in the construction of the Likupang SEZ. Namely contractors, consultants, government, investors, academics and the public. While the sampling in this study was done by stratified random sampling method. Stratified random sampling is a sampling process through the process of dividing the population into groups, selecting a simple random sample from each group, and combining them into a sample to estimate the population parameters. The samples in this study are Service Providers (Contractors and Consultants) who have handled projects with a sustainable construction approach, the Government (District, Province and Central), investors, academics and the community (around SEZ and the general public).

4. Results and Discussion

4.1 Validity and Reliability Calculation Results

The results of the calculation of validity were carried out using the convergent validity test (average variance extracted (AVE)). This technique is using to measure the AVE value with the measurement value must reach the value of each variable, which is ≥ 0.5. See Table 2.

Table 2. Average Variance Extracted Value

Variables	Cronbach's Alpha values	rho_A values	Composite reliability values	AVE values
X1 (Ec)	0.833	0.838	0.888	0.666
X2 (Env)	0.895	0.918	0.922	0.702
X3 (Soc)	0.784	0.702	0.859	0.604
X4 (Gov)	0.865	0.883	0.902	0.648
X5 (In)	0.857	0.874	0.902	0.697
X6 (Hr)	0.927	0.934	0.940	0.662
X7 (Cul)	0.791	0.816	0.877	0.704
Y (Sc)	0.831	0.832	0.887	0.664

The results of calculating the AVE value show that all the calculated factors have a value greater than 0.5. The result of calculating Discriminant Validity (Cross Loading) correlation between indicators and variables shows that the calculated value of the variable is greater than the arithmetic value of the variable indicator itself and is also greater than the calculated value of the other variables. Thus, it can be said that the validity of the indicators of each variable is valid. The results of the calculation show that the validity of the Discriminant Validity (Cross Loading) has met the requirements. The calculated reliability value (Composite Reliability and Cronbach's Alpha) is the efficacy of the instrument in measuring the indicator value.

The results of the Discriminant Validity (Fornell Lacker Criterion) calculation are carried out to determine the value of the Discriminant Validity (Fornell Lacker Criterion) validity, which is the correlation value between the variable itself and variables with other variables, cannot be smaller than other variables. The valid value must be greater between the value of the variable itself and other variables. The results are seen in table 3.

Table 3. Discriminant Validity (Fornell Lacker Criterion) Calculation Results

Variables	X1 (Ec)	X2 (Env)	X3 (Soc)	X4 (Gov)	X5 (In)	X6 (Hr)	X7 (Cul)	Y (Sc)
X1 (Ec)	0.816							
X2 (Env)	0.649	0.838						
X3 (Soc)	0.537	0.578	0.777					
X4 (Gov)	0.562	0.782	0.605	0.805				
X5 (In)	0.610	0.663	0.692	0.725	0.835			
X6 (Hr)	0.646	0.757	0.598	0.701	0.698	0.813		
X7 (Cul)	0.439	0.526	0.434	0.523	0.482	0.652	0.839	
Y (SC)	0.512	0.481	0.400	0.568	0.444	0.568	0.543	0.815

The results of the calculation show that the value of Counting Discriminant Validity (Fornell Lacker Criterion) has fulfilled the requirements of all the calculated variables. The validity value is also continued by calculating the value of Discriminant Validity (Cross Loading) which is a validity test between the indicator values that measure the variable itself and the indicator values of other variables. The validity value must be greater than the indicator with other variables. The calculated reliability value (Composite Reliability and Cronbach's Alpha) is the efficacy of the instrument in measuring the indicator value. Count Reliability Value (Composite Reliability and Cronbach's Alpha) must be > 0.7, can be seen in Table 4

Table 4. Reliability Calcualtion Results (Composite Reliability and Cronbach's Alpha)

Variables	Cronbach's Alpha values	rho_A values	Composite reliability values	AVE values
X1 (Ec)	0.833	0.838	0.888	0.666
X2 (Env)	0.895	0.918	0.922	0.702
X3 (Soc)	0.784	0.702	0.859	0.604
X4 (Gov)	0.865	0.883	0.902	0.648
X5 (In)	0.857	0.874	0.902	0.697
X6 (Hr)	0.927	0.934	0.940	0.662
X7 (Cul)	0.791	0.816	0.877	0.704
Y (SC)	0.831	0.832	0.887	0.664

The calculation results have shown that the calculated reliability value (Composite Reliability and Cronbach's Alpha) is greater than 0.7 so that it can be said that the instrument is reliable and effective for use in research.

4.2 Model Evalution

4.2.1. Model evaluation: Inner Model Test (R-Square)

The results of the model evaluation are carried out by calculating the following indicators.
The value of the Inner Model Test (R-Square) is a value that is only owned by the Y variable (Sustainability Construction). This value is a value that shows how much the independent variable (X1-X7) affects the dependent variable Y. The results show that the value of Y = 0.457 x 100% = 45.7% is influenced by X1-X7. See table 5.

Table 5. Inner Model Test (R-Square) Values

	R square	**R Square Adjusted**
Y (SC)	0.457	0.405

4.2.2. Model evaluation: Inner Model Test (Path Coefficients)

The value of the Inner Model Test (Path Coefficients) is a value that shows the direction of the relationship between positive or negative variables. The results of this calculation show the direction of the influence of each variable X on the Y variable (Continuous Construction). The acceptance value is at zero. If it is greater than 0 to 1, it means that it shows the direction of a positive influence, whereas if it is less than 0 to – 1, it means that it has a negative influence, it can be seen in table 6.

Table 6. Inner Model Test (Path Coefficients) Values

Variable	X1 (Ec)	X2 (Env)	X3 (Soc)	X4 (Gov)	X5 (In)	X6 (Hr)	X7 (Cul)	Y (Sc)
X1 (Ec)								0.256
X2 (Env)								-0.192
X3 (Soc)								-0.008
X4 (Gov)								0.403
X5 (In)								-0.131
X6 (Hr)								0.189
X7 (Cul)								0.264
Y (SC)								

The results of this study conclude the direction of the influence of the variables as follows:
X1 (Economy) has a Positive effect on Y (Sustanability Construction)
X2 (Environment) has a negative effect on Y (Sustanability Construction)
X3 (Social) has a negative effect on Y (Sustanability Construction)
X4 (Government) has a Positive effect on Y (Sustanability Construction)
X5 (Investment) has a negative effect on Y (Sustanability Construction)
X6 (Human Resources) has a Positive effect on Y (Sustanability Construction)
X7 (Culture) has a Positive effect on Y (Sustanability Construction)

These results indicate that the variables of the economy, government, human resources and culture show a positive influence on sustainable construction. While the environmental, social and investment variables show a negative influence.

4.2.3. Model evaluation: Inner Model Test (Significancy T-STATISTIC)

The result of calculating the value of the Inner Model Test Evaluation (Significancy T-STATISTIC) is a calculation result that shows the significance value of a variable. The acceptance value is the significance level used alpha = 0.05 or the T-Statistic value > 1.96 = SIGNIFICANT. See Table 7.

Table 7. Inner Model Test (Significancy T-STATISTIC) Values

	Original Sample (0)	Sample mean (M)	Standard Deviation (STDEV)	T Statistics (O/STDEV)	P Values
X1(Ec)->Y(Sc)	0.256	0.254	0.116	2.203	0.028
X2(Env)->Y[Sc)	-0.191	-0.160	0.197	0.972	0.331
X3(SocI->Y(Sc)	-0.008	0.004	0.166	0.047	0.963
X4(Gov)->Y{Sc)	0.403	0.376	0.199	1.016	0.043
X5[1n)->YiScI	-0.131	-0.130	0.143	0.916	0.360
X6[Hr)->Y(Sc)	0.189	0.193	0.175	1.078	0.282
X7(Cul)-> Y(Sc)	0.264	0.249	0.139	1.598	0.058

4.2.4. Model Evaluation: Inner Model Test (Predictive Relevance)

The results of the Inner Model Test (Predictive Relevance) are values to show how well the observations are made. The results of the Inner Model Test Predictive Relevance are calculated using blindfolding in PLS. Inner Model Test Predictive Relevance has an acceptance value of good observation value if > 0 it can be said that the observation is good. See Table 7. The results of the calculation show that the results of the Inner Model Test Predictive Relevance show the value so the conclusion of the observation is 0.237. This means that the observations that have been made in this study can be said that the observations are good.

Table 8. Inner Model Test (Predictive Relevance) Values

Variables	SSO	SSE	Q^2 (=1-SSE/SSO)
X1 (Ec)	324.000	324.000	
X2 (Env)	405.000	405.000	
X3 (Soc)	324.000	324.000	
X4 (Gov)	405.000	405.000	
X5 (In)	324.000	324.000	
X6 (Hr)	648.000	648.000	
X7 (Cul)	243.000	243.000	
Y (SC)	324.000	247.125	0.237

4.2.5. Model Evaluation: Inner Model Test (Model Fit)

The Inner Model Test (Fit model) is a value to show how well the model under study is. Acceptance value is measured using Value seen in NFI on PLS. See Table 9.

Table 9. Inner Model Test (Fit Model) Values

	Saturated Model	Estimated Model
SRMR	0.092	0.092
d_ULS	5.948	5.948
d_G	3.595	3.595
Chi_Square	1266.411	1266.411
NFI	0.568	0.568

The calculation results show that the Model Fit value is the NFI value = 0.568. This means that the NFI value shows that the built model is good. The percentage of the model built is obtained with NFI x 100%, NFI = 0.568 x 100% = 56.8% Fit model. This means that the sustainable construction model that has been built has 56.8% declared fit and can be implemented in sustainable construction in the Likupang SEZ. The results of this study have resulted in a model built to test whether the model that has been built is good or not. To measure the model, researchers used the Inner Model Test (Model Fit). See Figure 3.

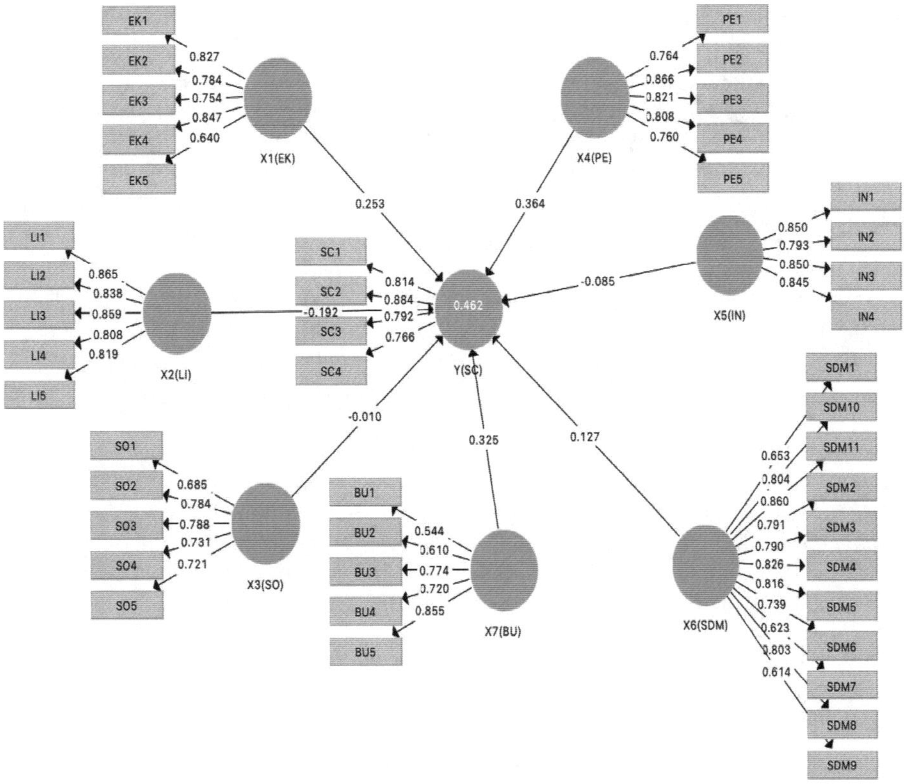

Figure 3. Sustainable Construction Fit Model

5. Conclusion

This study concludes that the resulting model has met the requirements as a good model and can be used in the development of sustainable construction. The proof results show that 56.8% of the model produced is good to use. The factors that influence the model have contributed to the development of the model. Several factors have been shown to have a positive and significant effect, such as economic and government factors, as well as human resource factors and cultural factors that have a positive effect. Meanwhile, environmental, social and investment factors have a negative influence on the sustainable construction model in the Likupang SEZ. The results of this research conclude that the model that has been produced can be implemented in the development of the Likupang SEZ directly by paying attention to the factors that have a positive influence and stakeholders need to develop a mature strategy in anticipating factors that have a negative influence and maximizing the factors that contribute positively.

References

[1] Durdyev S., S. Ismail, A. Ihtiyar, N. F. S. Abu Bakar, and A. Darko, "A partial least squares structural equation modeling (PLS-SEM) of barriers to sustainable construction in Malaysia," *J. Clean. Prod.*, **vol. 204**, pp. 564–572, 2018, doi: 10.1016/j.jclepro.2018.08.304.

[2] Ajibike W., A , Adeleke A. Q., F. Mohamad, J. A. Bamgbade, M. N. M. Nawi, and T. D. Moshood, "An evaluation of environmental sustainability performance via attitudes, social responsibility, and culture: A mediated analysis," *Environ. Challenges*, **vol. 4**, no. March, p. 100161, 2021, doi: 10.1016/j.envc.2021.100161.

[3] Araújo A. G, Pereira Carneiro A. M, and R. P. Palha, "Sustainable construction management: A systematic review of the literature with meta-analysis," *J. Clean. Prod.*, **vol. 256**, p. 120350, 2020, doi: 10.1016/j.jclepro.2020.120350.

[4] Lima L, E. Trindade, L. Alencar, M. Alencar, and L. Silva, "Sustainability in the construction industry: A systematic review of the literature," *J. Clean. Prod.*, **vol. 289**, p. 125730, 2021, doi: 10.1016/j.jclepro.2020.125730.

[5] Murtagh N, L. Scott, and J. Fan, "Sustainable and resilient construction: Current status and future challenges," *J. Clean. Prod.*, **vol. 268**, p. 122264, 2020, doi: 10.1016/j.jclepro.2020.122264.

[6] Udomsap A. D., and P. Hallinger, "A bibliometric review of research on sustainable construction, 1994–2018," *J. Clean. Prod.*, **vol. 254**, p. 120073, 2020, doi: 10.1016/j.jclepro.2020.120073.

[7] Sahadi S., and M. Agung Wibowo, "Ketepatan Klasifikasi pada Pengelompokan Manajer Proyek Konstruksi Berdasarkan Faktor Pengembangan Sumber Daya Manusia Menggunakan Analisis Diskriminan," *Media Komun. Tek. Sipil*, **vol. 20**, no. 2, 2015, doi: 10.12777/mkts.20.2.119-124.

Effect of eucalyptus pellita timber-pvd hybrid pile as a vertical drain on soft soil.

Suyadi[1] Tri Harianto[2] A B Muhiddin[2] A Arsyad [2]

[1]Doctoral Student of Civil Engineering, Hasanuddin University, Gowa, South Sulawesi, Indonesia, 92171
[2]Departement of Civil Engineering, Hasanuddin University, Gowa, South Sulawesi, Indonesia, 92171

E-mail: suyadi19d@student.unhas.ac.id

Abstract The innovative technique of a hybrid PVD timber pile as a vertical drain can be implemented to reinforce and improve soft soil. In this case, the acceleration of pore water dissipation could increase shear strength. The at soil of this study is to investigate the effectiveness of a hybrid eucalyptus pellita-PVD timber pile as a vertical drain. The test method applied is a laboratory model test, consisting of the installation of a hybrid pile on soft soil in a rectangular pattern with a spacing of 10 cm. Then by a static loading test was subjected on soft soil with an initial load of 10 kg and subsequent loads of 20 kg and 40 kg. Daily dial monitoring of the increase in pore water dissipation was performed during the preloading test. The results show that the hybrid pile could accelerate the process of pore water dissipation to the surface compared to using timber piles alone. The increase of soil density before and after the loading indicated that the hybrid piles enhanced the density by 3.9%, whereas the reinforcement using only timber pile just increased by 2.4%.

Keywords: hybrid pile, eucalyptus pellita, vertical drain.

1. Introduction

As well as the expansion of the construction sector, the need for developing land will continue to rise. However, some regions are unable to avoid conditions in which building sector development must be conducted on soft soil. Soft soil has a poor bearing capacity, a high soil plasticity index, low permeability, and a prolonged subsidence process. These qualities pose the most significant challenges for designers who wish to construct a building atop it.

Construction on soft soil requires many solutions to the problems of limited bearing capacity and extended consolidation period. And this issue can be resolved by improving the carrying capacity of the soil through micropile installation. [1] [2] [3] [4] [5] [6] [7] and constructing one-way vertical drain might accelerate changes in soft soil water content and the process of land subsidence. [8] [9] As for the conductivity issue, PVD can be utilized to speed up the process of pore water dissipation caused by loading processes. [10] [11] [12].

In Indonesia, micro-piles are constructed with timber piles since indigenous materials are permitted in several regions, and the availability of these materials is relatively high, making this a viable option in areas where the soil conditions have a limited bearing capacity.

As a breakthrough in soil improvement and reinforcement engineering, hybrid Eucalyptus pellita-PVD timber piles are anticipated to raise the soil's bearing capacity and hasten the consolidation process. This study intends to assess the effect of constructing a rectangle pattern of Hybrid piles of eucalyptus pellita-PVD timber piles as vertical drain on the soft soil's water release.

2. Materials and Testing Methods

2.1. Materials

Based on the result of laboratory tests, which included testing the physical and mechanical properties of soft soil samples, test data on the characteristics of the eucalyptus pellita timber material, and test data on the characteristics of the Prefabricated Vertical Drain (PVD) material, the initial research data were gathered. This information is necessary to establish whether the type of material used corresponds to the required design and description for the intended model test. This study utilized a soft soil sample from the Turikale, Maros Regency sub-district. The eucalyptus pellita timber originates from the Muting District of the Merauke Regency in South Papua Province. As indicated in Figure 1, the little timber pile has a length of 33 centimeters and a width of 1 centimeter. As shown in Figure 2, the Prefabricated Vertical Drain (PVD) length is 33 cm, and the width is 1 cm. Based on the shape of the two materials, two reinforcement models are created, the first of which consists of three connected timber piles. The second type uses wire, composed of three piles and one PVD in the center, as depicted in Figure 3.

Figure 1. Mikro pile eucalyptus pellita

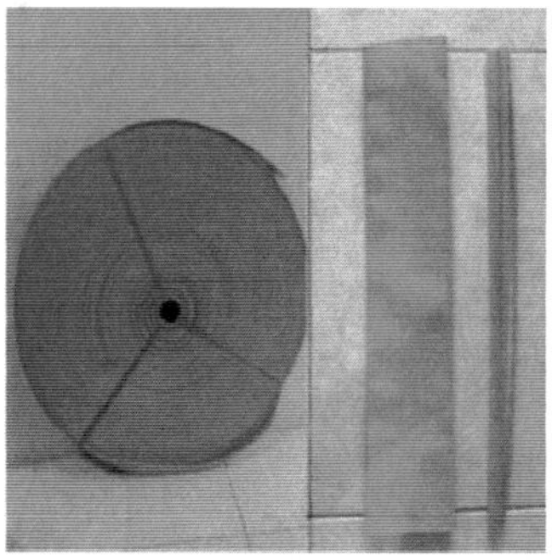

Figure 2. Prefabricated Vertical Drain (PVD),

Figure 3. Timber Pile and Timber pile+PVD (hybrid pile)

2.2. Testing Methods

The applied research design refers to the collected primary and secondary data. This study includes two test models: a one-way vertical drain test using a timber cone + PVD (Hybrid pile) and reinforcement using timber piles without PVD, installing a 10-cm-wide rectangular pattern. This layer will serve as a control point for the process of releasing water to the surface by evaluating the water content daily during the testing procedure, as depicted in Figures 4, 5, and 6. Subsequently, the preloading test was conducted with a progressive load, starting with 10 kg, followed by 20 kg, and finally 40 kilograms. By measuring the moisture content of the sand every day for 27 days, the process of water rising to the surface was observed and quantified. Figures 7 and 8 show that the test implementation model is shaped and sized according to the test dimensions.

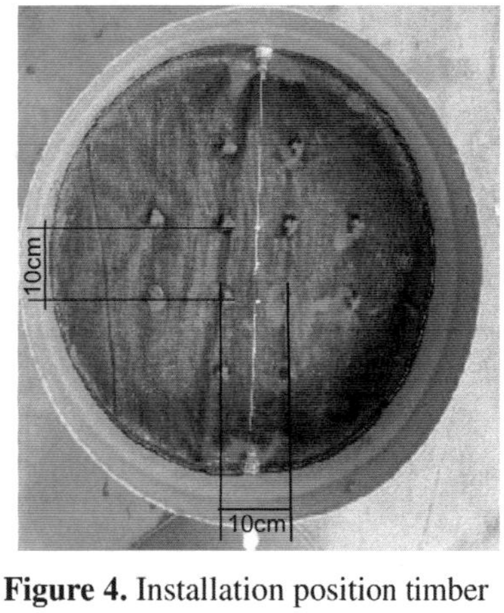

Figure 4. Installation position timber pile

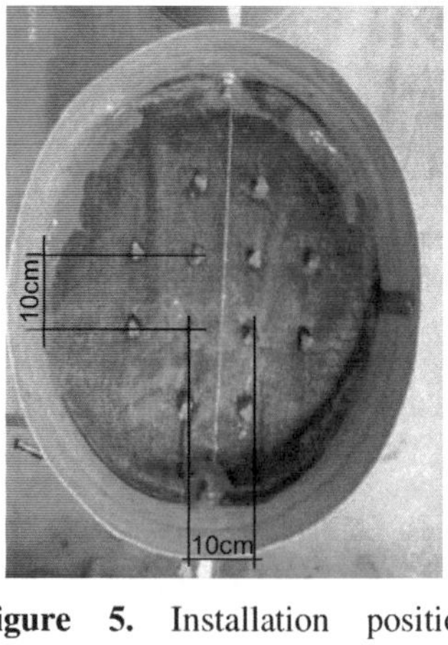

Figure 5. Installation position Hybrid pile

Figure 6. Water content sampling

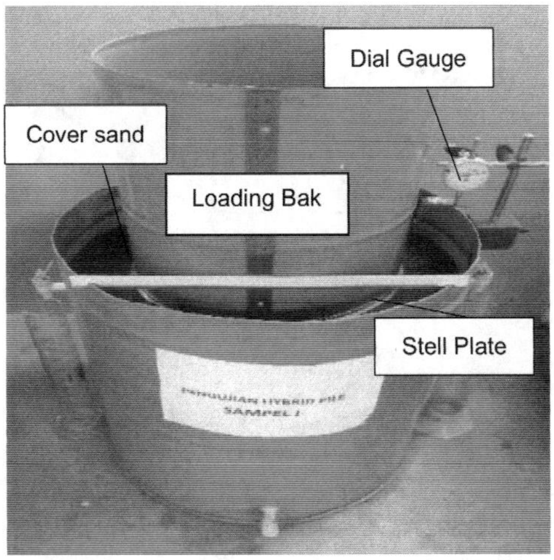

Figure 7. Test equipment installation position

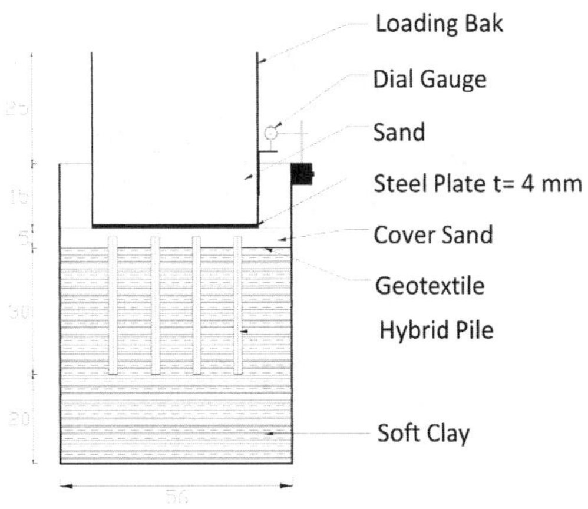

Figure 8. Model and size test dimensions

3. Result and Discussion

Based on the results of testing the soil characteristics, the type of soil utilized for model testing according to the USCS classification was high plasticity clay (CH), whereas the classification according to AASHTO was group A-7-6. The summary of the soil characteristics test may be found in Table 1.

Table 1. *Soil characteristics test results.*

No.	Kinds of Test	Unit	Test Result
Characteristic of Physical Properties			
1	Specific Gravity (Gs)	-	2,74
2	Water Content (w)	(%)	72.91
3	Atterberg Limit		
	shrinkage limit (S.L.)	(%)	13.88
	Plastic Limit (P.L.)	(%)	26.77
	Liquid Limit (L.L.)	(%)	82.75
	Plasticity Index (PI)	(%)	55.98
4	Grain size analysis		
	Gravel	(%)	0.00
	Sand	(%)	7.20
	Silt	(%)	25.50
	Clay	(%)	67.30
5	Permeability	(cm/s)	2.3×10^{-7}
Classification			
USCS			CH
AASHTO			A-7-6

Based on the outcomes of laboratory tests on Eucalyptus Pellita timber used as timber piles, this species is classified as a timber class I [13]. Based on the results of investigating the qualities of the timber material used for piles, the data presented in Table 2 were acquired. The characteristics of the Prefabricated Vertical Drain (PVD) material used in this investigation are of the CeTeau Drain CTD-812 type. This PVD has a permeability value of > 1 x 10-4 m/s, which is fairly high.

Table 2. *Eucalyptus Pellita test results*

No.	Kinds of Test	Unit	Test Result
1	Specific Gravity (Gs)	-	0,91
2	Water Content (w)	(%)	8.83
3	Flexural strength	Mpa	110.50
4	Compressive strength parallel to fiber	Mpa	60.15

Observations and testing of water content during the research period revealed that the use of a timber cone + PVD as a vertical drain on soft clay soils could accelerate the process of releasing surface water. At the initial time of loading, the increase in water content surface- cover sand and the graph repeats this pattern at the initial time of adding the second and third loads. Figure 9 depicts the graph of the relationship between the load and water content surface- cover sand.

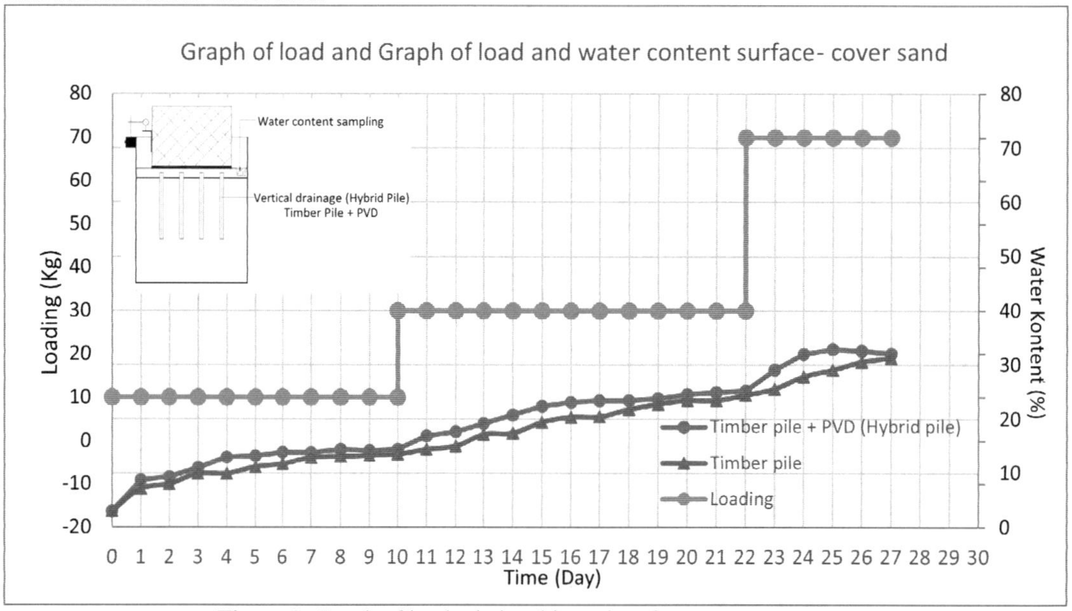

Figure 9. Graph of load relationship and surface water content.

Table 3. *The results of the water content and bulk density of the soil testing*

Test Model	Before loading			After Loading		
	γ_{wet} (gr/cm^3)	γ_{dry} (gr/cm^3)	W (%)	γ_{wet} (gr/cm^3)	γ_{dry} (gr/cm^3)	w (%)
Timber pile	1.49	0.82	82.69	1.51	0.84	81.07
Timber pile + PVD (hybrid pile)	1.49	0.82	82.20	1.53	0.85	79.93

Table 3 presents the results of a comparison between the soft soil test values with timber pile reinforcement and the soft soil test values with hybrid pile (PVD) reinforcement as vertical drain. The results of this preloading test indicate that the density of the soft soil has increased. The indicator for this rise in density is the change in the original value of dry density before loading and reinforcing system installation. The initial dry density value before loading was 0.82 kg/cm3 with a moisture content of 82%. However, following the application of timber pile reinforcement, the value increased to 0.84 kg/cm3 with a moisture content of 81.07%. Using a hybrid pile as vertical drain increased the soil's dry density to 0.85 kg/cm3 while decreasing the soil's water content to 79.93%.

This demonstrates that using timber piles and PVD (hybrid pile) can speed the soil compaction process. This situation emerges due to the PVD's function as a vertical drain system. Figure 10 depicts a comparison of the results of soil density tests. Figure 11 shows the variations in the moisture content of soft soil before and after loading. The results indicate that using Hybrid piles of Eucalyptus Pellita – PVD as vertical drain can enhance soil density by 3.9%, whereas the reinforcement that simply employs timber piles increases soil density by 2.4%.

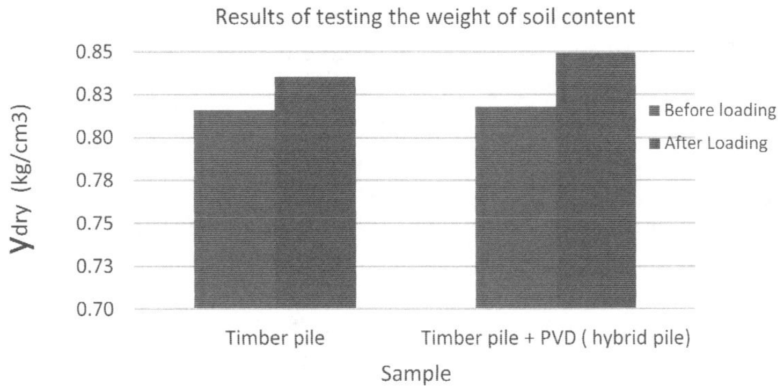

Figure 10. Dry density value before and after loading test

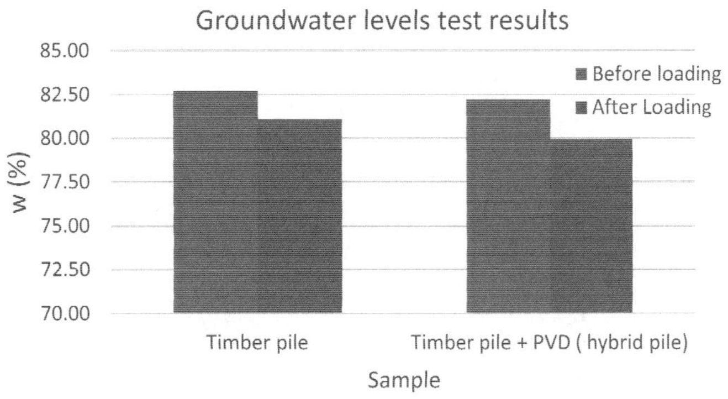

Figure 11. Comparing the results of water content tests

4. Conclusion

The conclusions that can be drawn from this research are:

1. Installation of vertical drain Hybrid piles of Eucalyptus Pellita – PVD timber piles are able to accelerate the process of removing water from the soil when compared to reinforcement that only uses timber cones.
2. The use of Hybrid pile piles of Eucalyptus Pellita – PVD as vertical drain was able to increase soil density by 3.9%, while the reinforcement that only used timber cones was 2.4%.

5. Suggestion

It is necessary to conduct further testing with different installation patterns and depth variations to obtain an effective design and depth used for the installation of hybrid piles as vertical drain.

6. References

[1] Harianto T., Djamaluddin R., Muhiddin A.B., Maricar I.& SitepuF 2012 *Characteristic'sand Behaviour Study on Raft and Pile Bamboo as a Soil Embankment Reinforcement (in Indonesia).(Indonesia:Trisakti University),* pp. G59-G65
[2] Gasruddin Ahmad., Harianto T., Arsyad Ardy 2016 *Reinforcementofthe Slopeof Road Surfacewith Micro Root Pile Typein Soft Soil Deposite.(Indonesia:Hasanuddin University),* Vol.5 pp. 80–85.
[3] Sandyutama, Y., Samang, L., Imran, A.M., Harianto, T., 2015 *Full Scale Model Test of Consolidation Acceleration on Soft Soil deposition with Combination of Timber Pile and PVD (Hybrid Pile).* (IJIRAE), Volume 2, pp. 23-28
[4] Moayed R.Z, Abolhassan Naeini S 2012 *Imrovement of Loose Sandy Soil Deposits using Micropiles.* KSCE Journal of Civil Engineering, pp. 334-340
[5] Mitrani .H and Madabhushi, S. P.G 2008 *Centrifuge modelling of inclined micro-piles for liquefaction remediation of existing buildings.* Geomechanics and Geoengineering An International Journal, pp. 245-256
[6] Harianto, T., Yunus,M and Walenna, M.A 2011 *Bearing Capacity Of Raft-Pile Foundation Using Timber Pile On Soft Soil, International Journal of GEOMATE, Vol.21(86)pp108-114.* doi:10.21660/2021.86.j2294
[7] Harianto T., Samang L., Suheriyatna and Sandyutama Y., 2016 *Filed Investigation of the Performance of Soft Soil Reinforcement with Inclined Pile, Proc. of the 5th Geotech. & Geophysic. Site Characteristics, Lehane, Acosta-Martinez & Kelly (Eds), pp. 1349-1352.*
[8] Abadi, T. C., 2004, *Uji Laboratorium Pemanfaatan Serabut Kelapa dan Ijuk Sebagai Bahan Drainase Vertikal Tanpa Filter*, (Indonesia:Institut Teknologi Nasional Bandung)
[9] Prastyo, B., Setiawan, B., Indrabaskara, R. H. D. H., 2017 *Pengaruh Kedalaman Drainase Vertikal Satu Arah Menggunakan Kolom Pasir terhadap Kecepatan Penurunan dan Perubahan Kadar Air Tanah Lunak*, (Indonesia:Universitas Sebelas Maret)
[10] Herwanto W 2012 *Studi Perbandingan Analisis Perbaikan Tanah Lunak Dengan MetodePrefabricated Vertical Drain Pada Berbagai Constitutive Model (Studi Kasus PT.Wilmar Nabati).*(Indonesia: *Institut Teknologi Bandung)*
[11] Juniarto 2011 *Analisa Waktu Penurunan Tanah Dengan KombinasiMetode Preloading dan Prefabricated Vertical Drain (PVD) Antara Pola Segitiga dan Persegi Pada PerbaikanTanah (Studi Kasus Landasan Pacu Bandar Udara Juwata Tarakan).(Indonesia; Lembaga PenelitianUniversitas Gunadarma)* ,pp. 91-99.
[12] Harianto, T. 2022 *Performance of Subbase Layer with Geogrid Reinforcement and Zeolite-Waterglass Stabilization, Civil Engineering Journal*, Vol. 8(20), pp251-262. doi: 10.28991/CEJ-2022-08-02-05.
[13] Yayasan Dana Normalisasi Indonesia, 1979. *Peraturan Konstruksi Kayu Indonesia NI-5 1961.* (Bandung: Yayasan Lembaga Penyelidikan Masalah Bangunan).

Acknowledgements

The authors would like to express their deepest gratitude for the financial support from the Education Fund Management Institute (LPDP), Ministry of Finance of the Republic of Indonesia.

Study on the Implementation of the Combination of 2 BIM and ERP Systems to Improve Project Cost Estimation Accuracy at the Auction Stage

Ispandi Pudael[1], Rusdi Usman Latief[2], Syarif Burhanuddin[3]

Civil Engineering, Hasanuddin University, Indonesia
Civil Engineering, Hasanuddin University, Indonesia
Civil Engineering, Hasanuddin University, Indonesia

email : ispandipudael@gmail.com,rusdiul@gmail.com,syarifbur9@gmail.com

Abstract. This study aims to obtain a process flow model of an organizational function that functions effectively in increasing the accuracy of project cost estimates at the auction stage by including implementation elements from a combination of 2 BIM and ERP systems in it. The location of the research was carried out in a state-owned contractor company, namely PT. Nindya Karya (Persero). This state-owned company has used ERP and BIM in its business processes. The methodology used in this study is a questionnaire-based primary data research where data relating to the development of process flow involving BIM and ERP elements in it have been collected from the Estimating department unit in the company PT. Nindya Karya. The questionnaire was developed to evaluate the importance of implementing BIM and ERP in improving the accuracy of cost estimation identified through responses obtained from the respondents to the questionnaire survey. The research data are then processed using Flow chart analysis techniques. The results of the study obtained that the process flow model of the estimating department of PT. Nindya Karya has currently implemented BIM in the tender calculation process, but has not involved ERP elements. The final result of drawing conclusions in this study is in the form of a workflow flowchart or process flow from the estimating department which in the process involves 2 BIM and ERP systems.

1. Introduction

The development of construction technology in Indonesia is increasingly advanced and growing rapidly, especially in the field of construction digitization. One application of construction digitization is BIM (Building Information Modeling) and ERP (Enterprise Resource Planning). BIM is a new innovation that is intended for working on a construction project, while ERP is a software system that is intended for systems within the company.

The concept of BIM imagines virtual construction before the actual physical construction, where we can simulate 3D building modeling, simulation of construction time and stages, simulation of cost estimation, simulation of energy analysis and operational cycles. ERP is an application integration program, with cross-functional parts such as marketing, production and HR, of a company organization, so that the information needed can be obtained instantly and up-to-date.

Seeing the benefits and advantages of BIM and ERP, it is very possible to apply them to help estimate the cost of the auction stage. In calculating the estimated cost of the initial stage of a project tender, there is often an increase in costs during the project implementation stage, this is due to the short time of preparation for the tender so that the cost estimate is less accurate.

Zainuddin and Fakhrizal (2015), in their research found that the factors that affect the accuracy of project cost estimates at the auction stage are the work method factor, the information factor, and the subcontractor factor. Seeing the potential of 2 BIM and ERP systems, these three risk factors can be minimized effectively by applying BIM and ERP technology in the cost estimation process at the auction stage.

In reality in the field ERP and BIM applications are separate application systems where ERP is more used to build enterprise systems and BIM is more used for a construction project cycle, each of which produces 2 separate outputs. This allows for potential errors in addition to the time lag. To overcome these problems, it is necessary to integrate 2 BIM and ERP systems in one platform.

Sarkar, et al, (2017) in their research results show that the proposed integrated BIM-ERP-IoT Module is a unique effort for real-time monitoring, asset management, inventory management, and improved enterprise revenue management for construction projects. This shows that the 2 BIM and ERP systems can be integrated.

The integration of 2 BIM and ERP systems in the cost estimation process at the auction stage can be started by designing the process flow of a company organizational function that includes elements of BIM and ERP implementation in it. The results of this flow process will later serve as a reference or guide for the company's IT department in its efforts to integrate 2 BIM and ERP systems into one digital platform.

Based on these problems, this study aims to design the process flow of an organizational function that functions effectively in increasing the accuracy of project cost estimates at the auction stage by including implementation elements from a combination of 2 BIM and ERP systems in it. The scope of this research focuses on corporations, namely the contractor company PT. Nindya Karya (Persero). This state-owned company has used ERP and BIM in its business processes, so it is very relevant to the theme of our research.

2. Literature review

2.1. The Concept of BIM

The concept of BIM envisions virtual construction prior to actual physical construction to reduce uncertainty, improve safety, solve problems, and analyze potential impacts (Smith, Deke 2007), Although BIM has been known since the 1970s, the term Building Information Modeling (BIM) is gaining popularity. after Autodesk released the paper "Building Information Modeling" in 2002. The term BIM reappeared in 2005 when the US General Service Administration (GSA) made the decision to build a new courthouse in Jackson, Mississippi (SIBIMA Construction, PUPR, 2019).

2.2. Understanding of ERP

ERP is an acronym for "Enterprise Resource Planning", which is the process of collecting and organizing business data through an integrated software suite. Basically, ERP is an application that automates business processes, controls, and collects input from various data. The data includes accounting, manufacturing, supply chain, sales, marketing and human resources (HR) data.

2.3. BIM + ERP integration

2.3.1. The Concept of BIM + ERP Integration

BIM is a design technology that creates 3D Visualizations of construction projects using a coherent system of computer models. This allows real-time collaboration across all departments rather than using separate sets of images. ERP is a business management solution that manages finances, operations and projects and enables companies to run their business, automate key processes and save costs. When combined, these create a powerful project management system that revolutionizes construction projects.

BIM creates a full-size three-dimensional building project inside a computer. ERP allows the user to determine the final costs, the time frame involved and where the equipment and subcontractors will come from. BIM is responsible for every system, channel, structural column, and pipeline. ERP schedules production and ensures project requirements are met. The bundled product ensures building projects are completed on time, with few errors and changes. Collaboration between all departments is seamless and a wide variety of tools are available immediately.

2.3.2. The Benefits of BIM + ERP Integration

Integrating BIM and ERP enables the integration of actions and controls through effective communication of information available to everyone from a central location. Always-on access to updated data will allow managers to gain transparency into the actual state of the project and then help make profitable business decisions. Of course, ERP and BIM integration is more than just collecting data. Let's take a look at other important benefits of integrating ERP and BIM in construction projects:

a. Financial management

Contract work is the norm in the construction industry. Tender bids are issued for large-scale projects. Accurate estimates of materials and labor are key in both situations. The combination of BIM + ERP can be used to get accurate estimates for each item in the BOM and for the efforts involved in project execution as well. This calculation is more precise and accurate. This helps achieve more profitable and predictable project resource planning.

b. Enabling Better Performance

Communication is an important factor in determining the success of a construction project. The integration of BIM and ERP provides a centralized source of information for everyone working on a project. These are fed into defined workflows so the right people get the right info when they need it. Updates are made and recorded in real-time, eliminating errors due to time lags in communication. What if scenarios could be analyzed quickly and their implications downstream This helped people and teams deliver better results and take responsibility for their efforts.

c. Avoiding contract conflicts

The solution helps in avoiding contractual conflicts with clients. This is done by making a baseline estimate and then tracking inventory, raw material requirements, labor, finances, and project budget against the forecast. Careful analysis of construction aspects leaves little or no room for misunderstanding and helps encourage more timely interventions if deviations become inevitable.

d. Simplify Data Several Projects

Handling multiple projects or several aspects of one project at a time will inevitably create confusion in communication.

2.4. Cost Estimation

Cost estimation is a process of determining or determining the costs involved for a construction for each existing project. According to the National Estimating Society-USA, estimation is the art of estimating (the art of approximating) the possible amount of costs required for an activity based on information available at that time. Cost estimation can also be defined as a cost evaluation of all project elements whose calculations are based on the approved scope of work (Phaobunjong, 2002). Cost estimation can also be interpreted as determining the possible construction costs of any given project. Many items (i.e., materials, labor, equipment, insurance, and overhead, as well as profit forecasts etc.) affect and contribute to the cost of a building project.

3. Research Method

Data collection for this research was conducted at the office of PT. Nindya Karya, especially the Estimating department, which is responsible for the process of estimating project costs at the auction stage in the DKI Jakarta area in June 2022. The data needed in this study consists of primary data and secondary data. Primary data obtained from interviews, focus group discussions (FGD) and filling out questionnaires, secondary data in the form of company documents, and literature studies. The selection of resource persons is based on purposive sampling with the intention that the selected sample has knowledge, expertise, and competence in the field studied in this study. Respondents consist of internal and external respondents. Internal respondents in this study were company managers. External respondents are other state-owned and private construction contractor companies.

The steps involved in analyzing and processing data in this research are: this research begins with viewing or photographing the process flow model run by the company PT Nindya Karya in the current Estimating department which has responsibility for the project cost estimation process at the auction stage. Furthermore, a detailed assessment of the risk factors that affect the accuracy of cost estimates at the auction stage is carried out. Then identify the ability of 2 BIM and ERP systems to minimize risk factors that affect the accuracy of cost estimates at the auction stage. The data analysis technique used the flow chart analysis technique. This technical analysis consists of three activities, namely data reduction, data display and drawing conclusions/verification. Data reduction is the simplification of data by removing unnecessary data, then displaying the data, namely displaying data in the form of tables, diagrams, or in the form of flow charts.

4. Result and Discussion

In early 2021, PT. Nindya Karya carried out an Organizational Restructuring, which was originally a territory-based business activity, turned into a Job Specialization with Centralized functions such as Supply Chain Management (SCM), Quality Health Safety and Environmental (QHSE) Information Technology & Human Capital (IT & HC). In the organizational structure of the Estimating Department led by a Senior Vice President, who oversees 5 units, namely Contract Administration, Estimating Infra 1, Estimating Infra 2, Estimating Building, and Engineering & BIM, each unit is led by a Vice President and under him several staff. Based on the mapping result of the Estimating department's organizational structure, it can be seen that PT. Nindya Karya has implemented BIM technology in the project cost estimation process at the auction stage, this can be seen in one of the unit units, namely engineering & BIM.

4.1 Identify The Current Estimating Department Process Flow Model.

From the results of interviews and data collection of company documents in the Estimating Department, the process flow model of the Estimating Department of PT. Nindya Karya can now be seen from the Estimating Procedure document. The current process flow model in the estimating department is known as the estimating procedure, there are 4 estimating procedures, namely (1) Contract Administration Procedures, (2) Tender Process Procedures, (3) BIM Modeling Procedures for Tenders, (4) Velue Engineering Procedures for tender.

The current process flow model of the estimating department has implemented BIM in the tender process, this can be seen from the BIM Modeling Procedure for tenders in Figure 1. In this procedure the BIM Levels applied are Level 3D and 4D, where the application of BIM in the redrawing process (major item 3D modeling), perform MS synchronization. Project and work methods, making construction sequence animations (Naviswork).

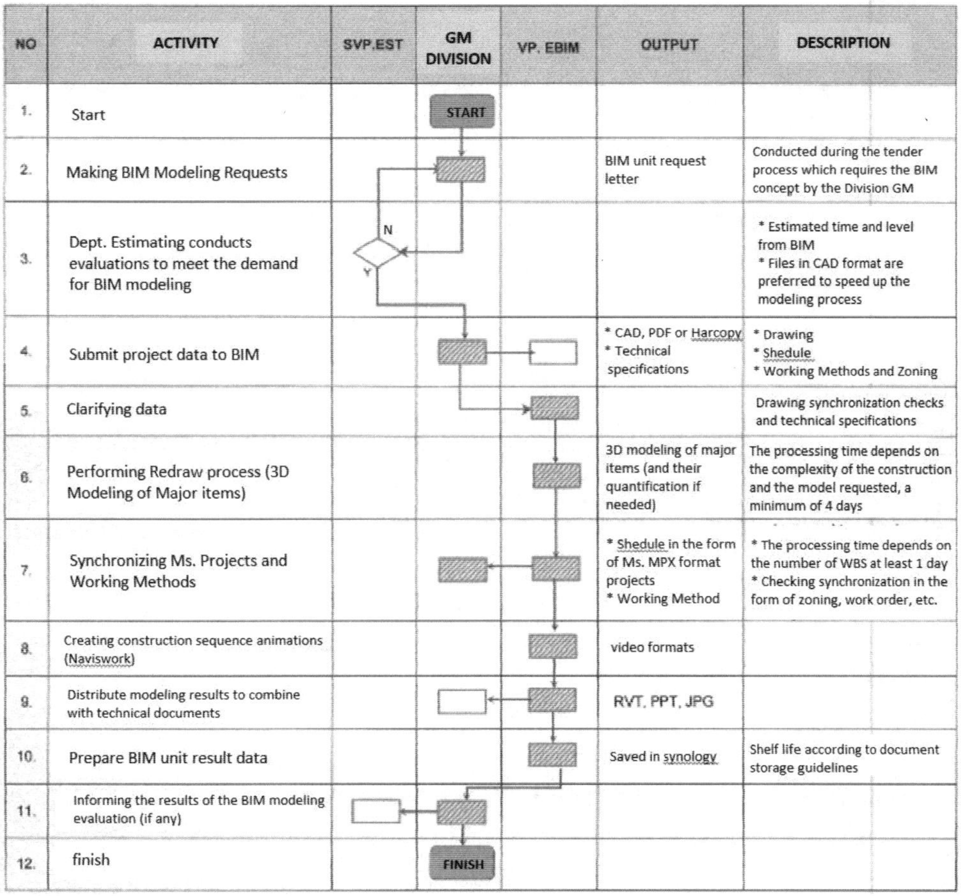

Figure 1. *Procedur BIM Modeling for tender*

Velue Engineering Procedure for Tender is a velue engineering process procedure from the volume of the main material work on the BOQ with modeling results using BIM software. This can be seen in Figure 2.

The current estimating department process flow model does not involve ERP elements, this is because the nindya ERP has not accommodated the Estimating department, the new ERP nindya accommodates the HR, Marketing, Finance, Production Departments, Efforts to accommodate the Estimating department in the ERP system are still in the development stage where later it is hoped that ERP can accommodate the estimating department in calculating or making online RAB, making work methods, tracking prices and vendors integrated in ERP.

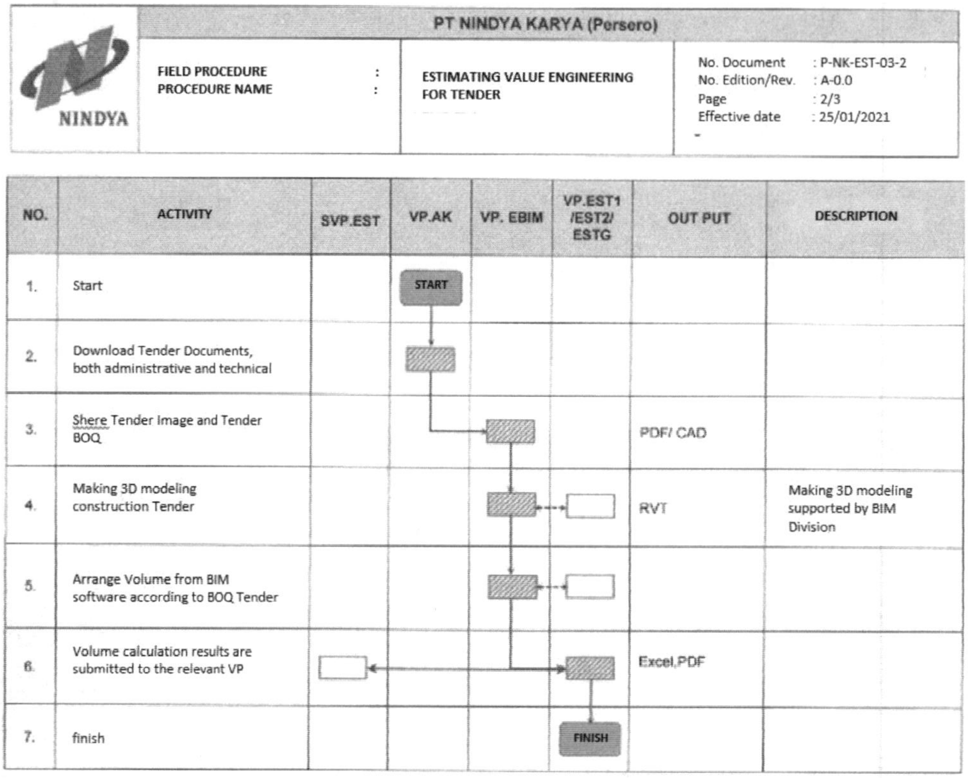

***Figure 2**. Value Engineering Procedure for Tender*

4.2 Designing a Process Flow Model For The Estimating Department Incorporating 2 Systems BIM and ERP

In designing the process flow of an organizational function that functions effectively in increasing the accuracy of project cost estimates at the auction stage by including the implementation element of a combination of 2 BIM and ERP systems in it, this study involved the Senior Vice President and the Vice Presidents of the Estimating Department as respondents to develop a process flow model in the estimating department that integrates 2 BIM and ERP systems. Data retrieval from the respondents was carried out by conducting FGDs, where each respondent answered each of the research questionnaire questions together regarding the development of a process flow model in the Estimating Department which integrates 2 BIM and ERP systems.

The final result of drawing conclusions is in the form of a workflow flowchart or process flow of an organizational function. Flowchart program using Microsoft Visio Software. The results of this flow process will later serve as a reference or guide for the company's IT department in its efforts to integrate 2 BIM and ERP systems into one digital platform.

5. Conclusion

PT. Nindya Karya has applied BIM technology in the project cost estimation process at the auction stage, this can be seen in the organizational structure of the estimating department, one of which is engineering & BIM. The current estimating department process flow model has implemented BIM in the tender process, this can be seen from the BIM Modeling Procedure for tenders and Velue Engineering Procedure for Tenders. The current estimating department process flow model does not yet involve ERP elements. Efforts to accommodate the Estimating department in the ERP system are still in the development stage. The final result of drawing conclusions in this study is in the form of a workflow flowchart or process flow from the estimating department which in the process involves 2 BIM and ERP systems. Flowchart program using Microsoft Visio Software. The results of this flow process will later serve as a reference or guide for the company's IT department in its efforts to integrate 2 BIM and ERP systems into one digital platform.

References

[1] Heryanto S, Subroto G and Rifa'ih 2020 *Study on the Application of Building Information Modeling (BIM) in the Indonesian Construction Services Industry*, Journal of Architecture Innovation ISSN 2549-080X

[2] Roland M, Simanjuntak A and Baskoro A 2020 *Review Of Project Financing Management In Implementation Of BIM on Building Projects*, ISBN: 978-602-51450-2-5

[3] Alianto H, and Wijaya S 2013 *Implementation of ERP System Implementation in Making Project Feasibility, Project Status and Project Monitoring in Contracting Companies*, **4** 582-7

[4] Chen Y and Tserng H 2017 *An Integrated Methodology for Construction BIM & ERP by Using UML Tool*, (Taiwan: University of National Taiwan University)

[5] Iskandar M and Krisnadi I 2020. *Application of Enterprise Resource Planning (ERP): Literacy Studies in Project Management.* Master of Electrical Engineering Postgraduate Mercu Buana University

[6] Alianto H and Wijaya S 2013, *Application of the Erp System in Making Project Feasibility, Project Status and Project Monitoring in Contracting Companies*, **4** 582

[7] Sarkar D, Pandya K, Dave B, Jha K and Dhaneshwar D 2021 *Development of an integrated BIM-ERP-IoT module for construction projects in Ahmedabad. Innov. Infrastruct. Solut.* 7, 50

[8] Zainuddin and Fakhrizal 2015. *Analysis of Factors Affecting the Occurrence of Underestimate and Overestimate Costs at Cost Accuracy Level*, **2** 22

Selected Mechanical and Physical Properties Concrete with Polypropylene Plastic Granule Aggregate

M Sofyan[1)*], H Parung[2)], M W Tjaronge[2)], and AA Amiruddin[2)]

[1)]Doctoral Student, Civil Engineering Department, Hasanuddin University, Indonesia
*sofyanm21d@student.unhas.ac.id
[2)] Civil Engineering Department, Hasanuddin University, Indonesia
parungherman@yahoo.com
[2)] Civil Engineering Department, Hasanuddin University, Indonesia
tjaronge@yahoo.co.jp
[2)] Civil Engineering Department, Hasanuddin University, Indonesia
a.arwinamiruddin@yahoo.com

Abstract. Amount of plastic use contribute significantly to the accumulation of plastic waste. The use of plastic waste in concrete is one of the world's partial solutions for dealing with plastic waste. Polypropylene plastic granules are used as a partial replacement for fine aggregate in this study. The purpose of this research is to examine the physical and mechanical properties of the chosen base concrete while considering the potential of polypropylene plastic granules as a sand substitute in the concrete mixture. Variables to be investigated incorporate compressive strength, split tensile strength, workability, and density. The results of this study show that using polypropylene plastic granules as a substitute for fine aggregate at 10%, 20%, 30%, and 40% rates slump values of 8.1 cm, 8.4 cm, 8.6 cm, and 8.8 cm, respectively. However, it does not perform well in terms of compressive strength or split tensile strength. The strength reduction achieved by using 40% PP plastic granule can reach 58.06 percent.

1. Introduction

The rapidly developing construction science and technology encourages stakeholders to focus on quality and productivity standards in order to fulfill quality and environmentally friendly construction products. Concrete technology is currently advancing at a rapid pace [1]. Waste is one option for producing concrete mix materials with unique properties [2]. The issue of accumulated plastic waste is currently in the spotlight. Plastic is a polymer composed of long carbon chains [3]. The world's population is rapidly increasing, which contributes to plastic waste. From 1950 to 2018, global plastic production was estimated to be 8.3 billion metric tonnes, with an annual increase of 5% (185 million tonnes) [4-6]. The overuse of synthetic plastics, combined with their inherent resistance to degradation, is a major environmental issue (Phuong et al., 2016b). Polypropylene (PP), Polyethylene (PE), polyethylene terephthalate (PET), polystyrene (PS), polyurethane (PUR) and spolyvinyl chloride (PVC) account for 90% of global plastic production [7].

In order of preference, plastic waste can be processed using three methods: recycling, incineration (often combined with an energy recovery process), and landfilling [8]. Several studies on the use of plastic waste in concrete mixtures have been conducted by researchers. The commonly reported result of incorporating plastic into concrete as an aggregate replacement is a reduction in workability, density, and mechanical performance of the resulting concrete mix, with no significant improvement in material durability [9-11]. Faraj et al, 2021 [12] investigated concrete using recycled PP plastic granules at percentages of 0%, 10%, 20%, 30%, 40% of the volume of fine aggregate, 20% fly ash, and 10% silica fume. The investigation's findings revealed that using PP plastic granules made the concrete lighter but reduced some mechanical properties. Frigione [13] used a w/c ratio of 0.45 and recycled PET bottles with a 5% weight basis replacement of natural fine aggregate. The compressive strength curves of reference concrete and PET concrete were found to be very similar. However, when compared to the reference concrete, PET concrete had higher ductility and toughness. Ridwan et al., 2014 [14] investigated the proportion of polystyrene granules calculated to the weight of cement and found an increase in the split tensile strength of concrete with percentage increases of 21.789 percent, 15.811 percent, 11.856 percent, and 10.329 percent. The compressive strength of the test object, on the other hand, decreased by 4.928 percent, 14.765 percent, 16.214 percent, and 22.826 percent.

Previous research found that concrete with plastic granular aggregate has a variety of physical and mechanical properties. The variables to be studied will be workability, volume weight, split tensile strength, and compressive strength of concrete using polypropylene plastic as an aggregate substitute. Workability is regarded as a parameter that contributes to concrete consistency. Meanwhile, good concrete initial strength is regarded as an important criterion in precast concrete fabrication.

Previous research has shown that Polypropylene plastic granules can be used as a concrete mixture with good mechanical properties; however, the size and shape of the plastic granules or pieces used are not uniform. In this study, a more uniform grain size and shape was used as a substitute for fine aggregate in concrete, with the volume of fine aggregate calculated. The workability, density, compressive strength, and split tensile strength of 7,28-day concrete were measured using Polypropylene plastic granules.

2. Materials and Methods

The specimen is a concrete cylindrical specimen with 10 cm in diameter and 20 cm in high, made up of 5 Variations of Test Objects, as shown in table 1. In this study, the mix method is dry mixing. All dry ingredients, including cement, aggregates, and additives, are mixed together in a mixer in dry mix applications. When the material is in the mixer, the material in the form of water is included. Each specimen will be tested at the ages of 7 and 28 days to determine the percentage of PP plastic granules based on the volume of fine aggregate. Plastic Granules are used in place of fine aggregates. Plastic granules are made of polypropylene (PP). Figure 1 depicts the physical appearance of the polypropylene granules used in this study. The cement used is portland cement type 2 from Indonesia, with a cement density of 3.01 gr/ml.

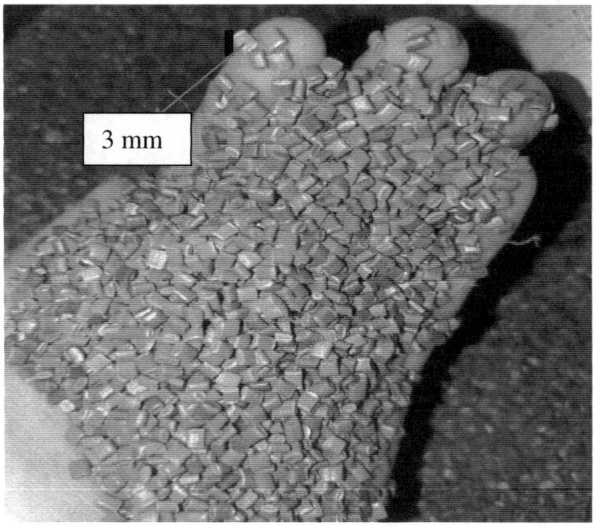

Figure 1. Physical appearance of Polypropylene Granule

IOP Conf. Series: Earth and Environmental Science 1117 (2022) 012014 doi:10.1088/1755-1315/1117/1/012014

Based on the physical material test results, the normal concrete mix design calculation for f'c = 20 MPa is performed, in accordance with SNI 7656 - 2012 [12]. SNI 7656 - 2012 [13] designates coarse aggregate gradation limits. Based on the results of the test, the unit weight of PP plastic granules is 547.93 kg/m^3. The aggregate physical data can be found in tables 2 and 3.

Table 1. Physical Properties Of Fine Aggregate

No	Sample	Constituent Material				
		water (kg)	cement (kg)	coarse aggregate (kg)	sand (kg)	Polypropylene granule (kg)
1	PPG0 (Normal Concrete)	11,611	21,501	69,918	51,565	-
2	PPG10(Polypropylene 10%)	11,611	21,501	69,918	46,408	2,997
3	PPG20 (Polypropylene 20%)	11,611	21,501	69,918	41,252	5,994
4	PPG30 (Polypropylene 30%)	11,611	21,501	69,918	36,095	8,991
5	PPG40 (Polypropylene 40%)	11,611	21,501	69,918	30,939	11,988
Total		58,055	107,505	349,59	206,259	29,97

Table 2. Physical Properties Of Coarse Aggregate

Physical Properties	Value
Water Content	2.8 %
Particle size	25 mm
Density	1405,825 kg/m^3
Fine Modulus	6.01
Sieve Analysis	Zone 3
Specific gravity	2.2

Table 3. Physical Properties Of Fine Aggregate (sand)

Physical Properties	Value
Water Content	4.6%
Specific gravity	2.29
Density	1640,78 kg/m^3
Fine Modulus	2.65
Sieve Analysis	Zone 2

To determine several key parameters of concrete quality, a compressive strength (fig. 2c) and split tensile strength test (fig. 2d) was performed. Concrete compressive strength is the maximum compressive strength per unit area. The nature, type of aggregate, type of mixture, workability, maintenance, and age of concrete all have an impact on its compressive and split tensile strength. The concrete is then cured by soaking it for 28 days.

After 28 days, the strength of the concrete can be determined. While the density can be calculated by weighing the specimen and dividing it by its volume. The slump test is used to determine the consistency or thickness of fresh concrete by measuring the height of the fresh concrete after the Abrahams cone is lifted.

a) b)

c) d)

Figure 2. a) Slump Test, b) Fresh Concrete, c) Compressive Strength test, d) Split Tensile Strength test

3. Results and Discussion

3.1. Density and Workability

Concrete density is one of the physical properties of concrete. Density is measured in kilograms per cubic meter. The dead load on the structure can be affected by density. Density is calculated by weighing the dry weight of the concrete and dividing it by the volume of the specimen. Figure 3 depicts the results of density measurements.

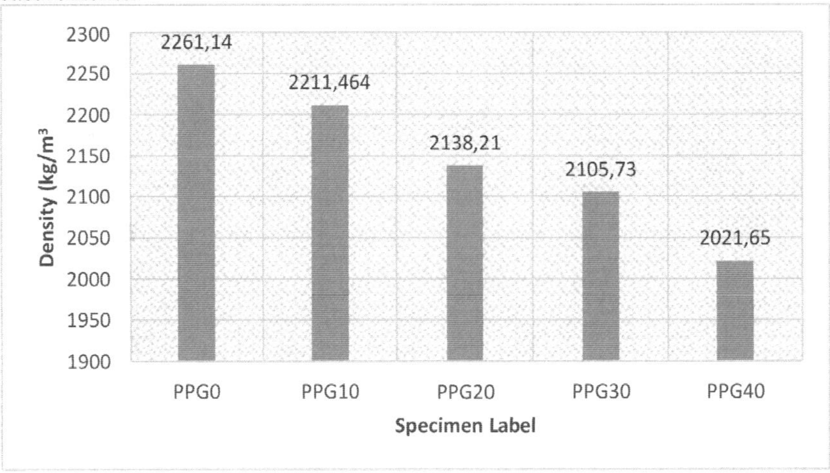

Figure 3. Density of Specimen

The density of each specimen is depicted in Figure 3. The addition of a percentage of polypropylene plastic granules can be seen to reduce the density of the concrete. The PPG0 specimen has the highest density at 2261.14 kg/m^3, while the PPG40 specimen has the lowest density at 2021.65 kg/m^3. The proportion of Plastic Granules in the PPG40 specimen is 40%. The addition of 10% Polypropylene Plastic Granules to the PPG10 decreased the density of PP-G0 by 2.2 percent. While using 40% plastic granules in PPG40 can reduce density by up to 10.59%. The percentage of PP plastic granules in concrete reduces its density. This is because plastic grains have a lower density than sand.

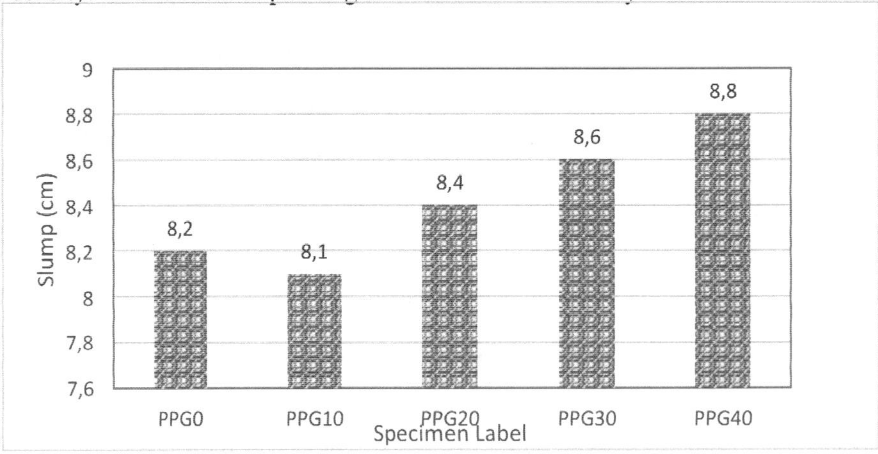

Figure 4. Slump Value

Workability Concrete demonstrates its consistency properties. A slump test is used to assess workability. Figure 4 depicts the results of the test object's Slump test. According to the slump test results, none of them are below or above the predetermined threshold of 7.5 cm to 10 cm. Concrete workability is expressed as a slump value, as shown in Figure 4. The slump value increases as the percentage of plastic granules increases. The increasing percentage of plastic granules is accompanied by an increase in slump value. However, PPG10 specimens behaved differently. Each specimen's mixing process was completed on a separate day. The cause could be the aggregate water content in each specimen's SSD condition.

3.2. Compressive Strength and Split Tensile Strength

Figure 5 depicts the specimen's compressive strength. The increase in the percentage of polypropylene in the concrete mixture resulted in a decrease in the compressive strength of both the 7-day and 28-day concrete. The use of 40% PP granules can reduce the strength of PPG0 control specimen concrete by up to 58.06 percent. The PPG40 specimen had the lowest compressive strength. The difference in compressive strength at 7 days for each specimen is not statistically significant. Similarly, the PPG40 specimen revealed that the strength of the concrete was nearly the same at 7 and 28 days. This phenomenon can occur for a variety of reasons, one of which is an imperfect compaction process. The use of a vibrator to compact fresh concrete during the molding process will yield better results.

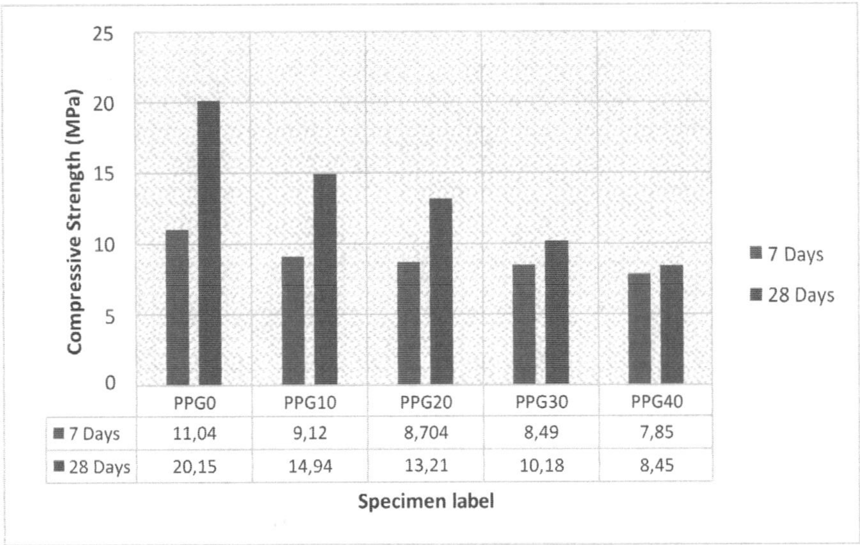

Figure 5. Compressive strength of specimen

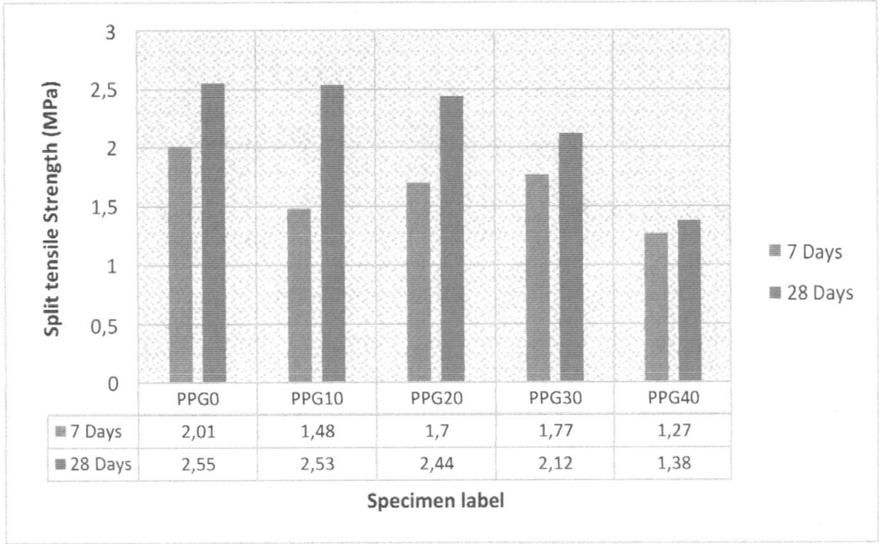

Figure 6. Split tensile strength of specimen

The split tensile strength (figure 6) of the 28-day specimen demonstrated the same behavior. At PPG40, the decrease in split tensile strength reached 45.9 percent. However, the percentage decrease in split tensile strength is not as large as the decrease in test object compressive strength. The behavior of 7-day concrete specimens differed. PPG10 strength was initially lower, but PPG30 strength increased by 16.4 percent. The PPG40 specimen has the lowest splitting strength. The non-uniform distribution of PP granules could be the cause of fluctuations in split tensile strength at 7 days' age, or an error occurred during the test object's molding. The split tensile strength trend differs from the compressive strength trend. It is caused by the matrix or uneven distribution of PP plastic granules in the mixture after the concrete has hardened.

4. Conclusion

It is possible to draw conclusions based on the results data and the discussion of several points.

- When applied to structural elements, the use of plastic granules can reduce the density of the concrete, lowering its self-weight.
- The increase in slump value is followed by an increase in the percentage of polypropylene granules. The addition of 40% polypropylene granules increased the test slump rate by up to 7.3%.
- The substitution of 40% polypropylene granules can reduce compressive strength by up to 58.06%.
- The percentage decrease in split tensile strength is not as large as the decrease in test object compressive strength. The distribution of plastic granules in the mixture can influence the trend of split tensile strength, which typically follows the trend of concrete compressive strength.

References

[1] Sofyan, M., Wicaksono, B., & Irlan, A. O. (2020). Effect of Combination of Steel Fibre and Silica Fume on the Mechanical Strength of Concrete. IOP Conference Series: Materials Science and Engineering, 771, 012048. https://doi.org/10.1088/1757-899x/771/1/012048

[2] D. A. Hamid, S. As, and E. Safitri, (2014). "TEKAN DAN MODULUS ELASTISITAS BETON BERKINERJA TINGGI GRADE 80 Agregat Alami dan Agregat Daur Ulang," vol. 2, no. 2, pp. 43–49,.

[3] Jakubowicz, I., 2003a. Evaluation of degradability of biodegradable polyethylene (PE). Polym. Degrad. Stab. 80, 39–43. https://doi.org/10.1016/S0141-3910(02)00380-4

[4] Jambeck, J.R., Geyer, R., Wilcox, C., Siegler, T.R., Perryman, M., Andrady, A., Narayan, R., Law, K.L., 2015. Plastic waste inputs from land into the ocean. Science 347 (6223), 768–771. https://doi.org/10.1126/science.1260352.

[5] Amobonye, A., Bhagwat, P., Singh, S., Pillai, S., 2021. Plastic biodegradation: frontline microbes and their enzymes. Sci. Total Environ. https://doi.org/10.1016/j. scitotenv.2020.143536.

[6] Geyer, R., Jambeck, J.R., Law, K.L., 2017. Production, use, and fate of all plastics ever made. Sci. Adv. 3 (7), e1700782. https://doi.org/10.1126/sciadv.1700782.

[7] Phuong, N.N., Zalouk-Vergnoux, A., Poirier, L., Kamari, A., Ch^atel, A., Mouneyrac, C., Lagarde, F., 2016a. Is there any consistency between the microplastics found in the field and those used in laboratory experiments? Environ. Pollut. 211, 111–123. https://doi.org/10.1016/j.envpol.2015.12.035.

[8] J. Gertsakis, H. Lewis,2003. Sustainability and the Waste Management Hierarchy, RMIT University, Melbourne, Australia. http://www.ecorecycle.vic.gov.

[9] W. Tang, Y. Lo, A. Nadeem, 2008. Mechanical and drying shrinkage properties of structural-graded polystyrene aggregate concrete, Cem. Concr. Compos. 30 (5) 403–409.

[10] M. Juki, M. Awang, M. Annas, K. Boon, N. Othman, M. Roslan, F. Khalid, 2013. Relationship between compressive, splitting tensile and flexural strength of concrete containing granulated waste Polyethylene Terephthalate (PET) bottles as fine aggregate, Adv. Mater. Res. 793, p356–359.

[11] S. Kou, G. Lee, C. Poon, W. Lai, 2009. Properties of lightweight aggregate concrete prepared with PVC granules derived from scraped PVC pipes, Waste Manage. 29 (2) p621–628.

[12] Faraj, R. H., Sherwani, A. F. H., Jafer, L. H., & Ibrahim, D. F..2020. Rheological behavior and fresh properties of self-compacting high strength concrete containing recycled PP particles with fly ash and silica fume blended. Journal of Building Engineering, 34, 1-28.

[13] M. Frigione, 2010. Recycling of PET bottles as fine aggregate in concrete, Waste Manag. 30 (6) 1101–1106.

[14] Ridwan, F., Subari, S., & YuIius, E.. Pengaruh Penggunaan Cacahan GeIas PIastik PoIypropyIene (Pp) Terhadap Kuat Tekan Dan Kuat Tarik Beton. 2014. Bentang, 2(1), 262524.
[15] Badan Standarisasi Nasional. SNI 7656-2012 Tata Cara Pemilihan Campuran untuk Beton Normal. BSN, Jakarta. 1-48. 2012
[16] ASTM-C33, 2003, *Standard Specification for Concrete Aggregates*, Annual Books of ASTM standards, USA.
[17] Badan Standardisasi Nasional. SNI 1972-2008. Cara Uji Slump Beton. Badan Standardisasi Nasional Indonesia, 1–5. 2008

Conceptual Framework For Supply Chain Risk Management Study on Breakwater Project in Indonesia Case Study : Makassar New Port Breakwater Development Project

Muhammad Nur Andry Djaja and Rosmariani Arifuddin

Civil Engineering, Hasanuddin University, Indonesia

andriedj78@gmail.com

Abstract. In carrying out its supply chain activities, a project often experiences problems starting from the procurement, production, to delivery processes. Likewise, the Makassar New Port Breakwater development project must consider its risks in planning, implementation and operations. This study aims to determine the risks that could potentially occur in supply chain activities at the Breakwater Makassar New Port development project in Indonesia. Furthermore, it will provide information on which risks fall into the high and low categories. Moreover, the research will provide the results of the mitigation options for each risk. An essential contribution of this research is that the results will give a lesson learned and initial data to determine future risks in the breakwater project. Based on the research conducted, the identified risks in the supply chain of the Makassar New Port Breakwater development project are 38 risks where the dominant risk factor in the material supply chain that influences the implementation of the Makassar New Port Breakwater development project is variable C6, namely the amount of production that does not match the target. In the moderate risk category. From the results of data analysis, it is known that the contractors and suppliers make efforts to reduce the dominant risk either preventively or curatively so that these risks do not interfere and cause a significant impact on the project objectives.

1. Introduction

The industry that produces construction services is a sector that plays an essential role in determining the pace of activity for the economy and advancing the other sectors. However, according to Ganesh (2021), usually managing construction projects, many project constraints are found, such as quality that is not achieved and the time to complete the project being late. In addition, the construction services industry is one of the most dynamic industries compared to other sectors due to the ever-changing market conditions, the relatively short construction period, and fluctuations in material prices which are very difficult to predict, as well as intense competition between service providers, subcontractors, suppliers and other parties [9].

The challenges in construction projects are increasingly complex, so increasing production results in terms of quality and completion time is very important for companies to continue to be competitive and survive. According to Briscoe and Dainty (2005), O'Brian et al. (2009) and Ahmed (2017) explained that supply chain management is essential to improve the performance of construction projects where in the construction supply chain management cycle changes are inevitable and may have an impact towards the project. Changes will continue to occur and may impact project timelines and cost overruns. According to Dainty (2007), supply chain management has proven strategic in managing projects involving many stakeholders, suppliers and materials.

In performing its supply chain activities, a project often encounters problems from the procurement, production and delivery processes. Similarly, the Makassar New Port Breakwater Development Project needs to consider the risks that will arise in its planning, implementation and operations. This development project has many risks because it uses heavy equipment in the supply of materials and involves a lot of human resources that require attention, especially to avoid troubles.

This study aims to determine what risks will be identified in the supply chain of the Makassar New Port Breakwater Development project. From this analysis, it is also possible to predict the dominant risks that will occur in the future and provide risk responses to reduce or anticipate these dominant risks.

2. Methodology

2.1 The project overview and Location

This research was conducted on the Makassar New Port Breakwater construction project, located on New Makassar Port Street, Makassar, South Sulawesi.

Figure 1. Research Site

2.2 Research Design

The research was conducted on the Makassar New Port Breakwater Development project with a qualitative descriptive research method. The qualitative descriptive method aims to make a systematic, factual and accurate description or description of a phenomenon or the relationship between the phenomena being investigated. The qualitative descriptive method used is the interview and survey method which aims to obtain opinions from experts and respondents regarding the risks that may occur in the supply chain activities of the Makassar New Port Breakwater Development project.

2.3 Conceptual Framework

The research was carried out in several stages, as shown in Figure 2.

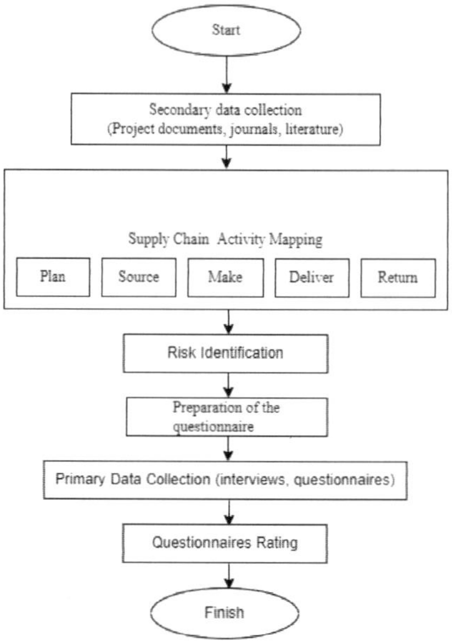

Figure 2. Conceptual Framework

The conceptual framework starts by collecting secondary data and mapping supply chain activities using the Supply Chain Operation Reference model (source, make, plan, deliver and return). Furthermore, identifying risks that may occur and potentially occur in supply chain activities, namely by collecting primary data through registering (listing) risks that may arise as much as possible through field surveys, interviews and questionnaires. The final stage is analysis and assessment to predict the dominant risks that will occur in the future and can provide risk responses to reduce or anticipate these dominant risks. Risk analysis is carried out using the Severity Index (SI) method. The goal is to get the results of a combination of probability assessment and risk impact on time and cost aspects. The formula calculates the Severity Index (SI):

$$SI = \frac{\sum_{i=}^{4} aixi}{a\sum_{i=0}^{4} xi} x100\% \quad (1)$$

2.4 Data Analysis

There are two types of data used in this study, namely:

1. Primary data is data obtained from distributing questionnaires and interviews to respondents who are involved in the Makassar New Port breakwater construction project.
2. Secondary data is general data on project implementation sourced from documents, journals, literature and previous related studies.

2.5 Respondents of the Research

Respondents in this study had a detailed and thorough understanding of the actual conditions of the project and had been directly involved in the Makassar New Port Breakwater Development Project (Package C). Respondents who were designated as samples in this study amounted to 20 people, where the respondents consisted of 3 service users (PT. Pelabuhan Indonesia IV (Persero), 12 service providers (implementing contractors), and 5 consultant work supervisors.

2.6 Instrument of the Research

A research instrument is a tool used to collect data and information from the questionnaire. The research instrument used was a questionnaire with a qualitative perception scale in the form of a scale ranging from 1 to 5. The scale used in the preparation of the questionnaire is the ordinal for the probability of a risk and the impact that will be caused. For example, supply chain activities could use the following rating scale:

- Probability/risk probability assessment scale (Duffel, 2003):
  Very Low = Rarely occurs, only under certain conditions 20%
  Low = Sometimes occurs in certain conditions >20–40%
  Moderate = Occurs under certain conditions >40-60%
  High = Often occurs in certain conditions >60-80%
  Very High = Always occurs in conditions >80-100%
- The risk impact assessment scale on implementation costs (Knight & Fayek, 2002):
  Very Low = <10 million rupiah
  Low = >10-25 million rupiah
  Medium = >25-100 million rupiah
  Height = >100 million – 1 billion rupiah
  Very High = >1 billion rupiah
- Risk impact assessment scale over time (Kerzner, 2006):
  Very Low = No impact on project schedule, 1 day project duration
  Low = There is a delay in the project schedule, >1-3 days project duration
  Medium = There is a delay in the project schedule, > 3-7 days project duration
  High = There is a delay in the project schedule, > 7-30 days project duration
  Very High = There is a delay in the project schedule > 30 days project duration

3. Result and Discussion

The results of the research describe that there will be risks that have the potential to occur in supply chain activities at the Makassar New Port Breakwater development project in Indonesia. Furthermore, it will provide information on which risks fall into the high and low categories. Moreover, the research will provide the results of the mitigation options for each risk. Finally, an essential contribution of this research is that the results of this study will provide a lesson learned and initial data to determine the risks that occur in the breakwater project in the future.

3.1 Risk Identification

Risk identification is determining what events affect the agency and documenting the risk's character. The risk identification results will be used for qualitative analysis and risk response processes. It could be identified based on direct observations in the field and brainstorming with related parties competent in providing input.

In this preliminary survey, there were 38 identified risks consisting of 8 risk variables in the Source activity, seven risk variables in the Plan activity, nine risk variables in the Make activity, eight risk variables in the Deliver activity and six risk variables in the Return activity. Identification of risks that occur in the Makassar New Port Breakwater Development Project. The identified risks can be described in table 1 as follows:

Tabel 1 Risk Identification

Activity	Code	Risk Identification
Source	A1	Error choosing supplier
	A2	Not evaluating supplier performance
	A3	Delay in picking up raw materials
	A4	Raw materials received do not match the quantity requested
	A5	The type/item of the raw material received does not match the request
	A6	There are raw materials that do not match the quality
	A7	There is no evaluation of raw materials from the company
	A8	Raw material pick-up schedule error
Plan	B1	Error planning quantity availability for raw materials
	B2	Incompatibility of raw material planning with financial planning
	B3	Error planning scheduling of taking raw materials
	B4	Production scheduling plan error
	B5	There is a change in the plan in the production system
	B6	Error in estimating the number of material requests
	B7	Incompatibility of distribution planning with production planning
Make	C1	Error in production scheduling
	C2	There was a delay in the production process
	C3	Insufficient raw materials for production
	C4	The production process is not according to the SOP
	C5	Error in product storage
	C6	The amount of production does not meet the target
	C7	Non-conformance of materials produced with customer orders
	C8	The production machine is broken
	C9	Not checking the quality of the products
Deliver	D1	No material quality check (before delivery)
	D2	Lack of material delivery capacity
	D3	Lack of means of transportation
	D4	Lack of material at the supplier
	D5	There was an error in the shipping process that damaged the material
	D6	The material error sent to the customer
	D7	Material delivery schedule error to customer
	D8	Delay in delivery of materials to customers
Return	E1	Material is returned to the supplier for some reason
	E2	Complaints from customers
	E3	Delay in the return process from supplier to customer
	E4	No returns for some reason
	E5	No management of inappropriate material results (return)
	E6	No handling is carried out for processes or material results that are not in accordance with (return)

3.2 Variable Risk Analysis

After conducting a preliminary survey and obtaining relevant risk variables related to the supply chain in the Makassar New Port Breakwater development project, this primary survey aims to determine the probability of risk that can occur and the impact of the risk that is caused when the risk occurs. This primary survey contains perceptions or points of view on each risk variable by determining the probability and impact value/score. Risk variable analyses were carried out to analyze the primary survey. The analysis is carried out on the risk probability and risk impact assessment. This analysis is calculated by the Severity Index (SI) method. The results of the probability assessment of supply chain activities for the Makassar New Port Breakwater development project are shown in Table 2.

Tabel 2 Probability Assessment

Code	1 VL	2 L	3 M	4 H	5 VH	Total	SI (%)	Category	Grade
A1	8	2	4	2	1	17	29,41%	Low	2
A2	6	6	2	3	0	17	27,94%	Low	2
A3	5	4	6	1	1	17	33,82%	Low	2
A4	4	4	4	5	0	17	39,71%	Low	2
A5	4	6	2	5	0	17	36,76%	Low	2
A6	5	4	2	5	1	17	39,71%	Low	2
A7	8	3	3	2	1	17	27,94%	Low	2
A8	6	5	4	1	1	17	29,41%	Low	2
B1	6	5	2	4	0	17	30,88%	Low	2
B2	7	5	1	4	0	17	27,94%	Low	2
B3	6	5	3	3	0	17	29,41%	Low	2
B4	6	5	3	3	0	17	29,41%	Low	2
B5	6	1	9	1	0	17	32,35%	Low	2
B6	6	4	6	1	0	17	27,94%	Low	2
B7	7	3	3	4	0	17	30,88%	Low	2
C1	8	3	3	3	0	17	26,47%	Low	2
C2	6	4	2	4	1	17	35,29%	Low	2
C3	7	4	3	2	1	17	29,41%	Low	2
C4	8	2	5	1	1	17	27,94%	Low	2
C5	8	2	4	2	1	17	29,41%	Low	2
C6	5	3	3	5	1	17	41,18%	Moderate	3
C7	5	6	4	1	1	17	30,88%	Low	2
C8	5	1	8	2	1	17	39,71%	Low	2
C9	9	2	3	2	1	17	26,47%	Low	2
D1	9	3	2	2	1	17	25,00%	Low	2
D2	6	3	7	1	0	17	29,41%	Low	2
D3	6	4	5	2	0	17	29,41%	Low	2
D4	6	4	4	3	0	17	30,88%	Low	2
D5	7	5	3	1	1	17	26,47%	Low	2
D6	5	7	4	0	1	17	27,94%	Low	2
D7	6	6	4	1	0	17	25,00%	Low	2
D8	5	6	2	4	0	17	32,35%	Low	2
E1	6	4	6	1	0	17	27,94%	Low	2
E2	5	4	6	1	1	17	33,82%	Low	2
E3	6	6	4	1	0	17	25,00%	Low	2
E4	6	3	8	0	0	17	27,94%	Low	2
E5	8	1	5	1	2	17	32,35%	Low	2
E6	9	1	4	0	3	17	30,88%	Low	2

Table 2 for probability assessment shows that 1 (one) Middle category contained in code C6, which is a Make activity with risk identification. The amount of production does not meet the target with an SI value of 41.18%. As for the identification of other risks in the Low category.

The following are the results of the risk impact analysis based on the cost aspect from the contractor's perspective on supply chain activities for the Makassar New Port Breakwater development project, which can be seen in Table 3.

Tabel 3 Assessment of Risk Impact on Cost Aspect

Code	1	2	3	4	5	Total	SI (%)	Category	Grade
	VL	L	M	H	VH				
A1	7	1	4	4	1	17	37%	**Low**	**2**
A2	7	2	3	4	1	17	35%	**Low**	**2**
A3	5	3	6	2	1	17	37%	**Low**	**2**
A4	5	3	5	3	1	17	38%	**Low**	**2**
A5	5	3	4	3	2	17	41%	**Medium**	**3**
A6	5	3	3	5	1	17	41%	**Medium**	**3**
A7	5	3	4	4	1	17	40%	**Low**	**2**
A8	4	6	4	2	1	17	35%	**Low**	**2**
B1	7	2	4	4	0	17	32%	**Low**	**2**
B2	6	3	3	2	3	17	40%	**Low**	**2**
B3	7	3	4	3	0	17	29%	**Low**	**2**
B4	5	5	2	5	0	17	35%	**Low**	**2**
B5	5	4	5	1	2	17	37%	**Low**	**2**
B6	6	4	1	2	4	17	41%	**Medium**	**3**
B7	6	2	4	5	0	17	37%	**Low**	**2**
C1	6	4	5	2	0	17	29%	**Low**	**2**
C2	5	4	6	1	1	17	34%	**Low**	**2**
C3	5	4	4	2	2	17	38%	**Low**	**2**
C4	5	4	2	4	2	17	41%	**Medium**	**3**
C5	6	4	2	3	2	17	37%	**Low**	**2**
C6	5	2	4	4	2	17	44%	**Medium**	**3**
C7	6	1	3	4	3	17	46%	**Medium**	**3**
C8	5	0	5	5	2	17	49%	**Medium**	**3**
C9	6	3	4	3	1	17	35%	**Low**	**2**
D1	8	1	5	2	1	17	31%	**Low**	**2**
D2	2	5	5	3	2	17	47%	**Medium**	**3**
D3	5	3	6	3	0	17	35%	**Low**	**2**
D4	6	4	4	2	1	17	32%	**Low**	**2**
D5	6	4	4	2	1	17	32%	**Low**	**2**
D6	6	1	6	0	4	17	43%	**Medium**	**3**
D7	5	4	4	1	3	17	40%	**Low**	**2**
D8	5	4	4	2	2	17	38%	**Low**	**2**
E1	5	3	4	4	1	17	40%	**Low**	**2**
E2	5	3	6	1	2	17	38%	**Low**	**2**
E3	6	4	6	1	0	17	28%	**Low**	**2**
E4	6	4	7	0	0	17	26%	**Low**	**2**
E5	5	3	6	3	0	17	35%	**Low**	**2**
E6	4	3	7	1	2	17	41%	**Medium**	**3**

Table 3 for assessing the risk impact on costs shows that for the Middle category with a value of 3, there are 10 (ten) codes A5, A6, B6, C4, C6, C7, C8, D2, D6, and E6. As for the Low category, with a value of 2, as many as 28 (twenty-eight) risk identification codes.

The following are the results of the risk impact analysis based on the time aspect from the contractor's perspective on supply chain activities for the Makassar New Port Breakwater development project, which can be seen in Table 4.

Tabel 4 Assessment of Risk Impact on Time Aspect

Code	1	2	3	4	5	Total	SI (%)	Kategori	Nilai
	VL	L	M	H	VH				
A1	6	3	5	0	3	17	37%	Low	2
A2	6	4	3	2	2	17	35%	Low	2
A3	5	2	3	4	1	15	40%	Low	2
A4	5	3	2	7	0	17	41%	Middle	3
A5	6	2	3	5	1	17	40%	Low	2
A6	5	3	4	3	2	17	41%	Middle	3
A7	6	4	2	4	1	17	35%	Low	2
A8	5	3	4	3	2	17	41%	Middle	3
B1	5	2	4	4	2	17	44%	Middle	3
B2	4	3	4	5	1	17	44%	Middle	3
B3	5	5	3	2	2	17	37%	Low	2
B4	6	2	5	2	2	17	38%	Low	2
B5	6	1	6	1	3	17	41%	Middle	3
B6	5	4	1	6	1	17	41%	Middle	3
B7	6	1	4	3	3	17	44%	Middle	3
C1	6	1	5	3	2	17	41%	Middle	3
C2	5	2	5	3	2	17	43%	Middle	3
C3	5	2	3	4	3	17	47%	Middle	3
C4	7	0	5	3	2	17	40%	Low	2
C5	6	5	2	2	2	17	34%	Low	2
C6	5	2	3	5	1	16	42%	Middle	3
C7	6	2	4	3	2	17	40%	Low	2
C8	4	2	2	6	3	17	53%	Middle	3
C9	5	4	3	4	1	17	38%	Low	2
D1	6	4	3	3	1	17	34%	Low	2
D2	4	3	6	4	0	17	40%	Low	2
D3	4	4	5	4	0	17	38%	Low	2
D4	4	4	4	4	1	17	41%	Middle	3
D5	4	5	4	3	1	17	38%	Low	2
D6	4	4	6	1	2	17	40%	Low	2
D7	5	4	3	4	1	17	38%	Low	2
D8	4	4	4	4	1	17	41%	Middle	3
E1	4	4	8	1	0	17	34%	Low	2
E2	4	3	6	3	1	17	41%	Middle	3
E3	6	4	4	3	0	17	31%	Low	2
E4	5	6	5	1	0	17	28%	Low	2
E5	3	5	5	4	0	17	40%	Low	2
E6	4	5	6	2	0	17	34%	Low	2

Based on table 4 for the assessment of the impact of risk on the time aspect shows that there are 22 (twenty-two) risk identifications in the Low category and 16 (sixteen) risk identifications in the Middle category.

IOP Conf. Series: Earth and Environmental Science 1117 (2022) 012015 doi:10.1088/1755-1315/1117/1/012015

3.3 Risk Assesment

The risk value is obtained by plotting the value on the probability and impact matrix. The categories of probability and impact consist of three categories: high, medium, and low. Then, risk analysis on cost and time is carried out by multiplying the results of the probability assessment (P) with the impact assessment (I) on the cost and time of each risk variable. The results of the calculation of the risk analysis on the cost and time aspects can be seen in table 5.

Table 5 Assessment of Risk Categories Against Cost and Time Aspects

Code	Significant risk assesment of cost				Significant risk assesment of time			
	Probability (P)	**Cost Impact (I)**	**P x I**	**Risk**	**Probability (P)**	**Time Impact (I)**	**P x I**	**Risk**
A1	2	2	4	Low	2	2	4	Low
A2	2	2	4	Low	2	2	4	Low
A3	2	2	4	Low	2	2	4	Low
A4	2	2	4	Low	2	3	6	Low
A5	2	3	6	Low	2	2	4	Low
A6	2	3	6	Low	2	3	6	Low
A7	2	2	4	Low	2	2	4	Low
A8	2	2	4	Low	2	3	6	Low
B1	2	2	4	Low	2	3	6	Low
B2	2	2	4	Low	2	3	6	Low
B3	2	2	4	Low	2	2	4	Low
B4	2	2	4	Low	2	2	4	Low
B5	2	2	4	Low	2	3	6	Low
B6	2	3	6	Low	2	3	6	Low
B7	2	2	4	Low	2	3	6	Low
C1	2	2	4	Low	2	3	6	Low
C2	2	2	4	Low	2	3	6	Low
C3	2	2	4	Low	2	3	6	Low
C4	2	3	6	Low	2	2	4	Low
C5	2	2	4	Low	2	2	4	Low
C6	3	3	9	Middle	3	3	9	Middle
C7	2	3	6	Low	2	2	4	Low
C8	2	3	6	Low	2	3	6	Low
C9	2	2	4	Low	2	2	4	Low
D1	2	2	4	Low	2	2	4	Low
D2	2	3	6	Low	2	2	4	Low
D3	2	2	4	Low	2	2	4	Low
D4	2	2	4	Low	2	3	6	Low
D5	2	2	4	Low	2	2	4	Low
D6	2	3	6	Low	2	2	4	Low
D7	2	2	4	Low	2	2	4	Low
D8	2	2	4	Low	2	3	6	Low
E1	2	2	4	Low	2	2	4	Low
E2	2	2	4	Low	2	3	6	Low
E3	2	2	4	Low	2	2	4	Low
E4	2	2	4	Low	2	2	4	Low
E5	2	2	4	Low	2	2	4	Low
E6	2	3	6	Low	2	2	4	Low

Based on table 5, the risk variables are categorized as medium and low. Therefore, one medium risk variable and 37 standard category risk variables were obtained in the analysis.

3.4 Risk Management

It is done by plotting the value on the probability and impact matrix to get the risk value and the categories of probability and impact consisting of three categories: high, medium, and low. Then, risk analysis on cost and time is carried out by multiplying the results of the probability assessment (P) with the impact assessment (I) on the cost and time of each risk variable. The results of the calculation of the risk analysis on the cost and time aspects can be seen in table 5.

From the analysis results from table 5, it is obtained that the risk variables have the dominant risk, namely the risk with a medium category that impacts cost and time. These risks are likely to have a significant impact on project implementation. The risk variable category analysis is carried out using the PMBOK risk matrix, where as shown in Figure 3, there is one medium category risk variable, code C6, with the type of risk, namely the number of producing results that are not on target. Table 6 shows some mitigation that can be done to overcome the moderate risk obtained.

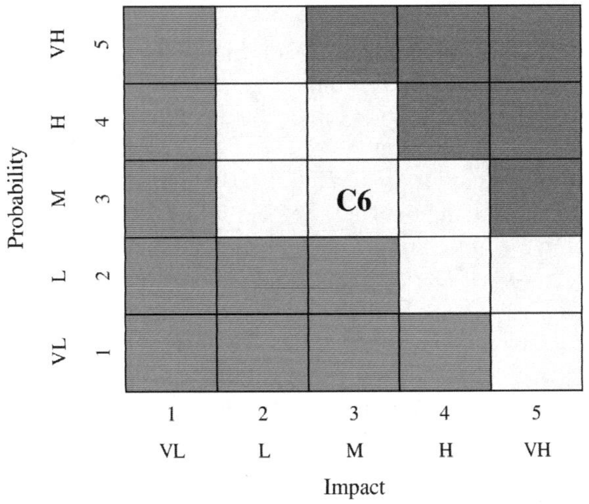

Figure 3 Risk Matrix Assessment (PMBOK)

Table 6 Management of the dominant risk of cost and time

Code	Risk Type	Risk Management
C6	The amount of production does not meet the target	Using previous historical data analysis to estimate production results to achieve the target
		Quality Control from both parties must also always be careful in checking the materials to be sent and those that have been sent
		Always communicate with the supplier to ensure the material to be produced is according to specifications
		Prepare a list of backup suppliers to anticipate suppliers who are lacking in sending materials

4. Conclusion

Based on the research and the results of the analysis carried out, it can be concluded:

- The identified risks in the supply chain of the Makassar New Port Breakwater development project are 38 risks consisting of 8 risk variables in the Source activity, seven risk variables in the Plan activity, nine risk variables in the Make activity, eight risk variables in the Deliver activity and six risk variables on the Return activity.
- The dominant risk factor in the material supply chain that affects the implementation of the Makassar New Port Breakwater development project is variable C6, namely the amount of production that is not on target with a moderate risk category.
- The risk response to the dominant risk is in the form of corrective actions, namely: using previous historical data analysis to estimate production results so that they can reach the target. Quality control from both parties must also always be careful in checking the materials to be sent and those that have been delivered. So sent, always communicate with suppliers to ensure the material to be produced is according to specifications, and prepare a list of backup suppliers to anticipate suppliers who are lacking in sending materials.

References

[1] Arifin, Wahyu Tenri, dkk. 2020. Analisa Rantai Pasok Material Usaha Jasa Konstruksi Kabupaten Tasikmalaya. *Jurnal Tekno Insentif*. Bandung.

[2] Tampubolon, Amos. 2018. Analisa Faktor-faktor Risiko Dominan Pada Pelaksanaan Proyek Konstruksi Dengan Metode Severity Index. *Tugas Akhir*. Universitas Sumatera Utara

[3] Cahyono Bintang Nurcahyo, I Putu Artama Wiguna. 2016. Analisis Risiko Rantai Pasok Beton Ready Mix pada Proyek Pembangunan Apartemen di Surabaya. *Jurnal Aplikasi Teknik Sipil* ISSN.1907-753X Volume 14, Nomor 2.

[4] Maditsaraga, Ganesh Adimas & Darmawan Pontan. 2021. Evaluasi Faktor Dominan Yang Mempengaruhi Rantai Pasok Pada Proyek Konstruksi Gedung Politeknik Negeri Bengkalis. *Prosiding Seminar Intelektual Muda* #5, Inovasi Keberlanjutan Binaan Melalui Riset dan Karya Desain.

[5] Rachmasari, Herlinda. 2019. Analisa Risiko Rantai Pasok Konstruksi Dengan Menggunakan Model Simulasi. *Tesis ITS*. Surabaya.

[6] Refdizalis, dkk. 2020. Studi Perilaku Rantai Pasok Konstruksi Pada Proyek Pembangunan Gedung. *Jurnal Talenta Sipil* pp.58-66.

[7] Salim, Indra Putra, dkk. 2021. Kajian Awal Manajemen Risiko Proyek Kereta Cepat Jakarta – Bandung. *Prosiding CEEDRiMS 2021*.

[8] Sandangan, Anggun, dkk. 2022. Analisis Risiko Dalam Sistem Rantai Pasok Pada Proyek Upgrade Trans Studio Mall Makassar. *Paulus Civil Engineering Journal (PCEJ) e-ISSN xJurnal Teknik Sipil UKIPaulus-Makassar* Volume 4 Issue 2.

[9] Sutoni, Akhmad., Dhendi Rukma Kurniadi. 2019. Analisis Risiko dalam Construction Supply Chain: Studi Kasus pada Proyek Renovasi Gedung Kantor VEDCA. Jurnal Media Teknik & Sistem Industri.

IOP Conf. Series: Earth and Environmental Science 1117 (2022) 012016 doi:10.1088/1755-1315/1117/1/012016

Compressive strength and microstructure of Self Compacting Concrete with nylon fiber substitution

MN Nurazuwa[1], **Munadrah**[2], **R Irmawaty**[3], **and A B Muhiddin**[3]

[1]Jamilus Research Centre, FKAAB, Universiti Tun Hussein Onn, Malaysia
[2]Virama Consulting, Makassar, Indonesia
[3]Civil Engineering Department, Universitas Hasanuddin, Indonesia

nurazuwa@uthm.edu.my

Abstract. Concrete is the most widely used construction material as building material because it has advantages such as easy workability, high compressive strength, and economical in terms of manufacturing and maintenance. However, there are some disadvantages obtained from the use of concrete, such as low tensile strength, easy cracking, and brittleness. Therefore, it is necessary to add fiber to the concrete to increase the ductility and use of SCC concrete to maintain workability after fiber substitution. The variation in the addition of nylon fiber is 0.5% and 1% to the weight of cement with a diameter of 0.35 mm and 0.65 mm, a length of 15 mm, and 20 mm. With the addition of 1% of 0.65 mm diameter and 15 mm length, the compressive strength of SCC concrete increased by 126%. The nylon fiber was also found to be evenly distributed in the concrete mixture and did not experience clumping and the volume weight of the concrete was reduced by 5-7% of the normal concrete. The SEM results show a clear interface between concrete and nylon fiber, and the slippery morphology of nylon fiber causes the bonding with concrete to become weak.

1. Introduction

At present, the development of science and technology is very rapid along with the adjustment of human needs that continue to increase. Development in the field of construction in the modern era shows significant developments, including in the construction of bridges, tunnels, housing, office buildings, hospitals, shopping centers, and so on. The most important part is the quality of concrete. Construction work for conventional concrete which has a large specific gravity and solidification or vibration of concrete is absolute to do. The goal of compaction is to reduce the amount of air trapped in fresh concrete, resulting in homogenous concrete with no voids. However, employing standard methods to obtain high-quality concrete typically entails more effort and cost.

Nylon fiber is a material with polymer fibers that have the properties of fibers, films, and plastics. Nylon is formed by a group associated with a repeat hydrocarbon unit of different lengths in a polymer having a specific gravity of 1.15. The main reason for adding fiber to concrete is to increase the energy absorption capacity of the mixed matrix, which means increasing concrete ductility. The addition of ductility also means the addition of concrete behavior to fatigue and impact. Nylon fibers are fibers produced with the fibers forming elements that are a long chain of synthetic polyamides, where less than 85% of amide bonds bind directly (- CO-NH -) two aliphatic groups. Polymers, namely linear polyamides, are referred to as nylon. There are two primary methods to produce nylon fo fiber applications. In the first method, molecules with an acid group (COOH) react with molecules containing amine groups (NH2), creates the material known as nylon, which is named based on the number of carbon atoms between two acid groups and two amine groups. A compound containing an amine on one side and an acid on the other side polymerized to form a chain with a repeating unit NH-[CH2] n-CO-) x. Since the early 1980s, synthetic micro or macro nylon fibres have been utilised to strengthen concrete shrinkage. In the case of micro, monofilament, and fibrillation forms, monofilament was used in lengths ranging from 13 to 38 mm, with a size of 20 mm being the most widely utilised. Because this fiber is so thin, the amount per weight (the amount of fiber) is in the range of millions per 1 kg of concrete. Concrete added with fiber in a certain volume percentage has more ductile properties compared to non-fiber concrete.

In the research of Suhana and Sugriana [1], the fiber used in the concrete mix was nylon. In general, monofilament is better known as nylon string. Monofilament nylon is a single-component product formed from a plastic that is melted into a braid through an iron mould. The strength of nylon monofilament varies depending on the size of the fiber diameter. In Gonzalo's study [2] in general, the increase in concrete strength caused by radiation fibers is 5, 10, 50, and 10 kGy depending on the percentage applied to the fiber and fiber content in the concrete. The fibers are combined with Portland cement, gravel, sand, and water in volumes of 1.5 percent, 2 percent, and 2.5 percent after their tensile strength is determined. The compressive strength of concrete is measured and compared to the findings of similar materials. The maximum compressive strength value is observed for 50 kGy fiber and the fiber volume of 2.0 % strength is 122.2 MPa, compared to 35 MPa for concrete without fiber. In the study of Aries Munandar [3], based on the test results obtained the average value of compressive strength of lightweight concrete gas with nylon fiber addition of 0%, 0.25%, 0.5%, 0.75%, 1% increase of 7.62%, 15.48%, 31.54%, 50.37% respectively, compared to lightweight concrete without fiber. With a fiber content of 1%, the splitting tensile strength raised to 46.04 %, and the modulus of elasticity rose to 63.18 %. SCC (self-compacting concrete) is defined by Tjaronge et al.[4] as concrete that, while still in the form of fresh concrete, can flow through the reinforcement criteria (passability) and fill the entire space in the mould in a solid manner without requiring manual compaction or mechanical vibration (criteria fillability).SCC is a plastic fresh concrete that easily flows because the weight itself fills the entire moulded concrete has properties for self-solidification, without the aid of a vibrator for compaction. A good SCC must remain homogeneous and cohesive, not in harmony, there is no blocking, and no bleeding. According to Okamura and Ouchi [5], one solution to obtain a durable concrete structure that is bound to the ability of construction work is the ability of the concrete itself to solidify, which is able to flow into the mould's nooks and crannies, due to its weight and no need any vibrators, with a composition of a mixture consisting of 50% coarse aggregate and 40% fine aggregate of the concrete volume and the water-cement ratio between 0.25 -0.40. Tjaronge, MW et al. [6] SCC is concrete which has high fluidity properties so that it can flow and fill spaces in the mould without compaction or only requires very little vibration to compact it. This can reduce the compaction time. When casting multi-story structures, the SCC can be lifted and moved through a pump to a higher level when there is a high amount of liquidity. One of the chemicals that affect the ability of SCC to flow is the superplasticizer. SCC has the advantages of being very thin, having a high slump for a long time (slump keeping admixture), not requiring manual compaction, being more homogeneous and stable, the compressive strength of concrete can be made for high / very high quality, being more impervious, having smaller porosity and shrinkage, being more durable, having better and smoother concrete surfaces, having less noise pollution, and requiring less labor. In Erniati's research [7] the SCC concrete microstructure analysis was performed by Scanning Electron Microscope (SEM) which was used to analyze the morphology of the SCC concrete and PCC cement. Also, the effect of seawater uses on SCC is analyzed.

This paper aims to analyze the compressive strength of SCC with the addition of nylon fibers, fiber distribution in concrete, and the microstructure of SCC concrete related to the morphology and interface between nylon fibers and mortar.

2. Experimental Program

2.1. Material

The materials used are Portland Composite Cement, aggregate, nylon monofilament (Figure 1), and superplasticizer. The source of aggregate material in this study is from Bili-Bili with aggregate characteristics that have an SSD density of 2.58 and modulus of fineness of 2.61 for fine aggregates whereas coarse aggregates have SSD density of 2.62 and modulus of smoothness of 6.00. The aggregate testing method refers to the Indonesian National Standard (SNI) and the American Society of Testing Materials (ASTM) [8].

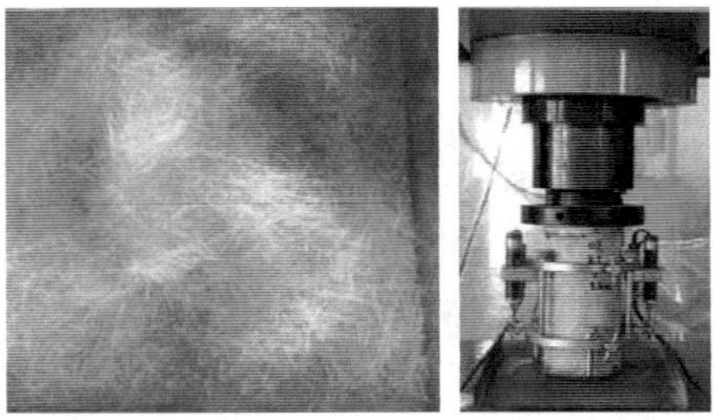

Figure 1. Nylon monofilament and Compression test

2.2. Specimen design

The cylindrical specimens used had a diameter of 100 mm and a height of 200 mm. Nylon fibers with diameters of 0.35 mm and 0.65 mm, lengths of 15 mm and 20 mm, and cement weights of 0.5 % and 1 % were added to SCC specimens to produce various SCC results. All materials (coarse aggregate, cement, fine aggregate and nylon) were mixed under dry conditions for 1 minute, then water containing SP was added and blended continuously for two minutes. This was followed by the manual mixing of the materials for one minute to scrap those trapped at the bottom and side part of the pan. The mixing was continued for two minutes until the concrete was homogeneous, before the fresh concrete was placed in the molds. Slump flow testing is performed using the Abrams cone, testing is carried out to determine the level of workability (ease in work) of the concrete mixture that has been made. The mixture design method used is *The European Federation of Specialist Construction Chemicals and Concrete System* (EFNARC) 2005 [9]. Then the treatment of test specimens is carried out aimed at keeping the surface of fresh concrete always moist until the concrete is considered to be hard enough. This humidity is maintained to ensure the hydration process of cement takes place perfectly. Test specimens are curing in water until they are tested.

2.3. Specimen testing

Testing for compressive strength is conducted at ages 7.14 and 28 days following SNI 1974: 2011 [10]. Then the concrete sample is split into two parts so that it looks inside the concrete. The distribution of the surface in the concrete was observed in the distribution and the amount of fiber in several parts which can be seen in Figure 2. Also tested the modulus of elasticity in concrete aged 28 days. Microstructure analysis was performed with the *Scanning Electron Microscope* (SEM) test, used to analyze the microstructure morphology of nylon fibers as well as the interface between concrete and fiber nylon.

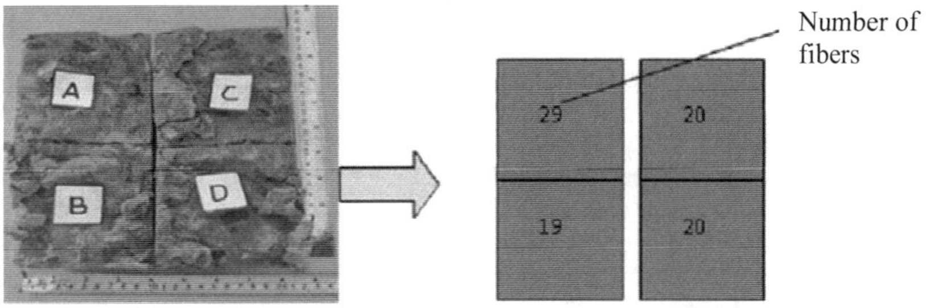

Figure 2. Sketch of the fiber mount calculation on a specimen

3. Result and Discussion

Concrete mixture planning with material weight ratio is carried out to determine the desired concrete strength, according to the methods of the European Federation of National Associations Representing Concrete (EFNARC) for normal concrete compositions, while for concrete with nylon fiber, the addition of nylon fiber is made with a variation of 0.5% and 1% of the weight of cement. The mixture composition shown in Table 1 is as follows.

Table 1. SCC mixture composition

No.	Materials	Normal Concrete	Nylon 0.5% D: 0.35 mm L : 15 mm	Nylon 0.5% D: 0.35 mm L : 20 mm	Nylon 0.5% D: 0.65 mm L : 15 mm	Nylon 0.5% D: 0.65 mm L : 20 mm	Nylon 1% D: 0.35 mm L : 15 mm	Nylon 1% D: 0.35 mm L : 20 mm	Nylon 1% D: 0.65 mm L : 15 mm	Nylon 1% D: 0.65 mm L : 20 mm
		kg / m³	kg / m³	kg / m³	kg / m³	kg / m³	kg / m³	kg / m³	kg / m³	kg / m³
1	Freshwater	175.00	175.00	175.00	175.00	175.00	175.00	175.00	175.00	175.00
2	Cement	500.00	500.00	500.00	500.00	500.00	500.00	500.00	500.00	500.00
3	Sand	723.08	720.57	720.57	720.57	720.57	717.48	717.48	717.48	717.48
4	Crushed stone 1-2	907.30	904.15	904.15	904.15	904.15	900.28	900.28	900.28	900.28
5	Nylon	-	2.50	2.50	2.50	2.50	5.00	5.00	5.00	5.00
6	Viscocreate 1.3%	6.50	6.50	6.50	6.50	6.50	6.50	6.50	6.50	6.50

3.1. Slump flow test

The slump flow test has the function to determine the SCC mixing ability, and the results of all variations of the SCC mix meet the slump flow requirements in a range of 550-650 mm.

3.2. Concrete volume weight

The SCC volume weight was carried out on cylindrical concrete aged 28 days. This test aims to determine the percentage change in weight of the SCC with nylon fiber to the normal weight of the SCC.

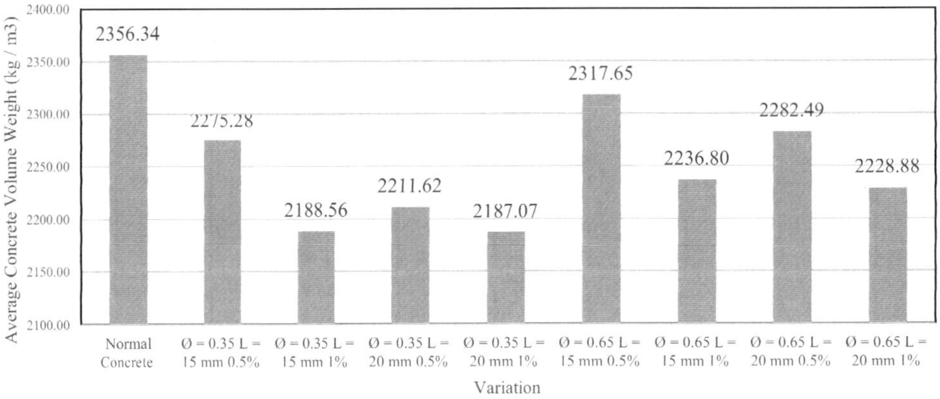

Figure 3. Average volume weight of SCC

Figure 3 shows the volume weight of SCC fiber concrete, where the greater the percentage of nylon fiber addition to SCC mixture, the volume weight of SCC fiber concrete will be smaller. Then SCC fiber concrete with 0.35 mm nylon fiber diameter has a smaller concrete weight than SCC fiber concrete with 0.65 mm nylon fiber diameter and normal concrete. This shows that the addition of nylon fiber to the SCC affects the value of the unit weight.

IOP Conf. Series: Earth and Environmental Science 1117 (2022) 012016 doi:10.1088/1755-1315/1117/1/012016

3.3. *Compressive strength*

At ages of 3 days, 7 days, and 28 days, compression test was conducted to evaluate the strength of normal SCC with SCC containing nylon fibre using a Universal Testing Machine (UTM) with a 1000 kN capacity. Figure 4 illustrates how the presence of nylon fiber affect the compressive strength result.

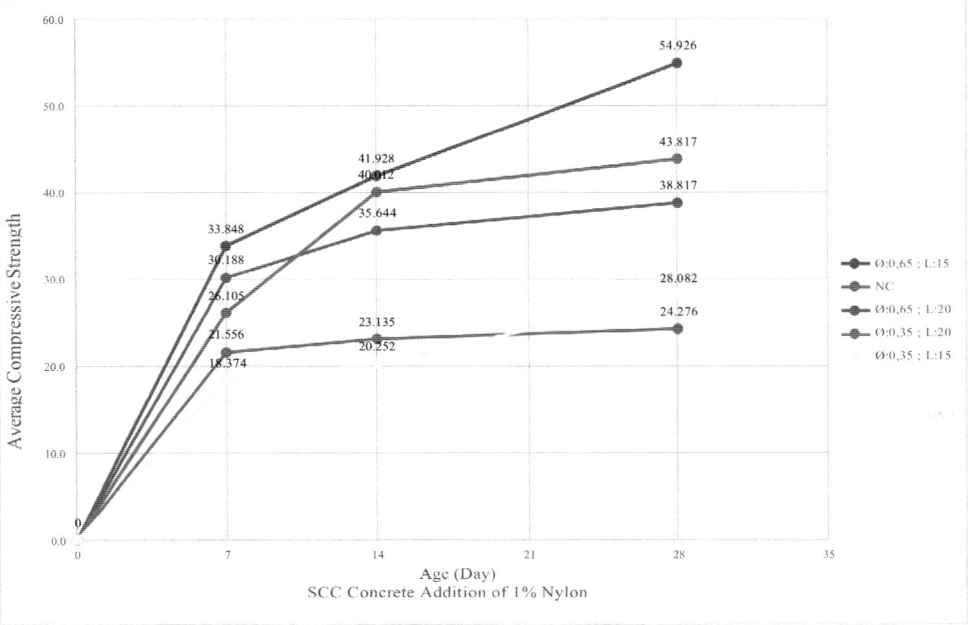

(a)

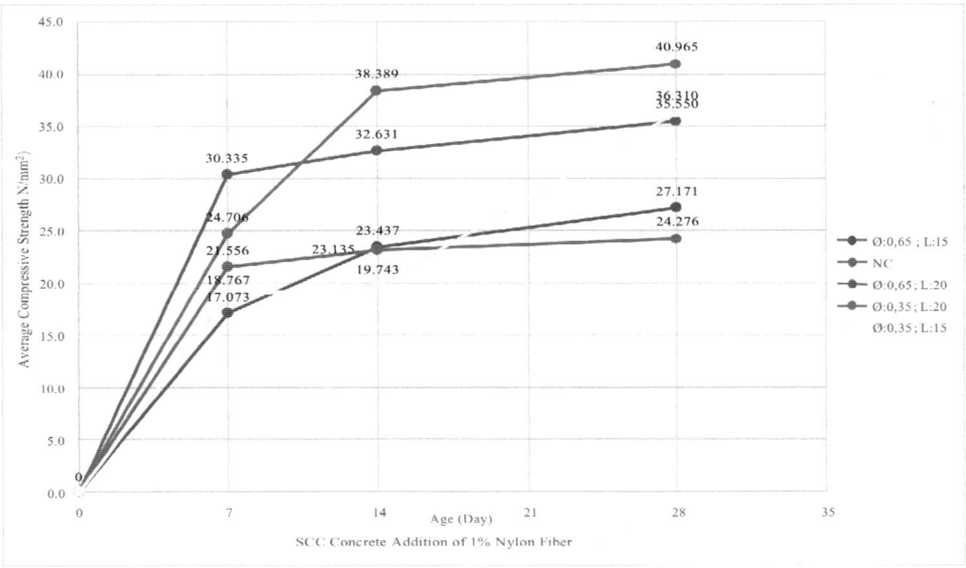

(b)

Figure 4. Average compressive strength of SCC concrete (a) 0.5% nylon fiber; (b) 1% nylon fiber

Figure 4 shows that SCC fiber concrete with the addition of 0.5% and 1% nylon fiber increased in compressive strength from the age of 7 days to 28 days. As the age of the concrete increases, the strength of the concrete also increases, as presented in figure 4.

Figure 5 depicts the highest compressive strength value of SCC fiber concrete with 0.5% nylon fiber addition achieved at 0.35 mm diameter variation with 20 mm fiber length with a compressive strength value of 40.97 MPa. Meanwhile, 0.65 mm diameter variation with 15 mm fiber length produces highest compressive strength value of 54.93 MPa for 1% nylon fiber. Compressive strength of SCC concrete increased by 68.75% with the addition of 0.5% nylon fiber (0.35 mm diameter and 20 mm length) and 126.26% with the addition of 1% nylon fiber (0.65 mm diameter and 15 mm length) at 28 days compared to normal SCC. Overall, SCC fiber concrete with 1% nylon fiber addition has a higher compressive strength than SCC fiber concrete with % nylon fiber addition and normal concrete.

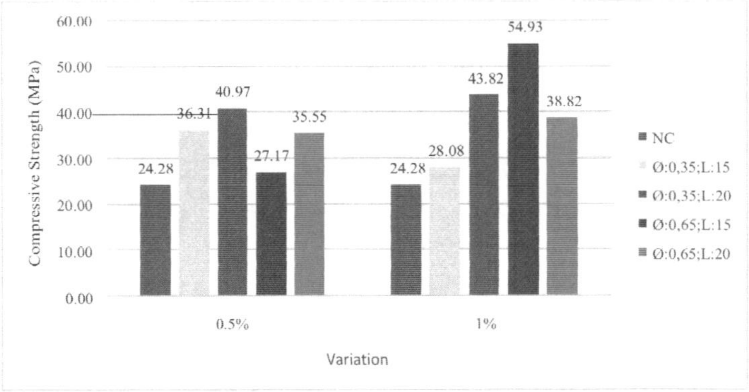

Figure 5. Average compressive strength of SCC at 28 days

3.4. Modulus of elasticity

The stress and strain relationship between the SCC concrete using nylon fiber with different variations are shown in Figures 6 and 7. The modulus of elasticity of SCC concrete is presented in Table 2. Form Table 2, it is clear that modulus of elasticity increases with increasing concrete compressive strength value.

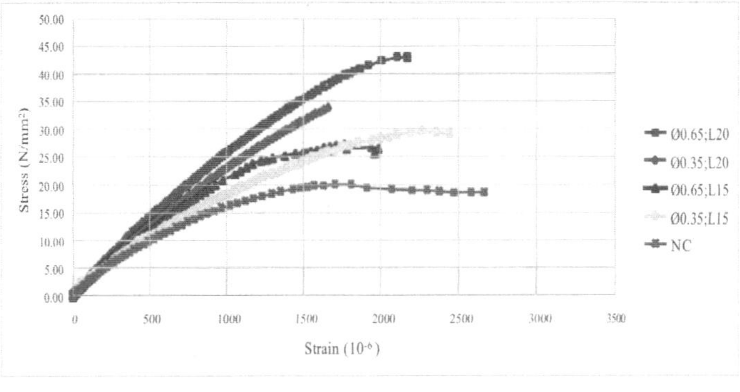

Figure 6. Stress-strain of 28-day-old SCC concrete at the addition of 0.5% nylon fiber

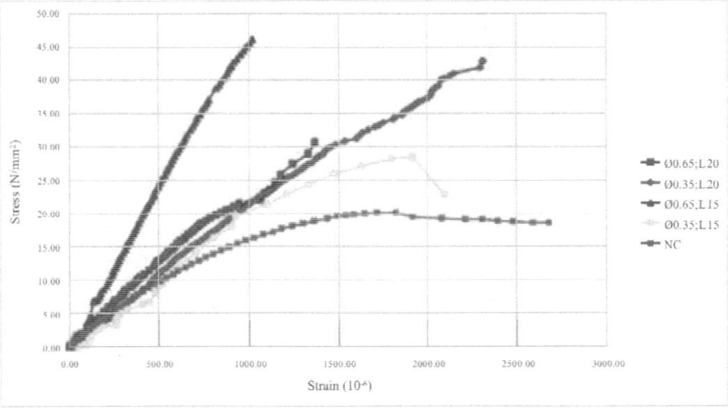

Figure 7. Strain-stress of 28-day-old SCC concrete at the addition of 1% nylon fiber.

Table 2. Modulus of elasticity of SCC

Variation	Elastic Modulus, Ec (MPa)	
	0.5% Nylon Fiber	1% Nylon Fiber
NC	19288.500	19288.500
Ø0.65; L15	23093.835	37665.11
Ø0.65; L20	27189.803	28431.315
Ø0.35; L20	22921.287	22968.397
Ø0.35; L15	21026.448	22229.046

3.5. Fiber distribution

Distribution test results on nylon fiber, show the effect of superplasticizers in improving concrete workability. In this test, each variation of content and type of nylon fiber is taken, each sample is divided into four segments, and the number of segmented nylon fibers is calculated. Observation of the distribution of nylon fibers is done by counting the amount of nylon fiber that is perpendicular to the concrete surface. The specimen is divided into two parts then marked and counted how many nylon fibers in the top half and the bottom half. The distribution of nylon fibers is presented in table 3 and Figure 8 show nylon distribution.

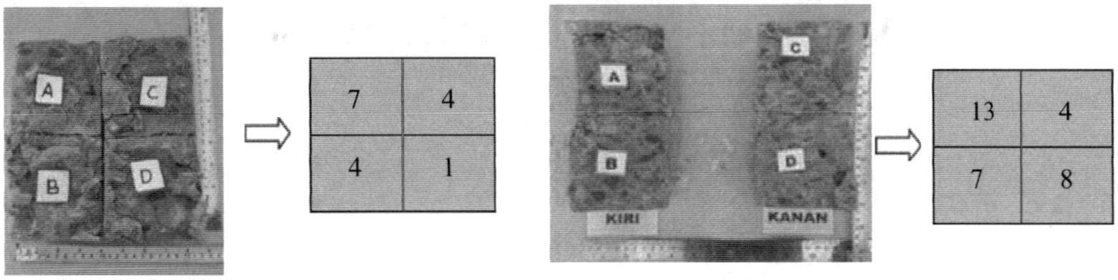

a. Ø=0.65 mm, L=15 mm, 0.5%

b. Ø=0.65 mm, L=15 mm, 1%

Figure 8. Nylon distribution testing

Table 3. Distribution of nylon fibers in SCC (%)

Variation	Fiber Count			
	A	B	C	D
Normal Concrete	0	0	0	0
Ø = 0.65 L = 15 mm 0.5 %	43.75	25.00	25.00	6.25
Ø = 0.65 L = 15 mm 1 %	40.63	21.88	12.50	25.00
Ø = 0.65 L = 20 mm 0.5 %	24.00	28.00	36.00	12.00
Ø = 0.65 L = 20 mm 1 %	26.00	34.00	18.00	22.00
Ø = 0.35 L = 20 mm 0.5 %	26.83	24.39	32.52	16.26
Ø = 0.35 L = 20 mm 1 %	22.87	23.94	30.32	22.87
Ø = 0.35 L = 15 mm 0.5 %	36.59	20.73	23.17	19.51
Ø = 0.35 L = 15 mm 1 %	28.07	21.64	24.56	25.73

With the addition of *superplasticizer*, it can improve workability on concrete helping to compact the concrete optimally and not hollow even though there is an addition of 0.5% nylon fiber and 1% of cement weight. This results in the nylon fibers in the SCC concrete being spread evenly, although some variations of the concrete on the top side of the concrete tend to be more fiber. This is caused by the effect of lighter nylon fiber density so that the fiber is lifted above the concrete section. The number of nylon fibers scattered in the SCC concrete increases with the percentage of fiber addition and length.

3.6. Microstructure of SCC with nylon fiber

From the results of the SEM test, the morphology of nylon fibers and the SCC concrete microstructure with the addition of nylon fibers can be determined. Figure 9 shows the morphology of nylon fibers where the nylon surface is slippery, which cause the bonding between nylon and mortar is weak. Even thought, cement paste can still adhere well.

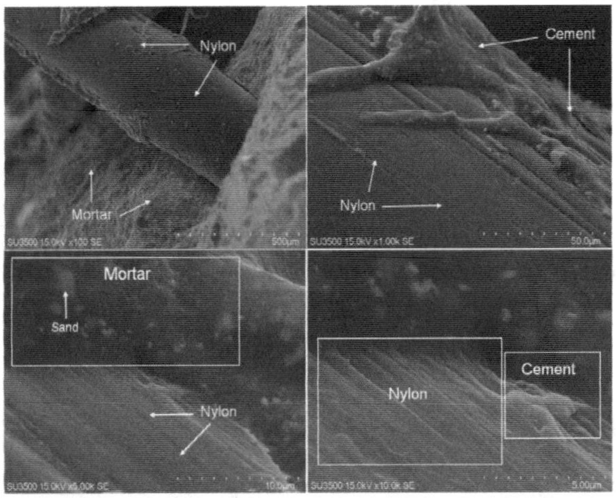

Figure 9. Morphology of nylon fibers and their bonding to cement paste

In figure 10, it is clear that there is a cavity between the nylon fiber and the mortar because the surface of the nylon fiber is slippery so that the mortar cannot blend with the nylon. Or in other words the bonding between nylon fibers and concrete is weak. SEM observations show good concrete density.

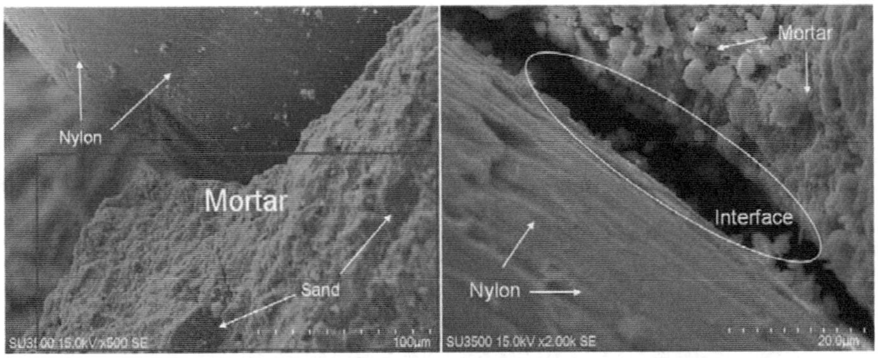

Figure 10. The interface between mortar and nylon fiber

4. Conclusion

In general, SCC concrete has an increase in compressive strength value along with an increase in age, percentage, diameter, and length of nylon fiber. The optimum compressive strength value is obtained at the addition of 1% fiber, 0.65mm diameter, and 15 mm fiber length with a compressive strength value of 54.93 MPa at 28 days. The distribution of nylon fibers in SCC is spread evenly without segregation, although in some variations there is a tendency for more fibers on the upper side. This is because of the specific gravity of the lightweight nylon fiber so that the fiber tends to the upper side of the concrete. The SEM test results clearly show that the surface of the nylon fiber is slippery, hence the bonding between the nylon fiber and concrete is weak, and the presence of interfaces in the two materials.

5. Acknowladgement

Thank you to the SCC research team for your cooperation, also the Nanoscience and Nanotechnology Research Center (PPNN) for your assistance in SEM testing.

Reference

[1] Suhana N et al 2016 Pengaruh Nilon Monofilament Pada Campuran Beton Terhadap Kuat Tekan dan Kuat Lentur. *Jurnal Rekayasa Infrastruktur* **3** 106-119.

[2] Martinez G B et al 2005 Concrete Reinforced with Irradiated Nylon Fibers *Cambridge Journals* **21** 484-490.

[3] Munandar A 2015 Pengaruh Penambahan Serat Nylon Pada Beton Ringan Dengan Teknologi Gas Terhadap Kuat Tekan, Kuat Tarik Belah, Dan Modulus Elastisitas *E-Jurnal Matriks Teknik Sipil Universitas Sebelas Maret Surakarta* **9** 750-758

[4] Tjaronge et al 2014 Compressive Strenght and Hydration Process of Self Compacting Concrete (SCC) Mixed with Sea Water, Marine Sand and Portland Composite Cement *Advanced Materials Research* **935** 242-246

[5] Okamura et al 2003 Self Compacting Concrete *Journal of Advanced Concrete Tecnology* **1** 5-15

[6] Tjaronge et al 2006 Slump Flow dan Kuat Lentur Self Compacting Concrete (SCC) dengan Kandungan Superpliticizer yang Bervariasi *Media Teknik Sipil* **6** 11-16.

[7] Erniati et al 2015 Compressive Strength and Slump Flow of Self Compactig Concrete Uses Fresh Water and Sea Water *ARPN Journal of Engineering and Applied Sciences* **10** 2373-2377

[8] ASTM C33-78 1978 Standard Spesification For Concrete Aggregates Philadelphia Pia

[9] EFNARC 2005 The European Guidelines for Self-Compacting Concrete, Specification, Production and Use UK

10] SNI 1974:2011 Cara Uji Kuat Tekan Beton dengan Benda Uji Silinder Jakarta

The Effect of Binder Types of Strengthening Panel for Shear Strengthening Reinforced Concrete Beams

R Irmawaty[1], Fakhruddin[1], and Muthmainnah[1]

1) Civil Engineering Department, Engineering Faculty, Universitas HasanuddinJl. Poros Malino KM. 06 Bontomarannu, Gowa, 92171, Indonesia

muthmainnah0105@gmail.com

Abstrac. This paper presents the shear behavior, crack patterns, and failure modes of reinforced concrete beams strengthened with geopolymer and Portland cement mortar panels. The specimens were prepared in the form of geopolymer and Portland cement mortar panels with a size of 300 mm x 700 mm, each consisting of 4 pieces. The reinforced concrete beams specimen were produced with dimension of 150 mm x 300 mm x 2300 mm and static monotonic loading test was carried out. To ensure the strength of the geopolymer and concrete used is in based on the target, a compressive strength test is carried out at the age of 28 days, with 3 samples each. According to the results of the loading test of reinforced concrete beams, the maximum load on the beam strengthed with geopolymer and Portland cement mortar panels increased by 34.98% and 29.54%, respectively, compared to the control beam. The addition of panels to reinforced concrete beams causes it's behavior to change, becoming more ductile. This also has an impact on the crack patterns and failure modes. CB beams experience shear failure, which is indicated by the appearance of diagonal cracks along the shear area to the point of loading, whereas mortar panel reinforced beams experience flexural failure.

Keywords: Strengthening, Mortar Panels, Geopolymer.

1. Introduction

In civil engineering, the term strengthening is known as a method of increasing the capacity of structural elements for loads. Strengthening of the structured elements are used in both new and old structures. Strengthening in new structures, was applied when a design failure occurs, or quality is not achieved during construction. While in the old structure, strengthening was applied to structural elements that were affected by environmental conditions, the age of the building, changes in building functions, and changes in standardization. Strengthening was needed if there is damage that causes degradation, resulting in non-fulfillment of technical requirements such as strength, stiffness, ductility, stability and resistance to certain performances [1].

There are several strengthening methods commonly used, such as Fiber Reinforced Polymer (FRP),Concrete Jacketing, and External Steel Plates. Each of these methods has its advantages and disadvantages [2]. Strengthening using mortar panels mounted on the sides of reinforced concrete beams using the Near Surface Mounted Technique, can cover the drawbacks of other reinforcement methods.

In general, cement is used as the main binder on mortar. However, the use of cement has a negativeimpact on the environment. Industrial cement can produce CO_2 gas which is known as greenhouse gas emissions [3]. The production of one ton of cement is about one ton of CO_2 into the atmosphere. In2016, world cement production produced around 2.2 billion tons of CO_2, or equal to the contributionof about 8% of the world's carbon dioxide (CO_2) gas emissions [4]. Therefore, another alternative is needed that can be used as a binder material.

Concrete without cement or geopolymer concrete is a geosynthetic concrete whose bonding reaction occurs through a polymerization reaction and not through a hydration reaction as in conventional concrete [5]. Geopolymer itself was first introduced by Davitdovits, and described as a non-cement material made of aluminosilicate as an alkali activator, including industrial waste (fly ash, blast furnace slag) or metakaolin at low calcination temperatures. One type of geopolymer comes from the main material of fly ash industrial waste which is rich in Alumina (Al) and Silica (Si) elements. Fly ash is one of the wastes generated from the coal combustion process at the Steam Power Plant (PLTU).

Studies on the use of geopolymers have been extended to various applications in structural elements, including reinforced concrete, columns, concrete-filled steel tube (CFT) columns, reinforcedconcrete wall panels, slabs, and precast railroad sleepers. Therefore, this research was conducted by exploring the use of geopolymer mortar as a shear reinforcement technique in beams using anchored geopolymer mortar (GM) panels with the addition of PVA (Poly-Vinyl Alcohol) fibers.

The study aimed to compare the shear capacity, crack patterns, and failure modes of beams reinforced with geopolymer mortar panels (GM) and Portland cement mortar panels (PC).

2. Materials and Methods

Ready mix concrete with compressive strength design of 20 MPa. Fly Ash obtained from PLTU Bosowa Jeneponto, South Sulawesi based on ASTM C618 [6]. NYCON-PVA RECS 15 type of PVA fiber with 8 mm in length, was used to increase the ductility of the Geopolymer mortar [7].

Table 1. Composition of Geopolymer and Portland Cement Mortar (kg/m^3)

Kode	kg per m3							PVA
	NaOH	Na_2SiO_3	Boraks	Pasir	Semen	*Fly Ash*	*plasticizer*	(%)
GM-0.6	156	234	10	780	-	390	7.8	0.6
PC-0.6	-	-	-	1360	636	-	-	0.6

The specimens in this study consisted of three types of beams with dimensions of 150 x 300 x 2300mm with compressive strength of f'c = 20 MPa, as shown in Figure 1. Normal beams as control beams are named with the code CB, beams with panel reinforcement geopolymer mortar is named with the code GM-W-200.

Table 2. Variation of specimens

No	Code	Strengthened materials	Number of the specimens
1	CB	-	1
2	GM-W-200	Geopolymer mortar + PVA Fiber	1
3	PC-W-200	Portland cement mortar + PVA Fiber	1
		Total Specimens	**3**

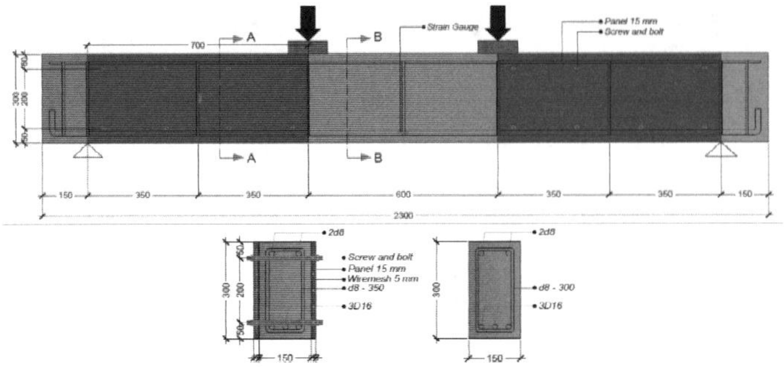

(a). Longitudinal and transverse section of GM-W-200 and PC-W-200

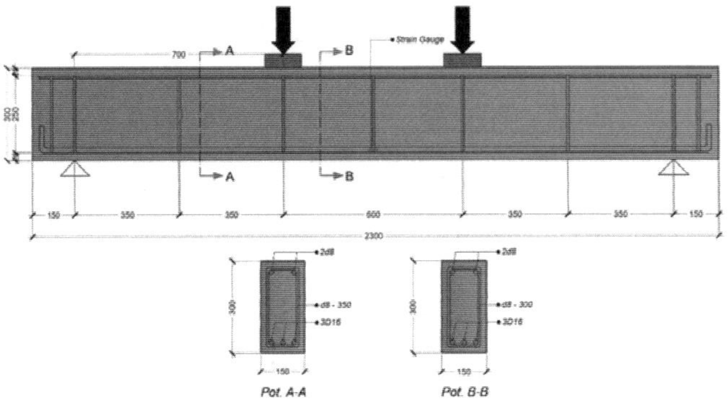

(b). Longitudinal and transverse section of CB

Figure 1. Specimens detail

3. Fabrication and Testing of specimens

This research was conducted in the Structure and Materials Laboratory, Department of Civil Engineering, Faculty of Engineering, Hasanuddin University. This type of research is an experimental laboratory study in the form of shear testing of reinforced concrete beams strengthened with geopolymer mortar panels and Portland cement mortar.

The fabrication of the specimens begins with the manufacture of geopolymer mortar and portland cement mortar panels. The manufacture of mortar panel test specimens begins with the manufacture of test specimen molds. After the preparation of the material (table 1), the mixing process is carried out using a mixer. At this stage, a sample of cubes measuring 50mm x 50mm x 50mm was also taken for testing the compressive strength of the mortar panel. After that, the panels are preserved by moist curing method.

During the panel maintenance process, the specimens of reinforced concrete beams were made, preceded by the manufacture of concrete and reinforcement. The manufacture of reinforced concrete beam molds is carried out simultaneously with the manufacture of panel mortar moulds. On the right and left sides of the reinforced concrete beam mold, holes are provided for the installation of anchor bolts.After that, the casting is done using ready mix concrete. Reinforced concrete beam specimens were also cured using the moist curing method. After the reinforced concrete panels and beams reach the age of 28 days, the panels are installed on the reinforced beams. Installation of mortar panels is done using an adjustable spanner (wrench) adjusted to the dimensions of the bolts used. After the reinforcement panel is inserted into the anchor of the test object, it is locked until the geopolymer mortar reinforcement panel is homogeneous with the reinforced concrete beam to be tested.

The specimen was loaded with a two-point load monotonically using a static test device with a capacity of 1500 kN. Loading uses deflection control with a loading speed of 0.2 mm/second. The instrumentation used is a strain gauge to measure the strain in concrete and steel, a Linear Variable Displacement Transducer (LVDT) to measure the deflection of the specimen, a load cell with a capacity of 200 KN to read the load that occurs, and a data logger to store all the data measured by instrument.

Strain gauges are installed in 4 locations, one on the top side of the concrete, one on the wiremesh, one on the shear reinforcement, and one on the flexural reinforcement. The strain gauge on the top side is used to measure the strain on the maximum compression side, while the strain gauge on the wiremesh is used to determine the maximum strain on the wiremesh panels and for the strain gauge onthe shear reinforcement and flexural reinforcement to determine the maximum strain on the shear and flexural reinforcement.

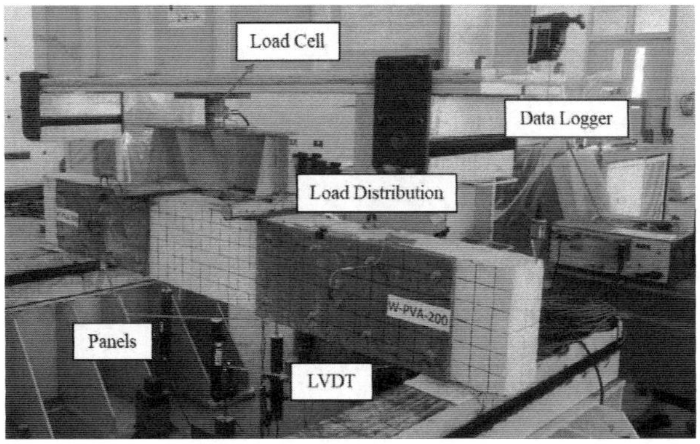

Figure 2. Set Up Specimen

4. Results and Discussion

4.1 Mechanical Properties

According to Table 4, the compressive and splitting tensile strengths of geopolymer mortar with PVA fibers at 28 days was 21.5 MPa and 2.1 MPa, respectively. Meanwhile, that of Portland cement mortar with PVA fibers for the same number of days was 23.2 MPa and 2.5 MPa. The results showed that the compressive and splitting tensile strengths of geopolymer mortar insignificantly different from that of PC. It needs to be noted that both binding materials applied the moist curing method.

Table 3. Compressive and splitting tensile strength at 28 days

Binding materials	Compressive strength (MPa)	Splitting tensile strength (MPa)
Geopolymer mortar + PVA	21.5	2.1
Portland cement mortar + PVA	23.2	2.5

4.2 Load – Deflection Relationship

In general, the behavior of the load-deflection relationship can be idealized into a trilinear form. However, the graph of the load and deflection relationship of the specimen in Figure 3 shows two forms of load-deflection relationship behavior, which is bilinear and trilinear. In the CB beam, the load-deflection relationship graph is bilinear which only consists of two phases. The first phase is the condition where the concrete has not cracked. The second phase is when the concrete begins to crack until it reaches its maximum load. Meanwhile, for the specimen with a strengthened panel, it consists of three phases. In the first phase, the concrete is still in an elastic condition (not cracked). The second is the phase where the concrete begins to crack and the third is the plastic phase where the steel reinforcement yields to the maximum load.

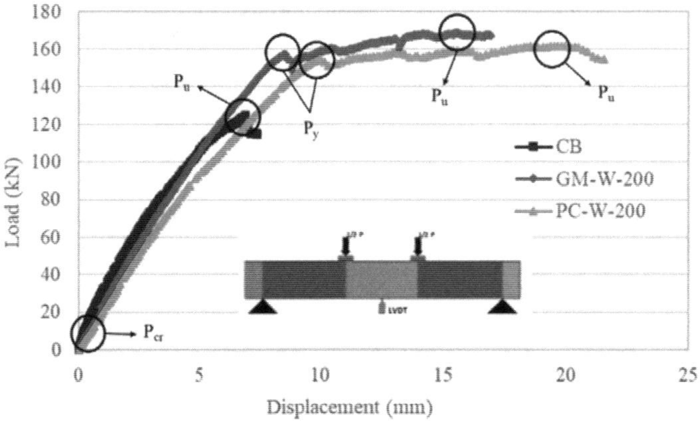

Figure 3. Load-deflection relationship

Figure 3 shows the load-deflection relationship of CB, GM-W-200 and PC-W-200 beams at mid- span. The effect of using geopolymer material on the reinforcement panels can be observed by comparing the load-deflection behavior of CB and GM-W-200 beams. Meanwhile, the effect of using Portland cement on the reinforcement panels can be observed by comparing the load-deflectionbehavior of GM-W-200 and PC-W-200 beams.

Based on Figure 3, the stiffness of the three beams had the same behavior at the initial crack. However, after the initial crack conditions reached a certain load, the CB beam experienced a decreasein stiffness until the beam was destroyed. The CB beam graph shows that after reaching the maximum load, the graph decreases suddenly, this indicates that the CB beam has shear failure.

However, in beams with strengthened panels, the loads are still increasing until it reaches the yield load, up to the maximum load. The PC-W-200 beam exhibits less rigidity than the GM-W-200 beam, this shows that both specimens suffer from flexural failure. The recapitulation of specimens is presented in Table 4, which consists of loads and deflections at initial crack conditions, yielding and maximum load.

Table 4. Load-deflection relationship recapitulation

No	Code	Load (kN)				Deflection (mm)			
		Pcr	$Py_{(Shear\ Rebars)}$	$Py_{(Tensile\ Rebars)}$	Pu	δ_{cr}	$\delta_{y(Shear\ Rebars)}$	$\delta_{y(Tensile\ Rebars)}$	δ_u
1	CB	10.06	-	-	124.75	0.26	-	-	6.89
2	GM-W-200	11.06	157.00	156.27	168.40	0.52	8.54	8.38	15.46
3	PC-W-200	11.06	153.34	154.87	161.60	0.61	10.17	10.00	19.65

In the control beam (CB), the reinforcing steel is fully elastic until the beam experiences initial cracks at a load of 10.06 kN with a deflection value of 0.26 mm. After the initial cracking occurs, the load continues to increase as the deflection increases. When the load reaches 124.75 kN, the load drops suddenly. This indicates that the beam has failed.

In the beam strengthened with geopolymer mortar panels, the initial crack occurred at a load of 11.06 kN with a deflection value of 0.52 mm. After initial cracking, the beam then shows the elastoplastic properties of the steel to yield reinforcement at a load of 157 kN with a deflection value of 8.54 mm for shear reinforcement with a strain value of 1411 με and 156.27 kN with a deflection of 8.38 mm for tensile reinforcement with a strain value of 2234 με. These results are in accordance with the tensile strength test value of steel, where the tensile reinforcement has a yield strain value of 1924 με while the shear reinforcement is 1684 με.After the reinforcement yielded, the load still showed an increase but was disproportionate. This can be seen from the relationship curve to be flatter than before the reinforcement yields. This happens until the beam collapses at an ultimate load of 168.4 kN with a deflection value of 15.46 mm. This shows that beams with geopolymer mortar panel reinforcement exhibit more ductile behavior than CB beams.

In the beam strengthened with Portland cement mortar panels, the initial crack occurred at a load of 11.06 kN with a deflection value of 0.61 mm. After the initial cracking, the beam then shows the elastoplastic properties of the steel to yield reinforcement at a load of 153.34 kN with a deflection of 10.17 mm for shear reinforcement with a strain value of 1905 με and 154.87 kN with a deflection of 10 mm for tensile reinforcement with a strain value of 1608 με. These results are in accordance with the tensile strength test value of steel, where the tensile reinforcement has a yield strain value of 1924 με while the shear reinforcement is 1684 με. Before yielding reinforcement, the load increases proportionally as happened in the GM-W-200 beam. After yield reinforcement, the beam still increased but disproportionately before the beam collapsed with a maximum load of 161.6 kN and a deflection value of 19.65 mm. This showsthat beams with geopolymer mortar strengthening panel exhibit more ductile behavior than CB beams.

The effect of Geopolymer mortar panel reinforcement can be observed by comparing the three CB beams, GM-W-200 and PC-W-200. The beams GM-W-200 and PC-W-200 show almost the same loaddeflection relationship behavior. It can be seen in the graph where the geopolymer mortar panel and portland cement mortar panel affect the behavior of the beam to be more ductile.

4.3 Maxsimum Load

Figure 4 shows the maximum load on each specimen. The maximum loads of CB, GM-W-200 and PC-W-200 beams are 124.75 kN, 168.39 kN and 161.6 kN, respectively. The maximum load of GM- W-200 and PC-W-200 beams experienced a significant increase compared to CB beams. GM-W-200 beam and PC-W-200 beam increased by 34.98% and 22.81% to the control beam

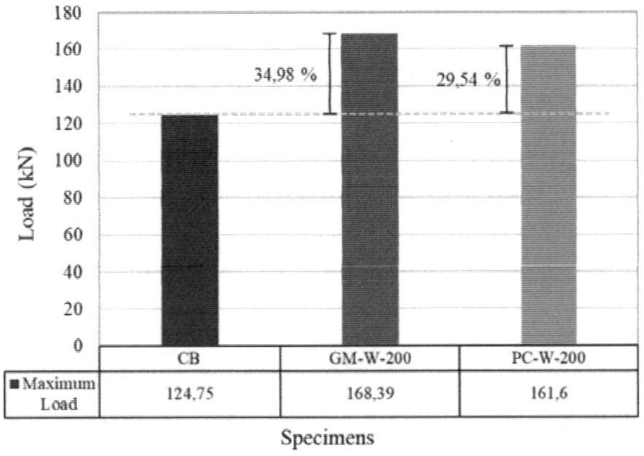

Figure 4. Maximum Load

4.4 Crack Pattern

Figure 5 shows the control beam (CB) crack pattern. The control beam (CB) has an initial crack whichis a flexural crack at a load of 37.3 kN. The crack occurs at the bottom left of the span and increases inlength. The initial shear crack occurs when the load is 56 kN. Shear cracks are characterized by the presence of diagonal cracks in the shear area of the beam. At the time of peak load, the concrete on theshear side crumbled, indicating that the beam had shear failure.

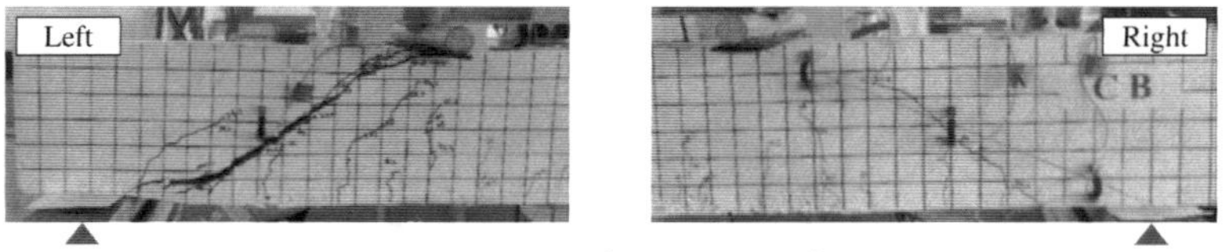

Figure 5. Crack pattern of CB

Figure 6 shows the GM-W-200 beam crack pattern. Based on observations, the initial crack which is a flexural crack in the GM-W-200 beam occurs at a load of 25.19 kN as shown in Figure 6. Meanwhile, the initial crack in the geopolymer mortar panel appears at a load of 57.38 kN. Cracks occur in the center of the panel. After the initial crack on the panel appears, other cracks start popping up randomly.

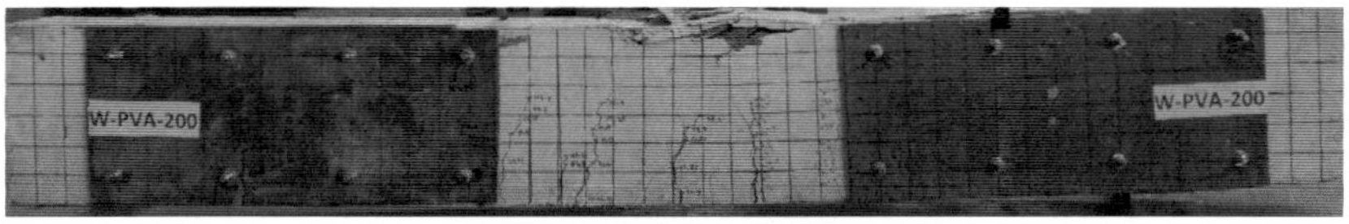

Figure 6. Crack pattern of GM-W-200

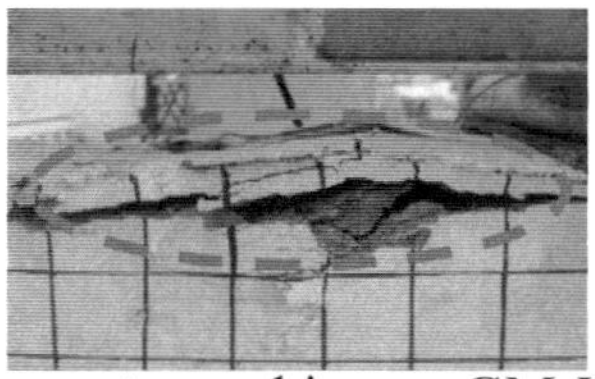

Figure 7. Concrete crushing on GM-W-200 beam

After reaching the maximum load, the concrete is destroyed on the compression side (Figure 7). This can also be seen in the graph of the load-strain relationship of concrete, where the strain of GM- W-200 beam concrete in the compression area is close to 3000 $\mu\varepsilon$.

Figure 8 shows the PC-W-200 beam crack pattern. Based on the observations, the initial crack which is a flexural crack in the PC-W-200 beam appears in the mid-span area when the load is 12.73 kN. After the initial crack appears, another flexural crack appears and then propagates towards the loading point. Initial cracks in the Portland cement mortar panel appeared at a load of 73.57 kN. Cracks occur in the upper area of the panel adjacent to the loading point After the initial crack in the panel, other cracks begin to appear randomly.

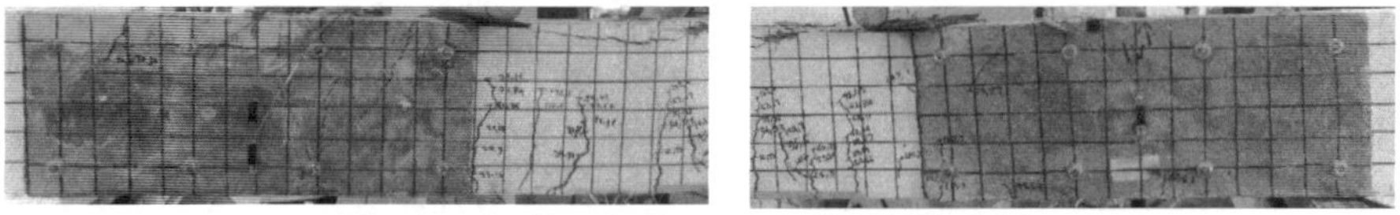

Figure 8. Crack Pattern of PC-W-200

Figure 9. Concrete crushing on PC-W-200 beam

After reaching the maximum load, the concrete is destroyed on the compression side. This can also be seen in the graph of the concrete load-strain relationship, where the concrete strain for PC-W-200 in the compression area reaches a value of 4313 με.

Based on observations on crack patterns and failure models on the GM-W-200 and PC-W-200 beams, it can be seen that both of them have similarities, including the type of initial crack and the number of cracks in each specimen. In addition, both variations of the beam were destroyed on the compression side. both specimens of the test object experienced flexural failure

5. Conclusion

Based on the results of research and discussion, it can be concluded that :

1. The additional of strengthened panels with geopolymer and Portland cement materials was able toincrease the loading capacity of reinforced concrete beams, respectively by 34.98% and 29.54% to the control beam, in the form of increasing the maximum load and changing the failure pattern from shear failure to flexural failure.
2. Crack patterns and failure modes on the GM-W-200 and PC-W-200 beams show similarities,both in terms of crack patterns and the number of cracks in each specimen. Both panel's variations show a flexural failure pattern while the control beam exhibits a shear failure.

References

[1] Triwiyono, A. (1998). Evaluasi Kinerja dan Perkuatan Struktur Gedung Guna Alih Fungsi Bangunan.Forum Teknik Sipil No. XVIII.

[2] Raza S, Khan M K I, dkk. 2019. Strengthening and Repair of Reinforced Concrete Columns byJacketing: State-of-the-Art Review. Sustainability MDPI Journal.

[3] Malhotra, V.M. 2002. Introduction: Sustainable Development and Concrete Technology. ACI ConcreteInternational. 24 (7), p. 22.

[4] Rodgers, L. 2018. Climate change: The massive CO2 emitter you may not know about. United StatesNawy, Edward G. 1990. Beton Bertulang – Suatu Pendekatan Dasar. Jakarta : Penerbit Erlangga.

[5] Davidovits, J., 2005, Green Chemistry and Sustainable Development Solutions, Perancis: GeopolymerInstitute.

[6] ASTM C618-93. 1991. "Standard Test Method for Fly Ash and Row or calcined Natural Pozzolan forUse as a mineral Admixture in Portland

[7] Zerfu, Kefiyalew. 2017. Effects of PVA Fiber on Bond Strength Improvement in Geopolymer Concrete.Departemen Teknik Sipil ITS.

[8] Zhang Y, Sun W, Li Z, Eddie, Chungkong C. Impact properties of geopolymer based extrudatesincorporated with fly ash and PVA short fibre. Constr Build Mater 2008; 22:37

Deterministic and probabilistic Seismic Hazard Assessments for Merangin Kerinci Sumatera

Ardy Arsyad, Akhmad Bakri Muhiddin, Farouk Maricar

Department of Civil Engineering, Faculty of Engineering, Hasanuddin University
Geo-Disaster Mitigation, Centre of Technology, Hasanuddin University

ardy.arsyad@unhas.ac.id

Abstract. This paper presents calculated ground motions for Merangin Kerinci in the Provinsi of Jambi Sumatera. The ground motions were calculated using a probabilistic seismic hazard analysis. The earthquake catalogue used in the analysis included all recorded earthquakes up to the end of July 2019, trimmed for magnitude completeness. Uncertainty in input parameters was considered using a logic tree, with branches including shallow crustal faults of the Great Sumatran Faults and Sumatera Megathrust Subduction. Ground motions are presented for the local site conditions, measured at those sites which can be classified as Site Class D/C (stiff soil to Soft Rock/Very Dense Soil). Ground motions, including Peak Ground Acceleration (PGA), and spectral accelerations at periods ranging from 0.1 s to 10.0s were calculated the site. Data are presented for return periods of 175 years, 475 years, 2,500 years, 5,000 years, and 10,000 years for PSHA, and at mean, 50th, and 84th percentile confidence levels for DSHA. Compared to the Indonesian Seismic Hazard Map (ISHM) 2017 (at DSHA at 5% damping and 84th Percentile), the deterministic PGA values calculated in this study are relatively closed, which is accounted for 0.58g found in this study compared to the range of 0.6g – 0.7 in the ISHM 2017. On the other hand, the probabilistic PGA values for a particular of 5,000 years return period (SEE), the PGA values found in this assessment is 0.73g which is lower than the range of 0.8g – 0.9g in the ISHM 2017.

1. Introduction

Merangin Kerinci is located in the Province of Jambi, Sumatera, where is in the proximity to the Great Sumatran Fault (Figure 1). The location is classified as "High" to "Very High" seismic hazard according to Indonesian Earthquake Hazard Maps [1], and must, therefore, be investigated to accommodate seismic ground motions resulting from earthquakes with an annual exceedance probability (AEP) of 1:2,500 to 1:10,000 for safety evaluation earthquake (SEE) and Maximum Credible Earthquake (MCE), and with AEP of 1:145 for operating basis earthquake (OBE).

Therefore, this study aims to utilize the magnitude-recurrence relations developed for shallow crustal faults (Great Sumatran Fault Segments), and Megathrust Subduction to assess the contributions to the seismic hazard at Merangin Kerinci from those regions; to develop several seismic source zone models for the Merangin Kerinci, based on seismotectonic considerations and patterns of historical seismicity, utilizing the earthquake catalogue to the end of July 2019; and to establish appropriate ground motion attenuation relations for the region; and to calculate earthquake ground motions at the locations for return periods of 200 years, 500 years, 2500 years, 5,000 years to 10,000 years, for bedrock-level local site conditions within addressing parameter uncertainty using a logic tree. Furthermore, probabilistic

IOP Conf. Series: Earth and Environmental Science 1117 (2022) 012018 doi:10.1088/1755-1315/1117/1/012018

and deterministic-based ground motions were provided based on a consistent methodology and considering the latest seismicity information. These ground motions could potentially be used for local and regional scale seismic hazard assessment on a consistent seismic hazard basis.

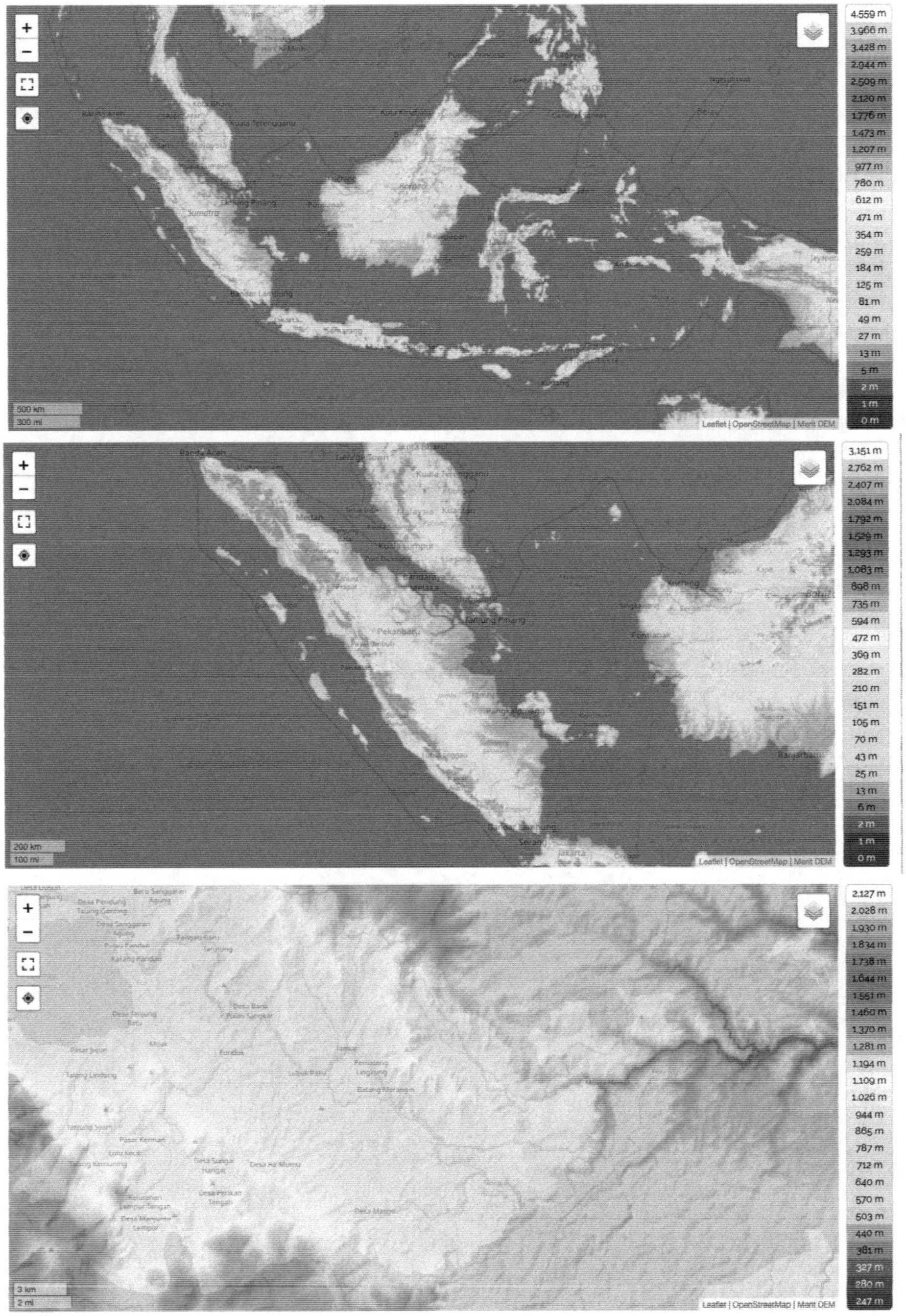

Figure 1. The study area of Merangin Kerinci, the Province of Jambi, Indonesia.

2. Main Input for Hazard Model

2.1. Geological and Seismotectonic Kerinci Merangin

Great Sumatran Fault is a main fault extending about 1,900 km in NW-SE along the western part of the island. The fault has been investigated by Katili [2], Bellier and Sebrier [3], and Sieh and Natawidjaja [4]. Propagation of this fault strongly relates to the tectonic development of the Sumatra, which is the result of continental collisions since Late Paleozoic until Late Mesozoic. Great Sumatran Fault is concentrated in the western part of the island. Sieh and Natawidjaja [4] have contributed substantial insight into the segmentation of Sumatran Fault. The Sumatran Fault can be recognized in 19 segments from Aceh to Sunda Strait with predominantly dilatational stepover. As a result, the formation of several pull-apart basins occurred, starting from Aceh to Kutacane Graben, then running along the west of Toba Caldera, moving to Tarutung Sarulla Basins and jumping to BarumunAngkola Segments. The trip continues to Lake Singkarak, Muaralabuh-Sungai Penuh Basins, Musi-Manna Segments, Lake Ranau, Suoh Valley, and eventually Sunda Strait. Active faults have existed in the Great Sumatran Fault, which is segmented into several faults (Figure 2).

Regarding historical earthquakes, the Merangin Kerinci is relatively closed to the Sungai Penuh, which is the city of the Regency Kerinci, the Provinsi of Jambi. The area has been experiencing at least two significant earthquakes, that is 1909 with 7.3 Mw and 1995 with 7.1 Mw. In the south, in 1952, a major earthquake occurred at 6.8 Mw. Along the Great Sumatran Fault, several major earthquakes occurred in 1943 with 7.6 Mw, 1926 6.8 Mw, 1942 7.3 Mw.

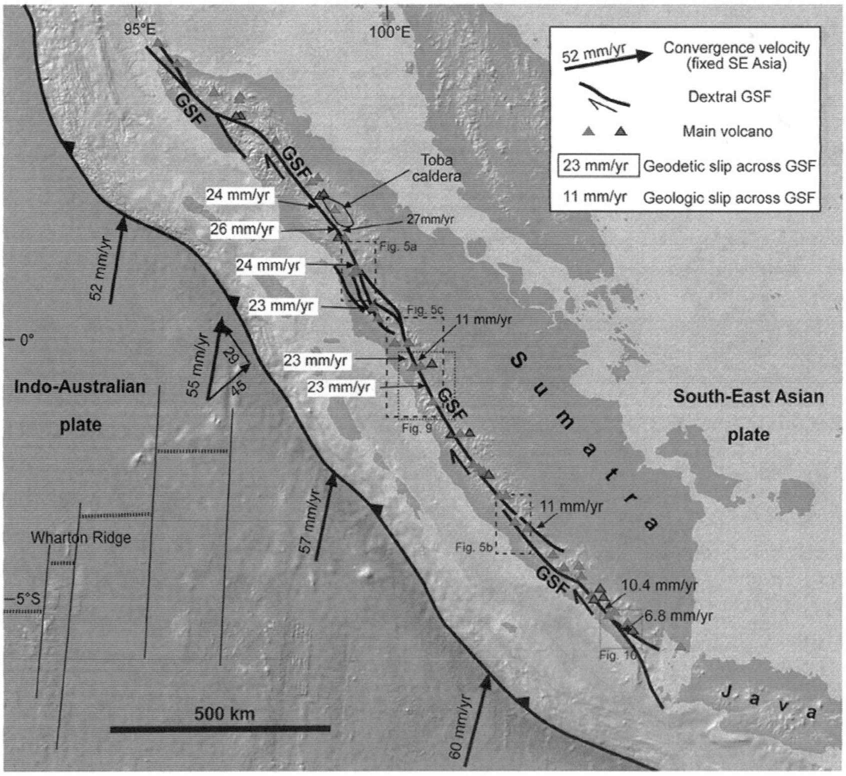

Figure 2. Active faults existed along the Great Sumatran Fault (after Sieh and Natawidjaja, 2000).

2.2. Seismic Source Model

2.2.1. Earthquake Catalogue

The earthquake catalogue used for this hazard analysis was the Seismic Hazard Earthquake Epicentre File obtained from the United States of Geological Survey (USGS) in July 2019. The catalogue includes all recorded earthquakes in the region, dating from before 1900 up to the end of July 2019 (Figure 3).

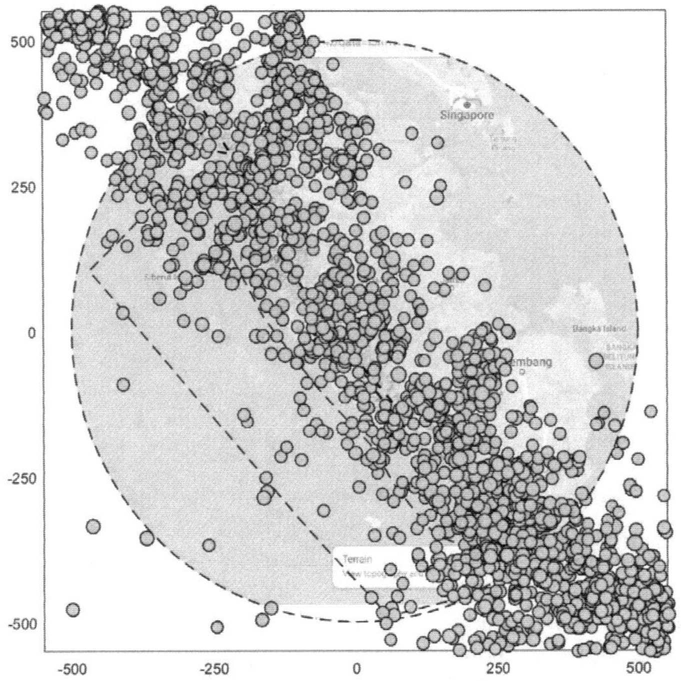

Figure 3. All recorded earthquakes were compiled in the study from 1900 – 2019.

2.2.2. Magnitude Completeness

The completeness of the earthquake record varies significantly with time. The magnitude and epicentre location for early historical earthquakes were estimated by calculating the Magnitude Completeness (Mc). As shown in Figure 4, Mc is accounted for Mw 4.7. It means that the data were compiled in the assessment considered as completed data. The attenuation relationships used in this hazard analysis are based on moment magnitude (M). The earthquake events in the USGS catalogue are in different magnitude scales, depending on the quality of record obtained and what was considered to be the best representation of the event. For this study, these magnitudes were converted to a consistent scale (moment magnitude) before calculating the magnitude- recurrence relations.

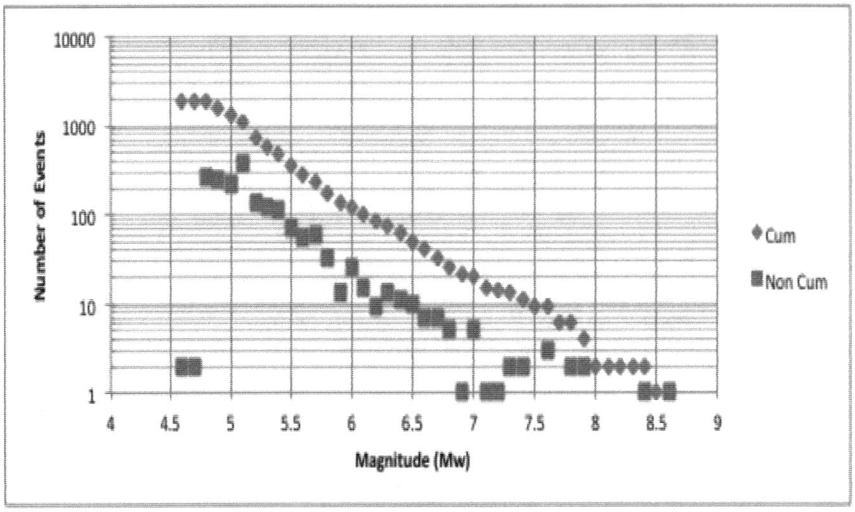

Figure 4. Estimation of magnitude completeness Mw 4.7 from the graph of cumulative count with increasing magnitude.

2.3. Ground Motion Prediction Equations (GMPEs)

Ground motion relations (attenuation relationships) describe the amplitude of a ground motion parameter, such as acceleration or velocity, as a function of earthquake magnitude, distance from the source, and spectral period. Based on the ground-motion evaluations, the attenuation relations of Atkinson & Boore 2008 [5], Campbell & Bozorgnia 2008 [6], and Campbell & Bozorgnia 2008 (CY-08) were used for this study on shallow crustal faults, and Atkinson & Boore 2006 (AB-06) and Youngs et al. 1998 [7] for megathrust subduction (Table 1 and 2), together with a recent California relation from the Pacific Earthquake Engineering Research Center (PEER) / Next Generation Attenuation (NGA) study. The PEER relations capture a wealth of empirical data from active tectonic regions around the world, and have been extensively studied. In the case of the deterministic approach, ground-motion attenuation relations were used, as illustrated in Table 3.

Table 1. GMPE for Shallow Crustal.

Ground-motion Attenuation Relation	Relative weight in Logic Tree
Atkinson & Boore 2008 NGA (AB 08)	0.33
Campbell & Bozorgnia 2008 NGA (CB 08)	0.33
Chiou & Young 2008 NGA (CY 08)	0.33

Table 2. GMPE for megathrust subduction.

Ground-motion Attenuation Relation	Relative weight in Logic Tree
Atkinson & Boore 2006 (AB 06) [8]	0.5
Youngs et al. 1998 (Y 98) [9]	0.5

Table 3 GMPE for Deterministic Seismic Hazard Analysis (DSHA).

Groundmotion Attenuation Relation	Relative weight
Abrahamson-Silva-Kamai 2014 [10]	0.22
Campbell & Bozorgnia 2014 [11]	0.22
Boore-Steward-Seyhan-Atkinson 2014 [12]	0.22
Chiou & Youngs 2014 [9]	0.22
Idriss 2014 [13]	0.12

For the deterministic approach, at least four faults in the vicinity of Merangin Kerinci were selected including Suliti, Siulak, Dikit and Ketaun (Table 4, and Figure 5). On the other hand, for probabilistic, calculated ground motions with spectral accelerations for 5% damping for horizontal and vertical directions were computed with shallow crustal earthquake and megathrust subduction earthquakes with annual exceedance probability (AEP) 1:175, 1:475, 1:2500, 1:5000, and 1:10000. The mean of the distribution of epistemic uncertainty was calculated from the logic tree (Figure 6). It must be stated that three segments of megathrust subduction (Megathrust South Sumatera, Megathrust Middle Sumatera 1, and Megahtrust Sumatera 2) were employed for the assessment (Table 5).

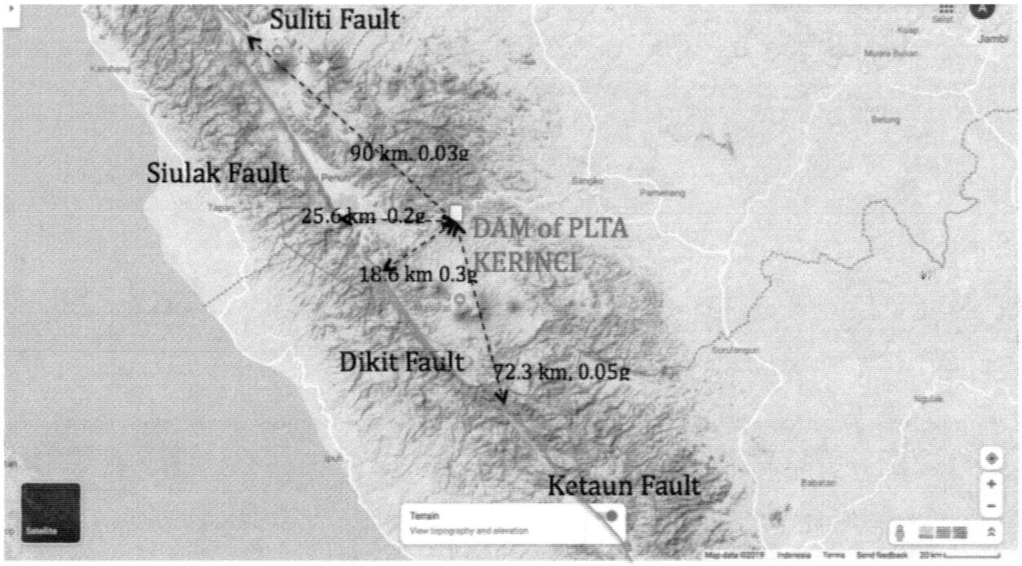

Figure 5. Seismic sources were accounted for Suliti Fault, Siulak Fault, Dikit Fault, and Ketaun Fault, utilized for DSHA.

Table 4 Shallow Crustal Faults for DSHA.

Faults	Estimated Max Mw Generated	Slip Rate (mm/year)	Closest Distance (km)
SULITI	7.2	14	90.0
SIULAK	7.1	14	25.6
DIKIT	7.1	11	18.6
KETAUN	7.2	11	72.3

Table 5 Megathrust Subduction for DSHA

Faults	Estimated Max Mw Generated	Closest Distance (km)
Megathrust South Sumatera	8.2	226.7
Megathrust Middle Sumatera 1	8.6	339.7
Megathrust Middle Sumatera 2	8.5	119.7

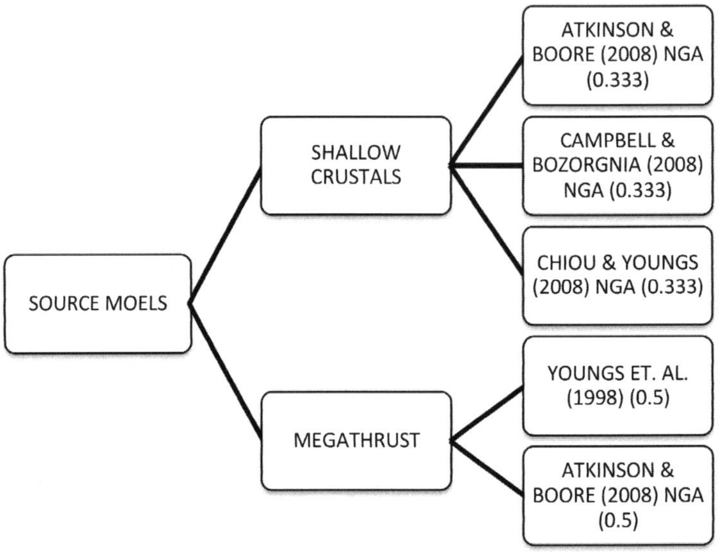

Figure 6. The hazard model logic tree.

3. Seismic Hazard Analysis Results

3.1. Deterministic Approach

By using the NGAWest2 software, the DSHA based PGAs in Merangin Kerinci were computed in which the target spectral accelerations were set to 50th percentile and 84th percentile. The spectral accelerations of the active crustal faults (Suliti, Seulak, Dikit and Ketaun) were generated based on GMPEs of the ASK-14, CB-14, BSSA-14, CY-14, and I-14. The generated spectral accelerations of the active faults can be seen in Figure 7. The GMPEs of CY-14 and ASK-14 seem to obtain higher accelerations than other GMPE for spectral period less than 3 sec. For the period above 3 sec, the I-14 provides higher spectral accelerations. Figure 8 shows that the Seulak and Dikit faults would be able to generate much higher seismic acceleration than the Suliti and Ketaun Faults.

The results are shown in Table 6, revealing that the Ketaun Fault deterministically predicted to generate 0.595g. Compared to the other faults (Suliti, Seulak, and Dikit), the PGA is relatively similar to the PGAs of Suliti, Seulak, and Dikit (0.580g – 0.594g). Since the distance of the Dikit fault to the Merangin Kerinci is about 18 km, this fault was selected as the controlling earthquake of the DSHA. Table 7 shows that the PGA with 5% damping at short period (0.2s) is 0.818g for 84th percentile and 0.94g for 50th percentile. On the other hand, the PGA at long period (1.0s) is 0.253g for 84th percentile and 0.158g for 50th percentile. The results of DSHA also indicate that the megathrust subduction zones just generate lower PGAs than the active shallow faults, as a high of 0.113g at Middle 2, and low of 0.026g at Middle 1.

Table 6 Horizontal Peak Ground Acceleration at 84th Percentile

No	Source	MCE	Distance to closest (km)	PGA (84th Percentile)					
A. MAJOR ACTIVE FAULTS									
	Segment			ASK-14 NGA	BSS-14 NGA	CB-14 NGA	CY - 15 NGA	I-14 NGA	Mean
1	Suliti	7.2	90.0	0.608	0.605	0.571	0.543	0.643	0.594
2	Seulak	7.1	25.6	0.591	0.605	0.542	0.530	0.649	0.583
3	Dikit	7.1	18.6	0.585	0.605	0.533	0.526	0.649	0.580
4	Ketaun	7.2	72.3	0.612	0.605	0.567	0.550	0.643	0.595
B. SUBDUCTION ZONE (SUMATERA MEGATHRUSTS)									
	Segment			AB-2003	Youngs-97	Zhao-2006			Mean
1	Middle 2	8.5	119.73	0.130	0.011	0.200			0.113
2	South	8.2	226.7	0.065	0.075	0.060			0.067
3	Middle 1	8.6	399.7	0.021	0.038	0.020			0.026

Table 7 Horizontal Peak Ground Acceleration of Dikit Fault at 84th and 50th Percentile

Period (sec)	PGA (g)		Source
	Fractile 0.5	Fractile 0.84	
	0.251	0.580	
0.01	0.216	0.355	Dikit fault, Mw = 7.1, 18.6 km from the Merangin Kerinci
0.02	0.219	0.361	
0.03	0.235	0.396	
0.05	0.282	0.498	
0.075	0.352	0.639	
0.1	0.407	0.732	
0.15	0.478	0.832	
0.2	0.494	0.818	
0.25	0.472	0.759	
0.3	0.434	0.689	
0.4	0.367	0.575	
0.5	0.313	0.491	
0.75	0.213	0.340	
1	0.158	0.253	
1.5	0.099	0.160	
2	0.070	0.114	
3	0.043	0.070	
4	0.030	0.048	
5	0.023	0.037	
7.5	0.013	0.021	
10	0.008	0.013	

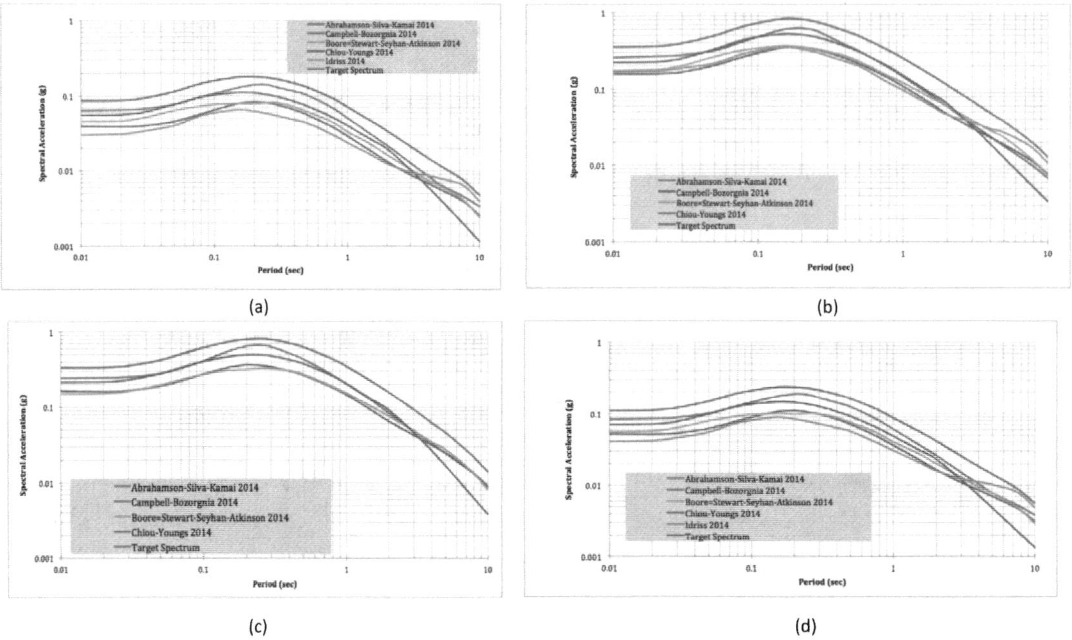

Figure 7. Deterministic spectral accelerations of the active faults of (a) Suliti, (b) Seulak, (c) Ketaun and (d) Dikit.

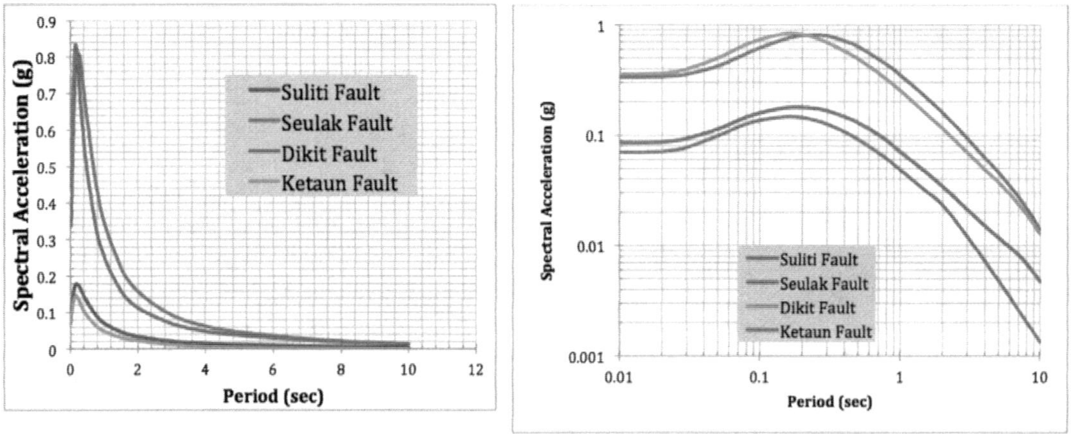

Figure 8. Comparisons of the deterministic spectral accelerations of the active faults.

3.2. Probabilistic Approach

It can be seen in Table 8 and 9 and Figure 10, the horizontal and vertical spectral accelerations obtained probabilistically would vary over return periods. As it would be expected that the accelerations would increase as the return period increases. For 175 years return period, the acceleration for short period would be 1.02g, while that for long period is 0.61g. In contrast, at 2500 years return period, the acceleration would be 1.62g, and 1.15g for short and long periods respectively. In terms of PGA, the value was found at 0.39g for 175 years return period, on the basis of two attenuation functions: the CB-08 and AS-08. While the value of PGA could be little higher with three attenuation functions of BA-08, CB-08, and AS-08. It must be noted that BA is too conservative, while CB and AS are quite similar. For

5,000 years return period, the PGA is at 0.73g. This result is bit lower compared to the similar PGA obtained through ISHM 2017 which is about 0.8g to 0.9g. In comparison to ISHM 2017 or SNI 1726-2017, the findings in this study seem to underestimate.

3.3. Magnitude – Distance Deaggregation

Deaggregation results indicating the magnitude and distances most significantly contributing to the calculated PGA are shown in Figure 9. These additional deaggregation calculations were performed to assist in selecting appropriate earthquake time-histories. It is evident that, the dominant contributions to the 1:10,000 PGA are from low to moderate magnitude earthquakes (Mw 5 – Mw 6) that occur close to the study area, less than 100 km. Dikit, Siulak, and Ketaun would become contributing fault whereas Dikit Fault is the most significant contributor with distance about 30 km - 100 km with Mw. 4.8 – Mw 6.2.

Table 8 Seismic Hazard Spectra for Merangin Kerinci at 5% Damping in various return periods – horizontal direction.

Period (sec)	Peak Ground Acceleration (g)				
	1:175	1:475	1:2,500	1:5,000	1:10,000
0.01	0.37	0.45	0.58	0.64	0.70
0.02	0.38	0.46	0.58	0.65	0.71
0.03	0.41	0.50	0.58	0.71	0.78
0.05	0.51	0.61	0.64	0.88	0.97
0.075	0.68	0.82	0.79	1.19	1.32
0.1	0.82	1.00	1.07	1.45	1.61
0.15	1.01	1.22	1.59	1.17	1.97
0.2	1.02	1.24	1.62	1.81	2.01
0.25	0.98	1.20	1.57	1.75	1.95
0.3	0.95	1.17	1.54	1.73	1.92
0.4	0.86	1.06	1.42	1.59	1.78
0.5	0.82	1.03	1.40	1.58	1.78
0.75	0.70	0.90	1.27	1.46	1.66
1	0.61	0.80	1.15	1.33	1.53
1.5	0.48	0.65	0.96	1.13	1.30
2	0.39	0.54	0.81	0.94	1.10
3	0.27	0.37	0.56	0.66	0.77
4	0.21	0.29	0.44	0.51	0.60
5	0.19	0.27	0.42	0.49	0.58
7.5	0.16	0.23	0.38	0.46	0.55
10	0.14	0.22	0.37	0.46	0.56

Table 9 Uniform Hazard Spectra for Merangin Kerinci at 5% Damping in various return periods – vertical direction.

Period (sec)	Peak Ground Acceleration (g)				
	1:175	1:475	1:2,500	1:5,000	1:10,000
0.01	0.25	0.30	0.39	0.43	0.47
0.02	0.25	0.31	0.39	0.43	0.47
0.03	0.27	0.33	0.39	0.47	0.52
0.05	0.34	0.41	0.43	0.59	0.65
0.075	0.45	0.55	0.53	0.79	0.88
0.1	0.41	0.50	0.54	0.73	0.81
0.15	0.51	0.61	0.80	0.59	0.99
0.2	0.51	0.62	0.81	0.91	1.01
0.25	0.49	0.60	0.79	0.88	0.98
0.3	0.48	0.59	0.77	0.87	0.96
0.4	0.34	0.42	0.57	0.64	0.71
0.5	0.33	0.41	0.56	0.63	0.71
0.75	0.28	0.36	0.51	0.58	0.66
1	0.24	0.32	0.46	0.53	0.61
1.5	0.19	0.26	0.38	0.45	0.52
2	0.16	0.22	0.32	0.38	0.44
3	0.11	0.15	0.22	0.26	0.31
4	0.08	0.12	0.18	0.20	0.24
5	0.08	0.11	0.17	0.20	0.23
7.5	0.06	0.09	0.15	0.18	0.22
10	0.06	0.09	0.15	0.18	0.22

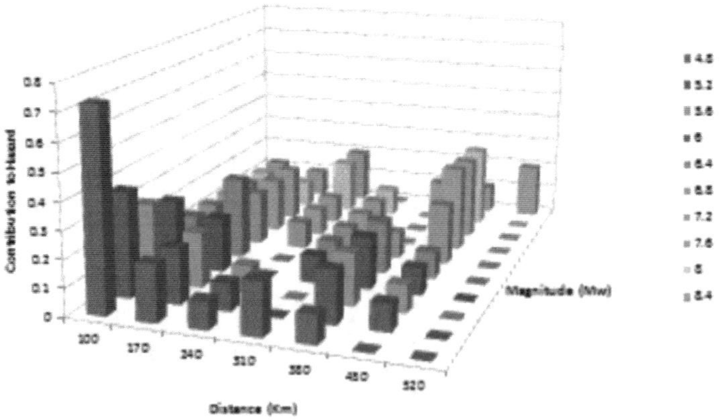

Figure 9. Magnitude – distance deaggregation in probabilistic seismic hazard of Merangin Kerinci.

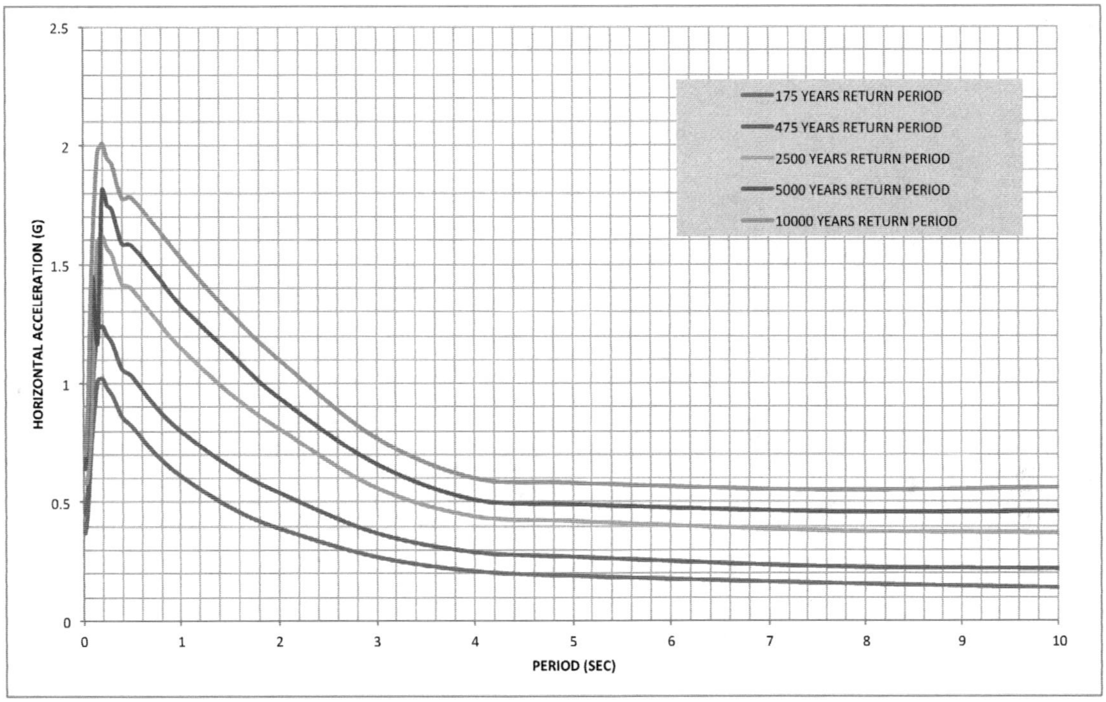

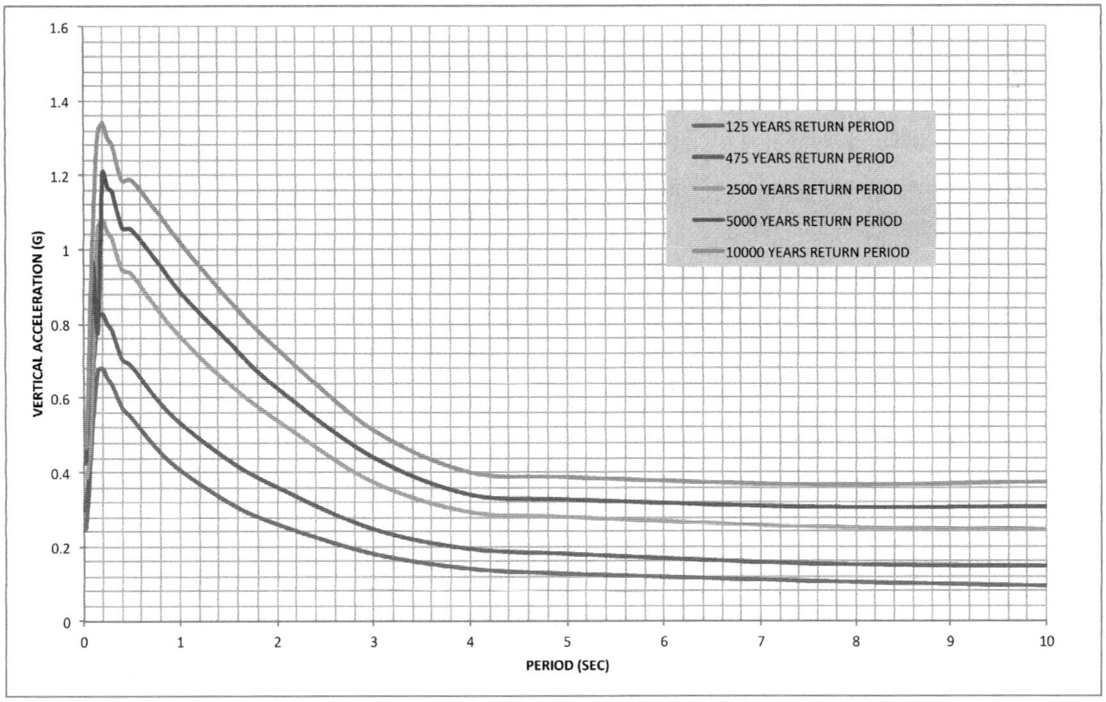

Figure 10. Horizontal and vertical spectral acceleration of Merangin Kerinci for return periods.

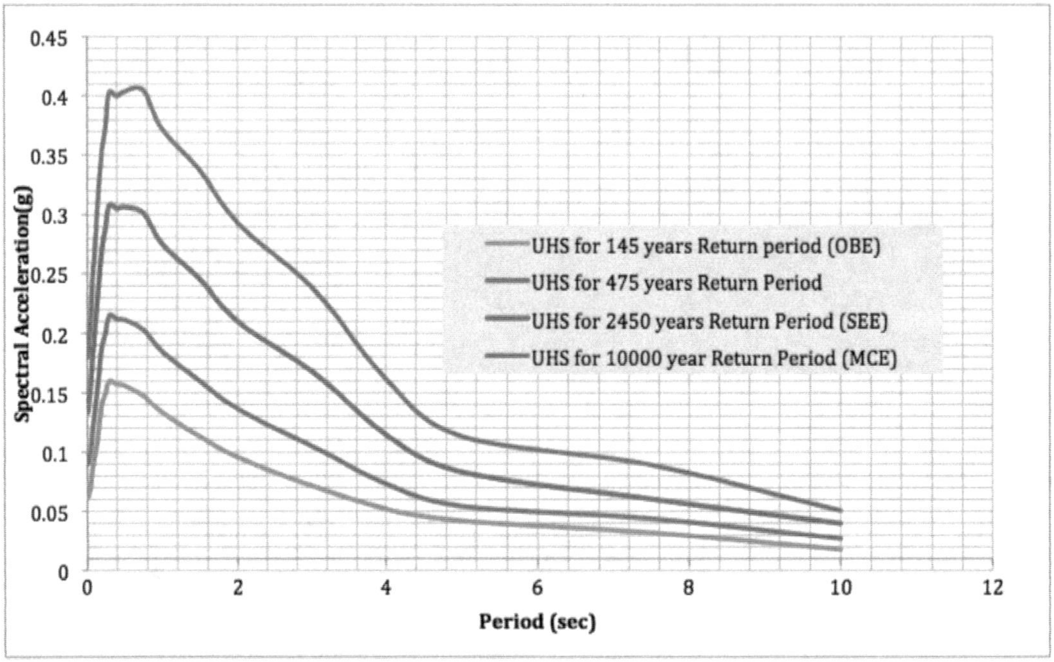

Figure 11. Uniform hazard spectral of Merangin Kerinci for various return periods.

3.4. Comparison to ISHM 1726-2017

Compared to the Indonesian Seismic Hazard Map (ISHM) 2017 (at DSHA at 5% damping and 84th Percentile), the deterministic PGA values calculated in this study are relatively closed, which is accounted for 0.58g found in this study compared to the range of 0.6g – 0.7 in the ISHM 2017. On the other hand, the probabilistic PGA values for a particular of 5,000 years return period (SEE), the PGA values found in this assessment is 0.73g which is lower than the range of 0.8g – 0.9g in the ISHM 2017.

Table 10 PSHA based PGA.

RETURN PERIOD (years)	PGA (g) Based on BA (2008), CB (2008), AS (2008)	PGA (g) Based on CB (2008) and AS (2008)	SNI 1726-2017 (National Seismic Hazard Map)
175	0.42	0.39	0.40 – 0.50
475	0.53	0.49	0.60 - 0.70
2,500	0.71	0.65	0.70 – 0.80
5,000	0.81	0.73	0.80 – 0.90
10,000	0.90	0.81	1.20 – 1.50

4. Conclusions

Seismic hazard assessment for site of Kerinci Merangin has been conducted. The site is situated at relatively closed to the Great Sumatran Fault. In particular to one of the fault segments, which is Dikit Fault, with 18.6 km away from the site. However, the study found that the deterministic PGA is 0.58g at 84th percentile. This is quite well agreement with the Indonesian Seismic Hazard Map (ISHM 2017) in deterministic approach with 0.6g – 0.7g. Meanwhile, the probabilistic PGA at SEE level (5,000 years return period) of this study is found at 0.73g. This seems to be much lower than that in the ISHM 2017 with 0.8g – 0.9g.

5. References

[1] Tim Pusat Studi Gempa Nasional-2017 (The 2017 PuSGen) (2017) Peta Sumber dan Bahaya Gempa Indonesia Tahun 2017. Jakarta: Kementrian Pekerjaan Umum dan Perumahan Rakyat (In Indonesian)

[2] Katili, J., and F. Hehuwat,1967 On the occurence of large transcurrent faults in Sumatra, Indonesia, J. Geoscience, Osaka City Univ., 10, 5- 17.

[3] Bellier, O., and M. Sebrier, 1994. Relationship between tectonism and volcanism along the Great Sumatran fault zone deduced by SPOT image analyses, Tectonophysics, 233, 215-231.

[4] Sieh, K. and Natawidjaja, D. 2000, "Neotectonics of the Sumatran fault, Indonesia", Journal of Geophysical Research: Solid Earth, Wiley Online Library, Vol. 105 No. B12, pp. 28295– 28326.

[5] Boore, D. M. and Atkinson, G. M., 2008. Ground-motion prediction equations for the average horizontal component of PGA, PGV, and 5%-damped PSA at spectral periods between 0.01s and 10.0 s, Earthquake Spectra, 24:1, 99–138.

[6] Campbell, K. W. and Bozorgnia, Y., 2008. NGA ground motion model for the geometric mean horizontal component of PGA, PGV, PGD and 5% Damped linear elastic response spectra for periods ranging from 0.01 to 10 s, Earthquake Spectra, 24:1, 139-171.

[7] Chiou B.S-J. and Youngs, R.R., 2008. An NGA model for the average horizontal component of peak ground motion and response spectra, Earthquake Spectra, 24:1, 173–215.

[8] Atkinson, G. M. and Boore, D. M., 2006. Earthquake ground-motion prediction equations for eastern North America, Bull Seism. Soc. Am, 96:6, 2181–2205.

[9] Youngs, R. R., S.-J. Chiou, W. J. Silva, and J. R. Humphrey. 1997. Strong ground motion attenuation relationships for subduction zone earthquakes, Seismol. Res. Lett. 68, no. 1, 58–73.

[10] Abrahamson, N. A., Silva, W. J., and Kamai, R., 2014. Summary of the ASK14 Ground motion Relation for Active Crustal Regions, Earthquake Spectra 30, 1025–1057.

[11] Campbell KW, Bozorgnia Y. 2014. NGA-West2 Ground Motion Model for the Average Horizontal Components of PGA, PGV, and 5% Damped Linear Acceleration Response Spectra. Earthquake Spectra. 30(3):1087-1115. doi:10.1193/062913EQS175M

[12] Boore, D. M., Stewart, J. P., Seyhan, E., and Atkinson, G. A., 2013. NGA-West2 Equations for Predicting Response Spectral Accelerations for Shallow Crustal Earthquakes, PEER Report No. 2013/05, Pacific Earthquake Engineering Research Center, University of California, Berkeley, CA, 134 pp.

[13] Idriss, I. M., 2014. An NGA-West2 empirical model for estimating the horizontal spectral values generated by shallow crustal earthquakes, Earthquake Spectra 30, 1155–1177.

IOP Conf. Series: Earth and Environmental Science 1117 (2022) 012019 doi:10.1088/1755-1315/1117/1/012019

Comparative Study of Delay Factors In Construction Projects In South Sulawesi (BUMN VS Private)

D Patasik[1], I R Rahim[2], and S Hamzah[3]

[1]Management Construction, Faculty of Engineering, Hasanuddin University,UKI Paulus Makassar

[2]Management Construction// Environmental Engineering, Faculty of Engineering, Hasanuddin University

[3]Management Construction, Faculty of Engineering,Hasanuddin University

degalpatasik@gmail.com

Abstract. *The development of the construction world is currently growing rapidly along with the times, where stakeholders in the construction world are required to be able to complete a project efficiently, but in reality there are still many obstacles that we encounter that result in a work being hampered. Referring to the Government Agency Performance Report (LKIP) of the Office of Highways and Construction Development of South Sulawesi Province (2018-2019) and the Office of Public Works and Spatial Planning of South Sulawesi Province (2020) there are still many jobs that are delayed. The purpose of this research is to analyze the factors that influence the completion of construction projects and to analyze the comparison of factors/variable delays that significantly affect construction project delays in the scope of government (BUMN) and private sector in South Sulawesi as well as alternative handling. The method used is Frequency Index Analysis, Influence Index, Interest Index, and risk analysis. The results of the research show that 27 of the 35 factors distributed are relevant in South Sulawesi and the comparison of the risk factors for delay is dominant. The first rank was for BUMN respondents. There was a design change by the owner (58 %) and the first rank was for Private. Delay in delivery of goods/materials (79.44 %). The alternative on the BUMN side is that there is a design change by the owner is to review and discuss together about the urgency to change items and improve design planning. The alternative for the private sector, namely delays in delivery/material goods, is to ensure that construction management conducts coordination meetings with contractors to evaluate the order mechanism so that it can be carried out early.*

Keywords: *Project Delay, Data Analysis, BUMN and Private, South Sulawesi*

1. Introduction

The development of the construction world is currently growing rapidly along with the times, where stakeholders in the world are required to be able to complete a project efficiently, but in reality there are still many obstacles that we encounter that result in a work being hampered[1].

Referring to the Government Agency Performance Report (LKIP) of the Office of Highways and Construction Development of South Sulawesi Province (2018-2019) and the Office of Public Works and Spatial Planning of South Sulawesi Province (2020) there are still many jobs that are experiencing delays in the scope of South Sulawesi [2] Abdullah's statement; et al. (2010) in Causes of Delay in MARA Management Procurement Construction Projects that over the years the problem of delays in construction projects has slowed down and was so significant that it slowed down implementation planning[3]. The same thing is also stated by Aziz (2013) in the Ranking of Delay Factors in Construction Projects After Egypt Revolution that delay is a frequent phenomenon and is related to almost all construction projects[4].

Construction project delays can basically be calculated but certain conditions that can occur outside the planning calculations can be caused by improper management, material, labor, equipment, and environmental problems that do not support project implementation[5]. Even Assaf and Al-Hejji (2006) in the Causes of Delays in Large Construction Projects say that the causes of delays can be seen in terms of materials, labor, equipment, costs, design changes, relations with related agencies, scheduling and control, supervision procedures, testing processes, environmental issues, contractual issues and the absence of professional consultants[6]. For delays, a special assessment is needed in the project implementation process (Ismail, 2013)[7]. The size of a project does not determine the level of complexity. Small projects can be more complex than larger projects. Complexity requires regulation and control in such a way that there are no problems in project implementation (Soeharto, 1999) [8].

The Problem of the reaserach, What are the factors that affect the completion of the project ?, What is the comparison of delay factors/variables that significantly affect construction project delays in the government (BUMN) and private sector in South Sulawesi? , What alternatives/strategies are there to reduce the impact of construction project delays? [9].

2. Methods and Literature Review

2.1 Location and Research Design

The design of this research is a comparative analysis research, where the subject in this study is a construction project that is currently underway or has been completed between the 2019-2021 timeframe which is reviewed from the BUMN and Private sector in the scope of South Sulawesi

2.2 Risk Management

The definition of risk is the possibility of something happening that will have an impact on the target, measured in terms of consequences and likelihood. What is meant by the consequence is the result of an event that is stated qualitatively or quantitatively, which is a loss, loss, or gain.

2.3 Validity and Reliability Test

Validity test is the accuracy of an instrument in measurement. while the reliability test is used to determine the consistency of the measuring instrument, whether the measuring instrument used is reliable and remains consistent if the measurement is repeated.

2.4 Analysis Frequency Index, Severity Index, and Importance Index

Analysis of the frequency index is known as the Frequency Index (F_I), showing the level of occurrence of factors that affect work performance, while the influence index is known as the Severity Index (S_I), showing the level of impact caused by each risk factor and Analysis of Importance Index (I_I) shows the results of the multiplication between frequency and the influence on the factors causing delays that have the most influence on contractor performance

$$F_I\,(\%) = \sum_{i=1}^{5} = 1\,\frac{(ai * ni)}{n} \times \frac{100}{5} \quad (1)$$

$$S_I(\%) = \sum_{i=1}^{5} = 1\,\frac{(ai * ni)}{n} \times \frac{100}{5} \quad (2)$$

$$I_I\,(\%) = \frac{FI\,(100\%) X SI\,(100\%)}{100} \quad (3)$$

2.5 Risk Analysis

Risk analysis is the development of an understanding of risk (SNI ISO 31000,2011). ISO 31000:2018 states that the purpose of risk analysis is to understand the nature of risk and its characteristics according to the level of risk.

2.6 Type of research

Metathesis study of previous research (national and international journals) as a reference for determining research variables and questionnaire results on respondents. The measurement uses the analysis of the Frequency Index, Influence Index and Interest Index, as well as risk analysis

2.7 Data Source

In this study, the respondents were contractors, both BUMN and private in the scope of South Sulawesi, with specifications for technical personnel in a project, minimum education level.

2.8 Measuring tools

2 stage questionnaire, Stage 1 (pre test) to validate the factors that have been determined based on the results of the metathesis study and stage 2 to determine the dominant factor in each respondent, both overall, BUMN, Private. as a reference in determining mitigation according to the level of occurrence and impact.

2.9 Research hierarchical process

Phase 1 questionnaire (pre-test) was conducted to validate the factors that have been determined based on the results of the metathesis study totaling 35 variables of delay for all respondents (BUMN and Private) then conducting a 2-stage questionnaire totaling 27 variables testing the occurrence and impact of the risk factors for project delays.

Stage 1 Stage II

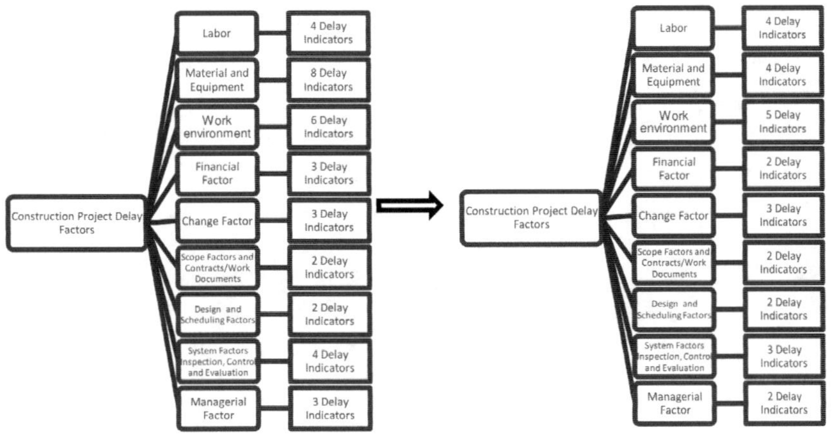

3. Result and Discussion

3.1 Identification of Risk Factors for Phase I Construction Project Delays

The first analysis is the validation of 35 delay factors to all respondents who occur in the implementation of construction projects from hierarchical process stage 1.

Figure 1. Results of Risk Factor Identification in the percentage of respondents' choice of project delays Phase 1 Questionnaire Construction (pre-test)

No	Delay Risk Variable	N	Indicator		P	Information
			Occur	Not Occur		
1	**Labor**					
X1	Unskilled workforce	**21**	**12**	**9**	57%	Relevant Occurring
X2	Shortage of labor	**21**	**14**	**7**	67%	Relevant Occurring
X3	Labor productivity	**21**	**13**	**8**	62%	Relevant Occurring
X4	Communication between the workforce and the chief handyman/foreman	**21**	**13**	**8**	62%	Relevant Occurring
2	**Material and Equipment**					
X5	Delay in delivery of goods/materials	**21**	**17**	**4**	81%	Relevant Occurring
X6	Lack of construction materials	**21**	**13**	**8**	62%	Relevant Occurring
X7	Material quality is not good	**21**	**6**	**15**	29%	Irrelevant
X8	Damage to materials in storage	**21**	**4**	**17**	19%	Irrelevant

X9	Material changes in form, function, and specifications	**21**	**5**	**15**	24%	Irrelevant
X10	Delay in delivery/equipment supply	**21**	**4**	**17**	19%	Irrelevant
X11	Equipment breakdown	**21**	**18**	**3**	86%	Relevant Occurring
X12	Availability of adequate equipment/as needed	**21**	**15**	**6**	71%	Relevant Occurring
3	**Work Environment**					
X13	Poor response from the surrounding community to the project/complaint	**21**	**11**	**10**	52%	Relevant Occurring
X14	Storage area for materials/materials away from the project site	**21**	**4**	**17**	19%	Irrelevant
X15	Access to the project site/ Different natural conditions from the time of survey	**21**	**13**	**8**	62%	Relevant Occurring
X16	Limited scope of work	**21**	**14**	**7**	67%	Relevant Occurring
X17	High rainfall intensity	**21**	**21**	**0**	100 %	Relevant Occurring
X18	The occurrence of unexpected things such as fire, flood, bad weather	**21**	**19**	**2**	90%	Relevant Occurring
4	**Financial Factor**					
X19	Material price	**21**	**17**	**4**	81%	Relevant Occurring
X20	Funding difficulties in contractors	**21**	**19**	**2**	90%	Relevant Occurring
X21	Payment difficulties by owner	**21**	**7**	**14**	33%	Irrelevant
5	**Change Factor**					
X22	There has been a design change by the owner	**21**	**21**	**0**	100%	Relevant Occurring
X23	Incorrect/incomplete planning (drawings/specifications)	**21**	**18**	**3**	86%	Relevant Occurring
X24	Changes in the scope of work at the time of implementation	**21**	**20**	**1**	95%	Relevant Occurring
6	**Scope Factors and Contracts/Work Documents**					
X25	Owner's delay in making decisions	**21**	**20**	**1**	95%	Relevant Occurring
X26	There is a request for changes to the work that has been completed	**21**	**14**	**7**	67%	Relevant Occurring
7	**Design and Scheduling Factors**					
X27	Work order plans that are not well structured/integrated	**21**	**14**	**7**	67%	Relevant Occurring
X28	Differences in the schedule of sub-contractors in project completion	**21**	**21**	**0**	100%	Relevant Occurring
8	**System Factors Inspection, Control and Evaluation**					
X29	The material sample approval process takes a long time by the owner	**21**	**15**	**14**	71%	Relevant Occurring
X30	Delays in the inspection and testing of materials	**21**	**17**	**4**	81%	Relevant Occurring
X31	A lot of work that must be repaired / redone due to defects / not correct	**21**	**16**	**5**	76%	Relevant Occurring
X32	The long process of evaluating work progress	**21**	**8**	**16**	38%	Irrelevant
9	**Managerial Factor**					
X33	Field manager experience	**21**	**8**	**13**	38%	Irrelevant
X34	Poor communication between the owner's representative and the contractor	**21**	**17**	**4**	81%	Relevant Occurring
X35	Poor communication between planners and contractors	**21**	**18**	**3**	86%	Relevant Occurring

IOP Conf. Series: Earth and Environmental Science 1117 (2022) 012019 doi:10.1088/1755-1315/1117/1/012019

3.2 Identification of Risk Factors for Phase 2 Construction Project Delays

Based on the results of the identification of risk of delays on figure 1, from 35 variables of risk of delays in construction projects to 27 variables of risk of delays in construction projects, due to the low level of selectability of risk variables below 40%, 8 variables of delay risk were eliminated based on the results of the phase I questionnaire.

3.3 Validity Test and Reliability Test

Validity test meets The value of r table for n = 21 is 0.4329, most of the calculated r values are greater than r The table of all indicators causes delays so that it can be used as an instrument in the questionnaire and Reliability Test meets because the value of the C-Alpha coefficient is greater than 0 ,60, then all categories of causes of delay are reliable and appropriate to be used as a questionnaire instrumen

3.4 Analysis Results Frequency Index (F_I) , Severity Index (S_I) and Importance Index (I_I)

Figure 2. Comparison of Highest Ranking Frequency Index (F_I)

No	Overall Respondents	BUMN contractor	Private Contractor
1	Communication between the workforce and the chief handyman/foreman	There has been a design change by the owner	Delay in delivery of goods/materials
2	Delay in delivery of goods/materials	Incorrect/incomplete planning (drawings/specifications)	Delay in delivery of goods/materials
3	Lack of construction materials	Changes in the scope of work at the time of implementation	Communication between the workforce and the chief handyman/foreman
4	There has been a design change by the owner	Labor productivity	Equipment breakdown
5	Labor productivity	Communication between the workforce and the chief handyman/foreman	Shortage of labor

Figure 3. Comparison of Highest Ranking Severity Index (S_I)

No	Overall Respondents	BUMN contractor	Private Contractor
1	Lack of construction materials	Funding difficulties in contractors	Lack of construction materials
2	Equipment breakdown	There is a request for changes to the work that has been completed	Delay in delivery of goods/materials
3	Delay in delivery of goods/materials	High rainfall intensity	Equipment breakdown
4	Unskilled workforce	Equipment breakdown	Shortage of labor
5	Shortage of labor	Lack of construction materials	Unskilled workforce

Figure 4. Comparison of Highest Ranking Importance Index (*Ii*)

No	Overall Respondents	BUMN contractor	Private Contractor
1	Lack of construction materials	There has been a design change by the owner	Delay in delivery of goods/materials
2	Delay in delivery of goods/materials	Changes in the scope of work at the time of implementation	Lack of construction materials
3	Communication between the workforce and the chief handyman/foreman	High rainfall intensity	Equipment breakdown
4	There has been a design change by the owner	Incorrect/incomplete planning (drawings/specifications)	Communication between the workforce and the chief handyman/foreman
5	Equipment breakdown	There is a request for changes to the work that has been completed	Shortage of labor

3.5 Alternative to reduce Impact of delay

Determination of alternativ for the dominant category based on the results of respondents' answers in the stage 2 questionnaire

Figure 5. Responses and Actions for the Treatment of Project Delay Risk Variables in South Sulawesi all respondents (top 5)

No	Delay Risk Variable	Risk Response Category	Risk Response and Action Risk Treatment Based on Questionnaire Results
1	**Labor**		
X4	Communication between the workforce and the chief handyman/foreman	*Avoidance*	Coordination meeting related to manpower, tools, working time
2	**Material and Equipment**		
X5	Delay in delivery of goods/materials	*Avoidance*	Ensure that construction management conducts coordination meetings with contractors to evaluate the order mechanism so that it is carried out early
X6	Lack of construction materials	*Avoidance*	Better evaluation of order mechanism and calculation of material requirements
X7	Equipment breakdown	*Avoidance*	Routine maintenance is carried out on equipment such as an excavator when the work is carried out when the equipment is damaged, the work is forced to wait for the equipment to be repaired
5	**Change Factor**		
X16	There has been a design change by the owner	*Avoidance*	Review and discuss together about the urgency to change items and improve design planning

4. Conclusion

1. From the results of this study, we can conclude that the dominant influence of the delay factor in construction projects on the BUMN side is more on the delay factor which is difficult to accommodate while on the private contractor side it is more on the delay factor that can be accommodated. Overall, respondents have 5 dominant risk factors for delays, among others, Lack of Construction Materials (56.67%). Delay in delivery/material goods (55,62 %) , Communication between the workforce and the chief handyman/ foreman (50,34 %) , There was a design change by the owner (50,23 %) , Equipment damage (49,34 %). In BUMN respondents, 5 dominant risk factors for delays were, among others, there was a design change by the owner (58%), Changes in the scope of work at the time of implementation (50.38%), Rain intensity (49.50%), Planning (drawing/specifications)) incorrect/incomplete (48.84%) , There is a request for changes to the work that has been completed (48.88%) while in the Private Contractor respondents 5 dominant risk factors for delays are, among others, delays in delivery of goods/materials (79, 44 %) , Lack of Construction Materials (79.33 %) , Equipment Damage (65 %) , Communication between the workforce and the head of the foreman (63.89 %). Lack of manpower (63.75%). Alternatives/strategies to minimize the impact caused by the risk of delays in construction projects are obtained after conducting risk mapping based on risk analysis at all levels of respondents, both overall respondents, BUMN respondents, and private respondents. Next, evaluate the risk based on the probability and consequence values for the occurrence of problems in the construction project so that the predicate/category of each variable risk of delay (Retentition, Reduction, Transfer, Avoidance) is found. In the final stage, the risk response is carried out for each predicate/category at each level of the respondent.

Bibliography

[1] Abdullah, M. R.; Rahman, I. A.; Azis, A. A. A. (2010), "*Causes of Delay in MARA Management Procurement Construction Projects*", Journal of Surveying, Construction & Property.

[2] Durdyev, S. ; Hoesseini, R. M. (2019), "*Causes Of Delays On Construction Projects: A Comprehensive List*" , Department of Engineering and Architectural Studies, Ara Institute of Canterbury, Christchurch, New Zealand, and School of Architecture and Built Environment, Deakin University, Geelong, Australia

[3] Hammadi, A. ; Nawab, S. M. (2016), "Study Of Delay Factors In Construction Projects", International Advanced Research Journal in Science, Engineering and Technology Vol. 3, Issue 4, April 2016

[4] ISO 31000. 2018. "Risk Management – Guildelines (ISO 31000:2018)". BSI Standards Limited 2018. Switzerland

[5] Kokou, R. ; Akogbe, M. T. K. ; Zho, J. (2013), "Importance And Ranking Evaluation Of Delay Factors For Development Construction Projects In Benin", Faculty of Infrastructure Engineering, Dalian University of Technology, Dalian, China

[6] Megha D.; Rajiv, Bhatt. (2013), "Causes Of Delay For Residential Construction Projects In Indian Context"; Institute of Technology, Gujarat, India

[7] Modares, T. (2015), "The Importance Of New Technology For Delay Mitigation In Construction Projects" , Sharif University of Technology, Tehran, Iran.

[8] Orangi, A.; Palaneeswaran.; Wilson J. (2011), "Exploring Delays In Victoria-Based Australian Pipeline Project", Faculty of Engineering and Industrial Sciences, Swinburne University of Technology, Australian.

Evaluating the Non-Erodible Cement-Bentonite Barrier Material Samples via Hole Erosion Test Apparatus

M A Walenna[1,2], A Royal[2], I Jefferson[2] and G Ghataora[2]

[1] The Graduate School, Hasanuddin University, Jalan Perintis Kemerdekaan Km. 10, Makassar 90245, Indonesia
[2] School of Engineering, University of Birmingham, Edgbaston, Birmingham B15 2TT, UK

mawalenna@unhas.ac.id

Abstract. Because of the damage that dams, levees, and other water-retaining structures cause, their failure has long been a source of concern. Despite the limitations of the testing apparatus (e.g., HET) used to study barrier material erodibility, they are still widely used. In previous studies, there was very little discussion of the hydraulic parameters (e.g., fluid pressure, flow rate, etc.) obtained from the HET test, particularly how they relate to the varying eroded hole. Therefore, the purpose of this paper was to look into that aspect of the relationship as well as the practical aspects of the HET apparatus that must be considered. Fluctuations in pressure and flow rate data collected during testing were the primary issue discovered. The noise from the pump machinery caused the data collected to fluctuate within a specific range, but the fluctuations were controlled and consistent. Throughout the test, the pressure recorded fluctuated ± 10 kPa below the target pressure, while the flow rate fluctuated approximately 0.3 L/min. The test was performed on a smooth-surfaced, uniform-diameter hole; however, this is not always the case for holes in eroded samples. The non-uniform surface can affect the friction between the eroded surface and the eroding fluid, which in turn affects the pressure loss across the erosion pipe. A non-invasive measurement method for determining the diameter of an eroded hole is suggested for future research, and this study provided an empirical relationship between pressure difference, flow rate, and eroded hole diameter. Despite the limitations, the results provided an overview of the expected fluctuation range from the apparatus as well as the relationship between hydraulic parameters.

1. Introduction

The failure of dams, levees, and other water-retaining structures has long been a source of concern due to the damage they cause [1][2]. The Teton Dam failure killed 11 people and caused $400 million in property damage [3]. Although vertical barriers such as the Cement-Bentonite Barrier (CBB) help to protect dams from internal erosion, previous research has shown that they are still vulnerable to failure due to crack formation and erosion [4]. This emphasizes the importance of investigating CBB's susceptibility to erosion and design considerations to ensure its performance to provide increased protection. Walenna et al. [5] conducted preliminary research into the erodibility behavior of CBB material using the Hole Erosion Test (HET) apparatus. It is worthwhile to look into the HET apparatus used in the previous study in order to gain a better understanding of the method for investigating the erodibility of barrier material.

2. Literature Review

The common laboratory-based tests for internal erosion are usually based on the Pinhole Erosion Test [6], the Hole Erosion Test [7], the Jet Erosion Tests [8] [9] and the Rotating Cylinder Test [10].

Both PET and HET simulate the erosion occurring due to fluid flow in a concentrated leak in the soil sample. If the objective is to investigate the hydraulic condition required to initiate the erosion on the sample with high resistance to erosion, PET is not suitable as the applied hydraulic heads at upstream are limited by the maximum height of standpipe; ranging from 50 mm to 1080 mm (e.g. [6][11]). Therefore, PET is only recommended to be used for dispersive material. Unlike PET, HET is more quantitative in investigating the erosion phenomenon. It is clear from the findings of the previous research that both PET and HET tests are limited in their scope, that perhaps they should be considered more akin to an index test rather than those investigating the inherent response of the materials under in situ conditions.

Marot et al. [12] suggested that the hydraulic head difference obtained from HET is more dependent on the energy loss due to change of section, approximately up to 75%, between upstream and/or downstream section to hole of sample (i.e. contraction and expansion) than on the friction loss due to eroding fluid-sample contact.

Despite these limitations, HET and PET are still widely used in estimating geomaterial erodibility. The discussion of the hydraulic parameters (e.g., fluid pressure, flow rate, etc.) obtained from the HET test was very minimum in the past studies, particularly how they relate to the varying eroded hole. Therefore, this paper attempted to investigate that aspect of the relationship as well as the practical aspects of the apparatus that must be considered.

3. Methodology

3.1. Sample Preparation

3.1.1. Material Preparation

Both bentonite and cementitious materials used (cement and GGBS) should be in a dry state. The bags used to store those materials were sealed to ensure that they were not affected by the laboratory environment's changing temperature and humidity levels.

The distilled water was sourced from the laboratory's filtered machine, which maintains its chemical consistency.

3.1.2. Mix Proportion

The mix was comprised of bentonite, cement, GGBS, and water. The following is the mix proportion used:

Table 1. Mix Proportion

Sample Type	GGBS Replacement (%)	Sample Composition			
		Water (mL)	Bentonite (gr)	Cement (gr)	GGBS (gr)
CB with GGBS	80%	1000	40	40	160

Because the goal was to create non-erodible samples, it was desired that the samples could withstand the hydraulic energy exerted by the eroding fluid and that no part of the sample eroded during the testing. As a result, the GGBS replacement of 80% (20% of total cementitious material) was chosen, as recommended by Opdyke and Evans [13] and Royal et al. [14], because it provided the greatest strength improvement.

3.1.3. Procedure

The sample production procedure used in this study is largely based on previous similar studies (e.g., [5][14]). The procedure is as follows:

- Mixing the dry bentonite with water for approximately 25 minutes and the bentonite slurry was left overnight (for 20-24 hours) at room temperature.
- For about 5 minutes, the bentonite slurry was mixed with cementitious material (cement and GGBS at the same time). The mixed cement-bentonite slurry was then transferred to individual molds, each with an acrylic rod in the center to create an eroded hole in the sample. The acrylic rod's diameter ranges from 3, 4, 5, 6, 8, 10, and 12 mm..
- The samples were vibrated for 10-15 minutes to allow entrapped air in the slurry to escape.
- All samples were cured in molds at 5 °C for 56 days, to simulate the average ground temperature in the UK (BGA, 2019).
- After the samples have completed the curing period, the mold cap and rod were removed. The samples were trimmed to the desired length of 100 mm (as seen in Figure 1).

Figure 1. CBB samples (with 3 mm hole diameter) after completing the curing period

3.2. Sample Testing

The non-erodible samples were tested via the Hole Erosion Test apparatus [7] (the schematic diagram of the apparatus is illustrated in Figure 2). The apparatus's main principle is to allow eroding fluid to flow from the upstream to the downstream side of the sample. When soil erosion has not yet occurred, there is a difference in pressure recorded at the upstream and downstream sides (pressure of the eroding fluid, at the downstream, dropped). When the pressure upstream is increased to the point where the erosion process begins, the differential pressure begins to decrease and continues to decrease as the hole enlarges due to the erosion that causes soil detachment. When used for erodible samples, the eroded soil should exit on the downstream side.

The testing procedure aims to ensure that both the main parameters of the eroding fluid were obtained, as this study focused on investigating the relationship between pressures, at the upstream and downstream sides, and flowrate of the eroding fluid; and how they could be correlated with the varying hole diameter. This testing was conducted in the following manner:

- The apparatus system was filled with water from the system's directly connected pump.
- The downstream valve was slowly opened, allowing the eroding fluid to flow from the upstream, and they were kept at low pressure and flow rate to ensure no particle detachment occurred. Both the flowmeter and the pressure transducer should be collecting data from the system by now.
- The test was initiated by gradually opening the upstream gate valve, and both the flow and pressure measurement tools were set to record data.

- The eroding fluid's pressure difference between the upstream and downstream sides was gradually increased in predetermined increments (target pressure). The maximum pressure could be generated by the system on the upstream side was approximately 350 kPa.
- Because the measurement tools used have a maximum data collection time of 10 minutes, the testing should be completed before that time.
- The test was repeated 2-3 times to ensure that the results obtained were reproducible.

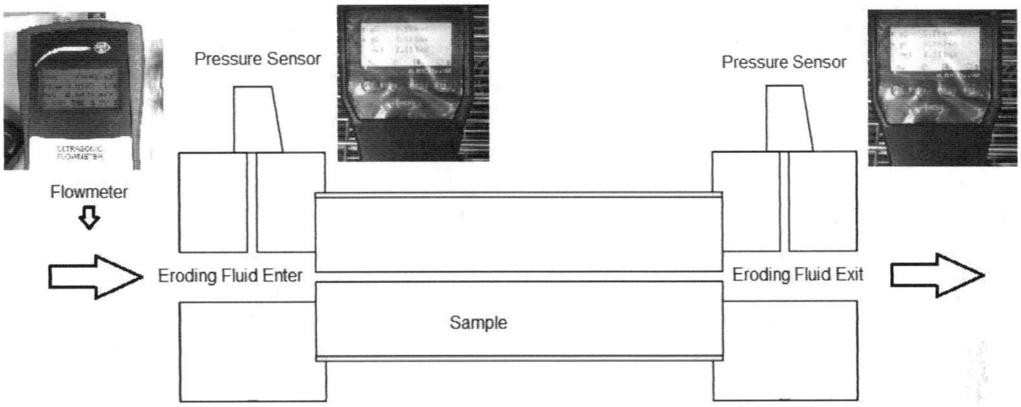

Figure 2. HET Apparatus Schematic Diagram

4. Results and Discussion

There were approximately 21 non-erodible CBB samples tested for this study. The fluctuations in pressure and flow rate data collected during testing were the primary issue discovered. This was primarily due to pump machinery noise, which caused the data collected to fluctuate within a certain range, but the fluctuations were controlled and consistent (as seen in Table 2). The pressure recorded during the test consistently fluctuate approximately below 10 kPa of the target pressure, while the flow rate fluctuated below 0.3 L/min throughout the course of the test.

Table 2. Average Fluctuation of Non-Eroding Fluid

Diameter (mm)	Pressure (kPa)		Flowrate (L/min)	
	Avg. Max	Avg. Min	Avg. Max	Avg. Min
3	9.3	-10.2	0.14	-0.15
4	6.95	-8.6	0.14	-0.14
5	7.59	-7.24	0.16	-0.16
6	7.1	-6.74	0.23	-0.2
8	6.7	-6.85	0.2	-0.24
10	6.99	-8.23	0.25	-0.29
12	8.14	-9.58	0.28	-0.26

4.1. Fluid Pressure Fluctuation

The pressure fluctuation of the eroding fluid slightly increased as the target pressure was increased (as seen in Figure 3).

The largest range of pressure reading fluctuation was recorded at the test of 3 mm hole, ranging from +9.3 to -10.16 on average. The rest were in the 6.95 to 8.14 range for the upper threshold and the -6.74 to -9.58 range for the lower threshold.

While it was not particularly evident, the bigger the eroded hole, the bigger the fluctuations. This suggests that the magnitude of fluctuations would affect the accuracy of the HET especially when the projected target pressure is bigger.

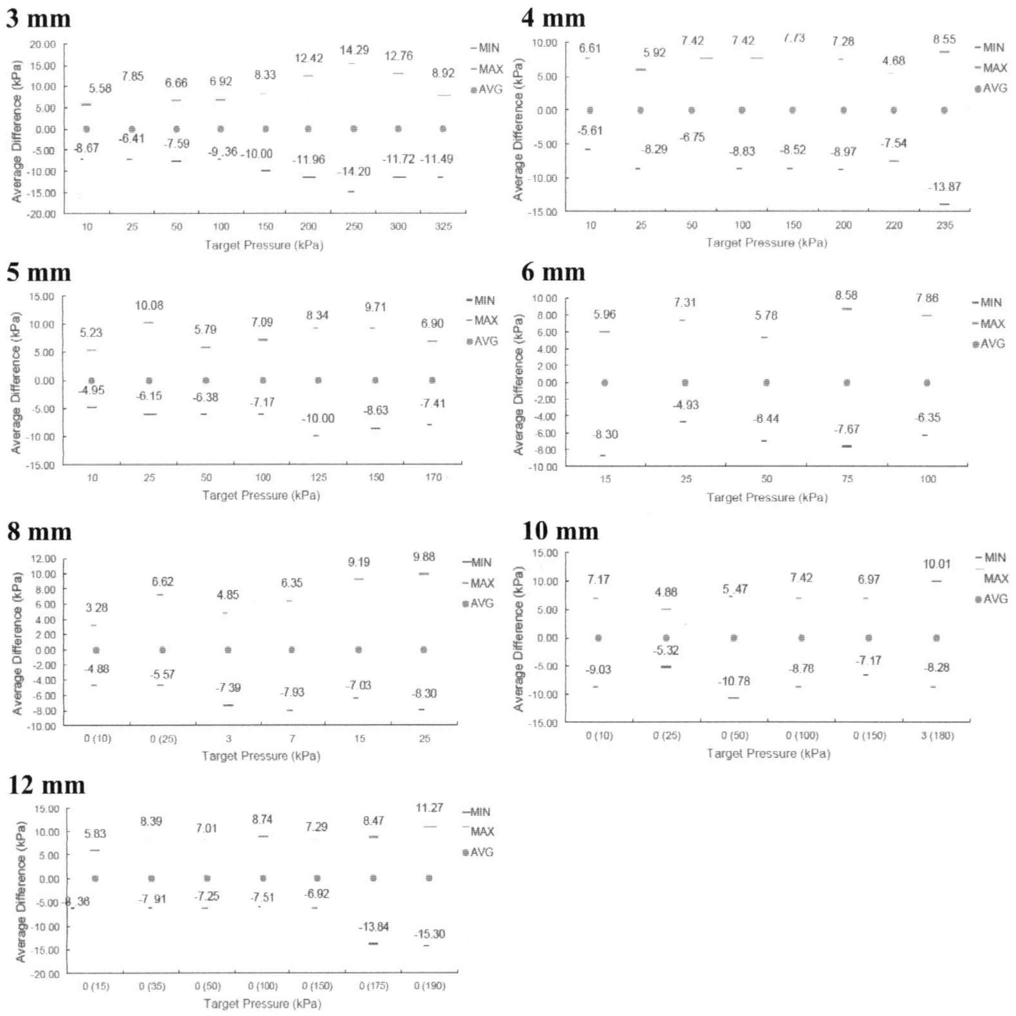

Figure 3. Eroding Fluid Pressure Fluctuation

4.2. Fluid Flow Rate Fluctuation

The fluctuations in the flow rate, like the fluctuations in the fluid pressure, increased as the target pressure increased, and the eroded hole grew larger (as seen in Figure 4). For 3 mm to 12 mm holes, the average upper limit range increased from 0.14 to 0.28, while the bottom limit expanded from -0.15 to -0.26. The increase in fluctuations was mainly linked to the increase in eroded hole size.

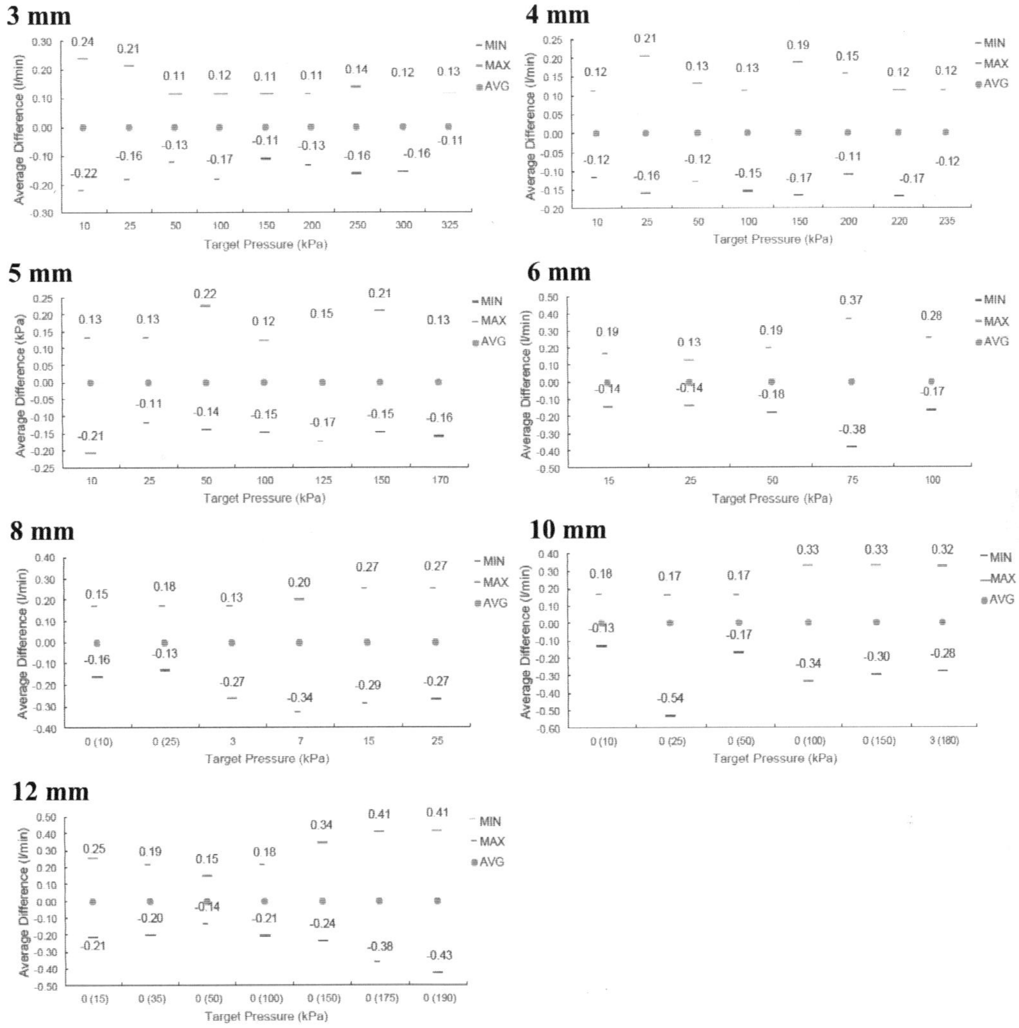

Figure 4. Eroding Fluid Flow Rate

4.3. The Relationship between Fluid Pressure and Flowrate

Figure 5-6 depicts the relationship between the Pressure Differential and the Flowrate. The measured flow rate from the flowmeter was correlated with the maximum pressure at which the pump capacity was set to the maximum for each diameter.

Due to pump noise, some differential pressures measured at 10 and 12 mm fell below zero. They did not indicate a negative pressure differential pressure, but it did imply that the pressures upstream and downstream were equal or that no pressure was lost across the pipe. The recorded pressure decreased as the hole diameter increased, while the flow rate increased in the opposite direction. This pattern is consistent with Bernoulli's principle.

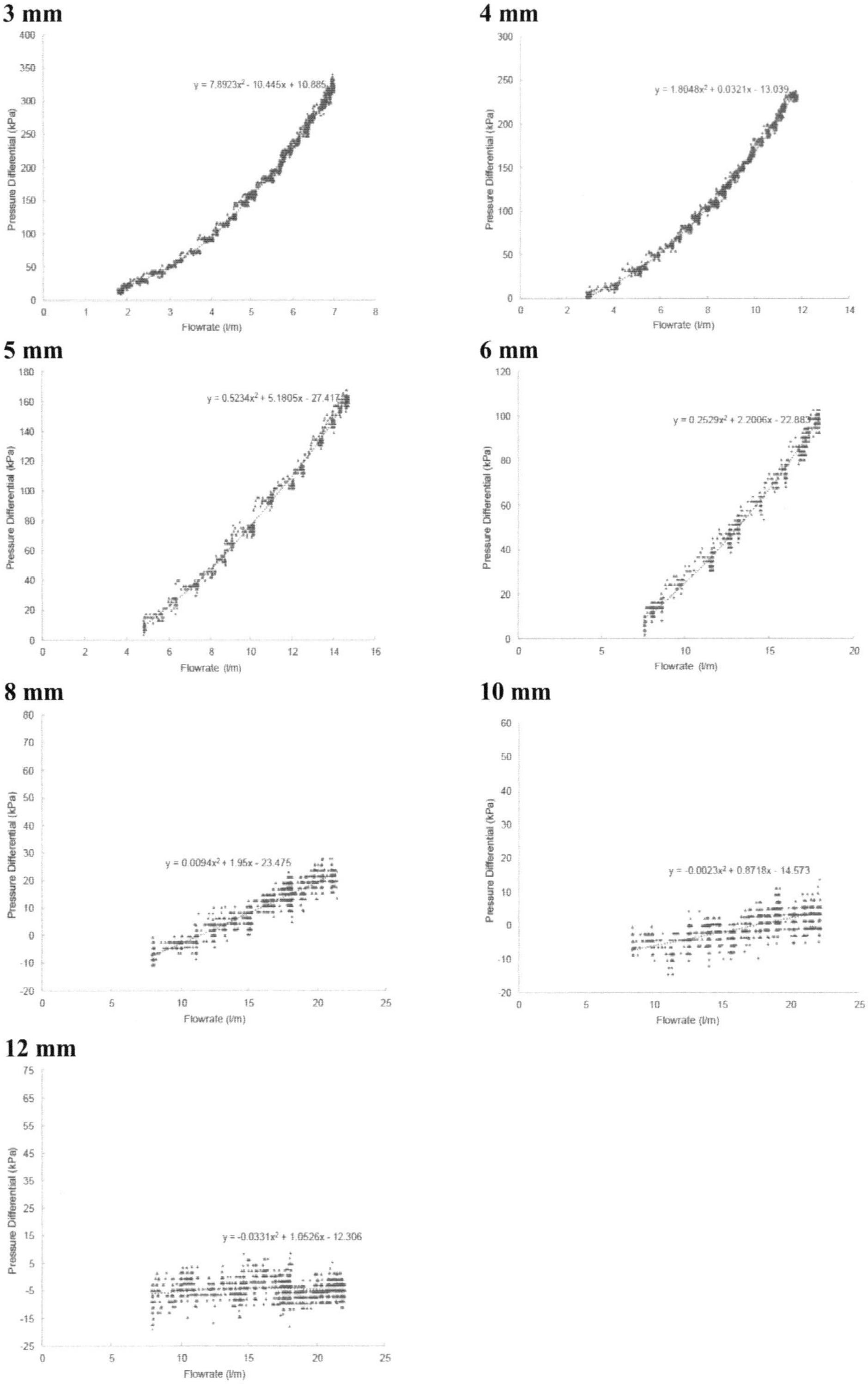

Figure 5. The Relationship between Eroding Fluid's Pressure and Flow Rate

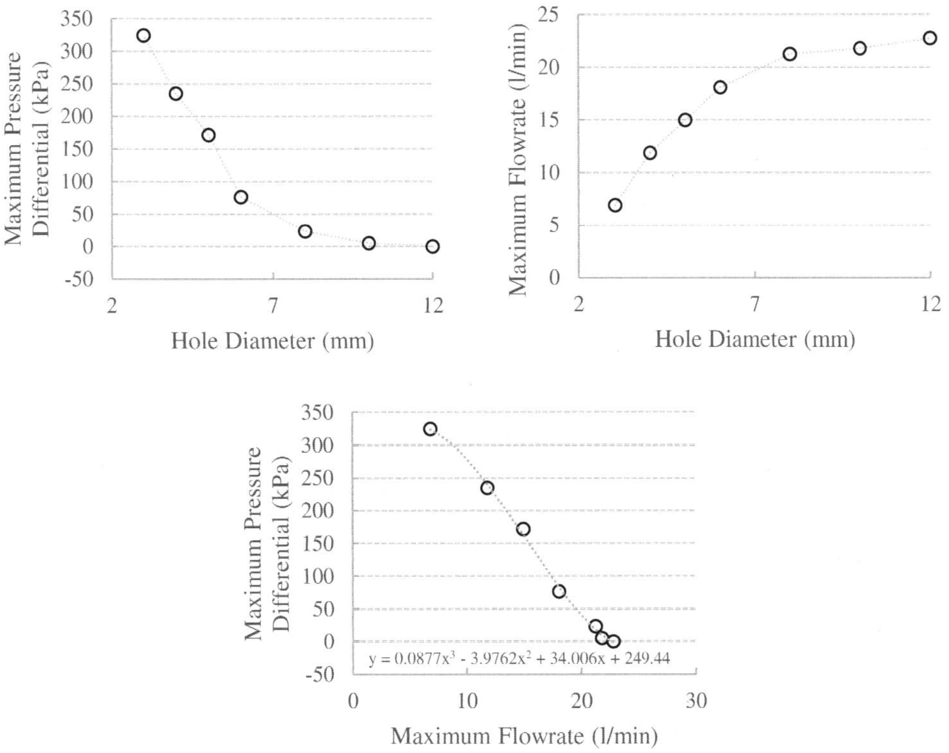

Figure 6. Maximum Pressure Differential and Flowrate

The test was performed on a hole with a smooth surface and uniform diameter; however, this is not always the case for eroded samples' holes. Because of the non-uniform response of the impacted surface to erosive force, which impacted the uniformity of the eroded diameter and shape, the eroded hole can vary in terms of surface roughness and symmetry. The non-uniform surface can influence the friction between the eroded surface and the eroding fluid, affecting the pressure loss across the pipe.

The technical difficulty encountered in producing the more representative hole was the method for producing samples with the desired hole diameter (i.e. non-uniform surface). Melted paraffin wax was poured into the hole to estimate its diameter in this study. The sample could not be used for the test again because it had to be split in half to retrieve the hardened wax. This suggested that the use of a non-invasive measurement method to determine the diameter of an eroded hole be investigated, and this study provided an empirical relationship between the pressure difference, flow rate, and eroded hole diameter. Despite the limitations, the results provided an overview of the range of data noise to be expected from the apparatus and the relationship between hydraulic parameters.

5. Conclusion

Investigating geomaterial erodibility has become a critical part of ensuring the long-term safety and reliability of dams or levees, especially when that infrastructure is expected to withstand a significant hydraulic load. Laboratory research on the erodibility of geomaterials has progressed [5][15], but the practical aspects of the apparatus have not been adequately discussed. With its limitations, this study attempted to contribute to the discussion and proposed the following:

- The primary issue discovered was fluctuations in pressure and flow rate data collected during testing. This was primarily due to noise from the pump machinery, which caused the data collected to fluctuate within a specific range, but the fluctuations were controlled and consistent. The pressure recorded during the test fluctuated around 10 kPa below the target pressure, while the flow rate fluctuated around 0.3 L/min throughout the test. It is suggested that the source of eroding fluid has a mechanism to reduce vibration caused by pump operation.

- The largest range of pressure reading fluctuation was recorded at the test of 3 mm hole, ranging from +9.3 to -10.16 on average. The rest were in the 6.95 to 8.14 range for the upper threshold and the -6.74 to -9.58 range for the lower threshold. The erratic flow rate, like the erratic fluid pressure, increased as the target pressure increased, and the eroded hole grew larger (as seen in Figure). The average upper limit range for 3 mm to 12 mm holes increased from 0.14 to 0.28, while the bottom limit increased from -0.15 to -0.26.
- The test was carried out on a hole with a smooth surface and uniform diameter; however, this is not always the case for holes in eroded samples. The non-uniform surface can influence the friction between the eroded surface and the eroding fluid, affecting the pressure loss across the pipe.
- The use of a non-invasive measurement method to determine the diameter of an eroded hole is suggested to be explored for future study, and this study provided an empirical relationship between pressure difference, flow rate, and eroded hole diameter. Despite the limitations, the results provided an overview of the data noise range to be expected from the apparatus as well as the relationship between hydraulic parameters.

References

[1] Bonelli S and Benahmed N 2011 Piping flow erosion in water retaining structures *Int. J. Hyd. Dams* 18 **3** 94-98
[2] Eddleston M. and Wan C F 2014 Use of basic soil test data in internal erosion risk assessments *Dams & Res.* 24 **1** 26-39
[3] USBR 2016 Teton Dam History and Facts *U.S. Dept. of the Inter. B. of Rec.*
[4] Rice J D and Duncan J M 2010 Deformation and cracking of seepage barriers in dams due to changes in the pore pressure regime *J. Geo. Geoenv. Eng.* 136 **1** 16-25
[5] Walenna M A, Royal A C D, Jefferson I, and Ghataora G S 2019 Investigating the erosion resistance of cement-bentonite barrier material using hole erosion test *Proc. of the 9th Int. Conf. Sco. & Er.* 103-109
[6] Sherard J L, Dunningan L, Decker R S, and Steele F 1976 Pinhole test for identifying dispersive soils *J. Geotech. Eng. Div.* **102** 69–85
[7] Wan C F and Fell R 2002 Investigation of internal erosion and piping of soils in embankment dams by the slot erosion test and the hole erosion test *UNICIV Report No R-412 UNSW, Sydney*
[8] Hanson G J 1991 Development of a jet index to characterize erosion resistance of soils in earthen spillways *Tran. of the ASAE* 34 **5** 2015-2020
[9] Henensal P and Duchatel F 1990 L'érodimètre à jets mobiles Bull. Lab. Ponts et Chaussées **167** 47-53
[10] Lim S S 2006 Experimental investigation of erosion in variably saturated clay soil *Unpublished PhD Dissertation UNSW Sydney*
[11] Fauzilah I, Zainab M, and Mukri M 2008 A study of the mechanism of internal erosion resistance to soil slope instability. *Elec. J. of Geotech. Eng.* **13** A 12
[12] Marot D, Reggazoni P, and Wahl T 2011 An Energy-Based Method for Providing Soil Surface Erodibility Rankings *J. Geo. Geoenv. Eng* 137 **12** 1290-1293
[13] Opdyke S M and Evans J C 2005 Slag-cement-bentonite slurry walls *J. Geo. Geoenv. Eng* 131 673-681

[14] Royal A C D, Opukumo A W, Qadr C S, Perkins L M, and Walenna M A 2018 Deformation and Compression Behaviour of a Cement-Bentonite Slurry for Groundwater Control Applications. *Geotech. Geol. Eng.* **36** 835-853

[15] Haghighi I, Chevalier C, Duc M, Guedon S, and Reiffstech P 2013 Improvement of Hole Erosion Test and Results on Reference Soils *J. Geo. Geoenv. Eng* 139 **2** 330-339

Improvement of Peat Soil Settlement Parameter Using Soybean Crude Urease Calcite Precipitation (SCU-CP) Method

M Yuza Augusti[1], H Putra[1], I Yudhistira[1]

[1] Department of Civil and Environmental Engineering, IPB University, Dramaga, Bogor, 16680, Indonesia

Correspondence author's : heriansyahptr@apps.ipb.ac.id

Abstract. The characteristics of peat soil with high organic and moisture content, also high compressibility can cause structural failure due to excessive settlement. Improvement of peat soil settlement parameters can be conducted using calcite precipitation method with soybean as bio-catalyst, namely Soybean Crude Urease-Calcite Precipitation (SCU-CP). SCU-CP is one of the grouting methods to improve soil consistency so that the bearing capacity of the soil increases. The purpose of this study was to analyze the increase in the parameters of the settlement parameter (R_{90}) and the time at 90% degree of compression (t_{90}). This study was conducted using a density of 90% Proctor standard compaction and the variation of untreated and treated samples. The optimum combination SCU-CP solution was used in the treated samples with a concentration of reagents (Urea and $CaCl_2$) of 1 mol/L and soybean concentration of 20 g/L. The results proved that using the optimum combination of SCU-CP method, it could significantly improved soil settlement parameter of R_{90} and t_{90}. The R_{90} values for untreated samples (UT) at each pressure of 2.0 kg/cm^2, 4.0 kg/cm^2, and 8.0 kg/cm^2 obtained 39 mm, 86.7 mm, and 162 mm respectively. The improvement on R_{90} values of treated samples (TE) obtained 16.4 mm, 28.1 mm, and 57.5 mm respectively after being treated with the SCU-CP method. In the treated samples (TE) the t_{90} value is lower than untreated samples which indicates that the time to reach 90% consolidation is getting faster.

1. Introduction

Peat soil is an organic soil that has a low bearing capacity, so that if it is used as the basis for building foundations, it will be a problem if soil stabilization is not carried out. Yulianto and Mochtar [1] stated that peat soil has poor physical properties, with high moisture content (500% - 900%), large void ratio (7 - 15) and low specific gravity (1.2 – 1.6), and low bearing capacity (5 - 7 kPa) and has uneven and large compression. Peat soil with high organic and moisture content so that it has a high level of compressibility and has a significant impact on settlement due to the secondary consolidation process [2]. Based on Kazemian *et al.* [3], high organic content can be an indicator of a high level of development and soil compatibility. As per Ali et al. [4], the highest peat soil moisture content can reach > 1000% depending on the composition, and peat soil has an initial void ratio of 5-15 so that the compression is high, and has a low bearing capacity with a value of 5-20 kPa. Pokharel and Siddiqua [5] stated, in dealing with development problems on peat soils are generally carried out using the preloading method and replacement of organic soil with granular material so that it will take a long time and high construction costs. The peat soil stabilization method using the preloading and surcharge method, which is applied to peat soil with a layer thickness of < 3 meters, can cause a very long settlement and allow carbon release from a decrease in the water content of the peat layer at the bottom under the embankment [6]. Based on these explanation, peat soil improvement can be done chemically by using lime stabilization [7]. One type of lime stabilization method that allows it to be sustainable and environmentally friendly is the precipitation of calcium carbonate ($CaCO_3$).

Precipitation of calcium carbonate ($CaCO_3$), especially in the form of calcite, can reduce permeability, reduce compressibility, and increase soil shear strength [8]. Calcite precipitation has been carried out by using bacterial activity, namely micro-crobial-induced calcite precipitation (MICP) and has been shown to increase the strength and compressibility of peat soils [9]. However, this MICP method has a fairly high level of complexity, so it has a high failure rate in its use. The usage of bacteria presents difficulties because it depends on factors such as temperature, pH, and the size of the bacteria [10]. As an alternative, calcite precipitation methods that do not use bacteria, such as Enzyme-Mediated Calcite Precipitation (EMCP) can be used. This method is done by using the enzyme urease to break down urea into carbonate ions, which will react with calcite ions in the formation of calcite [11]. Putra *et al.*[12] stated that the urease enzyme serves as a biocatalyst in the soil improvement process known as the calcite precipitation method, which hydrolyzes urea into ammonium ions and carbonate ions in the presence of calcium ions to produce calcium carbonate, also known as calcite. According to a research done by Putra *et al.* [13], strength enhancement evaluation of sandy soil using the EMCP method with the Unconfined Compression Strength (UCS) test obtained a range of values of 0.2 - 1.6 Mpa. The EMCP method, when applied on a large scale, will require injection in large volumes where the use of pure urease enzyme is less economical because the price of this enzyme can reach > 90% of the price of other materials [14]. Substitute for urease enzyme as a catalyst in the calcite precipitation method that has the potential without reducing the effectiveness of the EMCP method is the use of soybeans, namely Soybean Crude Urease Calcite Precipitation (SCU-CP). As per [13] Soybeans can be utilized as an alternate source of possible urease, with an enzyme reaction rate of up to 104 U/g. Previous research has been carried out regarding the replacement of the urease enzyme with soybean powder as a biocatalyst in calcite precipitation method [15–17]. Evaluation of increasing soil strength with UCS test using soybean as biocatalyst obtained a range of values of 65-870 kPa [16]. Based on a study conducted by Pratama *et al.* [18], regarding the application of the calcite precipitation method to an increase in the shear strength of peat soil, it resulted in an increase in the parameters of the peat soil shear strength in the treated sample with an increase in the cohesion value of 49 kPa and the internal friction angle within 34° to the control sample at a curing time of 14 days.

Based on the explanation of the bearing capacity of peat soil and several studies that have been carried out, grouting using SCU-CP solution will be used to improve the parameters of peat soil settlement. This study aims to analyze the effect of the SCU-CP method on the consolidation parameters, the settlement (R_{90}) and the compression rate at 90% (t_{90}) based on primary consolidation tests. The high initial permeability of peat soils causes the initial compression to occur very quickly. Primary compression occurs after the peat soil is loaded. Many studies related to peat soil compression have been carried out, including by Johari *et al.* [19] who conducted research on the compressibility behavior of peat soil using the log-time method regarding fiber effects on compressibility of Peat. In this study, peat soil compaction was carried out using the square root of time method (Taylor) under 90% density of Proctor standard compaction.

2. Materials and Method

2.1. Materials

The materials used in conducting the research consist of; peat soil passed through sieve no.16, reagents in the form of Urea ($CO(NH_2)_2$) with laboratory grade specifications (ACS, Reng. Ph Eur), and calcium chloride ($CaCl_2$) dihydrate laboratory grade specifications (ACS, Reag. Ph Eur), soybean powder with 100% purity Food Grade (Gasol) specifications, and distilled water as a solvent in SCU-CP solutions. Calcite formation reaction on SCU-CP method is shown in equation (1) – (3) and the materials used in this study are given in **Figure 1**.

$$CO(NH_2)^2 + 2H_2O \rightarrow 2NH_4^+ + CO_3^{2-} \quad (1)$$

$$CaCl_2 \rightarrow Ca^{2+} + 2Cl^- \quad (2)$$

$$Ca^{2+} + CO_3^{2-} \rightarrow CaCO_3 \downarrow \text{(precipitated)} \quad (3)$$

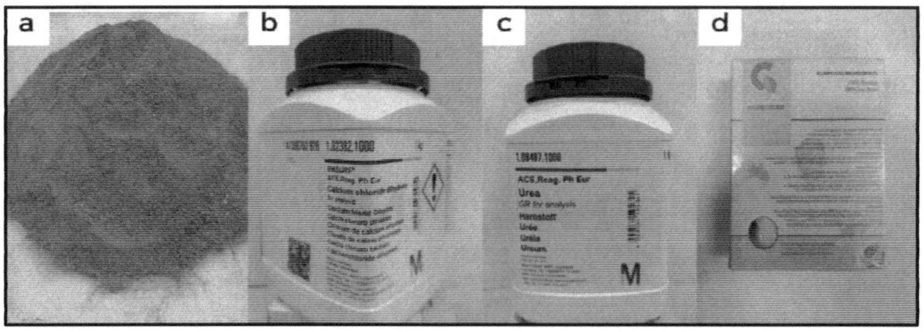

Figure 1. Peat soil *(sapric)* (b) Calcium chloride dihydrate (c) Urea (d) Soybean (Gasol)

Based on the results of the properties test of the peat soil sample, the ash and organic content of the peat soil sample was obtained at 83.41% and 16.55%, so based on ASTM D4427 [20], the peat soil sample was included in the High Ash classification (ash content >15%) and sapric peat (organic content <33%). The moisture content of the peat soil sample is 35.86%, so based on ASTM D4427 [20], it is classified as peat with a low absorption rate or slightly absorbent. The specific gravity (Gs) of the peat soil sample is 1.53, so based on Terzaghi [21], the soil is classified as organic or peat. Based on the test results, a low specific gravity value of peat soil will affect the bearing capacity based on the statement of Islam and Hashim [22] , where the low specific gravity (Gs) (1.2 – 1.6) will affect the compressibility of peat soil. Based on the liquid limit test (LL), the peat soil sample has an LL value of 69.58%, where the LL of the peat soil is directly proportional to the organic content. High water absorption in peat soil will cause a tendency to increase LL, but based on the results, the organic content was obtained <33% so the LL value was relatively lower than the LL value of peat soil in general (>100%) [23]. The condition of the gradation of the peat soil sample based on the test using a sieve shaker is given in **Figure 2.** Based on **Figure 2**, the condition of the soil sample grains is analyzed based on the uniformity coefficient (Cu) and the gradation coefficient (Cc) of D_{10}, D_{30}, and D_{60}, so that the Cu and Cc values are 5.45 and 0.80. According to Putra [24], a well-graded soil has a Cu value >15, so the peat soil sample has poor grain conditions (poor graded). Details of the properties of the peat soil sample are given in **Table 1.**

Table 1. Physical propesties of peat used in this study

Soil Properties	Value	Unit
Specific Gravity, Gs	1.53	-
Field Moisture Content	35.86	%
Organic Content	16.55	%
Ash Content	83.41	%
Liquid Limit, LL	69.58	%
Uniformity Coefficient, Cu	5.45	-
Curvature Coefficient, Cc	0.8	-
Soil Type	Peat	-

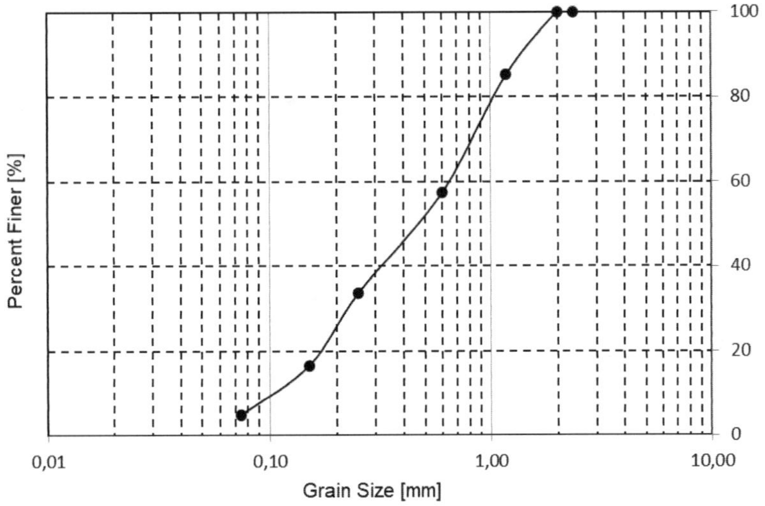

Figure 2. Grain-size distribution of peat used in this study

2.2. Preparation of SCU-CP Optimum Grouting Solution

The preparation of the SCU-CP solution refers to the research conducted by Putra *et al.* [25] using the optimum combination results from the precipitation test results on **Figure 3**. According to **Figure 3**, there is an increase in the precipitation ratio (%) on using soybean concentration as a catalyst in the range of 5 – 20 g/L. The results of the precipitation ratio reached a stable condition with the use of soybean concentrations in the range of 20 – 40 g/L, and decreased at the use of soybean concentration of 40 – 50 g/L for each variation of the reagent concentration (R) 0.5 – 1.5 mol/L. The results of the test precipitation show that the optimum combination was shown in the result of the highest precipitation ratio, by using the reagent (R) concentration (Urea and $CaCl_2$) of 1 mol/L and soybean concentration of 20 g/L. The preparation of the SCU-CP solution was carried out based on the results of the Proctor density test which was taken at 90% density of the maximum dry density value. The solution was made by stirring using a hot-plate stirrer for 5 minutes, and extracting the soybean using a sieve no. 200 then the three solutions were mixed and stirred. The schematic in making the SCU-CP solution is given in **Figure 4.**

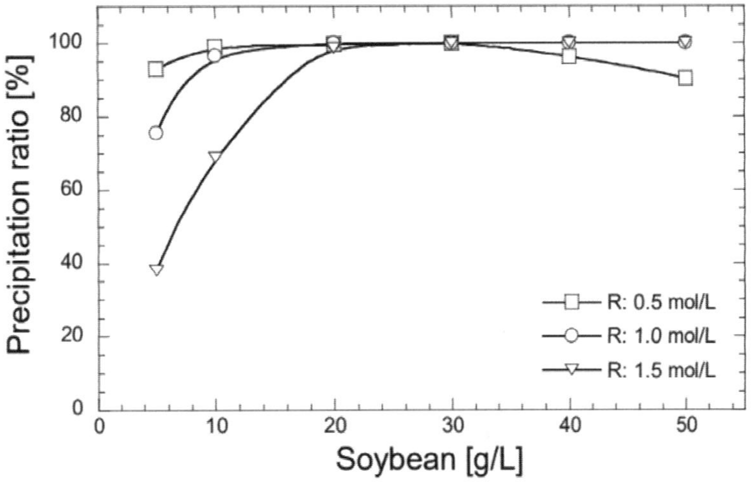

Figure.3 Precipitation test result with variations of soybean and reagent concentration [25]

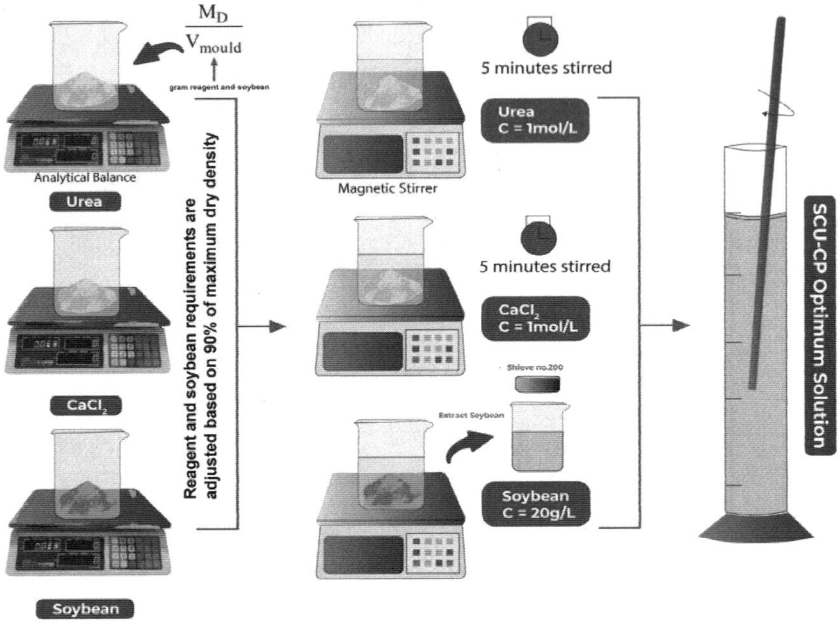

Figure.4 Preparation of optimum SCU-CP solution

2.3. Soil Sample Prepararion of 90% Standard Proctor Density

Preparation of soil sample with density of 90% Proctor standard refers to ASTM D698-07 [26] regarding Light Density Test Method for Soil. The test sample was prepared based on 90% Proctor density in the presence of treated (UT) sample by premixing method of SCU-CP solution and untreated (TE) sample. The density of 90% Proctor standard is based on the maximum dry density (MDD) value in the volume dimensions of the Proctor test mold. The volume requirement of the SCU-CP solution is based on the volume of the optimum water content (OMC) with the ratio of the volume of the solution for Urea: CaCl2: Soybean, which is 1: 1: 2. The 90% density of the standard Proctor obtained based on the Proctor test will be used as the density in the soil sample for consolidation testing in analyzing the effect of density in the presence of treated and untreated on the settlement parameters. The schematic in making the 90% density of Proctor standard is given in **Figure 5**.

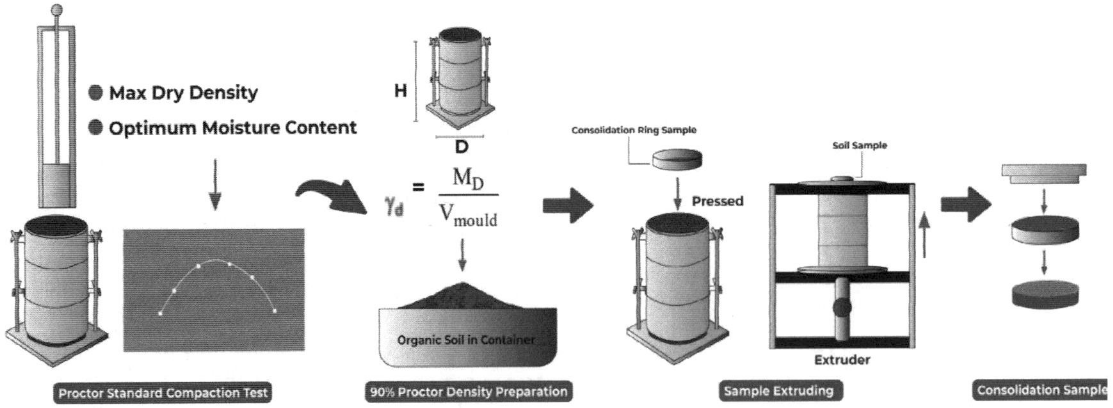

Figure.5 Consolidation Soil Sample Preparation with 90% Density of Proctor Standard Compaction

2.4. One-Dimensional Consolidation Test of Peat Soil Sample

The consolidation test used an oedometer to get the value of soil settlement when a load was applied to the soil sample. This consolidation test was carried out on soil samples with a density of 90% of the Proctor standard on untreated (UT) and treated (TE) samples that had been given a mixture of optimum SCU-CP solution to analyze the effect of the optimum SCU-CP grouting solution on the parameters of the settlement (R_{90}) and time to reach 90% degree of compression (t_{90}). This test refers to ASTM D2345M-11 [27] regarding Standard Test Methods for One-Dimensional Consolidation Properties of Soils Using Incremental Loading.

The consolidation test was carried out at every 90% density of the Proctor standard on untreated (UT) and treated (TE) samples with SCU-CP solution treatment, then loaded with loading steps starting from 0.5 kg to 8 kg then releasing the loading slowly back to 0.5 kg. Each loading was carried out for 24 hours. Then the settlement of soil sample data is taken and analyzed by the square root of time method (Taylor) . The process of soil sample consolidation test is given in **Figure 6.**

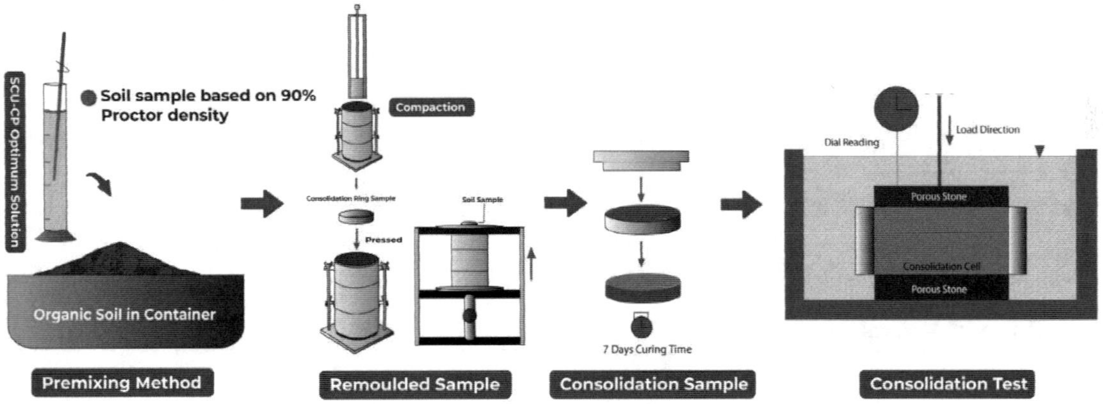

Figure 6. Consolidation test scheme on soil samples

3. Results and Dicussion

3.1. Compaction Test Result in Preparing Soil Sample with 90% Density of MDD

Proctor compaction test was carried out to obtain the optimum moisture content (%) and maximum dry volume weight (gr/cm3) in the manufacture of 90% Proctor density. Density of 90% Proctor was obtained based on the value of maximum dry density (MDD) based on the curve of the relationship between moisture content (%) and dry density (g/cm3). The results of the Proctor test is given in **Figure 7**. Based on **Figure 7**, the optimum moisture content (OMC) and maximum dry density (MDD) of the soil sample were obtained at 45.0% and 1.16 gr/cm3. The optimum water content value of the peat soil sample was obtained <100% unlike peat in general with the optimum moisture content value>100%. This is in accordance with the results of soil organic content testing where the soil sample has low soil organic content so that it has low water absorption. According to the statement of Desiani [28], high organic content will cause large water absorption. The maximum density value of the Proctor test is used as a reference density for sample preparation in the consolidation test to obtain a comparison of time to reach 90% degree of compression (t_{90}) and consolidation coefficient (Cv). Experimental condition of Peat sample in this study is given in **Table 2.**

Research conducted by Putri *et al.* [29] regarding the stabilization of peat soil using a cement mixture as cement resulted in an OMC value of <100% which is 84% with an MDD value of 0.480% in the control sample. The relatively high MDD value of peat soil in this study of 1.16 gr/cm3 correlated with a low OMC value, which was based on Sidhi *et al.* [30] a low volume weight value of peat soil correlates with a high moisture content value because the weight of the soil grains is replaced by water, where the water density is lower than the soil density. Furthermore, the low OMC value of peat soil in this study also correlates with research conducted by Ahmad *et al.* [31] regarding stabilization of peat soil using industrial by-product materials, in the control sample, the results of standard compaction treatment have an OMC value of 31.1% and an MDD value of 1.13 gr/ cm3. Based on this, the relatively high MDD value can be influenced by the physical parameters and the OMC value of the compaction treatment

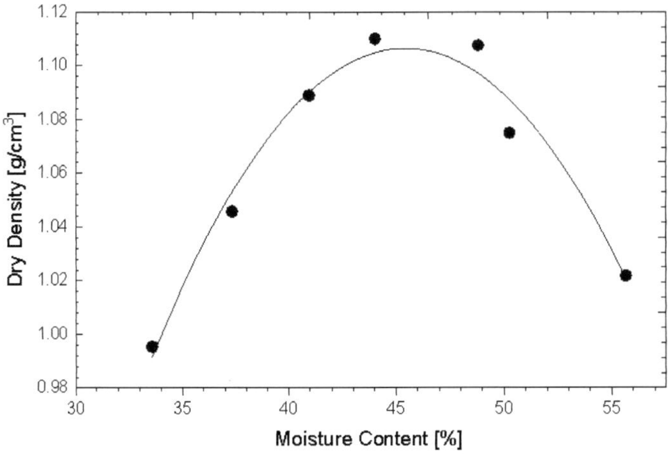

Figure 7. Proctor standar compaction test result of Peat soil used in this Study

Table 2. Experimental condition of Peat soil sample in this study

Case	Density	Dry Density	CaCl2	Urea	Soybean
	[% Proctor]	[g/cm^3]	[mol/L]	[mol/L]	[g/L]
UT90	90.0	1.16	-	-	-
TE90		1.16	1.0	1.0	20.0

3.2. Improvement in Time Parameter (t_{90}) to 90% Degree of Compression (U=90%)

Consolidation testing is carried out in measuring the parameters of the settlement speed that occurs in a structure. Determination of the rate of settlement and degree of consolidation was conducted based on the graph of settlement and time in the presence of reduced pore voids with saturation conditions due to loading. The consolidation test was carried out in comparing the settlement (R_{90}) and time parameter (t_{90}) to reach 90% degree of compression (U= 90%) on the treated soil with vertical drainage from the treated and untreated samples in analyzing the effectiveness of the calcite precipitation method to improve settlement of peat soil. Analysis of t_{90} parameter was performed based on the Root of Time (Taylor) method based on the relationship between settlement and time (t). The results of the consolidation test on untreated (UT) and treated samples (TE) samples are given in **Figure 8**. Based on **Figure 8**, the value of the consolidation settlement rate (t_{90}) at 2.0 kg/cm^2, 4.0 kg/cm^2, and 8.0 kg/cm^2 pressure, the t_{90} values are 8.41 minutes, 1.82 minutes, and 1.21 minutes respectively for untreated (UT) samples. As for the treated samples (TE), the t_{90} parameter improved with the use of the optimum SCU-CP grouting solution, where the lower value of t_{90} value indicates that the time to reach the 90% compression degree is getting faster. Based on **Figure 8,** the results obtained for the parameter t_{90} at a pressure of 2.0 kg/cm^2, 4.0 kg/cm^2, and 8.0 kg/cm^2 were 0.20 minutes, 0.36 minutes, and 0.56 minutes respectively.

Based on the concept of consolidation, where consolidation is the process of reducing the volume or reducing the pore voids of saturated soil due to loading. The high initial permeability of peat soils causes the initial compression to occur very quickly. Primary compression occurs after the peat soil is loaded. The deformation process due to the destruction of the structure in the peat soil fibers causes the compression behavior to be more complicated. Calcite formed from the use of optimum SCU-CP grouting solution in soil samples can bind between soil particles, limit movement, and fill pore spaces so that soil consistency changes from an increase in cohesion. A change in the level of soil consistency causes the time to reach 90% compression to be faster so that it can accelerate the consolidation process. Mackevičius *et al.* [32] stated, the bond between peat soil and calcite is one of the things that can increase the cohesion value in the soil. This is supported by the statement of Putra et al. [33] where the calcite crystals formed from precipitation can become adhesive between soil grains, so that movement between soil grains can be reduced and increase stiffness.

Research conducted by Bobet *et al.* [2] regarding the One-dimensional consolidation behavior of cement-treated organic soil using a variation of portland cement treatment in the range (8 – 20)% PC and the use of 50% and 100% PC by dry mass of soil, it was found that the higher the percentage of cement applied, the greater the stress. pre-consolidation as a result of a decrease in the value of the void ratio due to the hydration reaction of the cement which can close and bind the pores. A similar concept in this study using SCU-CP grouting solution, the calcite formed will fill the pore space and increase soil cohesion so that it can reduce the void ratio which will affect the compression process. This is in accordance with the statement from Canakci *et al.* [34], where the calcite formed in the pore spaces in organic soils can increase the bearing capacity of the soil.

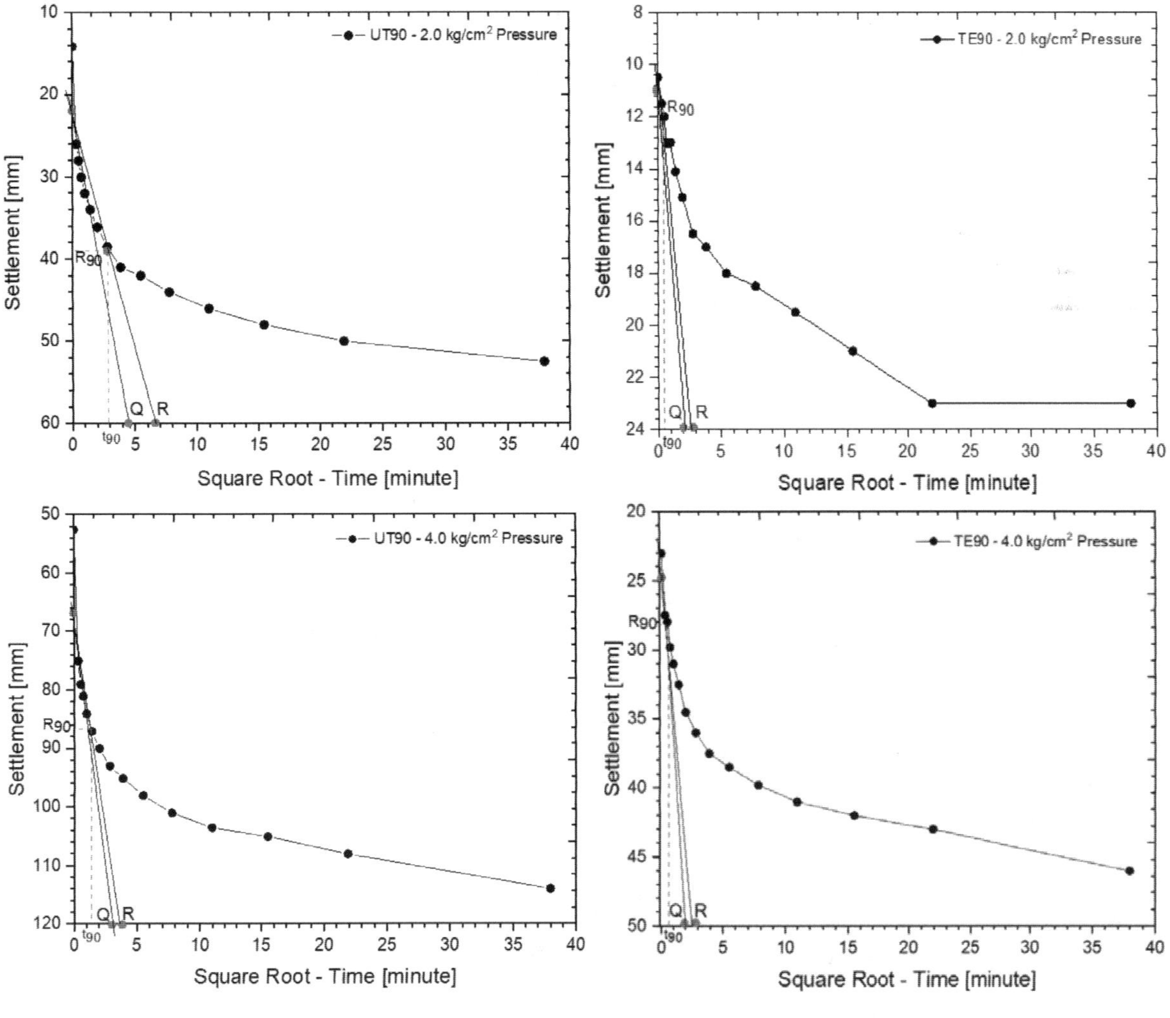

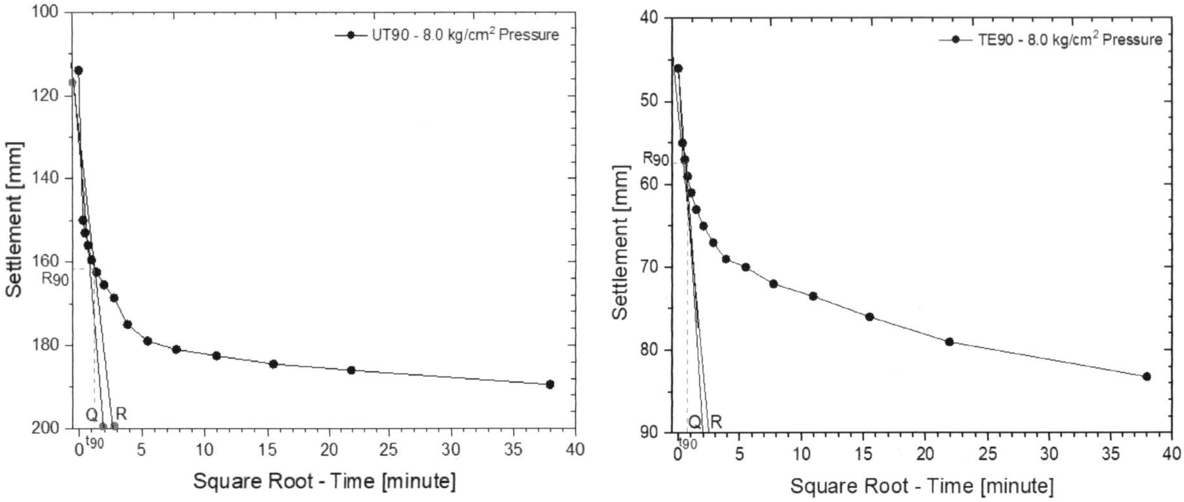

Figure.8 Comparison of t_{90} parameters in consolidated UT and TE samples

3.3. Improvement in Settlement Reduction at 90% Consolidation (R_{90})

The settlement parameter at 90% (R_{90}) consolidation can be determined by drawing a graph of settlement and time for a particular load. An increase in the settlement parameter resulted in the treated samples (TE) with the addition of the optimum combination SCU-CP solution. The results of the comparison of R_{90} parameter in the treated and untreated samples are depicted in **Figure 9**. R_{90} values in untreated (UT) conditions at 2.0 kg/cm², 4.0 kg/cm², and 8.0 kg/cm² pressure were obtained 39 mm, 86.7 mm, and 162 mm respectively. The improvement of the R_{90} parameter in treated samples (TE) for each pressure of 2.0 kg/cm², 4.0 kg/cm², and 8.0 kg/cm² obtained the R_{90} values of 16.4 mm, 28.1 mm, and 57.5 mm respectively. Improvement in R_{90} parameter indicates that the settlement is lower and decreases at 90% consolidation. The reduction in the settlement parameter shows that the improvement of peat soil using SCU-CP method reduce the level of compressibility at 90% consolidation.

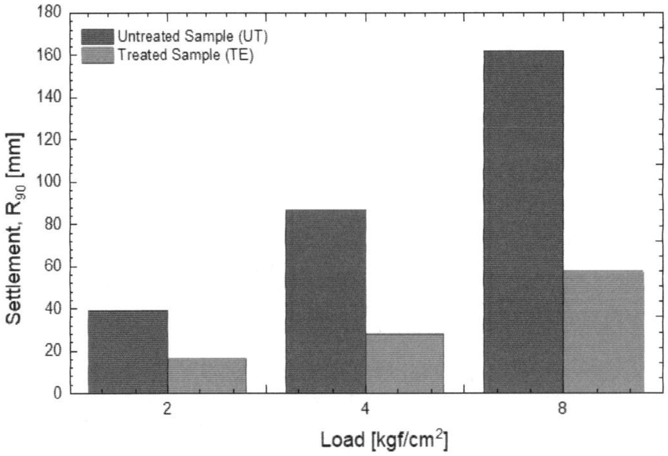

Figure 9. Settlement at 90% consolidation process

The reduction of settlement at 90% consolidation (R_{90}) value occurs due to the presence of calcite which fills the pore space and increases the soil cohesion so that the soil consistency will become stiff. This is in accordance with Yaushara *et al.* [35], calcite that settles in the soil can bind soil particles, limit movement, and improve soil properties. The precipitated calcite covers the pore space, reducing the porosity and permeability of the soil. Research conducted by Mochtar and Yulianto [36] regarding compression behavior of fibrous peat stabilized with admixtures of lime $CaCO_3$ and Rice Husk Ash (admix-1) also Lime $CaCO_3$ and Fly Ash (admix-2), resulted that the value of the compression rate was influenced by the type of mixture and The curing time, which in the use of admix-2, has a high compression rate because at the time of formation of $CaSiO_3$ in the pore space there is not enough water for reaction formation so that the volume changes during crystal formation. As for calcite percipitation method, an increase in the rate of compression occurs due to an increase in soil cohesion so that the soil has a rigid consistency so that the soil sample will reach the compression limit faster, but the effect of calcite formation is influenced by organic content. Based on Yunus *et al.* [37] lime stabilization is influenced by organic content, where with a high OMC value, the distance between the soil aggregate grains increases, thereby reducing the resulting binding level.

4. Conclusions

Based on the results of this study, it can be concluded that Soybean Crude Urease Calcite Precipitation (SCU-CP) method can improve peat soil settlement parameter based on settlement (R_{90}) and time parameter (t_{90}) to 90% degree of Compression (U=90%). Parameter t_{90} which indicates the time to reach 90% compression, the treated sample (TE) produces lower t_{90} values or t_{90} were getting faster at each pressure of 2.0 kg/cm^2, 4.0 kg/cm^2, and 8.0 kg/m^2, which resulting 0.20 minutes, 0.36 minutes, and 0.56 minutes respectively. The faster t_{90} parameter, the process to reach 90% consolidation is getting shorter. The increase in R_{90} values for each pressure of 2.0 kg/cm^2, 4.0 kg/cm^2, and 8.0 kg/cm^2 on the treated samples (TE), resulting 16.4 mm, 28.1 mm, and 57.5 mm respectively or lower than the untreated samples and it indicates that 90% consolidation were obtained with lower settlement process. Parameters t_{90} and R_{90} values are positively correlated, where in the treated samples (TE) the t_{90} value is lower so that the R_{90} value decreases which indicates that the time to 90% compression degree of consolidation (U=90%) is getting faster with low settlement.

Acknowledgments
The research described in this paper was financially supported by the Minister of Education, Culture, Research, and Technology of the Republic of Indonesia, grant number 082/E5/PG.02.00.PT/2022

References

[1] Yulianto FE, Mochtar NE. Mixing of Rice Husk Ash (Rha) and Lime for Peat Soil Stabilization Mixing of Rice Husk Ash (Rha) and Lime for Peat Soil Stabilization. Proc First Makassar Int Conf Civ Eng 2010:533–8.

[2] Bobet A, Hwang J, Johnston CT, Santagata M. One-dimensional consolidation behavior of cement-treated organic soil. Can Geotech J 2011;48:1100–15. https://doi.org/10.1139/t11-020.

[3] Kazemian S, Asadi A, Huat BBK, Prasad A, Rahim IBA. Settlement problems in peat due to their high compressibility and possible solution using cement columns. Forensic Eng Proc Congr 2009;41082:255–64. https://doi.org/10.1061/41082(362)26.

[4] Ali FH, Sing WL, Hashim R. Engineering properties of improved fibrous peat. Sci Res Essays 2010;5:154–69.

[5] Pokharel B, Siddiqua S. Effect of calcium bentonite clay and fly ash on the stabilization of organic soil from Alberta, Canada. Eng Geol 2021;293:106291. https://doi.org/10.1016/j.enggeo.2021.106291.

[6] Yulianto, F.E. dan Mochtar NE. Behavior of Fibrous Peat Soil Stabilized with Rice Husk Ash (RHA) and Lime, Proceedings of 8th International Symposium on Lowland Technology 2012.

[7] Sing WL, Hashim R, Ali FH. A Review on Chemical Stabilization of Peat. Geotech Geol Eng

2021;39:5429–43. https://doi.org/10.1007/s10706-021-01857-1.

[8] Burbank M, Weaver T, Lewis R, Williams T, Williams B, Crawford R. Geotechnical Tests of Sands Following Bioinduced Calcite Precipitation Catalyzed by Indigenous Bacteria. J Geotech Geoenvironmental Eng 2013;139:928–36. https://doi.org/10.1061/(asce)gt.1943-5606.0000781.

[9] Putra H, Yasuhara H, Kinoshita N, Hirata A. Optimization of enzymemediated calcite precipitation as a soil-improvement technique: The effect of aragonite and gypsum on the mechanical properties of treated sand. Crystals 2017;7:1–15.

[10] Soon NW, Lee LM, Khun TC, Ling HS. Factors Affecting Improvement in Engineering Properties of Residual Soil through Microbial-Induced Calcite Precipitation. J Geotech Geoenvironmental Eng 2014;140. https://doi.org/10.1061/(asce)gt.1943-5606.0001089.

[11] Neupane D, Yasuhara H, Kinoshita N, Ando Y. Distribution of mineralized carbonate and its quantification method in enzyme mediated calcite precipitation technique. Soils Found 2015;55:447–57. https://doi.org/10.1016/j.sandf.2015.02.018.

[12] Putra H, Yasuhara H, Kinoshita N, Hirata A. Optimization of enzyme-mediated calcite precipitation as a soil-improvement technique: The effect of aragonite and gypsum on the mechanical properties of treated sand. Crystals 2017;7. https://doi.org/10.3390/cryst7020059.

[13] Putra H, Yasuhara H, Erizal, Sutoyo, Fauzan M. Review of enzyme-induced calcite precipitation as a ground-improvement technique. Infrastructures 2020;5. https://doi.org/10.3390/INFRASTRUCTURES5080066.

[14] Zulfikar RA, Putra H, Yasuhara H. The Utilization of Milk as a Catalyst Material in Enzyme-Mediated Calcite Precipitation (EMCP) for Crack-Healing in Concrete. J Civ Eng Forum 2021;23:54–61. https://doi.org/10.9744/ced.23.1.54-61.

[15] Gao Y, He J, Tang X, Chu J. Calcium carbonate precipitation catalyzed by soybean urease as an improvement method for fine-grained soil. Soils Found 2019;59:1631–7. https://doi.org/10.1016/j.sandf.2019.03.014.

[16] Lee S, Kim J. An Experimental Study on Enzymatic-Induced Carbonate Precipitation Using Yellow Soybeans for Soil Stabilization. KSCE J Civ Eng 2020;24:2026–37. https://doi.org/10.1007/s12205-020-1659-9.

[17] Baiq HS, Yasuhara H, Kinoshita N, Putra H, Johan E. Examination of calcite precipitation using plant-derived urease enzyme for soil improvement. Int J GEOMATE 2020;19:231–7. https://doi.org/10.21660/2020.72.9481.

[18] Pratama EM, Putra H, Syarif F. Application of calcite precipitation method to increase the shear strength of peat soil. IOP Conf Ser Earth Environ Sci 2021;871:1–8. https://doi.org/10.1088/1755-1315/871/1/012058.

[19] Johari NN, Bakar I, Razali SNM, Wahab N. Fiber Effects on Compressibility of Peat. IOP Conf Ser Mater Sci Eng 2016;136:1–9. https://doi.org/10.1088/1757-899X/136/1/012036.

[20] ASTM International. Determining the Water (Moisture) Content, Ash Content, and Organic Material of Peat and Other Organic Soils 1. West Conshohocken, PA, USA,: ASTM; 2020.

[21] Terzaghi K. Theoretical Soil Mechanics. New York (USA): Wiley; 1943.

[22] Islam MS, Hashim R. Bearing capacity of stabilised tropical peat by deep mixing method. Aust J Basic Appl Sci 2009;3.

[23] Artiwi NP. Rekayasa Stabilisasi Tanah Gambut dengan Penambahan Terak/Klinker Ditinjau dari Nilai Permeabilitas. J JOSCE 2019;01:45–52.

[24] Putra H. Mekanika Tanah. Yogyakarta (ID): Gre Publishing; 2019.

[25] Putra H, Erizal, Sutoyo, Simatupang M, Yanto DHY. Improvement of organic soil shear strength

through calcite precipitation method using soybeans as bio-catalyst. Crystals 2021;11. https://doi.org/doi:10.3390/cryst11091044.

[26] ASTM International. ASTM D698-07 Standard Test Methods for Laboratory Compaction Characteristics of Soil Using Standard Effort. West Conshohocken, PA, USA,: ASTM; 2007.

[27] ASTM International. Standard Test Methods for One-Dimensional Consolidation Properties of Soils Using Incremental Loading. vol. i. West Conshohocken, PA, USA,: ASTM; 2010. https://doi.org/10.1520/D2435.

[28] Desiani A. Kajian Pengaruh Materi Organik pada Sifat Fisis Tanah Lunak. J Tek Sipil 2019;13:21–48. https://doi.org/10.28932/jts.v13i1.1427.

[29] Putri EE, Yuliet R, Harris LE, Makinda J. Stabilization of Rimbo Panjang peat soil using lightweight materials mixed with cement as subgrade for road pavement. Int J GEOMATE 2020;18:30–6. https://doi.org/10.21660/2020.66.9180.

[30] Sidhi K, Nuryanto H, Hartanto D. Kajian Karakteristik dan Kuat Geser Tanah Gambut Dengan Penambahan Semen Tipe I Sebagai Bahan perbaikan Tanah. Konfrensi Tek. Sipil, vol. 2, 2019, p. 2–9.

[31] Ahmad A, Sutanto MH, Ahmad NRB, Bujang M, Mohamad ME. The implementation of industrial byproduct in malaysian peat improvement: A sustainable soil stabilization approach. Materials (Basel) 2021;14:1–22. https://doi.org/10.3390/ma14237315.

[32] Mackevičius R, Sližyte D, Zhilkina T. Influence of Calcite Particles on Mechanical Properties of Grouted Sandy Soil. Procedia Eng 2017;172:681–4. https://doi.org/10.1016/j.proeng.2017.02.080.

[33] Putra H, Yasuhara H, Kinoshita N, . E, Sudibyo T. Improving Shear Strength Parameters of Sandy Soil using Enzyme-Mediated Calcite Precipitation Technique. Civ Eng Dimens 2018;20:91–5. https://doi.org/10.9744/ced.20.2.91-95.

[34] Canakci H, Sidik W, Halil Kilic I. Effect of bacterial calcium carbonate precipitation on compressibility and shear strength of organic soil. Soils Found 2015;55:1211–21. https://doi.org/10.1016/j.sandf.2015.09.020.

[35] Yasuhara H, Neupane D, Hayashi K, Okamura M. Experiments and predictions of physical properties of sand cemented by enzymatically-induced carbonate precipitation. Soils Found 2012;52:539–49. https://doi.org/10.1016/j.sandf.2012.05.011.

[36] Mochtar NE, Yulianto FE. Compression behavior of fibrous peat stabilized with admixtures of lime CaCO3+rice husk ash and lime CaCO3+Fly Ash. Int J Adv Sci Eng Inf Technol 2018;8:792–8. https://doi.org/10.18517/ijaseit.8.3.4317.

[37] Mohd Yunus NZ, Wanatowski D, Stace LR. The influence of chloride salts on compressibility behaviour of lime-treated organic clay. Int J GEOMATE 2013;5:640–6. https://doi.org/10.21660/2013.9.3143.

Shear Strength Characteristics of Stabilized Sedimentary Soil with Bacillus Subtilis Bacteria

T Harianto[1], M.A. Walenna[1], and M. C. P. A. Torano[1]

[1]Departement of Civil Engineering, Hasanuddin University, Gowa, South Sulawesi, Indonesia, 92171

E-mail: triharianto@unhas.ac.id

Abstract. Sedimentary soils are formed from deposits resulting from weathering of rocks through natural processes. One way to increase the bearing capacity of sedimentary soil is by stabilizing it. The purpose of this study was to determine the effect of variations in the mixture of Bacillus Subtilis bacteria on the characteristics of sedimentary soils and to determine the effect of the curing period on the shear strength of the sediment stabilized by Bacillus subtilis bacteria. Based on the results of physical properties testing, the original soil was classified as ML, which is silt with low plasticity based on the USCS classification system, and classified as soil group A-4, namely silty soil according to the AASHTO classification system. The addition of Bacillus Subtilis bacteria mixed with urea and CaCl2 was proven to increase the value of shear stress, cohesion, and shear angle of the soil. The addition of Bacillus Subtilis with a mixture of urea and CaCl2 to the sedimentary soil accompanied by a curing period was shown to increase the shear stress at the percentage of adding stabilizing agent 8% from 1.25 kg/cm2, 1.45 kg/cm2, and 1.77 kg/cm2 at the beginning to 1.73 kg/cm2, 2.09 kg/cm2, and 2.39 kg/cm2 at 28 days of curing, with an average increase of 35 percent. The cohesion value continued to increase from the initial value of 0.72 kg/cm2 to 1.57 kg/cm2 at 28 days of curing, resulting in an increase in cohesion value of 118 percent. For the value of the shear angle in the soil, there was an increase from the original soil shear angle value of 22◦ to 31◦ in a mixture of 8% bacteria with a curing period of 28 days so that there was an increase in the value of the shear angle by 40 percent.

Keywords: Dam, Sedimentary Soils, Soil Stabilization, Shear Strength, Cohesion Bacillus Subtilis.

1. Introduction

Soil is a very important component in the field of civil construction because it is a determining factor for the success of a construction work, but the availability of land for large-scale construction such as road embankment requires a large amount of land, while the supply of land with good bearing capacity is sometimes limited. One alternative to overcome these obstacles is to use waste soil (sedimentary soil) to become embankment soil.

Sedimentary soils are formed from deposits resulting from weathering of rocks through natural processes such as water, wind, and others. One example of sedimentary soil can be found at the Bili-bili Dam, Gowa Regency, South Sulawesi Province. Sedimentary soil at this location is quite abundant and has not been used in general, so that it can be investigated its use in stability against the influence of external forces on the bearing capacity of the sedimentary soil. One way to increase the bearing capacity of sedimentary soil is through stabilization. Stabilization can be carried out by adding a certain material or composite to increase the strength of the soil. One of the environmentally friendly stabilization methods is to use bacteria, for example Bacillus Subtilis bacteria. Bacillus belongs to the group of rods and cocci bacteria that form endospores with the characteristics of having a stem cell shape, motile with one flagellum, gram-positive, aerobic, forming endospores, having habitats in the soil, water, aquatic environment, as well as animal digestion. including humans), some species are pathogenic to humans and other animals [1]. Some of the stabilization measures are increasing soil density, adding inactive material to increase cohesion and/or shear resistance, adding material to cause chemical and physical changes in soil material, lowering the water table and replacing poor soil [2].

The purpose of soil stabilization is to increase the bearing capacity of the soil and to increase its stability. The basic principle of soil improvement is that poor soil capacity (in various aspects) can be improved through improving the properties of the soil, in accordance with the desired improvement objectives. If what is desired is to increase the bearing capacity and shear strength of the soil, then several soil parameters need to be improved, such as soil volume weight (γ), soil cohesion (c), angle of shear in the soil (Φ). In addition, the principles in sustainable development focus that development will only be sustainable and natural resources can also be utilized by future generations only if the aspect of protecting the environment remains a priority in every stage of planning, implementation, and operation of infrastructure. One of the soil improvement studies using Bacillus Subtilis bacteria on clay soils resulted in an increase in the value of the shear angle along with the addition of the volume of bacteria into the soil sample [3]. Investigated the shear strength characteristics of the soil using the biogrouting stabilization method of Bacillus Subtilis bacteria in which the clay sand soil sample was mixed with Bacillus Subtilis bacterial stabilizing agent [4]. This study analyzed the optimum stabilization of microorganisms with variations in solution and curing time. Bacillus Subtilis bacteria were injected into the soil with a variation of the solution 1x, 2x and 3x injection or 2 cc, 4 cc and 6 cc. The soil that had been injected with microorganisms was cured for 3 days, 7 days, 14 days, 21 days, and 28 days. The mechanical characteristics that were evaluated were the shear angle and the cohesion values derived from direct shear testing. The results obtained were an increase in the cohesion value of 297% to the original soil sample value and an increase in the value of 6.86% to the value of the shear angle in the original soil.

The application of soil improvement techniques must always be accompanied by considerations of environmental sustainability, so that the goal of soil stabilization is not only focused on achieving technical requirements but must also meet environmental safety requirements. Therefore, it is very important to look for alternative methods of stabilization that are environmentally friendly, namely by using microorganisms derived from microbes. In a previous study, namely a study of the mechanical characteristics of stabilized organic soils with Bacillus subtilis bacteria with curing times of 7, 21, and 28 days, it resulted in continuous soil shear strength parameter values, especially at 28 days of curing [5]. While in research using expansive soil samples stabilized by the bioremediation method, the compressive strength of the soil has increased by 430 times [6]. So that this study intends to examine the effect of Bacillus subtilis bacteria as a stabilizing agent for sedimentary soils on the shear strength of the soil.

2. Literature Review

2.1. Definition and Soil Classification

In a general technical sense, soil is defined as a material consisting of aggregates (granules) of solid minerals that are not cemented (chemically bonded) to each other and from solid particulate organic matter that has weathered accompanied by liquid and gaseous substances which fill the empty spaces between the particles. Soil is useful as a building material in various kinds of civil engineering work, besides that the soil also functions as a support for the foundation of the building [7]. Soil is a collection of minerals, organic matter, and relatively loose deposits, which are located on top of bedrock. The formation of soil from its parent rock can be a physical or chemical process. The process of soil formation physically changes rock into smaller particles, occurs due to the influence of erosion, wind, water, ice, humans, or the destruction of soil particles due to changes in temperature or weather [8].

The following is a grouping of soil types based on a mixture of grains according [9]:

- Coarse-grained soil is soil with large grains of soil in the form of sand and gravel.
- Fine-grained soils are soils where most of the soil grains are clay and silt.
- Organic soil is soil that contains quite a lot of organic matter.

2.1.1 Classification System by Texture2

In a general sense, what is meant by soil texture is the state of the soil surface. Soil texture is influenced by the size of each grain in the soil. Several classification systems based on soil texture have been developed long ago by various organizations to meet their own needs, some of these systems are still in use today such as the classification system based on soil texture developed by the United States Department of Agriculture (USDA).

This system is based on the boundary size of the soil grains as described by the USDA system, namely:

- Sand: granules with a diameter of 2.0 to 0.05 mm.
- Silt: granules with a diameter of 0.05 to 0.002 mm.
- Clay: granules with a diameter smaller than 0.002 mm.

2.1.2 Classification System by Usage

At present there are two more soil classification systems that are always used by civil engineers. Both systems consider the grain size distribution and Atterberg limits. These systems are the AASHTO Classification System and the Unified Classification System [10].

2.2. Clay Characteristics

Silts are generally formed from crushed quartz crystals that are the size of sand. Sometimes silt is also called dust in some literature. Deposits that float on the surface of the water or that sink are usually formed by silty soil material. Natural rock breaking involves chemical weathering of rocks and regolite as well as physical weathering through frost haloclasty. The main process involves abrasion, either solid (by glaciers), liquid (river deposition), or by wind [11].

2.3. Soil Stabilization

Soil stabilization in principle is to improve poor soil quality. If a soil found in the field is very loose or easily compressed, or if it has an inappropriate consistency index, too high permeability, or other undesirable properties that are not suitable for a civil construction project, then the soil must be stabilized. Stabilization may consist of any of the following measures [12]

- Increasing soil density.
- The addition of inactive material to increase the cohesion and frictional resistance that arises.
- The addition of materials to cause chemical and physical changes in the soil.
- Lowering of ground water level (soil drainage).
- Poor ground replacement.

2.4 Bacillus Subtilis Bacteria

Bacillus subtilis is a gram-positive, rod-shaped bacterium that is calatase-positive. At first this bacterium was named Vibri subtilis by Christian Gottfried Ehrenberg and its name was changed by Ferdinand Cohn to Bacillus subtilis in 1872 (subtilis is Latin for 'good'). Bacillus subtilis cells are usually rod-shaped, about 4-10 micrometers (μm) long and 0.25-1.0 m in diameter. Other bacteria from the genus Bacillus, namely Bacillus subtilis can form endospores, to survive in environmental conditions even from temperature and drying [13].

2.5 Direct Shear

Soil shear strength parameters are needed in analyzing soil bearing capacity, slope stabilization and thrust stress for retaining walls. Soil shear strength is the resistance force exerted by soil particles against pressure or pull. On the basis of this understanding, when the soil is subjected to loading it will be held back by soil cohesion and friction [14].

3. Methodology

3.1 Data Collection

Several types of data collection activities have been carried out on the test objects: the first stage of soil selection, and then visually identified and its availability factors in nature, then by identifying the characteristics of the soil to ensure its suitability with the stabilizing material to be used with the composition referring to the research objectives.

3.2 Material

The following are the soil investigated in this research:

3.2.1 Sedimentary Soil

The soil used in this study, as shown in Figure 1, is a sedimentary soil located in the Bili-Bili Dam, Gowa Regency, South Sulawesi. The color of the soil is gray and fine grained.

3.2.2 Bacillus Subtilis Bacteria

The bacteria used in this study were Bacilus Subtilis bacteria which were bred separately, and a bacterial growth culture test was carried out to determine the growth characteristics of a type of bacteria through a growth curve marked by turbidity in liquid media with the help of a shaker and incubator which can be seen in the Figure 2.

Figure 1. Sedimentary Soil from Bili-Bili Dam

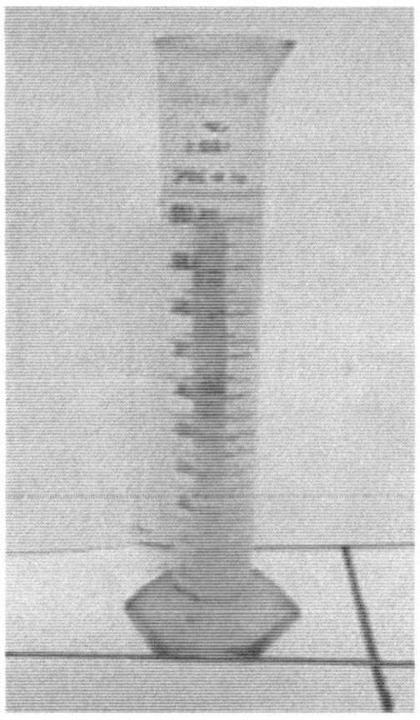

Figure 2. Bacillus Subtilis Bacteria

3.3 *Testing Standards*

The characteristics of the material used in this study were tested based on the reference to the standards issued by ASTM, as shown in Table 1:

Table 1. Testing Standards

Testing	Standards
Water content	D2216 – 98
Specific Gravity	D854 – 14
Atterberg Limits	D4318 – 05, D4943 – 08
Sieve Analysis and Hydrometer	D422 – 63
Compaction	D698 – 07
Direct Shear	D3080

3.4 *Sample Testing*

Sample testing is divided into 2 types of tests, namely testing of physical properties and mechanical properties of the soil:

3.4.1 *Physical Characteristics Testing*

Physical properties test aims to determine the index properties needed in the classification and determination of stabilizing materials and the composition of stabilizing materials and their effects on the soil used. In this study, the physical properties tests carried out on sedimentary soils were as follows: water content, specific gravity, sieve and hydrometer analysis, and Atterberg limits.

3.4.2 *Mechanical Characteristics Testing*

The purpose of testing the mechanical properties of the soil is to determine the behavior of the soil or the resistance of a soil that has been mixed with stabilizing material against a given pressure or load. The soil mechanical properties test is a parameter that can be seen to determine the bearing capacity of the soil. The mechanical properties test carried out in this study was a direct shear and compaction test.

4. Result and Discussion

4.1. *Compaction with a standard proctor apparatus*

4.1.1 *Water Content*

Based on the results of testing the water content value, the water content value for the sedimentary soil used in the study was 14.6%.

4.1.2 *Specific Gravity*

Based on the results of the specific gravity test for sedimentary soil, the density value for the Bili-Bili Dam sedimentary soil is 2.66. So that based on the specifications for the distribution of specific gravity based on soil type, the soil sample is classified as inorganic silt, with the specific gravity value being in the range of 2.62-2.68.

4.1.3 *Sieve Analysis and Hydrometer*

Tests for the percentage of soil grains can be done through sieve analysis and hydrometer. The results of this test can display the soil classification of the soil samples used. From the test results, the percentage of soil grains for gravel is 0.20%, sand is 2.60%, silt is 94%, and clay is 3.20%. From these results the soil used in this study is silty soil. The following is a graph of the results of the filter analysis and hydrometer analysis which can be seen in Figure 3.

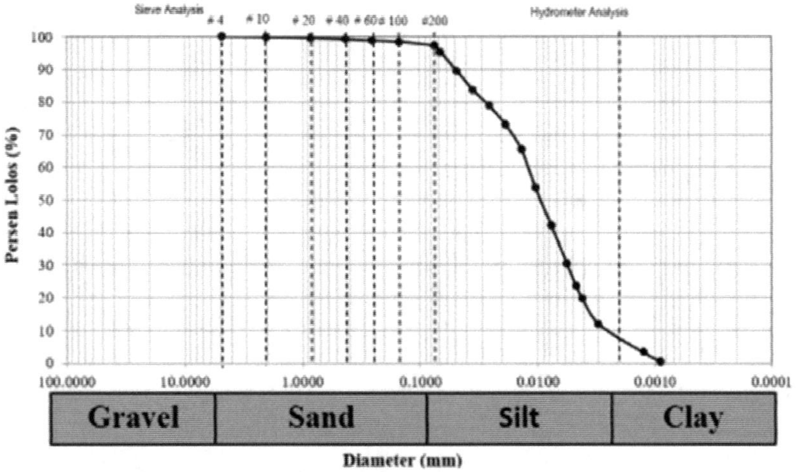

Figure 3. Graph of Sieve Analysis

4.1.4 Atterberg Limits

The Atterberg limits test was carried out to determine the nature of the soil to changes in the given water content. This test includes the Liquid Limit, Plastic Limit, and Shrinkage Limit testing. From the test, the following results were obtained:

- Plastic Limit

The plastic limit is the lowest water content at which the soil begins to become plastic. The plastic limit value obtained was 28.8%.

- Liquid Limit

The liquid limit is the water content at which the behavior of the soil changes from a plastic state to a liquid state. The liquid limit value obtained from the test results was 38.1%. The liquid limit graph can be seen in Figure 4

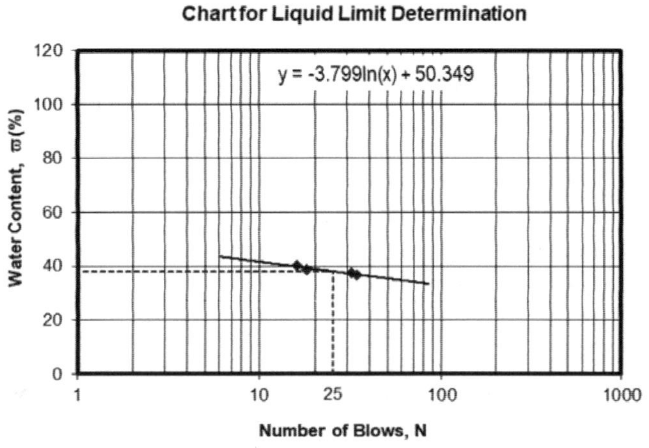

Figure 4. Liquid Limit Graph

- Shrinkage Limit

The shrinkage limit is the water content limit where the soil with a water content below the obtained water content does not cause a change in volume (shrinkage) again in the soil. The value of the shrinkage limit obtained was 8.6%.

- Plasticity Index

The plasticity index is the value of the difference between the liquid limit and the plastic limit. From the test results, the plasticity index value was 9.3%.

4.2 Soil Classification

From the test for the physical properties of the soil used in the study, the following classification is for the sedimentary soil of the Bili-bili Dam, Gowa Regency:

4.2.1 AASHTO

For sieve analysis, hydrometer and Atterberg boundary, the percentage of native soil grains that passed the No. 200 sieve was 97.2%. The liquid limit value was 38.1%, and the plasticity index was 9.3%.

From the values that have been obtained, it can be determined the type of soil based on the classification of the AASHTO system. The determination of the type of soil was found as a silty soil.

4.2.2 USCS

By connecting the plastic limit value and the plasticity index value, it is obtained that the soil type belongs to the ML (Milt with low plasticity) group, which means silty soil with low plasticity. There has been a relationship that shows the liquid limit value and the soil plasticity index to determine the classification of soil types. The relationship that shows these values can be seen in Figure 5.

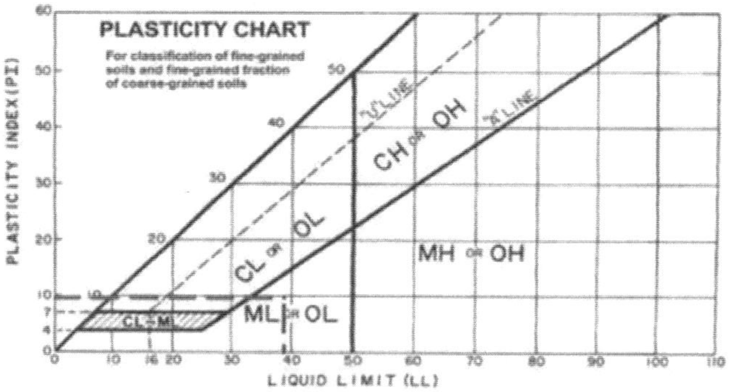

Figure 5. USCS Classification

4.3 Mechanical Characteristics

4.3.1 Compaction

In this test, compaction was carried out using the standard proctor compaction method, namely by inserting the soil into the mold and compacting it using a standard compactor. Compaction was carried out 25 times in every 3 layers of soil. The following is a graph of the relationship between water content and dry density in soil samples which can be seen in Figure 6.

Based on the results of the compaction test, the maximum dry density value was obtained, namely 1.34 g/cm3 which was achieved at the optimum moisture content of 30.25%. The purpose of compaction testing is to increase the volume weight of the soil, which means increase the strength of the soil to support the load on it, increase slope stability, and reduce soil compression [15]. It is indicated that the initial density of the natural soil is sufficient to apply as a stabilized soil material. Also, the moisture content on the origin soil indicated that the water content was set to the normal condition. The compaction value was important parameter in preparing the sample. All the sample were prepared at the mximum density and optimum moistute content.

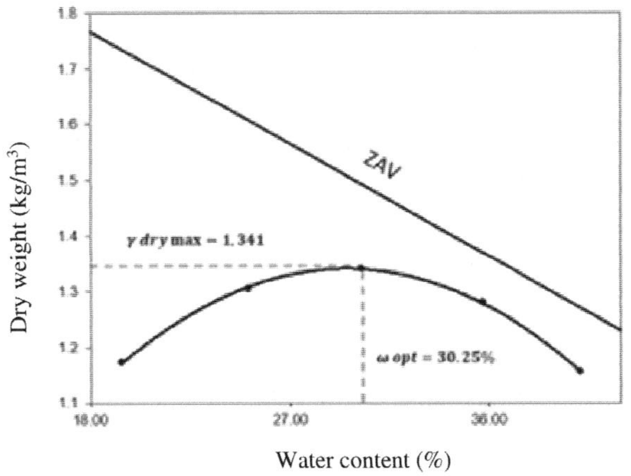

Figure 6. Graph of the relationship between moisture content and dry density of the compaction test results

4.3.2 Cohesion

From the test results, it can be seen that the cohesion value of the sedimentary soil mixed with bacteria with a curing period of up tp 28 days, the highest increase in cohesion value reached 1.57 kg/cm^2 with the addition of 8% bacteria with a curing period of 28 days. However, in the curing period of 0 day the mixed cohesion value was soil with 8% bacteria was 1.09 kg/cm^2 and without the addition of bacteria (untreated soil) was 0.72 kg/cm^2 which can be seen in Figure 7. It can be seen that the improvement soil by adding the bacteria was found optimum in bacteria content of 8% with maximum curing time was 28 days. Moreover, the improvement of cohesion for 28 day curing time was not significant with the percentage of the bacteria from 4% to 12%. The most improvement of the cohesion value was found for 3 and 7 days. This is due to the calcite precipitation mainly developed at the early stage of curing time. The curing time for 14, 21 and 28 days showed the impropvement of cohesion valus insignificant compared to the short curing time. In practical application, the bacteria content and 7 days of cutring time was sufficient value to apply in the field.

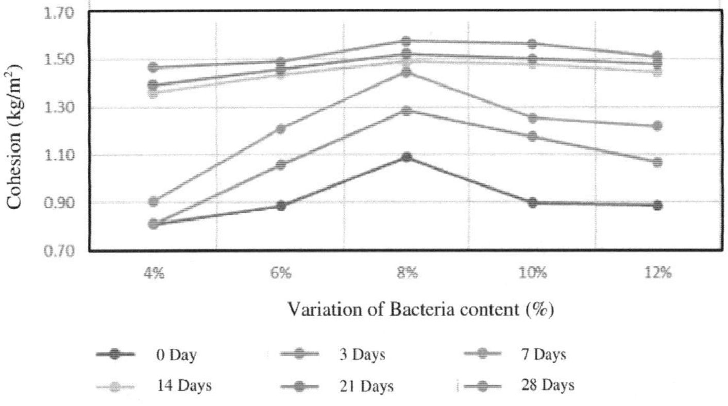

Figure 7. Variation of cohesion value with different curing time

5. Conclusions

Based on the results of physical properties testing, the sedimentary soil of the Bili-bili Dam is classified as ML, namely silt with low plasticity based on the USCS classification system, and classified as soil group A-4, namely silty soil according to the AASHTO classification system.
The addition of Bacillus Subtilis bacteria mixed with urea and CaCl2 was proven to increase the value of shear stress, cohesion, and shear angle in the soil.
From the test results, curing time of 0, 3, 7, 14, 21 and 28 days increased the value of shear stress, cohesion, and shear angle in the soil. In this case, the maximum value was achieved at 28 days of curing with the addition of 8% bacteria. For the value of shear stress increased by 35%. The cohesion value increased by 118%. For the value of the shear angle in the soil there is an increase from the original soil shear angle value of 22° to 31° so that there is an increase of 40%.

6. References

[1] Holt, John G., N.R. Krieg, P.H.A. Sneath, J.T. Staley dan S.T. Williams. 2000. Bergey's Manual of Determinative Bacteriology. Ninth Edition. Williams & Wilkins. Philadelphia.

[2] Bowles, J.E. (1993), Alih Bahasa Ir.Johan Kelana Putra Edisi Kedua, Sifat-Sifat Fisis Dan Geoteknis Tanah, Penerbit Erlangga, Jakarta.

[3] Iffah Fadliah (2013). Studi Eksperimental Stabilisasi Biogrouting Bacillus Subtilis pada Tanah Lempung Kepasiran. Program Studi Magister Teknik Sipil, Fakultas Teknik, Universitas Hasanuddin.

[4] Lynda, Angelina (2013), Karakteristik Kuat Geser Tanah Dengan Metode Stabilisasi Biogrouting Bakteribacillus Subtilis, Fakultas Teknik, Universitas Hasanuddin, Makassar

[5] Imelda Vera Tumanan (2014). Studi Karakteristik Mekanik Tanah Organik Terstabilisasi Bakteri Bacillus Subtilis. Jurnal Tugas Akhir. Program Studi Teknik Sipil, Fakultas Teknik, Universitas Hasanuddin.

[6] Rio Alvin Arfandy (2017). Stabilisasi Tanah Ekspansif dengan Metode Bioremediasi. Jurnal Tugas Akhir. Program Studi Teknik Sipil, Fakultas Teknik, Universitas Hasanuddin.

[7] Das, Braja M. 1995. Mekanika Tanah 1. Erlangga. Jakarta.

[8] Hardiyatmo Christady. (2001). Prinsip-prinsip Mekanika Tanah dan Soal Penyelesaian. Yogyakarta : Universitas Gadjah Mada

[9] Santosa, et al. (1996). Seri Diktat Kuliah : Dasar Mekanika Tanah. Gunadarma.Ankum, P., Koga, K., Segeren, W.A. and Luijendeijk, J., 1988. Lessons from 1200 years impoldering in the Netherlands. Proc. Int. Symposium on Shallow Sea and Lowland, Institute of Lowland Technology, Saga Univ. Saga: 102-108.

[10] Das, B. M. (1995). Mekanika Tanah (Prinsip-prinsip Rekayasa Geoteknik) Jilid I. Jakarta : Erlangga.

[11] Darwis, H. (2017). Dasar-Dasar Teknik Perbaikan Tanah. Yogyakarta : Pustaka AQ.

[12] Bowles, J. E. (1986). Sifat-Sifat Fisis Dan Geoteknis Tanah. (J. K. Hainim, Trans). Jakarta: Erlangga.

[13] Bergey, D.H. dan Boone, D.R. 2009. Bergey's Manual of Systematic Bacteriology. Second Edition. Volume Three: The Firmicutes. Springer. United States of America.

[14] Harianto, T., Yunus, M. and Walenna, M.A. (2021) Bearing Capacity of Raft-Pile Foundation using Timber Pile on Soft Soil. International Journal of GEOMATE, Vol. 21(86), pp. 108-114. doi:10.21660/2021.86.j2294.

[15] Harianto, T. (2022). Performance of Subbase Layer with Geogrid Reinforcement and Zeolite-Watergrass Stabilization. Civil Engineering Journal vol. 8(2), pp. 251-262. doi: 10.28991/CEJ-2022-08-02-05.

Geographic Information System (GIS-T) database design for transportation safety system (case study: spatial database for road transportation system in Indonesia)

Bambang Istiyanto [1], Yan El Rizal Unzilatirrizqi D [2], Mouli De Rizka Dewantoro [3], and Rahmat Ahmad [4].

Polytechnic of Bali Land Transport Indonesia [1,4], Polytechnic of Road Transport Safety Tegal Indonesia [2], and Jasa Tirta II Public Corporation [3].

bambang@poltradabali.ac.id

Abstract. This study aimed to develop a spatial database design on transportation safety system. Spatial databases is a fundamental factor that is very important in the utilization of GIS for transportation, so it is important that a good preparation of spatial databases. Preparation of spatial database will be effective if tailored to the needs. Factors affecting transportation safety is different from one region to another. Factors that affect transportation safety system in an area can be influenced by the transportation system infrastructure in the area. Database design for transportation safety in this study is expected to be a model of the spatial database according to the highway transportation system in a small town in Indonesia. Characteristics of the transport system in Indonesia was largely dominated by private vehicles and less of public transport. Spatial analysis has the advantage to combine data and analyze various factors based on spatial location. Spatial data that can be used to support the analysis include the basic data, thematic data path, data that can be supporting the highway transportation system. Spatial data and GIS - T can be used as an analysis tool and alternative approach to improve road transport safety.
Keywords: Spatial Database, GIS-T, Road Safety System

1. Introduction

Trends in the development of roads in Indonesia is often said to be directly proportional to the development of road users. The length of the road is said to be biased in accordance with the increase in motor vehicles continue to rise rapidly. Ease of purchase of motor vehicles facilitate people acquire the vehicle, the road will consequently diminishing the effectiveness of the use of the road.

Reduction in the effectiveness of the use of the road is shown by the way it was no longer able to accommodate vehicles that perform mobility. The number of vehicles using the road is getting more and more roads are not wide or increased, the result would be congestion on the roads. These factors make the management of road traffic becomes very important to do. Traffic management is not expected to inhibit the mobilization that use the road, but did not make the level of congestion is getting worse.

Road traffic management has been carried out using field data analysis utilizing the tabular analysis. Spatial approach with the use of maps and geographic information systems, have not been widely used. The use of spatial data actually allows because the road is a factor related to the phenomenon of spatial form of human movement. Everyday human movement is certainly a movement that originated from one location to the destination location, this is a phenomenon of the movement or the movement of people is a spatial phenomenon. Spatial or spatial approach will certainly help in the implementation of road traffic management.

An application of GIS technology is used for policy making occurred in the State of Miami, Kansas [8]. 12 fatal accidents in 11 months on a two-lane road along the 20 miles of U.S. Highway 169. Head of the police are very concerned that fatal incident in the street and ask for land information management office (LIMO) to create GIS maps that show the location of the accident and the attributes associated with the accident that can be accessed by the Governor and authorized personnel to quickly analyze the situation and take decisions where the accident occurred due to significant errors when the driver left the vehicle after leaving the road. After looking at the map, the Governor ordered to widen along the way. The police chief then focus area to reduce future accidents using crash data analysis based on other roads GIS [2].

Bob Thomson wrote the words of the head of the traffic safety bureau said that "GIS is a strategic revolution that could potentially improve the effectiveness of employment for police and policy makers in all parts of the State". The only potential strategies need to be explored. Utilize of GIS with dynamic map. The future of GIS to improve efficiency in every job with safety advantages of human life which is a potential asset [8].

Some explanation on how to use GIS technology showed in developed countries have started to be considered ranging from about 1 decades ago . It makes use of GIS technology and the advancement of information systems used in the State of Indonesia, which until now most of the movement of the future highway use.

This system is also suitable to be applied in Indonesia, GIS is a map-based analysis tool, as long as there are coordinates and digital maps can be implemented anywhere. The analysis is carried out with map layers that have data and information according to the theme. The results of the analysis are also in the form of a map that describes the purpose of the paper. For the appearance of the earth that is informative and can help transportation planners, consultants in carrying out urban and regional transportation planning.

1.1. GIS-Transportation

GIS is a computer-based system that has the ability to handle the data has geography references as data entry, data management, manipulation and analysis of data, and output as the final result [1]. While Burrough in 1986, defines GIS (geographic information system) as a computer-based system that is used to enter, store, manage, analyze and re-enable data have spatial reference for a variety of purposes related to mapping and planning. In the next phase, the GIS form and store it in rational tables at once connecting these elements and their attributes. As such attributes can be accessed through the site map elements, and vice versa map elements can be accessed based on the attributes [4].

Transportation safety systems, including services such as highway traffic safety systems and automatic location detection on the traffic system, is part of the purpose of integration between geographic information systems (GIS) and communications technology to develop this type of transportation services [9]. GIS-T is a holistic approach to the complex land use and transportation issues. GIS-T can also be used to reduce the lack linked between analysis and communication, makes a great public input into policy analysis, such as data selection, assumed models, and formulation of scenarios [6].

Geographic information system for transportation is the relationship between hardware, software, data, operators, organizations and institutions to collect, describe, analyze, and communicate various kinds of information about the earth. Information type is the transportation system, and the region is affected by the system [5]. Deploying GIS in transportation (GIS-T) at an enterprise level presents an opportunity to eliminate the traditional application- specific development pattern of information systems by providing a common data structure centered on transportation features [2].

2. Method

2.1. Data Inventory
Inventory data is needed in the preparation of GIS-T database design. Inventory Data is useful to know the different kinds of data to be entered into the database. Sorts the incoming data in the database is the data used in the implementation of the use of GIS for transportation. Inventory data contains the type of data, information from the data, and how the data acquisition of the data.

Table 1. GIS Data Inventory.

Data	Description	Data Acquisition
Traffic Infrastructure	Road segment, detail road information, object, etc.	Survey/GIS
Public Facilities	School, market, restaurant, fuel station, etc.	Survey/GIS
Thematic Data	River, land use, geology, boundary, etc.	External
Crash Data	Up to date data, prediction	Survey, study
Traffic Flow Arrangements	Traffic scenario	External
Hazardous Data	Flood, soil movements, etc.	External

Table 2. Tabular Data Inventory.

Data	Description	Data Acquisition
Road Metadata	History	Internal
Road Specification	Detail Information	Internal
Road Plan	Planning	Internal/external
Traffic Information	Traffic situation, scenario	Survey
General Information	Population, etc.	External
Other Data	Thematic information	Internal/Study

Table 3. Other Data Inventory

Data	Description	Data Acquisition
Real time Traffic Video	Road Monitoring	Telemetry
Real time Traffic Instrument	Road Monitoring	Telemetry

Inventory Data includes all the data that will be used in GIS-T database. Database design is critical to determining the foundation for the data entered in the database. Inventory data in this design were divided into three groups based on the type of data. The group of data is spatial data, tabular data, and other kinds of data, in this design is the data sourced directly from the telemetry tool in the field.

2.2. Relationship Design
The relationships between data with other data that is essential in the preparation of the data base. This study is concerned with how the GIS data and tabular data can be connected to the data that are complementary to the analysis.

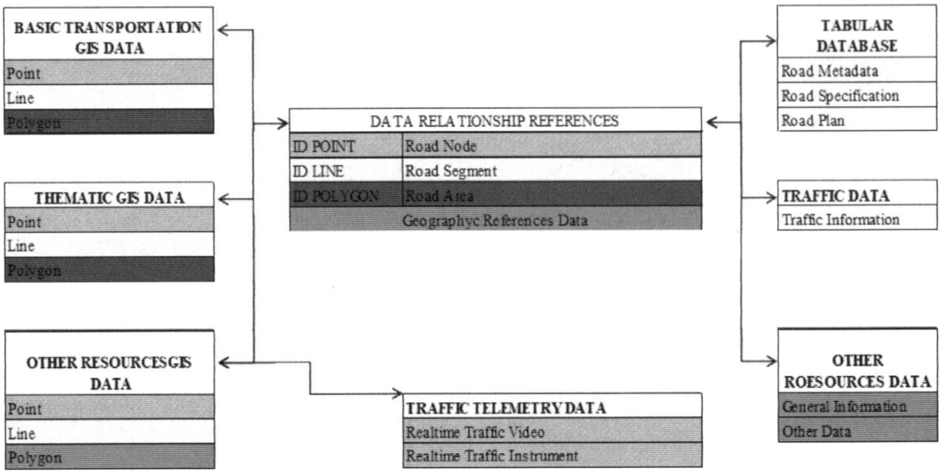

Figure 1. Database Relationship Design

Design relationships between databases using reference data relationships. Reference data is divided into four sections relations, namely the reference point, the reference line, the reference polygon, and geographical references. Reference points spatial location of an object contains data, reference data are used to connect the tabular data which is an attribute of the site and may be updated periodically. Data reference line connecting the data segments of roads and road attributes tabular data that can be updated regularly. Data detailed image reference polygon connecting paths can be combined with tabular data as attributes that can be updated regularly as well. Relationships that last reference data is based on the geographic location where they can relate to administrative boundaries, or polygons that are not raw like inundation polygon area, soil type, and others.

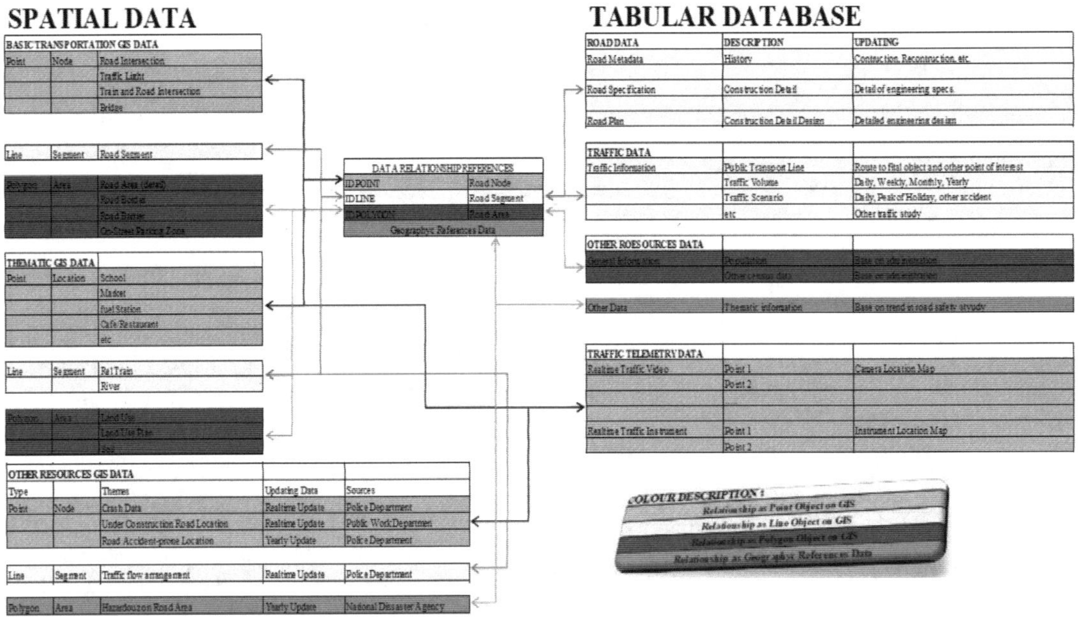

Figure 2. Detailed GIS-Transportation Database Relationship Design

2.3. *GIS-T Database Design*

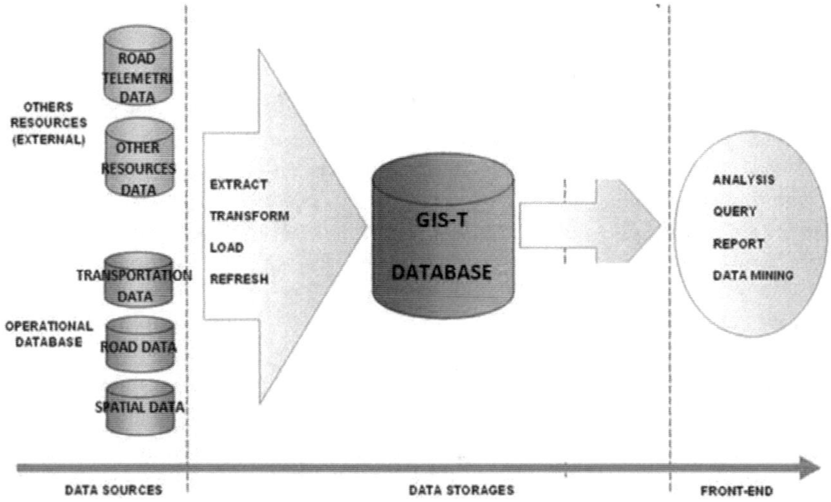

Figure 3. GIS-T Database Structure Design

The preparation of the database is expected to be used as a data center in decision making. Examples that can be explained from this database is the spatial data where every street has a street name and detailed description of the road tends to be static. The road segment data can be combined with traffic information data volumes on the roads which are updated every day. Daily data stored in the database will be analyzed to determine how the ability to pack a road, so it can be determined how the management or regulation of traffic on these roads.

The use of data sourced from other agencies, for examples geological data and the type of soil. This data can be used as input in the analysis of road maintenance prioritization. Prioritization of road maintenance can be performed by using the data of geology and soils are the foundation of the construction of the road, traffic volume, and type of vehicles that used the road. This analysis will also be used to determine how the scenario so that the road traffic arrangements can last longer.

The preparation of this database is also used to accommodate the specific phenomenon that has not been mentioned in the image database design details. Specific case in Indonesia is how the parking arrangements in the area of highway had not been set properly. Rules also not been properly enforced so that it may be necessary to monitor how the analysis of on-street parking areas. Other case is the number of spilled market in Indonesia. The spilled market can happen on certain days or every day. In the event of a long holiday as traffic peak season, spilled market data can be used as a reference for the selection or the nearest alternative route that does not result in a long traffic jam.

3. Result

Database design consists of an inventory of data, collate relationships among the data, and draw up into a series of databases for policy-making. Combination database compiled in this study is as shown in the figure below. Consists of two main types of data are sourced outside the operational and data sourced from daily operations. Data sourced from outside consists of telemetry data path, and other data such as population data, soil type, etc. Data into daily operations are divided into three databases namely road transport data, road data, and spatial data path. Overall the database and entered into the main database can be used for analysis of all the data that is designed using interconnected by reference.

The design shown in detail in the present study does not include all the data. This design is only used to illustrate how to use the tabular data that is already widely used and combined with spatial data.

Spatial data is best used for analyze road use which is a network that has a spatial reference. The use of tabular data with spatial data and connect it, will make monitoring the highway will be better and will hopefully facilitate decision makers in their policy. A good policy is expected to improve the safety of road transport.

4. Conclusion

This research is to design a database for transport safety, in this particular case road transportation in Indonesia, just designing how to accommodate between GIS and data transport. Connecting two data into a GIS database in transportation or that many called GIS-T database. The design is not include the all required data that used. This study focused on how to build a GIS-T database that could integrate data in consideration of transport policy-making, especially in Indonesia. The future research is needed to design more detail how the composition of each type of data and data types that are connected. All detail of data with better references and more systematics. It is expected that both the data structures in the database and make the data detail relationship between data in one integrated database can improve the effectiveness of the analysis.

References

[1] Aronoff, Stan. (1989). Geographic Information System a Management Perspective. WDL Publication : Ottawa-Canada.
[2] Brush, Trisha W. (1999). Governor Persuaded to Widen “Killer 169” by the Power of GIS. ESRI ArcNews, Fall 1999,p 22.
[3] Butler, J.A. and K. Dueker, (2001), Implementing the Enterprise GIS in Transportation Database Design. URISA Journal Vol. 13, No. 1 Winter 2001.
[4] Burrough, P.A. (1986). Principles of Geographic Information Systems for Land Resource Assessment. Monographs on Soil and Resources Survey No. 12, Oxford Science Publications : New York.
[5] Fletcher, D. (2000). Geographic information systems for transportation: A look forward in Transportation in the New Millenium: State of the Art and Future Directions. Washington, DC: Transportation Research Board.
[6] Harvey J. Miller and Shih- Lung Shaw. (2001). Geographic Information Systems for Transportation (GIS-T): Principles and Application. Oxford University Press.
[7] Mitra, Sudeshna. Enhancing Road Traffic Safety: A GIS Based Methodology to Identify Potential Areas of Improvement. Civil and Environmental Engineering Department California Polytechnic State University.
[8] Roche, Jerry. (2000). Geographic Information Systems-Based Crash Data Analysis And The Benefits To Traffic Safety. 2000 MTC Transportation Scholar Conference. Ames-Lowa.
[9] Souleyrette, R. R. and Strauss, T. R. (1999). Transportation. in S. Easa and Y. Chan (eds.) Planning and Development Applications of GIS. American Society of Civil Engineers, Reston, VA, 117-133.

Analysis of Motorized Vehicles Performance at Signalized Intersections Based on Micro Simulation

S H Aly[1], S Martono[2], and N M R Harusi[3]

[1]Department of Civil/Environmental Engineering, Hasanuddin University, Indonesia
[2]Department of Civil Engineering, Hasanuddin University, Indonesia
[3]Department of Environmental Engineering, Hasanuddin University, Indonesia

email: marni.hamidaly@gmail.com(corresponding author)

Abstract. Based on microsimulation, this study analyses the performance and behavior of motorized vehicles at the signalized intersection between the railway line and Ahmad Yani road-Jemursari road. For the survey of the volume of motorized vehicles, this intersection is divided into four intersections with eight directions of traffic movement and one direction of combined traffic movement. Microsimulation was carried out using Vissim software with nine parameters calibrated according to actual conditions during the peak hour observation period. The maximum queue length and motor vehicle delays for the peak hour observation period occur at intersections one in the north and two in the east. Performance of the intersection of Jl. Ahmad Yani-Jl. Jemursari is included in category F. The indicator F are traffic jams, low vehicle speeds, long queues, and vehicle delays. The following study of this paper is to make a geometric model of the intersection into a Fly Over model.

Keywords: intersection performance, plane intersection, geometric, flyover

1. Introduction

The performance and behavior of vehicles at an intersection are the main factors in determining the most appropriate handling to optimize the intersection function, especially at the intersection of a plot with the railway line [1]. Traffic conditions at high-density intersections, especially at signalized intersections and plots with railway lines, cause longer vehicle delays with long queues of vehicles. The problem at the lighted intersection is the occurrence of queues which result in vehicle delays at each arm of the intersection with different volumes.

Traffic conditions at the crossroad of Ahmad Yani road, especially at the meeting of Ahmad Yani road and Jemursari road as one of the accesses from the north (Surabaya) to the south (Sidoarjo). And vice versa from the south to the north (Surabaya city) or towards the east (Jemursari), there tend to be vehicle delays with Level of Service intersections already experiencing congestion and long queues of cars.

This study aims to analyze the performance of motorized vehicles in the form of queue lengths and vehicle delays using VISSIM software based on microsimulation models [2]. With VISSIM software, various parameters of motor vehicle behavior can be reviewed so that the analysis of queues and vehicle delays is more comprehensive.

IOP Conf. Series: Earth and Environmental Science 1117 (2022) 012024 doi:10.1088/1755-1315/1117/1/012024

2. Literature review

An intersection is a meeting of two or more road segments that serve to direct vehicles to change direction. Intersections are part of a road network with various points of conflict that occur [3], so they are critical in providing services to road users.

2.1 Intersection Type

According to Morlok (1988) [4], the types of intersections based on the arrangement method are divided into 2 (two) types, namely:

a. A crossing without a signal, that is, an intersection that does not use a traffic signal, and
b. A conjunction with a sign means road users can pass through the hub by the traffic signal operation.

2.2 Vehicle Delay at Intersection

Vehicle delays at the intersection are the real-time average obstacles experienced by vehicles when passing at the crossroad [5].

2.3 Vehicle Queue Length at Intersection

The queue length at the intersection is the distance in meters from the stop line where vehicles queue at the foot of the signalized intersection [6,7].

2.4 The conception of the Micro-simulation Model

The concept of a simulation model is often used in planning a transportation activity, especially one that is dynamic and very broad, such as traffic conditions and situations. Traffic conditions are very comprehensive and have a variety of characteristics and many parameters, so a simulation model approach is required to simplify a complex problem.

2.4.1 Traffic Simulation Micro Concepts Based on Vissim

Micro-simulation can simulate the behavior of individual vehicles in a predefined road network and is used to predict the possible impact of traffic patterns resulting from changes in traffic flow or changes in the physical environment around the road. In the concept of micro-simulation, the model used in the micro-simulation tool is explicitly known as the car following model. Car Following Model (CFM) is a model used to control the behavior of the driver or driver towards other drivers in the same lane, and the model was developed by Gipps (1981).

2.4.2 Traffic Micro-Simulation Parameters Based on Vissim

The vissim-based micro-simulation parameter is used in the calibration and validation process in traffic simulation modeling (Tutorial Ptv Vissim 8). The nine parameters selected in the modeling are 1) Average Standstill Distance, 2) Additive Part of Safety Distance, 3) Multiplicative Part of Safety Distance, 4) Minimum Headway, 5) Desired Position at Free Flow, 6) Overtake at Same Lane (left), 7) Overtake at Same Lane (right), 8) Minimum Lateral Distance (driving), 9) Minimum Lateral Distance (standing).

2.4.3 Calibration and Validity of Simulation Model

Validation on Vissim is a process of testing the correctness by comparing observations and simulation results. Validation is concerned with determining whether conceptually the simulation model can represent the model accurately [8]. The calibration and validation process is carried out based on the volume of traffic flow and queue length. The validation approach used is the Chi-square equation model and the statistical model according to Geoffrey E. Havers (GEH). The general formula for Chi-square (x2) can be seen in equation 1 [9]

$$x^2 = \sum_{i=1}^{k} \left| \frac{O_i - E_i}{E_i} \right| \quad (1)$$

Where: Oi = observation data, Ei = simulation data

The GEH equation model is a modified statistical formula of Chi-square by combining the difference between relative and absolute values. The test criteria are accepted when the Chi-square value is calculated Chi-square table with a significant level and the degree of confidence in the Chi-square test is 95% or = 0.05. The GEH equation is shown in equation 2 [10] and meets the requirements of the resulting error value, as shown in Table 1.

$$GEH = \sqrt{\frac{(q_{simulated} - q_{observed})^2}{0{,}5 \times (q_{simulated} + q_{observed}}} \quad (2)$$

where: q = traffic flow volume (vehicles/hour)

Table 1. Geoffrey E. Havers (GEH) Statistical Equation Model

Value of GEH	Describsion
GEH < 5,0	Accepted
5,0 ≤ GEH ≤ 10,0	warning: possible model error or bad data
GEH > 10,0	Rejected

3. Research Design

And holidays and Data analysis. This research was designed starting from conducting previous research at the study location at the signalized intersection of Jalan Ahmad Yani and Jalan Jemursari, the second stage of providing the equipment used, carrying out a traffic volume survey for the observation period in the morning, afternoon and evening respectively on weekdays, November, 2nd and 3rd 2020. The traffic volume data obtained by the Vissim program is the peak hour volume in the afternoon.

3.1 Location of Research

The research is located by the signalized intersection of Ahmad Yani road - Jemursari road in Surabaya city, as shown in Figure 1. Figure 2 shows the distribution and situation of the Ahmad Yani-Jemursari signalized intersection along with the direction of traffic movement, divided into four intersections with eight directions.

Figure 1. Research Location Ahmad Yani road - Jemursari Road

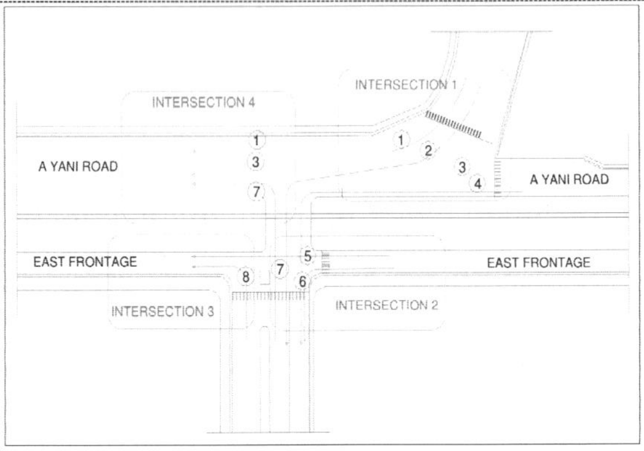

Figure 2. Division of intersection and direction of traffic flow

For more details on the direction of traffic flow at the four intersections, see Figure 3 and Table 2.

Tabel 2. Traffic Flow Movement at Four Intersections Jl. Ahmad Yani-Jl. Jemursari

No.	Name	Direction of Traffic Movement	Describsion
1	Intersection 1	P1	Moving from the south, turn around to the south, Ahmad Yani road (round the Dolog Roundabout)
		P2	Moving from the south of Ahmad Yani road, headed east to Jemursari road
		P3	Moving from the north Ahmad Yani road straight to the south of Ahmad Yani road
		P4	From the north, Ahmad Yani road turns left onto Jemursari road
2	Intersection 2	P5	Moving from the north of the East Frontage Road straight towards the south of the East Frontage Road
		P6	Moving From the north, East Frontage Road, turn left onto Jemursari road
		P7	From the east, Jemursari road turn left onto Ahmad Yani road
3	Intersection 3	P8	From the east, Jemursari road turn left towards East Frontage Road
4	Intersection 4	P1	Moving from the south of Ahmad Yani, turned back to the south of Ahmad Yani road (round the Dolog Roundabout)
		P3	Moving from the north Ahmad Yani road straight to the south of Ahmad Yani road
		P7	The movement from the east, Jemursari road turn left onto Ahmad Yani road

3.2 Micro-simulation Survey and Analysis Methods

The types of surveys conducted in this study are summarised in Table 3.

Table 3. Types of Surveys, locations, and tools used

Type of survey	Location of survey	Tools
Inventory	At the intersection	meter rollers, survey forms, and stationery
Vehicle Volume	Eight directions of movement at the intersection of Ahmad Yani road – Jemursari road	Handycam and tripod
Vehicle delay		
Vehicle Queue Length		

The micro-simulation analysis was brought off with PTV Vissim 8 to implement the micro-simulation model so that the analysis results match the reality in the field. The first step is to determine the parameter values in the simulation model, namely the calibration process using the trial and error method and the results of the GEH test analysis (Geoffrey E. Havers). The parameters that exist on PTV vissim and by the selected existing traffic flow, a calibration process is carried out on the volume of vehicles during the peak hour period.

4. Result and Discussion

4.1 Vehicle Volume at Intersection

The volume of total traffic flow at each movement is shown by the various types of vehicles crossing at the intersection in Table 4. The enormous volumes are at P3 and P1 at intersection one at 4,564 vehicles/hour and 4,502 vehicles/hour, and at P7 and P4 at intersection 2 of 3,190 vehicles/hour and 1,779 vehicles/hour. The type of vehicle is dominated by motorcycles by 71%.

Tabel 4. The volume of Vehicle Types in Peak Hour Period (vehicles/hour)

Vehicle Movement Direction	Volume Per Vehicle Type Peak Hour Period (vehicle/hour)												TOTAL (Veh./hour)
	Motorcycle	Big Motorcycle	City Car	Sedan Car	MPV	SUV	Truck	Box Car	Small Bus	Big Bus	1 Axel Truck	2 Axel Truck	
P 1	3,005	255	297	72	569	181	63	45	7	6	1	1	4,502
P 2	461	60	46	13	103	43	13	20	4	1	1	1	766
P 3	3,229	247	230	55	548	133	78	33	1	6	1	3	4,564
P 4	996	102	118	30	427	78	10	14	1	1	1	1	1,779
P 5	526	34	51	2	80	24	4	2	1	2	1	1	728
P 6	305	22	16	4	25	6	3	3	1	1	1	1	388
P 7	2,004	715	208	3	201	19	19	16	1	1	2	1	3,190
P 8	662	3	33	6	299	86	13	11	1	1	1	1	1,117

4.2 Micro-simulation of traffic at the intersection

The calibration values and results for the *nine* selected vissim parameters are shown in Table 5.

Tabel 5. Calibration values of 9 vissim parameters at the intersection

No.	Simulation parameter	Before Calibration (m)	After Calibration (m)
1	Average Standstill Distance	2	0.06
2	Additive Part of Safety Distance	2	0.06
3	Mutiplicative Part of Safety Distance	3	1
4	Minimum Headway	0.5	0.5
5	Desired Position at Free Flow	middle	any
6	Overtake at Same Lane (left)	off	on/yes
7	Overtake at Same Lane (right)	off	on/yes
8	Minimum Lateral Distance (driving)	1	0.3
9	Minimum Lateral Distance (standing)	1	0.5

Table 5 presents the nine selected vissim parameters with the calibration values for existing traffic behavior during the peak observation period. The calibration results show parameter values that differ before calibration (default) and after calibration.

4.3 Validation of micro-simulation model results

The volume of traffic flow observed and the results of the calibration of the micro-simulation model and Geoffrey E. Havers (GEH) test for each movement at the four intersections during the peak hour periods are presented in Table 6 and visually shown in Figure 3 and Figure 4.

Table 6. Traffic flow volume as a result of observation, model calibration, and GEH test

Intersection	The direction of Traffic Movement	Traffic Flow Volume		*GEH Test*	Result
		Observation	Simulation Model		
1	P 1	35	28	1.25	Accepted
	P 2	543	503	1.75	Accepted
	P3	1932	1763	3.93	Accepted
	P4	552	654	4.15	Accepted
	Total	3080	2948	2.40	Accepted
2	P 4	1101	1143	1.25	Accepted
	P 5	321	320	0.06	Accepted
	P 6	124	158	2.86	Accepted
	P 7	1328	1459	3.51	Accepted
	Total	2874	3080	3.78	Accepted
3	P 7	351	320	1.69	Accepted
	P 8	754	806	1.86	Accepted
	Total	1105	1126	0.63	Accepted
4	P 1	1698	1770	1.73	Accepted
	P 2	1093	1156	1.88	Accepted
	P 7	1340	1437	2.6	Accepted
	Total	4131	4363	3.56	Accepted

Table 6 shows that the traffic flow volume from the observation and micro-simulation model with the Geoffrey E. Havers test at the Ahmad Yani - Jemursari road intersection, has been well calibrated for the 4 (four) intersections in the eight directions of traffic flow movement. The simulation results with the GEH test can be accepted for all movements, both in the eight directions of movement and in total volume at the four intersections. GEH value < 5.00 or in the range 0.63-4.15.

Figure 3. Visualization of the micro-simulation model Before calibration

Figure 4. Visualization of the micro-simulation model After calibration

Changes in the volume of traffic flow and vehicle situation, which occur in Figures 3 and 4, are the results of the simulation-model calibration by changing the value of the driving behavior parameter according to the nine parameter values in Table 5. The traffic flow volume, before calibrated, has a queue of very regular vehicles in their respective lanes. Whereas, after calibration of the simulation model, the importance of traffic flow and vehicle conditions began to become irregular, the distance between vehicles followed the existing requirements, namely close together and some cars not in the lane. This indication shows traffic behavior at the intersection of Ahmad Yani road-Jemursari road.

4.4 Queue Length and Delay of existing micro-simulation vehicles

The results of the micro-simulation analysis of queue length and vehicle delay according to the nine actual parameter values (Table 5) for the four intersections showed in Tables 7, 8.9, and 10.

The tables show the maximum and minimum queue lengths and vehicle delays with the level of Service (LOS) from each direction of traffic flow during the peak hour observation period.

Table 7. Queue length and vehicle delays and LOS at intersection 1

Movement	Qlen-Min (m)	Qlen-Max (m)	Veh-Delay (det)	LOS
P1	146.93	161.28	138.97	F
P3	140.18	161.25	152.95	F
Total	143.56	161.28	173.51	F

Table 8. Queue length and vehicle delays and LOS at intersection 2

Movement	Qlen-Min (m)	Qlen-Max (m)	Veh-Delay (det)	LOS
P6	124.45	161.22	878.94	F
P7	142.84	161.23	164.83	F
P9 (P2 and P4)	80.21	100.80	1.63	A
Total	86.92	161.23	216.44	F

Table 9. Queue length and vehicle delays and LOS at intersection 3

Movement	Qlen-Min (m)	Qlen-Max (m)	Veh-Delay (det)	LOS
P8	132.04	161.25	0.58	A
Total	66.18	161.25	1.01	A

Table 10. Queue length and vehicle delays and LOS at intersection 4

Movement	Qlen-Min (m)	Qlen-Max (m)	Veh-Delay (det)	LOS
P7	137.81	160.85	9.71	A
Total	140.76	161.30	19.56	C

The performance and behavior of motorized vehicles from the microsimulation show that the signalized intersection of Ahmad Yani road – Jemursari road is already at LOS F. Where traffic is jammed, and there is a low speed, long queues, and enormous obstacles or delays. The maximum queue length occurs at P1 intersection 1 of 161.28 meters, and the minimum queue length occurs at Intersection 3 of 66.18 meters. The most considerable vehicle delay value occurred at P6 Simpang 2 of 14.65 vehicle minutes.

To provide optimal service, the signalized intersection of Ahmad Yani road - Jemursari road should be a non-level intersection with the Flyover model. So that the further analysis in this paper will examine the geometric model of the Ahmad Yani road-Jemursari road signalized intersection as a non-plane intersection with the Flyover model.

5. Conclusions

1. Analysis of the signalized intersection Jl. Ahmad Yani-Jl. Jemursari is carried out based on four intersections, eight directions of movement, and one movement of merged traffic flow.
2. Microsimulation of traffic volume is carried out by changing the nine values of the driver's behavior parameters according to the actual conditions at the intersection.
3. The traffic volume from the microsimulation model is validated with the Geoffrey E. Havers test and is accepted as the accurate traffic volume
4. The maximum queue length and motor vehicle delays for the peak hour observation period occur at intersections 1 in the north and 2 in the east, Jl. Ahmad Yani-Jl. Jemursari.
5. The performance of the Ahmad Yani-Jemursari signalized intersection is already at LOS F, where traffic is jammed, there is a low speed, long queues, and significant obstacles or delays.
6. Further study of this paper is analyzing the geometric model of the Ahmad Yani-Jemursari road signaled intersection into a non-plane intersection with the Flyover model.

References

[1] Asriandi R E P and Ramanda F 2018 *Green Time Optimization of Signal Intersection Through A Use of Ptv Vissim For Improving Intersection Performance* Vol 6 No.2 (BENTANG Jurnal Teoritis dan Terapan Bidang Rekayasa Sipil)
[2] ----------- 2016 *PTV VISSIM 8 User Manual* (PTV AG Karlsruhe Germany)
[3] ----------- 1997) *Manual Kapasitas Jalan Indonesia* (Direkotrat Bina Jalan Kota Jakarta)
[4] Morlok E K 1998 *Pengantar Teknik dan Perencanaan Transportasi* (Erlangga Jakarta)
[5] Tamin O Z 2000 *Perencanaan dan Pemodelan Transportasi* (ITB Bandung)
[6] Sutomo H 1992 *Appropriate Saturation Flow at Traffic Signals in Javanese Cities: A Modelling Approach* (PhD Thesis Institute of Transport Studies University of Leeds UK)
[7] Hoque M D 994 *The Modelling of Signalised Intersections in Developing Countries* (PhD

Thesis Department of Civil and Environment Engineering University of Southampton UK)
[8] Hellinga B and Hesham R 1996 *systematic Verification, Validation and Calibration of Traffic Simulation Models* (Presented at the 1996 Transportation Research Board Annual Meeting Washington DC)
[9] Hidayat A 2014 *Tutorial Rumus Chi Square Dan Metode Hitung* (Diakses Juni 2017 *https://www.facebook.com/anwar.hidayat1*)
[10] Putri N H dan Irawan M Z 2015 Mikrosimulasi Mixed Traffic Pada Simpang Bersinyal Dengan Perangkat Lunak Vissim (The 18th FSTPT International Symposium, Unila, Bandar Lampung) p 1-10

IOP Conf. Series: Earth and Environmental Science 1117 (2022) 012025 doi:10.1088/1755-1315/1117/1/012025

Early Age of Volume Weight, Indirect Tensile Strength and Tensile Elastic Modulus of Foam Concrete Containing Blended Cement

Mansyur[1], M W Tjaronge[2], R Irmawaty[2] and A A Amiruddin[2]

[1] Doctoral Student, Civil Engineering Department, Hasanuddin University, Jalan Perintis Kemerdekaan Km. 10, Makassar 90245, Indonesia
[2] Civil Engineering Department, Hasanuddin University, Jalan Perintis Kemerdekaan Km. 10, Makassar 90245, Indonesia

Mansyurusn14@gmail.com

Abstract. Portland composite cement is a type of blended cement produced by the Indonesian cement factories where fly ash as one of the ingredient and the production is based on efforts to reduce CO_2 emissions in the cement production process. This research is part of the sustainability infrastructure development which is carried out through the use of Portland composite cement as a cementitious material in the manufacture of foam concrete. The purpose of this paper was to contribute to the understanding of the early age volume weight, indirect tensile strength and tensile elastic modulus of the hardened foam concrete. In this study, two mix designs of foam to mortar ratio of 32.53%: 67.47% and 55.41%: 44.59%, by volume were used to produce foam concrete. The test results showed a good compatibility of Portland composite cement-superplasticizer based mortar in combination with foam can be achieved to determine the foam concrete slurry stability and can thus form hardened foam concrete. By using foam to mortar ratio of 32.53%: 67.47%, volume weight, indirect tensile strength and tensile elastic modulus achieved were 1508.3 kg/m^3, 0.83 MPa and 207.5 MPa at three days, and 1476.9 kg/m^3, 0.89 MPa and 254.3 MPa at 7 days, respectively. Tensile strength of foam concrete with foam content of 55.41%: 44.59% at 3 days and 7 days were 0.37 MPa and 0.46 MPa, respectively. The foam to mortar rasio of 55.41%: 44.59% produced foam concrete with tensile elastic modulus of 137 MPa and 200 MPa, at 3 days and 7 days respectively.

1. Introduction

In the last decades, sustainability is a major parameter in the technology development including the construction industry. In this regard, sustainability-based innovation in the production of cement, which is one of the most widely used materials in construction using concrete, continues to be carried out. By producing blended cement that allows the use of by products with contain large amounts of silica, it supports the reduction of CO_2 emissions, reduces the extraction of natural materials and reduces fuel consumption [1]. Fly ash is a by-product that contains large amounts of silica (SiO_2) and a number of materials that can be used as an ingredient in cement. In Indonesia, one of the blended cements that have been produced by national cement factories is Portland composite cement which contains fly ash from burning coal in power plants. Portland composite cement is produced based on

Indonesian National Standard [2]. Many studies have shown that Portland composite cement is able to produce concrete with good performance [3-7].

The use of lightweight materials in the tall building structures and the dwelling houses is part of continuous innovation based on sustainability. Lightweight materials usage can reduce the weight of the load borne by the structural elements so as to lessen the dimension of beam, column and foundation which have implications for reducing material use. Fuel consumption can be reduced on transport of precast lightweight element structures. One of the lightweight materials that continuously have been developed today is foam concrete. Foam concrete is made up of Portland cement based paste, filler or fine aggregate, water and foam without coarse aggregates. The foam content produces numerous microvoids in the hardened foam concrete hence the weight is lighter than the normal weight concrete, while in fresh state the foam concrete slurry has good consistency to flow freely with its own weight filling the mold [8-9]. Preliminary research on the use of Portland composite cement in the manufacture of foam concrete that has been performed by Tjaronge et al shows that the hardened foam concrete made of Portland composite cement has the same early compressive strength as foam concrete made from Ordinary Portland cement [10].

The ability to spread into the mold and self-compacted ability of foam concrete slurry can be evaluated by observing the shape and volume weight of the hardened foam concrete. The amount of foam content affects the volume weight while the volume weight of the hardened foam concrete is related to many physical properties including tensile strength and tensile modulus elasticity. One part of this research was to examine the effect of foam content on the early age volume weight of the hardened foam concrete. Knowledge of the 3 and 7 days volume weigth can be used to measure the load that will be created by 3 days and 7 days age lightweight foam panel if installed on the building structure or the dwelling house.

Solid materials such as rock consists of non-homogeneous compounds and containing numerous microcracks, consequently, the tensile and compression conditions will be responded differently hence there are two types of elastic modules: the compressive elastic module Ec and the tensile elastic modulus (Es). Much effort and attention have been placed on the tensile elastic modulus by researchers, where method to obtain tensile elastic modulus by using indirect tensile strength test (the Brazilian disc) has been systematically developed [11-12]. The hardened foam concrete which the ingredients are not homogeneous, and contains numerous microvoid so has physical similarities to the rocks. Indirect tensile strength and tensile elastic modulus at early age are important parameters with need to be studied in pursue to understand the bearing capacity of the hardened foam concrete to respond the tensile loads. Efforts to obtain indirect tensile strength and tensile elastic modulus of hardened foam concrete by using indirect tensile strength test with equipped with linear variable differential transformers (LVDTs) was reported in this study.

2. Materials and Methods

2.1. Portland Composite Cement

Table 1 displays the physical properties of Portland composite cement used in this study. The chemical compounds of Portland Composite Cement were obtained from X-ray fluorescence (XRF) test results. Chemical compound of the Portland composite cement used as cementitious material is dominated by the CaO and silica (SiO_2) elements, which are 61.79% and 18.39%. Whereas the other composed elements are 0.99% MgO, 1.81% SO_3, 5.15% Al_2O_3, 5.15% Fe_2O_3, and 4.61% LOI (Loss of Ignation), respectively.

Table 1. Physical Properties of Portland Composite Cement

Physical properties	Results
Fineness	382
Compressive strength	
a. 3 days	18.86 MPa
b. 7 days	26.81 MPa
c. 28 days	41.79 MPa
a. Setting time (Vicat test)	
b. Initial setting,	132.5 minutes
c. Final setting,	198 minutes
Hydration temperature 7 days,	65 cal/gr
Normal consistency	25.15 %
Specific gravity	3.03

2.2. Fine Aggregate

Table 2 shows the physical properties of fine aggregate. The fine aggregate in this research is silica sand taken from a local sand mine from a river in Pinrang Regency, South Sulawesi Province. This test was carried out based on the Indonesian National Standard regarding concrete requirements for structural buildings. It can be seen the physical properties of fine aggregates used in this study met the specifications of the Indonesian National Standard (SNI) for concrete materials required.

Table 2. Physical properties of fine aggregate

No.	Property	Value
1.	Spesific gravity	
	-Bulk density	2.58
	-Saturated surface density	2.61
	-Apparent density	2.65
2.	Water absorption	0.91 %
3.	Modulus of fineness	2.95
4.	Volume weight	
	-Loose	1.40 kg/lt
	-Dense	1.48 kg/lt
5.	Water content	3.59 %
6.	Organic content	No. 1

2.3. Mixtures Design of Foam Concrete

The foam concrete mixture design can be seen in Table 3.

Table 3. Mix design foam concrete (m^3)

No	Foam-mortar ratio (by volume)	Mortar			
		Portland Composite Cement (kg)	Water (kg)	Sand (kg)	Naphthalene based superplasticizer (kg)
1	32.53% : 67.47%	662.5	232	1325	16.6
2	55.41% : 44.59%	662.5	232	1325	16.6

In this study, 2 different mix designs were made, where mix designs 1 and 2 had a ratio of foam to mortar by volume of 32.53% : 67.47%, and 55.41% : 44.59%, respevtively. Meanwhile, the mortar composition was kept constant for all mix designs.

2.4. Production of Foam Concrete Specimens and Testing Slump Flow

Foam was prepared in a container. Foam agent was mixed with water in a ratio of 1:3.3 and stirred until stable foam was obtained. Meanwhile, Portland composite cement, fine aggregate, water and superplasticizer were mixed in a mixer to form mortar. The prepared foam was poured into mixer and stirred with mortar to form foam concrete slurry.

Foam concrete slurry was poured into the cylindrical mold with diameter of 10 cm and height of 20 cm without any compaction activity, and then stored in the room with temperature of 27.5 ± 5°C and humidity of 60 RH. After 24 hours, the hardened foam concrete removed from the mold and stored until the age of 3 days and 7 day.

2.5. Indirect Tensile Strength Test

Indirect tensile stregth test ITS testing was carried out on the 3 and 7 days specimens. Specimen was laid horizontally and subjected to a compressive load along the sides to produce tensile stress. The load speed is 5 mm/sec. A load cell was used to measure the given load. In the middle of the two sides, LVDTs were installed to measure the displacements that occurred. Load and displacement measurements were recorded automatically with a computerized data logger set. Furthermore, through the relationship of tensile stress and strain, tensile elastic modulus (Es) was calculated by the equation 1 [13]:

$$E = (\sigma/2)/\varepsilon \tag{1}$$

Where σ is the peak tensile stress; σ/2 is the stress in stress-strain curve which is half of the peak tensile strain; ε is the strain related to σ/2 in the stress-strain curve.

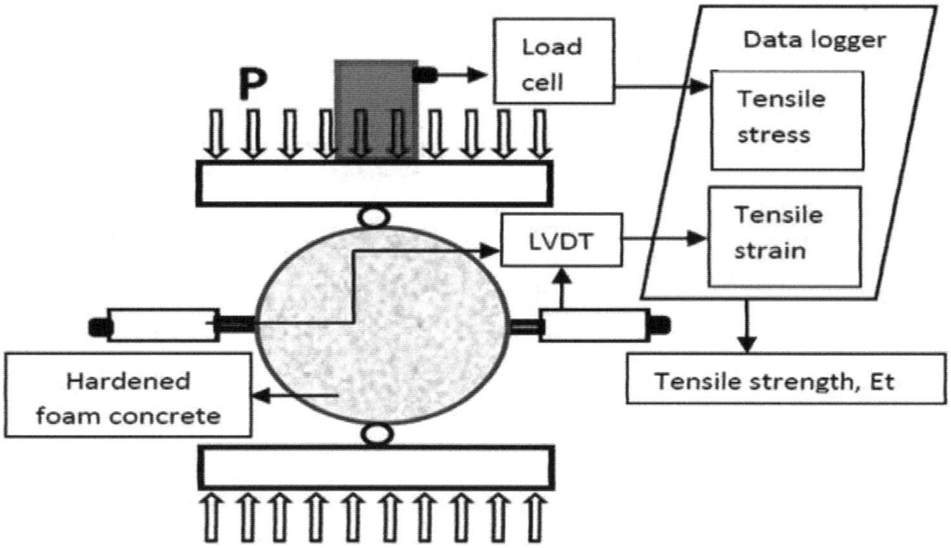

Figure 1. Indirect tensile strength test set up

3. Results and Discussion

Visual observation of the shape and appearance of the test object was carried out before measuring the volume weight of the specimens. All specimens had good shape and appearance without macrovoids, honey combs and no accumulation of fine aggregate at the bottom part. These results indicated that the composition of Portland composite cement superplasticizer mortar combined with foam had a good compatibility and can thus formed stable slurry foam concrete with good self-compaction ability and resistance to segregation during flowing and spreading into the mold.

Table 4 shows volume weight, indirect tensile strength and tensile elastic modulus of the specimens made based on a mix design with foam to mortar ratios of 32.53%: 67.47%, and 55.41%: 44.59%. Each value shown is the average value of the three specimens. At the mix design with the ratio of foam to mortar of 32.53%: 67.47% by volume, the volume weight, indirect tensile strength and tensile elastic modulus were 1508.3 kg/m^3, 0.83 MPa and 207.5 MPa, respectively at 3 days and 1476.9 kg/m^3. 0.89 MPa and 254.3 MPa, respectively at 7 days.

At the mix design with the ratio of foam to mortar of 55.41%: 44.59% by volume, the volume weight, indirect tensile strength and tensile elastic modulus were 1317.2 kg/m^3, 0.37 MPa and 137 MPa, respectively at 3 days and 1283.7 kg/m^3, 0.46 MPa and 200 MPa, respectively at 7 days. Knowledge of volume weight correlated with tensile strength and tensile elasticity at early agewill form the basis for structural applications and the wider usage of Portland composite cement based foam concrete.

Table 4. Volume Weight, Indirect Tensile Strength and Tensile Elastic Modulus

Mix Design	Foam to Mortar ratio	Volume Weight (kg/m^3)		Tensile Strength (MPa)		Tensile Elastic Modulus (MPa)	
		3 days	7 days	3 days	7 days	3 days	7 days
1	32.53%: 67.47%,	1508.3	1476.9	0.83	0.89	207.5	254.3
2	55.41%: 44.59%	1317.2	1283.7	0.37	0.46	137.0	200.0

4. Conclusion

This research produced foam concrete with two ratios of foam to mortar of 32.53%: 67.47%, and 55.41%: 44.59% by volume, respectively, while the composition of Portland composite cement, water, superplasticizer and fine aggregate was kept constant for all mixed designs. Based on the experimental results it can be concluded:

1. The composition of Portland composite cement-superplasticizer mortar in combination with foam established a good compatibility which determined stable foam concrete slurry and can thus form hardened foam concrete.
2. By using foam to mortar ratio of 32.53%: 67.47%, the volume weight, Indirect tensile strength and tensile elastic modulus were 1508.3 kg/m^3, 0.83 MPa and 207.5 MPa at 3 days, respectively and 1476.9 kg/m^3. 0.89 MPa and 254.3 MPa at 7 days, respectively.
3. Foam to mortar ratio of 55.41%: 44.59% by volume, resulted in volume weight, Indirect tensile strength and tensile elastic modulus of 1317.2 kg/m^3, 0.37 MPa and 137 MPa, respectively at 3 days and 1283.7 kg/m^3, 0.46 MPa and 200 MPa, respectively at 7 days

5. Acknowledgement

This work was supported by the Indonesian Lecturer's Excellence Scholarship Program (BPPDN). Most of the research works were conducted in the Eco Material and Concrete Laboratory, Civil Engineering Department, Hasanuddin University, Makassar-Indonesia.

References

[1] Jan Deja, Alicja Uliasz-Bochenczyk and Eugeniusz Mokrzycki. 2010. CO_2 Emissions From Polish Cement Industry. International
[2] Standard National of Indonesia. Portland Composite Cement. SNI 7064:2014
[3] Tjaronge M.W., Irmawaty R., Adisasmita S.A., Amiruddin A., and Hartini, Compressive Strength and Hydration Process of Self Compacting Concrete (SCC) Mixed with Sea Water,

Marine Sand and Portland Composite Cement. Advanced Materials Research 935, 2014, pp. 242–246.

[4] Erniati, M. Wihardi Tjaronge, Zulharnaha, and Ulva Ria Irfan, Porosity, pore size and compressive strength of self compacting concrete using sea water, The 5th International Conference of Euro Asia Civil Engineering Forum (EACEF-5), Procedia Engineering 125 (2015) 832 – 837

[5] M. A. Caronge, M. W. Tjaronge, H. Hamada and R. Irmawaty, Effect of water curing duration on strength behaviour of Portland composite cement (PCC) mortar, 2017 IOP Conf. Ser.: Mater. Sci. Eng. 271 012018

[6] M. A. Caronge, M. W. Tjaronge, R. Irmawaty, B. Bakri and S Hamuddin, Effects of vibration time on compressive strength and corrosion resistance of steel bars in concrete, 2020 IOP Conf. Ser.: Earth Environ. Sci. 419 012031

[7] Marewangeng, M.W.Tjaronge, A. R. Djamaluddin and Sumarni Hamid Aly, Mechanical characteristics of self compacting concrete using laterite stone as coarse aggregate, International Journal of GEOMATE, July, 2020, Vol.19, Issue 71, pp. 250 – 255

[8] Y.H. Mugahed Amran, N. Farzadnia, A.A. Abang Ali, Properties and applications of foamed concrete; a review, Constr. Build. Mater. 101 (2015) 990–1005.

[9] Valery Lesovik, Vasily Voronov, Evgeny Glagolev, Roman Fediuk, Arbi Alaskhanov, Y. H. Mugahed Amran, G. Murali and Andrey Baranov, Improving the behaviors of foam concrete through the use of composite binder, Journal of Building Engineering 31 (2020) 101414

[10] Y Sunarno, M W Tjaronge, and R Irmawaty, Preliminary study on early compressive strength of foam concrete using Ordinary Portland Cement (OPC) and Portland Composite Cement (PCC), 2020 IOP Conf. Ser.: Earth Environ. Sci. 419 012033

[11] Ye Jianhong, F. Q. Wu, J. Z. Sun, Estimation of the tensile elastic modulus using Brazilian disc by applying diametrically opposed concentrated loads. International Journal of Rock Mechanics & Mining Sciences 46 (2009) 568–576.

[12] J. Claessona and B. Bohlolib, Brazilian test: Stress field and tensile strength of anisotropic rocks using an analytical solution, International Journal of Rock Mechanics & Mining Sciences, 2002, 39(8): 991–1004.

[13] Feng Ming, Dongqing Li, Mingyi Zhang, Yu Zhang. 2017. A Novel Method for Estimating the Elastic Modulus of Frozen Soil. Cold Regions Science and Technology.

Structural Settlement Evaluation of Low-Crested Breakwaters Made of Geotextile Tubes

Dede M. Sulaiman, Ririn Rimawan, Mahdi Ibrahim Tanjung, I Putu Samskerta

Technical Implementation Unit for Hydraulics and Geotechnics, Directorate of Technical Development for Water Resources, DGWR, MPWH Jl. Ir.H.Juanda N0.193 Bandung 40135, Indonesia dedems@ymail.com; ririn.rimawan@pu.go.id; mahdi.ibrahim7@gmail.com; samskerta@gmail.com

Abstract. One of the efforts to improve successful in handling coastal erosion is to develop a coastal protection structure to be called as PEGAR, an acronym of Pemecah Gelombang Ambang Rendah or Low-Crested Breakwater (hereinafter referred to as LCB). The use of geotextile tubes filled with sand in LCB construction has been widely practiced since 2010. LCBs play a role in breaking waves and dampening their energy before reaching the coast. The breaking of the waves generates a rotating current behind the structure which results in the deposition of sediment supplied through the crest of the structure and longshore drift. Subsequent sedimentation processes produce sedimentary deposits and forms new shorelines (salient) and even forms tombolo. Through these two sediment supplies, natural sand filling and beach widening can take less than one season.

However, from the experience of applying LCBs in the field, there are several LCBs that experience structural degradation. In addition, it even subsided after several years of operation which resulted in a decline in its function as a coastal protection. Evaluations and suggestions for improving a more stable and durable low-crested structure are given in this paper.

1. Introduction

One of the efforts to increase the success of coastal erosion management is to develop a coastal protection structure called LCB, an acronym for Low-Crested Breakwater type. The use of sand-filled geotextile tubes in LCB construction has been widely practiced since 2010. LCBs play a role in causing waves to break, dampening their energy before reaching the coast. The breaking of the waves generates a rotating current behind the structure which plays a role in depositing sediment supplied through the crest of the structure and longshore drift from the surrounding coast. The subsequent sedimentation process produces sediment deposits and forms a new shoreline (salient) and even forms tombolos, sedimentary deposits that have merged with the structure. Through these two sediment supplies, the process of sand filling and natural widening of the beach can take place quickly[1].

LCB is designed to reduce, prevent, and control coastal erosion by accelerating the accumulation of sediment that forms new beaches. This LCB structure reduces wave energy by forcing the waves to break when in contact with the structure. Some of the energy is reflected and the rest is transmitted to the beach by transporting sediment[2]. A practical approach to produce an effective LCB as a coastal erosion control is to place the LCB structure in a position between the mean water level and the highest sea level. In addition, the distance of the LCB from the beach is proportional to the length of the structure[2]. A number of LCB prototypes have been implemented since 2010 at Pasir Putih Beach Anyer, Tanjung Kait Beach Tangerang (2011), Pisangan Beach Karawang (2012), Sari Pekalongan Beach (2012), Sigandu Batang Beach (2014 and 2015), Depok Batang Beach (2018 , and Sukaria Batang Beach (2019). The location of the PEGAR prototype is located on the conditions and characteristics of a sloping beach with relatively small wave and current conditions. The field application for strong wave conditions and coastal currents was carried out at Peisah Beach, Jembrana, Bali in 2018. The coastline response, both for relatively small wave conditions as well as for high waves and strong currents such as Pebuahan Beach, Bali, has given a positive response with the formation of a new beach in the form of a new coastline (salient) and sediment deposits that merge with the LCB (tombolo). The process of deposition of sediment until the formation of a new beach is less than one season[3].

Besides its advantages in restoring eroded beaches and making the beach wider, the weakness of the LCB structure of this geotextile tubes material is the stability of the structure and its service life.

IOP Conf. Series: Earth and Environmental Science 1117 (2022) 012026 doi:10.1088/1755-1315/1117/1/012026

Structural settlement, reduced sand volume, and tear damage have reduced its function as a wave absorber. This paper discusses how the LCB structure made from geotextile tubes can function durable through several suggestions for strengthening the foundation and the use of armor layer.

2. Materials and Methods

2.1 Low-Crested Breakwaters Concept

Offshore breakwater is a coastal protection structure that is placed parallel to the shoreline and functions as a damper for wave energy before reaching the coast. This type of coastal protection structure with the crest of the structure above the highest water level and not attached to the beach is very effective in restoring beaches affected by erosion. However, the use of conventional breakwaters, especially on Indonesian beaches, is still rare and less popular than other types of coastal protection structures such as groines, revetments, or sea walls.

The lack of use of this type of coastal protection is due to several factors, including this breakwater structure which is known as a very expensive coastal protection method and its impact is very aesthetically disturbing, especially for tourist beaches. One of the efforts to obtain effective and environmentally friendly coastal protection technology is to develop LCB structures. This LCB structure is an offshore breakwater structure with peak elevation slightly above the mean water level[1,2,3].

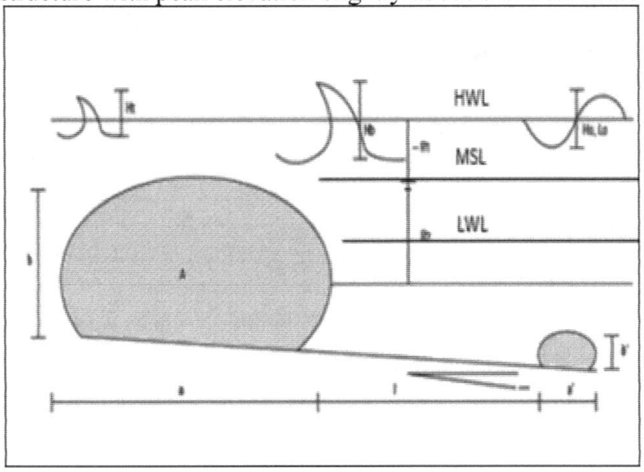

Figure 1. Typical Cross Section of Geotextile Tube LCB[3]

As the name implies, LCB is a coastal protection structure that is always overtopped by waves. The armor on the LCB is still being hit causing the waves to break[4]. The difference between LCB and conventional breakwater lies in the bending or wave diffraction which is more dominant in the LCB. Likewise, the reflection of waves towards the sea is more dominant in conventional breakwaters. Another difference is the amount of wave energy that can overtop the structure. The conventional breakwaters retain most of the wave energy, on the other hand the LCB releases some of the wave energy above it into the shadow region behind the structure [1,5].

This innovation to conventional breakwater structure is carried out through dimensional modifications, namely by reducing the peak elevation of the structure which is then directly proportional to the reduced volume and cost. Another innovation is by applying the use of geotextile tubes filled with sand so as to increase added value, namely reducing dependence on natural rocks which is starting to become scarce. The development of this LCB is based on a series of experimental results in the laboratory and in the field so as to produce a structure that has effective protection with relatively cheaper construction costs than conventional breakwaters. Research and development of the LCB structure began in 2009 through physical model testing at the Coastal Engineering laboratory in Ciparay, Bandung Regency [3,6].

2.2 Geotextile Tubes

LCB-forming materials can be used ranging from natural rocks, concrete blocks, to innovative materials such as geotextiles. Geotextile is the most popular material as a coastal protection structure material. The geotextile tube was chosen because of its stability and ability to reduce wave energy, and also because of its flexibility to adapt to dynamic coastal media. In addition, the adaptation of geotextile tobe to coastal morphological responses is fast and the installation and maintenance costs are relatively more economical [1,6].

The use of geotextile sacks provides several advantages, namely 1) types of geotextile materials available in the market with various sizes; 2) the geotextile material is relatively light which allows the transportation cost to the location to be relatively cheap; 3) fill material such as sand can be obtained at the project site; 4) response to flexible waves; and 5) can be installed using local power without the need for heavy equipment and special skills[6]. Although economically cheaper than natural stone or concrete materials, in terms of durability and service life, geotextiles are largely determined by scouring, impact, and tearing which are mostly caused by human activities. Geotextile material as a form of LCB has shown the restoration power of beaches affected by rapid erosion with the accumulation of sediment on the coast. The new coastline due to the LCB will then act as a beach protector that will dampen the waves before they reach the coast.

2.3. As Generator of Circular Currents And Natural Sand Nourishment

The main function of this LCB structure is to reduce wave energy through wave breaking. Waves coming from the deep sea, either perpendicular or at an angle to the coast, when passing through the LCB structure will break and produce a rotating current behind the structure[7]. The current generated by the breaking of the waves tends to rotate, weakening each other and encouraging the formation of sediment deposits on the coast. This current circulation phenomenon causes the formation of salient and tombolos that make the beach wider, as well as restore the previously eroded beaches.

LCB is also a method of protecting, growing, and restoring beaches affected by erosion through wave overtopping which contains various sediments that through the sedimentation process settle behind the structure. Because it is effective in depositing sediment and forming a new beach behind it, this low-crested structure can be considered as a natural sand nourishment technology [1,3,8].

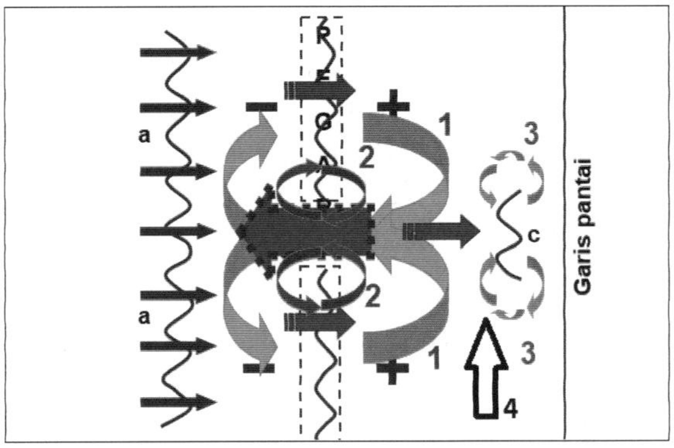

Figure 2. Illustration of the formation of circular current and sedimentation processes behind LCB[3]

2.4 Locations of LCB Prototypes

The prototype of the LCB field made of geotextile tubes was carried out on three different types of coasts: sandy beach with a steep slope (Pasir Putih Beach, Anyer- Banten Province)), sloping beach (Tanjung Kait Beach, Tangerang), and muddy coast (Pisangan, Karawang). Finally, a field test for steep coastal conditions with high waves and dominant tidal currents was carried out at Pebuahan Beach, Jembrana, Bali. The results of monitoring three months after installation showed that the LCB was able to deposit 185.8 m3 of sediment in coastal conditions with a longshore current speed of 66 cm/s, thus forming a new beach in a short time.

The rapid use of geotextiles as a coastal structure material has a direct impact on the increase in LCB construction made from geotextile tubes. This positive trend has contributed to the increasing length of coastal protection structures in Indonesia. The local governments of Pekalongan City and Batang Regency, Central Java, have built geotextile LCBs of 2,500 m and 2,280 m, respectively. Therefore, the length of coastal structures made of geotextiles up to 2018 is 4780 meter. This significant increase in beach length has enriched the repertoire of new beach structure types with the emergence of LCB structures.

Tabel 1. Length of National coastal protection structure Until 2018[3]

Types of Coastal Structure	Length (m)	Percentage
Breakwaters	1.351,45	0,74%
Groins	5.39,07	3,00%
Jettty	429,92	0,24%
Revetment	65.843,66	36,30%
Seawall	80.183,81	44,17%
Submerged Breakwaters/LCB	4.780,00	2,91%
NCICD Seadike	6.000,00	3,30%
Total Length	**163.985,91**	100 %

3. Results and Discussions

3.1 Settlement of LCB Geotextile Tubes Along Depok Beach

Project implementation is generally carried out after the disbursement of the budget, both state and regional budget, and is carried out at the end of the East monsoon towards the end of the year. The installation was carried out in a simple way, some companies deploy a layer of geotextile as a scouring apron, some directly place the LCB geotextile tubes directly at a certain distance from the beach without any foundation or scouring apron. The installation of the LCB at the end of the east monsoon period is, in fact, the laying of the LCB structure on top of the sediment that was accreted during the east monsoon. These sediments are unstable sediments, easily eroded and carried by longshore sediment transport during the next wind season. In the west wind season (January-June) with the direction of sediment transport along the coast from west to east, the sediment that has settled during the east monsoon is eroded again and transported by longshore drift that leads to the east. Therefore, the LCB structure that is installed above the east monsoon sediment heap, starting at the beginning of the west monsoon, has decreased or settled. The results of monitoring in April 2019, which is 3 months since it was installed in December 2018, the LCB Geotextile tubes structure at Depok Beach has decreased by about 50 cm from the position at the time of installation. The peak elevation of LCB only appears to the surface at low tide (LWL), whereas at the beginning of the installation the top of the structure was installed at the mean sea level (MSL). The phenomenon of structural subsidence is dominantly caused by seasonal changes as shown in the illustration in Figure 3. In addition, some geotextile tubes experience a decrease in elevation because the volume of sand is reduced by leakage[9].

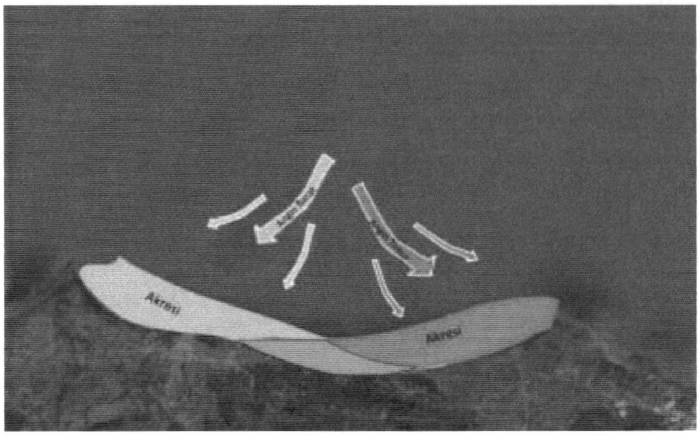

Figure 3. Illustration of changing accretion location depending on season [9]

3.2 Settlement of LCBs Along Sukaria Beach

Sukaria Beach is one of the Depok beach segments which has developed into a tourist beach area following the opening of a coastal highway to the location of the Power Steam (PLTU) of Batang in Ujung Negoro. The beach is narrow and experiencing severe erosion after a storm surge that destroyed almost all of Depok Batang Beach in December 2018. In order to disseminate LCB technology to prevent coastal erosion, a training for personnel of the human resources of the Ministry of Public Works and Housing was conducted in April 2019. The selection of LCB technology made of geotextile tubes in this training refers to the successful of the technology was used to deal with coastal erosion in Batang Regency in 2014. As the location for the application of LCB, Sukaria beach was chosen which is a tourist beach and has coastal erosion problems that need to be addressed immediately. Planning and implementation of the installation of the LCB structure by 30 trainees who all went to the field. The length of the LCB structure from geotextile tubes is 60 m consisting of three different types of geotextiles, namely woven, non-woven, and composite geotextiles with a length of 20 m each and a height of 1.2 m. The LCB position was installed at a depth of 100 cm (MSL), the structure height was 120 cm, the distance from the beach was about 60 m. Monitoring of shoreline changes has been carried out in July 2019, December 2020, and December 2021. The results of monitoring in December 2021 the beach formed by the presence of the LCB was about 350 m length with a volume of deposited sand, was 6480 m3. The layout of the LCBs, wave conditions, tides, sediment types, and current patterns formed by the breaking of waves on the LCB contributes greatly to the formation of new beaches through sediment deposition on the coast. The phenomenon of structural decline and the rotary current pattern behind this low-threshold structure is a concern and demands an in-depth discussion.

Figure 4. Geotextile Tube LCB shortly after installation (left) and after 3 years(right) [10]

3.3 Efforts to make LCB more stable and durable

3.3.1 Geotextile Tube as a Core Covered with Armor

The limited service life and durability of geotextile tubes is one of the weaknesses in the use of geotextiles as LCB material. Therefore, it is necessary to strengthen the outer layer structure of the LCB with armor to make it more durable. The armor on the outer layer can be made of concrete cubes, natural rocks, tetrapods, or other types of armor and the geotextile tube is placed as the core layer. The addition of this outer layer does not change the character of the low-crested structure and the function of the LCB remains as a wave absorber that causes waves to break and overtop above it [10].

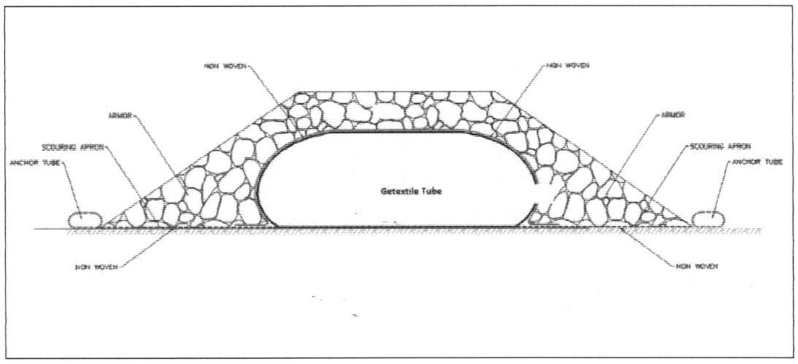

Figure 5. Proposed Armored LCB Geotextile tube[10]

3.3.2 Foundation Reinforcement and Scour Protection

Foundation stability means that the bearing capacity of the soil must be able to withstand structural loads without decreasing above the required tolerance threshold. For this reason, according to the planning and implementation guidelines for geotextile tubes on sandy beaches using geotextiles or woven bamboo (Figure 6). On the other hand, for the muddy coast, it recommends to use a bamboo raft combined with the installation of bamboo stick underneath (Figure 7). For both sandy beach and muddy coast an additional protection to reduce local scour, scoure aprons must be installed both in front and behind the LCBs. Scour aprons and toe protection must be fitted prior to filling the LCBs to minimize pulling, pressing, folding, or creasing [11].

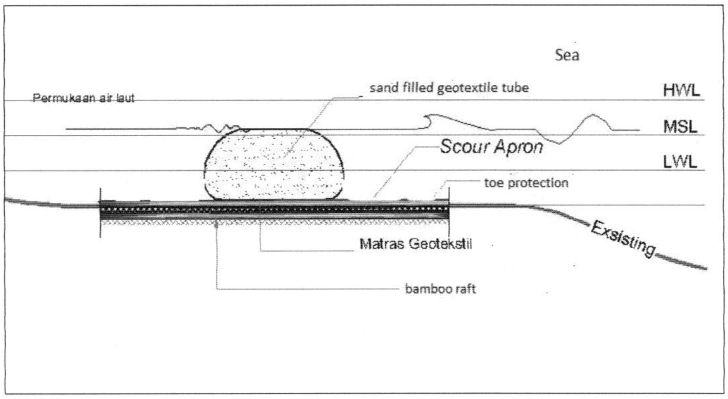

Figure 6. Typical crossection and scouring apron for sandy beach [11]

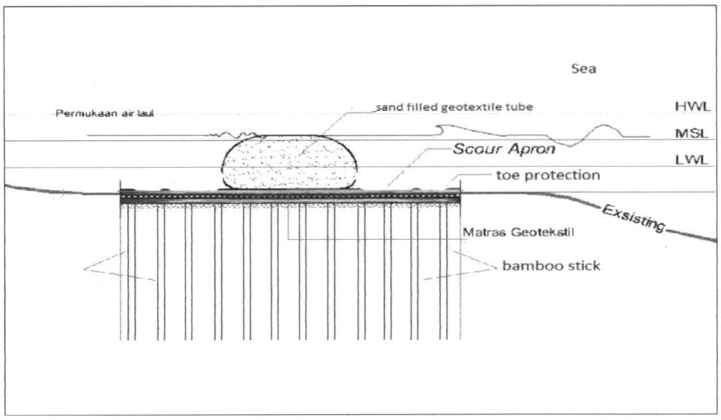

Figure 7. Typical Crossection and scour apron for mud coast[11]

4. Conclusion

The main function of the LCB is to reduce wave energy through wave breaking. Waves coming from the deep sea, either perpendicular or at an angle to the coast, when they pass the crest of the LCB will break and produce a rotating current behind the structure. The circular currents are in opposite directions and play a role in weakening longshore sediment transport which causes sediment to settle behind the LCB to form new beaches.

The decline in the geotextile tubes LCB structure that occurred at Sigandu Beach, Batang was not absolutely caused by extreme conditions and the occurrence of large waves at the end of December. The decrease in structure was mostly due to the placement of the LCB on a stretch of accreted sediment at the end of the East monsoon. Once the season changes to the west wind season, the direction of sediment transport along the coast also changes, so the sediment that forms the basis of the LCB is also eroded and the entire body of the LCB structure undergoes settlement or decline.

LCB layout planning, both vertically and horizontally is very important and determines its successfull in protecting and rehabilitating the coast. The vertical layout is related to the position of the LCB peak elevation relative to the mean sea level. While the horizontal layout is related to the distance from the beach, the length of the structure, and the width of the gap.

5. Recommendation

In order to avoid massive structural degradation, especially for the northern coast of Java, it is recommended that the installation be carried out during the transition season, more precisely April or June, when net transport is in normal condition, the coast is not in an accreted condition.

In order to avoid subsidence of the structure and to extend the service life of the LCBs, on muddy coasts with soft materials, it is necessary to install bamboo woven and bamboo rafts as suggested in the LCB Planning and Implementation Guidelines. The LCB structure will experience local scouring on the front and behind of the structure, therefore it is necessary to install toe protection both in front and behind the it. With the application of LCB technology, it is expected to be able to solve the problem of coastal erosion that occurs throughout Indonesia and can increase the width of the coast significantly in a fast tempo

Acknowledgments

The authors would like to thank both individuals and institutions for the data, information, and materials allowing this paper to be prepared. My special thanks go to Director of Technical Development for Water Resources, Directorate General for Water Resources, Ministry of Public Works and Housing; and Chief of Technical Implementation Unit for Hydraulics and Geotechnics, for the opportunity and support to complete this paper.

References

[1] Sulaiman DM. 2017. *Pegar Bercelah Bangunan Pelindung dan Penumbuh Pantai*. Deepublish. Yogyakarta.. 237 hal.

[2] Burcharth HF, Hawkins SJ, Zanuttigh B, Alberto L. 2007. *Environmenta Desogn Guidelines for Low Crested Coastal Atructures*. First edit. New York: Elsevier BV;. 397 hal.

[3] Sulaiman DM. 2021 Teknologi Pegar Untuk Penanggulangan Erosi dan Abrasi Pantai. Deepublish. Yogyakarta.. 42 hal.

[4] Pascual CV, Mera FL, Rodriguez IL. 2007. Breakwaters Stability Analysis of Low Crested and Submerged Rubble Breakwaters: Relationship between Flow Characteristic. *Proceedings of the 5th International Conference on Coastal Structures 2007*.. hal. 939–50.

[5] Buccino M, Calabrese M. 2003. On The Hydraulic Response of Low Crested Breakwaters. *Dept Hydraul Environ Eng Univ Naples "Federico II" Via Claudio 21 80125* Naples, Italy.; hal.1–14.

[6] Oumeraci H, Recio J. . 2010. *Geotextile Sand Containers for Shore Protection.* Handbook of Coastal and Ocean Engineering. ed YCK im. World Sci Publ;Chapter 21: hal. 553–600.

[7] Kerry Black and Shaw Mead. 2009. Wave Rotation for Coastal Protection. *Reef Journal*. hal.301-311.Vol.1 (1). Raglan. New Zealand..

[8] Álvarez E, Rubio R, Ricalde H. 2006. Shoreline restored with geotextile. *Geosynthetics*. 24(3):hal. 1–8.

[9] Sulaiman DM, Azhar RM, Jufri RG, Hartawan IP, Fernando D, Sugiatmo. 2019. Proses Sedimentasi dan Penurunan Struktur PEGAR di Pantai Sigandu Batang Jawa Tengah. *Prosiding Pertemuan Ilmiah Tahunan HATHI XXXVI Kupang*, NTT. hal. 1–8.

[10] Sulaiman DM, Hendra Hidayat, Doyo L.Dwiarso. 2019. Pengisian pasir dan Perlebaran Pantai SecaraAalami Dengan Teknologi Pegar *Prosiding Pertemuan Ilmiah Tahunan HATHI XXXIX Makassar*,.

[11] Ministry of Public Works and Housing, 2017. Draft Pedoman Perencanaan dan Pelaksanaan Pemecah Gelombang Ambang Rendah (PEGAR) Berbahan Geotekstil, Rapat Konsensus, Bandung,

DISASTER MITIGATION OF SLUM-SCALE FIRES IN JONGAYA VILLAGE AT TAMALATE

M Arifin[1], S Wunas[1], I Sastrawati[1], W Osman[1], G Lakatupa[1], J Mandasari[1], and Umar[1]

[1]Department Urban and Regional Planning, Engineering Faculty, Hasanuddin University, Makassar, South Sulawesi, Indonesia. 92111

mimiarifin@yahoo.com, shirly_wunas@yahoo.co.id, isfa.sastrawati@gmail.com, w_wahidahosman@yahoo.com, lakatupa@gmail.com, jayanti.mandasari@unhas.ac.id , accessumar08@gmail.com

Abstract. There are exactly about 834 fire cases occurred in Makassar City during 2016 to 2020. Tamalate Subdistrict being one of several subdistricts get the highest number of fire cases which was about 108 cases. The worst fire cases occurred in Jongaya Village around 2021. The fire caused the damage of 110 houses. Mitigation efforts are needed to prevent similar events from occurring. The purpose of this study is to analyze the level of risk fire disaster and formulate fire disaster mitigation directions that are in accordance with the characteristics of the residential environment at the research site. The analytical techniques used are qualitative and quantitative descriptive analysis, spatial, and scoring. The results showed that the study site was included high density with a density 92.24 buildings per hectare. The level of fire disaster risk is dominated by buildings with a classification of high fire risk (67%). Mitigation directive efforts are focused on location directions for the placement of portable fire pumps and fire engines, directions for the location of water filling of fire trucks, as well as directions for circulation of evacuation routes and temporary gathering points.

Keywords: Disaster Mitigation, Fire Disasters, Dense Settlements, Jongaya Village.

1. Introduction

The settlement density as well as environmental degradation accompanied by inadequate infrastructure, led to the formation of slums [1]. Slums are areas that have poor environmental conditions, densely populated, dirty, and limited space for ventilation, air, light, open fields, sanitation, and environmental conditions and the damage of the human life [2].

The high population activity and the increase of urban residents forming new settlements will lead to a greater chance of urban fire disasters [3]. Areas that are vulnerable to disaster risk are characterized by densely populated residential areas, one of the disasters that often occurs is the danger of fire. Dense settlements are one of the problems of fire disasters which frequently occur in Makassar City [3]. It is undeniable that the problem of fire is a case that occurs a lot in dense settlements [4].

Tamalate has an area of 21.47 Km^2 with a population growth rate of 180,824 people and a population density, which is 8,947 people / Km^2. Jongaya Village has the third highest density (31,878 people/ Km^2) which identifies that Tamalate has a relatively high level of activity [5].

During 2016 to 2020 there have been 834 fire cases where Tamalate is at the highest fire cases with 108 fire cases. Seeing the magnitude of the impact of this fire disaster, it is necessary to handle and pay attention from all parties in settlements, including the local government, especially in Tamalate. The results of the site review showed that the fire case in RT 04 RW 02 Jongaya Village at Tamalate was the largest fire case for the past five years in Makassar City, there were 110 houses, the density of buildings, narrow road access and construction of building materials that were vulnerable to fire and spread high fires make this location need to be applied mitigation to the danger of residential fire disasters. The purpose of this study was to analyze the level of fire disaster risk and formulate fire disaster mitigation directives that are in accordance with the characteristics of the residential environment at the research site.

2. Literature review

2.1 Slums and Fire Disasters

Slums are related to fire disasters, judging from the condition of buildings related to aspects of age, density, and regularity of residential buildings. The cause of fire disasters and damage to buildings can be seen from one of the factors such the age of the building [6]. It can be known that the area becomes shabby because the size of the building that is below the standard and the density of buildings [7].

Apart from these aspects, the type of building material can also be the thing that determines how much the level of resistance of building materials to fire disasters. Slums basically use combustible building materials, this is because the materials from the building use combustible materials such as wood, zinc, cloth, and bamboo [8]. Another cause that can affect the quality of slums is the condition of environmental roads. Road concessions and garbage disposals that exist in slums often have poor conditions [9].

According to Indonesian Law Number 4 of 1992 concerning housing and settlements, one of the main basic means for the functioning of a residential environment is the road network. The existence of a road network in settlements can create space or distance between buildings. Therefore, it can prevent fire propagation during a fire disaster [9]. Meanwhile, waste management can be an indicator of slums because slums usually do not have waste management facilities that are in accordance with applicable technical requirements. Incompatibility of waste management in slums can result in environmental pollution and the threat of fire disasters [10].

Indonesian Ministry of Public Works and Housing Regulation Number 14/PRT/M/2018 concerning Prevention and Quality Improvement of Slum Housing and Slums, to identify slum areas, criteria are used. The determination of the criteria for slum areas is carried out by considering various aspects or dimensions such as, buildings, environmental roads, drinking water supply, environmental drainage, wastewater management, waste management and fire protection.

2.2 Fire Disaster Risk Level

According to The Disaster Mitigation Center of the Institute Technology of Bandung, the risk of disasters can be seen from the hazards that meet with vulnerability and the absence of resilience (capacity). In the following model, disaster risk can be reduced by increasing the value of resilience and minimizing existing vulnerabilities, each factor, both potential sources of danger, vulnerability, and regional resilience, is used as a benchmark for high or low disaster risk in the region.

Disaster Risk = Hazard + Vulnerability – Capacity

Figure 1. Disaster Occurrence Model According to the ITB Disaster Mitigation Center
Source: Sagala et al, 2013

In this case, the disaster management formula model used is a Crunch Model based on the ITB Disaster Mitigation Center [11], namely that disasters (D) are the product of vulnerability (V) to meet a hazard (H).

Disaster Risk = Hazard + Vulnerability – Capacity

Using the formula above, it provides an overview of the relationship between Hazard (H), Vulnerability (V), and Capacity (C) easily.

2.3 Fire Disaster Mitigation

The Mitigation is a form of disaster management efforts. Based on Indonesian Law Number 24 of 2007 concerning Disaster Management, a mitigation is a series of efforts to reduce disaster risk, both through physical development and awareness as well as increasing the ability to face disaster threats [12]. Mitigation functions in reducing the impact of natural disasters is caused by human negligence in a nation or community.

Mitigation efforts are divided into 3 sections which are pre-disaster, during disasters, and post-disasters. Pre-disaster is a preparedness or effort to provide understanding to the population in order to anticipate disasters through providing information, increasing preparedness in the event of a disaster there are measures to reduce disaster risk. At the time of the incident, it is an emergency response, namely efforts made immediately at the time of the disaster to overcome the impact caused, especially in the form of saving victims, property, and evacuation. Post-disaster is an effort to restore rehabilitation and development. Mitigation can be categorized into structural and non-structural mitigation [13].

2.3.1 Structural Mitigation. Structural mitigation is an effort to reduce vulnerability to disasters by strengthening buildings and infrastructure to hazards through building codes, engineering design, and construction implementation for the development of building resistance to hazard threats. Disaster-resistant buildings are buildings with structures that are planned in such a way that the building is able to survive or suffer less severe damage if the disaster in question occurs. Technical engineering is a procedure for designing building structures that have taken into account the action characteristics of a disaster.

2.3.2 Non-Structural Mitigation. Non-structural mitigation is an effort to reduce the impact of disasters through regulations or policies such as making laws and regulations. The Disaster Management Law is an example of non-structural efforts in disaster mitigation. Another example is the creation of urban spatial planning, community capacity building, and even reviving various other activities that are useful for strengthening community capacity, as well as part of this mitigation. This is all done for and by the communities living around disaster-prone areas. Non-structural policies include regional planning, legislation and insurance.

Non-structural policies are more concerned with policies aimed at avoiding unnecessary and destructive risks. Therefore, it is necessary to identify the risks in advance. Physical risk assessment includes the process of identifying and evaluating possible disasters and possible impacts. In addition, non-structural mitigation also includes the provision of training and education, public education, evacuation planning, institutional buildings, and warning systems. Rules related to mitigation, both structural and non-structural, must support each other. The use of technology to predict, anticipate and reduce the risk of a disaster must be balanced with the creation and enforcement of adequate regulatory tools supported by appropriate spatial plans.

The purpose of this study was to determine the model of fire propagation in urban dense areas, using a different computational model than the model in general. The research method used in this study is through physics-based knowledge, where the spread of fire is depicted by predicting the fire behavior of each building in conditions affected by exposure to thermal radiation from surrounding buildings that are burning. Prediction of fire behavior in each building using numerical techniques based on the one-layer zone model. The variables used are the equation of mass, energy and chemical type of building components calculated simultaneously with the process of temperature increase, concentration of types of building materials and furniture materials in buildings.

3. The Method

The research location is in RT 04 RW 02, Jongaya Village, Tamalate, Makassar City (Figure 1). Qualitative descriptive analysis is used to provide information to determine the existing condition of the research area related to the characteristics of residential buildings which are then presented in the form of maps so that the information conveyed is easier to understand using spatial analysis.

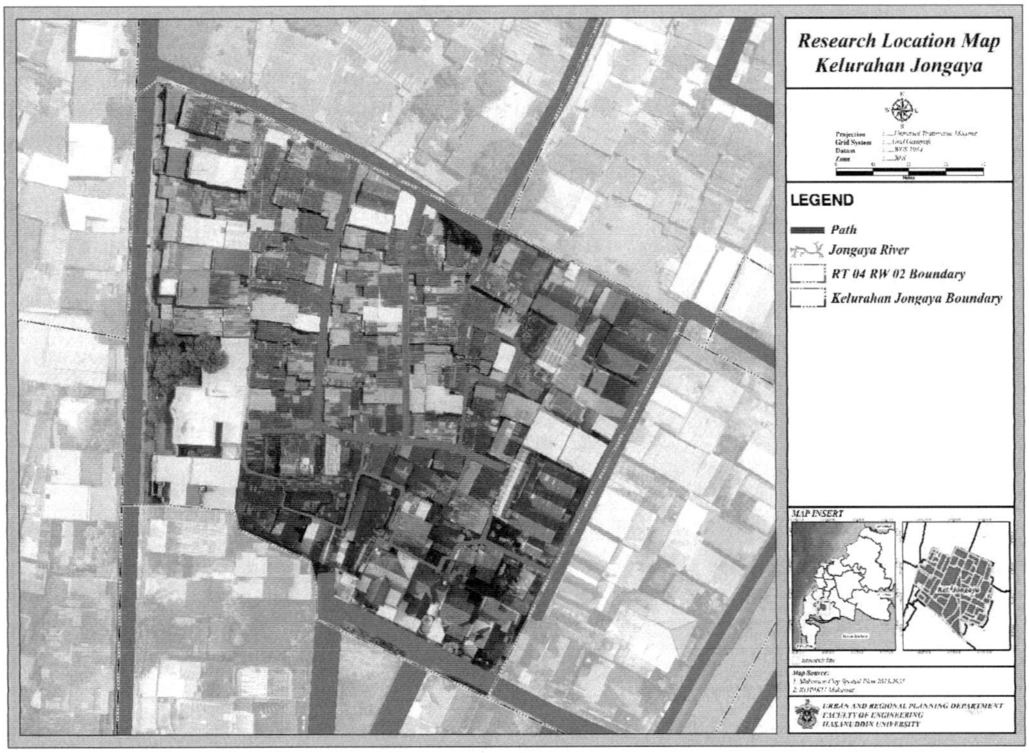

Figure 2. The Research Location Map

In assessing the risk level of this fire disaster, using the Crunch Model formula based on the ITB Disaster Mitigation Center.

Disaster Risk (R) = Hazard (H) + Vulnerability (V) – Capacity (C)

Details:
(H) = Hazard Score Value
(V) = Vulnerability Score Value
(C) = Capacity Score Value

The level of fire risk uses a risk analysis method by scoring on each research variable. The benchmark in assigning score values to each variable is given based on Regulation of National Disaster Management Agency Number 02 in 2012 concerning General Guidelines for Disaster Risk Assessment.

Table 1. Variable Value Benchmark

Benchmark	Evaluation		Maximum Variable Value	Minimum Variable Value
	Standard	Not Up to Standard		
Threat	0	1	4	0
Vulnerability	0	1	10	0

Benchmark	Evaluation		Maximum Variable Value	Minimum Variable Value
	Standard	Not Up to Standard		
Resistance	1	-1	6	-6
Fire Risk			20	-6

Source: Regulation of National Disaster Management Agency Number 02 in 2012

The difference between the maximum score and the minimum score is then divided by the desired number of classes to get the interval value and used to classify the risk class for the residential fire disaster which is the location of the research case study. The class interval is at 8.66. So that the results of the calculation of fire risk classes are classified based on high, medium, and low fire levels with their respective class intervals making it easier to determine each fire-prone building unit by paying attention to each variable and indicator in weighting or scoring the determination of the level of fire disaster risk. The fire risk assessment based on research variables and indicators can be seen in table 3.

Table 2. Fire Disaster Risk Class

Class Interval	Fire Disaster Risk Level
-6 to 3	Low Fire Disaster Risk Level
4 to 13	Moderate Fire Risk Level
14 to 23	High Fire Disaster Risk Level

Table 3. Fire Risk Assessment Based on Threat, Vulnerability and Resilience Factors

No	Variable	Indicator	Evaluation	
			In accordance	It is not in accordance with
1	Hazard	Electric current	0	1
		Large/small industry	0	1
		Seller of oil, gasoline and LPG gas	0	1
		Live fire ignition	0	1
Total Hazard Score			0	4
2	Vulnerability	building materials	0	1
		Building size	0	1
		Number of floors of the building	0	1
		Building density	0	1
		Building density	0	1
		KDB	0	1
		Resident income	0	1
		Vulnerable age population	0	1
		Wind speed and direction	0	1
		temperature	0	1
Total Vulnerability Score			0	10
3	Capacity	Water sources	1	-1
		Presence of a hydrant	1	-1
		Presence of RTH	1	-1
		The width of the road	1	-1
		Firefighting distance and time	1	-1
		Water reservoir	1	-1
Total Capacity Score			6	-6

The direction of disaster mitigation follows the existing conditions of the research location relating to the determination of evacuation points, evacuation routes, hydrant location points, water pumps. Furthermore, using spatial analysis to outline the location points that are the focus of fire disaster mitigation. This disaster mitigation directive is an effort to reduce the risk of residential fire disasters based on the results of the identification of the level of fire disaster risk and through a literature study.

4. Result And Discussion

4.1 Building Characteristics

The results showed that the study site was included in the high density with a density 92.24 buildings per hectare dominated by permanent buildings consist of 52 units (39%) and 37 units (28%) non-permanent.

Table 4. Number of Buildings Based on Materials in RT 04 RW 02 Jongaya Village

Building Type	Buildings (units)	Persentage
Wood	37	28%
Wood and Cements	43	33%
Cement	52	39%
Total	132	100%

Figure 3. Wood Material House

Figure 4. Wood and Cement Material House

Figure 5. Cement Material House

Building materials at the study site is further classified into flammable, non-flammable and fireproof buildings. Furthermore, the welding of buildings based on their flammability is assessed based on their construction materials starting from the use of roofs, roof frames, walls and floors. Based on the construction criteria of the building, buildings with a classification of highly flammable are assessed based on the use of roofing with zinc material, roof frame from wooden beams, wooden materialized walls, and using wooden floors (wooden floors are only assessed on buildings with 2 floors). For the classification of non-flammables is judged based on the use of roofing from zinc, roof from wooden beams, material cement walls, and cement floor. While the fireproof classification building is assessed based on the use of cement material roofs, roof frames of iron, mild steel or concrete, and cement materialized walls.

Table 5. Number of Buildings Based on flammability

Building Type	Buildings (units)	Persentage
Flammable	50	38%
Non-flammable	41	31%
Fireproof	41	31%
Total	132	100%

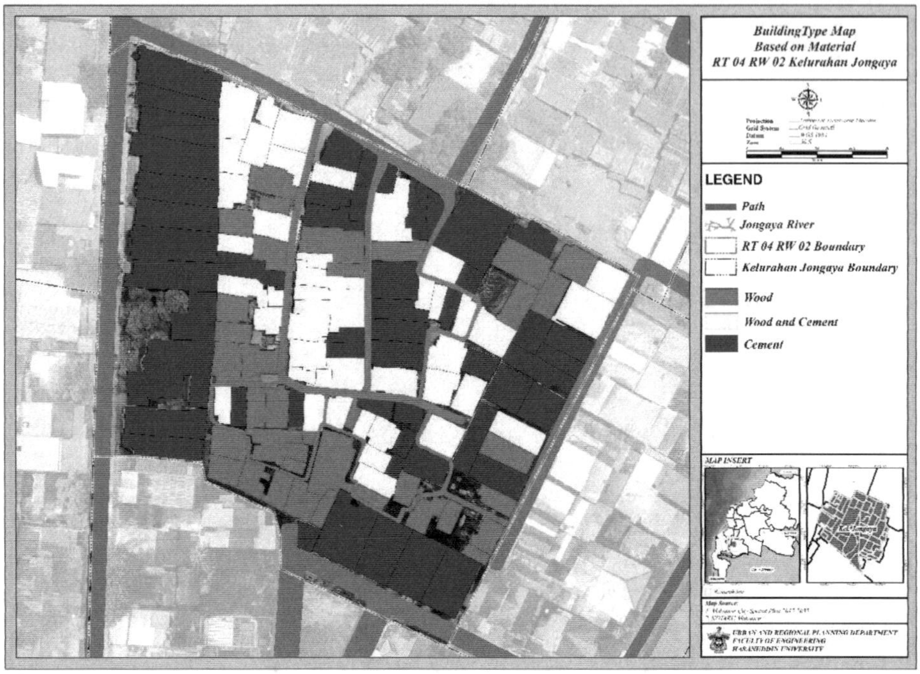

Figure 6. Building Type Map Based on The Material

In this study, building density is a comparison between the number of buildings and the area (ha). In this regard, density and density level are interrelated with each other. The higher the building density of a residential area, the tighter the distance between the buildings. If the distance between buildings is tighter, the chances of a fire are also higher. This is what happened in RT 04 RW 02 Jongaya Village which is a residential area located in the downtown area. The density of buildings in RT 04 RW 02 Jongaya Village can be seen from the following calculation results: The number of buildings in RT 04 RW 02 Jongaya Village is 132 units with an area of 1.43 ha.

Overall RT 04 RW 02 Jongaya Village has an area of 1.43 hectares and consists of 132 which means it consists of 92 buildings/ha, including a very high building density which shows that this location is prone to fire disasters. The results of the identification of the type of material, the size of the building, the height, the density of the building show that the characteristics of the settlements at the study site are very vulnerable to fire disasters. This is due to the tight condition of the building and the absence of open space in RT 04 RW 02 Jongaya Village so that the road conditions to the location are also narrow.

4.2 Fire Disaster Risk Level

Determining the level of fire disaster risk in RT 04 RW 02 Jongaya Village, Makassar City, an assessment was given based on risk variables, namely in the form of threat variables, vulnerability variables and resilience variables. In assessing the level of risk of this fire disaster, using the Crunch Model formula based on the ITB Disaster Mitigation Center. From this formula, scoring is then carried out according to the results assessment of each risk variable. The results assessment of risk variables can be seen in the following table.

Table 6. Risk Variable Results

Variable	Score
Hazard	4
Vulnerability	10
Capacity	-3

Variable	Score
Total Risk Variable	15

The results in the assessment table of each variable and indicator, calculations are then carried out based on the crunch model formula. Here is the calculation:

Table 7. Fire Disaster Risk Classification

Classification	Buildings (units)	Percentage (%)
Low Fire Risk Classification	0	0
Medium Fire Risk Classification	44	33
High Fire Risk Classification	88	67
Total	132	100

The results show spatial analysis level of fire risk. It can be seen that buildings in the north, south and east of the study site are classified as high risk of fire. Buildings with moderate fire risk are predominantly in the western part while buildings with low risk of fire are not found at the research site.

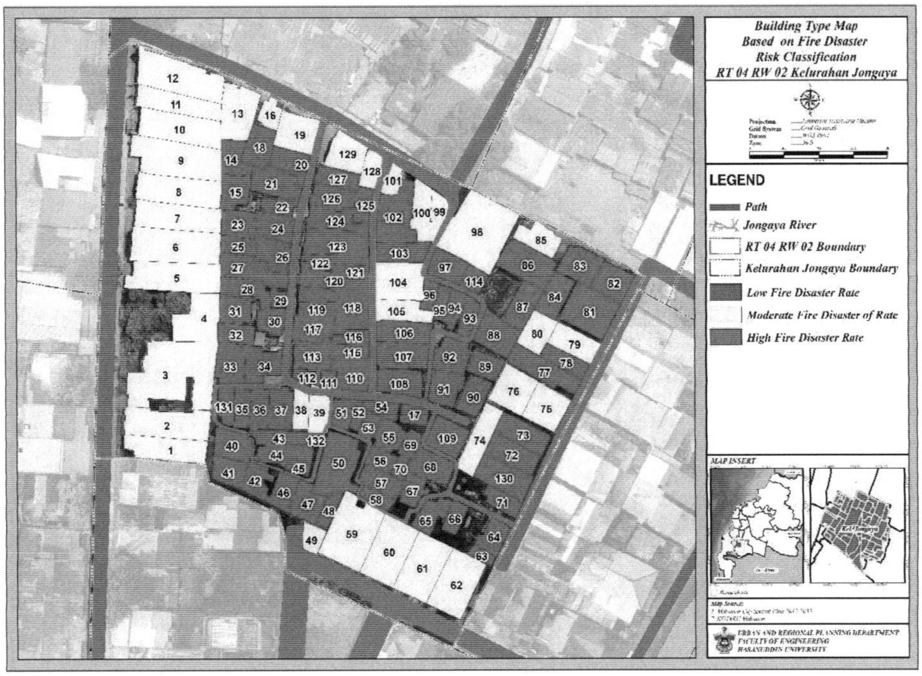

Figure 7. Fire disaster risk classification

4.3 Fire Disaster Mitigation Directives

The mitigation directives in this study leads to structural mitigation directives that focus on the conditions before the occurrence of a fire disaster so this can be a preventive measure to avoid fire disasters as much as possible and become an effort to reduce the risk of fire disasters. There are also several directions as an effort to reduce the risk of fire disasters described in the following points.

4.3.1 Electrical Installation Check. Building units or houses that already have electrical installations older than 15 years should check and replace the electrical installation by contacting PLN.

4.3.2 Using An electric current safety device. This is a tool that can be used to disconnect automatically if an electrical installation experiences disturbances such as excessive loads, short circuits, sparks, and others. At the research site, there are still building units or residents' houses that do not use automatic heating fuses when an electrical short circuit occurs. Hence, preventing the occurrence of an electrical short circuit that can cause a fire, should use an MCB (Mini Circuit Breaker) which has almost the same function as a fuse (**Figure 9**). The advantage of using an MCB is that it can break the connection of all phases of a 3-phase circuit even if there is a short circuit in one of its phases only, if there is a short circuit or more load then the MCB can be reused, and has a good level of safety or quick response in the event of a short circuit or more load.

4.3.3 Using Fireproof Building Materials. Types of buildings including vertical buildings with highly flammable materials can use fireproof building materials, for example lightweight brick walls. The materials are more efficient and fireproof than adobe, stone material walls are more fireproof than wood, gypsum walls are more fireproof than plywood, aluminum sills are more fireproof than wooden sills, and others. Build fire-proof dividing walls or partition walls between buildings and must be built simultaneously with individual floors (there should be no wall openings).

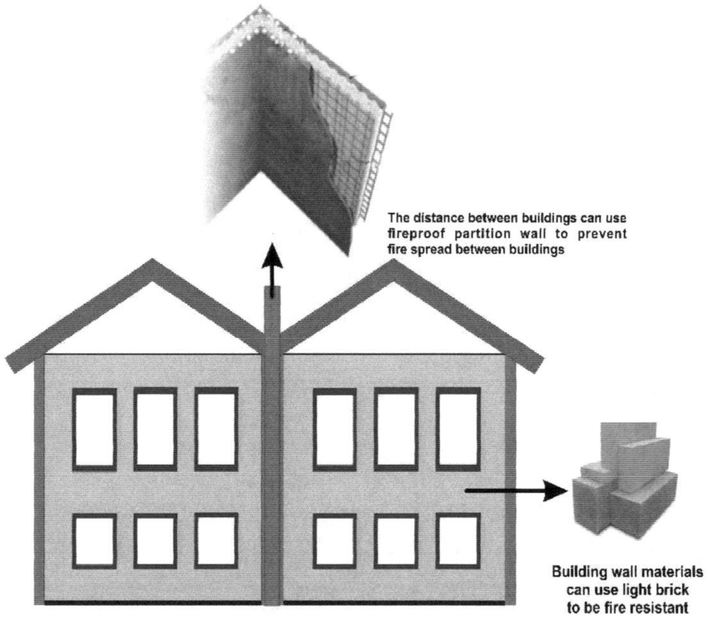

Figure 8. Fireproof Partition Wall Illustration

4.3.4 Safe Kitchens design. One of the places which is prone to fire disasters is the kitchen. Many objects are prone to cause fires in the kitchen such as gas cylinders, kerosene, and other kitchen materials that are flammable, so a special design is needed so that the kitchen becomes safer and less prone to fire, especially in houses where there are activities in it. The kitchen should be designed by keeping the stove away from combustible materials, using non-wood material walls using fireproof walls such as cement, light brick (hebel) or gypsum covered with emergency exits.

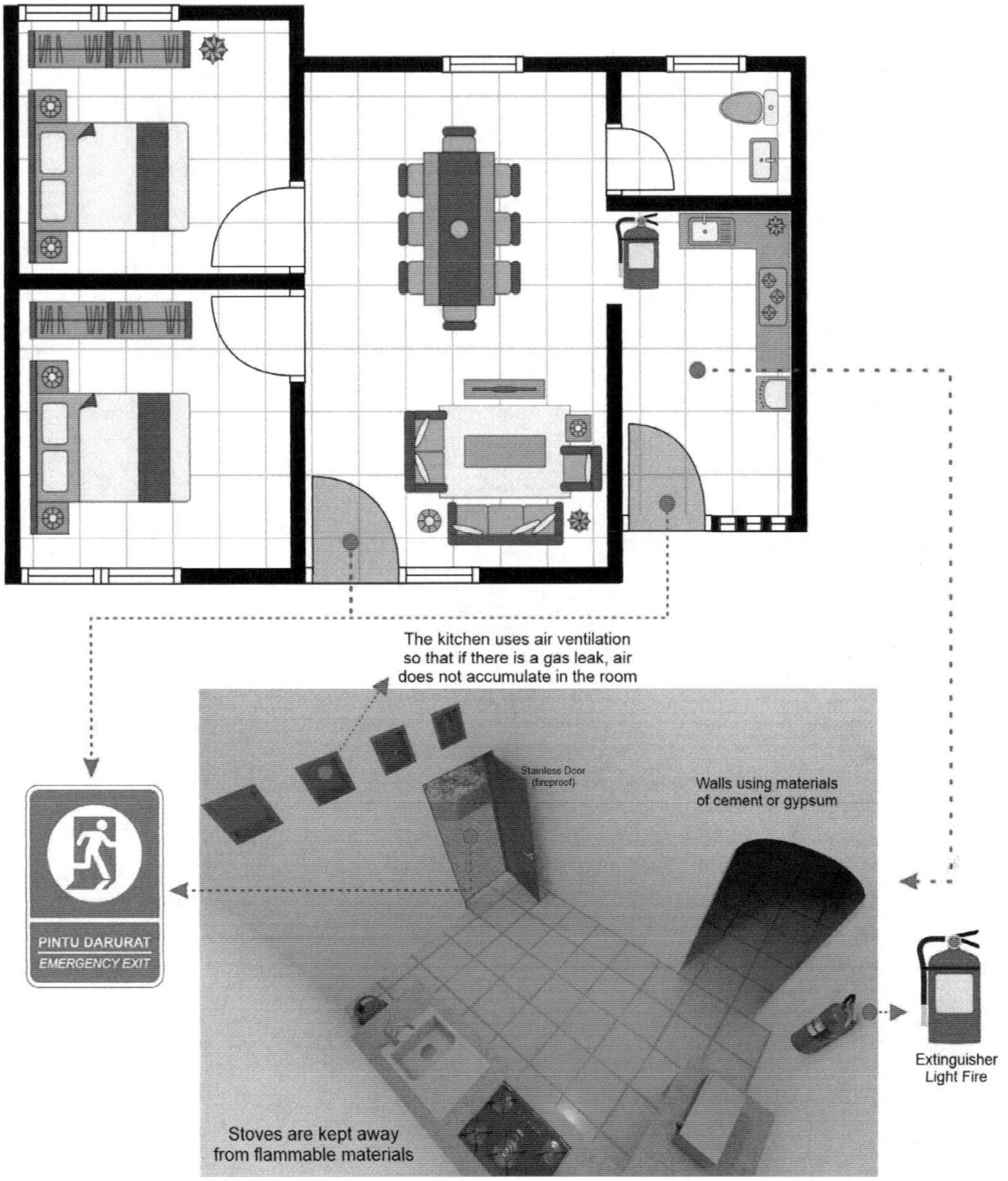

Figure 9. Design illustration for safe kitchen

4.3.5 Provide Fire Engine motor. This fire engine motor can be used to reach the location or route which is difficult to be entered by fire trucks on routes that have a width below 4 meters. A fire engine motor that has a width of 1.5 meters being able to reach locations that have a route below 4 meters with a tank capacity of 200-300 liters, and pump suction power of 6 meters, a hose of 20 meters with a spray pump thrust of 26 meters vertical can reach building units or burning houses. Provision of portable water pumps at locations or routes cannot be reached by fire engines and environmental hydrants can reach the research location at RT 04 RW 02 Jongaya Village with a service radius of 200 m.

Figure 10. Hydrant Location Directives in RT 04 RW 02 Jongaya Village

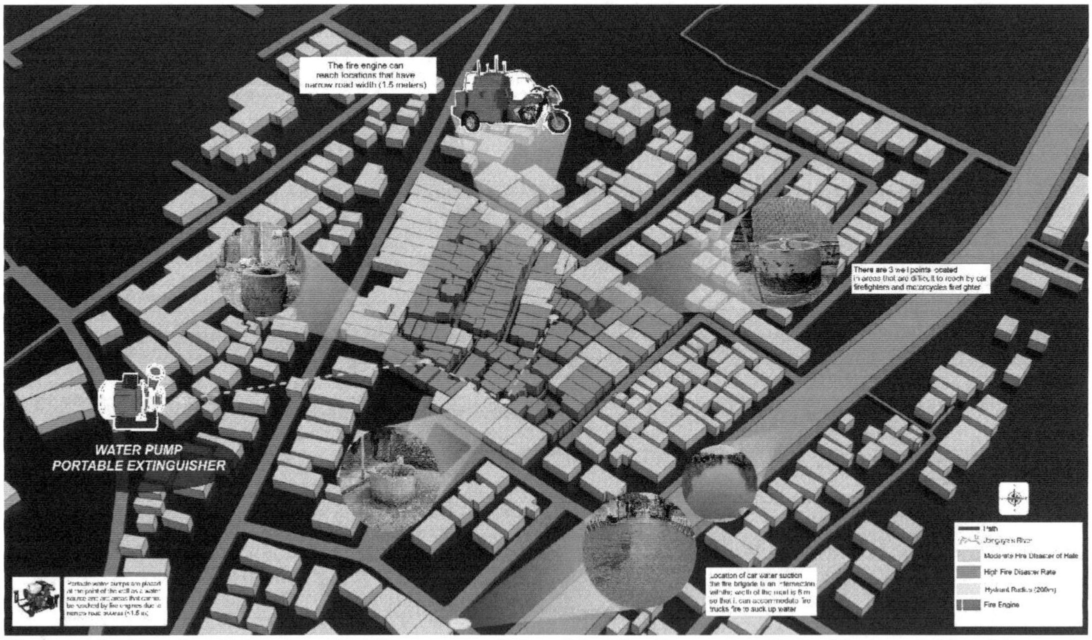

Figure 11. Location Direction for Placement of Portable Fire Pumps and Fire Extinguishing Motors in RT 04 RW 02 Jongaya Village

IOP Conf. Series: Earth and Environmental Science 1117 (2022) 012027 doi:10.1088/1755-1315/1117/1/012027

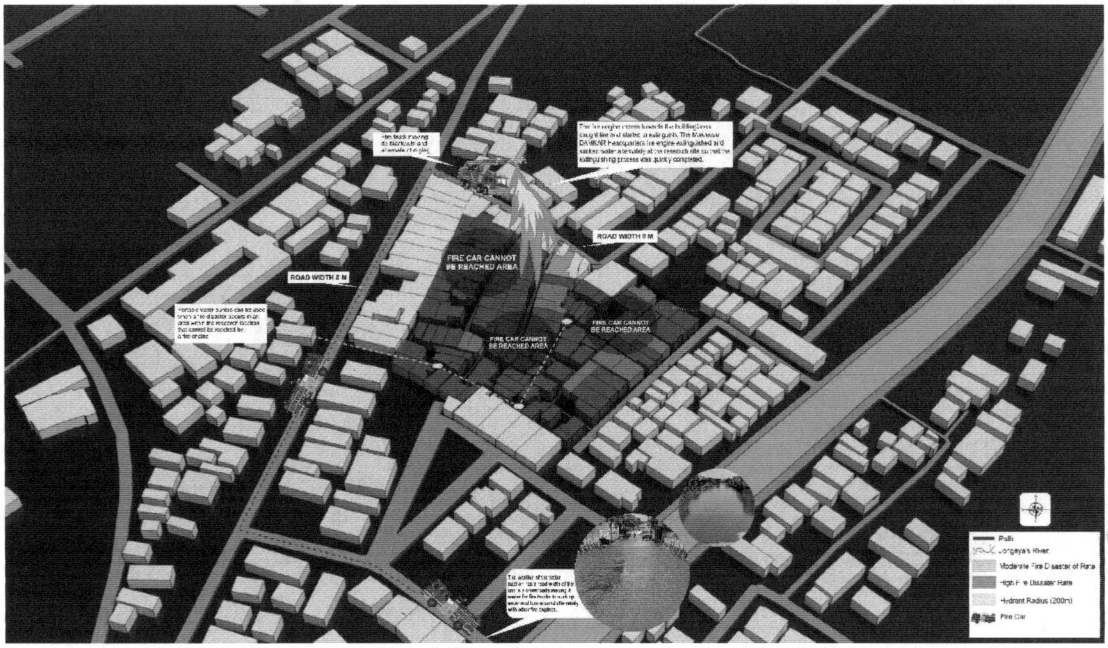

Figure 12. Direction for Fire Truck Water Filling Location in RT 04 RW 02 Jongaya Village

4.3.6 *Provide Fire Extinguishing*. Based on the result of interviews at the research site, it is identified that 78% of the people in study case needed the help of fire extinguishers from the government, 11.9% needed material assistance from the government for home repairs and 9.2% needed financial assistance for home and electrical repairs. Therefore, efforts that can be made to take action to anticipate fire disasters are by providing light fire extinguishers. This fire extinguisher can be placed in certain houses or buildings, for example in houses as bases or LPG sales points or places to sell gasoline as a quick response effort in the event of a fire ignition, it can also be placed in buildings with public facilities functions such as security posts or at the house of the rt chairman or other community leaders who have easy access for local residents.

4.3.7 *Provide Evacuation Routes*. Furthermore, to avoid casualties, it is necessary to provide evacuation routes and temporary evacuation sites. This direction is also carried out based on the existing conditions of the research site by taking into account accessibility conditions. It is necessary to provide disaster information boards containing information about disaster-prone areas or their dangers, information on disaster events that have occurred and/or have the potential to occur as well as the location of temporary gathering places or refugee camps (National Disaster Management Agency Number 07 of 2015). Such information boards can be both evacuation paths and marker information.

IOP Conf. Series: Earth and Environmental Science 1117 (2022) 012027 doi:10.1088/1755-1315/1117/1/012027

Figure 13. Direction for Circulation of Evacuation Routes and Temporary Gathering Points in RT 04 RW 02 Jongaya Village

Figure 14. Information Board Directives on Evacuation Routes in RT 04 RW 02 Jongaya Village

5. Conclusion

The results showed that the study site was included in the high density with a density 92.24 buildings per hectare dominated by permanent buildings consist of 52 units (39%) and 37 units (28%) non-permanent. The level of fire disaster risk is dominated by buildings with a classification of high fire risk (67%) with building sites mostly located in the northern, eastern, and southern parts of the study site.

Mitigation directive efforts are focused on structural mitigation such as location directions for the placement of portable fire pumps and fire engines, directions for the location of water filling of fire trucks, as well as directions for circulation of evacuation routes and temporary gathering points. Moreover, to avoid any risk of fire disaster the local community used to design a safe home kitchen and used fireproof materials building.

6. References

[1] Sutanti, N., Tjahjono, B. and Syaufina, L., 2020. *Analisis Risiko Bencana Kebakaran di Kecamatan Tambora Kota Administrasi Jakarta Barat. TATALOKA*, *22*(2), pp.162-174.

[2] Mononimbar W, Supardjo S, Wiarni S. (2018). Analisis Tingkat Kekumuhan Kawasan Permukiman di Kecamatan Kotamobagu Timur. Jurnal Spasial. 5(1), 61-70.

[3] Widiantoro, B.A., 2016. *Analisis Tingkat Resiko Bencana Kebakaran Di Kecamatan Mariso Kota Makassar Berbasis Sistem Informasi Gegrafis (SIG). Plano Madani: Jurnal Perencanaan Wilayah dan Kota*, *5*(1), pp.94-102.

[4] Haq, F., 2017. *Studi Kerentanan Kawasan Pemukiman Padat terhadap Bencana Kebakaran dan penanggulangannya (Studi Kasus: Kecamatan Tallo Kota Makassar)* (Doctoral dissertation, Universitas Islam Negeri Alauddin Makassar).

[5] Badan Pusat Statistik Kota Makassar. 2020. Kecamatan Tamalate Dalam Angka Tahun 2020.

[6] I. Akmal. 2012. Membetulkan dan Merawat Rumah. Jakarta: Gramedia Pustaka Utama.

[7] Komarudin. 1997. Menelusuri Pembangunan Perumahan dan Permukiman. Jakarta: Yayasan Realestatet Indonesia.

[8] Surtiani, E.E., 2006. *Faktor-faktor yang mempengaruhi terciptanya kawasan permukiman kumuh di kawasan pusat kota (studi kasus: kawasan pancuran, salatiga)* (Doctoral dissertation, program Pascasarjana Universitas Diponegoro).

[9] Firda, D., Sobirin, S. and Indra, T.L., 2017, July. Sensitivitas Permukiman Kumuh Terhadap Bencana Kebakaran di Kota Bekasi. In *Prosiding Industrial Research Workshop and National Seminar* (Vol. 8, pp. 600-605).

[10] SNI 19-2454-2002 tentang Tata Cara Teknik Operasional Pengelolaan Sampah Perkotaan.

[11] Sagala, Saut., Adhitama, Praditya., Sianturi, Donald G., 2013. "*Analisis Upaya Pencegahan Bencana Kebakaran di Permukiman Padat Perkotaan Kota Bandung, Studi Kasus Kelurahan Sukahaji*". Resiliance Development Initiative. Bandung.

[12] Undang-undang Republik Indonesia Nomor 24 Tahun 2007. Penanggulangan Bencana.

[13] Trinofansyah. 2011. *"Identifikasi Tingkat Risiko Bencana Kebakaran Di Pemukiman Padat (Studi Kasus: Pemukiman Padat Lebakgede Coblong Bandung)"*. UNIKOM. Bandung.

IDSL for Tsunami Early Warning System in Indonesia

S Husrin[1,3], A Annunziato[2], G S Prasetya[3], R Hidayat[4,3]
[1, 3] Research Centre for Geological Disasters – The National Research and Innovation Agency - Indonesia
[2] Joint Research Centre - European Commission (JRC-EC) - Ispra - Italy
[3] Indonesia Tsunami Expert Society (IATsI) - Jakarta - Indonesia
[4] Coordinating Ministry of Maritime and Investment Affairs- Indonesia

semeidi@gmail.com

Abstract. The Inexpensive Device for Sea Level measurement (IDSL) has been implemented for tsunami monitoring and early warning in the Mediterranean regions. Following the events of 2018 Palu and the Sunda Strait Tsunami, The MMAF, JRC-EC, IATsI, and BMKG worked together and installed eight (8) units of IDSL in Indonesia starting from January 2019. This paper discusses the concept of IDSL, the installation and the operation in the last 18 months and its potential use as tsunami early warning system in Indonesia. We concluded that IDSL and other similar systems could strengthen the existing TEWS in Indonesia considering its reliability in delivering the alerts in real-time with the characteristics of short distance tsunami and the presence of small islands, low costs in installation and maintenance that may guarantee the sustainability of the system.

1. Introduction

After the 2004 Indian Ocean Tsunami, the Government of the Republic of Indonesia immediately established the Tsunami Early Warning System (TEWS) supported mostly by the Government of Germany and other countries/international's institutions [1]. The TEWS known as Ina-TEWS primary based on the seismometer sensors with the system that is capable of generating parameters for tsunami model and immediately combine with pre-calculated tsunami model results database for providing alerts or early warnings [2][3]. This capability should allow the system to identify the possible tsunami events and deliver the early warnings in less than 5 minutes to the target communities [4][5]. However, the events of the 2018 Palu and Sunda Strait tsunamis have shown that the Ina-TEWS is still not adequate due to the facts that those two events were unable to be detected by the existing system [6]. The 2018 Palu tsunami was generated by submarine landslides triggered by 7.8 Mw large strike-slip earthquakes [7][8]. Meanwhile, the 2018 Sunda Strait tsunami was also generated by the submarine landslides due to flank collapsed during the eruption of Mount Anak Krakatau [9][10]. Without any warnings, the 2018 Palu and Sunda Strait tsunamis devastated and claimed thousands of lives. It is clear that the existing Ina-TEWS was not designed to detect tsunamis generated by non-tectonic or by near shore events [6][10].

Following the 2018 Sunda Strait Tsunami, in early January 2019, the head of the Indonesian National Disaster Management Agency or BNPB summoned all Indonesian institutions with technical capabilities to establish immediate emergency tsunami early warning system in the Sunda Strait due to the presence of strategic industries and critical infrastructures for Indonesia. One of the proposals was the Inexpensive Device for Sea Level measurement (IDSL), a tide gauge with capabilities of collecting high quality data, transferring data in near real-time and delivering alerts instantly based on the anomalies in the sea levels [10][11]. IDSL for TEWS was first developed by the Joint Research Centre – The European Commission (JRC-EC) in the Mediterranean Sea to support the Member States of the North East Atlantic and Mediterranean Tsunami Warning System (NEAMTWS) in collaboration with the UNESCO [11].

For Indonesia, the concept of using tide-gauge networks for TEWS was not new because it was also mentioned in the technical documents of Ina-TEWS [1][2]. However, the release of alerts from the tide gauges was not possible due to real-time tide-gauges had not been available yet [10][12]. Therefore, recalling previous successful joint research collaboration in Palu [7], the Marine Research Centre of MMAF, the Indonesian Tsunami Experts Society (IATsI) and JRC-EC proposed the implementation of IDSL for emergency tsunami early warning system in the Sunda Strait. The proposal was also later supported by BMKG as the sole authority for tsunami early warning system in Indonesia to be expanded for other regions across Indonesia. The paper describes about the first implementation of IDSL in Indonesia since January 2019 and discusses the key performance of IDSL to strengthen the existing TEWS in Indonesia.

2. Methodology

The hardware of IDSL consists of the sea level radar-based sensor, the CCTV Camera, communication box (GSM modem), the computer, the power supply (solar panel and batteries) and the metal frame structures. The computer as the main engine of IDSL is the Raspberry Pi consists of components such as USB, HDMI, Ethernet port, video card, sound card, and other busses that can be connected with external devices [11]. The radar sensor of IDSL measures the sea levels in near-real time mode and when the anomalies are detected, the in-built algorithm (Kalman filter) will automatically deliver alerts to the dedicated emails and phone numbers (for SMS or text messages) [10]. The CCTV will capture more frequent images (e.g., every 2 minutes) for visual validation. The speed of transmitting the data relies on the GSM network in the area (with 4G network, the transmission or the latency time can be in less than 20 seconds). Therefore, with these capabilities, the authority will have enough time to deliver tsunami alerts in less than 5 minutes. The key performance of IDSL for TEWS focuses on the following parameters:

- *Data quality:* The first six-months data of IDSL were analysed and compared with other devices such as the nearby BIG tide gauge and the Tidemaster from Valeport
- *Latency time:* The speed of data transfer from the device to the server is very crucial in order to timely provide alerts to the community if the tsunami occurred.
- *Delivering Alerts*: The capability of delivering alerts when the anomalies of the water levels were detected. This distinguishes IDSL for TEWS from other ordinary tide gauges
- *CCTV images*: the quality of CCTV images is important for investigating visually the conditions in the surrounding area such as assuring the anomaly of the water levels (e.g., drawdown of the sea level)

Considering that tsunami in Indonesia is a near-field tsunami (tsunami arrival is less than an hour), a tide gauge like IDSL for TEWS remains challenging. However, since the system had a capability of near-real-time in delivering the measurements and alerts instantly based on the anomalies of sea levels, these could provide an immediate information of the impending tsunamis. Therefore, based on the monitoring of 8 units IDSL installed in Indonesia in the last 20 months, the above parameters were assessed to draw a conclusion on using IDSL to strengthen the TEWS in Indonesia. Moreover, the advantage and disadvantage of a tide gauge for TEWS will also be discussed based on the characteristics of the tsunami, budgets, infrastructures availability, socio-economic conditions and general tsunami mitigation measures.

3. Results and Discussion

In general, a tide gauge can be used as a tsunami early warning system if it can provide *high quality data (an error less than 0.5 cm), short time interval (~ 5 seconds), near real-time (small latency time), and capability to deliver the early warning* [10][11]. Those requirements should also be supported by sufficient number of installed tide gauges, secured communication infrastructures, independent power supplies and robustness of each hardware components against hostile marine environments (strong winds, corrosive, etc). For the case of Indonesia, the installation and operational costs of the system is also very important to secure sustainablity and to cover such wide area. Moreover, the involvement of coastal local communities plays a major role for the sustainability of the system due to high occurrence of vandalism activities to the remote monitoring devices in the oceans [4].

3.1. Distribution and Installation of IDSL

As of November 2020, 8 (eight) units of IDSL has been in operation in Indonesia. The first two units of IDSL were installed in January 2019 in the Sunda Strait waters (Sebesi Island and Marina Jambu) followed by the installation in Sadeng Port (DI Yogyakarta), Pangandaran (West Java), Pelabuhan Ratu (West Java), Prigi (East Java), Bungus Port (West Sumatera) and Tua Pejat-Mentawai (West Sumatera) (Table 1, Figure 1 and Figure 2). Most of the devices were installed in the fisheries ports and other port facilities (Bungus, Tua Pejat, Prigi, Pelabuhan ratu dan Sadeng). IDSL in Pangandaran is the only one installed outside the ports and port facilities because it was installed in a BIG (The National Geospatial Agency) platform specifically built for housing a tide gauge measurement (Figure 2). The installation of IDSL was conducted in cooperation with the local governments, Port authorities, The Agency of Meteorology, Climatology and Geophysics (BMKG), The National Geospatial Agency (BIG) and the National Agency of Disaster Management (BNPB) and the remote support of the JRC-EC.

Table 1. Installation of IDSL in Indonesia since January 2019.

No	Device ID	Location's name	Lat (°)	Long (°)	Installation's date	Town/Province
1	IDSL-302	Marina Jambu	-6.189322	105.841088	29.01.2019	Pandeglang/Banten
2	IDSL-301	Sebesi	-5.936047	105.512106	30.01.2019	Lampung Sel/Lampung
3	IDSL-306	Sadeng	-8.190314	110.799053	26.10.2019	DIY Yogyakarta
4	IDSL-307	Pangandaran	-7.748359	108.501402	29.10.2019	West Java
5	IDSL-305	Teluk Bungus	-1.029661	100.396191	03.07.2020	Padang/West Sumatera
6	IDSL-304	Pel. Ratu	-6.989507	106.541573	05.08.2020	Sukabumi/West Java
7	IDSL-308	Prigi	-8.291638	111.731428	15.08.2020	Tul. Agung/East Java
8	IDSL-303	Tua Pejat	-2.037654	99.585453	16.09.2020	Mentawai/W. Sumatera

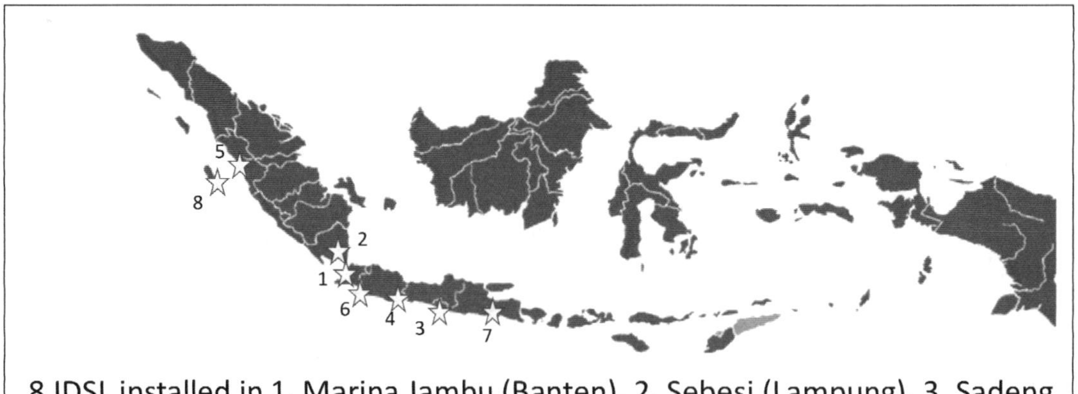

Figure 1. Installation of IDSL in Indonesia (see table 1).

The data of IDSL in Indonesia can be viewed in the website of JRC and the raw data can be freely downloaded (Link: https://webcritech.jrc.ec.europa.eu/TAD_server/Home?group=Indonesia). In each individual site (e.g. for IDSL in Sebesi Island of Lampung: https://webcritech.jrc.ec.europa.eu/TAD_server/Device/206), the IDSL webpage contains information about the characteristics of the device and the collected data such as the water level elevation data, air temperature, battery voltage, latency time, alerts, and the CCTV images. The following link can view the data of all 8 IDSL within the last 5 minutes (http://bit.ly/idn_idsl).

Figure 2. The installation of IDSL in Indonesia (see table 1 and figure 1)

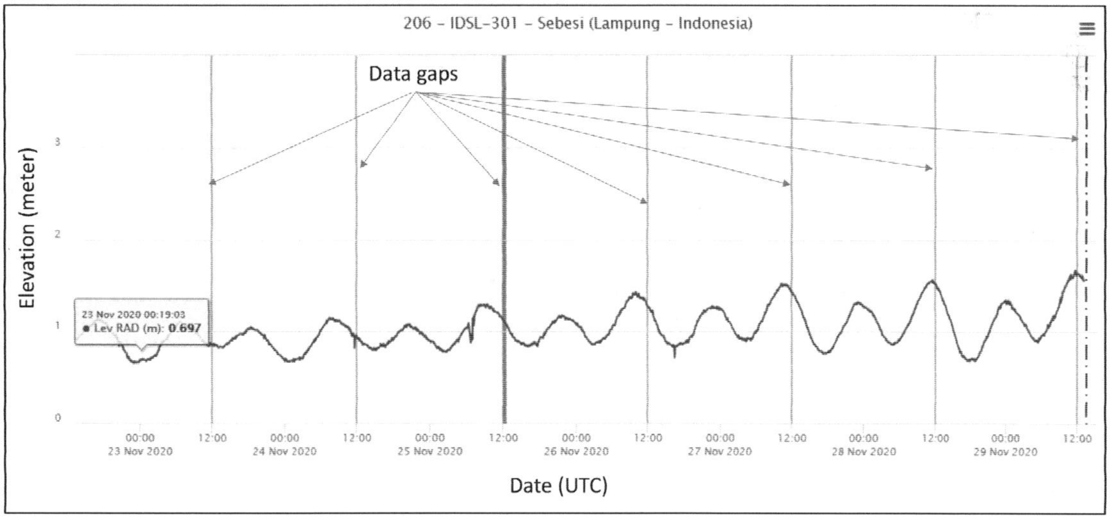

Figure 3. Example of data gaps during reset time at 12 UTC

3.2. Data quality of IDSL

The first six-month data of IDSL in Marina Jambu and Sebesi were analysed and compared to the data of tide gauges belong to BIG and the data collected by TideMaster (a commercial tide gauge produced by Valeport, link: https://www.valeport.co.uk/products/tidemaster-portable-tide-gauge/). Those three devices were installed at the same location. By using *t_tide* routine, a matlab-based harmonic-tide analyses tool [13], all data provided similar tidal magnitudes and constituents. Results also showed that the collected data from the three devices are almost similar and provide small variabilities which can be considered as negligible. The correlation of the data collected by IDSL, BIG and TideMaster are between 95%-99% [12]. The small differences in correlation values were resulted from the different of measurement intervals. The collected data of IDSL successfully maintained the time interval of 5 seconds while the others had time interval of 2 minutes (BIG) and 1 minutes (Tidemaster) [12].

Another important aspect on the quality of the data is the percentage of missing data or data gaps. Data gaps can be detected as the missing data due to a large latency time, poor communication network and communication lost (Figure 3). Throughout the measurements, poor communication network and communication lost were caused by operational management of the device (e.g. periodical reset, monthly internet payment and general maintenance) and the quality of communication (internet) infrastructures. The only aspect that may reduce the performance of IDSL is the large latency time occurring during some occasions (data gaps, see Fig. 3). This mainly occurred in the range of 30 second up to 5 minutes. Excluding the aspects of the operational management, the percentage of data gaps for IDSL were less than 9 % [12][14].

3.3. Latency time

Latency time is the time interval from the measurement to the availability online (i.e. it indicates the speed of data transfer from the device to the server). For the purpose of TEWS, the latency time should be as small as possible or near real-time. For the case of IDSL, *the latency time is kept to be less than 30 seconds* depending on the quality of the communication networks. Most of the time, *the latency time was less than 10 seconds*. Larger values on latency time will influence the numbers of missing data as already discussed in the previous sub section (data gaps). Based on the analyses of latency time for IDSL in Pangandaran with a stable communication network, the IDSL's latency time under 30 seconds reached 99% [14]. This means, IDSL with a stable communication network has reached a near real-time performance.

3.4. Delivering alerts

The capability of delivering alerts immediately when detecting sea level anomalies is a key for the device to be used for TEWS. A more detailed information on the algorithm of IDSL for the prediction of sea levels and detecting the anomalies of sea levels can be found in [10]. For the case of 8 IDSL installed in Indonesia, there were no tsunami events. However, several anomalies were detected, and alerts were delivered timely to the server, dedicated email addresses and dedicated phone numbers with SMS texts. The sea level anomalies detected by IDSL have proved that the alert system has been working properly.

It should be noted that whenever an 'alert' is identified, the system activates a frequent image collection from the CCTV (one image every 2 min instead of 15 to 60 min during non-alert mode). The role of CCTV images to visually check whether the tsunami events occurred (or not) was very important. Based on the alerts delivered by the system and the quality of the CCTV images, the following aspects influence the alert system of IDSL:

- Bad weather, especially during heavy rain with strong wind will trigger the system to deliver alerts due to detection of "waters" (rain waters) in the air. CCTV images helped the recognition of the alerts due to bad weather events.
- The sway of the metal frame will cause "anomalies". This event may be caused by strong wind or human activities in the metal frame structure.
- The presence of boats under the sensor. This is the most common alerts found so far because IDSL mostly are located in the fisheries ports. This sometimes can be easily recognised by the CCTV images.
- Data gaps may also trigger alerts because of sudden reading in the elevation data. Alerts due to data gaps were easily recognised visually.

3.5. CCTV images

The main objective of the CCTV images is to detect visually the water lines as the sign for the tsunami event. The CCTV images were set to capture image for every 15 minutes during normal time and every 2 minutes during alert's time. In case of alerts, the CCTV images proved to be helpful in analysing the on-going event. Several detections of alerts were easily explained based on the CCTV images such as those caused by bad weathers, strong wind and the presence of boats under the sensor. However, there are some conditions that the CCTV images were not clear enough for the observation of the waterlines:

- During night-time where the lights in surrounding area are poor (or no lights at all). The CCTV images were mostly dark, and the identification of the waterlines was impossible. The CCTV in Prigi and Bungus Ports are those perform 100% clear even during night-time due to excellent lighting of the port and surrounding area.
- During bad weather where identification of the water line was also very difficult.

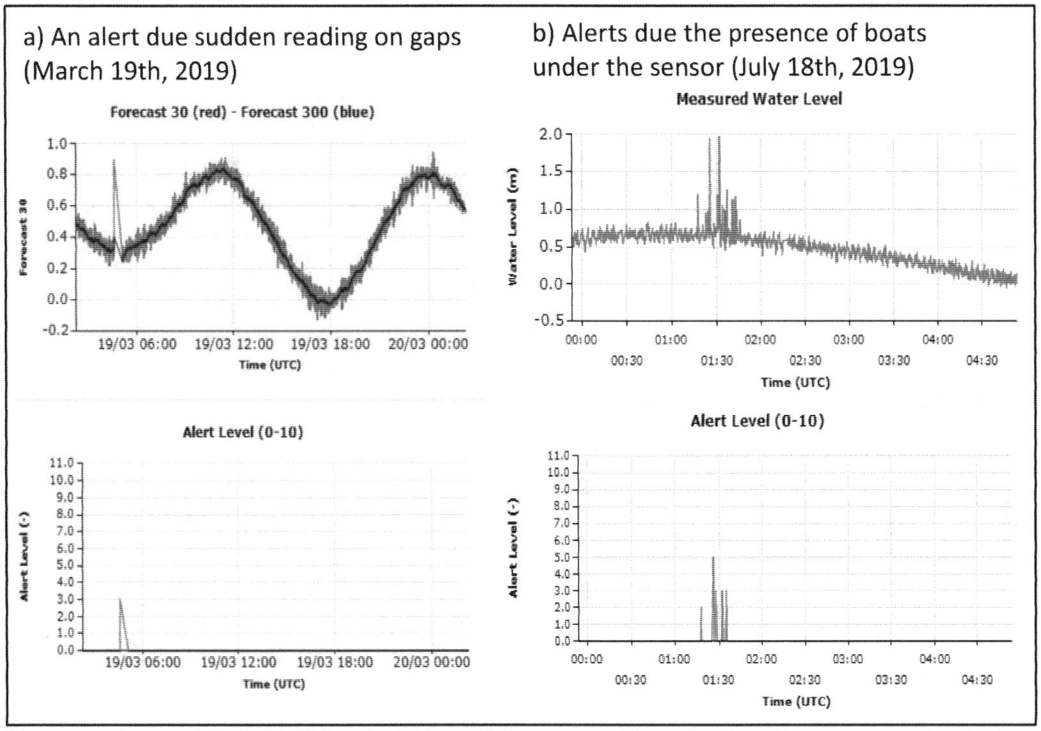

Figure 4. Examples of alerts detection by IDSL

Besides the above key parameters, other aspects are also very important for the consideration of using IDSL for the tsunami early warning system (TEWS) in Indonesia:

- The construction costs of the device are much cheaper compared to the other existing system. IDSL can be fabricated with initial coast less than 2000 USD (Rp. 25 million) including the metal frame. The monthly cost for internet data was 5 USD per month (Rp. 68.000,-)
- The installation method is very simple and so anyone without specific skills can participate in the installation processes
- The maintenance and the operation of IDSL are also simple. The only action that was required more often is to switch on the reset button
- The endurance of the electrical components is excellent. In the last 20 months, the parts that need a replacement are the fuse. The sensor surface was found to be dirty at two locations. Cleaning process was conducted, and the devices were back to normal afterwards.
- The involvement of the coastal community in monitoring and maintaining the operational of IDSL is very important. The IDSL team has worked together with the coastal community to assure the normal operation of IDSL

- The use of IDSL or other similar systems are clearly needed and more units should be installed. In order to get an optimum additional number of IDSL/Tide gauge and ideal locations, and assessment using scenario bases of tsunami modelling can be utilised. Ideally, IDSL should be placed as close as to the potential sources of tsunami. Considering of nearfield tsunami from megathrust source - most of the coastal areas along the Indonesian Archipelago are closed to the source.

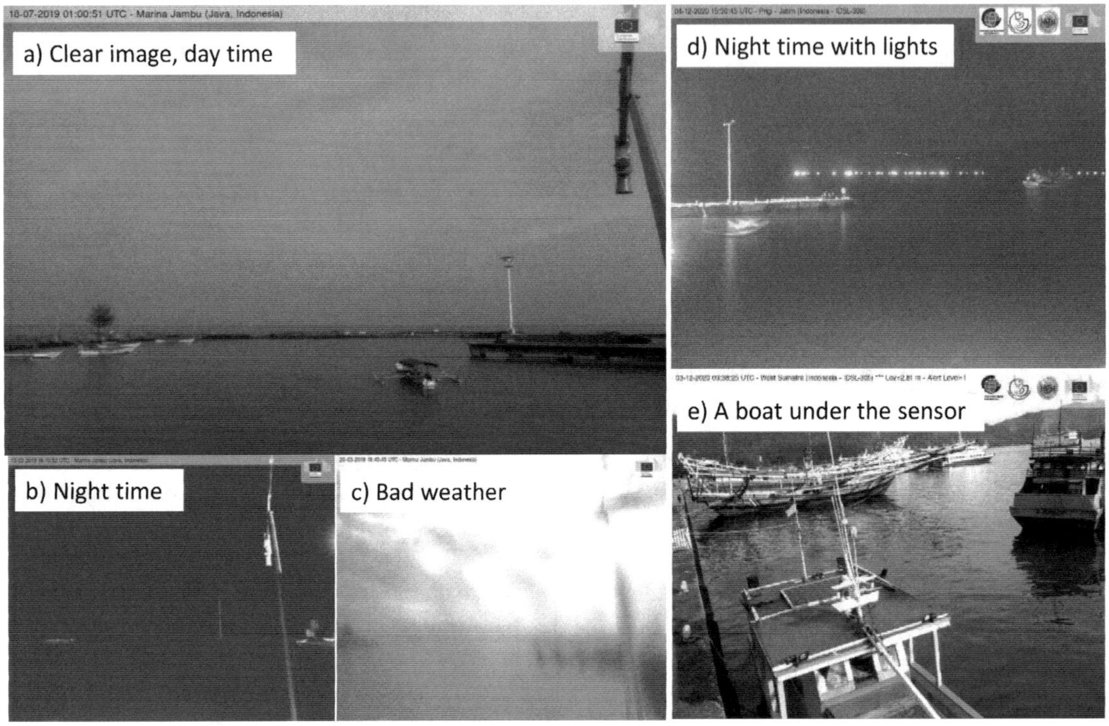

Figure 5. The quality of the CCTV images can explain the alerts

4. Conclusion and Recommendation

Based on the discussion above, the use of IDSL to strengthen the existing TEWS in Indonesia is strongly recommended considering *the system is near real-time, provide high quality data, low cost in construction and maintenance, capable of delivering alerts, and has additional features for double check (the CCTV images)*. Most of the weaknesses of IDSL also persist in other system, especially those related to the quality of communication network. To increase the performance of IDSL, the following steps can be taken:

- Increase the numbers of units across Indonesia and place them as near as possible to the potential sources of tsunami
- For the CCTV images, the quality can be improved by adding better quality of CCTV devices or adding additional lights in the surrounding area
- Improve the quality of communication infrastructures
- Integration with the existing systems within TEWS. This will help quick response in case the real tsunami events occurred.

Acknowledgments

We would like to thank the Marine Research Centre (KKP) for funding IDSL research based on APBN-DIPA 2019/20 financial budget. We also would like to thank the Joint Research Centre – European Commission for the donation of 8 units of IDSL for Indonesia and all the technical supports. The supports from other institutions such as the Indonesian Tsunami Expert Society (IATsI), BIG, Coordinating Ministry of Maritime Affairs, IOC-UNESCO Jakarta, & BMKG are highly appreciated. All authors contributed equally in preparing, analysing the data, writing and reviewing the manuscript. We also would like thank "Rumah Program Kebencanaan Geologi – BRIN 2022".

References

[1] BMKG 2010 *InaTEWS Indonesia Tsunami Early Warning System Konsep dan Implementasi*, link: https://www.gitews.org/tsunami-kit/id/E2/sumber_lainnya/InaTEWS%20-%20Konsep%20dan%20Implementasi.pdf, accessed on 23 April 2020

[2] BMKG 2012 *Buku Pedoman Pelayanan Peringatan Dini Tsunami InaTEWS – Edisi Kedua.* Jakarta: BMKG.

[3] BMKG 2014 *Peta jaringan akselegraf seluruh Indonesia sampai dengan tahun 2015*, link: https://www.bmkg.go.id/seismologi-teknik/?p=jaringan-accelerograph-intensity-meter&lang=ID, accessed on 23 April 2020

[4] Lauterjung, J., Letz, H. (Eds.) 2017 10 Years Indonesian Tsunami Early Warning System: Experiences, Lessons Learned and Outlook, Potsdam: GFZ German Research Centre for Geosciences, 68 p. DOI: http://doi.org/10.2312/GFZ.7.1.2017.001

[5] Lauterjung, J., Münch, U., Rudloff, A., 2010 The challenge of installing a tsunami early warning system in the vicinity of the Sunda Arc, Indonesia. *Natural hazards Earth System Sciences*, 10, pp. 641 – 646.

[6] BMKG 2019 Indonesian Master Plan for Tsunami Early Warning System and Inclusion of Atypical Tsunami, A keynote speech by the Head of BMKG, the *International Symposium on the Lesson Learnt from the 2018 Tsunamis in Palu and Sunda Strait-Jakarta 2019.*

[7] Omira, R., Dogan, GG., Hidayat, R., Husrin, S., Prasetya, G., Annunziato, A., Proietti, C., Probst, P., Paparo, MA., Wronna, M., Zaytsev, A., Pronin, P., Giniyatullin, A., Putra, PS., Hartanto, D., Ginanjar, G., Kongko, W., Pelinovsky, E., Yalciner, AC., 2019 The September 28th, 2018, tsunami in Palu-Sulawesi, Indonesia: a post-event field survey. *Pure Appl Geophys* 176(4):1379–1395. https://doi.org/10.1007/s00024-019-02145-z

[8] Husrin, S., Azzahra, FY., Prihantono, J., Tussadiah, A., Abida, AF. 2020 Hydrodynamics of Palu Bay during the Event of 2018 Palu Earthquake and Tsunami, Bulletin of The Marine Geology, Vol 35, No 1 (2020) DOI: http://dx.doi.org/10.32693/bomg.35.1.2020.658

[9] Borrero, J.C., Solihuddin, T., Fritz, H.M. *et al.* 2020 Field Survey and Numerical Modelling of the December 22, (2018). Anak Krakatau Tsunami. *Pure Appl. Geophys.* 177**,** 2457–2475 (2020). https://doi.org/10.1007/s00024-020-02515-y

[10] Annunziato, A., Prasetya, G. and Husrin, S. 2019 Anak Krakatau volcano emergency tsunami early warning system, *Science of Tsunami Hazards*, 2019, ISSN 8755-6839 (online), 38 (2), p. 68-95

[11] Annunziato A., Galliano D., Bonaita M. 2016 IDSL Sea Level Measurement Devices, Technical report JRC-EC. doi:10.2788/470647

[12] Husrin, S., Fauzah, S., Safi'i, AN., Patriabekti, YDH. 2021 Sistem Peringatan Dini Tsunami, book chapter: Manajemen Bencana dan Pembangunan Berkelanjutan di Pesisir Selat Sunda, Editor (Solihuddin, et. al), IPB Press

[13] Pawlowicz, R., Beardsley, B., and Lentz, S. 2002 Classical tidal harmonic analysis including error estimates in MATLAB using T_TIDE", Computers and Geosciences 28 (2002), 929-937

[14] Husrin S, Novianto D, Bramawanto R et al. 2021 Analisa Kinerja IDSL/PUMMA untuk Peringatan Dini Tsunami di Pangandaran, Jurnal Kelautan Nasional Vol. 16. No. 2 (2021), DOI: http://dx.doi.org/10.15578/jkn.v16i2.9846

Point load test method for prediction strength of sustainable mortar made from blended cement and fly ash

A Maysyurah[1] and M A Caronge[2*]

[1]Magister Student, Civil Engineering Departement Univercity Hasanuddin, Indonesian
[2]Lecturer, Civil Engineering Departement Univercity Hasanuddin, Indonesian

E-mail: caronge_eng@yahoo.co.id

Abstract. Fly ash (FA) is a by-product of power plants where a number of serious efforts in the field of sustainable building materials have spawned a recycling method in the form of cementitious material containing FA as raw material, including blended cement. The contribution of blended cement to producing sustainable mortar is presented in this study. The FA at a percentage of 10% to 30% by weight of cement is used instead of blended cement in the manufacture of mortar. The results of the point load strength index (PLSI) at the age of 7 days the percentages of FA-0, FA-10, FA-20, and FA-30 were 0.58, 0.46, 0.41, 0.41 N/mm^2 and at the age of 28 days were 0.88, 0.85, 0.79, and 0.78 N/mm^2 respectively, indicating that the mortar produced with blended cement and FA is carried the point loads well because it has stiffness and adequate robustness. The contrast test with available methods verifies that PLSI offers an alternative to compressive test and it has high accuracy. The point load test method can be used to predict the strength of mortar produced of blended cement and FA.

Keyword: Fly ash, blended cement, point load test, sustainable, mortar

1. Introduction

The use of coal continues to increase which is used to supply power plants to be able to provide stable electrical energy to support economic growth. One of the by-products produced from burning coal is FA. The use of coal that continues to increase has resulted in the number of FA that has accumulated in the storage ponds, which has a negative impact on the surrounding environment.

Those working in the building materials sector have deep concerns about the environmental impact caused by the increasing number of FAs. In this regards, various sustainable technologies have been developed with the aim of recycling industrial by-products including FA. FA is a pozzolanic material that is rich in alumina-silica elements so that it has the ability to replace type I Portland cement in the manufacture of concrete and mortar [1]. Blended cement is one type of hydraulic cement of which FA is one of its constituents [2]. Based on SNI 15-7064-2004 pozzolanic materials such as FA in blended cement ranges from 15-30%. The compressive strength value of blended cement is relatively lower in the water and air curing test. Although FA has been absorbed by cement factories to be combined with Portland cement clinker in the blended cement process, there is still a lot of FA that accumulates in the dumpily pond. In connection with this problem, this study uses blended cement as the main binder in the manufacture of mortar and also includes FA as a partial substitute for blended cement in the manufacture of blended cement-mortar. The introducing of FA as a mortar additive for bonding and plastering nature appears to be a feasible and useful method for recycling the amount of FA disposal.

2. Materials and testing method

2.1 Fine aggregate

The aggregate used for this research obtained from river sand which is washed and sifted with a maximum size that passes through the #4 sieve. The fineness modulus used for mortar is 2.96%, water absorption and specific gravity in the SSD sand state are 2% and 2.37, respectively.

2.2 Cement and *Fly ash*

The blended cement used is composite Portland composite cement (PCC) which is available in local market and produced based on SNI 7064-2014 [3-7]. Table 1 shows physical properties cement. FA comes from Barru, South Sulawesi, Indonesia. Fly ash is used to replace cement at rate of 10%-30%.

Table 1. Physical Properties cement

Specific gravity (gr/cm^3)	Bulk density (kg/liter)	Air of mortar (%)	setting time (min)		Fineness, specific surface		Compressive strength (MPa)		
			Initial	Final	Air perm. test (m^2/kg)	Residu on 45 mic	3 days	7 days	28 days
3.15	1.1	8.97	130	225	370	16.99	15.7	22.7	31.1

2.3 Mix proportion

All test specimens are designed with a flow value of 190 mm. In order to meet the designed flow, the water content of each mixture is adjusted based on the number of FA used to replace the cement content % by weight. The variations of fly ash used were 0% (FA-0), 10% (FA-10), 20% (FA-20) and 30% (FA-30) of the cement weight, as shown in Table 2. After mixed evenly and reached a flow value of 190 mm, the fresh mortar was added and compacted in a cube mold measuring 50 mm x 50 mm x 50 mm. then, allowed to stand for 24 hours in a room with a temperature of 25°C until the day of testing [8].

Table 2. Mix Proportion *fly ash* mortar (m^3)

Mix designation	Replacement of cement by FA (%)	Water/Binder ratio	Quantities (kg)			
			Water	FA	Blended cement	Sand
FA-0	0	0.59	284.61	0	484.03	1331.07
FA-10	10	0.65	284.06	48.31	434.78	1328.50
FA-20	20	0.76	291.71	95.33	381.31	1310.77
FA-30	30	0.87	291.15	141.72	333.02	1308.28

2.4 Point load strength index

PLSI has the advantage that it can be applied to test specimen with irregular shapes and test objects with a certain size. Formulas 1, 2 and 3 are used to obtain mortar point load index I_{50} (N/mm2), P is the failure load (N) and Dx diameter (mm)[9]. The ASTM D5731-16 axial test was applied to a mortar specimen with a length/diameter ratio between 0.3 and 1 following [10] which performed a point load test on rock specimens with an almost cube-like shape.

$$I_{S(50)} = \frac{P}{De^2} \quad (1)$$

$$D_e = \sqrt{4DW/\pi} \quad (2)$$

$$F = \sqrt{\frac{De}{50}} \quad (3)$$

PLSI testing on 50 mm equivalent size samples [11,12]. For the 50 mm nonstandard test, the correction factor is calculated using the F approximation.

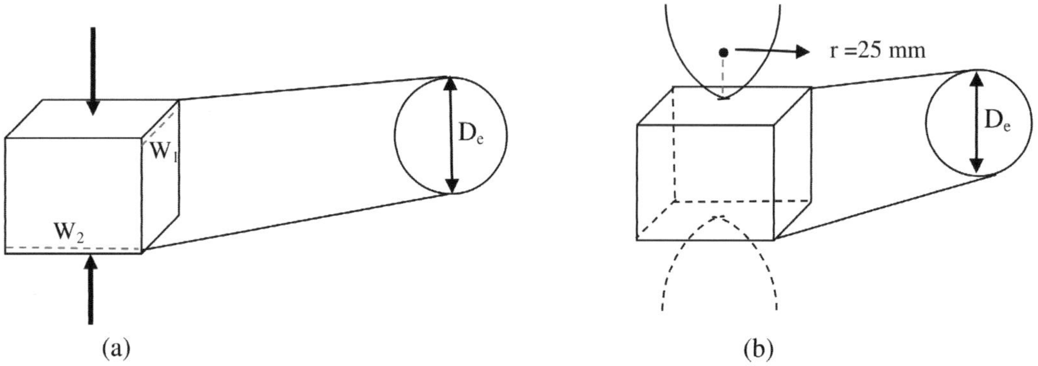

Figure 1. Cube specimens a) diameter of load cube, b) conical plate diametrically

For the diametral test the sample length should be greater than or equal to the diameter, and the load should be applied (location of the conical platens) at least 0:5D from the ends of the sample[13].

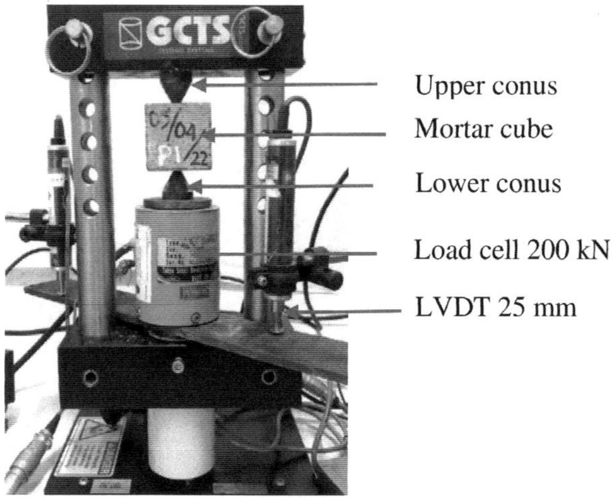

figure 2. Point load index testing

The standard radius of the cone plate of a point load testing machine is 5 mm. Figure 2 present in detail the PSLI testing device equipped with a load cell 2 LVDTs. The load and penetration of the truncated cone of the PSLI tool were carefully measured using load cells and LVDTs. A set of monitoring and recording devices consisting of a set of data loggers and a computer connected to the load cell and LVDTs.

3. Results and discussion

3.1 Characteristic *fly ash*

The chemical properties of PCC and FA obtained by X-ray fluorescences (XRF) test is shown in Table 3. The SiO2 content of PCC in the ranged of 18 – 20% which is fulfilled the SNI 7064-2014 for blended cement. The total SiO_2+Al_2O_3+Fe_2O_3 of the FA more than 70% and CaO content < 10% satisfied the requirement for Class F according to ASTM C618-12 [14].

Table 3. Chemical properties of PCC and FA

Material	SiO_2	Al_2O_3	Fe_2O_3	CaO	MgO	SO_3	Ins. residu	Free lime	LOI
cement (PCC)	18.16	5.10	3.40	61.24	0.93	1.46	2.08	1.58	6.20
FA	51.88	7.95	34.32	3.18	0.57	4.81	-	-	-

3.2 Point load strength index (Is_{50})

Table 4 shows at the age of 7 and 28 days with each Is(50) using 3 samples, Is(50) decreased with increasing in the amount of FA to replace blended cement. According to ASTM D5731 the whole mixture has a medium strength or Is(50), hence can be used as a binder and plaster.

Table 4. Point load strength index 7 and 28 days

Days	variation Specimen	$Is_{(50)}$ (N/mm^2)	CoV (%)	Strength Classification (ASTM D5731)
7	FA-0	0.58	4.00	Medium
	FA-10	0.46	2.79	Medium
	FA-20	0.41	8.18	Medium
	FA-30	0.41	8.33	Medium
28	FA-0	0.88	3.85	Medium
	FA-10	0.85	6.02	Medium
	FA-20	0.79	7.85	Medium
	FA-30	0.78	8.09	Medium

Table 4 shows the CoV obtained for each variation of the fly ash mortar specimen is below 10%, hence this finding indicates that good mixing of blended cement and FA can be achieved so that Is(50) of all specimens is almost similar at each testing day.

3.3 $Is_{(50)}$ and peak penetration value

Table 5 shows the average peak penetration at the age of 7 and 28 days using 3 test samples, respectively, where the penetration depth of the truncated cone plate of the PLSI equipment on the PLSI peak is measured that the peak penetration value of the mortar is affected by the FA content in the mortar. The

Table 5. Peak penetration rate

Specimen Name	Penetration (mm)			
	7 days	CoV (%)	28 days	CoV (%)
FA-0	3.37	8.91	3.48	6.87
FA-10%	2.95	7.19	3.32	11.25
FA-20%	2.76	9.92	3.02	10.73
FA-30%	2.81	7.57	2.96	8.78

Peak penetration value was observed to decrease with a decrease in fly ash mortar content at curing ages of 7 and 28 days. At 7 days, the peak penetration value of the control mortar was 3.36 mm, where the penetration of mortar with fly ash 10% (FA-10), 20% (FA-20), and 30% (FA-30) was 2.94, 2.75 and 2.80 mm. At 28 days of age, the peak penetration value of the control mortar was 3.47 mm, where the mortar with fly ash 10% (FA-10), 20% (FA-20), and 30% (FA-30) had penetrations of 3.31, 3.01 and 2.96. mm. Overall, the achieved penetration values clarify and support the findings obtained from the discussion of the relationship PLSI and FA content which shows that all specimens without FA or with FA content up to 30% have good stiffness so that the specimens are able to accept cone plate penetration to a depth that causes failure.

4. Concluding remarks

Based on the value of PLSI (I_{50}) and peak penetration the specimens with FA and without FA have similar behaviour which is allowed cracks to occur at point loads approaching the peak. Overall, the test results show that FA and blended cement can form mortar with ideal stiffness.

Acknowledgement

The authors express their deepest gratitude for their support and assistance during specimen testing. We also thank the Department of Civil Engineering, Hasanuddin University for granting permission to use the tool and publish the PLSI fly ash mortar used in this study and for providing materials and instruments for this research.

Reference

1. M.L. Wang, V. Ramakrishnan, Evaluation of blended cement, mortar and concrete made from type III cement and kiln dust, Construction and Building Materials, Volume 4, Issue 2, 1990, Pages 78-85. https://doi.org/10.1016/0950-0618(90)90005-L.
2. SNI 7064:2004. 2004. Portland Composite Cement. Indonesian National Standard (SNI). Jakarta, Indonesia. (in Bahasa)
3. A. Marewangeng, M.W. Tjaronge, A.R. Djamaluddin, & S. H. Aly (2020), Mechanical Characteristics of self compacting concrete using laterite stone as coarse aggregate. International journal of GEOMATE. http://doi.org/10.21660/2020.71.357214.
4. A. Adnan, Parung, Herman, Tjaronge, M. D. Rudy (2020). Bond between Steel Reinforcement Bars and Seawater Concrete. Civil Engineering Journal. http://doi.org/10.28991/cej-2020-SP(EMCE)-06
5. Djamaluddin, A.R., Caronge, M.A., Tjaronge, M.W. et al. Abrasion resistance and compressive strength of unprocessed rice husk ash concrete. Asian J Civ Eng 19, 867–876 (2018). https://doi.org/10.1007/s42107-018-0069-5.
6. Caronge, M. A., Tjaronge, M. W., Hamada, H., & Irmawaty, R. (2017). Effect of water curing duration on strength behaviour of portland composite cement (PCC) mortar. IOP Conference Series: Materials Science and Engineering, 271(1). https://doi.org/10.1088/1757-899X/271/1/012018.
7. Caronge, M. A., Tjaronge, M. W., Rahim, I. R., Irmawaty, R. & Lapian, F. E. P. (2022). Feasibility study on the use of processed waste tea ash as cement replacement for sustainable concrete production. Journal of Building Engineering, 52, 104458.
8. SNI 03-6825-2002. 2002. Method of testing mortar. Indonesian National Standard (SNI), Jakarta, Indonesia. (in Bahasa)
9. Mehdi Koohmishi, Assessment of strength of individual ballast aggregate by conducting point load test and establishment of classification method, International Journal of Rock Mechanics and Mining Sciences, Volume 141, 2021, 104711. https://doi.org/10.1016/j.ijrmms.2021.104711.
10. ASTM D5731-16. 2016. Standard Test Method for Determination of the Point Load Strength Index of Rock and Application to Rock Strength Classifications

11. Yang, Suhang & Zhou, Junwen & Bian, Zhengning & Xu, Zhifeng. (2020). Point-Load Test Method for Estimation of In Situ Masonry Mortar Strength. Journal of Materials in Civil Engineering. http://doi.org/10.1061/(ASCE)MT.1943-5533.0003347.
12. Yin, Jian-Hua & Wong, Robina & Chau, Kt & Lai, David & Zhao, Si. (2017). Point load strength index of granitic irregular lumps: Size correction and correlation with uniaxial compressive strength. Tunnelling and Underground Space Technology. http://doi.org/10.1016/j.tust.2017.09.011.
13. Broch, Einar and Franklin, J.A.(1972), 669-676, The point-load strength test. International Journal of Rock Mechanics and Mining Sciences & Geomechanics Abstracts. https://doi.org/10.1016/0148-9062(72)90030-7.
14. ASTM C618-12a. 2012. Standard Specification for Coal Fly Ash and Raw or Calcined Natural Pozzolan for Use in Concrete, ASTM International.

The Strength of Reinforced Concrete Slabs of 30% Candlenut Shell Variation Using Destructive Testing Methods

Nasruddin and P. Mushar

Laboratory of Building Material, Structure, and Construction, Architecture Department, Faculty of Engineering, Hasanuddin University

Corresponding author: nas_junus@yahoo.com

Abstract. The compressive strength of concrete is one of the parameters used to control the quality of concrete. Concrete compressive strength testing method which is considered the highest level of reliability is a destructive testing method using a compression testing equipment. Based on the case, a testing tool is needed to measure or determine the compressive strength of concrete known as the Universal Testing Machine (UTM). Experimental research is conducted with a scientific approach using several of variables. The purpose of this research is to get the compressive strength value of concrete slabs variation of 30% candlenut shells. From the test results using a UTM so that it can be obtained the quality of reinforced concrete. The research variable are the 28 days age of concrete and the curing methods (wet curing and dry curing). The number of specimens of reinforced concrete slabs is 4 pieces measuring 50cm x 25cm x 12cm. The results of the study are the compressive strength of concrete slabs 30% variation of candlenut shells using UTM shows that the strength of reinforced concrete slabs reaches 75 MPa in wet curing and 59 MPa in dry curing or equivalent to the quality of K400 to K500 concrete at 28 days. The results also show the compressive strength of concrete in wet curing is higher than in dry curing.

1. Introduction

Reinforced concrete construction methods are frequently used to create many buildings. Even though some structures, like monuments or historic buildings, have a lengthy lifespan, the strength of those structures decreases significantly with age, for instance in reinforced concrete buildings. The hypothesis states that although we are aware that concrete achieves its maximum age at 28 days, beyond that point, there is a slight increase in strength. However, the strength of concrete will decline after 50 years if it is still in place. One of the criteria used to regulate the quality of concrete is its compressive strength. The most reliable way to measure the compressive strength of concrete is by a destructive test utilizing a compressive testing machine like the Universal Testing Machine (UTM).

The primary material used to make buildings is concrete. Because of its benefits, such as its ease of shaping to fit the needs of building, ability to support enormous loads, resistance to high temperatures, and low maintenance requirements, concrete is used extensively. Concrete's evolution included the use of substitute ingredients to lessen the brittleness of the material. Substitution material is a material that can replace concrete material either coarse aggregate, fine aggregate, or cement with other materials, namely Portland Cement with steel slag, split stone (coarse aggregate) with pumice rock, and others. One benefit of the concrete material replacement process is the use of organic waste. Residues from manufacture and use, such as candlenut shells made from candlenut waste, can result in organic waste. Candlenut shell is a novel potential that may be expanded upon and exploited even further. Naturally, this might increase the value of candlenut shell, which was previously only thought of as waste from the candlenut business. The usage of candlenut shell waste in the Kemirie neighborhood of Parepare City, South Sulawesi Province, is not yet at its best. But as reality demonstrates, there is even more that can be done with the candlenut shell's potential. The candlenut shells are made up of CaO, SiO2, Al2O3, MgO, H2O, and Fe2O3. When everything has reacted, any remaining SiO2 will form a reaction with the CSH-2 gel to create a denser CSH-3 gel, which will increase the cement paste and aggregate [1]. It is quite legitimate to say that used candlenut shells can be substituted for in concrete compositions given the aforementioned justification. In this topic, the problem is the value of the concrete's compressive strength when different candlenut shells are put to the concrete slab in a destructive way using UTM, per the description.

2. Literature Reviews

Mulyono [2] claims that concrete may generally be split into the following groups: 11 - Concrete, classified according to its quality and class, is: (a) Class I concrete. This category of concrete is used for non-structural projects. This concrete doesn't require specialized knowledge to be implemented. Its quality control is minimal, consisting just on modest ingredient quality monitoring. It is not necessary to inspect the compressive strength. Bo denotes class I concrete; (b) denotes class II concrete. It is a type of concrete used for basic structural projects. Its execution must be carried out with the necessary skills and under the supervision of professionals. The standard qualities of Class II concrete are B1, K125, K175, and K225. B1's definition of quality control solely includes the supervision of ingredient quality. It is not necessary to inspect the compressive strength. It is necessary to regularly examine the concrete compressive strength in the K125 and K175 using the test specimen results; (c) Class III concrete. It belongs to a higher category of structural concrete than K225. Its implementation must be done with specialized knowledge and under the guidance of professionals. It needs a concrete laboratory with all the necessary tools, staffed by professionals who can perform ongoing quality control of the concrete.

Table 1. Conversion of Concrete Quality in Kg/cm2 to MPa

QUALITY K OF CONCRETE (kg/cm2)	QUALITY f_c OF CONCRETE (MPa)
K100	8,3 MPa
K-150	2.35 MPa
K-175	14.53 MPa
K-200	16.60 MPa
K-225	18.68 MPa
K-250	20.75 MPa
K-275	22.83 MPa

IOP Conf. Series: Earth and Environmental Science 1117 (2022) 012030 doi:10.1088/1755-1315/1117/1/012030

QUALITY K OF CONCRETE (kg/cm2)	QUALITY f_c OF CONCRETE (MPa)
K-300	24.90 MPa
K-350	29.05 MPa
K-400	33.20 MPa
K-450	37.35 MPa
K-500	41.50 MPa

Source: SNI 03-2847-2002 [3]

The concrete quality is divided into three quality classes, namely class I concrete, class II concrete, and class III concrete, based on Table 1 above. Class I concrete, which is represented by K-100, K125, K-150, K-175, and K-200, is used for non-structural work; Class II concrete, which is represented by K-225, K-250, and K-275, is used for structural work, such as floors, roads, foundations, floors, columns, etc.; and Class III concrete, which is represented by K-325, K-350, K-375, K-450, and K-500, is special concrete.

According to ASTM C 494 / C494M - 05a, there are seven different categories of chemical additives:

1. Water-reducing additives
2. Retarding additives
3. Accelerating additives
4. Water-reducing and retarding additives
5. Water-reducing and accelerating additives
6. Water-reducing and high range additives
7. Water-reducing, high range and retarding additives

A concrete compression test that is adapted to the concrete's hardening period provides the compressive strength of the material. The concrete compression test can be conducted for up to 28 days according to the law.

The ability of the test object to endure the compressive force or the test specimen's greatest capacity in resisting the force that causes destruction is known as the crushing compressive strength of an individual. The following elements affect the compressive strength of concrete:

- Cement Quality and Type
- Aggregate Surface Texture and Type
- Curing (Treatment)
- Temperature

The tensile stress and compressive strength of a material or materials are tested using the universal testing machine. As seen in Figure 1, a testing machine is also frequently referred to as a universal tester, materials testing machine, or materials test frame. This testing apparatus has demonstrated its capability to carry out a variety of common tensile and compression tests on various materials, parts, and buildings. Applying a compressive or tensile force to the material being tested is how you use the universal testing machine. We require another specimen in order to conduct compressive or tensile testing on the material. The specimen is fixed to a testing device with compressive and tensile forces that gradually grow in size, pressing and tugging on the rod until it changes length from short to long.

In order to properly achieve the intended standardization, Universal Testing Machine will supply information about how much measurement will be tested on the material. (www.alatuji.com, accessed on October, 2019) [4]

Figure 1. Universal Testing Machine (UTM)
Source: http://indonesian.tensiontestmachine.com

Reinforced concrete plates are frequently used in construction, whether as a building's floor, roof floor, bridge floor, or pier floor. Typically, the load operating on the plate is measured against the gravity load (dead load and / or live load). The load results in a bending moment (as in the case of the beam) [5].

According to Ali Asroni in the book Beam and Reinforced Concrete Plates, (2010) [6] what is meant by reinforced concrete slabs is thin structures made of reinforced concrete with fields perpendicular to the plane of the structure. The reinforced concrete plate is particularly rigid and horizontal in direction, therefore in structures, these plates serve as diaphragms or stiffening elements in a structure. The thickness of the plate plane is relatively tiny when compared to the length and width of the plane.

3. The Method

The test's methodology involves using UTM equipment to destructively measure the concrete's compressive strength value. A reinforced slab measuring 50 cm by 25 cm by 12 cm and having two layers of 8 mm reinforcement serves as the test specimen. Figure 2 depicts the kind of universal testing machine used in this study. Wet curing and dry curing techniques are used to treat the specimens. 28 days later, concrete is compacted tested with a sample size of two pieces for each variation and treatment. Thirty percent of the overall coarse aggregate is made up of various candlenut shells.

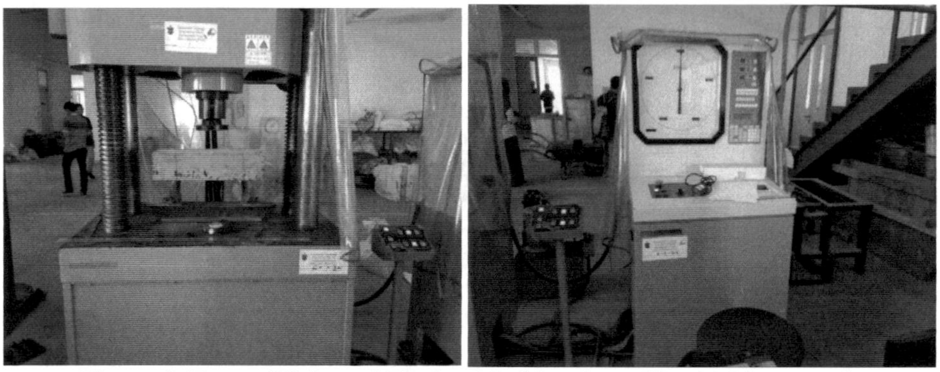

Figure 2. Type of Universal Testing Machine that utilize in this research

4. Discussion

4.1 Previous Research

The concept of concrete material products with discarded candlenut shells as a gravel substitute and as a new building material for self-help housing construction was developed in 2018 based on earlier research on cylindrical specimens. The final cylindrical concrete demonstrated that a modification of candlenut shells of 30% had stronger concrete compressive strength and attained the concrete quality of K200-K225 [7]. The results of these data can be used to produce a 30% candlenut shell variation when applied to specimens that act as a slab structure system. Figure 3 shows the case where candle nut shell material is used as a replacement material, the slump of conventional concrete, and the slump of the variation of 30% candle nut shell.

(a) (b) (c)

Figure 3. (a) Candlenut shell material; (b) Normal Concrete Slump (0%); (c) 30% Concrete Variation Slump

The concrete slabs specimens are casting on mold as shown in Figure 4. After 24 hours, the mold was opened and the specimen was cured in two conditions, namely dry curing and wet curing. In the wet curing method, the test specimen is placed in the pool in the curing room as shown in Figure 5. The setup of concrete slab specimen on the Universal Testing Machine is shown in Figure 6.

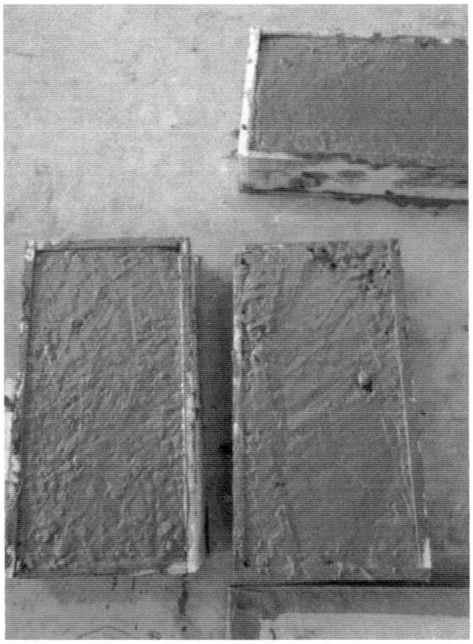

Figure 4. The concrete slabs specimens are casting on mold

Figure 5. Concrete slab specimens in the wet curing methods

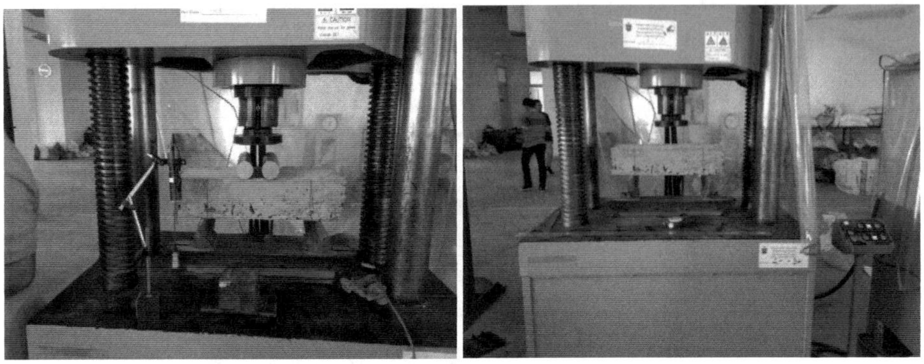

Figure 6. The setup of concrete slab specimen

4.2 Concrete Slab Specimens Testing Result

According to the previous study [8], a concrete slab with a candlenut shell variation weighs less than a slab with a regular concrete variation of 0%. Due to the substitution of candlenut shell, the weight difference is between 10-15%. Table 2 displays the characteristics of concrete slab specimens.

Table 2. Properties of Concrete Slab Specimens

Test Specimes	Weight (Kg)	Length (Cm)	Width (cm)	Height (cm)
D-1-30	32.52	50	25	12
D-2-30	30.94	50	25	12
W-1-30	31.73	50	25	12
W-2-30	30.61	50	25	12

Table 3. Results of destructive concrete compression testing with UTM

D-1-30	D-2-30	W-1-30	W-2-30
95,30	23,50	86,70	64,58
59,4		**75,64**	

NB. D = Dry Curing
W = Wet Curing

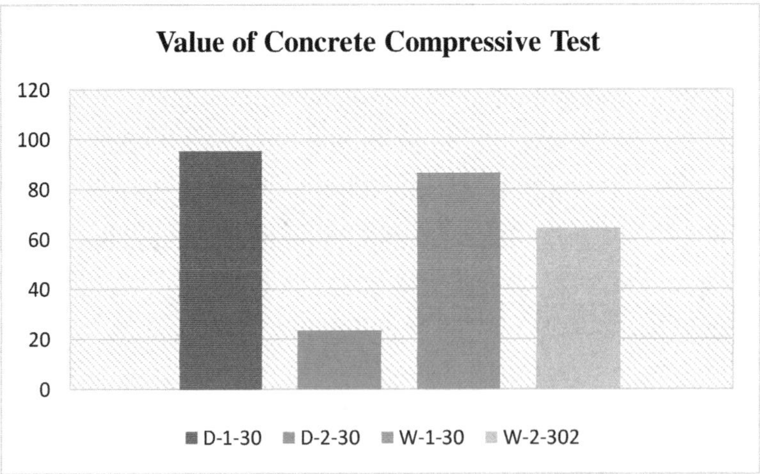

Figure 7. Value of Compressive Strength of Concrete Slab Specimens by Using UTM

IOP Conf. Series: Earth and Environmental Science 1117 (2022) 012030 doi:10.1088/1755-1315/1117/1/012030

Wet curing at 28 days increases the compressive strength of reinforced concrete slabs with 30% shell variation compared to dry curing, according to the data and figure above (Table 3 and Figure 7). The testing of the tool suffering bolt looseness on the test stand resulted in a significantly lower compressive strength rating for dry treatment D-2-30. Therefore, based on this instance, it is conceivable for the specimens that are dry curing to have a greater compressive strength value than those that are wet curing.

5. Conclusion

Results of the study demonstrate that the strength of reinforced concrete plates achieves 75 MPa in wet curing and 59 MPa in dry curing, or similar to the quality of K400 to K500 concrete at 28 days, utilizing 30% variation of candlenut shells and UTM. The findings also demonstrate that concrete's compressive strength is stronger when it is wet dried than when it is dry dried.

6. References

[1] Goldberd H.D Sinaga, 2013, The Effect of Mixed Skin Powder and Candlenut on the Properties and Strength of Concrete, Physics Department, Medan State University
[2] Mulyono, Tri, 2005, Concrete Technology, C.V Andi Offset, Yogyakarta.
[3] SNI 03-2847-2002 Procedure for Calculating Concrete Structure for Buildings
[4] www.alatuji.com, accessed in October, 2019
[5] http://indonesian.tensiontestmachine.com
[6] Ali Asroni .2010.Beam and Reinforced Concrete Plates
[7] Nasruddin, V Sampebulu, Pratiwi Mushar. "Value of concrete compressive strength with variation of candlenut shell applicated to plate on a non-destructive test using UPV", IOP Conference Series: Earth and Environmental Science, 2020
[8] Nasruddin, V Sampebulu, P Mushar. "Value of Concrete Compressive Strength on 28 days with Variation of Candlenut Shell Applicated to Plate on a Non-destructive Test Using Hammer Test", IOP Conference Series: Materials Science and Engineering, 2020

IOP Conf. Series: Earth and Environmental Science 1117 (2022) 012031 doi:10.1088/1755-1315/1117/1/012031

Analysis of confined masonry with strengthening around the opening

I A M Budiwati[1], M Sukrawa[1], and I G A D Wirasukma[2]

[1]Department of Civil Engineering, Universitas Udayana, Bali, Indonesia
[2]Graduate Student, Department of Civil Engineering, Universitas Udayana, Bali, Indonesia

idabudiwati@unud.ac.id

Abstract. Analysing confined masonry (CM) with strengthening around the opening (analytical model) was conducted numerically. The opening variations were centric and eccentric opening doors with an opening ratio of 16.93% and 31.75%; centric and eccentric opening windows with opening ratios of 10.16% and 19.05%. This analysis aims to apply the procedure for analysing the CM models by using layered shell elements and also comparing the behaviour and performance of each model. The masonry and frames were modelled using layered shell elements in SAP2000 software and analysis were done by using nonlinear static pushover analysis. This analysis starts by making a validation model based on other experimental work and comparing the behaviour of the structure modelled in the form of CM without openings and CM with openings. The results show that both models have load-displacement diagrams similar to the experimental results. By using the same modelling method applied to the analytical model, it was found that CM with large opening ratios have less strength than small opening ratios, and CM with centric openings was more robust than the eccentric openings.

1. Introduction

The Confined wall structure (CM) is a building technology that utilizes masonry walls as the main structure to withstand vertical and lateral loads, with reinforced concrete beams and columns functioning as binding members. The CM can accept the load due to the earthquake well [1]. In reinforced concrete buildings, the frame structure is not only filled with full brick walls but also brick walls with openings (holes) that function for windows, doors, ventilation and others. Reducing the area of the wall through the addition of openings and the location of the openings themselves will affect the behavior of the CM.

There have been several studies on the structure of confined walls, one of which is a laboratory study by [2] who analyzed the CM model without openings and with openings in the middle of the wall. Lateral cyclic loads are applied to the beam-column top joint. The experimental results show that on both walls the crack pattern is predominantly in the diagonal direction. The effect of the opening does not significantly affect the load carrying capacity and ductility due to the reinforcement in the form of wood frames in the opening.

In evaluating the seismic capacity of reinforced concrete buildings based on standards issued by the Building Research Institute, Japan [3] the effect of brick walls with openings is neglected in the calculations assuming the walls do not make a significant contribution to the supporting frame structure. However, a number of researchers [4] and [5] state that brick walls with openings have an influence on the lateral strength of the frame structure but the magnitude depends on the ratio of the opening area to the wall area.

An analysis of masonry with different size openings subjected to cyclic loading to determine the effect of opening size on wall crack patterns and lateral strength of the structure was conducted by [6]. The test results show that the stiffness and strength of the walls decrease with increasing opening area under cyclic loading. The type of opening (window, door, etc.) and the location of the opening in the wall have an important role in the location of the crack in the wall and the potential location of the crack. The value of the lateral strength of walls with the same shape and the same opening area is different and for walls with centric openings stronger than eccentric openings.

From the description above, a study was conducted to examine the behaviour of perforated CMs with variations in the location and ratio of wall openings using laboratory test data by [2] and with the addition of reinforcement around the opening in the form of reinforced concrete columns and beams (lintel band). Considering that the lintel around the opening in the research by [2] is made of wood frames, while in this study the lintel will be modelled with reinforced concrete so that it will later affect the failure pattern

2. Material

2.1. *Resistance of Confined Walls to Seismic Strength*

Based on Chilean regulations at the National Information Center of Earthquake Engineering [7], the drift ratio of buildings with the CM concept must be less than or equal to 0.002 (0.2%). The controlled stresses on the walls are tensile stress, compressive stress, and shear stress. The expected compressive stress does not exceed the quality of the wall pair so that cracks do not occur in the wall. Based on ACI 530-02 [8], the allowable tensile stress due to bending for walls made of clay and concrete, for solid units with a diagonal crack direction is 0.414 MPa. While the allowable shear stress should not exceed 0.827 MPa.

2.2. *Modelling with Layered Shell*

In the CSI Analysis Reference Manual for SAP2000 MV14 [9] it is explained that layered shell is a feature where shell elements can behave nonlinearly. The type of nonlinearity that occurs is material nonlinearity. Non-linear material originates from a stress-strain relationship that is disproportionate, both in the elastic region and outside the elastic region (elasto-plastic or perfectly plastic). The material can be isotropic, uniaxial, or orthotropic. The behavior of the material depends on the behavior of the material components selected for the layer.

2.3. *Non-Linear Static Analysis*

According to SNI Earthquake 1726-2019 [10], the static analysis of the thrust load (pushover) is a static nonlinear analysis, in which the influence of the planned earthquake on the building structure is considered as a static load on the center of mass of each floor, the value of which is gradually increased until exceeds the load causing the first to occur (plastic hinge) in the building structure, then with a further increase in load undergoes a large post-elastic deformation until it reaches the expected transition target or until it reaches the plastic condition.

3. Method

3.1. *Data Structure Model Validation*

The data structure of the validation model was taken from the experiment conducted by [2]. The compressive strength of concrete (f_c') and masonry (f_m') is 15.32 MPa and 2.33 MPa, respectively. For reinforcing steel material, at D10 it has f_y = 384.9 MPa, at D8 it has f_y = 350.9 MPa. The structure is a one-span and one-story portal measuring 3m x 3m which is restrained by reinforced beams and columns with a cross-sectional size equivalent to 15x15 cm2 and in MV2 there is an opening with a size of 80x120 cm2 in the middle of the wall. The size of the wooden frame is 5x10 cm.

The validation model was made to determine the suitability of the confined wall modeling with the results of laboratory tests. The load used is a monotonic load that is loaded on the upper right of the beam and column meeting as shown in Figure 1. The shell element does not need to be wider than the wall thickness.

(a) Model validation without opening MV1 (b) Model validation with opening MV2

Figure 1. Model validation

3.2. *Modelling on SAP2000*

The assumption of bonding between concrete and masonry is using full shell where the concrete forms a monolithic relationship with the brick wall. Layered-shell elements are used where shell elements can behave nonlinearly for all structural elements. The nonlinear type that occurs is material nonlinearity based on the material stress strain curve, Figure 2. In this modeling the material used is isotropic material for beam, column and wall elements and uniaxial for reinforcement.

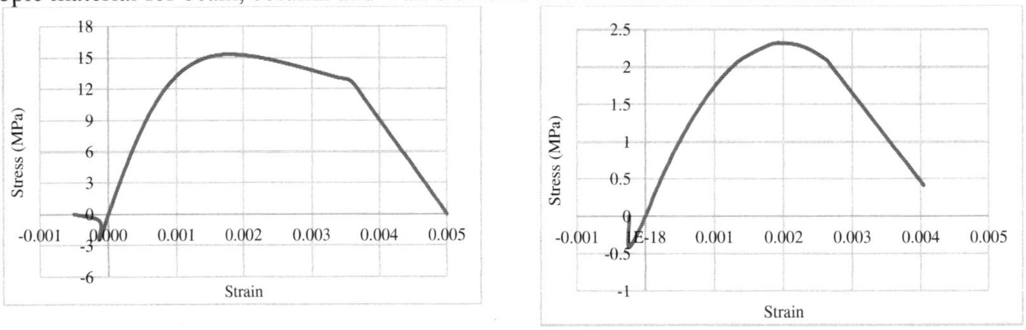

(a) Concrete (b) Masonry

Figure 2. Stress strain curves

In the model, the concrete and reinforcement elements are modeled with different layers so that the behavior of the material can be well simulated. Each layer is modeled at a different thickness. In the concrete layer, the stiffness reduction of the structure is carried out as stated in SNI 2847-2019 [11], the value of the crack reduction factor for each structure is 0.7 for the column, 0.35 for the beam, and 0.35 for the wall. Reduction is done on the concrete layer plate on the shell thickness. As for the reinforcement material, the shell thickness for the 2D model can be calculated by dividing the area of one reinforcement installed against the distance between the bars multiplied by the number of reinforcements in the same thickness direction.

In masonry and concrete elements will behave in the direction of the material angle 0° at this angle the stress component that behaves nonlinearly is S22. For uniaxial materials such as reinforcement, where the behavior is determined only in the direction of the 90° material angle with a stress component that behaves nonlinearly along S11.

The model that has been made is then analyzed by pushover analysis. The displacement target data is inputted according to the experiment, namely the largest deviation applied to the structure is 3.5 percent or equivalent to a peak shift of 105 mm. Pushover analysis is carried out in two stages of loading, namely the first stage, the building structure is loaded by gravity loads, namely a combination of dead loads with a loading coefficient of 1.0. The second stage, the analysis is continued by providing a pattern of lateral thrust loads on the structure.

3.3. *Data Structure Analytical Model*

The analytical model uses structural data based on experimental data. It is planned that the CM Analytical model is perforated with reinforced concrete columns and beams (lintel bands) around the opening with lintel cross-sectional dimensions adjusted to the wall thickness of 100x100mm using 10 MPa concrete. Analytical models were made of eight models, namely centric perforated walls in the form of centric doors with an opening area ratio of 16.93% (PS16), 31.75% (PS31) and centric windows with a opening area ratio of 10.16% (JS10), 19.05 % (JS19), eccentric doors with an opening area ratio of 16.93% (PE16), 31.75% (PE31), and eccentric windows with a opening area ratio of 10.16% (JE10), 19.05% (JE19) such as in Figure 3.

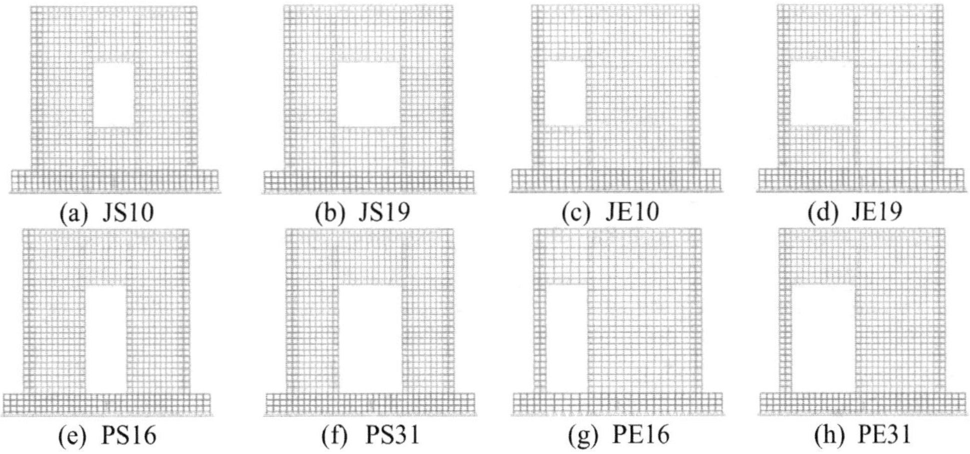

Figure 3. Analytical models

4. Results and discussion

4.1. *Force-Displacement Curve*

Validation Model. To compare the force-displacement curve of the validation model to the experiment, the curves of the two models are displayed in a Cartesian plane with a displacement limit of 105 mm and a force limit of 60 kN as shown in Figure 4. The model displacement-force curve shows that the initial stiffness of MV1 and MV2 are close to the results experiments on the linear section. The peak load of MV1 is in accordance with the experimental results while the peak load on MV2 tends to be smaller than the experimental results, so the model becomes safer and more realistic than the experimental V2 model where the wall strength is not significantly reduced by the presence of openings. Based on these results, the analysis with SAP2000 is quite in accordance with the experimental results so that the modeling method with layered shell elements in SAP2000 can be used for analytical models.

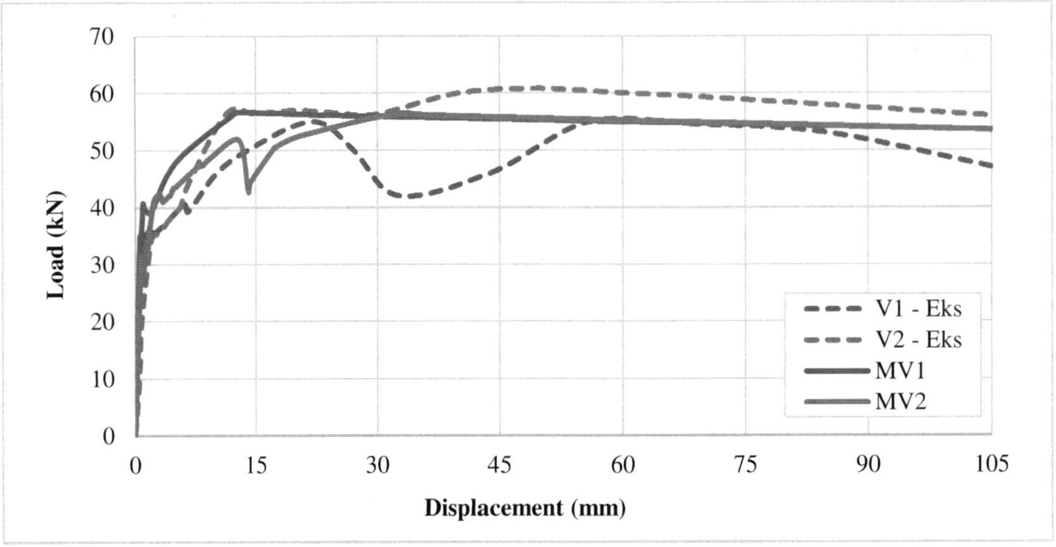

Figure 4. Load-displacement curves of model validation

Analytical Model. To The capacity curves from the pushover analysis results for all analytical models are displayed in one graph with a displacement limit of 30 mm and a force limit of 100 kN as shown in Figure 5. To make it easier to see the comparison of the perforated CM model with opening variations, the curve is displayed according to the compared variables as in Figures 6 to 8.

Based on the load-displacement curve in Figure 6, JS10 has a maximum base shear force of 85,214 kN which is about 40% higher than MV2 with a maximum base shear force of 53,499 kN and MV1 with a maximum base shear force of 53.552 kN.

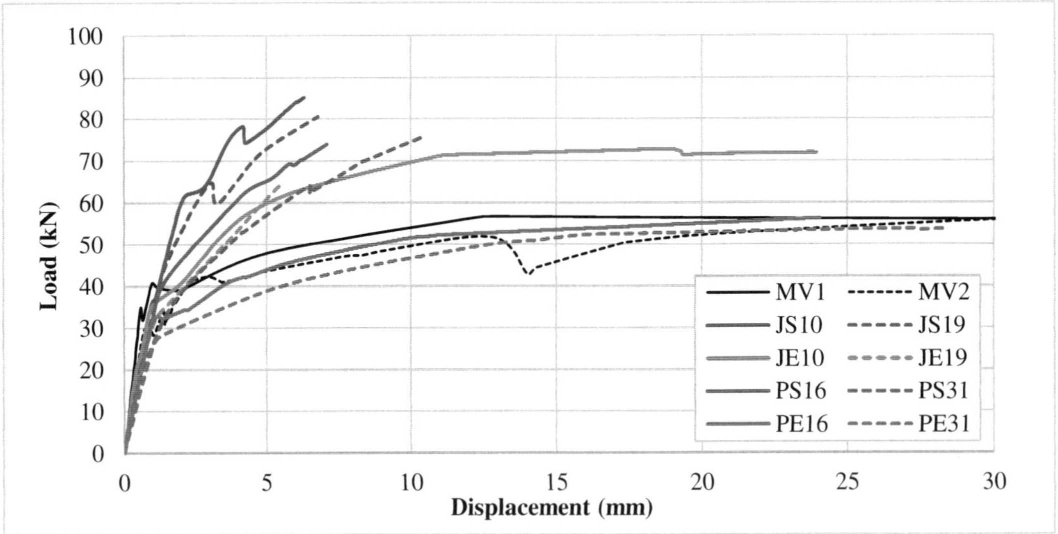

Figure 5. Load-displacement curves of model validation

Based on the load-displacement curve in Figure 7. The large opening ratio in the slotted CM has less strength than the small opening ratio. The same thing happened to the CM with eccentric openings, the larger the aperture ratio, the lower the CM power. From the load-displacement curve in Figure 8. The centric opening has greater strength than the eccentric opening.

According to Yi Shuan, et al (2014) eccentrically perforated CM will produce different results if it is subjected to a monotonic load from the opposite direction. Then a model is made with different loading directions on the entire analytical model to prove this. In this analysis the load is applied to the upper corner on the left of the wall model.

There are differences in the behavior of the structure when loading is carried out from different directions. The eccentric model which is loaded from the left tends to be more rigid than the eccentric model which is loaded with a monotonic load from the right. On the other hand, in the centric model, there is no difference in behavior in the model that is loaded from the right or left as can be seen in Figure 9.

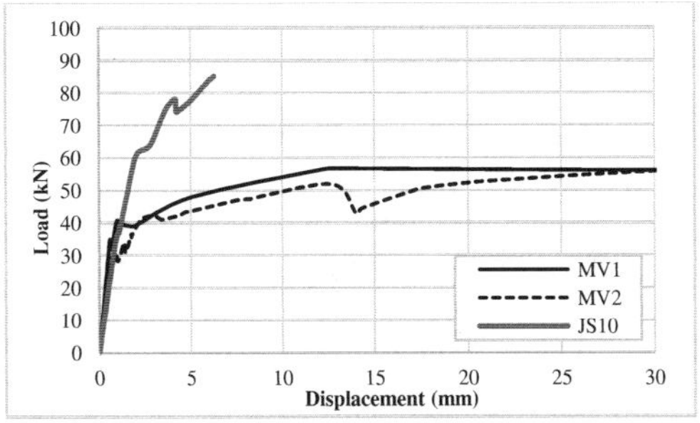

Figure 6. Load-displacement curves of CM with opening with and without strengthening around the opening

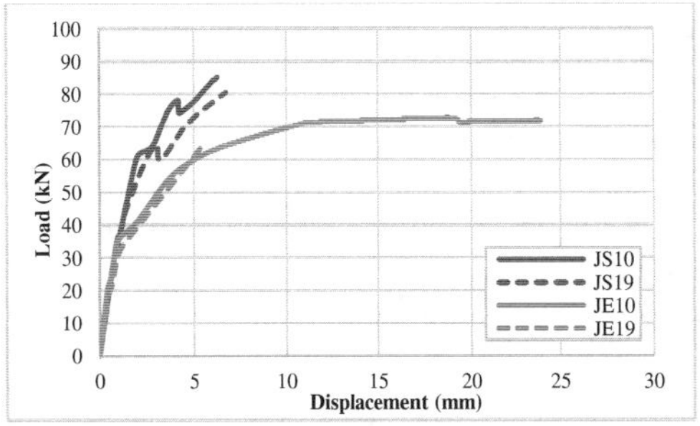

Figure 7. Load-displacement curves of CM with centric and eccentric window opening

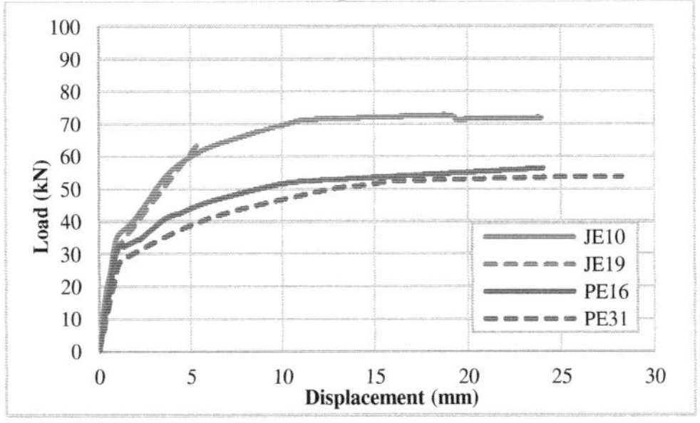

Figure 8. Load-displacement curves of CM with eccentric window or door opening

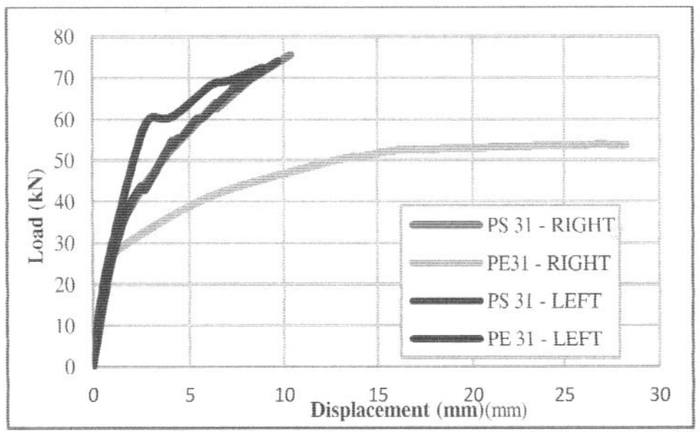

Figure 9. Load-displacement curves of CM with centric and eccentric door opening that applied with opposite load direction

4.2. *Form of Structural Collapse*

The behavior of MV1 and MV2 analyzed by SAP2000 is quite in accordance with the experimental results. MV1 and MV2 show a typical development of a diagonal crack pattern, which then develops into a strut and tie mechanism between the wall and the tie column for the lateral load-bearing mechanism. The main difference is that in MV2 compression struts cannot form a complete straight line because they are disturbed by window openings. However, the pattern of cracks and deformation of the window frames during loading indicates that the window frames resist the forces from compression struts and act as connecting elements that fill gaps in the window openings, so that the compression strut supports can withstand and transfer lateral loads from the ends of the walls to the foundation.

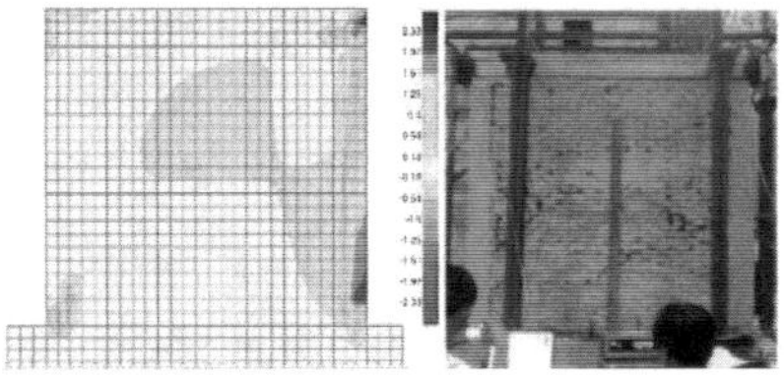

(a) SAP2000 (b) Experiment

Figure 10. Failure of MV1

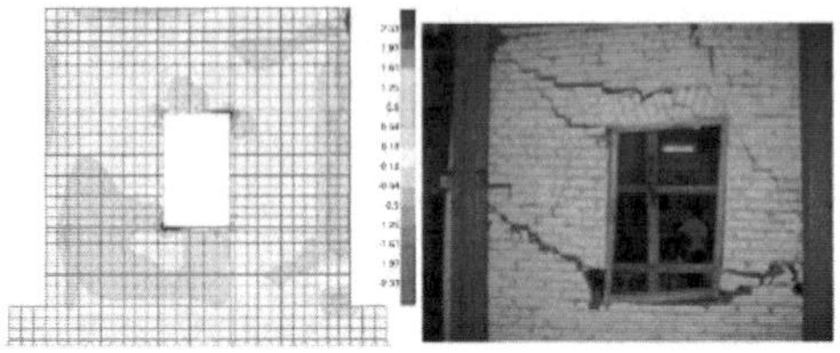

(a) SAP2000 (b) Experiment

Figure 11. Failure of MV2

4.3. *Comparison of stresses on the wall*

Validation Model. When the base shear force is 37 kN, the stress that occurs in MV2 is 16.8% greater than that of MV1, Figure 12. So, it can be concluded that the presence of an opening in the CM causes an increase in the stress that occurs in the CM. MV1 and MV2 have started to crack where the tensile stress has exceeded the tensile strength, which is 0.414 MPa, while the compressive and shear stresses have not exceeded the allowable stress

(a) MV1 (b) MV2

Figure 12. Stress (MPa) on the walls of model validation at load of 37 kN

Analytical Model. When the basic shear force is 38 kN, the stress that occurs in the JS10 model is smaller than the MV2 which is loaded with a load of 37 kN. It can be concluded that the reinforcement in the form of beams and columns of reinforced concrete lintels can withstand greater lateral loads than reinforcement with wooden frames.

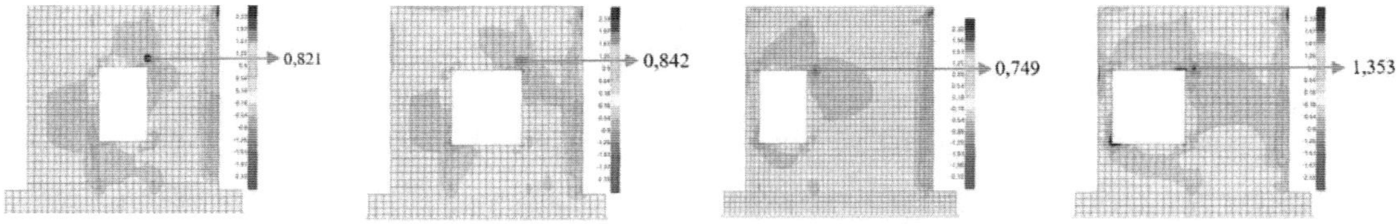

(a) JS10; 38 kN (b) JS19; 32 kN (c) JE10; 32 kN (d) JE19; 32 kN

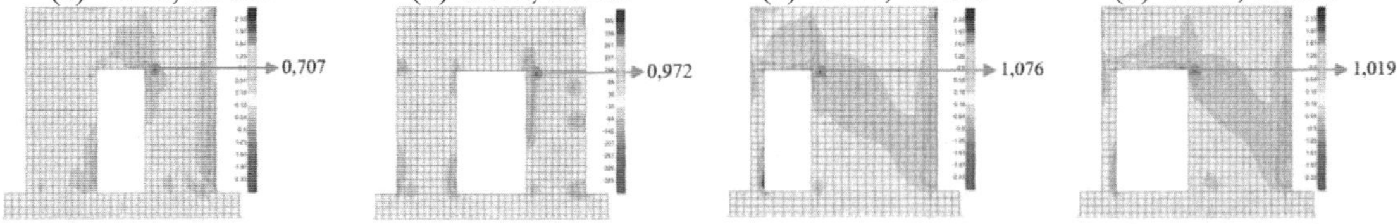

(e) PS1; 30 kN6 (f) PS31; 38 kN (g) PE16; 34 kN (h) PE31; 30 kN

Figure 13. Stresses (MPa) on the wall of analytical models

Meanwhile, based on Figure 13, it can be concluded that the stress that occurs in the perforated CM analytical model increases with the increase in the *opening* ratio in the CM. The voltage that occurs in the JS10 model which is loaded with 38 kN is smaller than the stress that occurs in the JS19 model which is loaded with 32 kN. Likewise at the door opening, the stress that occurs in the JS10 model which is loaded with 38 kN is almost the same as the stress that occurs in the PS16 model which is loaded with 30 kN. The position of the *opening* also affects the stress of the slotted CM. The stress that occurs in the JS10 model which is loaded with 38 kN is almost the same as the stress that occurs in the JE10 model which is loaded with 32 kN. So it can be concluded that the centric *opening* has a smaller stress than the eccentric *opening*. Likewise at the door opening, the stress that occurs in the PS16 model which is loaded by 30 kN is smaller than the stress that occurs in the PE16 model which is loaded by 30 kN.

4.4. Comparison of stresses *in reinforcement*

Validation Model. Experiments conducted by [2] did not examine the tensile strength of reinforcing steel on reinforced concrete sections. However, in this study, an examination of the tensile strength of reinforcing steel in reinforced concrete sections will be carried out to determine and compare the strength of the reinforcing steel loaded with lateral loads from each model. The maximum reinforcement stress in both validation models has not exceeded the yield strength limit of reinforcing steel, where the elastic limit of the stress is 384.9 MPa. In Figure 14, it can also be seen that the opening in the CM affects the tensile strength of the reinforcing steel in the tie column, where MV2 has a tensile stress of reinforcing steel that is 29% greater than MV1.

(a) MV1 (b) MV2

Figure 14. Stress (MPa) on the reinforcement of model validation

Analytical Model. Based on Figure 15, it can be seen that in the JS10 model the maximum reinforcement stress is at a value of 335.030 MPa when the basic shear force is 57.702 kN, while in MV2 with the same opening ratio, the maximum reinforcement stress is at a value of 344.971 MPa when the base shear force is equal to 53,552 kN. So, it can be concluded that JS10 has a maximum stress on reinforcement which is smaller than MV2. So that with the reinforcement in the form of reinforced concrete lintel beams and columns, it can increase the strength of the structure so that it can withstand collapse

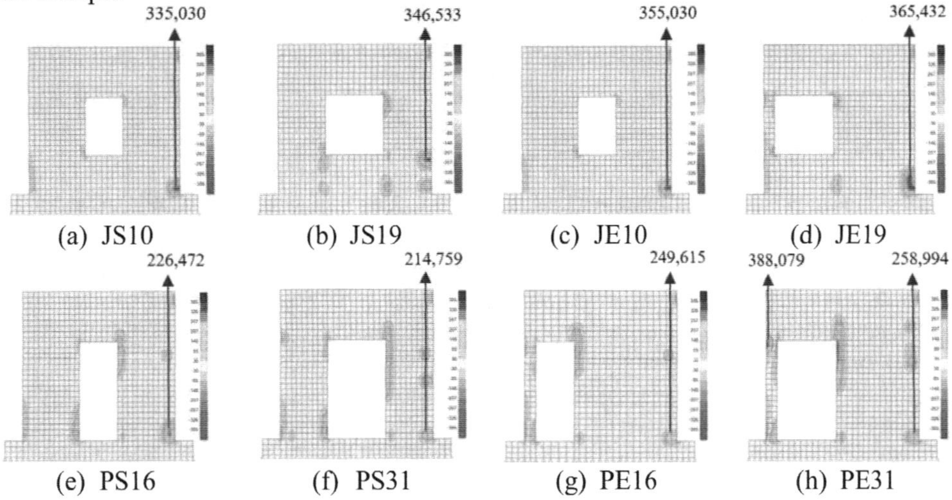

(a) JS10 (b) JS19 (c) JE10 (d) JE19

(e) PS16 (f) PS31 (g) PE16 (h) PE31

Figure 15. Stresses (MPa) on the reinforcement of analytical models

CMs with eccentric openings have yield stresses in the bonding column which are higher than those with centric openings. In the models JS10, JS19, JE10, JE19, PS16, PS31 and PE16 yielding has not occurred in all the reinforcement, while in PE31, yielding begins to occur in the middle of the reinforced concrete binding column. So prone to tensile failure in openings with a large ratio in the eccentric position.

To prevent the occurrence of tensile collapse in the cross section of the column, this study tried to widen the size of the column on the PE31 eccentric door model to 300x100mm where yielding occurs in the reinforcement, the column thickness still adjusts to the wall thickness, as shown in Figure 16(a).

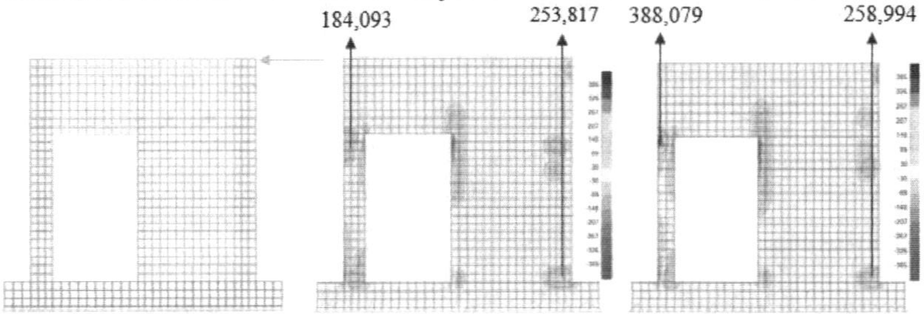

(a) Model (b) Streses after column widen (c) Streses before column widen

Figure 16. Modeling of CM with column widen

Based on Figure 16, it can be seen that by widening the size of the binding column in the eccentrically perforated CM, it can withstand yielding that occurs in the reinforcing steel of the tie column. The stress on the bonding column reinforcement decreased until the PE31 model decreased to 57.64%. So it can be concluded that by widening the size of the tie column, the CM with a large ratio in the eccentric position can withstand a greater stress than before.

4.5. Performance Evaluation

Validation Model. In order to meet the performance requirements of the ultimate limit of the structure, the Chilean regulation states that the drift ratio of buildings with the CM concept must be less than or equal to 0.002 (0.2%) times the height of the structure (6 mm). In the validation model a drift ratio of 0.002 is achieved when the base shear force is 49.28 kN on MV1 and 45.32 kN on MV2. Based on these results it can be concluded that MV1 can withstand a load of about 8.1% greater than MV2 before failure.

Analytical Model. Perforated CM with reinforced concrete lintel beams and columns (JS10) has better performance than the CM model without openings (MV1) and CM with openings (MV2) using frame reinforcement before failure (0.002 drift). JS10 (84.27 kN) can withstand a base shear force 41% greater than MV1 (49.28 kN) and 46% greater than MV2 (45.32 kN). While in the JS19 model a drift ratio of 0.002 is achieved when the base shear force is 78.51 kN. On the model JE10 at 63.86 kN, JE19 at 65.23 kN, PS16 at 70.27 kN, PS31 at 64.02 kN, PE16 at 46.24 kN, and PE31 at 40.88 kN. So it can be concluded that the base shear force that can be resisted by the perforated CM before failure (0.002 drift) decreases as the opening ratio increases. The position of the opening also affects the stress of the slotted CM. The centric opening can withstand a greater base shear force than the eccentric opening.

5. Conclusion

From the results it can be concluded that models using layered shell elements can simulate the testing results of CM. This means modeling with layered shells can be used to model CMs with opening.

The behavior CM model with opening reinforcement (JS10) is stronger than that the model without opening (MV1). The beams and columns around the opening effectively increase the strength of the CM. CM with small opening ratio is stronger than CM with large opening ratio. CM with centric opening is stronger than eccentric opening. CM with eccentric openings and with large opening ratio (PE16 and PE31) are weaker than CM without openings (MV1), while CM with centric and eccentric openings and with a small opening ratio are still stronger than CMs without openings. This means that the effect of reinforcement on eccentric opening is not as effective as in the case of centric opening. .

It is suggested that the opening with a large ratio (door opening) should not be installed on the edge of the wall and if necessary, the column dimensions should be enlarged to prevent tensile failure in the column.

References

[1] Meli R, Brzev S, Astroza M, Boen T, Crisafulli F, Dai J, Farsi M, Hart T, Mebarki A, Moghadam AS, Quiun D, Tomazevic M, Yamin L 2011 *Seismic Design Guide for Low-Rise Confined Masonry Buildings* (A Project Of The World Housing Encyclopedia, EERI & IAEE, Committee Of International Experts)

[2] Suarjana M, Kusumastuti D, Pribadi KS, Rildova 2012 Experimental Study on the Behaviour of Plastered Confined Masonry Wall under Lateral Cyclic Load *15th World Conference on Earthquake Engineering*

[3] Building Research Institute (BRI) 2005 *Standard for Seismic Evaluation of Existing Reinforced Concrete Buildings* (Building Research Institute, Japan)

[4] Mondal G and Jain SK 2008 Lateral Stiffness of Masonry Infilled Reinforced Concrete (RC) Frames with Central Opening *Indian Institute of Technology Kanpur*

[5] Surendran S and Kaushik HB 2012 Masonry Infill RC Frames with Openings: Review of In-Plane Lateral Load Behaviour and Modeling Approaches *The Open Construction and Building Technology Journal*

[6] Tabbakhha M and Modaressi A 2014 Influence of Size of Opening on Performance of Unreinforced Masonry Walls under Cyclic Loads *9th International Masonry Conference*

[7] Brzev S 2007 *Earthquake-Resistant Confined Masonry Construction* (National Information Center Of Earthquake Engineering Indian Insitute of Technology Kanpur)

[8] MSJC 2002 *Building Code Requirements for Masonry Structures* (ACI530-02/ASCE 5-02/TMS 402-02 of American Concrete Institude, American Society of Civil Engineers, and The Masonry Society). Masonry Standards Joint Committee

[9] CSI 2011 *Structural Analysis Program*, Ver.14.2.4. Computers and Structures, Inc. Berkeley, California, USA

[10] BSN 2019a *Tata Cara Perencanaan Ketahanan Gempa Untuk Struktur Bangunan Gedung dan Non-gedung* (1726:2019). Badan Standardisasi Nasional Indonesia

[11] BSN 2019b *Persyaratan Beton Struktural Untuk Bangunan Gedung* (SNI 2847:2019). Badan Standardisasi Nasional Indonesia

Analysis of Response and Performance of Bugis-Makassar Stilt House Structures Using Pushover Analysis

A. Aryadi, M. A. Kahar, and R. Mardiana

Civil Engineering Department, Hasanuddin University, Indonesia

Corresponding Author : civilarchitects.aryasi@gmail.com

Abstract. South Sulawesi is one of the areas prone to earthquakes. The Walanae Fault extends from the Makassar Strait to the Bone Bay. This fault stretches along the Walanae River and passes through the districts of Pinrang, Sidrap, Bone, to the Gulf of Bone. The structure of the stilt house is one that is earthquake resistant because it does not directly receive earthquake energy. Traditional techniques on the structure of houses on stilts in Bima district for earthquake resistance researched by Agus Hariyanto, Sugeng Tiyadi, and Andy Widyowijatnoko, stated that the local structural technique, namely pa'a which was empirically developed by the community over the years can reduce the vulnerability of houses. against earthquakes. This technique turns out to have similarities with the framework of the Bugis-Makassar stilt structure, which is classified as a rigid joint truss system. The purpose of this study was to determine the response and performance of the Bugis-Makassar stilt house structure in the Bone region which represents a high level of earthquake risk. The analysis was carried out on the structure of a Bugis stilt house in Bone using the ETABS v.20 program to obtain the response and performance of the structure. The results of the structural response from the study indicate the value of the lateral deviation of the floor. The Bone region has a floor lateral deviation of 3.64 mm and 6.45 mm for the X and Y directions. Based on the pushover analysis, the structure performance level has a B performance level for the Bone Region.

Keywords : Earthquake, Stilt House, Structural Performance, Structural Response, Pushover Analysis

1. Introduction

Indonesia is an archipelagic country that has a high level of risk of earthquakes. The Indonesian archipelago is located between four active tectonic systems, namely the Eurasian plate boundary, the Indo-Australian plate, the Philippine plate, and the Pacific plate. South Sulawesi is one of the areas prone to earthquakes. The Walanae Fault extends from the Makassar Strait to the Bone Bay. This fault stretches along the Walanae River and passes through the districts of Pinrang, Sidrap, Bone, to the Gulf of Bone. The structure of the stilt house is one that is earthquake resistant because it does not directly receive earthquake energy.

Nurjannah Baharta, Servie O. Dapas, and Ronny Pandaleke, (2005) suggested an evaluation of the structural design of a traditional wooden house produced by Woloan Village, which was carried out technically, both in terms of materials and structural systems, indicating that the structure of the wooden house is capable of bearing the planned earthquake load [1] . Hartawan, B. Suhendro, E.

Pradipto, A. Kusumawanto, stated that the results of tracing the development of the structural system of Bugis houses led to the discovery of the sequence of structural systems used by the Bugis people. Testing all these systems with the mappasituppu principle shows an effort to increase the stiffness of the structural system [2]. Traditional techniques on stilt house structures in Bima district for earthquake resistance researched by Agus Haiyanto, Sugeng Tiyadi, and Andy Widyowijatnoko (2020), stated that the local structural technique, namely pa'a which was empirically developed by the community over the years can reduce the vulnerability of houses to earthquakes [3]. This technique turns out to have similarities with the framework of the Bugis-Makassar stilt structure, which is classified as a rigid joint truss system.

The configuration of the structure is one of the factors that affects how well it can support the load. A regularly configured building is better at resisting earthquake forces, preventing structural damage [4][5]. As a result of the earthquake force being absorbed by the building's base, the magnitude of the earthquake force also influences how the structure reacts [6][7]. The response and performance of the structure in each Indonesian region, which represents regions with low to high earthquake risk, can be calculated from the earthquake force applied to the structure. SNI 1726-2019 [8] is the regulation that is used in earthquake calculations, and it contains an earthquake coefficient value that is used in structural planning. The Bone Regency area is considered one of South Sulawesi's seismically active regions for the purposes of this study. The goal of this study is to evaluate how the Bugis-Makassar stilt house structure in the Bone region responds and performs in accordance with SNI 1726:2019 regulations and the FEMA 440 method.

2. Research Methodology

The ETABS v.20 program is used in this study's computational simulation method, which uses the finite element method. The Bone Regency region of Indonesia used the following information on the structure of wooden stilt houses to calculate the response and performance of the structure:

Building function	=	Residential home
1 floor height	=	2,64 m
2nd floor height	=	2,6 m
Roof height	=	3,35 m
Long	=	6 m
Wide	=	9,3 m
Wood Quality	=	E13 and quality class A, E = 13771.83 mpa, Emin = 6436.65 mpa

Literature review

The initial step in this research's development was the collection of data relevant to the study's topic through literature review. Table 1 lists the guidelines on which the calculations in this study were based.

Table 1. The Rules Used

Calculation	Regulation
Loading	SNI 1727:2020
Earthquake Load	SNI 1726:2019
Timber Structure	SNI 7973:2013
Structure Performance	FEMA 440

Preliminary Design

Planning the dimensions of structural elements, such as beams, columns, floor beams, and roof trusses, according to SNI 2847:2019 is known as preliminary design. The ETABS v.20 program is then used to model the planning results in accordance with the information obtained through the preceding steps:

1. Create a data grid to make it simpler to use the structural components that will be described.
2. Define Material, which involved entering Material Information to match the Material

Specifications.

3. Defining structural elements entails describing beams, columns, floor beams, and roof trusses in terms of their dimensions and the caliber of the materials they are made of.
   Figure 1 depicts the structure that the ETABS v.20 program used to model.

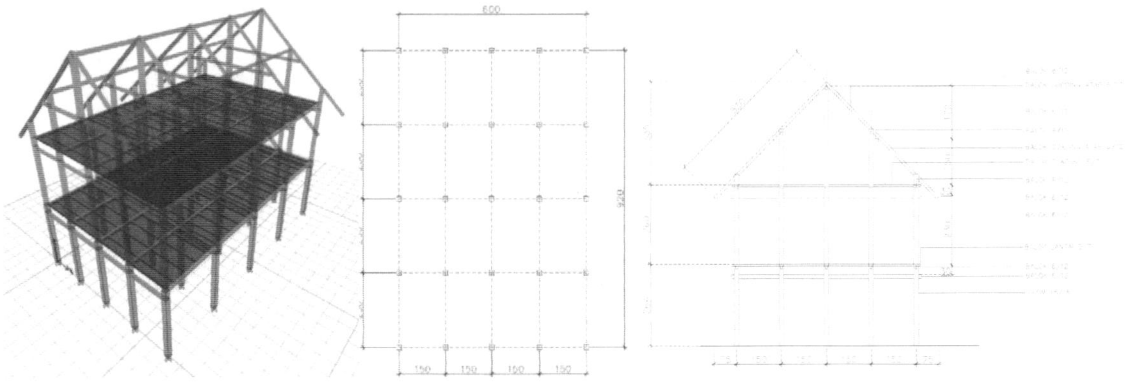

Figure 1. Structural Model Used

Structure Data Input

Following modeling, the necessary structural data input, including structural element dimension data, structural element material data, and load data used, is carried out at this stage. The stilt house structure will accept the loads that have been applied in order to generate the response and structural performance of the building. Dead loads, live loads, and earthquake loads are among the loads deployed. Live load and earthquake are calculated using SNI 1727:2020 and SNI 1726:2019, while dead load is calculated using PPPURG 1987 and SNI 1727:2020 [9]. The weight of all the installed building materials is known as the "dead load" [10]. The load created by users and occupants of the building is known as the "live load" [9], whereas the earthquake load requires careful analysis because it is difficult to predict the magnitude and timing of earthquakes [11]. The seismic coefficient for the region is required when calculating the earthquake load and can be found on the PU Cipta Karya website. Calculating the seismic parameters used to describe the spectrum response curve requires this seismicity coefficient. Table 2 displays the seismicity coefficient for each area examined.

Table 2. Seismicity Coefficient

No.	Region	Site Class	S_s	S_1
1.	Bone	Medium Ground (SD)	0,775	0,2870

Additionally, the loads are combined from all potential outcomes in accordance with SNI 1726:2019 regulations. There are the following load combinations:

1. 1,4 DL
2. 1,2 DL + 1,6 LL
3. 1,2 DL + 1 LL + 1,3 RX1 + 0,39 RY0
4. 1,2 DL + 1 LL + 1,3 RX1 - 0,39 RY0
5. 1,2 DL + 1 LL - 1,3 RX1 + 0,39 RY0
6. 1,2 DL + 1 LL - 1,3 RX1 - 0,39 RY0
7. 1,2 DL + 1 LL + 0,39 RX0 + 1,3 RY1
8. 1,2 DL + 1 LL + 0,39 RX0 - 1,3 RY1
9. 1,2 DL + 1 LL - 0,39 RX0 + 1,3 RY1
10. 1,2 DL + 1 LL - 0,39 RX0 - 1,3 RY1
11. 0,9 DL + 1,3 RX1 + 0,39 RY0

12. 0,9 DL + 1,3 RX1 - 0,39 RY0
13. 0,9 DL - 1,3 RX1 + 0,39 RY0
14. 0,9 DL - 1,3 RX1 - 0,39 RY0
15. 0,9 DL + 0,39 RX0 + 1,3 RY1
16. 0,9 DL + 0,39 RX0 - 1,3 RY1
17. 0,9 DL - 0,39 RX0 + 1,3 RY1
18. 0,9 DL - 0,39 RX0 - 1,3 RY1

Linear and Non-linear Analysis

After modeling and loading the buildings that are examined in each area, the analysis stage is completed. There are two stages to the analysis process: response spectrum analysis in the first stage, and non-linear analysis in the second (pushover analysis).

To determine whether the structure being used is safe or not, linear analysis calculates the fundamental shear force and the deviation between floors. The SNI 1726: 2019 regulation, which specifies the coefficients that must be determined in order to calculate the basic shear force, is referenced in the calculation of the basic shear force. Following are the steps involved in calculating the fundamental seismic shear force:

1. Determine the building risk category which can be seen in table 3 of SNI 1726:2019.
2. Determine the seismic design category in each region based on tables 8 and 9 of SNI 1726:2019.
3. Determine the reduction factor based on table 12 of SNI 1726:2019, the results of which are: $R = 8$; $\Omega_0 = 3$; $C_d = 5{,}5$.
4. Determine the fundamental period, where the equation is as follows :

$$T_a \leq T_x \leq C_u\, T_a$$

$$C_t \text{ x } h_x{}^n \leq T_x \leq C_u\, T_a \quad (1)$$

Value T_x must be smaller than $C_u\, T_a$, if T_x greater than $C_u\, T_a$ the value T used is $C_u\, T_a$

5. Determine the seismic design coefficient (C_s) based on article 7.8.1.1. SNI 1726:2019.

$$C_{s\,min} \leq C_s \leq C_{s\,maks}$$

$$0{,}44\, S_{DS} \text{ x } I_e \leq C_s \leq \frac{SD_1}{T(\frac{R}{I_e})} \quad (2)$$

Value C_s must be smaller than $C_{s\,maks}$, if C_s greater than C_s maks so value C_s used is $C_{s\,maks}$.

6. Determine the basic seismic shear force with the following equation :

$$V = C_{s\,min} \times W \quad (3)$$

$$W = \text{Self Weight Structure + Dead Load + 30\% Live Load} \quad (4)$$

Linear analysis is the calculation of the deviation between floors. In calculating the value of the deviation between floors, it takes the value of the lateral deviation caused by the given earthquake load. The value of the lateral deviation can be displayed through the ETABS v.20 program which is then included in the calculation of the deviation between floors. The drift between floors is calculated according to article 7.8.6. SNI 1726:2019. The drift between floors is the displacement that occurs at the center of mass above or below the floor under consideration. The following are the steps for calculating the deviation between floors:

1. Determine the value C_s, I_e,and hs_x.
2. Displaying lateral deviation (δ_e)
3. Calculate the enlarged deviation with the following formula:

$$\delta_I = \frac{C_d \, x \, \delta_e}{I_e} \quad (5)$$

4. Determine the deviation between the permit floors, where the calculation used is based on table 20 SNI 1726:2019 which is determined by the type of structure and risk category. In this study, the building used is included in risk category IV and the following calculations are obtained:

$$\Delta_{izin} = 0{,}015\, hs_x \quad (6)$$

5. Calculate the deviation between floors, where the value should not exceed the deviation between the permitted floors. For ground floor:

$$\Delta_1 = \delta_1 \leq \Delta_{izin} \tag{7}$$

For 2nd floor onwards:

$$\Delta_2 = (\delta_{e2} - \delta_{e1})\, C_d / I_E \leq \Delta_{izin} \tag{8}$$

6. If the deviation between floors (Δ_i) smaller than the allowable floor drift (Δ_{izin}), then the structure can be categorized as safe. However, if the deviation between floors (Δ_i) greater than the allowable interfloor deviation (Δ_{izin}), then the structure can be categorized as unsafe because the state of the structure has collapsed.

Buildings that have been declared safe, based on linear analysis will then be analyzed non-linearly (Pushover Analysis). This analysis is important because when a large earthquake occurs, some structural elements such as beams and columns will experience plastic hinge yielding [12].

Pushover analysis is carried out by applying a static load gradually until the structure collapses. The load placement in this analysis is carried out horizontally which is a combination of earthquake loads, live loads and dead loads [13]. In this study, calculation analysis [14]. The FEMA 440 method is an extension of the FEMA 356 (displacement coefficient method) [15]. The FEMA 440 calculation procedure begins by determining the effective vibration time (T_e) [15]. To determine the value of the natural vibration time of a structure, an effective vibration time value from the pushover curve is needed [16]. This solution is done by modifying the elastic response of the single degree of freedom system which is equivalent to the coefficient factors C_0, C_1, C_2, and C_3 so that the largest value of global displacement in both elastic and inelastic conditions is obtained, which is called the displacement target (δ_T) [15]. The formula used in determining the displacement target (δ_T) is as follows:

$$\delta_T = C_0\, C_1\, C_2\, C_3\, S_a \frac{Te^2}{4\pi^2} \tag{9}$$

3. Research Results and Discussion

Load Calculation Results

In giving the load to the house on stilts, the dead load and live load are given in accordance with the standard loading for residential houses. Meanwhile, the earthquake load is given a value according to the calculation of the earthquake load in the area. In this case the seismic load was analyzed using the response spectrum method [17]. The following is presented in Table 3 regarding the calculation of the spectrum response design in the area under review.

Table 3. Seismicity Parameters

Region	SDS (g)	SD1 (g)	T0 (s)	Ts (s)
Bone	0,61	0,39	0,13	0,64

From the parameters SDS, SD1, T0, and Ts, the spectrum response design curve for each region under review can be drawn which is presented in Figure 2.

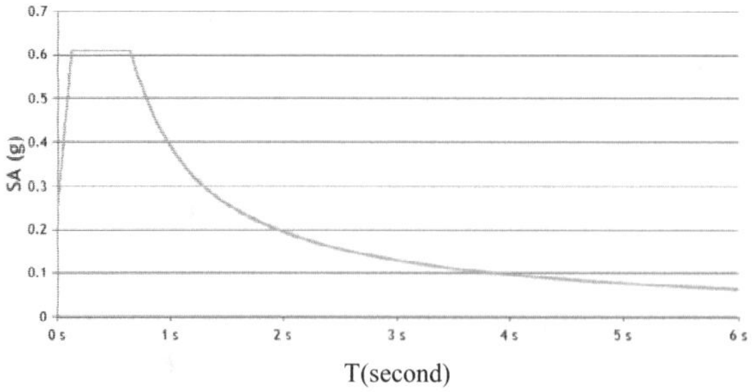

Figure 2. Response Spectrum Design

Results of Calculation of the Basic Seismic Shear Force

To get the calculation of the seismic base shear force, it takes the value of the fundamental period of the structure, the seismic design coefficient, and the weight of the structure in the calculation. The following is presented in Tables 4 and 5 in the form of the calculation results of the fundamental period of the structure, the seismic design coefficient, the weight of the structure, and the seismic base shear force.

Table 4. Period Fundamental Structure

Region	S_{D1}	C_u	C_t	x
Bone	0,394	1,4	0,0488	0,75

Table 5. Period Fundamental Structure (Advanced)

Region	T_a (s)	C_uT_a	$T_{a\ use}$ (s)
Bone	0,243	0,338	0,338

Table 6. Seismic Design Coefficient (Cs)

Region	$C_{s\ min}$	C_s	$C_{s\ maks}$	$C_{s\ use}$
Bone	0,0271	0,0769	0,1293	0,0769

Table 7. Total Weight of Structure and Base Seismic Shear Force

Region	C_s	W (ton)	V (ton)
Bone	0,0769	128,8313	9,9012

From Table 7, it can be seen that the Bone region has a base shear force value of 9,9 kN. The magnitude of the basic shear force depends on the magnitude of the earthquake force in the area. This is because the Bone region has a greater seismic force compared to other regions, so the seismic coefficient obtained is greater and can affect the value of the seismic base shear force calculation.

Result of Interfloor Deviation Calculation

The drift between floors is calculated based on the lateral displacement of the floors that occur, where the base shear force also affects the value of the drift between floors [18]. The deviation on the floor occurs due to the force on each floor of the building [7]. The base shear force which is distributed to each floor will produce a larger lateral force according to the height of the structure used. The following is presented in Tables 8 and 9 of the lateral floor forces taken from the results of the ETABS v.20 program.

Table 8. X Directional Floor Deviation

Lt	H_x (m)	**Bone**	
		δxe (mm)	**δxi (mm)**
2-Roof	8,42	3,637	-
1	2,6	1,417	-
Base	0	-	-

Table 9. Y Direction Floor Lateral Deviation

Lt	H_x (m)	**Bone**	
		δye (mm)	**δyi (mm)**
2-Roof	8,42	6,45	-
1	2,6	1,95	-
Base	0	-	-

From the calculation of the deviation between floors presented in Tables 8 and 9, the largest deviation between floors is on the 2nd floor, this is due to the distribution of forces that affect the height of the building.

Structural Performance Analysis Results
To determine the level of structure performance, the calculation of the displacement target in the area is carried out. The amount of the displacement target can be seen in Table 10.

Table 10. Migration Target in Bone Region

Koef	Bone	
	X	Y
T_e	0,56	0,56
T_s	0,4	0,4
C_0	1,4	1,4
C_1	1,0	1,0
C_2	1,0	1,0
C_3	1,0	1,0
S_a	0,19	0,19
$g (m/s^2)$	9,81	9,81
δ_T (m)	0,0021	0,0021

After determining the displacement target, then the results of the pushover analysis will be evaluated to determine the level of performance of the structure that is determined based on the load case analysis of the pushover. The capacity curves in the X and Y directions from the results of the pushover analysis in the Bone region can be seen in Figures 3 and 4.

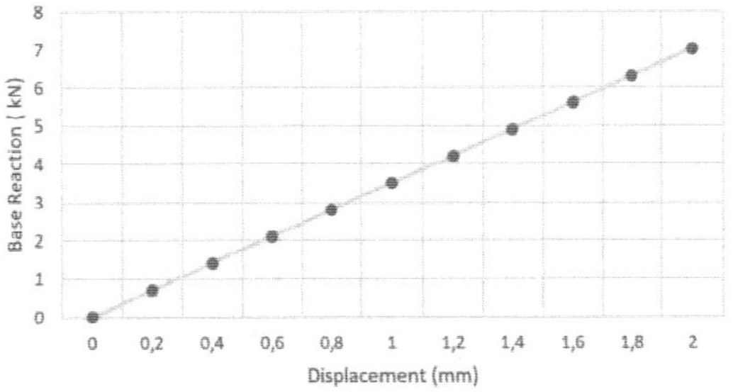

Figures 3. X Direction Capacity Curve

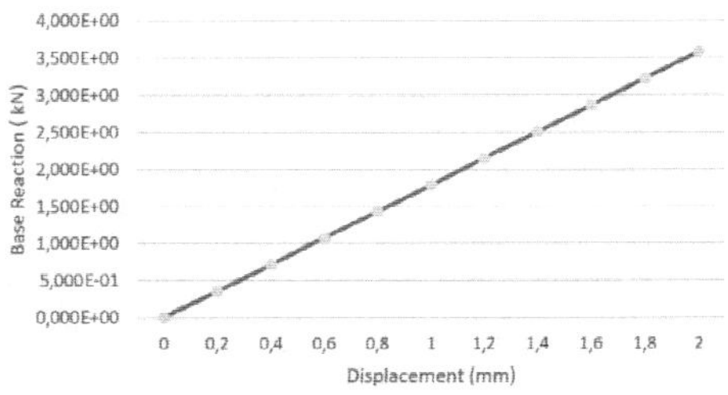

Figures 4. Y Direction Capacity Curve

From the load case pushover analysis, in the X and Y directions there is no plastic hinge condition, and an elastic condition occurs. The plastic hinge criterion is a condition where there is rotation in a structure that takes place continuously before finally reaching failure caused by external loading. This does not occur in the plastic hinge of the Bugis-Makassar Stilt House structure, because the cross section does not experience rotation. And in the Y direction, elastic conditions are also obtained, and there is a decrease in elastic conditions. To determine the level of performance of the structure, the condition of the load case pushover analysis is needed when the condition of the displacement target is achieved. With the achievement of the displacement target, it can be determined the performance points obtained in the area. The recapitulation of the results of the pushover load case analysis on the condition of achieving the performance point can be seen in Table 11.

Table 11. Pushover analysis load case results
Performance Point

Region	Direction X		
	Step	*Base Force* (kN)	*Disp* (mm)
Bone	8	7,01	2,00

Table 12. Pushover analysis load case results
Performance Point (Advanced)

	Direction Y		
	Step	*Base Force* (ton)	*Disp* (mm)
Bone	8	3,58	2,00

From the recapitulation of the results of the pushover analysis load case in Table 11, it can be determined the level of performance which is presented in Table 12.

Table 13. Structure Performance Level

Region	Performance Level	
	Direction X	Direction Y
Bone	B	B

From Table 13, it can be seen that the Bugis-Makassar Stage House in the Bone region is at the B performance level where the performance level in the X and Y directions does not occur either structural or non-structural damage. This is because the house on stilts behaves elastically which relies on the flexibility of the connection elements of the house structure to reduce earthquake energy.

4. Conclusion

The results of linear analysis show that the largest shear force of the Bugis-Makassar Stage House in the Bone region is 9,9 kN. The magnitude of the basic shear force in this structure will affect the value of the lateral drift and the drift between floors, so that the value of the lateral drift and the drift between floors in the structure is directly proportional to the value of the basic shear force. The results of the structural performance analysis based on pushover analysis in the Bone region have a performance level B, both in the X and Y directions. Therefore, the Bugis-Makassar Stage House Bone area does not experience damage to structural or non-structural elements.

5. Acknowledgement

Thank you to Indonesia Endowment Funds for Education (LPDP) and The Center for Education Financial Service (PUSLABDIK) for providing research funding assistance for the Doctoral Program (S3), so that it can be used to finance this research.

References

[1] Baharta, Nurjannah, Servie O. Dapas, and Ronny Pandaleke. 2019. Technical Evaluation of the Strength of Traditional Wooden House Structures produced by Woloan Village against Earthquakes in accordance with Indonesian National Standards (SNI). Journal of Civil Statistics. Faculty of Civil Engineering, Sam Ratulangi University, Manado

[2] Hartawan, B. Suhendro, E. Pradipto, A. Kusumawanto, 2015. Development of the Bugis House Building System in South Sulawesi. Proceeding The 5th Annual Engineering Seminar (AES 2015) Free Trade Engineers: Opportunity or Threat. Faculty of Engineering UGM. Yogyakarta 12 February 2015 A-(51-60

[3] Hariyanto, A. Dwi, T. Sugeng, W. Andry, 2020. Traditional Techniques on Stilt House Structures in Bima Regency for Earthquake Resistance. Journal of the Built Environment. Space Space. ISSN: 2355-570X

[4] A. S. Elnashai and L. Di Sarno, *Fundamentals of Earthquake Engineering*. John Wiley andSons, 2008.

[5] A. J. Pierre and I. Hidayat, “Seismic performance of reinforced concrete structures with pushover analysis,” *IOP Conf. Ser. Earth Environ. Sci.*, vol. 426, no. 1, p. 012045, Feb. 2020.

[6] F. P. Hsiao, Y. Oktavianus, and Y. C. Ou, “A pushover seismic analysis method for asymmetric and tall buildings,” *J. Chinese Inst.Eng. Trans. Chinese Inst. Eng. A*, vol. 38, no.8, 2015.

[7] R. Cornelis, W. Bunganaen, and B. H. U. Tay, "Comparative Analysis of Level Shear Force, Level Shear Force and Interstory Deviation Due to Earthquake Load Based on Earthquake Regulations SNI 1726-2002 and SNI 1726-2012," Tek. Civil, vol. III, no. 2, pp. 205– 216, 2014.

[8] National Standardization Agency, "SNI 1726:2019 Procedures for Planning Earthquake Resistance for Buildings and Building Structures." 2019. National Standardization Agency, “SNI 1727:2020 Minimum Design Loads and Associated Criteria for Buildings and Other Structures.” 2020.

[9] N. H. Crista, T. N. Widorini, and L. N. Anggraini, "Comparison of Flexural Reinforcement in the Faculty of Psychology, University of Semarang by Comparing the 2012 Earthquake SNI Regulations with the 2019 Earthquake SNI," J. Pengemb. Engineering and Technol., vol. 16, no. 2, pp. 143–162, 2020.

[10] B. Nasution and A. Purqon, "Test Response of Multi-storey Building Structures to Earthquakes Using the Finite Element Method," in Proceedings of SNIPS 2016, 2016.I. Samsya, "Evaluation of Base Isolation Applications in the Grand Keisha Building Using Pushover Analysis," Ten November Institute of Technology, Surabaya, 2017.

[11] B. Nasution and A. Purqon, "Test Response of Multi-storey Building Structures to Earthquakes Using the Finite Element Method," in Proceedings of SNIPS 2016, 2016.I. Samsya, “Evaluation of Applications Using Base Isolation in Grand Keisha Building Using Pushover Analysis,” Institut Teknologi Sepuluh November, Surabaya, 2017.

[12] R. N. Pujianto, "Analysis of Non-Linear Static Pushover Earthquakes Using the Fema 440 Method for Evaluation of Building Structure Performance," in Proceedings of the National Seminar: Public Housing Infrastructure Reliability for the Progress of the Nation, 2019, pp. 129–142.

[13] FEMA, “Improvement of Nonlinear StaticSeismic Analysis Procedures,” *FEMA 440, Federal Emergency Management Agency, Washington DC*, vol. 440, no. June. p. 392,2005.

[14] Tavio and U. Wijaya, Performance Based Design, 2nd ed. Andy, 2018.

[15] D. Purnamasari, "Performance Analysis of Reinforced Concrete Structures with Variations in Shear Wall Configuration Using Pushover Analysis," Sriwijaya University, 2019.

[16] W. A, R. Suryanita, and Z. Djauhari, "Predicting the Response of Building Structures Based on the Response of the Indonesian Earthquake Spectrum Using the Artificial Neural Network Method," J. Sainstek STT Pekanbaru, vol. 4, no. 1, pp. 1–15, 2016.

[17] B. Junita, "Response of Tall Building Structures with Variations in Outrigger Placement to Lateral Loads (Case Study: Tower A Building ST. Moritz Panakukang)," Jakarta State University, 2018

Study Of Airport Development Through Smart Airport Concept Approach In Supporting National Tourism Strategic Area In South Sulawesi Province

Milawaty Waris[1], Sakti Adji Adisasmita[2], and Muhammad Isran Ramli[3]

[1]*Post Doctoral Student of Civil Engineering Department, Hasanuddin University, Jalan Poros Malino KM-6 Gowa, Telp 085242692297, email:* milawaty.waris@gmail.com
[2]*Lecturer of Civil Engineering Department, Hasanuddin University, Jalan Poros Malino KM-6 Gowa, Telp 08179842957, email:* adjiadisasmita@yahoo.com
[3]*Lecturer of Civil Engineering Department, Hasanuddin University Jalan Poros Malino KM-6 Gowa, Telp 081242441126, email:* isranramli@unhas.ac.id

Abstract.
The aim of the research is to analyze a study of quality of the development of Sultan Hasanuddin International airport through the concept of the Smart Airport. The methodology of research using descriptive quantitative and qualitatifve research. The descriptive quantitative analysis in this research use the scale of likert and path analysis with the help of SmartPLS software. The results of the research showed that the quality and performance level of services of Sultan Hasanuddin International airport Makassar based on the quality and performances services attributes, namely reliability, responsiveness, assurance, empathy and tangible and the level of customer satisfaction. the quality of airport development services indicate a significant indirect or direct significant effect on customer satisfaction. The value of r-square on the performance of airport development services is 0.823 > 0.70 at the level of customer satisfaction 0.6278 > 0.70, it can be concluded that the variables and indicators are reliable because they fulfill of requirements > 0.70. In this case the T-value < 1.96 and P-value < 0.05. Thus, it can be concluded that the X4 variable has not significant effect on the Y variable with its indicators. Based on the results of the research, it can be concluded that the quality and performance of airport development services through a conceptual approach Smart Airport significant effect on the level of customer satisfaction. Hoped that in the future airport development will be further improved considering that the Sultan Hasanuddin International Airport in Makassar will be planned to go to Makassara erocity with the aerotropolis concept and Smart Airport concept as a supported of technological service innovation that provides basic infrastructure so that all activities it can be well connected.

Keywords :*Quality services, Performance services, Airport, Smart Airport. customer satisfaction*

1. Introduction

One of the gateways to a country or city is an airport. The development of air transportation, both passengers and goods, which is increasingly congested, requires a touch of technology, as well as development of technology in this *Society 5.0* era wiyh various sectors, including management of aviation and airport supporting tourism development [1] In the tourism industry, airports are access to tourist destinations. Three elements of the tourism ecosystem program is access one of launched by the Ministry of Tourism of the Republic of Indonesia, namely accessibility, amenities, and attractions. [13] South Sulawesi Province is the gateway to the eastern region of Indonesia which has great potential for tourism, culture and business. Sultan Hasanuddin International Airport is the center of economic development in eastern Indonesia which has a very positive influence on economic development. In supporting the National Tourism Strategic Area in South Sulawesi Province, Sultan Hasanuddin International Airport is connected surrounding at the airports. Currently, the management of Sultan Hasanuddin International Airport was taken over by PT. (Persero) Angkasa Pura I under the auspices of the Ministry of BUMN and the Ministry of Transportation. Airport management which is commercial and non-commercial is one of the business sectors run by PT. (Persero) Angkasa Pura I, which plays an important role in bridging the obligations and authorities in the field of transportation, especially air transportation and as a medium to support the success of development programs from various sectors within the government. Planning for the development and construction of an airport ideally has a large capacity with the hope of being able to serve air transportation traffic activities and the development of the number of passengers which tends to grow rapidly in the long term, but which is faced with various obstacles so that it becomes an inhibiting factor. The biggest obstacle faced is the budget for development funds and the limited amount of available regional space. Although there are limitations, the service provided to passengers must improve the quality of service[2]. Digital technology as a driving force for the presence of the smart airport concept that allows the creation of collaborative relationships between airport facilities, data facilities, and other applications that can encourage airport service levels so that they will be more effective and efficient in serving flight service users. With this concept, it can support the experience of flight service users in facilitating travel[3]. Sultan Hasanudin International Airport in Makassar is the largest airport in eastern Indonesia. The airport operates domestic and international flights. This airport is managed by PT. Angkasa Pura I and is 30 km from downtown Makassar. Air traffic movements experienced a significant increase in March 2022. In February 2022 the number of passengers arriving and departing was 830,929, an increase of 40% from 589,566 passengers [18] Since 2019 until now, the Sultan Hasanuddin International Airport development project has been implemented. Where in the development of this airport, the process is at the stage of expanding the passenger terminal area which includes the expansion of a capacity of 11 thousand visitors to 30 thousand to 40 thousand visitors, existing, and the development of the main road to and from the terminal [18] As one of the gateways for tourists, the existence of Sultan Hasanuddin International Airport is expected to increase visitor/tourists both from domestic and from abroad. It is hoped that through the development of this airport, the existence of attractive and unique tourism destinations will become one of the attractions for visitor/tourist arrivals in increasing economic income in South Sulawesi Province. In this regard, it is necessary to study the quality and performance of services in airport development through a conceptual approach smart airport on visitor/customer satisfaction.

2. Literature Review

2.1. Air Transportation

Transportation as one of youngest sub-sectors, the development of air transportation routes in this regard is growing very rapidly. Advances in aviation play an active role in supporting aspects of economic development, tourism and other sectors at large. The construction and development of airports must be planned in the long term so that the airport's capacity can accommodate air traffic movements in the long term[2]. In the implementation of air transportation activities, there are two systems that are the focus of attention, including the existence of airports as managers of supporting facilities and infrastructure and airline companies that operate aircraft as their fleet. Air transportation is very important in the economic development of a region. From a historical point of view, the development of transportation is the main reason for travel and tourism. Transportation and tourism have a reciprocal relationship. Tourism has a number of transportation requirements, especially on the capacity and frequency of transportation, on distance and speed, safety and comfort, and last but not least, the essence of the transportation service itself which can improve the quality of a service and the performance of air transportation. [6]

2.2. *Airport*

An airport is legally defined as a land and/or sea area with predetermined boundaries that is used as a landing and take-off place. It is a place for aircraft, passenger loading and unloading, cargo loading and unloading, intramodal and intermodal transportation, with aviation security facilities, basic facilities and other supporting facilities. Airport infrastructure services and facilities not only affect airport operations, but also contribute to tourism development. [2] The Role of Airports in Ministerial Decree No. 116 of 2019 on National Airport Regulations to Support Tourism. In supporting tourism in South Sulawesi, Sultan Hasanuddin International Airport is supported by surrounding airports. Apart from being a supporter of economic growth, air transportation services can also be a driving force for the growth of tourism and other sectors. [9] The Regulation of the Minister of Transportation of the Republic of Indonesia in PM No. 39 of 2019, describes the Regulation of the National Airport Order regarding the Airport System. This regulation regulates airport planning at the national level based on the benefits of urban spatial planning, economic sector growth and regional advantages of a region. Degree of integration in geographical and natural conditions, intra- and intermodal transport integration, environmental protection, flight safety and security, and other areas of development. Includes roles, functions, uses, hierarchies, and a general classification of airports. And the national airport master plan [8] Airports need to optimize limited terminal space while making more space available for commercial activity. To achieve operating cost reductions, airlines can take advantage of adapting the latest technology. Convenient and smooth passenger handling is a top priority throughout the terminal process.

2.3. *Smart Airport Concept*

In the era of Society 5.0, Concept Smart Airport present to improve services at the airport. This concept is a sub-system of the development concept *smart city* [11] This concept is developing very rapidly throughout the world by utilizing integrated technology with other sectors as its support. Definitionsmart airportclosely related to the definitionsmart city.Smart Airportis a subsystem ofsmart citythat applies technology to urban life to create a more comfortable and sustainable environment. Passengers expect all terminal facilities without any interruption to the journey. [3] Travelerssmart airportno need to wait long like in conventional airports. Sensors connect to IoT to provide information about shortest paths, parking lots, and self-checkable baggage. The effectiveness and efficiency of improving services for passengers through the development of airport digitization with the concept of "the future of airport through advanced of high technology must be supported by airport service users [3][4] [13], The ability of technological innovation through the smart airport concept enables new nuances in the business world, including better integration with a wider ecosystem of urban-based companies and organizations that can interact with surrounding airports. The form of the future of airport will result in an increase in the portion of airport revenue [13]

2.4. *Quality services of Airport*

Based on the Regulation of the Minister of Transportation Number 178 of 2015 concerning Service Standards for Airport Service Users, the service standards regulated in this regulation include service standards for prospective passengers and passengers at airport passenger terminals. This service standard starts from entering the service area as an airport service user in the departure area until leaving the service area in the arrival area. [12] Service standards include facilities used in the process of departure and arrival of passengers, facilities that provide comfort to passengers and the capacity of airport terminals to accommodate during peak hours. One of the special concerns in developing airports is to pay attention to the quality of the service system in various sectors. Service innovation should be given special attention by the airport management. This is very important for the satisfaction of consumers who use airport services. The implementation of services at an international airport requires coordination between airport managers and interrelated parties such as Customs, Immigration, Quarantine, other agencies related to intelligence and security, airlines, aviation support companies (including ground handling services, refueling aircraft, ramp handling services, aircraft loading walkaways, aircraft cleaning and catering services), airport activities supporting companies (including porter services, greeting services, taxi services, cleaning services, concessionaires and others). [12] If only one of the service companies experiences a problem, of course it will affect the quality of airport services as a whole. The level of service perceived by consumers cannot be separated from the influence of management in which management makes and improves service quality and knows the desires and needs of consumers and always makes changes on an ongoing basis in improving service quality based on consumer desires and expectations. In order for the level of quality service to be achieved, management and consumers should work together. The management must know the extent of the services that have been provided and what services must be provided in order to improve service quality, provide customer satisfaction and know what consumers want at Sultan Hasanuddin International Airport Makassar. In measuring service quality, it is not only based on consumers but also on the management side. Here, consumers can know and feel how big the level of service provided by the airport so that it can lead to a level of customer satisfaction. The higher the level of service provided, the higher the level of customer satisfaction. Airport managers as the organizers of the air transportation sector are required to be able to provide facilities and always provide optimal services included in airport services consist of punctuality, flight comfort and safety, to meet these expectations, airport services must be supported by good facilities, innovation and reliable services, including check-in area, airline tickets, inspection x -ray procedures, waiting room, parking and the other facilities. [4] [15]

2.5. *Customer Satisfaction*

Satisfaction is closely related to customer perceptions or the performance of a product. Furthermore, service performance will be compared with standards that represent the service performance expected by customers. Customer satisfaction is considered a global evaluation, which includes the overall evaluation of the product or service of the customer. Indeed, when customer satisfaction is analyzed in relation to other factors [17]. There are three general components to building customer satisfaction. First, customer satisfaction is a summary of emotional reactions of varying intensity. Second, the answers relate to a specific focus, such as product selection, purchase, or end use. Finally, reactions occur at specific times. This depends on the situation, but generally the time is limited. These three aspects provide a framework for context-specific operational definitions. They describe customer satisfaction as a collection of emotional responses of varying intensity, with well-defined timing and finite duration, directed at local aspects of product acquisition and/or consumption.[6]

IOP Conf. Series: Earth and Environmental Science 1117 (2022) 012033 doi:10.1088/1755-1315/1117/1/012033

3. Methodology Of Research

This research was conducted at Sultan Hasanuddin International Airport Makassar as a review in applying the conceptual approachSmartwhich is expected to support the National Tourism Strategic Area in South Sulawesi Province.

3.1. *Data Collection and Processing Techniques*

The collection of data in this study consisted of two, namely secondary and primary data collection. Secondary data are documents contained in various related agencies or institutions and from the results of previous research. Primary data were obtained through direct observation, completion of questionnaires by respondents or resource persons and interviews at the research site [17] Respondents or resource persons in this study came from experts in air transportation, the air transportation service and the tourism office, as well as passengers/visitors/tourists at Sultan Hasanuddin International Airport. The filling out of the questionnaire is done in two ways, namely manually and through filling out an online questionnaire that contains questions about the identity and experience of the respondent, the respondent's assessment of what if the concept Smart Airport implemented in the development of the airport as a supporter of the national tourism strategic area of South Sulawesi. .

3.2. Data Analysis Techniques

In this research is using non-experimental research with qualitative and quantitative descriptive approaches. After all the data from the required data collection has been collected, then data analysis is then carried out so that the primary data obtained from the survey and secondary data collected from various sources become focused and systematic and ready to be analyzed. From the results of the survey, a number of parameters were calculated and analyzed which were used through validity and reliability tests. [17] Calculating the number of population and samples In this study, to calculate the number of samples needed that can represent the population at the research location airport. In this study the determination of the number of populations and samples used the formula Slovin

$$n = \frac{N}{1+N\,(e)^2} \tag{1}$$

With information n is the sample size / number of respondents, N is the population size and eis the level of sampling error that can still be tolerated; e=0.1 In the Slovin formula, it has provisions if the value of e = 0.1 (10%) for a large population, and if the value ofe=0.2 (20%) for small population 2) Determination of Research Variables Determination of the variables in this study using a scalelikertto assess how much influence the quality of service has on airport development with a concept approach smart airporton passenger satisfaction. The variables in this study were adjusted to the measurement parameters for the quality and performance of airport development services on visitor or passenger satisfaction. The variables in this study were divided into 10 variables and 52 indicators.

Table 1. Variable Of Research

No	Variable	Notation	Indicator
1	*(Reability)* (X1)	Real1	Service coverage through a smart airport concept approach, making it easier for passengers to do activities
2		Real2	Digital facilities are available at the passenger departure and arrival terminals
3		Real3	Immigration check processing services for international passengers

No	Variable	Notation	Indicator
4		Real4	Easy information obtaining of passengers
5		Real5	Speed of check-in and levies PJP2U
6		Real6	Service at the airport is provided right from the moment the passenger arrives
7		Real7	Service at the airport is implemented as promised
8		Resp1	Clarity of information submitted from the airport information center
9		Resp2	Officers are trying to help passengers who have problems at the airport
10		Resp3	Reliability of airport facilities and equipment
11		Resp4	Airport staff are not too busy so they respond quickly to passenger requests/needs.
12	*(Responsiveness)* (X2	Resp5	Airport officers quickly service X-Ray customs
13		Resp6	Security and safety system checks through smart cyber security
14		Resp7	Airport officers carry out security surveillance on the air and land side (Security Surveillance)
15		Resp8	Service available Biometric system Baggage handling system
16		Resp9	Smart CCTV facial recognition service and identification of lost passengers' belongings is available
17		Resp10	Adequate skills and knowledge of airport officers in serving users/passengers
18		As1	The pasengger feel interacting comfortable at the airport
19		As2	The officers of an airport are always friendly to passengers
20		As3	The officers of an airport can be trusted for providing a sense of security for passengers
21		As4	The quality of information provided (sound system sound and speed of wifi connectivity) at the airport terminal is adequate
22	*(Assurance)* (X3)	As5	The employess of airport have experience and responsibility as well as honesty at work
23		As6	The employess of airport have mastered technology skills
24		As7	The employess of airport are motivated and reliable at work
25		As8	The operating of an airport times are suitable/suitable for passengers
26		As9	The officers of an airport are fair (it is not discriminate against passengers) when providing services
27		As10	Security services on parking area
28		Em1	The staff airport pays personal attention to passengers
29		Em2	The Officers airport put the interests of passengers first
30		Em3	The operating of an airport times that are suitable/suitable for passengers
31	*(Empathy)* (X4)	Em4	The Officers of an airport understand the specific needs of consumers
32		Em5	Information display at airport terminal is availibility
33		Em6	The Officers airport are fair (do not discriminate against passengers) when providing services

No	Variable	Notation	Indicator
34		Tang1	Biometric services (Face recognition, Eye recognition, Digital fingerprint recognition)
35		Tang2	Intelligent self service for passengers through self drop baggage machines, self check-in machines, self boarding pass, self boarding gates, domestic and international terminals
36	(*Tangible*) (X5)	Tang3	Availability of parking space/parking area and automatic parking service system (Self Parking)
37		Tang4	Digital lounge Airport equipped with wi-fi network
38		Tang5	Waiting room is clean and comfortable
39		Tang6	Available Places for shopping/malls, office facilities, restaurants, hotel facilities/micro hotels, condominiums
40		Tang7	Security and safety system check through smart cyber security
41		Y1	Satisfied with the services provided through the Smart Airport concept approach
42		Y2	The suitability of real conditions with expectations if the Smart Airport concept approach can be applied in airport services
43	Airport Development Performance (Y)	Y3	More value obtained when using airport services
44		Y4	The existence of airport services if applying the Smart Airport concept approach will be very beneficial for users
45		Y5	Feel comfortable with the available service facilities
46		CS1	Satisfied when using self service at Kiosk (Swalayan) can provide added value for customers
47		CS2	satisfied that the advanced X-ray machine safety check can add value to customers
48		CS3	Satisfied with the use of social media at the airport
49	Customer Satisfaction **(Z)**	CS4	Satisfied with using available hotels around the airport that can provide added value for customers
50		CS5	Satisfied with good airport security checks which increase customer value
51		CS6	Satisfied with the good organization of the station terminal thereby increasing customer value
52		CS7	Satisfied with the accessibility service of Sultan Hasanuddin International Airport

3.3. Techniques of Analysis data

In analyzing the data, software SPSS 25 was used to test the validity and reliability with the statistical analysis technique of Pearson's Product Moment Parametric correlation. The data collected were grouped using a Likert scale, with a very influential distribution of influence, less influential, not influential and very infuential. Determination of categorization of weight (value) between numbers 1 to 5, where number 5 is the highest value and number 1 is the lowest value. [7] [10]

3.3.1 Data Testing Method

The validity test used in this study is *(Construct Validity)*. because itinvolves many procedures including content validity and criterion validity. Validity test used correlation formula *Pearson Product Moment* as follows Where;*rxy* is the correlation coefficient of an item/item, N is the number of subjects, X is the score ofan item/item and Y is the total score.

$$r_{XY} \frac{n(\sum XY)-(\sum X).(\sum Y)}{\sqrt{\{n.\sum X^2-(\sum Y)^2\}.\{n.\sum Y^2-(\sum Y)^2\}}} \quad (2)$$

The next stage is the reliability test with the interpretation of the coefficient value of *r* then consulted with *r-table* (*r-critical)*. The item is valid, when *r-count* from the formula is greater than *r- table*, .

3.3.2 Hyphothesis

Furthermore, The validity test obtained by comparing the values of *r*-calculate and *r*2-sided test table with significant 0.05. The data are valid, if *r*-count *r-table*, the score total of instrument questions is significantly correlated with the test criteria.. The data is invalid if *r*-count < *r-table*, the instrument is not significantly correlated with total scores. Additionally, the obtained data were analyzed by *Structural equation modelling* (SEM) methods using with SmartPLS software. With the steps of Development of theoretical models, The development of path diagrams and conversion of path diagrams into equations. Test the validity of the PLS method, including the calculation of convergent validity, discriminant validity, and testing the inner structural model. *Convergent Validity* is a measurement of the validity of the variable measuring indicator which can be seen from the *outer loading* of each variable indicator. An indicator is said to have good reliability, if the value of *outer loading* above 0.70. While the value of *outer loading* still can be tolerated to 0.50 and under the value of 0.50 [14] [16]. *Discriminant validity* is done by looking at the cross-loading value of the depth construct. Calculating of the inner structural model, is evaluated by looking at the *r-square* (indicator reliability)for the dependent variable and the *t-statistics* values from path coefficient analysis which assumption that the higher the value of R means better than prediction model of the proposed on research model. Score *path coefficients* indicated by the value of tstatistics must be above 1.96 (for the two-tailed hypothesis *–two tailed*) and greater than 1.64 (for the one-tailed hypotheses *–one tailed*).

Both values are used with an alpha level of 5% and power of 80%[6]. The parameters shown on Table 2:

Table 2. Rule of thumb for measurement model evaluation

Validity and Reliability	Parameter	Rule of thumb
Convergent validity	Loading factor Cross loading	> 0.5 < 0.5
Discriminant validity	Average Variations Extracted (AVE)	> 0.5
Reliability	Composite reliability Cronbach's alpha	> 0.7 > 0.7

4. Results and Discussion

Determination of the population and sample in this research study are included to category of large numbers so that the value of e used is 0.1 (10%). the number of samples of the data adequacy test taken in this study were 100 respondents:

$$n = \frac{1.797.572}{1 + 1.797.572\ (0,1)^2} = 99,994 \infty 100 \text{ respondent sample}$$

Furthermore, the data obtained in this study were analyzed using the SEM method with thehelp of SmartPLS software. [12][14]. (1) Development of theoretical models; (2) Path diagram development; (3) Convert the path diagram into an equation:

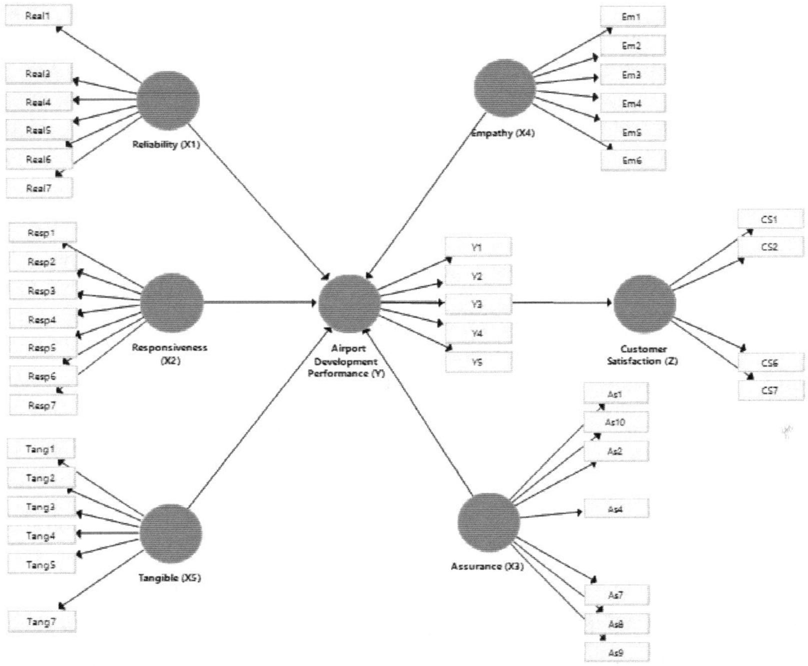

Figure 1. Research of Path Analysis Diagram

Table 3. Results of Calculation of Discriminate Validity and Reability

	Cronbach's Alpha	rho_A	Composite Reliability	Average Variance Extracted (AVE)	Keterangan
Assurance (X3)	0,914	0,925	0,930	0,625	Reliable/Valid
Empathy (X4)	0,884	0,896	0,909	0,627	Reliable/Valid
Airpoert Development Service PerformanceY)	0,859	0,870	0,899	0,642	Reliable/Valid
Reliability (X1)	0,934	0,938	0,948	0,751	Reliable/Valid
Responsiveness (X2)	0,908	0,912	0,927	0,644	Reliable/Valid
Tangible (X5)	0,943	0,947	0,953	0,744	Reliable/Valid
Customer Satisfaction (Z)	0,838	0,846	0,891	0,673	Reliable/Valid

Table 4. Outer Model Test Results (Measurement Model)

Outer loading

Variable	Reliability (X1)	Responsiveness (X2)	Assurance (X3)	Empathy (X4)	Tangible (X5)	Airpoert Development Service Performance (Y)	Customer Satisfaction (Z)
Real1	0,896						
Real3	0,841						
Real4	0,844						
Real5	0,855						
Real6	0,888						
Real7	0,872						
Real1	0,896						
Resp1		0,802					
Resp2		0,753					
Resp3		0,769					
Resp4		0,835					
Resp5		0,817					
Resp6		0,791					
Resp7		0,846					
As1			0,782				
As2			0,734				
As4			0,866				
As7			0,804				
As8			0,724				
As9			0,821				
As10			0,884				
Em1				0,726			
Em2				0,796			
Em3				0,751			
Em4				0,71			
Em5				0,868			
Tang1					0,855		
Tang2					0,871		
Tang3					0,879		
Tang4					0,826		
Tang5					0,814		
Tang6					0,908		
Tang7					0,88		
Y1						0,777	
Y2						0,83	
Y3						0,88	
Y4						0,763	
Y5						0,747	
CS1							0,794
CS2							0,857
CS6							0,773
CS7							0,854

In table 4. There are variables that are not included, namely Real 2 variables, As3, and As5 variables as wellas CS3, CS4 and CS5 variables. This is because the variable does not meet the requirements of < 0.70.

4.1. *Inner Model Test (Structural Model)*

Table 5. Analysis of Variant (R^2) or Test of Determination

Variable	R Square	R Square Adjusted
Airport Development Service Performance (Y)	0,823	0,813
Customer Satisfaction (Z)	0,678	0,675

Analysis of Variant (R^2) orDetermination test In table 5 is carried out to determine the magnitude of the influence of the independent variable on the dependent variable, the R-square value on the airport development service performance obtained 0.823 > 0.70 and at the passenger satisfaction level 0.6278 > 0.70. the conclusion can be shows reliable because its not fulfill of requirements qualify > 0.70.

Table 6. Model Summary of fit

	Saturated Model	Estimated Model
SRMR	0,089	0,094
d_ULS	7,529	8,386
d_G	6,487	6,681
Chi-Square	2614,539	2642,830
NFI	0,554	0,549

Table 7. hypothesis testing

Hipotesis	Original Sample (O)	Simple Mean (M)	Standart Deviation (STDEV)	T statistics (O/STDEV)	P Values	Status
H1. (Y –> Z)	0,823	0,827	0,029	28,105	0,000	Significant
H2. (X3 –> Y)	0,394	0,382	0,146	2,695	0,007	Significant
H3. (X2 –> Y)	0,328	0,323	0,085	3,855	0,000	Significant
H4. (X5 –> Y)	0,305	0,296	0,129	2,358	0,019	Significant
H5. (X4 –> Y)	0,045	0,02	0,045	0,988	0,324	Not Significant
H6. (X1 –> Y)	-0,065	-0,041	0,145	0,450	0,653	Not Significant

In table 7. There are 4 hypotheses in this study that were accepted. because T-value and P-value fulfill the requirements, T-value > 1.96 and P-value > 0.05, namely hypothesis H1, H2, H3 and H4. There are 2 hypothesis in this study cannot be accepted, namely the H5 and H6 hypothesis, because the T-value and P value its not fulfill the requirements. In this case the T-value < 1.96 and P-value < 0.05. Thus it can be concluded that the X4 variable has not significant effect on the Y variable with its indicator.

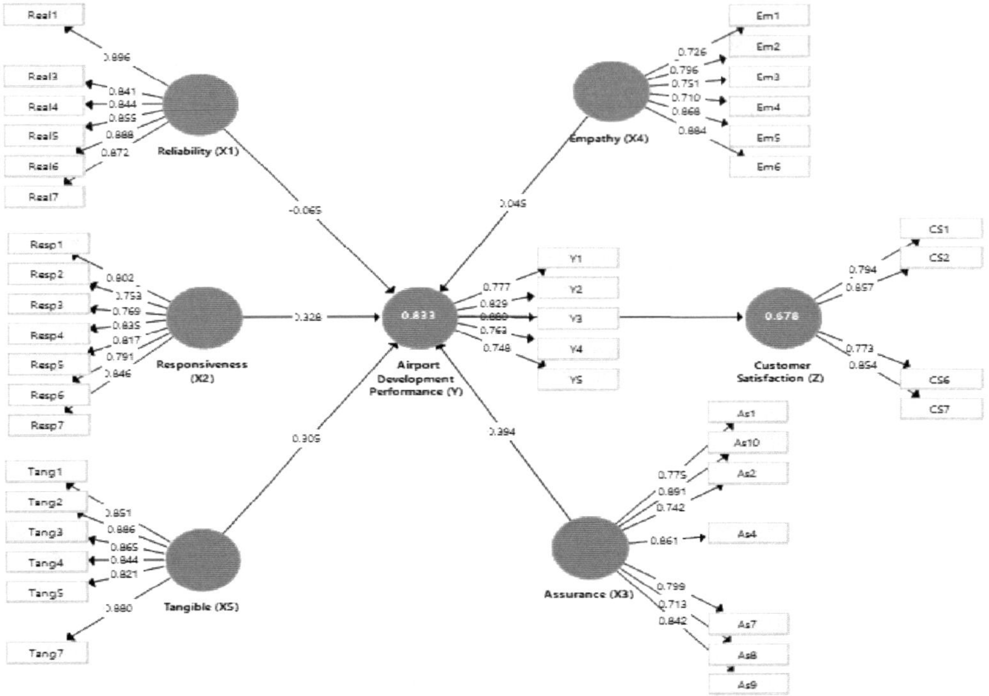

Figure 2. The final result model of data analysis using SmartPLS

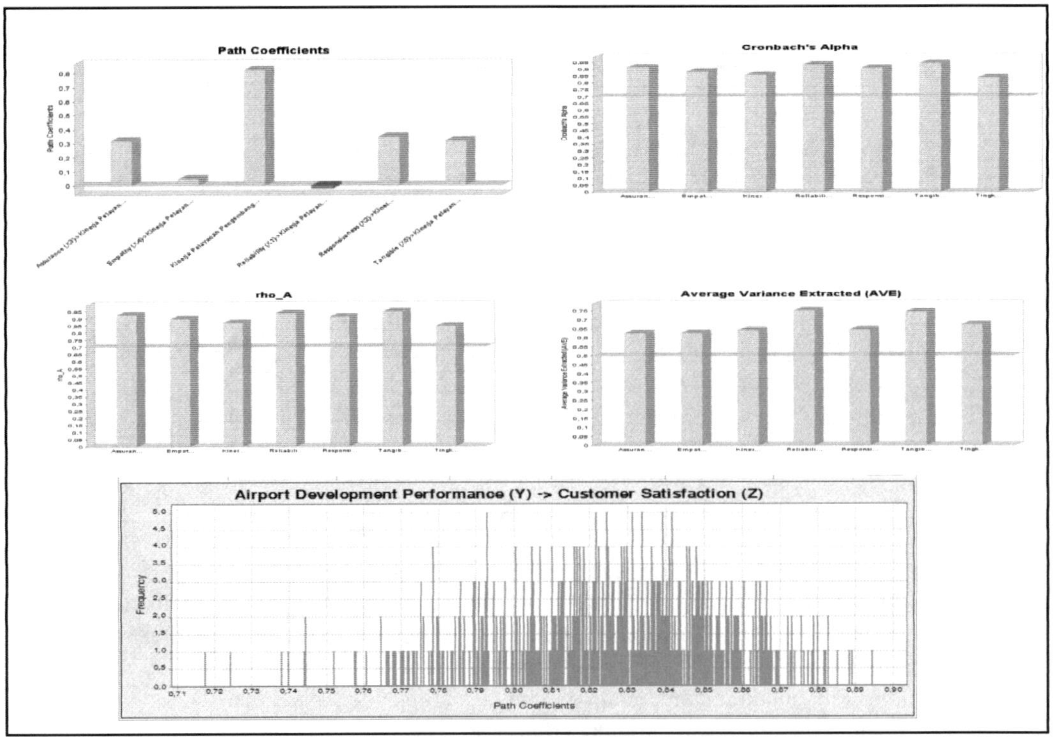

Figure 3. Graph of Path Coefficient, Cronbach's Alpha, rho_A, AVE and histogram imagesAirport development service performance with Customer satisfaction level

Conclusion

1. The results of the analysis of this study indicate that the quality and performance of airport development services have a significant indirect or direct significant effect on visitor satisfaction. The value of r-square on airport development service performance is 0.823 > 0.70 and at the level of passenger satisfaction 0.6278 > 0.70, it can be concluded that the variables and indicators are reliable because they are eligible > 0.70.
2. Analysis of Variant (R^2) orDetermination test In table 6 is carried out to determine the magnitude of the influence of the independent variable on the dependent variable, the R-square value on the airport development service performance obtained 0.823 > 0.70 and at the passenger satisfaction level 0.6278 > 0.70. the conclusion can be shows reliable because its not fulfill of requirements qualify > 0.70.
3. There are 4 hypotheses in this study that were accepted because they met the requirements for the T-value > 1.96 and P-value > 0.05, namely hypotheses H1, H2, H3 and H4. And there are 2 hypotheses in this research cannot be accepted, namely hypotheses H5 and H6, the T-value and P value not eligible. In this case the T-value < 1.96 and P-value < 0.05. Thus, it can be concluded that the X4 variable has not significant effect on the Y variable with there indicators.

Acknowledgments

Based on the results of the study, it can be concluded that the quality of airport development services through a conceptual approach *Smart Airport* significant effect on the level of visitor satisfaction. It is hoped that in the future the performance of airport development services will be further improved considering that the Sultan Hasanuddin International Airport in Makassar will be planned to go to Makassar *aerocity* with the aerotropolis concept and concept approach *Smart Airport* as a supporter of technological service innovation that provides basic infrastructure so that all activities in it can be well connected.

References

[1] Adisasmita, S. A., Hamzah, S., Ramli, M. I., Asdar, M., & Pati, S. (2017). Pengembangan Infrastruktur Bandar Udara menuju Bandar Udara Masa Depan dengan Konsep Airport City: (Studi Kasus: Bandar Udara Internasional Sultan Hasanuddin). *IPTEK Journal of Proceedings Series*, *3*(5), 93–100. https://doi.org/10.12962/j23546026.y2017i5.3119

[2] Adisasmita, S.A (2019). Tatanan Bandar udara nasional, Yogyakarta: Graha Ilmu

[3] Alabsi, M. I., & Gill, A. Q. (2021). A Review of Passenger Digital Information Privacy Concerns in Smart Airports. *IEEE Access*, *9*, 33769–33781. https://doi.org/10.1109/ACCESS.2021.3061425

[4]_Faisal, F., Fatihah, A. N., Aprilia, S. N., & Wiliyanda, W. (2020). Smart Airport Model Implementation as A Tourist Attraction in West Java International Airport Kertajati. *Digital Press Social Sciences and Humanities*, *4*(Ihtf), 00008. https://doi.org/10.29037/digitalpress.44355

[5]_ Ghozali, Imam, and Fuad. (2011). *Structural Equation Modeling: Teori Konsep dan Aplikasi Dengan Program Lisrel 8.0.*

[6] Gúčik, M., Vetrákova, M., & Marcic, M. (2018). The Role of Slovak Airports in Tourism Development. *MATEC Web of Conferences*, *236*, 4–11. https://doi.org/10.1051/matecconf/201823602008

[7] Hair, JR.JosephF., (1998) Anderson,Rolph E.,Tatham, Ronald L.and Black, William.C. 1998. Multivariat Data Analysis. *Prentice-Hall International, Inc.*

[8] Keputusan Menteri Perhubungan No.39 Tahun 2019. *Tatanan Kebandar udaraan Nasional,* Jakarta

[9] Keputusan Menteri Perhubungan No.116 Tahun 2019. *Tatanan Kebandar udaraan Nasional,* Jakarta

[10] Latan, H and I. Ghozali, *Partial least squares konsep, teknik dan aplikasi menggunakan program SmartPLS 3.0 untuk penelitian empiris*. Semarang: Badan Penerbit Universitas Diponegoro, 2015

[11].Mohamed, M., Gomaa, H., & El-Sherif, N. (2018). Exploring the Potentiality of Applying Smart Airport Technologies in Egyptian International Airports. *International Journal of Heritage, Tourism and Hospitality*, *12*(2), 122–129. https://doi.org/10.21608/ijhth.2019.31984

[12] Peraturan Menteri Perhubungan Nomor 178 tahun 2015 tentang Standar Pelayanan Pengguna Jasa Bandar Udara

[13] Rajapaksha, A., & Jayasuriya, D. N. (2020). Smart Airport: A Review on Future of the Airport Operation. *Global Journal of Management and Business Research*, *20*(March), 25–34. https://doi.org/10.34257/gjmbravol20is3pg25

[14] Sarwono J, Budiono, H. S (2012). Statistik Terapan Aplikasi untuk riset skripsi, tesis dan disertasi menggunakan SPSS, AMOS dan Excel, PT.Elex Media Komputindo, Kompas Gramedia, Jakarta

[15] S. Subekti and S. H. Winahyu, "Strategi Pengembangan Bandar Udara Komodo Labuan Bajo," *War. Penelit. Perhub.*, vol. 27, no. 5, p. 289, 2019, https://doi:10.25104/warlit.v27i5.800

[16] Stevens, J. J. (1995). Confirmatory factor analysis of the Iowa Tests of Basic Skills. Structural Equation Modeling: *A Multidisciplinary Journal*, 2(3), 214-231.

[17] Sugiyono. (2013). *Metode Penelitian Pendidikan Pendekatan Kuantitatif, Kualitatif, dan R&D*. Bandung: Alfabeta

[18]https://www.antaranews.com/berita/2317146/plt-gubernur-perluasan-bandara-sultan-hasanuddin-selesai-mei-2022

Performance of High Early Strength Concrete (HESC) using Different Superplasticizer

Y Sunarno[1], M W Tjaronge[2], R Irmawaty[2], A B Muhiddin[2]

[1]Doctoral Student, Civil Engineering Department, Universitas Hasanuddin, Makassar, Indonesia.
[2]Civil Engineering Department, Universitas Hasanuddin, Makassar, Indonesia.

email: yohanssunarno@gmail.com

Abstract. High Early Strength Concrete (HESC) is one type of high-performance concrete. HESC production requires high-quality materials and additives, namely chemical admixture type C (superplasticizer) to get high performance. This study aims to obtain a HESC mix design using a superplasticizer with the best performance. The 3 variations of HESC are made by adding a superplasticizer to a mixed concrete with the same dosage but from 3 different manufacturers, namely HESC-1, HESC-2, and HESC-3. Specimens performance consisted of tests for both fresh concrete and hardened concrete. The fresh concrete was tested by the slump, and the hardened concrete was tested by density and compressive strength tested at the age of 12 hours, 14 hours, and 3 days on cylindrical specimens measuring 100 mm x 200 mm which have been cured by water curing. Based on the test results, it was found that the average slump values in HESC-1, HESC-2, and HESC-2 were (cm): 64.0, 69.5, and 70.5. This shows that all HESC specimens are categorized as Self Compacting Concrete. According to the results of the density test on all specimens ranging from 2382 to 2436 kg/m3, all specimens are included in normal-weight concrete. With the initial compressive strength value of HESC at the age of 14 hours in the range of 21.0 to 24.9 Mpa, it can be concluded that all HESC specimens meet the requirements of High Early Strength Concrete, where HESC-3 is the best performance one.

1. Introduction

High early strength concrete (HESC) is one type of high-performance concrete that can achieve the quality of structural concrete with a compressive strength of > 21 MPa at 24 hours after site pouring [1]. High early strength concrete (HESC) reaches the designed strength in a short time. High strength at an early age is desirable for high-speed cast-in-place construction, cold weather construction, quick repairs of rigid pavements to reduce traffic downtime, fast-track pavement, and other uses. HESC can develop by using ordinary concrete constituents, chemical admixtures, and specific concrete engineering. Some methods used to achieve HESC are as follows: using high early strength cement (cement type III), increased cement content, low water-cement ratio, added chemical admixture, and steam curing [2] [3].

Chemical admixture on concrete has an important function to improve the performance of concrete. Its use can increase the characteristics of fresh concrete, assurance the quality of concrete at the time of mixing, delivery, casting, and curing, as well as overcome certain crises that may arise during the construction process, while reducing its production costs. ASTM C494 [4] provides standards for seven types of chemical admixture that differ according to their characteristics. One of which is chemical admixture type F (superplasticizer) which has a function as water-reducing and high-range admixtures

Superplasticizers are used to increase the fluidity of concrete without adding water to the mixed concrete. These molecules physically separate the cement particles by counteracting their attractive forces by steric and/or electrostatic forces. So that concrete is easier to work on at the time of pouring concrete. The workability of concrete can be maintained for a long period of time (1-2 hours or more), thus facilitating the process of casting, from transportation, pouring, pumping, compaction, and casting [5]. The availability of superplasticizers on the market today is very much, but it is still necessary to conduct research and development to meet the need for more complex concrete technologies [6][7]. Some of the most widely used type superplasticizers related to the improvement of water reducing ability are (a) Lignosulfonate (LS), with a limited water reduction ability of about 10%, used mainly in formulations to improve workability retention in ready mix applications. (b) Sulfonated naphthalene formaldehyde condensates (PNS), also known as poly naphthalene sulfonates (PNS), have a weak interaction with clay minerals with a water reduction ability of up to 30%. (c) Polymelamine sulfonates (PMS) and sulphonate formaldehyde condensate (PMS) with the ability to reduce water content by more than 20-30%. (d) Synthetic polymers, such as polycarboxylates and acrylic copolymers (PCEs), have a versatile chemical structure and can achieve water reduction of up to 40%, but have a low tolerance to clay minerals [7].

This study aims to get a mixed design of HESC which the best performance. HESC is produced by adding a superplasticizer with the same dose but comes from 3 different factories, namely HESC-1, HESC-2, and HESC-3. HESC performance consisted of tests for both fresh concrete and hardened concrete. The fresh concrete was tested by the slump, and the hardened concrete was tested by density and compressive strength tested on cylindrical specimens measuring 100 mm x 200 mm which have been cured by water curing. Testing of hard concrete is carried out at the age of 12 hours, 14 hours, and 3 days.

2. Material and Methods

2.1 *Chemical and Physical Characteristics of Ordinary Portland Cement (OPC)*

Testing the physical characteristics of ordinary portland cement (OPC) is carried out to obtain its physical properties as a reference in the manufacture of concrete mix-design. Chemical characteristics testing was carried out to analyze the chemical composition of OPC. The chemical characteristics testing carried out is in the form of XRF (*X-Ray Flourence*) testing which aims to determine the chemical content of the material being tested in the form of elements that make up the material used to produce concrete [8] [9]. Table 1 shows the tested results of OPC's physical and chemical characteristics.

Table 1. The Chemical and Physical Characteristics of OPC

PROPERTIES	Unit	Test Methode	Result	Specification
Chemical Properties				
MgO	%	ASTM C114	1.49	Max. 6.0
SO_3 if $C_3A \leq 8.0$	%	ASTM C114	2.29	Max. 3.5
Loss of Ignition	%	ASTM C114	2.47	Max. 5.0
Insoluble residue	%	ASTM C114	1.24	Max. 3.0
Equivalent Alkali (Na_2O+0.658 K_2O)	%	ASTM C114	0.38	-
Physical Properties				
Fineness/Blaine meter	m^2/kg	ASTM C204	338	Min. 280
Autoclave expansion	%	ASTM C151	0.23	Max. 0.80
Time of setting (Vicat test)				
- Initial Set	Minute	ASTM C191	112	Min. 45
- Final Set	Minute	ASTM C191	256	Max. 375
Compressive strength				
- a. 3 days	kg/cm^2	ASTM C109	196	Min. 135
- b. 7 days	kg/cm^2	ASTM C109	251	Min. 215
- c. 28 days	kg/cm^2	ASTM C109	364	Min. 300
False Set, final penetration	%	ASTM C451	84.02	Min. 50
Air Content	% volume	ASTM C185	3.30	Max. 12

Tested result of the physical and chemical characteristics of cement was carried out to determine the feasibility of cement used as a binding material in this study. The quality of OPC greatly affects the quality of the concrete produced [10]. Based on the test results of the physical characteristics of OPC in Table 1 shown, it can be seen that the OPC used meets the specifications of ASTM C150 [11].

2.2 *Physical Characteristics of Aggregate*
Table 3 shows the physical characteristics of coarse aggregates and fine aggregates. The coarse aggregate used is crushed stone which is produced by a stone crusher that uses raw materials derived from Bili-bili, Gowa Regency, South Sulawesi. It can be seen that the physical characteristics of the coarse aggregate are used to meet the specification of ASTM C33 [12]. The fine aggregate used is silica sand from the Pinrang river, South Sulawesi Indonesia. From table 3 it can be seen that the physical characteristics of the fine aggregates used to meet the specification of ASTM C33 [12].

Table 2 Physical Characteristics of Aggregate

Properties	Coarse Agg	Fine Agg
Colloid Content [%]	0.76	4.80
Fineness Modulus	6.99	3.01
Water absorption [%]	1.20	3.02
Moisture Content [%]	0.9	6.20
Specific Gravity (SSD)	2.61	2.58

2.3 *Characteristics of Superplasticizer*
The chemical admixture used in this study is a type of chemical admixture type F (superplasticizer) which has a function as a water-reducing and high-range admixture. This study uses a type of superplasticizer with 2 different product bases: Carbolic Acid Copolymer and Modified polycarboxylate in water from 3 different factories, namely superplasticizer-1, superplasticizer-2, and superplasticizer-3. Table 3 shows the characteristics of the superplasticizer used.

Table 3 Characteristics of Superplasticizer

Properties	Carboxylic Acid Copolymer	Modified polycarboxylate in water	
	Superplasticizer-1 (Sp-1)	Superplasticizer-2 (SP-2)	Superplasticizer-3 (Sp-3)
Density [kg/l]	1.10 ±0.05	1.06 ± 0.01	1.08 ± 0.02
Water reducing [%]	-	up to 25	up to 40
Dosage [%] of binder	0.3 - 2.0	0.8 - 2.0	0.8 – 2.0
Appearance	Dark brown liquid	Turbid yellowish liquid	Brown liquid

2.4 *Mixed Design*
The mixed design is made consisting of 3 variations of superplasticizer added to the mixed concrete. HESC-1 uses superplasticizer-1 (Sp-1), HESC-2 uses superplasticizer-2 (Sp-2), and HESC-3 uses superplasticizer-3 (Sp-3). The method of adding admixture is by mixing the liquid admixture into the water, stirring until homogeneous then adding it to the wet concrete mix. The dosage of admixture according to the technical data sheet of each product (table 3) is based on the percentage of the total weight of binder (cement). The HESC mixed design made in this study is in Table 4 below:

Table 4. HESC Composition (per 1 m^3)

Code	OPC (kg)	CA (kg)	FA (kg)	Water (kg)	Sp-1 (ltr)	Sp-2 (ltr)	Sp-3 (ltr)
HESC-1	580	740	1020	135	6.96	0	0
HESC-2	580	740	1020	135	0	6.96	0
HESC-3	580	740	1020	135	0	0	6.96

2.5 *Fresh Concrete Testing*

Workability testing was carried out by slump testing on fresh concrete according to ASTM C1611 standard [13], as shown in Figure 1.

Figure 1 Slump test on fresh concrete.

Figure 2 Cylindrical spesimen 100x200 mm

The weight and compressive strength tests were carried out on hard concrete at the age of 12 hours, 14 hours, and 3 days in the form of cylindrical specimens 100 mm x 200 mm which had been treated with water curing as Figure 2 according to ASTM C31 [14].

2.6 *Compressive Strength Test of Concrete*

The test method of compressive strength of cylindrical test refers to ASTM C39 [15] including the preparation and testing. At the preparatory stage follow the following procedure: (a) The test object is taken from the soaking bath. (b). Cleaning adhering dirt with a wet cloth. (c). Determine the weight and size of the test object. (d). Coating the upper and lower surfaces of the test object with sulfur mortar/capping (e). the test object is ready to be tested. After that, it is continued with the testing stage with the following procedure: (1). Placing the test object on the press machine in a centrist manner. (2). Running the press machine with an additional load between 2 to 4 kg /cm^2 per second. (3). Carry out loading until the test object becomes failure. (4). Record the maximum load that occurs during the inspection of the test object. (5). Drawing/documenting the shape of the damage to the test object. (6). Record the state of the test object. (7). Calculate the compressive strength of concrete, and the magnitude of the load of the unity of the area according to equation 1.

$$\text{Fc} = P/A \qquad (1)$$

Fc is the compressive strength of concrete (MPa)
P is the maximum load (N)
A is the surface area of the test object (mm^2)

IOP Conf. Series: Earth and Environmental Science 1117 (2022) 012034 doi:10.1088/1755-1315/1117/1/012034

3. Result and Discussion

3.1. Fresh Concrete Behaviours

From Table 5 can be observed the values of the results of slumps tested of all the mixed concrete. Based on the data from the slump test results on each mixed concrete, the average slump values of HESC-1, HESC-2, and HESC-3 were 64.0 cm, 69.5 cm, and 70.5 cm. This indicates the behavior of all the resulting HESC mixtures categorized as SCC (Self Consolidating Concrete) [16] [17]. HESC-3 has the highest slump value compared to other specimens because based on the admixture characteristics (table 3), the water reducer ability of Sp-3 is the highest, which is up to 40%. So with the same water content HESC-3 has the highest slump. From visual observations, it can be seen that the fresh concrete mixture from each mixture is adhesive: no segregation or bleeding occurs.

Table 5 Slump Tested Results

Specimen	Slump (Cm)		
	Max	Min	Average
HESC-1	65.0	63.0	64.0
HESC-2	70.0	69.0	69.5
HESC-3	72.0	69.0	70.5

3.2. Density and Visual Observation of Hard Concrete.

Based on Figure 3 shows that the density of HESC ranges from 2382 to 2436 kg/m^3. HESC-1 has the smallest density value with values (kg/m^3) at the age of 12 hours, 14 hours, and 3 days, respectively, namely 2382, 2383, and 2389. While in HESC-2 the density value (kg/m^3) is 2420, 2421, 2424. And the density value of HESC-3 (kg/m^3) is 2424, 2427, and 2436. Based on the density value, all HESC specimens are categorized as normal-weight concrete typically. According to ACI 318-11 [18], normal-weight concrete is concrete that has a density of 2155 kg/m^3 to 2560 kg/m^3. The density value increases with the age of concrete.

Based on visual observation of the hard concrete of all HESC specimens, it was found that the surface of the HESC specimen was smooth, flat, and shiny which showed that all types of superplasticizers used were well compatible with OPC so that the hardening process was good, in addition, there was no bladdery appearance in the specimens even though in the casting there was no vibrator used, this shows that Self Consolidating Concrete (SCC) behavior applies to all HESC specimens.

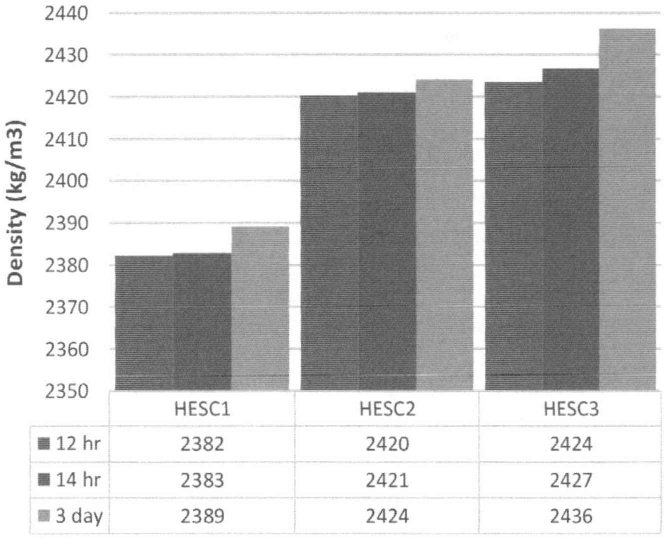

Figure 3 Density

IOP Conf. Series: Earth and Environmental Science 1117 (2022) 012034 doi:10.1088/1755-1315/1117/1/012034

3.3. Compressive Strength Results

Figure 4 showed compressive strength results at the age of 12 hours, 14 hours, and 3 days in each specimen variation. The compressive strength test results for HESC-1 are (MPa): 17.4, 21.0, and 34.0. In HESC-2 is (MPa): 21.8, 24.2, and 43.2. And for HESC-3 is (MPa): 21.6, 24.9, and 39.1. HESC-1 had the lowest average compressive strength value compared to other HESCs at each age, while HESC-2 specimens had the highest average value. An initial compressive strength test value of 14 hours on all HESCs above 21 Mpa shows that all HSCE specimens meet the requirements of High Early Strength Concrete (HESC) [1]. The highest compressive strength at the age of 14 hours HESC-3, this is influenced by the characteristics of the admixture used (Sp-3) which is the highest water reduction ability.

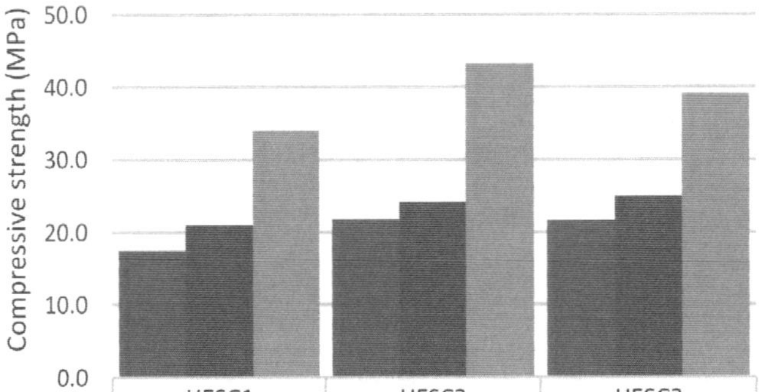

	HESC1	HESC2	HESC3
12 hr	17.4	21.8	21.6
14 hr	21.0	24.2	24.9
3 day	34.0	43.2	39.1

Figure 4. Compressive strength value of specimens against curing age

Figure 5 shows the percentage of initial compressive strength at the curing age of 12 hours, 14 hours, and 3 days against compressive strength at the age of 3 days in each specimen variation. The percentage values of compressive strength on HSCE-1 are (%): 51.2, 61.9 and 100.0. And HSCE-2 are (%): 50.4, 56.0, and 100.0. Then on HSCE-3 are (%): 55.3, 63.7 and 100.0. Based on Figure 5, it can be seen that the initial compressive strength value of the HSCE-1 is the lowest both at each concrete age, while HESC-2 has the highest percentage.

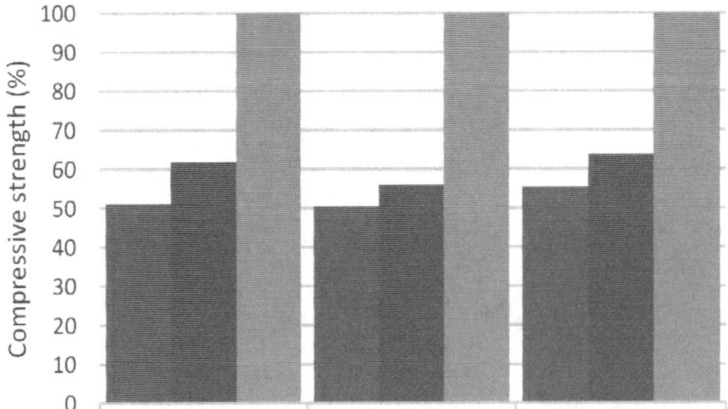

	HESC1	HESC2	HESC3
12 hr	51.2	50.4	55.3
14 hr	61.9	56.0	63.7
3 day	100.0	100.0	100.0

Figure 5. Percentage of compressive strength against 3 days age

4. Conclusions

Based on the density test results, it can be concluded that the density of all specimens can be categorized as normal-weight concrete typically according to ACI 318-11. Meanwhile, the results of slump testing on fresh concrete show the behavior of the HESC produced is included in the category of SCC (Self Consolidating Concrete). And based on the compressive strength results of all HSEC at a lifespan of 14 hours greater than 21 Mpa, all HESC specimens meet the requirements of the High Early Strength Concrete (HESC) category. HESC-3 has the highest compressive strength results at 14 hours of age. Visual observation of hard concrete on HESC shows that surface of specimens has a smooth, flat, and shiny surface indicating that all superplasticizers used in this study have good compatibility with OPC.

References

[1] Yasin, A.K., R Bayuaji, R., Susanto, T.E., 2017. *A review in high early strength concrete and local materials potential*, IOP Conf. Series: Materials Science and Engineering 267 (2017) 012004. doi:10.1088/1757-899X/267/1/012004EPRA

[2] Reddy, B. N., Naqash, J.A., 2020. *Review on Early Strength Concrete*. International Journal of Multidisciplinary Research (IJMR) -Volume: 6 | Issue: 12. DOI: 10.36713/epra2013

[3] Ananyachandran P., Vasugi, V. 2017. *Review on High Early Strength Concrete*. International Journal of Civil Engineering and Technology (IJCIET) Volume 8, Issue 7, July 2017.

[4] ASTM C494, *Standard Specification for Chemical Admixture for Concrete*, 2017

[5] Gelardi, G., Flatt, R.J., 2016. *Working mechanisms of water reducers and superplasticizers. Science and Technology of Concrete Admixtures*. Elsevier, (Chapter 11), pp. 257e278.

[6] Yaroshevskiy V. V., Kononenko A. M., Rodionov A .V., Saginov A. S. and Kwon S. S., 1986 *Complex admixture for early strength filling mixture*, Cert. of Auth. No. 1234380

[7] Perfilov V. A, 2005. *Complex admixture for early strength mixture*, RF Invent. Pat. No. 2258049

[8] Sunarno Y., Tjaronge M. W. and Irmawaty R. 2020. *Preliminary study on early compressive strength of foam concrete using Ordinary Portland Cement (OPC) and Portland Composite Cement (PCC)*. IOP Conf. Ser.: Earth Environ. Sci. 419.

[9] Sunarno Y., Tjaronge M. W., Irmawaty R. and Muhiddin A.B. 2022. *Unit Weight of Foam Concrete Containing Blended Cement and Fly Ash*. Jurnal Teknik Sipil & Perencanaan 24 (1) (2022) p 36 – 41.

[10] Rangan P.R., Tumpu M., Caroles M., Mansyur. 2021. *Compressive Strength of High-strength Concrete with Cornice Adhesive as a Partial Replacement of Cement*. IOP Conference Series: Earth and Environmental Science, Volume 871, Issue 1, id.012006, 7 pp

[11] ASTM C150, *Standard Specification for Portland cement*, 2012

[12] ASTM C33, *Standard Specification for Concrete Aggregate*, 2012

[13] ASTM C1611, *Standard Test Method for Slump of Self Consolidating Concrete.* 2021

[14] ASTM C31, *Standard Practice for Making and Curing Concrete Test Specimens in the Field*, 2019.

[15] ASTM C39, *Standard Test Method for Compressive Strength of Cylindrical Concrete Specimens*, 2012

[16] Okamura H., Ouchi M., 2003. *Self Compacting Concrete. Journal of Concrete Technology*. Vol 1 No.1, 5-15

[17] ACI 237R-07, *Self-Consolidating Concrete*, 2007.

[18] ACI 318-11, *Building Code Requirements for Structural Concrete and Commentary*, 2011.

Uniform Deflection in PCI Girder with Diaphragm Installation

Andi Indianto[1], Aprilia Rossa Ariesta[2], Erlina Yanuarini[3], Rikki Sofyan Rizal[4], Sukarman Sukarman[5]

[1,2,3,4,5]Civil Engineering Department, Politeknik Negeri Jakarta, Prof. DR. G.A. Siwabessy Street, Indonesia, 16424

andi.indianto@sipil.pnj.ac.id, apriliarossa3@gmail.com, erlina.yanuarini@sipil.pnj.ac.id, rikki.sofyanrizal@sipil.pnj.ac.id, sukarman@sipil.pnj.ac.id

Abstract. Keramasan Bridge 2 is a bridge with an age of more than 30 years, so it is necessary to review its feasibility and serviceability. Based on visual observations, there are signs of cracks in the transverse and longitudinal directions on the bridge floor. Based on the results of the loading test, it is stated that the deflection on the girder has exceeded the allowable deflection, and the cause of the longitudinal floor crack is caused by the non-uniform vertical movement of the girder. Based on this, it is necessary to strengthen the superstructure first before repairing the bridge floor. The strengthening of the superstructure is carried out by installing a diaphragm between the girders. The diaphragm is an angled steel profile of 100 mm x 100 mm with a thickness of 10 mm. To determine the number and position of diaphragms, two alternatives were analyzed. The first alternative is to add three diaphragms at every 6.275 m distance and the second alternative is the diaphragm at every 3 m distance. The results of the analysis show that the deflection that occurs between the girders after the installation of the diaphragm is uniform. The installation of diaphragms with a distance of 6.275 m and 3 m can increase the uniformity of the girder structure up to 8.12 times and 6.60 times respectively. Whereas the installation of diaphragms with a distance of 6.275 m and 3 m can increase the stiffness of the girder structure up to 10.96 times and 11.26 times, respectively. This shows that adding a diaphragm can provide rigidity and compactness between the girders.

1. Introduction

Keramasan 2 Bridge is one of the bridges in Palembang, which was built to connect Jalan Mayjen Yusuf Singedekane - Lintas Sumatra Street with a span of 25.10 m across the railway crossing below. Based on the observations that have been made, it signifies that there are cracks on the bridge floor. Based on the Report on the Feasibility and Serviceability Study of the Palembang Keramasan Bridge in 2021, the crack of the bridge floor indicates that the deflection in the girder structure is not uniform [1]. Non-uniform girder deflection occurs when the load on the girder shifts.

Based on the description above, the purpose of this study is to justify the girder reinforcement so that the girder movement becomes uniform, justify the girder reinforcement so that the permit deflection of the girder is fulfilled, and determine the type of damage repair to the bridge floor.

1.1. Theories

The following theories below are used to support this research.

1.1.1. Prestressed Concrete Bridge

A prestressed concrete bridge is a bridge structure consisting of reinforced concrete that is provided with additional steel cables (tendons) to provide additional strength to the structure [2].

1.1.2. Cracks in Concrete

Based on the Ministry of Public Works and Public Housing of Republic Indonesia module regarding bridge reinforcement, concrete cracks are divided into structural and non-structural cracks [3]. Structural cracks are cracks caused by the moment and transverse forces, while non-structural cracks are cracks on the surface.

1.1.3. Epoxy Injection

Crack repair using epoxy material cannot be used for leaking concrete structures or water seepage that has occurred in crack gaps [4].

1.1.4. Structural Reinforcement with Diaphragm Installation

The diaphragm on the bridge is needed so that the bridge that carries the load of the bridge floor and certain truck loads above it is safe against bending and lateral bending [5].

1.1.5. Bridge Loads

The bridge load used in this study refers to SNI 1725:2016 regarding Loading for Bridges.

1.1.6. The relationship between deflection and stress

Deflection is closely related to the load, the greater the load, the greater the deflection. While the load will cause a moment on the element. Furthermore, the moment on the element will cause stress in the cross-section. So that if the load is greater, the moment will increase, as a result the stress on the cross-section will be even greater. As shown in the following equation:

$$\text{Service deflection} = \frac{P_{LL}.l^3}{48EI} + \frac{5Q_{LL}.l^4}{384EI} \quad \text{(Equation 1)}$$

P_{LL} : traffic point load (kN)

Q_{LL} : traffic distributed load (kN/m)

l : length (m)

E : modulus of elasticity (Mpa)

I : inertia moment (m^4)

2. Method

This research can be explained by the following flowchart below.

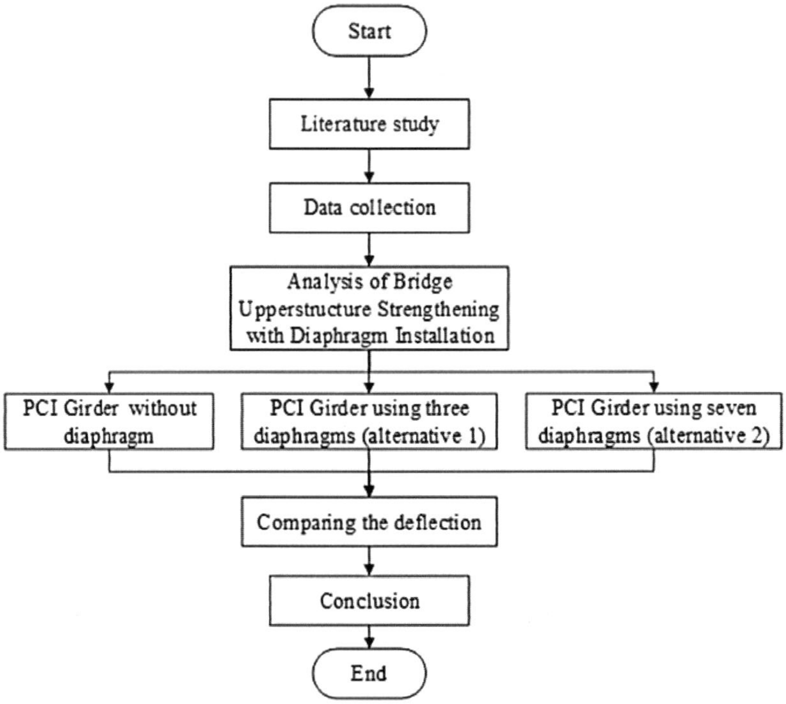

Figure 1. Flowchart.

The method used in this research is analytical. The structure analysis software calculated the deflection of three models. The first model was PCI Girder without a diaphragm (Figures 2 – 3), the second model was PCI Girder using three diaphragms (Figures 9 – 10), and the third models were PCI Girder using seven diaphragms (Figures 15 – 16). The girder was supported by simply supported. At the end of one support, U2, U3, R2, and R3 were restricted, then the other supported U1, U2, U3, R2, and R3. Then, comparing deflections based on analysis results using SAP2000 Software.

3. Result

In the superstructure of the bridge under review, the deflection of the structure has a large difference between the girders when the load on the girders moves. One way to overcome this is by installing a diaphragm that aims to stiffen the girder. The following is an analysis of the deflection of the girder, which was not previously installed with a diaphragm, and after the diaphragm was installed.

3.1. *Analysis Of The Superstructure Bridge Before Installation Of The Diaphragm*

The superstructure of Keramasan Bridge 2 can be seen in Figure 2 and 3D in Figure 3. The superstructure is then loaded with moving truck traffic loads (TLL).

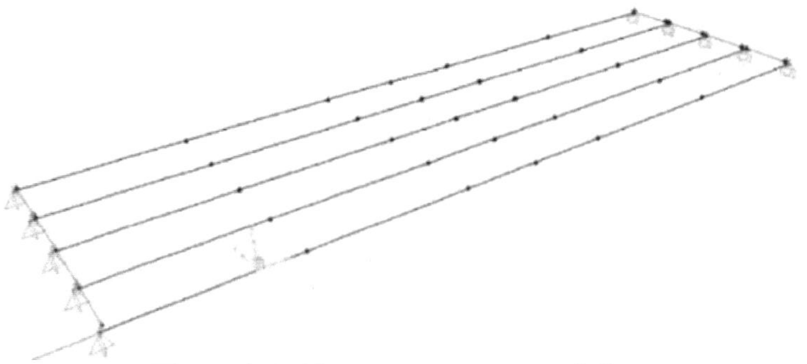

Figure 2. Bridge superstructure modeling.

Figure 3. Bridge superstructure modeling in 3D.

Based on SNI 1725:2016, truckloads can be calculated by the following equation [6].

TLL_{back} = 112.5 kN x 1.3
= 146.25 kN
TTL_{front} = 25 kN x 1.3
= 32.5 kN

The deflection caused by the moving truckload can be seen in Table 1 and Figures 4 – 7.

Table 1. Deflections due to moving truck loads.

Girder	Deflection (mm)				Max Deflection (mm)[a]	Min Deflection (mm)[b]	a-b (mm)
	TLL_1	TLL_2	TLL_3	TLL_4			
1	37.5113	2.2145	0.1694	0.0131	37.5113	0.0131	37.50
2	35.6109	35.4608	2.0584	0.0023	35.6109	0.0023	35.61
3	2.0697	35.4497	35.4548	2.0695	35.4548	2.0695	33.39
4	0.1582	2.0583	35.4659	35.611	35.6110	0.1582	35.45
5	0.0131	0.1693	2.2147	37.5112	37.5112	0.0131	37.50
Average							**35.89**

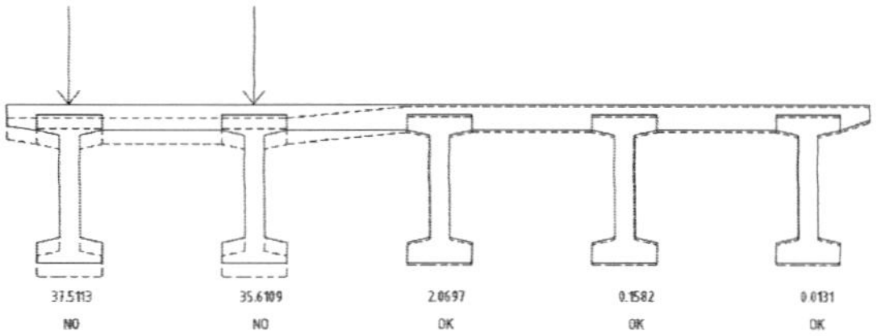

Figure 4. Girder deflection with no diaphragm caused by TLL_1.

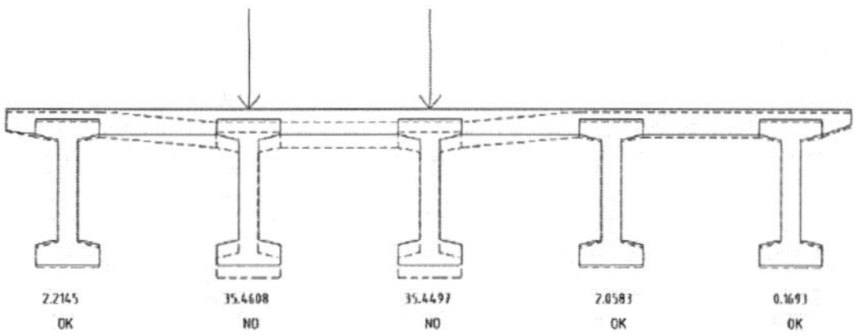

Figure 5. Girder deflection with no diaphragm caused by TLL_2.

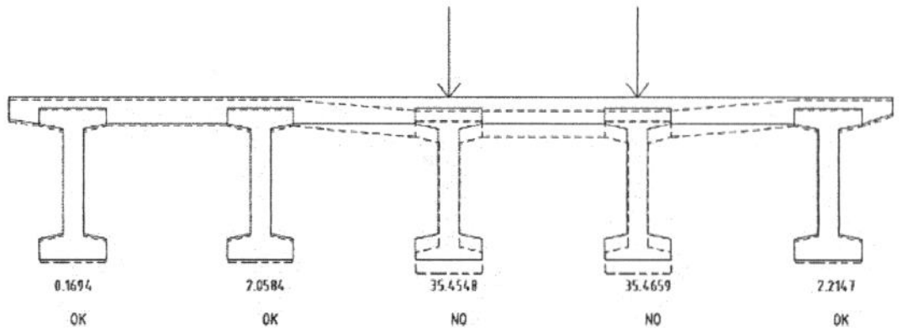

Figure 6. Girder deflection with no diaphragm caused by TLL_3.

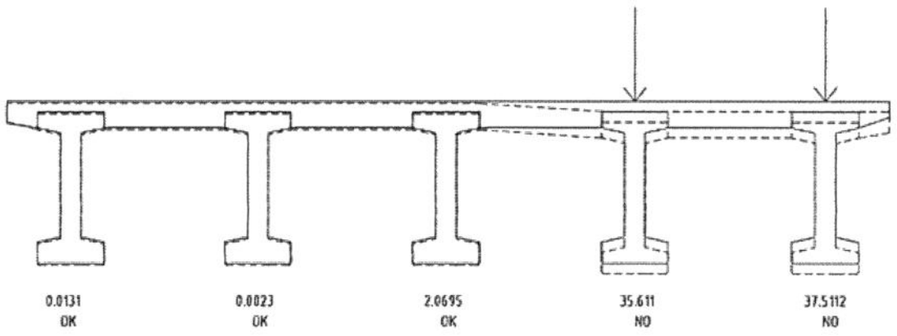

Figure 7. Girder deflection with no diaphragm caused by TLL_4.

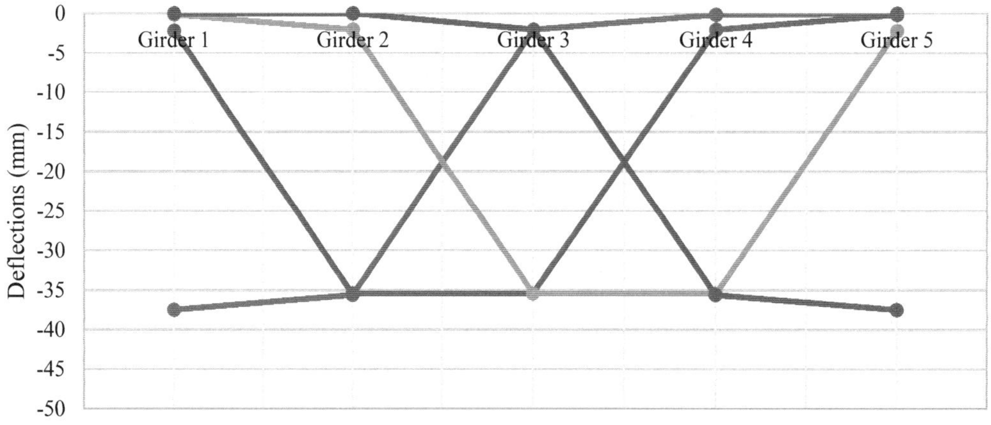

Figure 8. Deflections before diaphragm installation.

Based on Table 1 and Figure 8, the analysis results show a striking non-uniformity of the deflection of the girder when the bridge structure is given moving loads. Referring to the explanation above, it can be seen that the difference between the maximum and minimum loans is very high at 35.89 mm. So that the structure is categorized as non-uniform. As a result of the non-uniform deflection, the bridge's superstructure must be strengthened so that the girder deflection becomes uniform because no diaphragm can cause the structure not to be stiff [7]. In order to control and minimize distortion stresses due to torsional behavior when an eccentric live load is applied, it is necessary to install an intermediate diaphragm [8]. Intermediate diaphragms are used in box girders not only to prevent premature distortion under eccentric loading conditions but also to improve the distribution of live loads [9].

3.2. Analysis of the Superstructure Bridge After Installation of the Diaphragm

There are two alternatives for diaphragm installation in this study. The following is an analysis of each of these alternatives with a preliminary design, as shown in Figure 9.

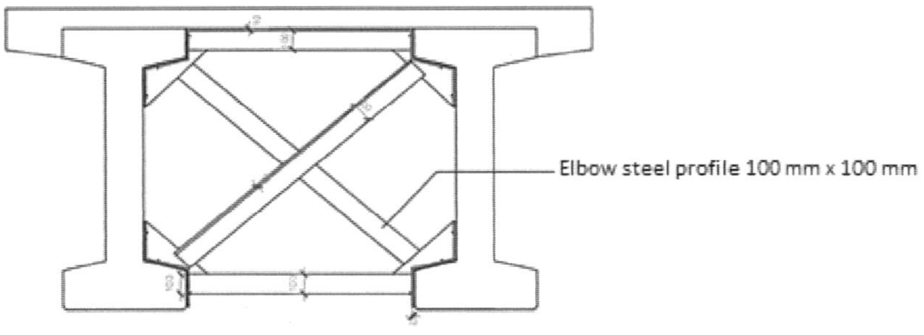

Figure 9. Preliminary design.

Diaphragms on bridges are often neglected and are considered ordinary beams that function only as longitudinal girder stiffeners and should not carry the load, therefore when 3D analysis the diaphragm is analyzed so that it does not carry the load [10].

3.2.1. Alternative 1

It is planned for three bridge diaphragms every 6.275 m with angled steel 100 mm x 100 mm and 10 mm thick, as shown in Figure 10 and Figure 11. The steel material used in this diaphragm is JISG3101 SS540.

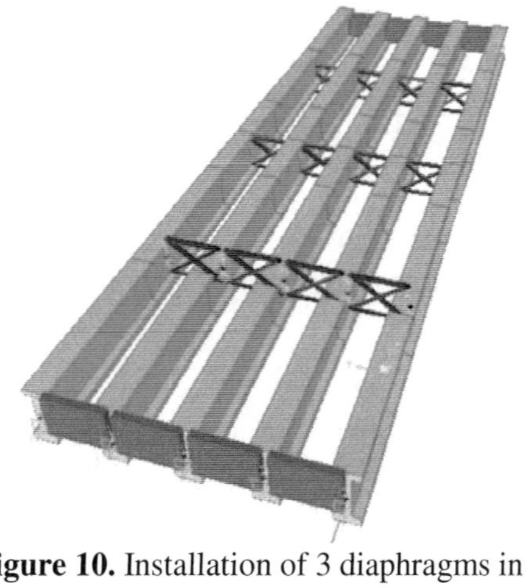

Figure 10. Installation of 3 diaphragms in 3D.

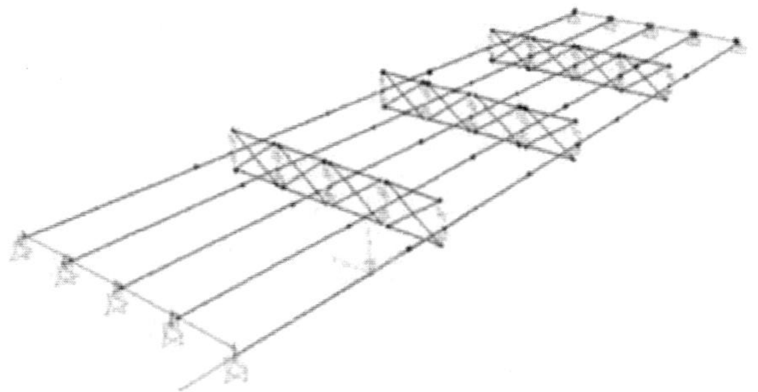

Figure 11. Bridge superstructure modeling with 3 diaphragms.

The bridge structure is loaded with truck loads from TLL_1, TLL_2, TLL_3, and TLL_4 loads. The truckload produces a deflection that can be seen in Figure 12 – Figure 15 and Table 2.

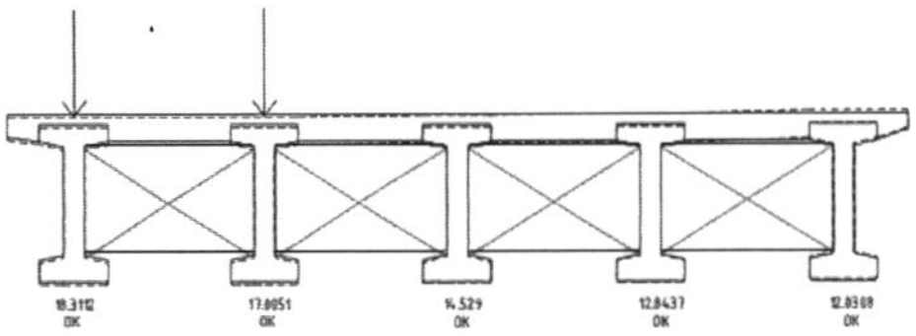

Figure 12. Girder deflection after installing 3 diaphragms caused by TLL_1.

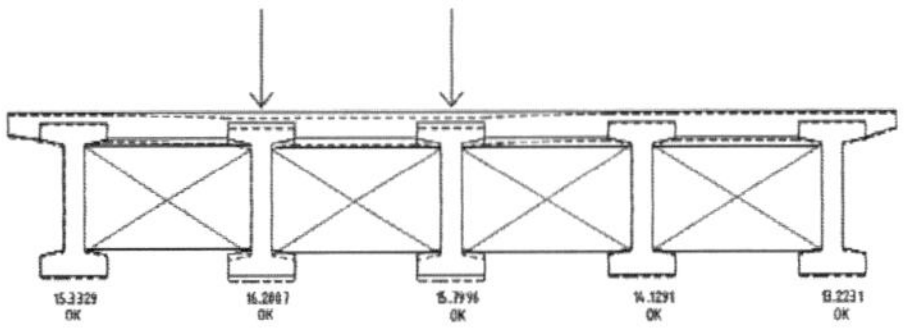

Figure 13. Girder deflection after installing 3 diaphragms caused by TLL_2.

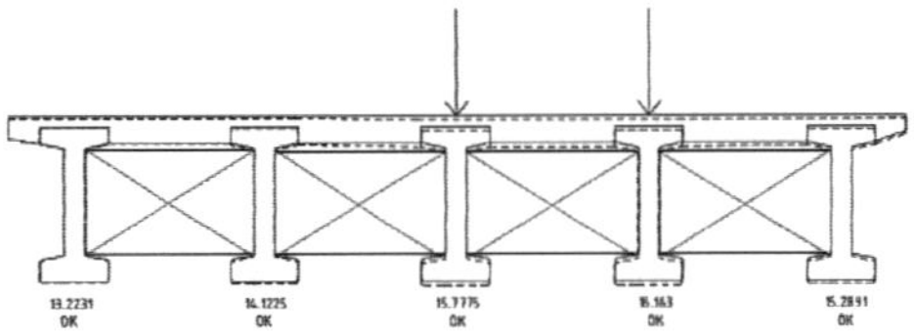

Figure 14. Girder deflection after installing 3 diaphragms caused by TLL_3.

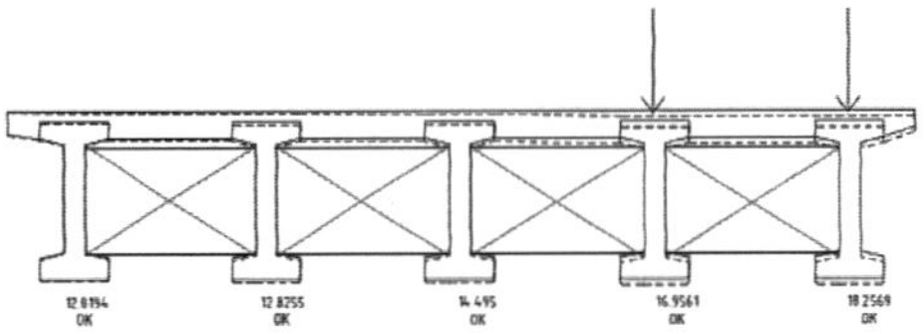

Figure 15. Girder deflection after installing 3 diaphragms caused by TLL_4.

Table 2. Deflections due to moving truck loads after installing 3 diaphragms.

Girder	Deflection (mm)				Max Deflection (mm)[a]	Min Deflection (mm)[b]	a-b (mm)
	TLL_1	TLL_2	TLL_3	TLL_4			
1	18.3112	15.3329	13.2211	12.0194	18.3112	12.019	6.29
2	17.0051	16.2007	14.1225	12.8255	17.0051	12.826	4.18
3	14.5290	15.7996	15.7775	14.495	15.7996	14.495	1.30
4	12.8437	14.1291	16.163	16.9561	16.9561	12.844	4.11
5	12.0308	13.2231	15.2891	18.2569	18.2569	12.031	6.23
Average							**4.42**

IOP Conf. Series: Earth and Environmental Science 1117 (2022) 012035 doi:10.1088/1755-1315/1117/1/012035

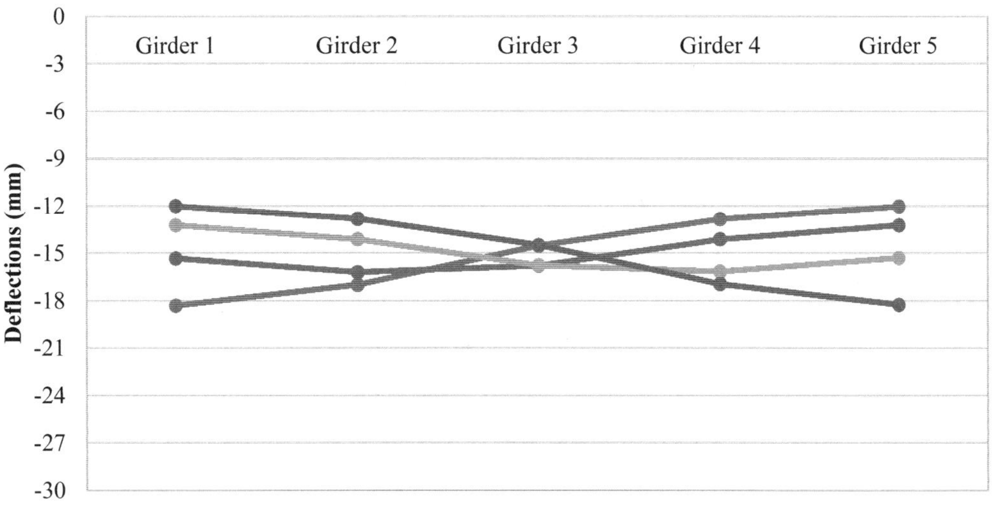

Figure 16. Deflections after installing 3 diaphragms.

Based on Figure 16, the installation of 3 diaphragms can improve cooperation deflection between girders. Furthermore, the difference between the maximum and minimum deflection for all girders already installed with 3 diaphragms is 4.42 mm (Table 2). Whereas the difference between the maximum and minimum deflection for all girders without diaphragms is 35.89 mm (Table 1). So that the increase in girder uniformity after installing 3 diaphragms is 812% or 8.12 times (as shown in equation 2).

$$\text{Increased Uniformity} = \frac{(a-b)_{without\ diaphragm}}{(a-b)_{with\ diaphragm}} x100\% \quad \text{(Equation 2)}$$

Furthermore, the increase in stiffness after installing 3 diaphragms that can be seen in Table 3.

Table 3. The increase in stiffness after installing 3 diaphragms.

Girder	Deflection (mm) a-b without diaphragm (mm)	Deflection (mm) a-b with diaphragm (mm)	Increased Stiffness (%)
1	37.50	6.29	596
2	35.61	4.18	852
3	33.39	1.30	2568
4	35.45	4.11	863
5	37.50	6.23	602
Average			**1096**

Based on Table 3, it can be seen that the increase in stiffness after installing 3 diaphragms is 1096% or 10.96 times (as shown in equation 3).

$$\text{Increased Stiffness} = \frac{(a-b)_{without\ diaphragm}}{(a-b)_{with\ diaphragm}} x100\% \quad \text{(Equation 3)}$$

3.2.2. Alternative 2

In the second alternative, a steel diaphragm profile of 100 mm x 100 mm with a thickness of 10 mm is used, which is placed at a distance of 3 m as shown in Figure 17, and 3D modeling in Figure 18.

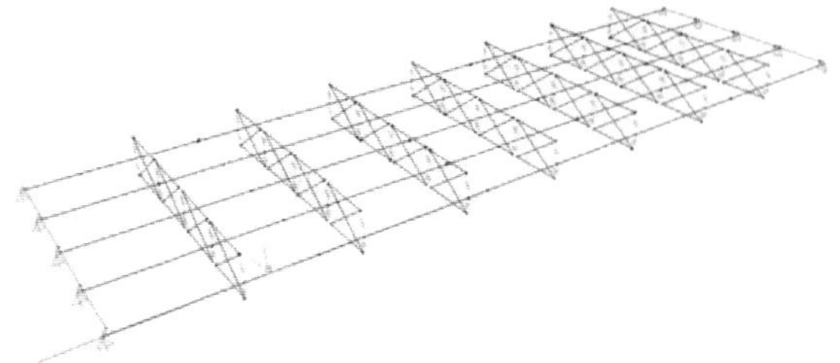

Figure 17. Bridge superstructure modeling with diaphragms at every 3 m distance.

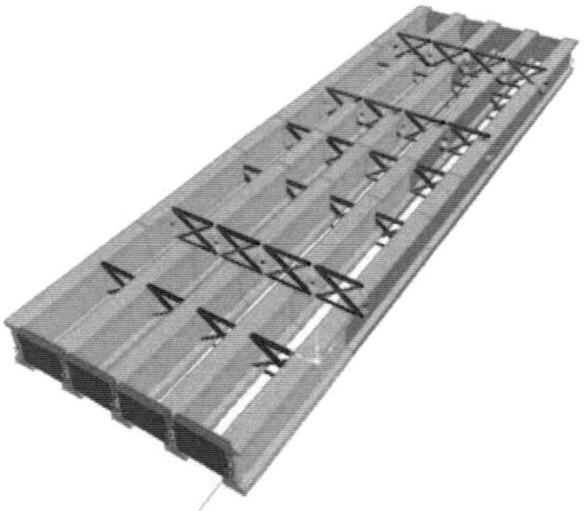

Figure 18. Installation of diaphragms at every 3 m distance in 3D.

The bridge structure is loaded with running loads, namely truck loads starting from TLL_1, TLL_2, TLL_3, and TLL_4 loads. The truckload produces a deflection which can be seen in Figure 19 – Figure 22 and Table 4.

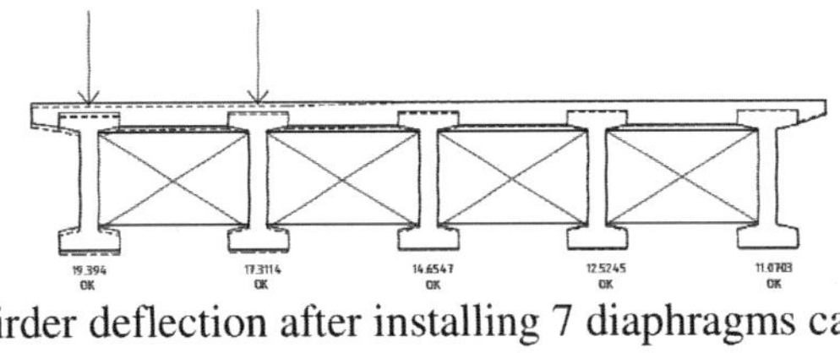

Figure 19. Girder deflection after installing 7 diaphragms caused by TLL_1.

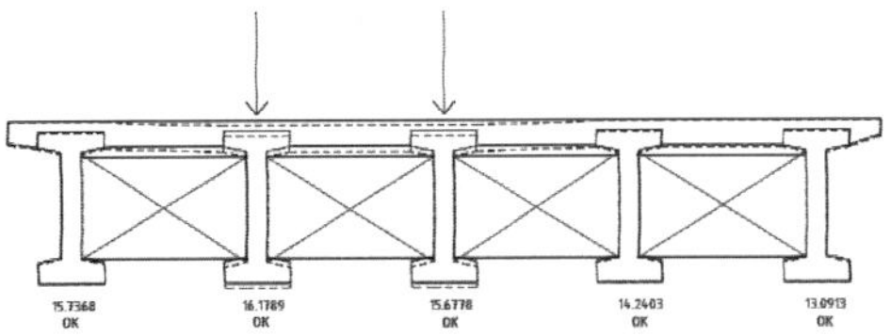

Figure 20. Girder deflection after installing 7 diaphragms caused by TLL_2.

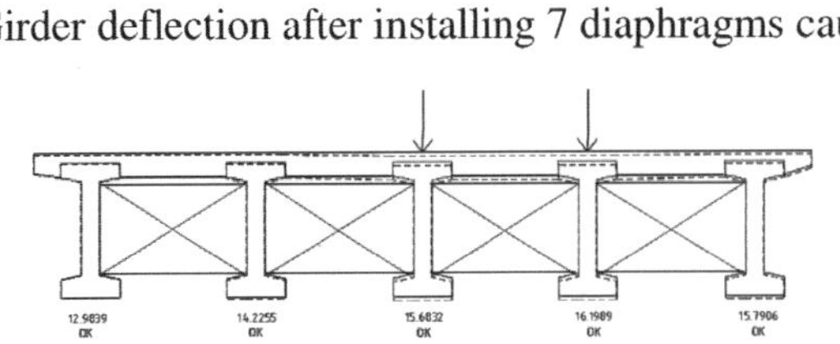

Figure 21. Girder deflection after installing 7 diaphragms caused by TLL_3.

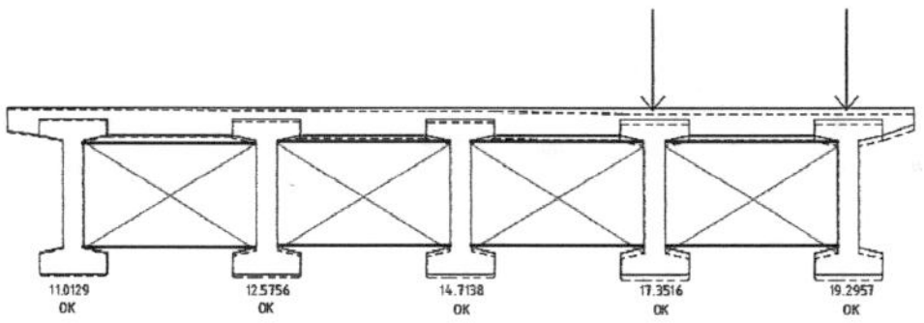

Figure 22. Girder deflection after installing 7 diaphragms caused by TLL_4.

Table 4. Deflections due to moving truck loads after installing 7 diaphragms.

Girder	Deflection (mm)				Max Deflection (mm)[a]	Min Deflection (mm)[b]	a-b (mm)
	TLL_1	TLL_2	TLL_3	TLL_4			
1	19.3942	15.7368	12.9839	11.0129	19.3942	11.013	8.38
2	17.3114	16.1789	14.2255	12.5756	17.3114	12.576	4.74
3	14.6547	15.6778	15.6832	14.7138	15.6832	14.655	1.03
4	12.5245	14.2403	16.1989	17.3516	17.3516	12.525	4.83
5	11.0703	13.0913	15.7906	19.2857	19.2857	11.070	8.22
Average							**5.44**

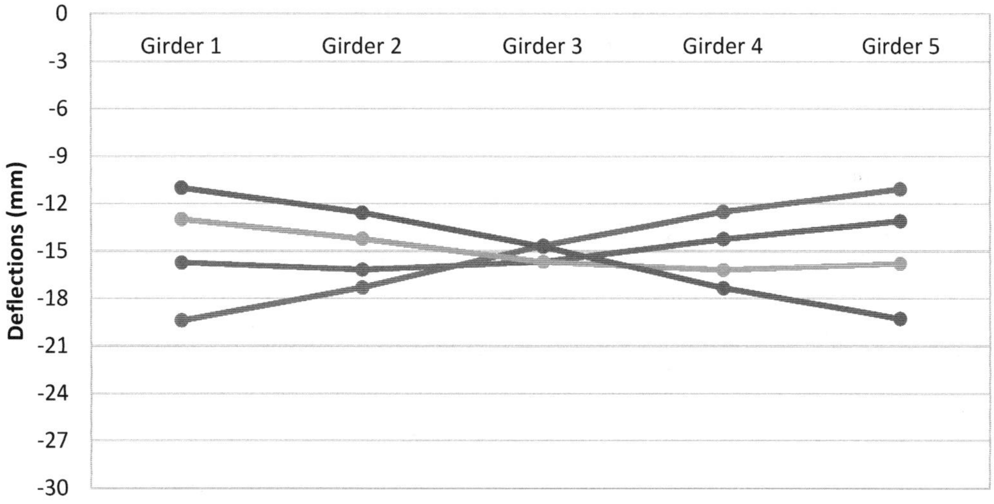

Figure 23. Deflections after installing 7 diaphragms.

Based on Figure 23, the installation of 7 diaphragms can improve cooperation deflection between girders. Furthermore, the difference between the maximum and minimum deflection for all girders already installed with 7 diaphragms is 5.44 mm (Table 4). Whereas the difference between the maximum and minimum deflection for all girders without diaphragms is 35.89 mm (Table 1). So that the increase in girder uniformity after installing 7 diaphragms is 660% or 6.60 times (as shown in equation 4).

(Equation 4)

$$\text{Increased Uniformi ty} = \frac{(a-b)_{without\,diaphragm}}{(a-b)_{with\,diaphragm}} x100\%$$

Furthermore, the increase in stiffness after installing 7 diaphragms that can be seen in Table 5.

Table 5. The increase in stiffness after installing 7 diaphragms.

Girder	Deflection (mm)		Increased Stiffness (%)
	a-b without diaphragm (mm)	a-b with diaphragm (mm)	
1	37.50	8.38	447
2	35.61	4.74	751
3	33.39	1.03	3241
4	35.45	4.83	734
5	37.50	8.22	456
Average			**1126**

Based on Table 3, it can be seen that the increase in stiffness after installing 7 diaphragms is 1126% or 11.26 times (as shown in equation 5).

$$\text{Increased Stiffness} = \frac{(a-b)_{without\,diaphragm}}{(a-b)_{with\,diaphragm}} x100\% \quad \text{(Equation 5)}$$

4. Conclusion

Based on the results of the analysis that has been carried out, it can be concluded that: a) the installation of diaphragms with a distance of 6.275 m and 3 m can increase the uniformity of the girder structure up to 8.12 times and 6.60 times respectively; b) the installation of diaphragms with a distance of 6.275 m and 3 m can increase the stiffness of the girder structure up to 10.96 times and 11.26 times, respectively.

5. References

[1] Feasibility Study and Serviceability Report of the Keramasan Bridge, 2021.
[2] D. Margan and P. Saelan, Study on Analysis of Partial Prestressed Beams on Workload, vol. 5, no. 2, 2019.
[3] Ministry of Public Works and Public Housing. Bridge Reinforcement Module, Bandung, 2015.
[4] Bina Marga, *Crack Repair With Epoxy Material*, 2017.
[5] A. Jamal, Analysis of the Effect of the Diaphragm on Lateral Bending in Longitudinal Bridges, Medan, 2007.
[6] National Standardization Agency, *SNI 1725:2016*. Indonesia, 2016. www.bsn.go.id
[7] J. M. Guci, The Effect of the Diaphragm on the Behavior of the Bridge Structural System, vol. 6, 2017.
[8] Lee J., Kim S., and Kang Y.J., Effect of cross-sectional rigidity on intermediate diaphragm spacing of steel-box girder bridges, Journal of Constructional Steel Research, vol 188, 2022. https://doi.org/10.1016/j.jcsr.2021.107006
[9] Quang-Viet Vu, Duc-Kien Thai, Seung-Eock Kim, Effect of intermediate diaphragms on the load – carrying capacity of steel – concrete composite box girder bridges, Thin-Walled Structures, vol 122, 2018. https://doi.org/10.1016/j.tws.2017.10.024
[10] Selfas A., Priyono P., and Dewi I. C., Influence of Using Diaphragm in Girder I Against Capability of High Grade Bridge Class A Capability Using Steel Construction, Jember, 2018.

Preference of Urban Commuters on Implementation of Rail-based Public Transportation Plan in Mamminasata Area

M I Sabil[1], M I Ramli[2] and S A Adisasmita[2]

[1] Graduate School of Civil Engineering, Hasanuddin University , Indonesia ,
[2] Department of Civil Engineering, Hasanuddin University, Indonesia.

sabil.ikhsan@gmail.com

Abstracts . The density of population and the increasing needs of Makassar city will affect the movement of society to fulfill their necessary. Public acceptance on rail transportation is one of the important keys in implementing sustainable transportation policies in the Mamminasata area. The aim of this study is to evaluate public acceptance regarding the planned public transportation. The survey in this study used the methods of questionnaire and interview that included the South Sulawesi Provincial Office with a total of 1002 civil servants who were also the users of commuters, then the data analysis was carried out by conduct the research about characteristics of the respondent's acceptance on rail-based transportation. The method of data collection was conducted with respondent chooses their perspective and their acceptance with likert scale. The results of this study show the perception of acceptance is the main factor that affect the policy of implementing rail-based public transportation in the Mamminasata area.

Keywords: Preference, Commuting, Public Transportation, Mamminasata

1. Introduction

The population numbers are continuing increase every year, it makes the population in the city of Makassar increasingly so fast. The population density in the city of Makassar pushes more demands in order to move and find the needs of society's life. Today, individual tourism activities in urban areas of developing cities in Asia, such as Makassar, Indonesia, are increasing too. Tourism requires the provision of infrastructure such as modes of transportation and transportation networks. By the needs of tourism activities in the city of Makassar, Makassar is one of the capital city of South Sulawesi province. [1] This is supported by the increasing number of official governments travel to and from nearby cities such as Maros, Gowa and Takalar, but this study only limited to Makassar. [2]

Transportation is a necessity of human life that cannot be separated. There is a close relationship between place and human activity. The importance of transportation is shown by the increasing need for transportation services to move people and goods to every corner of the country or even abroad. [3] From the perspective of the urban transportation system, public transportation is an integral and important part of the urban transportation system and have a very fundamental role. The road network is a node also link the other modes of transportation such as ports, airports, etc. [4]

Makassar City, which is the gateway to Eastern Indonesia (KTI) is one of capital city of South Sulawesi Province with number population around 1,526,677 people (Makassar City in Figure, 2020). The population density in the city of Makassar in 2012 reached 8,686 people/km2. The population density in the 15 sub-districts is quite diverse, with the highest population density are in Makassar

District with a total of population around 33,935 people/km2 and the lowest are in Tamalanrea District at 3,638 people/km2. [5] From the fairly large population density disparity, it can be seen that the uneven distribution of the population in Makassar City has an impact on the accumulation of community activity centers in several designated areas. Indirectly, people's travel demand will also increase.

Currently the city of Makassar has one means of transportation, namely transportation within the city. Many people use urban transportation. However, the government as a service provider are less able to meet the needs of the community and society. The phenomenon such reckless drivers of public transportation and unqualified vehicles are common sight in Makassar City. According to the data from Makassar City Transportation Service. Currently Makassar City has 25 traffic lanes within the city with a total of 5,069 vehicles. These numbers of traffic among existing city transportation because many routes are similar. Moreover, the use of private vehicles is increasing day by day.

Based on one of the problems mentioned above, through the Directorate General of Railways based on the National Railway Master Plan in an effort to accelerate national development in accordance with the RPJMN 20115-2019, it is planned to gradually plan the railway infrastructure in the development of the Mamminasata regional rail network and services [6] as one of the solutions to those problems. transportation problems in the Mamminasata area, with this plan it is necessary to analyse the preferences for rail-based transportation.

2. Method

2.1 Research Location

This research was conducted in the MAMMINASATA urban area with data collection in Makassar City with the target respondents are Civil Servants who work in the South Sulawesi Provincial office who work as office commuters every day.

2.2 Data Collection

The data used in this study is from the results of the questionnaire distribution. The distribution of the questionnaires was carried out in the offices and government agencies of the South Sulawesi Province with a total of 30 agencies that became the target. This are the following names of the agencies that are became the targets of questionnaire distribution.

Table 1. Office/Agency of Respondents

No	Office/Agency	sample
1	Department of energy and mineral resources	20
2	Department of culture and tourism	21
3	Department of Forestry	97
4	Department of Marine Affairs and Fisheries	20
5	Department of Health	91
6	Department of Food Security	19
7	Department of Cooperatives Small and Medium Enterprises	19
8	Department of Women Empowerment	6
9	Department of Community and Village Empowerment	19
10	Department of Youth and Sports	21
11	Department of Investment	20
12	Department of Education	77
13	Department of Environment	6
14	Department of Trade	25
15	Department of Transportation	42
16	Department of Industry	7
17	Department of Library	24

No	Office/Agency	sample
18	Department of Land, Housing & Urban Development	19
19	Department of Livestock and Animal Health	8
20	Department of Public Works	115
21	Department of Social Affairs	66
22	Department of Food Crops	49
23	Department of Manpower	25
24	Department of Demographics and Civil Records	17
25	Department of Public Information and Communications	13
26	Inspectorate of South Sulawesi Province	7
27	Regional Revenue Office	67
28	Regional Development Planning Agency	41
29	Regional Personnel Agency	11
30	General Bureau	30
	Total	1002

The population in this study was urban commuters. Given the unknown population, the researchers took samples with a minimal sample calculation and used a sampling technique, namely quota sampling. Quota sampling is one of technique that identify the sample from a certain population which is carried out to determine the classification according to certain characteristics until the required quota is reached.

The commuter perception variable is regarding approval of the LRT plan in the Mamminasata urban area and is measured by using a Likert scale. [7] [8] The Likert scale is a scale that used to measure attitudes, opinions and perceptions of a person or group of people regarding certain social phenomena. [9] In this case, the Likert scale is used to measure perceptions of the LRT plan by providing answers in the available measurement scales, as follows:

1 = Strongly Disagree
2 = Disagree
3 = Neutral
4 = Agree
5 = Strongly Agree

Determine the lowest level of perception and the highest level of perception of all alternative answers based on the characteristics of the respondents. The method used to determine the level of perception of the LRT is the number of respondents and the percentage of respondents, then the next step was conduct the process of identification [10] Furthermore, a significance test was conducted to determine the characteristics that have a significant effect on the perception of the LRT planning.

Table 2. Categories of Characteristics of Respondents

Characteristics	Categories of Characteristics
Gender	a. Male b. female
Age (Years)	a. 18 – 25 e. 50 - 57 b. 26 – 33 f. 58 - 65 c. 34 – 41 g. 66 - 73 d. 42- 49
Education	a. Senior High School e. Master b. Diploma (D4) f. Doctor

Characteristics	Categories of Characteristics
	c. Bachelor
Income (IDR)	a. 500,000-1,500,000 d. 3,500,000 - 4,500,000 b. 1,500,000 – 2,500,000 e. 4,500,000 – 5,500,000 c. 2,500,000 – 3,500,000 f. > 5,500,000
Vehicle Ownership	a. Not Know d. Car b. Motorcycles e. All Vehicles
Motorcycle Ownership	a. 0 d. 3 b. 1 e. 4 c. 2
Car Ownership	a. 0 d. 3 b. 1 e. 4 c. 2 f. 5
Transportation Mode	a. Microbus d. Car b. Taxicar e. Motorcycle c. Taxibike
Travel Cost (IDR)	a. <5,000 d. 25,000-35,000 b. 5,000-15,000 e. 35,000-45,000 c. 15,000-25,000 f. 45,000-55,000
Travel Distance (Meters)	a. < 2,000 e. 10,000 - 12,000 b. 2,000 - 4,000 f. 12,000 - 14,000 c. 4,000 - 6,000 g. 14,000 - 16,000 d. 6,000 - 8,000 h. > 16,000 e. 8,000 - 10,000

Table 1 shows that there are categories of questions that asked to respondents, both questions about the characteristics of respondents and questions about the attributes of commuting trips for Civil Servants.

3. Results and Discussion

3.1 Result

According to gender, age, education level, income, travel distance, travel cost and mode of transportation urban commuters in Makassar City have the following socio-economic characteristics as indicated in the table below:

Table 3. Characteristics of Respondents

Characteristics	Categories	Frequency	Percentage (%)
Gender	Male	488	48.70
	female	514	51.30
Age (Years)	18 - 25	98	9.78
	26 - 33	220	21.96

IOP Conf. Series: Earth and Environmental Science 1117 (2022) 012036 doi:10.1088/1755-1315/1117/1/012036

Characteristics	Categories	Frequency	Percentage (%)
	34 - 41	256	25.55
	42- 49	207	20.66
	50 - 57	198	19.76
	> 58	23	2.30
Education	Senior High School	113	11.28
	Diploma (D4)	39	3.89
	Bachelor	618	61.68
	Master	211	21.06
	Doctor	21	2.10
Income (IDR)	500,000-1,500,000	28	2.79
	1,500,000 – 2,500,000	88	8.78
	2,500,000 –3,500,000	162	16.17
	3,500,000 - 4,500,000	212	21.16
	4,500,000 – 5,500,000	176	17.56
	> 5,500,000	336	33.53
Travel Distance (Meters)	< 2,000	118	11.78
	2,000 - 4,000	129	12.87
	4,000 - 6,000	138	13.77
	6,000 - 8,000	138	13.77
	8,000 - 10,000	144	14.37
	10,000 - 12,000	123	12.28
	12,000 - 14,000	65	6.49
	14,000 - 16,000	92	9.18
	> 16,000	55	5.49
Travel Cost (IDR)	< 5	80	7.98
	5-15	351	35.03
	15–25	274	27.35
	25–35	151	15.07
	35-45	59	5.89
	45–55	87	8.68
Transportation Mode	Microbus	36	3.59
	Taxicar	93	9.28
	Taxibike	210	20.96
	Car	213	21.26
	Motorcycle	450	44.91

IOP Conf. Series: Earth and Environmental Science 1117 (2022) 012036 doi:10.1088/1755-1315/1117/1/012036

Table 2 shows the number and percentage of survey respondents, from these results it can be seen the number of the largest and smallest respondents and the percentage of respondents on each question category.

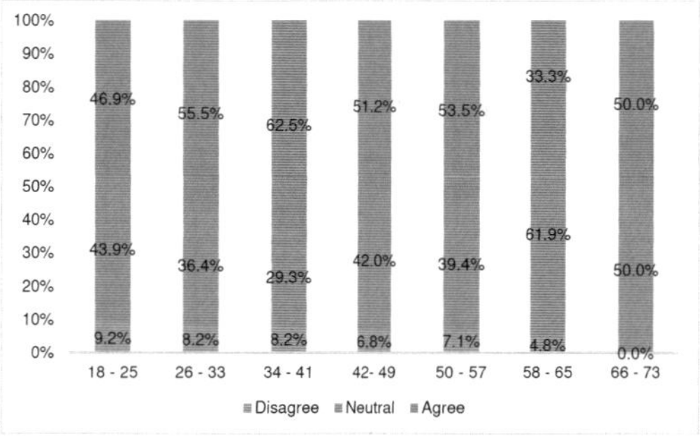

a. Preferences of respondents based on Age

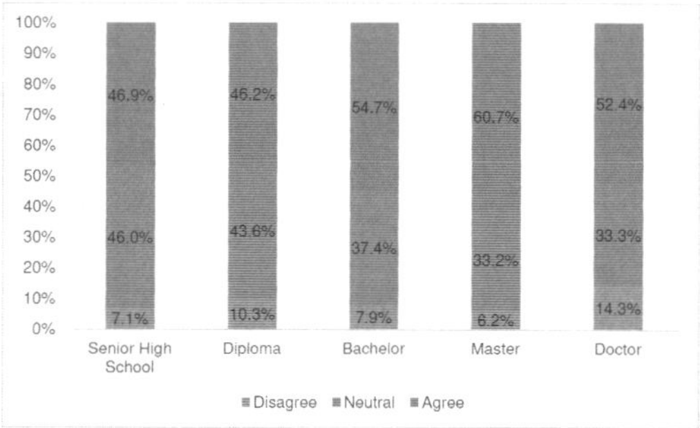

b. Preferences of respondents based on Education

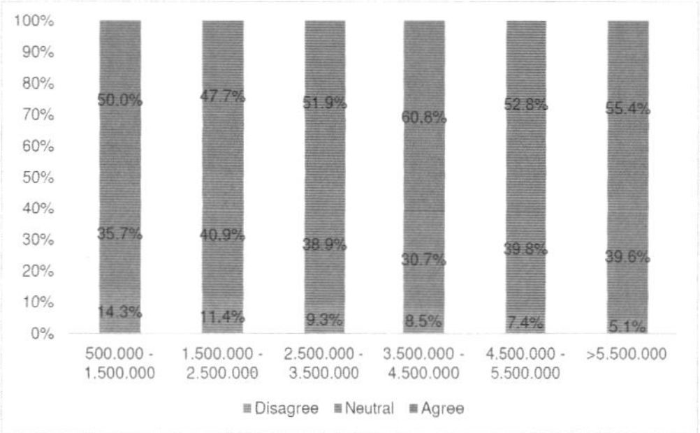

c. Preferences of respondents based on Income

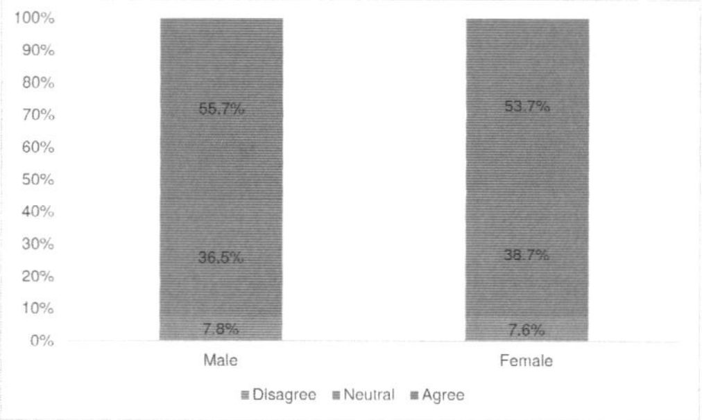

d. Preferences of respondents based on Gender

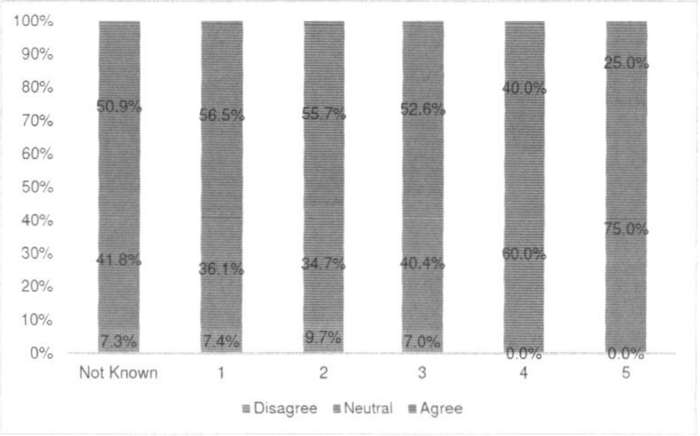

e. Preferences of respondents based on Car Ownership

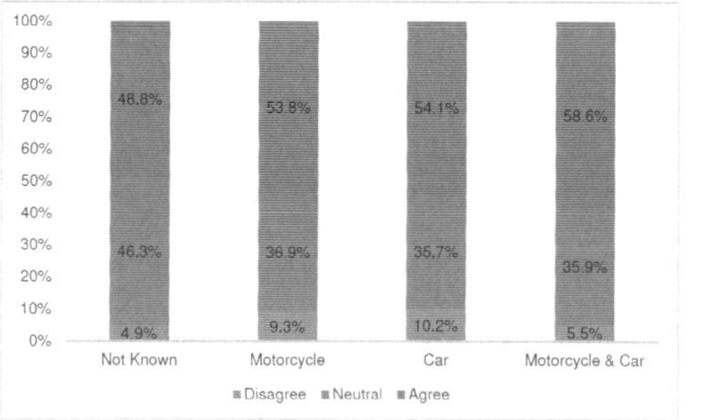

f. Preferences of respondents based on Vehicle Ownership

Figure 1. Characteristics of Individual Commuter

Based on Figure 1.a, it shows that the age range between 34-41 years has the largest percentage of 62.5% who agree about the upcoming LRT plan, while the smallest percentage is 58-65 with 33.3%. On the other hand, in the preference that states neutral, the largest percentage is at the age of 58-65 and the smallest is at the age of 34-41 with 61.9% and 29.3%, respectively, and in the disagree preference, all age ranges have a percentage below 10%.

Respondent's preference based on the last education as shown in Figure 1.b explains respondents in Master education has the largest percentage of approval preferences in the LRT plan with a percentage of 60.7% and the smallest percentage at the diploma level with a percentage of 46.2%, while the preference for stated that the largest percentage was neutral in the level of Senior High School with a percentage of 46% and the smallest was at the Masters with 33.2% and in the preference that stated Disagree with all levels of education below 15%.

Figure 1.c shows that income in the range of 3.5 million – 4.5 million has the largest percentage that agrees with 60.8% while the smallest is on income of 1.5 million – 2.5 million with a percentage of 47.7%. stated that the largest percentage was neutral at 1.5 million – 2.5 million with 40.9% and the smallest was in level income from 3.5 million – 4.5 million with 30.7%. Meanwhile, in the disagree preference, all income has a percentage below 15%.

On the gender characteristics as shown in Figure 1.d, there is no significant difference in the percentage of preference between men and women towards the LRT plan, either in terms of agree, neutral, or disagree.

Based on car ownership as shown in Figure 1.e, the largest percentage who chooses to agree with LRT for ownership of 2 cars is 55.7% and the smallest percentage is ownership of 5 cars with a percentage of 25%, in contrast to the preference which states neutral, the largest percentage is ownership of 5 cars and the smallest in ownership of 2 cars are 75% and 34.7%, respectively. In the preference that states Disagree, all car ownership variables are stated with a percentage below 10%.

Respondents' preferences based on the type of vehicle they have in Figure 1.f, respondents who choose agree with LRT with the largest percentage on the type of motorcycle and car with a percentage of 58.6% and the smallest percentage of respondents not owning a vehicle as much as 48.8% and vice versa on the preference that states neutral, the largest percentage of respondents who do not have a vehicle with 46.3% and the smallest percentage of respondents who have a car with 35.7%, while the preference that states disagree is less than 10% for all variables of the type of vehicle owned.

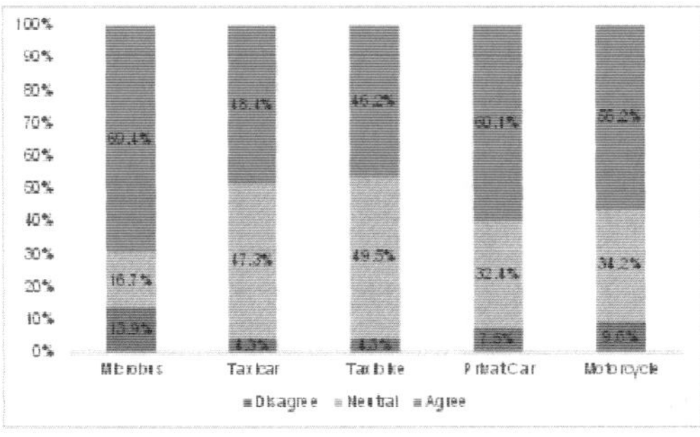

a. Preferences of respondents based on mode transportasi

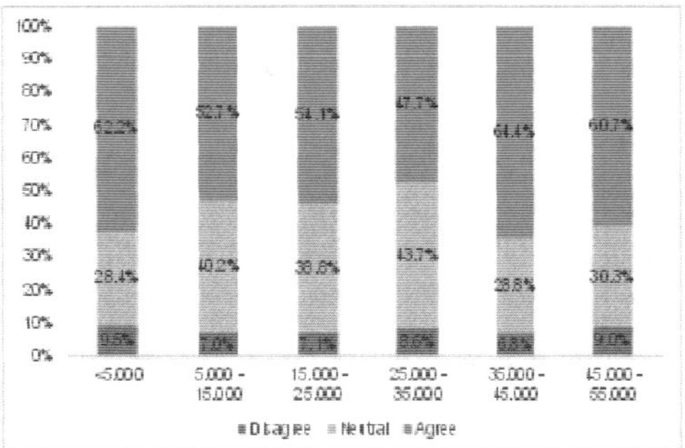

b. Preferences of respondents based on Traavel cost

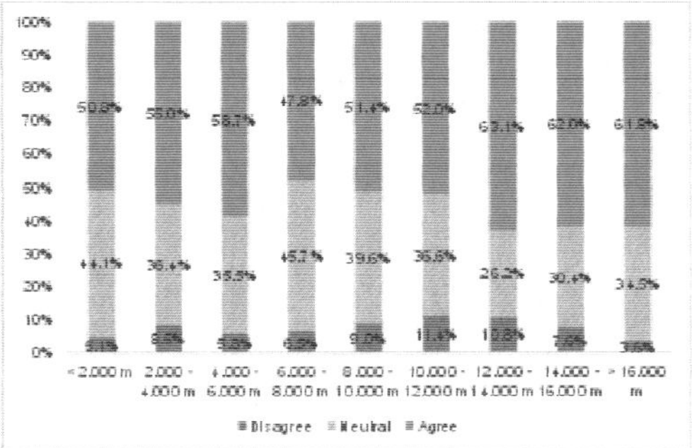

c. Preferences of respondents based on travel time

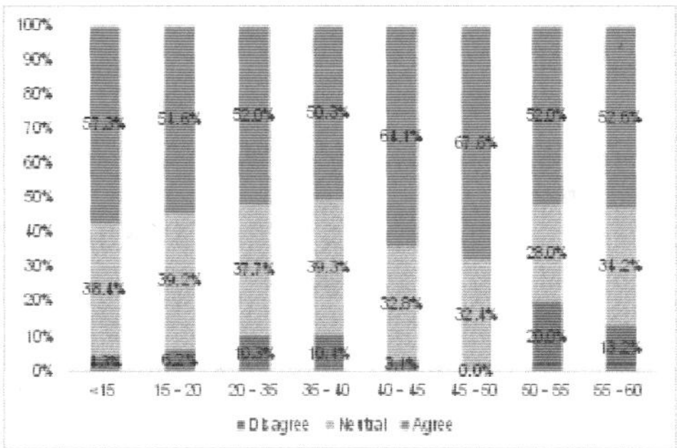

d. Preferences of respondents based on Travel Distance

Figure 2. Travel Characteristics of Commuter

Figure 2.a shows the preferences of respondents based on the mode of transportation, it can be seen that the most respondents who agreed were in the microbus mode (*pete-pete*) with a percentage of 69.4% and the smallest was in the taxi bike mode with 46.2%, while the preference was neutral on the plan. LRT has the largest percentage on taxi bike mode with 49.5% and the smallest on microbus with a percentage of 16.7%. In the preference that states disagree, all modes of transportation are stated with a percentage below 15%.

In the characteristics of the trip, the cost of the trip is shown in Figure 2.b. The cost of the trip is Rp. 35,000 – Rp. 45,000 has the largest percentage with 64.4% and the smallest on travel costs of Rp. 25,000 – Rp. 35,000 with a percentage of 47.7% for preferences who agree, and on the contrary, the largest percentage of respondents who say neutral is seen as a travel cost of Rp. 25,000 – Rp. 35,000 with 43.7% and the smallest on travel costs < Rp. 5,000 with a percentage of 28.4%. In the preference that states disagree, all travel cost ranges, respondents choose below 10%.

Based on the respondent's travel time as shown in Figure 2.c, the largest percentage who chose to agree with the LRT on travel time with a percentage of 45-50 minutes was 67.6% and the smallest percentage was 35-40 minutes with a percentage of 50.3%, and vice versa on the preference stated that the largest percentage was neutral at 35-40 minutes and the smallest at 45-50 minutes with 39.3% and 32.4%, respectively. In the preference that states disagree, the average travel time variable states with a percentage below 15% except for travel time of 50-55% which has the largest percentage with 20%.

Respondents' preferences based on travel distance are as shown in Figure 2.d respondents who chose to agree with LRT with the largest percentage in the range of travel distance of 12,000 – 14,000 meters with a percentage of 63.1% and the smallest percentage on travel distance of 6,000-8,000 m as much as 47.8% and on the other hand, in the preference which states that the largest percentage is neutral, the percentage of respondents who travel distance is 6.000-8.000 m with 46.7% and the smallest percentage is on the respondent's travel distance of 12.000-14.000 as much as 26.2%, while the preference which states disagree is less than 12% for all types of variables.

Table 4. Significance of Respondents

Variable	LRT Preferences	
	Coef.	P(Value)
Gender (Male)	0.0172	0.671
Age (Years Old)	-0.0010	0.614
Education	-0.0259	0.530
Income	0.0253	0.076**
Car Ownership	0.0668	0.023*
Motorcycle Ownership	-0.0017	0.944
Travel Cost	0.0026	0.867
Travel Time	-0.0270	0.040*
Distance (Meters)	0.0124	0.189
Cons	2.3447	0.000
Number of obs		1002
MSE root		0.629
Pseudo R2		0.0146

Notes:
* Significance level is 95%
** Significance level is 90%

Based on the table above, it describes how the relationship between individual characteristics and trips to community preferences in this case commuter urban travel is related to the existence of a rail-based transportation (LRT) development plan in the Mamminasata area. In the significance-based analysis, there are 3 variables that have a significant effect on respondents' preferences, namely Income, Car Ownership and travel time variables.

The Income variable has a significant effect of 90% as seen from the p value of 0.076, a positive value on the coefficient indicates the greater the income, the greater the preference to agree to the LRT plan, as well as the car ownership variable with a significant influence of 95% with a p value of 0.023, the higher number of car ownership, the greater also preference for agreeing to the rail-based transportation plan, which is indicated by a positive value in the coefficient variable. In the travel time variable, the results obtained are inversely proportional to other influential variables which state that the greater the travel time, the smaller the preference to agree or to the LRT plan with a negative coefficient.

3.2 Discussion
Based on the results of the relationship of the characteristics above, it is more in-depth to explain that productive age 26-49 years prefer to agree with the LRT plan, on the characteristics of education, it can be seen that education with a minimum undergraduate background states the largest percentage to agree with LRT compared to undergraduate education and below. more people choose to be neutral, while the number of vehicle ownership is directly proportional to the preference for LRT , the more the number of cars, the greater the preference that chooses to be neutral compared to agree.

On the income characteristics of the respondents, it can be seen that preferences are more dynamic from the level of each income, in gender the preferences are relatively the same between men and women, in the ownership of car vehicles, it shows that the greater number of cars owned, the fewer respondents who agree and vice versa more who choose to buy a car, as well as the type of vehicle ownership respondents who have 2 types of vehicles, namely motorcycles and cars which show a greater level of agreement with LRT.

On the travel characteristics of private vehicle users, there are more preferences that agree with LRT than respondents who use online transportation, also on the high cost of travel which makes the majority agree with rail-based transportation, as well as the average travel time of respondents are choose to agree from each time span by travel and on the long distance travelled also have an impact on the level of consent by the respondents.

4. Conclusion
This study describes the reference to the LRT Rail-based transportation plan and its effect on the characteristics of commuter trips for Civil Servants in the South Sulawesi. As seen in the results of the discussion above, there is an effect on the characteristics of income, car ownership, and travel time. In accordance with the results of the significance test, it was concluded that commuting trips with all income ranges the majority chose to agree also on car ownership which had a significant effect even though it was inversely proportional to the preference for LRT approval. Meanwhile, the travel cost is directly proportional to the preference which states that the greater the travel time, more people choose to agree with the plan.

Reference
[1] Ramli MI and Hustim M 2003 *Study on influence passenger dwelling process of para transit at area of foot intersection to performance of signal intersection in Makassar city* Proc. of Annual National Symp. 6th of Transportation Studies Forum inter-University (Indonesia: Makassar South Sulawesi)
[2] JICA 2007 *Study on Mamminasata Metropolitan Area (MMA) Development Plan* Final Report of JICA & South Sulawesi Province Collaboration Project
[3] Tamin, OZ 2000. Transportation Planning and Modeling. Second Edition. Publisher Institute of Technology Bandung, Bandung.

[4] Hardyanti Putri Harsono, N., Syafri, S., & Jufriadi, J. (2020). The Effect of Makassar New Port Development on Socio-Economic Viewed From Public Perception. Journal of Urban Planning Studies, 1(1), 042-049.

[5] Makassar City Central Bureau of Statistics. 2021. Makassar City in Figures 2021: Makassar City Statistics Center. Makassar.

[6] National Development Planning Agency (Bappenas). 2014. National Medium Term Development Plan (RPJMN) 2015-2019. Jakarta.

[7] Rahadian J and Sardjito 2015 *Criteria for public transport services based on people's preferences in West Jakarta (case study: Jalan Daan Mogot corridor)* ITS Journal of Engineering Vol. 4, No. 1.

[8] Oktariansyah, Damayanti R, Usman B and Putra AE 2015 *Analysis of the Quality of Public Transport Services (Transmusi) Through Performance Against Community Satisfaction in Palembang City.* Sriwijaya Journal of Management and Business Vol.15 (1).

[9] Gustina SB 2016 *Public Perception of Public Transportation in Jababodetabek* Proceedings of the IPLBI Scientific Meeting Bandung Institute of Technology

[10] Pratiwi NMF, Indrayani L and Suwena KR 2020 *Public Perception of Trans Sarbagita Public Transportation in Bali Province* Journal of Economic Education Volume 8 Number 1.2020 .

Risk management in the regency road improvement projects in Buleleng regency

A A D P Dewi[1], D K Sudarsana[1], I G B W Mulyaputra[1]

Engineering Faculty of Udayana University, Badung, Bali, Indonesia

anakagungdewi@unud.ac.id

Abstract. The largest district in Bali Province is Buleleng Regency with an area of approximately 24.25 percent of the total area of Bali Province. To facilitate the activities of the Buleleng community in all fields, adequate road construction is needed. The Public Works and Spatial Planning Office of Buleleng Regency carries out a road improvement project every year to improve road conditions from an unstable condition to a stable condition. The implementation of road improvement projects can cause various risks so it is necessary to conduct research for risk management with the stages of identifying risks, formulating dominant risks, formulating risk mitigation actions, determining risk ownership and analyzing the impact of dominant risks on potential cost and time losses in district road improvement projects in Buleleng Regency. The research method was carried out by compiling and distributing questionnaires, brainstorming, and focus group discussions. Respondents were selected using a purposive sampling method, namely the respondent determination method with the consideration that the respondent had the competence and experience in the district road improvement project in Buleleng Regency. The number of risks identified in this study were 40 risks with 31 (77.50%) risk being the dominant risk. The results of the formulation of mitigation actions against the dominant risk obtained 47 mitigation actions. Ownership of the most dominant risk (28 dominant risks) is the responsibility of the contractor. The risk with the dominant category is for example poor material quality that affects the quality of work with mitigation measures, namely ensuring that the material sent by the supplier matches the sample material that meets the required specifications with the risk ownership being the contractor. The impact of the dominant risk on the potential cost increase is IDR. 707.569.114,60 and the potential for additional time is 11 days.

1. Introduction

The position of Buleleng Regency in the northern part of Bali Island. Buleleng Regency is a regency that has the largest area in Bali Province with an area of about 24.25 percent of the total area of Bali Province or about 136,588 hectares [1]. To facilitate the activities of the Buleleng community in all fields, adequate road construction is needed. The Department of Public Works and Spatial Planning of Buleleng Regency (DPUTR Buleleng) carried out a road improvement project in order to improve road conditions from unstable to stable conditions. Road improvement is an activity to handle road sections so as to be able to improve the condition of roads that are not stable or in a damaged condition to a steady condition (good and moderate) in accordance with the age of the plan [2].

DPUTR Buleleng Regency carries out regular district road improvements every year so that the implementation should be carried out at the right cost, on time, and produces quality in accordance with the required technical specifications. However, in the implementation according to the data available in the Buleleng DPUTR almost every year there is a delay in implementation and according to the supervisory consultant report that there are various risks at each stage of work in the implementation of the district road improvement project in Buleleng Regency. Based on the description above, this study aims to carry out risk management with the stages of identifying risks, formulating dominant risks, formulating risk mitigation actions, determining risk ownership and analyzing the impact of dominant risks on potential cost and time losses in district road improvement projects in Buleleng Regency. so that it can be used as a reference for the implementation of road improvement projects in the following year so that it can run on time, with right quality and in accordance with the existing budget.

2. Literature Review

2.1 Risk management

Everyone should be concerned with risk management, because every construction project has risks due to uncertainty with potentially damaging consequences or consequences, the success or failure of the project depends on the approach to risk [3. The construction industry is more susceptible to risk and uncertainty than many other industries [4]. All construction projects are associated with risks from the beginning to the end which result in project delays and increased costs [5] Risk management is a systematic process for tackling, avoiding, and minimizing the occurrence of risk [6]. According to [3], although we cannot eliminate all uncertainties in a project, with risk management we can increase our chances of completing the project on time, cost, and produce quality according to the required technical specifications.

2.2 Risk Identification

This stage is the initial stage of the risk management process and is very important because poor identification of a risk will lead to further risks. Risk identification is done by identifying the source and type of risk [4]. Risk identification can be done based on the Work Breakdown Structure (WBS) so that for each type of work the risks that occur can be known [7] WBS is a hierarchical breakdown of the total scope of work to be carried out by the project party so that the project objectives are achieved. Each descending level of the WBS represents an increasingly detailed definition of project work [8]. Risk identification based on WBS can be done by using Brainstorming technique. According to [3], one of the risk identification techniques is to do Brainstorming. This technique is effective if it is carried out for a maximum of 2 (two) hours and is carried out in groups with a maximum of 5 (five) people.

2.3 Risk Assessment and Risk Acceptance

According to [3] risk assessment is a stage of risk management that aims to analyze the effects of identified risks so that they can be grouped into dominant/major risks and minor risks as the level of risk influence. Likelihood is the high and low chance of occurrence of losses that have the potential for project failure, the level and scale of possibility can be seen in Table 1. Consequences is a value that explains the size of the effect on the occurrence of an event that becomes a risk. The level and scale of consequences can be seen in Table 2.

Table 1. Likelihood Level and Scale

Scale	Level	Description
1	Very rarely	The percentage of possible occurrence or frequency is not significant, which is up to 20%
2	Rarely	The percentage of possible occurrence or frequency is not significant, whisch is 20 to 40%
3	Moderate	The percentage of possible occurrence or frequency is sufficient, which is more than 40% to 60%.
4	Often	The percentage of possible occurrence or high frequency, which is more than 60% to 80%
5	Very often	The percentage of possible occurrence or frequency is very high, which is more than 80%.

Table 2. Consequences Level and Scale

Scale	Level	Description		
		Cost	Time	Quality
1	Very Low	Cost increase is not significant	The increase in time is not significant	Quality decrease is not significant
2	Low	Cost increase < 10%	Time increment < 5%	Decrease in quality can still be fixed
3	Moderate	Cost increase 10%-20%	5-10% increase in time	Quality reduction needs approval
4	High	Cost increase 20%-40%	10-20% increase in time	Quality reduction is unacceptable
5	Very High	Cost increase >40%	Increase in time 20%>	Final work project cannot used effectively

The result of multiplying the consequences with the likelihood is a way to get the level of risk acceptance. After obtaining the level of risk acceptance, the risk acceptance scale can be determined as in Table 3 (Risk Acceptance Scale Table). The level of acceptance of unacceptable and unexpected risks is the dominant or major risk. Minor risk is an acceptable and negligible level of risk acceptance [9].

Table 3. *Risk Acceptance Scale*

Indicator	Scale
Unacceptable	$X > 12$
Undesirable	$5 \leq X \leq 12$
Acceptable	$2 < X < 5$
Negligible	$X \leq 2$

2.4 Risk Mitigation

This stage is the stage of the risk management process carried out to avoid, reduce, transfer, and contain the identified risks. Risk mitigation in risk management is grouped into four actions, namely risk transfer, risk reduction, risk avoidance, and risk retention. Risk transfer is risk mitigation carried out to transfer risk to third parties such as insurance. Risk reduction is risk mitigation that is carried out to reduce risk even though it cannot be completely reduced. Risk avoidance is risk mitigation that is carried out to avoid risk because the impact of the risk is unacceptable or very large. Risk retention is risk mitigation that is carried out on risks that have a small impact so that they are still acceptable. Each dominant or major risk requires mitigation action [10].

2.5 Risk Ownership

According to [4], risk ownership is a stage of the risk management process by allocating risk responsibilities to parties who are able to manage risk or carry out risk mitigation actions, using the principles of risk allocation as following:

1. The actor who has the best control over the event that creates the risk
2. Actors who can mitigate the risk if the risk arises/occurs
3. Actors who have responsibility if the risk is not controlled
4. If the risk is not controlled by all actors, then it is assessed as a joint risk

2.6 Quantitative Analysis

Quantitative risk analysis is a process carried out to analyze the impact of risk by means of risk variables that have been identified given a rate in the form of numbers [11.] Quantitative analysis was conducted to analyze the impact of dominant risk on potential cost and time losses in the implementation of the district road improvement project in Buleleng Regency. The dominant risk analyzed is risk originating from technical risk and project risk. This analysis was carried out through a Focus Group Discussion (FGD) involving parties who have competence and experience in the implementation of district road improvement projects so that the amount of potential additional costs and time can be calculated.

2.7 District Road Improvement

According to [12], district roads are the status of roads whose determination as district roads is carried out by the Regent through the Regent's decree. District Road Improvement is an activity to handle the condition of district roads to improve the condition of roads that are not stable or in a damaged condition to a stable condition (good and moderate) in accordance with the age of the plan [1]. Road improvement works in the form of structural improvement or strengthening work, increased traffic capacity in the form of widening traffic lanes and upgrading from dirt roads to gravel roads and to asphalt roads or from gravel roads to asphalt roads.

3 Research Method

3.1 Research design

The research design uses a qualitative descriptive method with the aim of describing the dominant risk, mitigating the dominant risk, the party responsible for carrying out the mitigation action, as well as the potential loss of cost and time accurately and systematically.

3.2 Research Location and Time

The research location is the existing district road improvement projects in the DPUTR Kab. Buleleng. The time of the research was carried out during the implementation of district road improvement projects in Buleleng Regency in 2021.

3.3 Determination of Data Source

This study uses primary data sources, namely respondents who have competence and experience in district road improvement projects in Buleleng Regency. The stages that require primary data are risk identification, risk assessment, dominant risk mitigation, dominant risk ownership, assessment of potential cost and time losses. The secondary data source of this research is literature in the form of research journals on risk management in construction projects. The stage that requires secondary data is the risk identification stage to determine the risks that have been identified in the construction project and their sources of risk.

3.4 Research Instruments

In this study using a questionnaire as a research instrument. Questionnaires are used at the risk assessment stage to obtain a frequency/probability scale and a consequence scale from the respondents. The quality of the research instrument is measured by the validity and reliability of the instrument. The questionnaire is declared valid if it is able to measure the desired research variables (probability scale and consequence scale). The questionnaire is declared reliable if the questionnaire produces relatively the same data at the time of re-measurement on the same object. The measurement scale in the questionnaire uses a Likert scale. This scale is used to measure the respondent's opinion about the likelihood and consequence scale of the identified risks by assigning a scale of 1 (one) to 5 (five). Respondents were selected using the Purposive Sampling Method. Purposive Sampling method is a technique for determining respondents with the consideration that respondents have competence and experience in road improvement projects so that they can guarantee the quality of the data obtained. The validity test in this study uses the Construct Validity Test. This test is carried out by giving questions to the respondents, then a correlation test is carried out with the value/score of each question with the total score/score of the questionnaire. The technique for conducting the Construct Validity Test is the Product Moment Correlation Technique. The correlation value of the product moment obtained must be compared with the critical number. The correlation table value of r, with degrees of freedom (df) = N (number of samples) - 2, with a significance level (alpha value) of 5%. If the calculation results obtained that the value of r count is greater than r table then the measurement is valid, but if r count is smaller than r table then the measurement is invalid. To perform the Product Moment Correlation Technique using the IBM SPSS V.26 computer program. The reliability test in this study used the IBM SPSS V.26 program on the reliability function. The minimum reliability that must be met by the questionnaire based on the established criteria is 0.7 [13].

3.5 Data collection
Data collection at the risk identification stage uses literature study and brainstorming methods. The literature used is research journals on risks in construction projects. Risk identification is continued by conducting brainstorming techniques based on the work breakdown structure. Data collection at the risk assessment stage uses a questionnaire. This questionnaire contains the identification of risks as well as the scale of possibility and the scale of consequences of the risk. Collecting data at the stage of risk mitigation, risk ownership, and calculating the impact of risk on potential cost and time losses using the FGD method involving experienced and competent parties in road improvement projects.

3.6 Data Analysis
Data analysis was carried out after collecting data from all respondents or other data sources. Data analysis activities consist of performing calculations, tabulating data, grouping data, presenting data so that they can answer the problem formulation [14]. The risks identified from the literature study and brainstorming based on the Work Breakdown Structure are grouped according to the source of the risk. After obtaining the probability scale and consequence scale data, calculations are carried out to obtain the risk acceptance scale by multiplying the likelihood scale and the consequence scale, the value of this multiplication is used to determine the risk acceptance indicator using Table 3 (Risk acceptance scale table). Mitigation actions obtained from the FGD are then grouped into actions to contain risk, reduce risk, transfer risk, or avoid risk. Data analysis at the risk ownership stage was carried out to determine the party responsible for tackling the dominant risks that might occur in the implementation of the district road improvement project in Buleleng Regency. The determination of the responsible party was carried out by FGD with parties who have competence and experience in the district road improvement project in Buleleng Regency with reference to the principles of risk allocation. Analysis to determine the impact of dominant risk on the potential for additional costs and additional time is carried out through FGDs. The results of the FGD are the estimation of additional processing time and additional volume of materials due to risk. Potential cost losses are calculated by multiplying the estimated increase in the volume of materials multiplied by the unit price of the material and the addition of processing time multiplied by the unit price of labor wages. The potential loss of time is calculated by the estimated additional processing time which is converted into days.

4 Result and Discussion

4.1 Identification of Risks in the District Road Improvement Project in Buleleng Regency
There are 40 (forty) risks identified in the district road improvement project in Buleleng Regency, consisting of 20 (twenty) risks from previous studies and 20 (twenty) risks from brainstorming results. The risk from the source of project risk has the largest percentage of 25%. The source of risk in project risk is the most because the research is carried out at the time of project implementation so that the risks regarding resources, quality control, and implementation standards are something that plays a very important role in achieving the expected project results. The percentage of total risk based on risk sources can be seen in Table 4.

Table 4. Percentage of Total Risk Based on Risk Source

No	Risk Source	Risk Amount	Persentage
1	Environment	6	15,00
2	Planning	2	5,00
3	Economy	1	2,50
4	Nature	4	10,00
5	Project	10	25,00
6	Technical	5	12,50
7	Man	8	20,00
8	Criminal	1	2,50
9	Safety	3	7,50
	Total	**40**	**100,00**

4.2 Dominant Risks in the District Road Improvement Project in Buleleng Regency

Risk assessment with the result equal to or more than 5 (five) is the dominant risk. Dominant risk control must be carried out because the presence of dominant risk has the potential to hinder project implementation. Based on the results of this study, there are 31 (thirty one) risks out of the 40 (forty) identified risks that are dominant risks, this indicates that the district road improvement project in Buleleng Regency is a project that has a high risk. Based on Table 5. (Risk Acceptance Distribution Table for Each Risk Source) the risk with an unacceptable risk acceptance level (unacceptable) is 10 (ten) risks with a percentage of 25.00%. Risk with an undesirable level of risk acceptance is 21 (twenty one) risks with a percentage of 52.50%. Risks with an acceptable level of risk are 7 (seven) risks with a percentage of 17.50%. Risks with a negligible level of risk acceptance are 2 (two) risks with a percentage of 5.00%. Unacceptable risk can be seen in Table 6 (Table of Risk with Unacceptable Risk), the undesirable risk can be seen in Table 7 (Table of Risk with Undesirable Risk))

Table 5. Distribution of Risk Acceptance for Each Risk Source

Risk Source	**Risk**	**Level of Risk Acceptance**			
	Identification	***Unacceptable***	***Undesirable***	***Acceptable***	***Negligible***
	(%)	**(%)**	**(%)**	**(%)**	**(%)**
Politic	0	0	0	0.00	0
Environment	15	2.50	7.50	5.00	0
Environment	5	2,50	2.50	0	0
Planning	0	0	0	0	0
Economy	2.5	0	2.50	0	0
Nature	0	0	0.00	0	0
Project	10	7.50	2.50	0	0
Technical	25	5	15.00	5	0
Man	12.50	2.50	10.00	0	0
Criminal	20	5	12.50	2.5	0
Safety	2.50	0	0.00	0	2.5
Safety	7.50	0	0.00	5	2.5
Persentage	**100.00**	**25.00**	**52.50**	**17.50**	**5**

Table 6. *Unacceptable Risk*

Risk Value	Risk Source	Risk Identification	Risk Acceptance
20	Project	Poor material quality that affects the quality of work	*Unacceptable*
16	Environment	There are traffic disturbances that hinder work due to no other alternative road for road users	*Unacceptable*
16	Planning	There are utilities that hinder work due to the difficulty of coordination related to relocation	*Unacceptable*
16	Nature	Delayed work due to rain	*Unacceptable*
16	Nature	Work is hampered due to difficult road access to the project site	*Unacceptable*
16	Nature	The quality of work is difficult to achieve due to the steep and winding job location	*Unacceptable*
16	Project	The delay in the progress of work due to the slowness of testing the quality of work/materials	*Unacceptable*
16	Technical	Work is hampered due to low equipment productivity due to inadequate equipment maintenance or aging equipment	*Unacceptable*
16	Man	The occurrence of work delays due to low labor productivity	*Unacceptable*
16	Man	Delayed work approval due to slow administration by service providers	*Unacceptable*

Table 7 *Undesirable Risk*

Risk Value	Risk Source	Risk Identification	Risk Acceptance
12	Environment	The delay in work is due to the community not allowing drainage of water on their land	*Undesirable*
12	Economy	Increased work costs due to rising material prices	*Undesirable*
12	Project	Incompatibility of the tools used with the requirements that affect the quality of work	*Undesirable*
12	Man	Results of work that do not meet technical specifications due to lack of skilled labor	*Undesirable*
9	Project	Delay in equipment mobilization due to equipment still being used in other locations	*Undesirable*
9	Technical	Difficulty in placing work materials due to narrow roads	*Undesirable*
9	Man	Worker fatigue due to a lot of asphalt laying work carried out at night so that it affects the quality of work	*Undesirable*

8	Environment	Delays in work due to land boundary conflicts with the community	*Undesirable*
8	Man	Delays in work due to land boundary conflicts with the community	*Undesirable*
8	Man	Test samples that do not conform to technical specifications due to lack of supervisor's knowledge of specification	*Undesirable*
6	Environment	The delay in work is due to the presence of community buildings on the shoulder of the road	*Undesirable*
6	Planning	There is a change in design due to suggestions from agencies or the community	*Undesirable*
6	Nature	Decreased work quality due to rain	*Undesirable*
6	Project	The quality of work is not measurable due to the lack of equipment to measure the quality of work	*Undesirable*
6	Project	Lack of lighting on work done at night so it affects the quality of work	*Undesirable*
6	Project	Decrease in quality due to non-conformance of repair methods on quality defects	*Undesirable*
6	Project	The difference in the calculation of the volume of work due to the calculation method that is not clear in the specifications	*Undesirable*
6	Technical	The discrepancy in the volume of work on BQ and field conditions due to the volume of work that cannot be calculated with certainty when MC 0	*Undesirable*
6	Technical	There is a change in design due to changes in field conditions from the conditions at the time of planning	*Undesirable*
6	Technical	Inefficient use of materials due to bumpy road surfaces	*Undesirable*
6	Man	Lack of quality of work due to inadequate supervision	*Undesirable*

4.3 Risk Mitigation

Risk mitigation against the dominant risk aims to reduce the impact of the dominant risk. Mitigation actions are formulated by conducting FGDs involving parties who have competence and experience in road improvement projects. The process of formulating risk mitigation actions begins with determining the categories of the main causes of dominant risks and the main causes of dominant risks using the Ishikawa Diagram, then formulating risk mitigation actions through FGDs. Based on the results of the study, 18 mitigation actions were obtained for unacceptable risks and 29 mitigation actions for unexpected risks. Risk mitigation measures for unacceptable risks consist of fourteen risk reduction actions, three risk avoidance actions, and one risk transfer action. Risk mitigation measures for undesirable risks consist of seventeen risk reduction measures and twelve risk avoidance measures.

4.4 *Risk Ownership*
The risk ownership stage is a stage of the risk management process by allocating risk responsibilities to parties who are able to manage dominant risks in the implementation of district road improvement projects in Buleleng Regency, these parties have the best control over events that cause risks or can mitigate risk if the risk it happened. These parties are the project owner (Service Users from DPUTR Buleleng Regency), implementing contractors, planning consultants, and supervisory consultants. Risk ownership for risks with unacceptable categories, namely:

- The contractor is responsible for 10 (ten) risks
- The Service User is responsible for 3 (three) risks
- The Supervisory Consultant is responsible for 2 (two) risks

Risk ownership for risks with the undesirable category, namely:

- The contractor is responsible for 18 (eighteen) risks
- The Supervisory Consultant is responsible for 11 (eleven) risks
- The Service User is responsible for 7 (seven) risks

4.5 *Quantitative Analysis of the Impact of Dominant Risk*
The dominant risk analyzed is risk originating from technical risk and project risk. This analysis was carried out by conducting FGDs involving parties who have competence and experience in implementing district road improvement projects. The dominant risk originating from the project and technical has an impact on the implementation of the district road improvement project in Buleleng Regency in the form of a potential cost increase of Rp. 707,569,114.60 and a potential extension of 11 days. If the contractor does not anticipate the potential extension of time, the project implementation time will exceed the contract implementation period so that it is potentially subject to a fine of 1/1000 of the contract value after deducting the value added tax (VAT).

5 Conclusions and Suggestions
5.1 *Conclusion*
Based on the research and analysis that has been done, it can be concluded:

1. In the implementation of the district road improvement project in Buleleng Regency, 40 (forty) risks were identified. The identified risks include 6 (six) environmental risks, 2 (two) planning risks, 1 (one) economic risk, 4 (four) natural risks, 10 (ten) project risks, 5 (five) technical risks, 8 (eight)) human risk, 1 (one) criminal risk, 3 (three) safety risk. The results of the analysis of risk assessment and acceptance of the identified risks show that there are 10 (ten) risks that are unacceptable, 21 (twenty one) risks are undesirable, and 21 (twenty one) risks are acceptable.) obtained 7 (seven) risks, and risks that can be ignored (negligible) obtained 2 (two) risks.
2. Risks including major risk are 31 (thirty one) risks consisting of 10 (ten) unacceptable risks and 21 (twenty one) undesirable risks. The dominant risks in the district road improvement project in Buleleng Regency include poor material quality that affects the quality of work, traffic disturbances that hinder work due to no other alternative road for road users, utilities that hinder work due to difficulty in coordination related to relocation, delays in work due to rain, as well as work being hampered due to difficult road access to the project site.
3. Risk mitigation actions for risks with unacceptable risk categories are carried out 18 (eighteen) mitigation actions, one of which is mitigation actions for the risk of poor material quality so that it affects the quality of work, namely the service user instructs the supervisory consultant personnel to make a presentation regarding technical specifications of materials that meet the requirements and how to control materials to meet technical specifications at the time of work before starting the work, the supervision consultant ensures that the materials used in the project meet the specifications by comparing the materials in the field with samples of materials that meet the specifications, if different, the material must be rejected or cannot be used, and the contractor ensures that the material sent by the supplier is in accordance with the sample material that meets the required specifications. Risk mitigation actions for risks with undesirable

risk categories are carried out 29 (twenty nine) mitigation actions, one of which is mitigation actions for the risk of work delays due to the community not allowing drainage water disposal on their land, namely service users and contractors conducting socialization and negotiations to land owners by involving village/kelurahan officials and making minutes of land use approval.
4. The risk ownership of the parties involved in the district road improvement project in Buleleng Regency are Service Users (Buleleng Regency PUTR Office), planning consultants, contractors, and supervisory consultants. For risks in the unacceptable category, the contractor has responsibility for 10 (ten) risks, service users have responsibility for 3 (three) risks, and the supervisory consultant has responsibility for 2 (two) risks. For risks with the undesirable category, contractors have responsibility for 18 (eighteen) risks, supervisory consultants have responsibility for 11 (eleven) risks, and service users have responsibility for 7 (seven) risks. The ownership of the dominant risk is the responsibility of the contractor because the risk identification is mostly a risk that occurs at the implementation stage.
5. The impact of the dominant risk on the potential cost increase and time extension of the district road improvement project in Buleleng Regency has a major impact on the technical and project aspects. The total potential cost increase is Rp. 707,569,114.60 and the total potential time extension due to risks originating from the project and technical for a minimum of 11 days, if the contractor does not anticipate this potential extension of time, the project implementation time will exceed the contract implementation period so that it is potentially subject to a fine of 1/1000 of the value contract after deducting the value of value added tax (VAT).

5.2 Suggestion

Suggestions that can be submitted based on the research and analysis that have been carried out are:

1. All parties involved in the implementation of the district road improvement project in Buleleng Regency, especially service users, contractors, and supervision consultants must work together to take action to ensure that the materials used in the project meet the specified technical specifications so that the quality of work in accordance with the requirements is achieved.
2. Before starting the implementation of the work, the service user must hold a coordination meeting with all parties or agencies that own the utility in order to obtain certainty regarding the existing utilities and a clear plan regarding the relocation of the utility.
3. The contractor as the owner of the most risk in the district road improvement project in Buleleng Regency before making a bid must conduct a field survey so that he bids according to the work method, especially for work locations with access roads that are steep and winding and high rainfall.
4. The results of this study are expected to be a guide for the parties involved in the implementation of the district road improvement project in Buleleng Regency and can be a reference for further research on risk management of road construction projects.

References

[1] BPS Buleleng Regency 2018 *Kabupaten Buleleng Dalam Angka 2018* Badan Pusat Statistik Kabupaten Buleleng.
[2] Ministry of Public Work 2010 *Petunjuk Teknis Penggunaan Dana Alokasi Khusus Bidang Infrastruktur* Jakarta.
[3] Godfrey P S and Halcrow S W1996 *Control of Risk. A Guide to the Systematic Management of Risk from Construction. Westminster London* : CIRIA, .
[4] Flanagan R and Norman G 1993 *Risk Management and Construction* In Wiley-Blackwell.
[5] Luu VT, Kim S Y, Tuan N Van and Ogunlana S O 2009 Quantifying Schedule Risk in Construction Projects Using Bayesian Belief Networks. *Int. J. of Project Management* **27 39-50**

[6] Jaya I N M, Sudarsana D K and Wiratni G A K I 2019. Manajemen Risiko Terhadap Pelaksanaan Proyek Konstruksi Hotel Di Kawasan Sarbagita. *Jurnal Spektran*, 7(1): 51–57.

[7] Purba E A, Kurnia M A, Dwi J U and Kistiani F 2015 Analisa Manajemen Risiko Pada Proyek Pdam Semarang (Studi Kasus: Proyek Pemipaan Ipa Kaligarang Semarang Barat). *Jurnal Karya Teknik Sipil*, 4: 274–282.

[8] Project Management Institute 2013 *A guide to the project management body of knowledge (PMBOK® guide)* 5th Ed. Newtown Square Pennsylvania 19073-3299 USA: Project Management Institute Inc.

[9] Norken I N, Purbawijaya I B N and Suputra I G N O 2015 *Pengantar Analisis dan Manajemen Risiko pada Proyek Konstruksi* Udayana University Press.

[10] Norken I N, Astana I N Y and Manuasri L K A 2012 Manajemen Risiko Pada Proyek Konstruksi Di Pemerintah Kabupaten Jembrana *Jurnal Ilmiah Teknik Sipil,* 16(2): 202-211.

[11] Imansari A 2019 Analisis Risiko Berdasarkan Aspek Waktu Dengan Metode Monte Carlo Pada Proyek Gedung Baru Di Universitas Brawijaya.*Naskah Terpublikasi Teknik Sipil.*

[12] Pemerintah Republik Indonesia 2006 *Jalan* Jakarta.

[13] Ghozali I 2021 *Aplikasi Analisis Multivariate dengan Program IBM SPSS 26* 10th Ed. Semarang: Badan Penerbit Universitas Diponegoro.

[14] Sugiyono 2015 Metode Penelitian Pendidikan (Pendekatan Kuantitatif, Kualitatif, Dan R&D) In *Alfabeta* (21st Ed.) Bandung: Alfabeta.

IOP Conf. Series: Earth and Environmental Science 1117 (2022) 012038 doi:10.1088/1755-1315/1117/1/012038

Demand Estimation for Non-Working Trips: Forecasting Passengers regarding the New Open of Railway Line Maros-Barru

Hajriyanti Yatmar[1], Panguriseng[1], Eka Buqra[1], and Ricky Rinaldi[1]

[1] Civil Engineering Department, Hasanuddin University, Makassar, Indonesia

hajriyantiyatmar@unhas.ac.id

Abstract. This study aims to identify the demand estimation of passengers regarding the operational railway of Maros-Barru line. The new railway is expected to attract passengers using public transport from origin to destination. The existing condition of Maros-Barru majority used private vehicle as the choice and minibus with fragmented line. The new construction railway provided by the government services the passengers through Maros-Barru. The demand for the railway plays an important role in describing the potential passengers of each region. The non-working trips with flexible time for the trip are the main focus of this study. The potential passengers using the origin-destination (O/D) matrix are used. The analysis uses the growth factor of the population and forecasts demand according to OD data. The high trip based on OD estimation result showed that Pangkep was at the first position followed by Maros and ended with Barru. This result indicates that Pangkep has a high demand according to the forecasting passenger which is located in the middle regency instead of Maros and Barru that close to the Capital area. The high trip occurred for the total not only for the Origin trip but also Destination trip. Hence, the population growth of the district on Pangkep is the most compared to other districts in other regions.

1. Introduction

In the urban planning concept, transportation plays an important role to serve people for travel. The growth of the area simultaneously affects the activity of citizens. Public transport as the main mode to serve people to travel is supposed to be integrated with other modes. Makassar-Parepare line is the first line to service the railway in South Sulawesi. The line through five regencies from Makassar as the Capital Metropolitan, Maros, Pangkep, Barru, and ends with Parepare. The line is planned to be open, especially for the Maros-Barru section. The infrastructure of the railway was constructed in 2018 and the line for Maros-Barru is planned open in December as the pilot line [1].

The demand for passengers becomes the main issue in investigating the forecasting of passengers [2]. Maros-Barru line section passing region of Maros-Pangkep-Barru. The region has different populations and characteristics. The forecasting due to these constraints is investigated to estimate the number of future passengers. Methods of forecasting are utilized by railroad companies to anticipate potential challenges and outcomes that may affect the company in the coming weeks, months, and years. The central goal of the planning and management of railway transportation is to bring the supply of transportation closer to the demand for railways. The key to successfully identifying and analyzing issues that are currently associated with railway traffic is to have a comprehensive understanding of the existing pattern of railway customers. For the purpose of developing demand forecasting and prediction models, detailed data on both the existing pattern and the volumes of railway traffic are required. Within the context of the process of long-range railway transportation planning, the forecasting of future demand is an essential task for the purpose of determining strategies for meeting anticipated requirements.

This study focused on non-working trips with the flexible time of people to perform trips. The number of non-working trips has significantly increased over the past few years, and it has even surpassed the number of trips that are related to work each day. Several studies show that high accessibility leads people to more travel. The non-work trips are more accessible than workers to be investigated especially in the Maros, Pangkep, and Barru regions. The forecasting demand of trips with identifying the potential passenger to respond to the new line opening of Maros-Barru becomes the baseline of the railway operator to provide services.

2. Literature Study

The demand for railroad transportation can be segmented into passenger and freight demand. Intercity travel is one component of passenger demand, while commuting is another. The demand for intercity travel stems from either commercial or recreational pursuits. The same is true for commuting, however, all of the components of commuting demand have similar characteristics, which is why there is no difference between business and leisure in terms of commuting demand. The need for freight services is virtually always a component of the manufacturing process. It is heavily impacted by factors such as pricing policy, seasonality, the type of goods being sold, as well as geographical and socio-economic parameters [4].

Both quantitative data and qualitative observations can be used in the forecasting methods that are used for modeling the demand for railway services. The fact that the experiences of qualified executives and employees are used as the primary source of data is the primary benefit of using qualitative research methods [5]. Quantitative methods of forecasting, on the other hand, base their projections on the reliability of previously collected data. There are only two types of forecasting techniques, quantitative and qualitative methods. This study uses qualitative methods based on the origin and destination trips of non-work passengers [6].

2.1. OD Matrix

The places of Origin and Destination (O-D) Matrix is used to describe the movement of people in a particular area. The development of a reliable network of public transportation requires the utilization of an O-D matrix. OD matrix has the principle of origin and destination trips. To obtain the OD matrix, it is necessary to interview people with the historical trip. The total trips were then plotted into an OD matrix and estimated using several methods [7]. One of the famous estimation methods uses Gravity Model with constraints [8]. The constraints that were used to estimate the OD matrix used some parameters such as socio-demography, population, and accessibility that are affected by distance. The simple formulation with gravity model concept.

$$T_{ij} = \frac{\alpha PiPj}{dij^2} \quad (1)$$

where Pi and Pj are the sample population of origin and destination, and dij^2 is the distance between origin and destination. The gravity model is then used with the following equation.

$$T_{ij} = O_i D_j A_i B_j f C_{ij} \quad (2)$$

Here, Tij is the trip from zone origin (i) to zone destination (j), Oi is the total people in zone i, Dj is the number of destinations in zone j, while the f (cij) is the deterrence function, and Ai and Bj is the balancing factors or convergent value.

2.2. Forecasting

In the context of railway demand forecasting, one way to categorize the various forecasting methods is according to thc amount of time in the future that they cover. The five to the ten-year horizon is

typically used in long-term forecasting. Short-term forecasts involve predictive intervals within six to eighteen months, whereas medium-term forecasts cover a time period ranging from two to five years into the future [9]. The time horizon influences the method of forecasting that is selected due to the availability and relevance of historical data, the amount of time that is available to create the forecast, the cost that is involved, the severity of errors, and the amount of effort that is considered worthwhile.
The use of a forecasting model and historical data in a structured manner is at the heart of quantitative forecasting techniques. The model provides an official summary of the patterns found in the data and expresses a statistical relationship between the values of the variable in the past and those in the present. Then, the model is used to make predictions about the future based on the patterns that are present in the data. The econometric model and the general time series model are the two primary types of forecasting models that are utilized today [10]. The relationship between the variable of interest and one or more other variables that are related to it can be used as a predictor in econometric models.

3. Methodology

The data collection was performed by interviewing the citizens, especially in the three regions as the sample. The line of Maros-Barru with the stations is described in the following Figure 1. The construction railway line planned started from station Makassar to Parepare. In addition, the Maros-Barru as the part of the Main Line (Makassar-Parepare) is about to open first (blue), continue with the green line, and last red line sequencely.

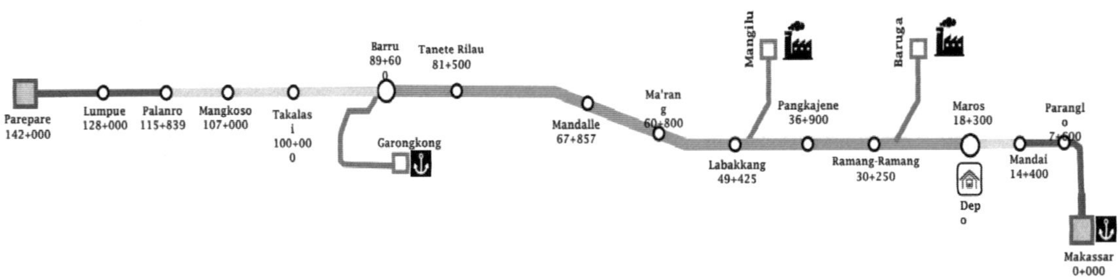

Figure 1. Makassar-Parepare Railway Line

The Makassar-Parepare line passed five region, Makassar, Maros, Pangkep, Barru, and Parepare. The line Maros-Barru is soonly opened and became the first operation line. This study has interviewed citizens in Maros, Pangkep, and Barru with a questionnaire. The questionnaires recorded the socio-demography characteristics including gender, occupation, age, income, zone address, and region. This used to be a part of OD matrices, especially determining the OD matrix and forecasting the demand of passengers with the growth factor population of each region.

3.1 Data Collection

This study interviewed citizens in Makassar-Parepare line with recording their non-work trip history and socio-demography condition. The total sample (N= 473) was then clustered according to the variable and category (Table 1). The variable gender showed the majority of respondents were female, the occupation with housewife, the age with 15 – 22 years old, income with non-fixed income (none), zone address in the urban area, and most respondents in Makassar.
The total sample is then distributed according to their demography condition or area. The total distribution is in Figure 2. The distribution is tabulated according to the address zone of respondents. The district area is then used to determine the zone of the station that already exists. The determined zone from district to zone of the station used the approaching of the close area from the station of each respondent.
The statistical data for the population growth factor is available for each district and used to forecast the demand of passengers using the railway based on the station. The important thing from the

clustered zone is to estimate the demand using OD matrices. The data that supported the OD matrix is the trip recorded from origin to destination of each district that connects to the station.

Table 1. Individual Characteristics of Non-Work

No	Variable	Category	Unit (%)
1	Gender	Female	60,5%
		Male	39,5%
2	Occupation	Civil Servants	13,7%
		Army	0,6%
		State-Owned Enterprise	1,5%
		Entrepreneur	23,0%
		Students	18,0%
		Outsourcing	1,3%
		Retired	1,9%
		Housewife	27,3%
		Others	12,7%
3	Age	15 - 22	19,9%
		23 - 30	15,0%
		31 - 38	18,6%
		39 - 46	18,8%
		47 - 54	16,5%
		55 - 62	8,9%
		63 - 70	2,3%
4	Income (IDR)	None	33,6%
		< Rp1.000.000	17,3%
		Rp1.000.000 - Rp3.000.000	26,6%
		Rp3.000.001 - Rp5.000.000	18,6%
		Rp5.000.001 - Rp10.000.000	3,6%
		> Rp10.000.000	0,2%
5	Zone Adress	Urban	57,9%
		Rural	42,1%
		Industrial	0,0%
6	Regency	Makassar	22,4%
		Maros	16,7%
		Pangkep	19,0%
		Barru	19,9%
		Pare Pare	22,0%

Figure 2 showed the distribution of respondents, especially in the Maros-Barru line through the station location. Most of the respondents according to the district, dominated in the Barru with Barru (7.98%) and Tanete Rilau (6.84%). The total respondent of each district then clustered into the station. The passengers that come from each district are determined at the close station. The clustered is used to determine the OD matrices based on the trip to the railway station.

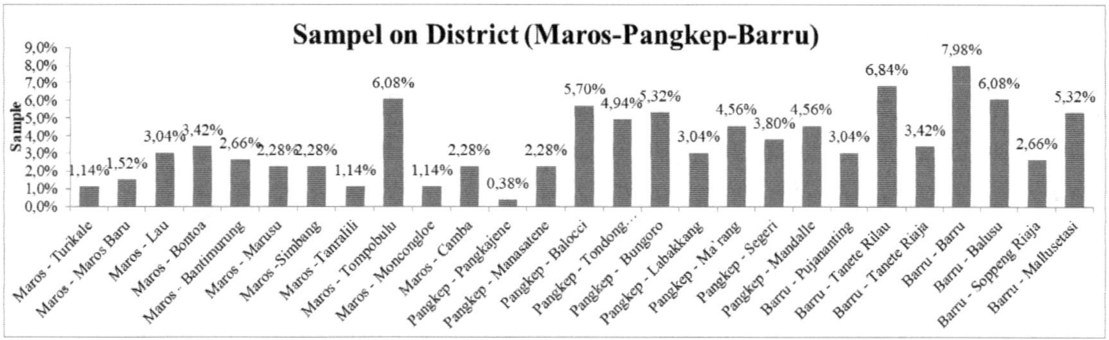

Figure 2. Sample Distribution based on Districts in Maros, Pangkep, and Barru

3.2 Data Analysis

The OD matrix is then pre-determined using the station as the origin and destination. The total respondents were then investigated and plotted into the trip OD matrices and estimated. The result of the OD matrix then uses the gravity model with a distance of each station as the constraints and deterrence function. The iteration process is used to obtain the value of the convergent balancing OD matrix.

The OD matrix of the non-work trip is obtained (Table 2). The OD estimation matrix is then used to forecast the demand for passengers using the growth factor of the population. The growth factor that used based on stations from Maros-Barru; Maros (0,22), Rammang-rammang (0,20), Pangkajene (0,91), Labbakkang (0,89), Ma'rang (0,77), Mandalle (0,07), Tanete Rilau (0,67), and Barru (0,85).

Table 2. OD Matrix based on Station in Maros-Barru line

O/D Matrix (Station)	Origin	Barru	Tanete Rilau	Mandalle	Ma'rang	Labakang	Pangkajene	Ramang"	Maros	Oi
Destination		**1**	**2**	**3**	**4**	**5**	**6**	**7**	**8**	
Barru	**1**	0	12	11	11	11	10	10	10	**75**
Tanete Rilau	**2**	24	0	23	22	21	21	21	20	**153**
Mandalle	**3**	9	10	0	10	9	9	9	9	**64**
Ma'rang	**4**	18	19	21	0	20	18	18	18	**131**
Labakang	**5**	16	16	17	18	0	18	17	16	**119**
Pangkajene	**6**	29	30	31	32	33	0	35	32	**222**
Ramang	**7**	18	19	19	20	20	22	0	21	**140**
Maros	**8**	22	23	23	23	24	25	26	0	**166**
Dj		**138**	**128**	**145**	**135**	**139**	**124**	**136**	**126**	**1070**

The growth factor was obtained from the statistical center of local government for the specific districts of each region along Maros, Pangkep, and Barru. The growth factor was then used to estimate the forecasting passengers using the OD matrix.

4. Result and discussions

This session showed the result of forecasting passengers of each station. The total number of passengers as a result of forecasting is shown in Table 3.

Table 3. Passengers Forecasting of Railway Maros-Barru

Station	Years					Total
	2023	**2024**	**2025**	**2026**	**2027**	
Barru	337	560	966	1704	3050	6617
Tanete Rilau	436	704	1167	1966	3354	7628
Mandalle	279	405	629	1033	1762	4108
Ma'rang	337	463	688	1093	1823	4404
Labakang	336	471	708	1129	1883	4527
Pangkajene	564	951	1640	2869	5062	11086
Ramang"	349	482	719	1147	1921	4618
Maros	489	853	1531	2795	5153	10820
Total	**3127**	**4889**	**8048**	**13736**	**24007**	**53808**

The forecasting of passengers of the railway is investigated. The high forecasting value for as long as five years came from Pangkajene followed by Maros with over 10.000 passengers. The trendline of forecasting is shown in Figure 3. The figure shows the total forecasting of passengers for 5 years in the Maros-Barru line. The trendline of each station is quite similar to the R2 values. For example, the station that is close to the center activity that represents each region has a quite similar value of R square with linear time series. Maros with R2 = 0,8908, Pangkep R2 = 0,9033, and Barru R2 = 0,8976.

Pangkep as the middle region between Maros and Barru has the high forecasts than Maros and Barru which is the buffer city of the capital (Makassar and Parepare). These may be indicated the trip occurred not only from the Maros-Barru but also inter-zone at Pangkep and also the population growth factor of Pangkep that indirectly affect the forecasting. The OD estimation of passengers showed the trip and calibrated with the distance as the deterrence function. From the total stations, Pangkep has four stations while Maros and Barru only have two stations.

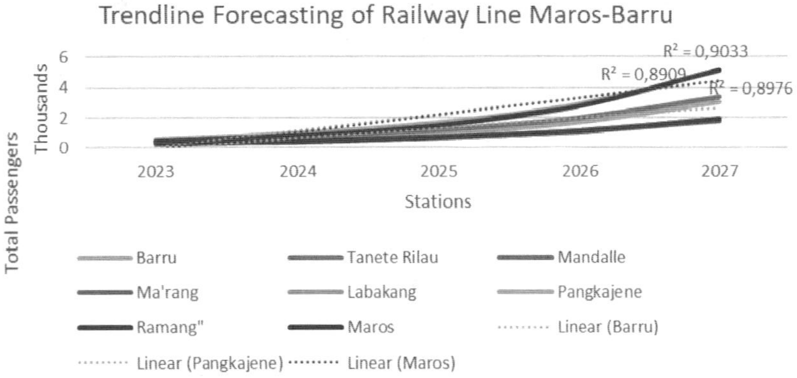

Figure 3. The Trendline Forecasting of Maros, Barru, and Pangkajene Stations

Pangkep with geographical conditions covers most of the central activities close to the stations. These situations have increased the accessibility of people who want to perform trips along the region. The spatial condition of the Pangkep region allows people to take short trips, especially non-work trips.

5. Conclusions

This study described the forecasting passengers due to the railway demand of the new open line Maros-Barru. The most passengers forecasting came from Pangkep followed by Maros then Barru. Pangkep also has four stations instead of Maros and Barru. The geographical conditions of the Pangkep area support the activity of citizens to travel. This is expected to attract people to use the railway, especially the line Maros-Barru, especially for non-work trips with a flexible schedule. The operator of the railway line has to be concerned not only for the Maros and Barru near the port city (Makassar and Parepare) but also for the potential of high trips of Interzone in Pangkep.

References

[1] Hajriyanti Yatmar et. all., 2021 *The Impact of Socio-Demographic and Activity-Travel Participation Variables on Mode Choice Preference of Sub-urban Commuters: A Case Study on the New Railway Operation in Maros-Makassar Line* Proceedings of the International Conference on Civil, Offshore and Environmental Engineering ICCOEE 2021: ICCOEE2020 pp 945–955 Springer Link

[2] Wenxiang Li et. all., 2018 *Exploring the Relationships between the Non-Work Trip Frequency and Accessibility Based on Mobile Phone* Transportation Research Record Sage pub

[3] Milos Milenkovic and Nebojsa Bojovic 2016 *Railway Demand Forecasting* Handbook of Research on Emerging Innovations in Rail Transportation EngineeringChapter: 5Publisher: IGI GlobalEditors: Umesh Rai

[4] Valtteri Helve 2015 *Demand Forecasting in a Railway Revenue Management System* Aalto University School of Business

[5] Jonathan Preston 1991 *Demand Forecasting for New Local Rail Stations and Services* Journal of Transport Economics and Policy Vol. 25, No. 2 (May 1991), pp. 183-202

[6] I Ekowicaksono, F Bukhari, and A Aman *Estimating Origin-Destination matrix of Bogor City Using Gravity Model* IOP Conf. Series: Earth and Environmental Science 31 (2016) 012021

[7] Rosa Basagoiti et. all., 2012 *Origin Destination matrix estimation and prediction in vertical transportation* (Conference: 2nd Symposium on Lift and Escalator technologies)

[8] Zhanhong Cheng, Martin Trepanier, and Lijun sun 2022 *Real-time forecasting of metro origin-destination matrices with high-order weighted dynamic mode decomposition* Transportation Science of Cornell University
[9] M I Ramli, D Runtulalo, H Yatmar, and A Mangessi 2019 *An Estimation of Origin-Destination Matrices for a Public Transport Network in Makassar using Macrosimulation Visum* IOP Conf. Series: Materials Science and Engineering 875 (2020) 012027
[10] Wenxiu Cao, Shaolong Sun, and Hongtao Li 2021 *A new forecasting system for high-speed railway passenger demand based on residual component disposing of* Measurement Volume 183, October 2021, 109762 Elsevier

IOP Conf. Series: Earth and Environmental Science 1117 (2022) 012039 doi:10.1088/1755-1315/1117/1/012039

Behavior of Building Structure Subject to Tsunami Load

Gede Pringgana[1], Ida Ayu Made Budiwati[2], Putu Ayu Ratih Yustinaputri[3]

[1,2]Civil Engineering Study Program, Faculty of Engineering, Udayana University
[3]Formerly Graduate Student in Civil Engineering, Faculty of Engineering, Udayana University

gedepringgana@unud.ac.id

Abstract. This study aims to analyze of the structural behavior of two-story building under the combination of earthquake forces and tsunami hydrodynamic forces, with model variations included the percentage of wall's opening and thickness. The 3-D modeling of two-story buildings were conducted in this study using software SAP 2000; the first model consisted of wall panels with the percentage of openings of 0% (PD0), 25% (PD25), 35% (PD35), and 50% (PD50), while the second model included buildings with 150 mm and 250 mm wall thickness. In addition, the hydrodynamic force predicted by numerical simulations based on *Smoothed Particle Hydrodynamic* (SPH) was compared to that of predicted by semi-empirical design equations. The results of the wall panel modeling show that the greater the percentage of wall's opening, the smaller the hydrodynamic load that can be carried out by the structure and hence, the lesser the stresses on the wall. The increase of building's wall thickness from 150 mm to 250 mm under the combination of earthquake and tsunami loads decreased the stress on walls, moments and transverse forces. Conversely, the tensile and shear stresses on the walls were increased due to the combination of tsunami loads. The comparison of hydrodynamic force shows that the maximum force predicted by SPH was generally higher to that of predicted by the semi-empirical design equations.

1. Introduction

The earthquake disaster and the following tsunami in 2004 that hit Aceh Province and several regions in the surrounding countries, caused great material losses and fatalities [1]. Recently, the tsunami disaster in 2018 due to the eruption of Mount Anak Krakatau caused death toll to reach 437 people [2]. Since then, the government and researchers pay more attention to the concept of tsunami-resistant building to prevent loss of lives during disaster. The tsunami-resistant building allows the building to be used immediately, only with minor repairs, after the tsunami disaster. Research related to the interaction of tsunami waves with building structures has been widely published, both experimental and numerical research. In addition to conducting experimental and numerical research, there are also literatures that contain formulas for predicting tsunami loads, including tsunami hydrodynamic loads or forces. The formulations available in literature to date, assume the hydrodynamic force as a quasi-static force, whereas in reality, the hydrodynamic force is dynamic, i.e., the values change over time (*time-dependent*). In order to represent the behavior of actual tsunami waves, one of possible solutions is by performing numerical modeling of tsunami waves using *software* DualSPHysics [3]. Based on differences in assumptions between quasi-static and dynamic behavior of tsunami load characteristic, this study will compare the hydrodynamic forces resulting from the SPH modeling with the semi-empirical design formulas. The hydrodynamic force from the SPH modeling used in this study is the hydrodynamic force from the *software* by [4], while the semi-empirical design formula used as a comparison is the formula proposed by [5, 6, 7, 8, 9].

The tsunami-resistant structure has been widely studied [10, 11]. It was found that the openings of the wall were able to reduce the force or load caused by the tsunami waves. Therefore, in this study, structural modeling was performed, referred to as the wall structure, with variations in the percentage of wall openings to determine the effect of the percentage of wall openings on the structure under tsunami loading. Moreover, to survive tsunami that is generated by earthquake, the performance of structure to resist earthquake loading that occurs earlier is critical. Referring to research by [12], one of the key parameters for assessing the seismic performance of reinforced concrete structures with infilled walls was wall thickness. The results showed that an increase in wall thickness caused a decrease in wall stress and an increase in structural stiffness leads to less potential of failure or collapsed of the structure due to earthquake loads. Based on finding by [12], this study performed structural modeling of building with variations in wall thickness subjected to earthquake loads and tsunami hydrodynamic loads.

In this study, the tsunami hydrodynamic force applied to the numerical models of wall panel and two-story building structure was the hydrodynamic force taken from the SPH modeling output conducted by first author and published in [4] because it was found that the SPH modeling prediction was closed to the testing result. The numerical finite element models were then performed using *software* SAP2000. The wall panel modeling was performed to analyze stress on the walls, while the modeling of two-story building was conducted to analyze the structural behavior of building, including internal forces, displacements, and wall stresses. The type of tsunami in this study is a *far-field tsunami*, where the epicenter is far from the coast (e.g., in the middle of the ocean), so that earthquake shaking is assumed not to damage buildings significantly prior tsunami impact. With this assumption, the structural model in this study does not experience a reduction in strength before the tsunami load applied.

The hydrodynamic force is applied on structure when the water flows around structure. The hydrodynamic force is a function of the density of the moving fluid, the flow velocity, and the geometry of the structure [9]. The hydrodynamic force which is one of the forces that must be taken into account in the design of tsunami-resistant buildings can be calculated based on proposed equations as follow:

[5] $$F = \frac{1}{2}\rho g h_b^2 + \rho h_b v_j^2 \quad (1)$$

[6] $$F = 1{,}3\rho h_j v_j^2 \quad (2)$$

[7] $$F = \frac{1}{2}\rho g h_b^2 + \rho h_j v_j^2 + \rho g^{1/3}\left(h_j v_j\right)^{4/3} \quad (3)$$

[8] $$F_{dyn} = \frac{1}{2}C_d \rho V^2 A \quad (4)$$

[9] $$F_d = \frac{1}{2}\rho_s I_{tsu} C_d b(h_e u^2) \quad (5)$$

where:

ρ = density of water of 1000 kg/m^3, while 1025 kg /m^3 for seawater
ρ_s = density of water including sedimentation (1.1 x 1025 kg/m^3 = 1128 kg/m^3)
g = gravity (9,81 m/s^2)

h_b , h_e	=	height of the water
v_j , u , V	=	velocity of water
h_j	=	the height of the water flowing over the flat reef
I_{tsu}	=	the tsunami importance factor (can be seen in Table 2.2)
C_d	=	coefficient of drag (can be seen in Table 2.3)
b	=	width of the structural member perpendicular to the current
B , A	=	structural area in the normal plane to the wave direction

Due to difference between the dimensions of the SPH numerical model and the structural finite element model using SAP2000, it is necessary to adjust the hydrodynamic pressure of the resulting SPH numerical model before being applied to the SAP2000 for structural modeling. The scaling of dimensions and pressures of a model is based on Froude scaling [13] as presented in Equation (6) where the dimension for length, *L*, of the building is directly proportional to the dimension of pressure, *P*.

$$n_p = n_L \tag{6}$$

where:

n_p	=	pressure scaling factor
n_L	=	length scaling factor

The inter-story drift is a critical parameter on the performance of building structure and its value should not exceed the requirements. The drift between floors is need to be controlled to ensure the performance of buildings that is related not only to the safety of building and but also the occupants during excessive force includes earthquake and tsunami. More information about the limit of inter-story drift can be found in Indonesian building code namely SNI 1726:2019 article 7.8.6 or 12.1 for all levels.

The ability of wall structures to support loads can be controlled through stresses that occur on the wall. The resulted stress on wall includes compressive stress, tensile stress, and shear stress. The compressive stress that occurs should not exceed the strength of material used to build the wall to prevent excessive cracks that often lead to failure. At the present time, there is no standard or regulation in Indonesia regarding the tensile stress and allowable shear stress for walls, so this study used the [14] as a reference. The allowable tensile stress due to bending in the wall can be seen in Table 1.

Table 1. The allowable stresses of the wall

Type of stress	Stress values (MPa)	Remark
Tensile stress	< 0.414	Solid unit with mortar type of Portland cement
Compressive stress	< 3.000	Depend on the quality of wall
Shear stress	< 0.103	-

2. Methodology

In this study, the tsunami force determined from the numerical SPH tsunami wave pressure on wall using Equation (1) as presented in [4] was used as input load on two-story building in the finite element modeling utilizing SAP2000. The magnitude of tsunami force profile in [4] was then scaled up so that the initial height of tsunami wave in SPH model (0,6 m) equivalent to the height of two-story building (8,4 m) which was used as tsunami-impact object in the finite element model. The tsunami force was not a quasi-static as assumed by Equation (1) through Equation (5), but dynamic in nature as shown by tsunami force profile in Figure 1. Based on SPH modeling output in [4], the tsunami force values were varied along the width of impacted structure as measured by probes within the same height, however the differences were not so significant thus, for convenient, the mean value was used to represent the force acting at a certain height along the width of structure. The tsunami force value was also varied with time along the simulation time.

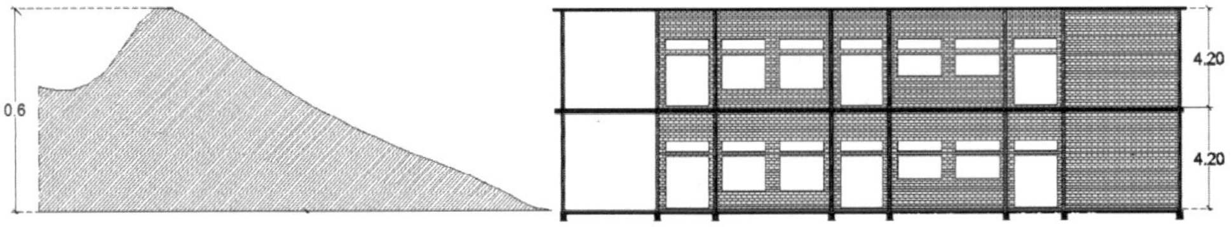

Figure 1. The tsunami force profile acting on two-story building structure in the SAP2000 model.

The percentage of opening and the thickness of wall were used as parameters in the finite element modeling in this study. Regarding wall's opening, four different percentage were evaluated as can be seen in Figure 2, included wall without opening which is denoted as PO-0, wall with 25%, 35%, and 50% opening which are denoted as PO-25, PO-35, PO-50, respectively. Moreover, in regard with the wall's thickness, two variations of thickness were used, i.e.,150 mm and 250 mm which denoted as Var1 and Var2, respectively. The numerical building model has the same structural element dimensions and materials. The building structure is assumed to be a two-story public-school building located in Banda Aceh, Indonesia. The concrete material used for the structural elements has a peak compressive strength ($f'c_f$) of 20 MPa, while the compressive strength of the brick wall ($'_m$) is 3 MPa. The slab thicknesses were designed to be 120 mm and 110 mm for the floor slab and roof slab, respectively.

Table 2. Dimensions of structural elements

Structural elements	Dimension (mm)	Structural elements	Dimension (mm)
Column at 1st and 2nd floor (K)	350 x 350	Joist at 2nd floor (BAL)	200 x 300
Column of staircase (K2)	300 x 300	Main beam at roof (BIA)	200 x 300
Main beam at 2nd floor (BIL)	300 x 400	Joist at roof (BAA)	150 x 200
Beam at 2nd floor (BL)	300 x 400	Sloof (SL)	200 x 300

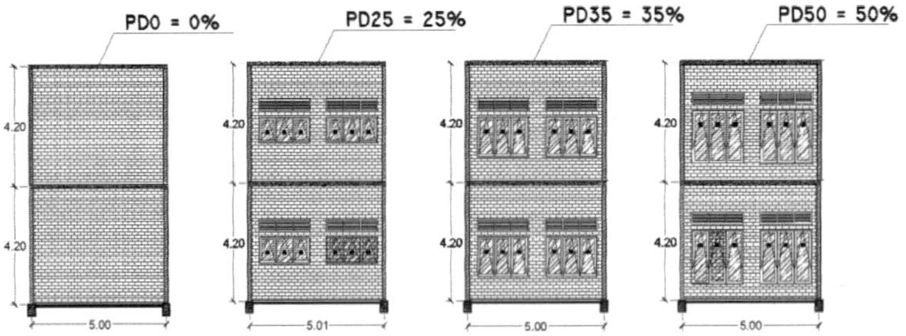

Figure 2. Wall panel model

The dead load of the structure was determined automatically by SAP2000, while the additional dead load was based on the Loading Planning Guidelines for Houses and Buildings (1987). Table 4-1 of SNI 1727:2013 used as reference in determining the live load of structures. The procedure and parameters for calculating earthquake loads were based on the SNI 1726:2019, with ground acceleration data of site class D (moderate soil) was obtained from the Puskim website (http://puskim.pu.go.id). In determining the earthquake loads, the structure was assumed to be a special moment-resisting reinforced concrete frame. The dimensions of the structural elements are presented in Table 2, where the dimension of the structural elements was ensured to be adequate (not overstressed) in resisting load combinations including earthquake load. The plane frame and the wall's opening variation of building model are shown in Figure 3.

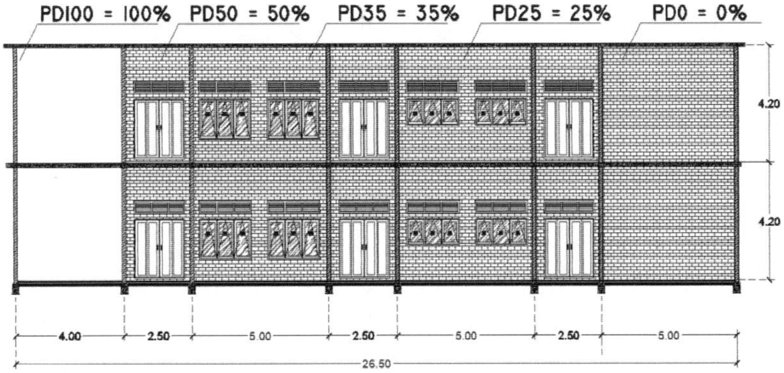

Figure 3. Building model

3. Results and Discussion

Figure 4 shows the hydrodynamic force predictions given by several proposed equations versus the SPH model prediction as reported in [4]. In Figure 4, the hydrodynamic force from the SPH numerical modeling results referred to as SPH. Furthermore, the hydrodynamic forces determined by the design equations referred to as Cross (Equation 1), Fujima (Equation 2), Robertson (Equation 3), FEMA CCM & CCH and (Equation 4), and ASCE 7-16 (Equation 5). In Figure 4, the force determining from SPH numerical modeling by Cunningham et al (2014) was measured during 20 second of time simulation with time increment of 0.01 that resulted in 200 time-steps. The magnitude of the SPH force prior to tsunami impact on structure (before t = 15.3 second) were constant which indicated the force (about 1000 N) as the results of water pressure of 1 m-depth still water measured by the probes.

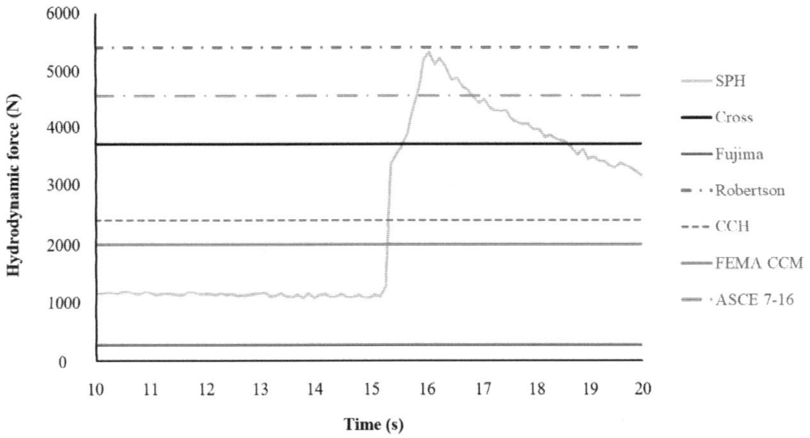

Figure 4. The prediction of the SPH hydrodynamic forces versus quasi static approach

The interaction of tsunami bore with structure is depicted by the graph of SPH force that is suddenly increase until reaching the peak and then followed by gradual decrease. This dynamic force profile could not be predicted by the quasi-static approaches of semi-empirical equations. The transient load of tsunami as showed by SPH prediction can help the structural engineer to design tsunami-resistant structures based on actual dynamic load instead of the static load condition. Different accuracy in predicting the tsunami hydrodynamic force as shown by semi-empirical design equations in Figure 4, depends on certain assumptions and specific experimental testing conditions the proposed design equations were based on.

The numerical analyses regarding the effect of the percentage of wall's opening on stress under tsunami load combinations were carried out via four 3-dimensional structural models, namely PD0, PD25, PD35, and PD50. Figure 5 shows the tensile, compressive and shear stress values decrease as the percentage of wall's openings increase. The maximum tensile and shear stresses which were occured at 0.9 seconds correspond to hydrodynamic force profile. In Figure 5(b), it can be seen no compressive stress occurred on wall within the time span of 0,9-2,0 seconds due to applied load.

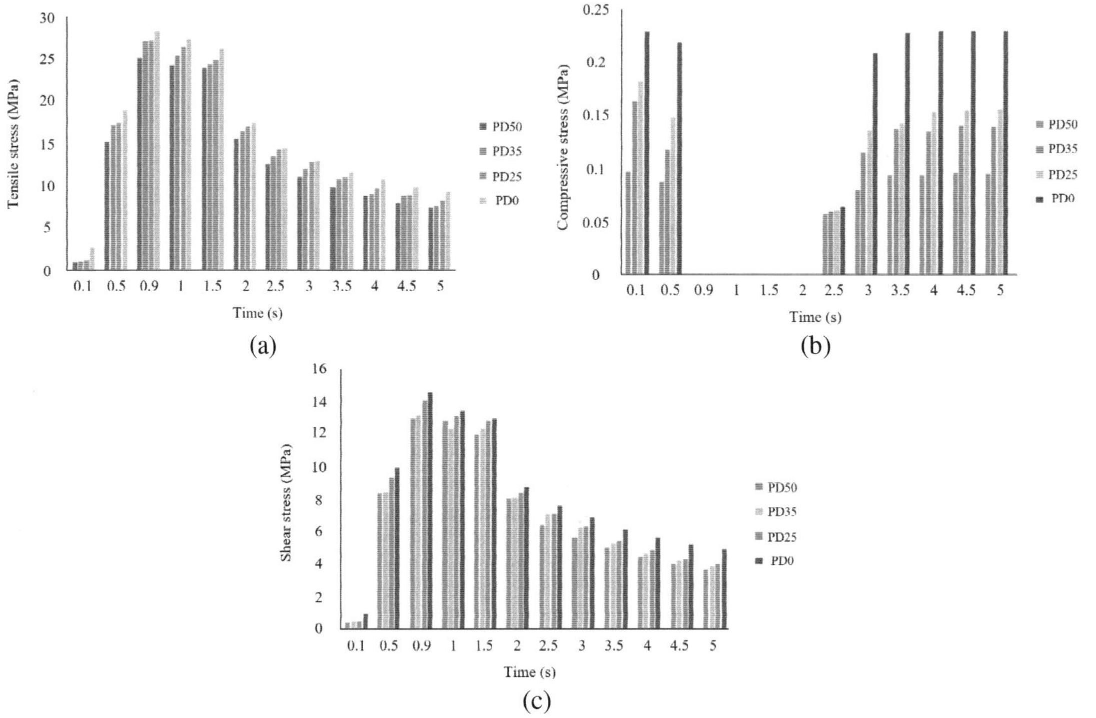

Figure 5. Stress on wall: (a) tensile stress, (b) compressive stress, (c) wall shear stress (in MPa)

The behavior of structural elements including columns, beams, and sloof can be seen from responses in the form of internal forces due to applied loads, where in this case no structural elements were in over-stressed condition. As a result of the increased of wall thickness from 150 mm to 250 mm, the structural elements experienced a decrease in the moment and transverse forces in the range of 9,2%-47,6% and 17,4%-94,4%, respectively. Only sloof whose moment and transverse force values thst were still remained. The axial force of columns at the 1st and 2nd floor increased by 1,4%-2,5%.

Due to applied loads, the structure also experienced lateral displacement as part of structure's response where the displacement was related to the stiffness of the structure. The maximum value of lateral displacement then used to calculate the drift ratio between floors, where the resulting drift ratio is 0.09% which occurs at the 2nd floor level in the X-axis direction. The drift ratio was smaller than the drift ratio limit (1%), so the structure is categorized safe from sudden collapse due to applied lateral loads. The increase of wall thickness to 250 mm caused lateral displacement decrease in the X-direction by 98,9% and 98,0% measured at the 2nd floor and roof levels, respectively. Similarly, lateral displacement decreased by 43,1% and 47,6% in the Y-direction measured at the 2nd floor and the roof, respectively.

In addition to internal forces and displacements, the tensile, compressive, and shear stresses on the walls were also analyzed. From the comparison of wall stresses, Var1 and Var2, it can be found that the maximum wall's compressive stress was less than the allowable compressive stress. In contrast, the tensile stress and shear stress exceed the allowable tensile stress and shear stress. The location of the maximum wall stress is shown in Figure 6. It can be seen from Figure 6 that the maximum stress occurred around the opening, with higher stress occurred at the corner of openings. This finding is in accordance with findings published in [15] which stated that the maximum stress of the infilled wall with opening occurred at the corner of the opening. For walls without openings, the maximum stress occured at the corner of the wall, which is the connection area between the wall and the column. The results of this study are in accordance with the research by [16] which stated that a wall without openings has a maximum stress at the corner of the stressed wall. The tensile, compressive, and shear stresses were decreased following the increased of the wall thickness. The tensile, compressive, and shear stressed decreased by 78,4%; 65,4%; and 51,3%, respectively.

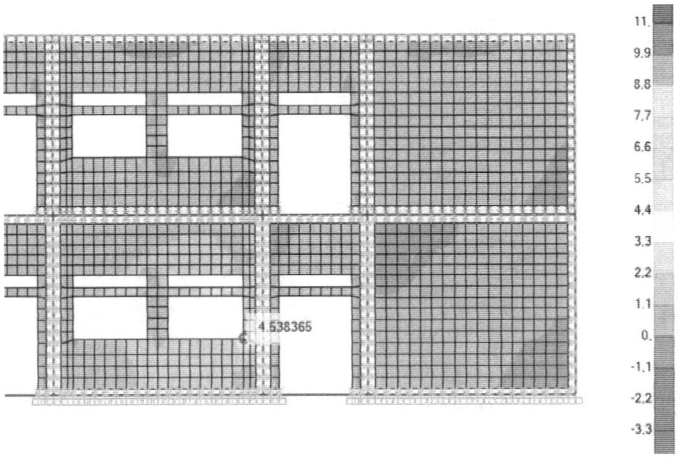

Figure 6. The tensile stress of Var1 model due to the combination of earthquake loads (in MPa)

The hydrodynamic force applied to the structural model in this study was within the time range of 15-20 seconds to obtain the most significant influence given by the tsunami wave profile. The behavior of building's structural members changed following the increased of wall's thickness. As a result of wall thickening, the 1st floor columns, 2nd floor beams, and roof beams experienced a decrease in moment ranging from 0,1%-4,5%, and only 2nd floor columns experienced increase in moment up to 4,9%. Considering the resulting shear forces, the columns at the 1st and 2nd floor experienced decrease in shear forces up to 3,5%, while the 2nd floor beams and roof beams experienced an increase in the shear forces up to 11,4%. The addition of wall thickness caused the strength of the column increased to carry out applied load in the direction perpendicular to the column, resulting in a decrease in the shear force of the column. This was in contrast with the case of beams that supported heavier loads due to the increase of wall thickness, which contributes to the increase in beam cross-section. The normal force of the 1st and 2nd floor columns has increased between 0,6%-11,0%, because the addition of wall thickness has an influence in increasing the load carried by the column. The greater the load carried by the column, the greater the normal force acting on the column. In addition to the internal forces, the base shear forces of the structure were also compared due to the combination of earthquake loads and the maximum combination of tsunami loads as presented in Table 3. The base shear forces due to the combined tsunami loads were greater than the earthquake loads.

4. Conclusion and Future Work

4.1. Conclusion

1. The hydrodynamic force prediction given by SPH numerical model was overestimated the predictions of semi-empirical design equations, except the one by [7]. The quasi-static approach of hydrodynamic force doesn't reflect the dynamic behavior of impact event where force values vary over time.
2. The greater the percentage of wall's opening, the smaller the hydrodynamic force on wall and hence, the smaller the stress on the wall. Due to the combination of earthquake and tsunami loads, for the

walls with and without openings, the maximum stress occurred at the corner area of the openings and at the corner of the wall, respectively. The results of this study reaffirm that the existence of opening is effective in reducing tsunami loads on wall.

3. The applied load and the increase of wall thickness (from 150 mm to 250 mm) influenced the member's inertial forces, building lateral displacement, and stress on wall.
   Due to earthquake load combinations and the increase of wall thickness, the member's inertial forces including maximum moment, shear and axial of structural members decreased by 9,2% to 47,6%; 17,4% to 94,4% and 1,4% to 2,5%, respectively. The lateral displacements of structure in x-direction decreased by 98% measured at both the 2nd floor and roof levels, while lateral displacements in y-direction decreased by 43.1% and 47.6% measured at the 2nd floor and roof level, respectively. The tensile, compressive and shear stresses on walls decreased up to 78,4%; 65,4%; and 51,3%, respectively.
   Due to tsunami load combinations and the increase of wall thickness, the members' moment decreased by 0,1% to 4,5%; shear of columns decreased by 0,9% to 3,5%; in contrast the shear of columns increased by 9,2% to 11,4%; axial of members increased by 0,6% to 11%. The lateral displacements of structure decreased by 8,5% to 52,9% and 8,1% to 21% measured at the 2nd floor and roof levels, respectively. The tensile, compressive and shear stresses on walls decreased up to 56,9%; 34,6%; and 64,5%, respectively.

4.2. Future Work
Further studies are needed considering not only the hydrodynamic aspect but also all types of tsunami loads. In addition, it is also necessary to take into account the effect of scouring to the building foundation and the near-field tsunami that might significantly reduce the stiffness of building prior to the tsunami impact. Regarding the excessive stress concentration occured on walls with opening as found in this study, the application of of lintel bands around the opening is worth to be evaluated.

5. References
[1] Levin, B., Nosov, M. 2009. Physics of Tsunamis. In Journal of Chemical Information and Modeling (Vol. 53).
[2] Heidarzadeh, M., Ishibe, T., Sandanbata, O., Muhari, A., Wijanarto, A.B. 2020. Numerical Modeling of the Subaerial Landslide Source of the 22 December 2018 Anak Krakatoa Volcanic Tsunami, Indonesia. Ocean Engineering, 195(June 2019): 106733.
[3] Pringgana, G. 2015. "Improving Resilience of Coastal Structures Subject to Tsunami-like Waves." PhD Thesis, The University of Manchester.
[4] Cunningham, L.S., Rogers, B.D., Pringgana, G. 2014. Tsunamiwave and Structure Interaction: An Investigation with Smoothed-Particle Hydrodynamics. Proceedings of the Institution of Civil Engineers: Engineering and Computational Mechanics, 167(3): 106–116.
[5] Cross, R.H. 1967. Tsunami Surge Forces. Journal of the Waterways and Harbors Division, 201–231.
[6] Fujima, K., Achmad, F., Shigihara, Y., Mizutani, N. 2009. Estimation of Tsunami Force Acting on Rectangular Structures. Journal of Disaster Research, 4(6): 404–409.
[7] Robertson, I.N., Riggs, H.R., Paczkowski, K., Mohamed, A. 2011. Tsunami Bore Forces on Walls. Engineering, 1–9.
[8] FEMA 55 CCM. 2011. Coastal Construction Manual. Washington, DC: Federal Emergency Management Agency.
[9] ASCE 7-16. 2016. Minimum Design Loads and Associated Criteria for Buildings and Other Structures. Reston, Amerika Serikat.
[10] Lukkunaprasit, P., Ruangrassamee, A., Thanasisathit, N. 2009. Tsunami Loading on Buildings with Openings. 28(5): 303–310.
[11] Wang, T., Meng, T., Zhao, H. 2015. Analysis of Tsunami Effect and Structural Response. Technical Gazette, 22(6): 1365–1371.

[12] Khan, K., Defence, A., Academy, F., Raza, S. 2015. Seismic performance assessment of masonry infilled reinforced concrete (RC) frame structures (August).
[13] Mccormick, M.E. 2010. Ocean Engineering Mechanics with Applications. (United States Naval Academy and Johns Hopkins University).
[14] ACI Committee 530. 2002. Building Code Requirements for Masonry Structures (ACI 530-02).
[15] Sukrawa, M., Pringgana, G., Yustinaputri, P.A.R. 2019. Modelling of Confined Masonry Structure and its Application for the Design of Multi-Story Building. 1034.
[16] Dorji, J., Thambiratnam, D.P. 2009. Modelling and Analysis of Infilled Frame Structures Under Seismic Loads. The Open Construction and Building Technology Journal, 3(2): 119–126.

Knowledge Development on Urban Public Transportation Concepts: A Literature Study in Bibliometric Analysis

Efendhi Prih Raharjo

Politeknik Transportasi Darat Bali

Sri Sarjana

Politeknik Transportasi Darat Indonesia - STTD

srisarjana@gmail.com

Abstract. Urban public transportation (UPT) is the development of public transportation systems in urban areas by optimizing resources and infrastructure that are integrated with each other by adopting environmentally friendly technologies to create a sustainable city. Urban public transportation is important to be developed through the differentiation of knowledge and technology which refers to development of metropolitan cities as a benchmark in supporting sustainable development. Qualitative research using bibliometric analysis as a meta-analysis study was developed to analyze the research theme. The scientific literature analyzed came from journal publications published from 1966-2020. The results of analysis state that significant increase the themes studied occurred in the last two periods with priority topics including sustainable transportation, service quality, air pollution, bus rapid transit, and light rail transit. New topics that have not been widely researched need to be strengthened in their analysis through the development of knowledge and technology to produce a sustainable transportation system.

Keywords: urban public transportation, sustainable city, urban area, knowledge development, bibliometric analysis

1. Introduction

Public transportation has an important role in urban development, especially the support of an environmentally friendly transportation system in order to create sustainable city. Urban public transportation (UPT) is important topic in the development of urban infrastructure in various regions today. An important key to realizing sustainable city is to develop convenient urban public transportation and provide space for pedestrian which causes the city center to grow rapidly [1]. Various problems must be faced in creating a sustainable city including efforts to overcome traffic congestion [2][3][4], reducing the rate of traffic accidents [5][6][7]. In addition, other obstacles that still occur in the use of public transportation are low comfort, accessibility and flexibility [8], fare mismatches and route gaps [9], dirty and unsafe public transportation [10], inefficient public transportation [11][12][13], it is even possible that there is a risk of disease transmission while on public transportation [14]. The occurrence of disruptions and delays in the departure schedule of public transportation due to tight scheduling [15][16][17]. Efforts to produce efficient transportation travel routes need to be addressed, including efforts to overcome the problem of transit routes in urban areas [18]. There is a need for an innovative and simple fast charging infrastructure for charging energy storage systems for electric buses to support public transportation [19]. The problem of a surge in passenger flow that has potential to be a threat to public safety is important to consider in public transportation system [20][21][22].

To overcome the problems that occur in the development of public transportation systems in urban areas, appropriate solutions are needed, including the need for effective route management through modular transit system [8][23], it is necessary to consistently select design criteria in improving transit network connectivity in order to increase comfort for passengers transit users [24][25], it is possible to have transportation services and routes analyzed using traveling salesman problem modeling [26]. In addition, it is also necessary to apply stochastic and real time data in improving the efficiency of transportation system [27], it is important to make multi-criteria decisions as alternative evaluation step in solving public transportation problems [28][29], need to reduce the number of vehicles for personal transportation through system dynamics simulations for the development of transportation systems in the future [30], utilizing six sigma methodology which is able to identify public transportation routes and to reduce risk as effort to improve public services [6][31], need for advanced dynamic traffic analysis capabilities through the use of light-weight dynamic traffic assignment to integrate transportation systems and improve travel efficiency comprehensively [32]. Coupled with Covid-19 pandemic, it is necessary to use a face mask to avoid disease transmission while on public transportation [14][33]. The implementation of big data in public transportation is able to revolutionize the service concept, management mechanism, and business model of public transportation industry [34], and is equipped with traffic information analysis that can be done using big data and artificial intelligence methods [35][36][37].

UPT is an effort to meet the increasing demand for travel in urban areas and at the same time to reduce the level of traffic congestion [38], part of public transportation system as a tool to improve travel standards in urban areas and to increase the efficiency of public transportation [39]. Public transportation is one of the systems in transportation, especially in big cities and with high density level [40]. Transportation is effort to move goods or person from one place to another [41] [42], and plays an important role in urban development by providing access to work, education, markets, recreation, health, and other services in effort to meet the needs and customer demand [10][43]. The development of urban transportation is way to increase the utilization of transportation resources in order to overcome congestion and create harmonious society [44]. Public transportation is very important in strategic development to increase the utilization of mass transit systems [45][46][47].

2. Method

Mapping UPT theme is an expected goal in this study as an effort to improve understanding of the concept of public transportation systems developed in various urban areas in the world. Knowledge mapping is applied to develop various required study topics according to the level of urgency [48]. Qualitative research is applied in this study through a review of scientific literature originating from journal publications [49]. Research analysis uses bibliometric analysis that examines various scientific literatures as main data source [50]. Scientific literature review has relevance to the topic under study that utilizes bibliometric software [51]. Monitoring the evolution under study can be used to see current research trends. So that the latest issues relevant to research topics need to be monitored and evaluated to support the development of knowledge and technology. The qualitative review in detail describes scientific journals that have made a major contribution to the development of UPT theme through various current topics that are very likely to be further improved along with the development of science and technology to create a sustainable public transportation system. The qualitative approach is implemented through the development of a meta-analysis to obtain novelty according to the specified period [52]. VOSviewer application is used to display the visualization of results for data analysis from collection of scientific journals according to the research theme under study [53]. VOSviewer is able to analyze research data on scientific journals that are presented in an attractive and interactive visualization form [54][55].

The data collected by researchers was obtained through systematic search in October 2021. The database from collection of various scientific journals with UPT theme was obtained with publication time span 1966 - 2020. The database is obtained through search process by entering keyword "urban public transportation" to obtain data that matches. In the final process of selecting scientific journals, 1499 journal articles were published and detected in titles, abstracts, and main articles of each selected journal article. Based on the obtained journal publication time span, it is divided into three time periods to facilitate data analysis. Analysis of the findings of each period can be presented referring to the categorization of data collected so that the progress of each publication period can be known. Issues that have novelty that refer to developmental patterns in scientific evolution can be used as new studies to strengthen knowledge development and support sustainable development [49][56][57].

3. Result and Discussion

The UPT concept is described in detail through several visualizations described in terms of network, density and data classification. Data obtained by entering the keyword "urban public transportation" to get journal articles. After the process of searching for journal articles, 1499 papers were obtained from various journal publications with three periods based on the time span of journal publication. The search for journal articles with UPT theme begins with journal publications published in 1966 until 2020. From journal articles collected in this study, we get 24012 the number of citations within a span of 75 years of journal publication. To obtain the density and network level, database obtained is adjusted to research theme used by utilizing VOSviewer.

Table 1. Urban Public Transportation Publication Period Classification

Clasifications	**1966-2000 (I)**	**2001-2010 (II)**	**2011-2020 (III)**
Citation years	55	10	10
Number papers	149	480	870
Citations	6580	9150	8282
Cites/year	119.64	457.50	828.20
Cites/paper	44.16	19.06	9.52
Authors/paper	1.62	2.12	2.87
h-index	31	39	43
g-index	80	89	70
hA-index	7	10	19

The keywords proposed in the process of searching for journal articles on theme under study are divided into three periods according to the time span of publication. The 10-year time span is carried out on journal publications published after 2000, while before 2000 it is calculated based on 55-year time span. Table 1 shows that the number of journal publications on themes studied in three periods experienced significant increase in the number of publications. However, for the number of article citations, the second period has higher number of citations of 9150 than other periods. The number of citations in period 3 is not optimal because the journal publication time is not too long so that many researchers have not received or recognized it. The third period has largest citations per year, which are 828.20 because the trend of increasing research on themes studied is very rapid. Meanwhile, the citation value per paper in period 1 was 44.16 because the number of journal articles published was the least and it was used as citation divider. Meanwhile, with the authors per paper value, period 3 with the value of 2.87 are the largest because the majority of journal publications have more than 1 author and collaborate with each other in the same research field. To determine the quality of journal publications, it can be determined by the value of h-index, g-index, and hA-index where this assessment refers to number of citations and number of publications produced so that the greater index value obtained, better the journal publication value. Efforts to compare the three periods of journal publication on UPT theme are expected to show the current trend of research topic developments.

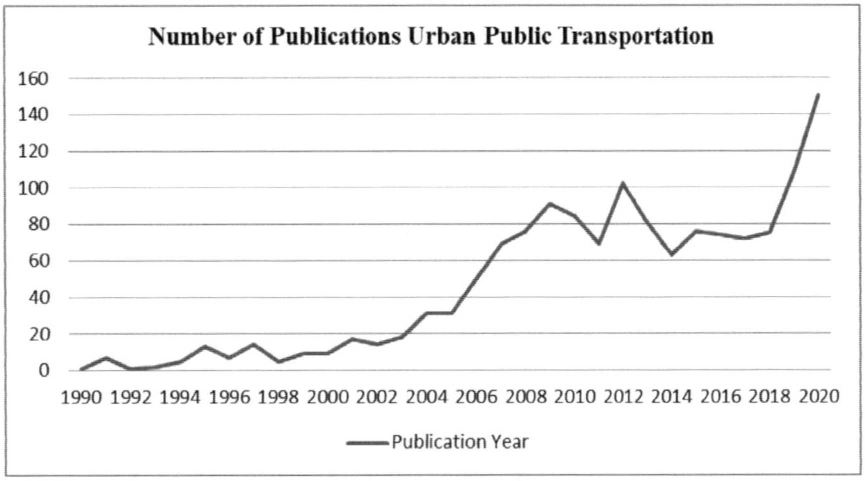

Figure 1. Urban Public Transportation Trends

The trend of journal publications occurs in UPT theme as shown in Figure 1, where in the last two years of third period of journal publication time span under study, trend increased significantly. Publications in 2020 are 150 journals and there are 108 journals in 2019, while in 2018 there were 75 published journals so that increasing trend experienced growth. Journal publications before 2000 had not experienced rapid development, time span from 1990 to 2000 the number of journal publications only ranged from 1-14 publications. However, starting in 2004, the increase in publications occurred in second period with 31 journals and peaked in 2012 with 102 journal publications. The development of journal publications at the end of third period was triggered because many cities in various parts of the world adopted smart city and sustainable city concepts. The concept developed in this research theme is also implemented in Jakarta and several other big cities in Indonesia.

Figure 2. Overlay visualization

In the analysis of journal publications conducted through overlay visualization in third period, there are several topics that are widely used by researchers in that time span. The topics that were widely used in third period were mostly in 2012-2018 range, which are presented in Figure 2. Urban planning, service quality, transportation engineering, sustainable transportation, bus rapid transit, and air pollution were some of the topics that developed in third period. At the end of third period, air pollution, service quality and land use were the most discussed and researched topics published in reputable journals.

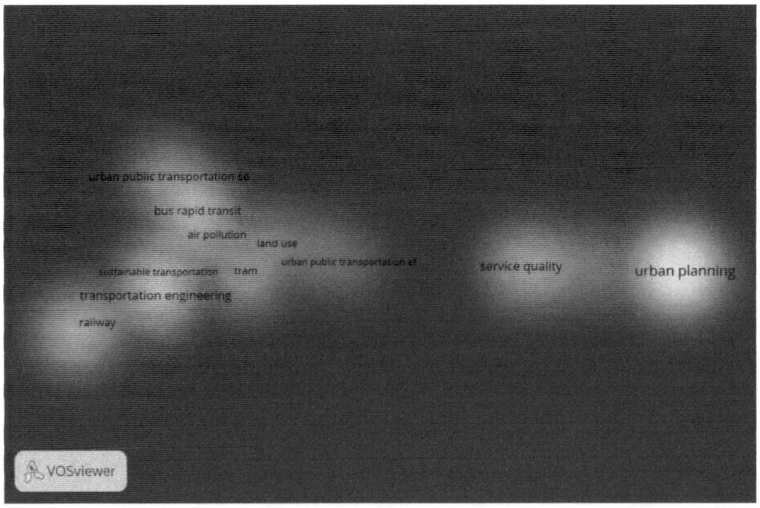

Figure 3. Density Visualization

Figure 3 shows density visualization on the topic under study based on yellow color density shown. The brighter yellow color shown, topic studied in that color has a large level of density. Urban planning shows the clearest yellow color compared to other topics, so it can be seen that there are a large number of journal publications on topic of urban planning. For research in the next period, it is expected to take advantage of research topics that have a low level of visualization for density or an unclear yellow color so that the potential for the topics studied to be published has large enough opportunity, such as air pollution, service quality, and sustainable transportation.

Table 2. Ranked in The Top Ten Journal Publications 2011-2020

Rank	Authors	Title	Source Journal	Year	NP	APY
1	Y Geng, Z Ma, B Xue, W Ren, Z Liu, T Fujita [58]	Co-benefit evaluation for urban public transportation sector–a case of Shenyang, China	Journal of Cleaner Production	2013	102	12.75
2	T Mayer, C Trevien [59]	The impact of urban public transportation evidence from the Paris region	Journal of Urban Economics	2017	78	19.50
3	TM Fenta [60]	Demands for urban public transportation in Addis Ababa	Journal of intelligent transportation and urban planning	2014	22	3.14
4	A Psaltoglou, E Calle [61]	Enhanced connectivity index–A new measure for identifying critical points in urban public transportation networks	International Journal of Critical Infrastructure Protection	2018	10	3.33
5	Y Davidich, Y Kush, A Galkin [62]	Improving of urban public transportation quality via operator schedule optimization	Journal of Urban and Environmental Engineering	2019	2	1.00
6	J Frez, N Baloian, J Pino, G Zurita, F Basso [63]	Planning of urban public transportation networks in a Smart City	JUCS-Journal of Universal Computer Science	2019	4	2.00
7	W Tu, R Cao, Y Yue, B Zhou, Q Li, Q Li [64]	Spatial variations in urban public ridership derived from GPS trajectories and smart card data	Journal of Transport Geography	2018	81	27.00
8	G Blanco Ramírez [65]	Many choices, one destination: Multimodal university brand construction in an urban public transportation system	International Journal of Qualitative Studies in Education	2016	14	2.80
9	Y Xue, H Guan, J Corey, H Wei [66]	Evolutionary game model of private capital investment in urban public transportation	KSCE Journal of Civil Engineering	2018	5	1.67
10	M Cao, S Ma, M Huang, G Lü, M Chen [67]	Effects of free-floating shared bicycles on urban public transportation	ISPRS International Journal of Geo-Information	2019	3	1.50

TC=Number of citations; APY= Average citation per year

Data related to the ranking of top ten journal publications in third period on UPT theme by researchers can be shown in table 2. Journal publications in the first rank with article title "Co-benefit evaluation for urban public transportation sector–a case of Shenyang-China", with authors Yong Geng, Zhixiao Ma, Bing Xue, Wanxia Ren, Zhe Liu, and Tsuyoshi Fujita published in Journal of Cleaner Production in 2013 with 102 citations and an average yearly citation of 12.75. While in third place with an article written by Thierry Mayer and Corentin Trevien in 2017 with article title "The impact of urban public transportation evidence from the Paris region" published by Journal of Urban Economics with 78 citations and 19.50 an average citations per year. Furthermore, the third rank was published by Journal of intelligent transportation and urban planning with article title "Demands for urban public transportation in Addis Ababa" where Tilahun M Fenta was the author with total number of citations of 22 and average citation of 3.14 per year. Ten journals with article publications with highest citations, nine of them were indexed by Scopus, while one other journal was indexed by Web of Science, included Journal of intelligent transportation and urban planning. Several journal articles that are ranked in top ten still have minimal number of citations because time of publication of articles is still under five years so that publication of articles is still limited and not widely known by researchers.

Table 3. Largest Number of Citation Data 1966-2020

TC	APY	Cite Scores	SJR	Title	Source Journal	Authors	Publisher	Year
1348	42.13	6.6	1.470	Gasoline consumption and cities: a comparison of US cities with a global survey	Journal of the American planning association	PWG Newman, JR Kenworthy [68]	Taylor & Francis	1989
1081	83.15	9.5	2.161	Network hub location problems: The state of the art	European Journal of Operational Research	S Alumur, BY Kara [69]	Elsevier	2008
1018	78.31	6.6	3.886	Why do the poor live in cities? The role of public transportation	Journal of urban Economics	EL Glaeser, ME Kahn, J Rappaport [70]	Elsevier	2008
816	45.33	2.7	0.451	Economic small-world behavior in weighted networks	The European Physical Journal B	V Latora, M Marchiori [71]	Springer	2003
561	70.13	9.5	2.161	A review of urban transportation network design problems	European Journal of Operational Research	RZ Farahani, E Miandoabchi, WY Szeto [72]	Elsevier	2013
440	11.00	-	-	Personal and occupational factors associated with carpal tunnel syndrome	Journal of occupational medicine	LJ Cannon, EJ Bernacki, SD Walter [73]	Wolters Kluwer	1981
413	9.83	-	-	Carpal tunnel syndrome and selected personal attributes.	Journal of occupational medicine	TJ Armstrong, DB Chaffin [74]	Wolters Kluwer	1979
397	11.03	13.8	11.771	Conceptual developments in the economics of transportation: an interpretive survey	Journal of Economic Literature	C Winston [75]	American Economic Association	1985
351	8.56	9.5	2.161	Evaluation and optimization of urban public transportation networks	European Journal of Operational Research	CE Mandl [76]	Elsevier	1980
349	20.53	9.5	2.161	A DEA approach for evaluating the efficiency and effectiveness of urban transit systems	European Journal of Operational Research	MG Karlaftis [77]	Elsevier	2004

TC=Number of citations; APY= Average citation per year; SJR= Scimago Journal Rank

Table 3 shows data on the largest number of citations in journal publications with time span 1966-2020. The largest number of citations was 1348 citations, 42.13 average citations per year, citation scores of 6.6 and 1,470 in the ranking of journal scimago in article title "Gasoline consumption and cities: a comparison of US cities with a global survey" with source Journal of the American Planning Association by Taylor & Francis publishers in 1989. The second place in number of citations was obtained by Sibel Alumur and Bahar Y. Kara with 1081 citations, 83.15 citations per year average, 9.5 citation values, and 2.161 in Scimago Journal Ranking published in article entitled "Network hub location problems: The state of the art" published in European journal of operational research in 2008. While in third place was published in Journal of urban economics with article title "Why do the poor live in cities? The role of public transportation" written by Edward L. Glaeser,

Matthew E. Kahn, and Jordan Rappaport in 2008 with 1018 citation counts, 78.31 citation averages per year, with citation score of 6.6, and 3.886 in schimago journal rating. Ten articles with the highest number of citations, eight journal articles of which have high citescores and scimago journal rankings are indexed in scopus, while two other journal articles are currently discontinued on scopus so that citescores and scimago journal rankings are not detected.

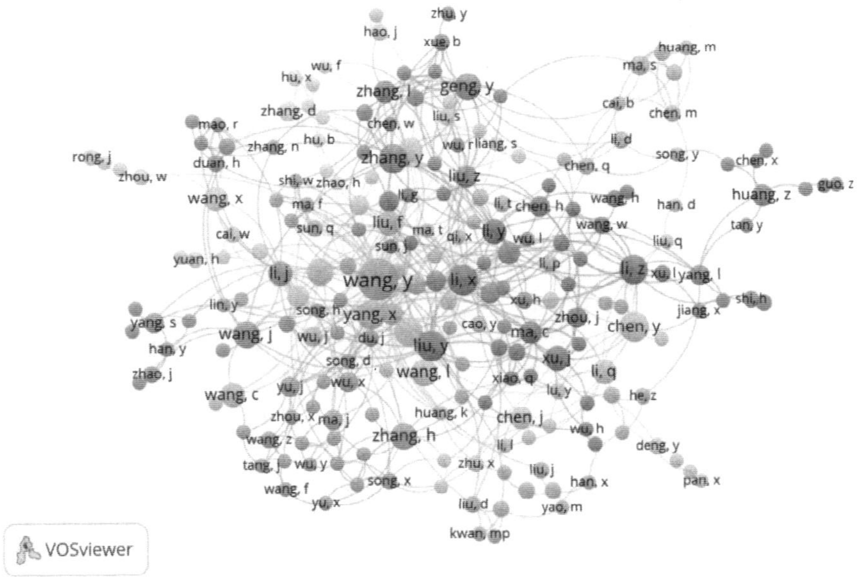

Figure 4. Network visualization 2011-2020

Network visualization of authors published in journal articles with UPT theme shown in Figure 4 shows the relationship between researchers. An author is displayed in the form of round node with a certain color and dimension. Nodes with large or small dimensions indicate how many journal publications each author has, more journal publications, larger the node and vice versa. While the color of node shows the clustering of topics for published journal articles so that nodes with the same color indicate that nodes are in the same cluster. The line that connects two nodes indicates the existence of connection and connectedness of research topics published in journals. The vertices with the largest and most dominant dimensions are identified by author's name, Yihui Wang (Wang, Y), while other authors with large vertex dimensions include Yi Zhang, Yuqi Liu, Yong Geng and Yihsu Chen. If you look at the description of authors' names in journal publications with the theme of UPT, it can be seen that the dominance of country or region where the authors are located, shows that the development of urban development in authors' home region is quite at the forefront in accordance with the development of their research.

4. Conclusion

The development of research topics related to development of public transportation systems in various urban areas is currently very much needed as reference and benchmark in order to design a city development that is better suited and environmentally friendly. The results of this study stated that the use of UPT theme in various studies published in scientific journals had significant developments based on publications each year. In 2019-2020 in period three and 2006-2008 in period two there was an increase in research related to themes studied. The concept of UPT has a rapid research development by researchers because it has a goal to develop public transportation system in urban areas through the use of various resources and infrastructure to integrate with each other by adopting environmentally friendly technologies to create sustainable city. Various current topics related to UPT concept need special attention and need to be researched further through optimizing the latest knowledge and technology including sustainable transportation, service quality, air pollution, bus rapid transit, and light rail transit. The exposure of relationships between topics studied in the form of network visualization, overlays and density becomes an interesting analysis material so that it can be seen which topics have been frequently researched and which are still rarely studied, as well as researchers who are dominant in scientific publishing journals. In addition, it can also be seen which journals have a rating or have a large number of citations so that their reputation can be known. Most of highly rated and citation journals are published by well-known publishers including Elsevier, Taylor & Francis, and Springer. For this reason, it is important for researchers to know that in order for journal articles to have good reputation, they need to be published in various journal publishers that also have good quality.

References

[1] R. Matsunaka, T. Oba, D. Nakagawa, M. Nagao, and J. Nawrocki, "International comparison of the relationship between urban structure and the service level of urban public transportation — A comprehensive analysis in local cities in Japan , France and Germany," *Transp. Policy*, vol. 30, pp. 26–39, 2013, doi: 10.1016/j.tranpol.2013.06.008.

[2] K. M. Almatar, "Traffic congestion patterns in the urban road network: (Dammam metropolitan area)," *Ain Shams Eng. J.*, no. 101886, pp. 1–9, 2022, doi: 10.1016/j.asej.2022.101886.

[3] M. A. Fattah, S. R. Morshed, and A. Kafy, "Insights into the socio-economic impacts of traffic congestion in the port and industrial areas of Chittagong city , Bangladesh," *Transp. Eng.*, vol. 9, no. 100122, pp. 1–12, 2022, doi: 10.1016/j.treng.2022.100122.

[4] L. C. Reyes, C. A. O. O. Zezzatti, C. G. Santillan, P. H. Hernández, and M. V. Fuerte, "A Cultural Algorithm for the Urban Public," *Hybrid Artif. Intell. Syst.*, vol. II, pp. 135–142, 2010, doi: 10.1007/978-3-642-13803-4_17.

[5] H. Ospina-mateus, S. B. Garcia, L. Q. Jimenez, and K. Salas-navarro, "Dataset of traffic accidents in motorcyclists in," *Data Br.*, vol. 43, no. 108461, pp. 1–8, 2022, doi: 10.1016/j.dib.2022.108461.

[6] Ü. Kuvvetli and A. R. Firuzan, "Applying Six Sigma in urban public transportation to reduce traffic accidents involving municipality buses," *Total Qual. Manag. Bus. Excell.*, vol. 30, no. 1–2, pp. 82–107, 2017, doi: 10.1080/14783363.2017.1297198.

[7] C. H. Wu, S.-C. K. Kao, and C.-C. Chang, "A knowledge elicitation approach to traffic accident analysis in open data: comparing periods before and after the Covid-19 outbreak," *Heliyon*, no. 101541, pp. 1–40, 2022, doi: 10.1016/j.heliyon.2022.e10302.

[8] Z. Zhang, A. Tafreshian, and N. Masoud, "Modular transit: Using autonomy and modularity to improve performance in public transportation," *Transp. Res. Part E Logist. Transp. Rev.*, vol. 141, no. January, p. 102033, 2020, doi: 10.1016/j.tre.2020.102033.

[9] Herabudin, "Role of transportation department governance urban public transport in solok city," *J. Kelola J. Ilmu Sos.*, vol. 2, no. 2, pp. 11–20, 2019, doi: 10.15575/jk.v2i2.5439.

[10] R. Hanny and S. F. Almassawa, "Community Perception of the Usage of Urban Public Transport and Online Application Based Transportation in South Tangerang," *J. akuntansi, Manaj. dan Ekon.*, vol. 23, no. 1, pp. 29–36, 2021, doi: 10.32424/1.jame.2021.23.1.4108.

[11] D. Ushakov, E. Dudukalov, L. Shmatko, and K. Shatila, "Artificial Intelligence as a factor of public transportations system development," *Transp. Res. Procedia*, vol. 63, pp. 2401–2408, 2022, doi: 10.1016/j.trpro.2022.06.276.

[12] T. Tamaki, H. Nakamura, H. Fujii, and S. Managi, "Efficiency and emissions from urban transport : Application to world city-level public transportation," *Econ. Anal. Policy*, vol. 61, pp. 1–9, 2016, doi: 10.1016/j.eap.2016.09.001.

[13] H. T. Khanh-Van, T. V. Ha, T. Asada, and M. Arimura, "Assessing transportation system efficiency in its relationship with urban housing: A data envelopment analysis," *Asian Transp.*

Stud., vol. 8, no. 100065, pp. 1–13, 2022, doi: 10.1016/j.eastsj.2022.100065.
[14] A. Tirachini and O. Cats, "COVID-19 and public transportation: Current assessment, prospects, and research needs," *J. Public Transp.*, vol. 22, no. 1, pp. 1–34, 2020, doi: 10.5038/2375-0901.22.1.1.
[15] B. Amberg, B. Amberg, and N. Kliewer, "Robust efficiency in urban public transportation: Minimizing delay propagation in cost-efficient bus and driver schedules," *Transp. Sci.*, no. Articles in Advance, pp. 1–24, 2018, doi: 10.1287/trsc.2017.0757.
[16] E. Vitvitsky, E. Galaktionova, and N. Yuryeva, "Optimization of the Schedule for Road Transportation of 'Tails' Recycling of Solid Municipal Waste," *Transp. Res. Procedia*, vol. 61, pp. 185–190, 2022, doi: 10.1016/j.trpro.2022.01.030.
[17] M. LaRoche-Boisvert, R. Dimitrakopoulos, and J. A. Ferland, "Simultaneous production scheduling and transportation optimization from mines to port under uncertain material supply," *Resour. Policy*, vol. 73, pp. 1–12, 2021, doi: 10.1016/j.resourpol.2021.102150.
[18] L. Ahmed, C. Mumford, and A. Kheiri, "Solving urban transit route design problem using selection hyper-heuristics," *Eur. J. Oper. Res.*, vol. 274, no. 2, pp. 545–559, 2019, doi: 10.1016/j.ejor.2018.10.022.
[19] F. Ortenzi, M. Pasquali, P. P. Prosini, A. Lidozzi, and M. Di Benedetto, "Design and validation of ultra-fast charging infrastructures based on supercapacitors for urban public transportation applications," *Energies*, vol. 12, no. 2348, pp. 1–13, 2019, doi: 10.3390/en12122348.
[20] H. Wang, L. Li, P. Pan, Y. Wang, and Y. Jin, "Early warning of burst passenger flow in public transportation system," *Transp. Res. Part C*, vol. 105, pp. 580–598, 2019, doi: 10.1016/j.trc.2019.05.022.
[21] M. Mohammed and J. Oke, "Origin-destination inference in public transportation systems: A comprehensive review," *Int. J. Transp. Sci. Technol.*, pp. 1–14, 2022, doi: 10.1016/j.ijtst.2022.03.002.
[22] S. Sun, Y. D. Wong, X. Liu, and A. Rau, "Exploration of an integrated automated public transportation system," *Transp. Res. Interdiscip. Perspect.*, vol. 8, no. 100275, pp. 1–9, 2020, doi: 10.1016/j.trip.2020.100275.
[23] J. Wu, B. Kulcsár, Selpi, and X. Qu, "A modular, adaptive, and autonomous transit system (MAATS): A in-motion transfer strategy and performance evaluation in urban grid transit networks," *Transp. Res. Part A Policy Pract.*, vol. 151, pp. 81–98, 2021, doi: 10.1016/j.tra.2021.07.005.
[24] M. Owais, A. S. Ahmed, G. S. Moussa, and A. A. Khalil, "Integrating underground line design with existing public transportation systems to increase transit network connectivity: Case study in Greater Cairo," *Expert Syst. Appl.*, vol. 167, pp. 1–21, 2021, doi: 10.1016/j.eswa.2020.114183.
[25] P. H. da Silva, S. Kaplan, and P. W. Gonzales-Taco, "Service climate antecedents of transit passenger complaints and temporary opt-out: The case study of Brasília," *Res. Transp. Bus. Manag.*, vol. 41, no. 100671, pp. 1–10, 2021, doi: 10.1016/j.rtbm.2021.100671.
[26] C. Puchongkawarin and K. Ransikarbum, "An Integrative Decision Support System for Improving Tourism Logistics and Public Transportation in Thailand," *Tour. Plan. Dev.*, vol. 0, no. 0, pp. 1–16, 2020, doi: 10.1080/21568316.2020.1837229.
[27] R. Elbert and J. Rentschler, "Freight on urban public transportation: A systematic literature review," *Res. Transp. Bus. Manag.*, vol. 40, 2021, doi: 10.1016/j.rtbm.2021.100679.
[28] S. Seker and N. Aydin, "Sustainable Public Transportation System Evaluation: A Novel Two-Stage Hybrid Method Based on IVIF-AHP and CODAS," *Int. J. Fuzzy Syst.*, vol. 22, no. 1, pp. 257–272, 2020, doi: 10.1007/s40815-019-00785-w.
[29] J. Zak, "Multiple-Criteria and Group-Decision Making in the Fleet Selection Problem for a Public Transportation System," *Transp. Res. Procedia*, vol. 27, pp. 43–52, 2017, doi: 10.1016/j.trpro.2017.12.102.
[30] W. Sardjono, E. Selviyanti, and W. G. Perdana, "Modeling the relationship between public

transportation and traffic conditions in urban areas: A system dynamics approach," *J. Phys. Conf. Ser.*, vol. 1465, no. 1, pp. 1–9, 2020, doi: 10.1088/1742-6596/1465/1/012023.

[31] Z. Cserdi and Z. Kenesei, "Attitudes to forced adoption of new technologies in public transportation services," *Res. Transp. Bus. Manag.*, vol. 41, no. 100611, pp. 1–11, 2021, doi: 10.1016/j.rtbm.2020.100611.

[32] L. Tong, Y. Pan, P. Shang, J. Guo, K. Xian, and X. Zhou, "Open-Source Public Transportation Mobility Simulation Engine DTALite-S: A Discretized Space–Time Network-Based Modeling Framework for Bridging Multi-agent Simulation and Optimization," *Urban Rail Transit*, vol. 5, no. 1, pp. 1–16, 2019, doi: 10.1007/s40864-018-0100-x.

[33] A. Shabani, A. Shabani, B. Ahmadinejad, and A. Salmasnia, "Measuring the customer satisfaction of public transportation in Tehran during the Covid-19 pandemic using MCDM techniques," *Case Stud. Transp. Policy*, pp. 1–11, 2022, doi: 10.1016/j.cstp.2022.05.009.

[34] X. Ma and X. Chen, *Public Transportation Big Data Mining and Analysis*. Elsevier Inc., 2019.

[35] G. Xiong *et al.*, "applied sciences Building Urban Public Traffic Dynamic Network Based on CPSS : An Integrated Approach of Big Data and AI," *Appl. Sci.*, vol. 11, no. 1109, pp. 1–14, 2021, doi: 10.3390/app11031109.

[36] V. V. Okrepilov, B. B. Kovalenko, G. V. Getmanova, and M. S. Turovskaj, "Modern Trends in Artificial Intelligence in the Transport System," *Transp. Res. Procedia*, vol. 61, pp. 229–233, 2022, doi: 10.1016/j.trpro.2022.01.038.

[37] I. Evgrafova, L. Egorova, A. Marchenko, and A. Tarasov, "Ethical problems of practical interaction between strong artificial intelligence and humans in the urban transport environment and legal proceedings," *Transp. Res. Procedia*, vol. 63, pp. 2094–2098, 2022, doi: 10.1016/j.trpro.2022.06.234.

[38] M. Brons, P. Nijkamp, E. Pels, and P. Rietveld, "Efficiency of urban public transit : A meta analysis," *Transportation (Amst).*, vol. 32, pp. 1–21, 2005, doi: 10.1007/s11116-004-0939-4.

[39] K. Solecka and Ż. Jacek, "Integration of the urban public transportation system with the application of traffic simulation," *Transp. Res. Procedia*, vol. 3, pp. 259–268, 2014, doi: 10.1016/j.trpro.2014.10.005.

[40] M. Nassereddine and H. Eskandari, "An integrated MCDM approach to evaluate public transportation systems in Tehran," *Transp. Res. Part A Policy Pract.*, vol. 106, pp. 427–439, 2017, doi: 10.1016/j.tra.2017.10.013.

[41] E. P. Raharjo, S. Sarjana, and M. Safitri, "Transportation infrastructure planning in supporting disaster mitigation : Case study in Mount Gamalama," *Jàmbá - J. Disaster Risk Stud.*, vol. 14, no. 1, pp. 1–11, 2022, doi: 10.4102/ jamba.v14i1.1123.

[42] S. Sarjana, N. Khayati, L. Warini, and P. Praswiyati, "The Role of Transportation Management in Optimizing Supply Chain Management at Industrial Estate," *J. Transp. Multimoda*, vol. 18, pp. 1–14, 2020.

[43] R. Goel, R. Maini, and S. Bansal, "Vehicle routing problem with time windows having stochastic customers demands and stochastic service times: Modelling and solution," *J. Comput. Sci.*, vol. 34, pp. 1–10, 2019, doi: 10.1016/j.jocs.2019.04.003.

[44] Z. Xinlan and W. Tao, "Active Development Strategy of Urban Public Transportation Based on the Marketing Management Theory," in *International Conference on Transportation Engineering 2007*, 2007, pp. 2765–2770, doi: 10.1061/40932(246)454.

[45] R. A. Scott, B. T. George, and V. R. Prybutok, "A Public Transportation Decision-Making Model within a Metropolitan Area," *Decis. Sci.*, vol. 47, no. 6, pp. 1048–1072, 2016, doi: 10.1111/deci.12203.

[46] S. Tabassum, S. Tanaka, F. Nakamura, and A. Ryo, "Feeder Network Design for Mass Transit System in Developing Countries (Case study of Lahore, Pakistan)," *Transp. Res. Procedia*, vol. 25, pp. 3129–3146, 2017, doi: 10.1016/j.trpro.2017.05.343.

[47] Y. Tang, Y. Jiang, H. Yang, and O. A. Nielsen, "Modeling and optimizing a fare incentive strategy to manage queuing and crowding in mass transit systems: Modeling and optimizing a

fare incentive strategy to manage queuing and crowding in mass transit systems," *Transp. Res. Part B Methodol.*, vol. 138, pp. 247–267, 2020, doi: 10.1016/j.trb.2020.05.006.

[48] S. Sarjana, M. A. A. Najib, and N. Khayati, "Dissemination of Economic Resilience in Scientific Covid-19 Pandemic through Bibliometric Analysis," *JDE (Journal Dev. Econ.*, vol. 7, no. 1, pp. 142–155, 2022, doi: 10.20473/jde.v7i1.33796.

[49] S. Sarjana, J. R. Widokarti, H. Fachri, and D. Pranita, "Hybrid Energy to Drive Renewable Energy Diversity in Bibliometric Analysis," *Int. J. Energy Econ. Policy*, vol. 12, no. 1, pp. 500–506, 2022, doi: 10.32479/ijeep.11956.

[50] R. R. Pai and S. Alathur, "Bibliometric Analysis and Methodological Review of Mobile Health Services and Applications in India," *Int. J. Med. Inform.*, vol. 145, no. 104330, pp. 1–11, 2021, doi: 10.1016/j.ijmedinf.2020.104330.

[51] S. K. Aski, R. Akbari, S. Hantoushzadeh, and F. Ghotbizadeh, "A bibliometric analysis of Intrauterine Growth Restriction research," *Placenta*, vol. 95, pp. 106–120, 2020, doi: 10.1016/j.placenta.2020.03.010.

[52] S. Sarjana and N. D. Djajasinga, "Scientific Evolution of Knowledge Management in Meta-Analysis Studies," *Cent. Asia Caucasus*, vol. 23, no. 1, pp. 3409–3419, 2022, doi: 10.37178/ca-c.23.1.242.

[53] S. Sarjana, M. A. A. Najib, I. K. Dewi, and N. Khayati, "Knowledge transformation on the economic recovery in post-pandemic," *J. Innov. Bus. Econ.*, vol. 5, no. 2, 2021.

[54] S. Sarjana, M. A. A. Najib, and N. Khayati, "Bibliometric Analysis to Encourage the Development of Digital Economy Scientific Studies," in *Advances in Economics, Business and Management Research*, 2021, vol. 198, pp. 14–21, doi: 10.2991/aebmr.k.211207.002.

[55] S. Sarjana, "Perspektif Urban Public Transportation dalam Kajian Meta-Analysis," *Media Komun. Tek. Sipil*, vol. 27, no. 2, pp. 277–287, 2021, doi: mkts.v27i2.40635.

[56] J. Raimbault and D. Pumain, "Trade-offs between sustainable development goals in systems of cities," *J. Urban Manag.*, vol. 11, pp. 237–245, 2022, doi: 10.1016/j.jum.2022.05.008.

[57] J. Schulte and S. Knuts, "Sustainability impact and effects analysis - A risk management tool for sustainable product development," *Sustain. Prod. Consum.*, vol. 30, pp. 737–751, 2022, doi: 10.1016/j.spc.2022.01.004.

[58] Y. Geng, Z. Ma, B. Xue, W. Ren, Z. Liu, and T. Fujita, "Co-benefit evaluation for urban public transportation sector. A case of Shenyang, China," *J. Clean. Prod.*, vol. 58, pp. 82–91, 2013, doi: 10.1016/j.jclepro.2013.06.034.

[59] T. Mayer and C. Trevien, "The impact of urban public transportation evidence from the Paris region R," *J. Urban Econ.*, vol. 102, pp. 1–21, 2017, doi: 10.1016/j.jue.2017.07.003.

[60] T. M. Fenta, "Demands for Urban Public Transportation in Addis Ababa," *J. Intell. Transp. Urban Plan.*, vol. 2, no. 3, pp. 121–128, 2014, doi: 10.18005/itup0203002.

[61] A. Psaltoglou and E. Calle, "Enhanced connectivity index – A new measure for identifying critical points in urban public transportation networks," *Int. J. Crit. Infrastruct. Prot.*, vol. 21, pp. 22–32, 2018, doi: 10.1016/j.ijcip.2018.02.003.

[62] Y. Davidich, Y. Kush, A. Galkin, N. Davidich, and I. Tkachenko, "Improving of urban public transportation quality via operator schedule optimization," *J. Urban Environ. Eng.*, vol. 13, no. 1, pp. 23–33, 2019, doi: 10.4090/juee.2019.v13n1.023033.

[63] J. Frez, N. Baloian, J. A. Pino, G. Zurita, and F. Basso, "Planning of Urban Public Transporation Networks in a Smart City," *J. Univers. Comput. Sci.*, vol. 25, no. 8, pp. 946–966, 2019.

[64] W. Tu, R. Cao, Y. Yue, B. Zhou, Q. Li, and Q. Li, "Spatial variations in urban public ridership derived from GPS trajectories and smart card data," *J. Transp. Geogr.*, vol. 69, no. 3688, pp. 45–57, 2018, doi: 10.1016/j.jtrangeo.2018.04.013.

[65] G. Blanco Ramírez, "Many choices, one destination: multimodal university brand construction in an urban public transportation system," *Int. J. Qual. Stud. Educ.*, vol. 29, no. 2, pp. 186–204, 2016, doi: 10.1080/09518398.2015.1023233.

[66] Y. Xue, H. Guan, J. Corey, and H. Wei, "Evolutionary Game Model of Private Capital Investment in Urban Public Transportation," *KSCE J. Civ. Eng.*, vol. 22, no. 9, pp. 3620–3632, 2018, doi: 10.1007/s12205-018-1998-y.

[67] M. Cao, S. Ma, M. Huang, G. Lü, and M. Chen, "Effects of free-floating shared bicycles on urban public transportation," *ISPRS Int. J. Geo-Information*, vol. 8, no. 323, pp. 1–15, 2019, doi: 10.3390/ijgi8080323.

[68] P. W. G. Newman and J. R. Kenworthy, "Gasoline consumption and cities: A comparison of U.S. cities with a global survey," *J. Am. Plan. Assoc.*, vol. 55, no. 1, pp. 24–37, 1989, doi: 10.1080/01944368908975398.

[69] S. Alumur and B. Y. Kara, "Network hub location problems: The state of the art," *Eur. J. Oper. Res.*, vol. 190, pp. 1–21, 2008, doi: 10.1016/j.ejor.2007.06.008.

[70] E. L. Glaeser, M. E. Kahn, and J. Rappaport, "Why do the poor live in cities? The role of public transportation," *J. Urban Econ.*, vol. 63, pp. 1–24, 2008, doi: 10.1016/j.jue.2006.12.004.

[71] V. Latora and M. Marchiori, "Economic small-world behavior in weighted networks," *Eur. Phys. J. B*, vol. 32, no. 2, pp. 249–263, 2003, doi: 10.1140/epjb/e2003-00095-5.

[72] R. Z. Farahani, E. Miandoabchi, W. Y. Szeto, and H. Rashidi, "A review of urban transportation network design problems," *Eur. J. Oper. Res.*, vol. 229, no. 2, pp. 281–302, 2013, doi: 10.1016/j.ejor.2013.01.001.

[73] L. J. Cannon, E. J. Bernacki, and S. D. Walter, "Personal and occupational factors associated with carpal tunnel syndrome," *J. Occup. Med.*, vol. 23, no. 4, pp. 255–258, 1981, doi: 10.18005/ITUP0203002.

[74] T. J. Armstrong and D. B. Chaffin, "Carpal tunnel syndrome and selected personal attributes," *J. Occup. Med.*, vol. 21, no. 7, pp. 481–486, 1979.

[75] C. Winston, "Conceptual Developments in the Economics of Transportation: An Interpretive Survey," *J. Econ. Lit.*, vol. 23, no. 1, pp. 57–94, 1985.

[76] C. E. Mandl, "Evaluation and optimization of urban public transportation networks," *Eur. J. Oper. Res.*, vol. 5, no. 6, pp. 396–404, 1980, doi: 10.1016/0377-2217(80)90126-5.

[77] M. G. Karlaftis, "A DEA approach for evaluating the efficiency and effectiveness of urban transit systems," *Eur. J. Oper. Res.*, vol. 152, pp. 354–364, 2004, doi: 10.1016/S0377-2217(03)00029-8.

Coastline changes monitoring induced by man-made structures using synthetic aperture radar: A new simple approach

Ni Nyoman Pujianiki

Civil Engineering Department, Faculty of Engineering, Udayana University, Bukit Jimbaran Campus, Badung Regency, Bali, Indonesia.

pujianiki@civil.unud.ac.id

Abstract. This study presents a new simple approach to detect the coastline changes due to the construction of man-made structures. The new approach uses low-pass filter to reduce the speckles errors and Otsu thresholding method the create binary image. Then, a closing morphological operation was conducted to improve the binary image. In the final step, the canny edge detection is used for coastline delineation. The coastlines changes analysis was done by using Digital Shoreline Analysis System (DSAS). In general, the processing steps of this new approach is semi-automatic processes, it does not require manual digitization to extract the coastline. This method was tested using two independent synthetic aperture radar (SAR) satellite images, i.e., ALOS-1 and Sentinel-1 SAR data. The results show that this new approach produced better results than other well-known methods. This new approach detects clearly the man-made structure and its impact to the coastal zone, i.e., abrasion and accretion. The study area is located in Pengambengan Port, Jembrana Regency, Bali, Indonesia. In addition, the result of this study can be used as evaluation parameters for future development of Pengambengan port or re-design of present structures.

1. Introduction

The monitoring of coastal zone evolution is an essential works and it provides critical information for government or decision makers. The investigation of coastline changes is widely used as an important parameter in coastal area management [1]. The long-term of coastline changes dataset provides useful information in order to evaluate the coastal health. Coastline monitoring can be conducted using direct surveys or remote sensing technique, such as aerial photograph or artificial satellite (The Earth Observation satellite) [2]. Direct survey is well-kwon that this method is time consuming and it takes huge labor cost. On the other hand, the remote sensing technique is commonly known as an efficient method to detect coastline changes with high accuracy and periodical measurements [3].

In general, there are two types of sensors to obtain the image by The Earth Observation Satellite, i.e., optical and radar sensors. The optical sensor is a passive remote sensing method. It uses the reflected energy from the sun to collect the image of the Earth surface. It provides an optical photograph and easy to be analyzed. However, because it uses the energy of the sun, the optical sensor cannot obtain the data in the night time observation and in the cloudy condition

The radar sensor is an active remote sensing method. It transmits the electromagnetic wave at microwave frequency to Earth surface and receives the reflected electromagnetic wave from the Earth surface [4]. It is independent from the sun illumination, can collect the data at night time observation, and independent from weather conditions [4]. However, unlike optical photograph, the radar image requires advanced knowledge to understand and interpret the physical meaning. It is because radar image contains several distortions, such as radiometric distortion and geometrical distortion.

It is not easy to define the coastline of radar image using human eye or manual digitization. It requires advanced pre-processing, main-processing and post-processing steps. Therefore, this study developed a new simple approach to extract coastline from radar image. The pre-processing is done using The Sentinel Application Platform (SNAP), main-processing is using our owned code written in Python and Quantum Geographic Information System (QGIS), and the post-processing is done in ArcGIS with Digital Shoreline Analysis System (DSAS) add-on. The results of this method are visually assessed using temporal images from Google Earth Pro. In addition, the results of this method also compared with automatic coastline delineation developed by [5].

2. Study area and satellite datasets

2.1. Study area

The study area is located in the Pengambengan fisheries port, Jembrana regency, Bali Province, Indonesia (Fig.1a). This port was built in 1976. The first development was conducted in 2000 and 2003 by creating break water, road, and jetty (Fig. 1b-c). The second development was done in 2014 by construction of ship dock and groynes (Fig. 1d-f).

Fig.1c show a sedimentation was occur in this port. It may be one of the reasons of construction of groins. However, these groins produced a sedimentation area in the south-west part of the port (Fig. 1e-f).

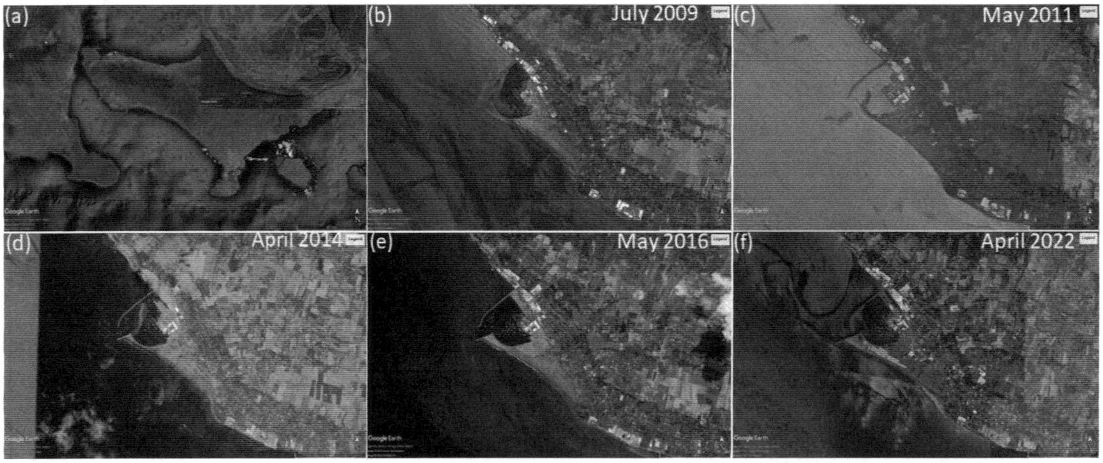

Figure 1. Location of Pengambengan fisheries port and several photos of their developments: (a) location of Pengambengan fisheries port by Google-Earth, (b) Condition of Pengambengan fisheries port in July 2009, (c) in May 2011, (d) in April 2014, (e) in May 2016, and (f) in April 2022.

2.2. Satellite dataset

This study uses radar images provided by the Advanced Land Observing Satellite (ALOS-1) and the Sentinel-1 satellite. Both radar images were downloaded by free-of-charge from Alaska Satellite Facility (https://search.asf.alaska.edu/#/). The ALOS-1 satellite is operated by Japan Aerospace Exploration Agency (JAXA), while the Sentinel-1 satellites are operated by European Space Agency (ESA). Both satellites are the Earth Observation satellite that used radar wavelength (microwave) to collect the information of the Earth surface. The detail information of those radar datasets is presented in Table.1.

IOP Conf. Series: Earth and Environmental Science 1117 (2022) 012041 doi:10.1088/1755-1315/1117/1/012041

Table 1. Radar dataset specification

Satellite name	Frequency	Polarization	Data level	Spatial resolution	Observation date and time (dd-mm-yyy hh:mm:ss)
ALOS-1	L-Band (1.27 GHz)	Dual (HV)	Geo-referenced (Level 1.5)	10x10 meter	02-07-2009 15:16:05 UTC
Sentinel-1	C-Band (5.405) GHz	Dual (VH)	Ground range detected (GRD)	10x10 meter	13-04-2022 10:41:35 UTC

Fig. 2 show the comparison of optical photos and radar images. Both independent datasets are taken at the same date. Optical image in Fig. 2a was taken at the same day with radar image in Fig 2c. It shows that the groins were not build yet. The groins were detected by optical and radar images that were taken on 13 April 2022, as shown by Fig. 2b and 2d.

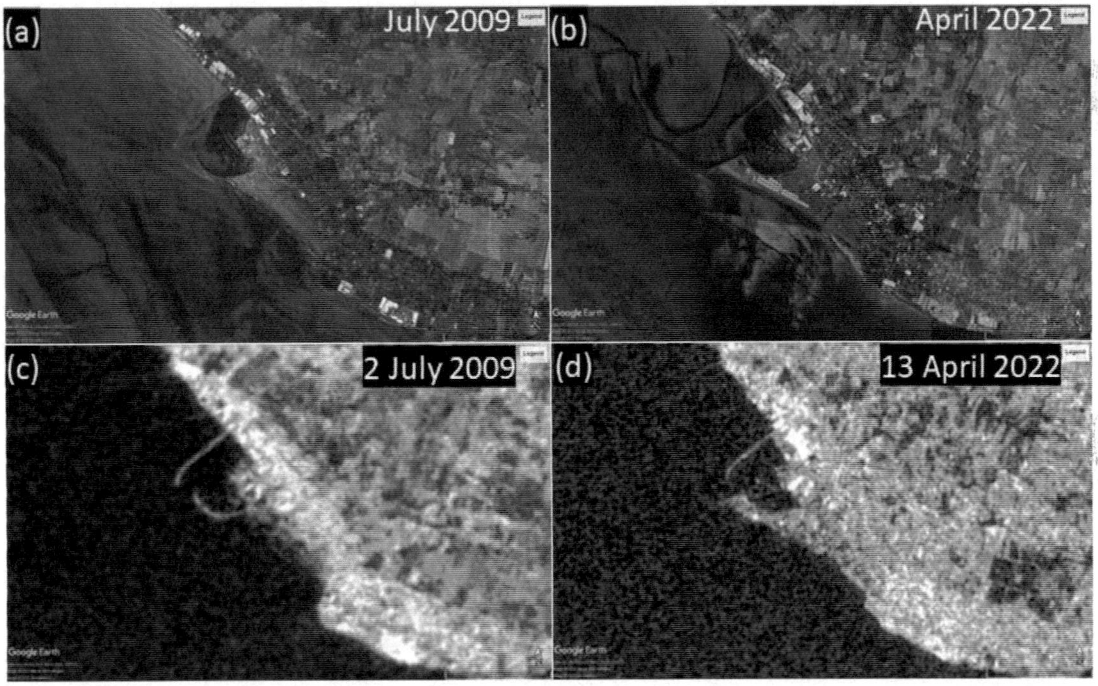

Figure 2. Comparison of optical and radar images: (a) Optical image on 2 July 2009, (b) Optical image on 13 April 2022, (c) Radar image on 2 July 2009, and (d) radar image on 13 April 2022.

3. Method

The processing steps of the proposed method is presented in Fig. 3. The pre-processing of radar images mostly consists of radiometric calibration and geometrical correction or geocoding (transform from radar coordinate to geographical coordinate system). The low-pass filtering was done by kernel size 3x3 in SNAP application software.

The Binarization (Otsu Thresholding) and Morphological Operation calculation were conducted in Python. The QGIS software is used for edge detection process to extract the coastline. The tidal corrections and the calculation of coastline changes were conducted in ArcGIS software.

The tidal data is obtained from http://wxtide32.com/. It provides archived and real time tidal data. The date and time of the tidal data should be same with the date and time of SAR satellite observation. The SAR satellite provides the observation time in Universal Time Coordinated (UTC). Then, a time conversion from UTC to local time is conducted. The tidal data during SAR satellite observation is presented in Tabel 2.

The beach slope is required in order to conduct tidal correction from detected coastline by SAR image. The beach slope data (β) is obtained from Indonesian Digital Elevation Model and Bathymetry data (DEMNAS). The tidal correction calculation is done by:

$$r = \frac{X - MSL}{\beta} \tag{1}$$

Where X is the tide hight (in meter) at SAR satellite observation and MSL is mean sea level. While r is correction value for tidal correction process. If the tide condition during SAR satellite observation is in high tide, then the original detected coastline is moved to sea direction. Conversely, If the tide condition during SAR satellite observation is in low tide, then the original detected coastline is moved to land direction.

Table 2. Tidal condition at study area during SAR satellite observation

Satellite date and time (UTC) (dd-mm-yyyy hh:mm:ss)	Local date and time (GMT+8) (dd-mm-yyyy hh:mm:ss)	Tide	Condition
02-07-2009 15:16:05	02-07-2009 21:16:05	1.79 m	High tide
13-04-2022 10:41:35	13-04-2022 18:41:35	0.93 m	High tide

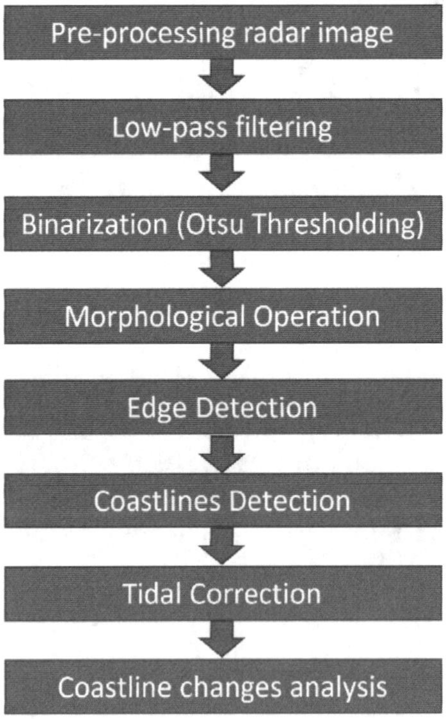

Figure 3. The processing steps of the proposed method to obtain the coastlines and calculate the coastline changes

4. Results and Discussion

The effectiveness of low-pass filter method to increase the quality of the pixel value distribution (histogram) was tested by author in [6]. The good histogram for detection of coastline shows there is a gap or separation between pixel from water and land. This is the first process to improve the final result of detected coastline. Fig. 4 show the effectiveness of low-pass filter to increase the quality of histogram. Fig. 4a show before low-pass filtering and Fig. 4b show after low-pass filtering.

Fig. 4 show the comparison the proposed method and the method developed by [5] (in this case it is called method-1). The comparison of proposed method and method-1 [5] is presented in Table 3. Based on the result in Fig. 5, it seems that the proposed method provides better results than method-1. Method-1 removes the real coastline that created due to construction of break water and groins Fig. 5c. It is maybe due to strong effect of the Median-filtering by kernel 5x5. The other possible reason also the effect of opening morphological operation that used in Method-1.

The proposed method shows very clear the shape of coastline due to construction of break water and groins (Fig. 5d). It is also agreed well visually with the optical image in Fig. 5a. This detected binarization image (Fig. 5d) becomes one of the key accuracies of the final detected coastline in this area.

Table 3. Comparison of the proposed method and method-1 developed by [5]

Proses or calculation	Proposed Method	Method-1
Despeckling filter	Low-pass filtering by kernel 3x3	Median-filtering by kernel 5x5
Binarization	Otsu Thresholding	Otsu Thresholding
Morphological operation	Closing operation	Opening and closing
Edge detection	Canny Edge Detection	Canny Edge Detection

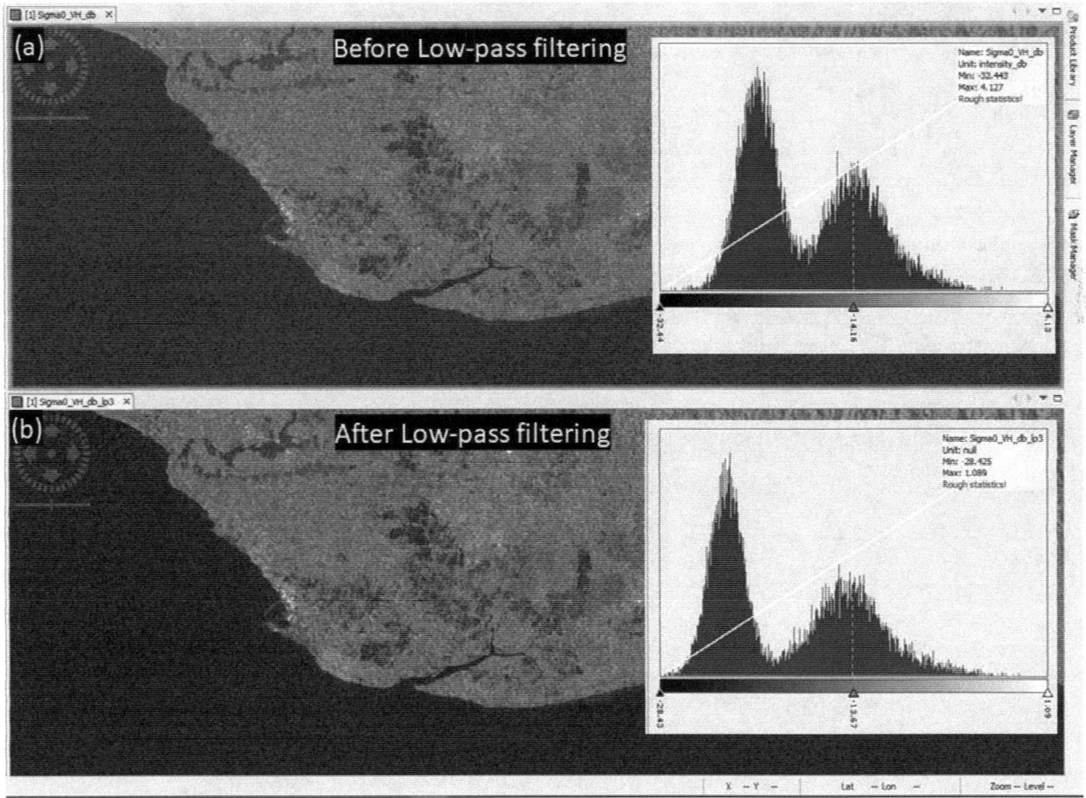

Figure 4. The effectiveness of the low-pass filter method to increase the quality of the pixel value distribution (histogram), (a) before low-pass filtering, and (b) after low-pass filtering.

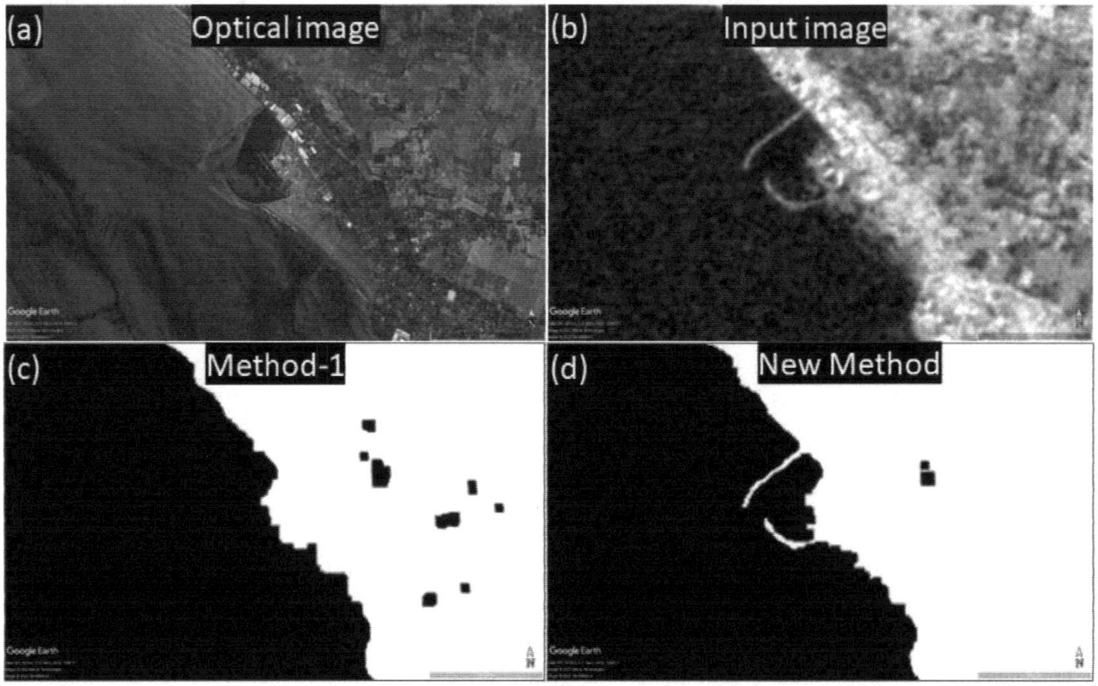

Figure 5. Comparison of binarization images produced by proposed method and method-1: (a) optical image by Google-Earth as a reference, (b) input image, (c) binarization by Method-1, and (d) binarization by New Method.

Fig. 6 shows the effect of tidal correction for each detected coastline. The Blue line indicates before tidal correction and red line indicated after tidal correction Fig. 6a shows the coastlines before and after tidal correction for ALOS-1 (2 July 2009). The base photo in Fig. 6a was taken at same date with the acquisition date of ALOS-1. The detected coastline by Sentinel-1 satellite on 13 April 2022 is presented by Fig. 6b (before and after tidal correction).

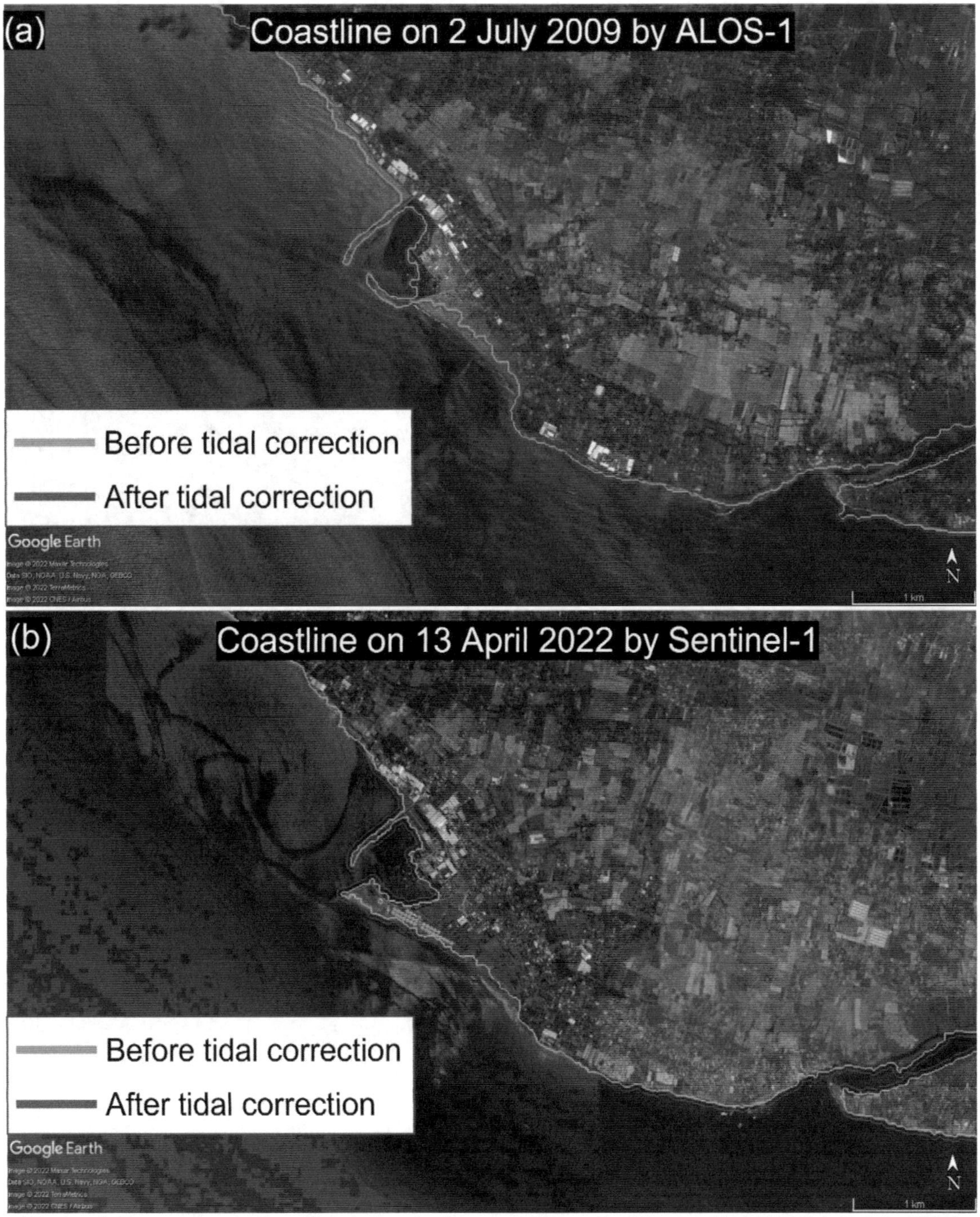

Figure 6. Detected coastlines for each radar image: (a) coastline detected by ALOS-1 (2 July 2009) and (b) coastline detected by Sentinel-1 (13 April 2022).

The final coastline changes calculation is presented by Fig. 7. In general, the coastline changes are presented in Net Shoreline Movement (NSM) and End Point Rate (EPR). The NSM is the total movement between the two shoreline positions. It is calculated based on the distance between the oldest and the youngest shorelines. The EPR is the result of distance calculation of shoreline movement by the time elapsed between the two shorelines. The unit of EPR is meters per year with positive values indicating accretion and negative values indicating erosion.

Fig. 7 shows the NSM of Pengambengan Port and its surrounding area. The maximum value of NSM is 90 to 300 meters and it occurs mostly in the south-east area of Pengambengan port. The minimum value is located in north-west part of Pengambengan part with -95.73 to -50.00 meters.

The EPR of Pengambengan port and its surrounding are is presented in Fig. 8. The positive value (accretion) is mostly located in south-east of the Pengambengan port with range 7.01 to 23.00 meters per year. The negative value (erosion) is dominant occur in north-west of the Pengambengan port, -7.49 to – 4.00.

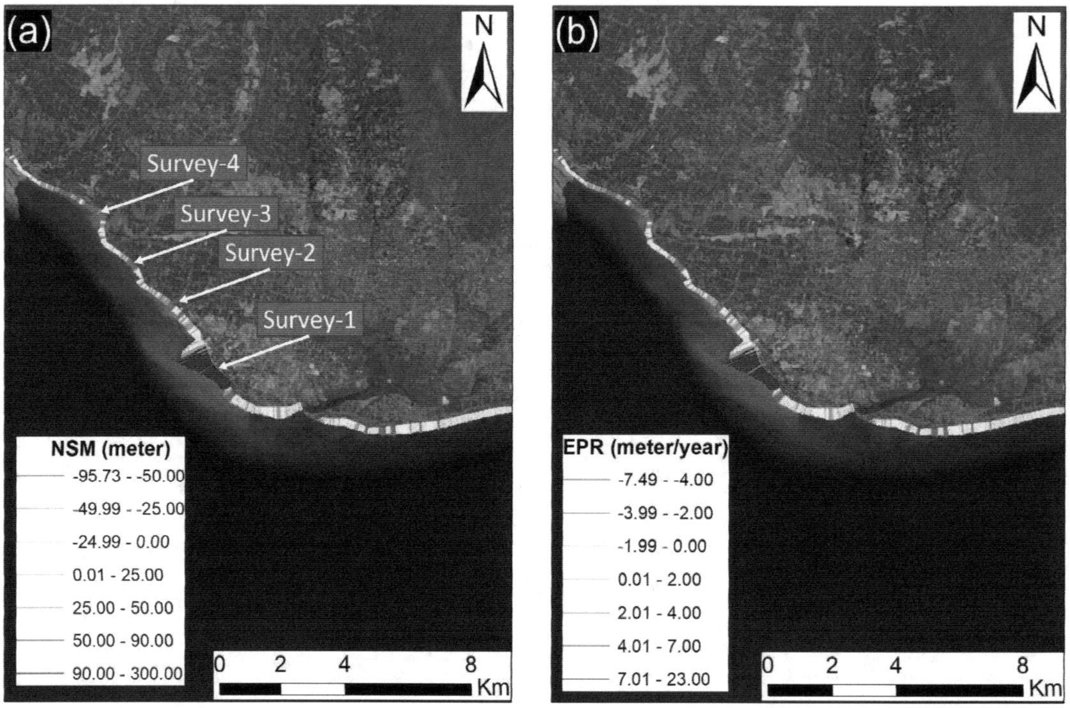

Figure 7. The result of coastline changes calculation: (a) Net Shoreline Movement (NSM) in meter, and (b) End Point Rate (EPR) in meter/year.

In order to verify the results of this study, several ground surveys were conducted at selected locations as shown in Fig. 7a. The survey-1 is located in the Pengambengan Port, it is confirmed that this area is accretion area (Fig. 8a). The second survey, survey-2, is located in Cupel beach. This area is erosion area and it is confirmed by Fig. 8b. The survey-3 is located near Baluk beach. Based on the SAR results, this area is erosion area and it is confirmed by Fig. 8c. The most eroded area is located in Pebuahan beach (Survey-4). The current condition of Pebuahan beach is shown by Fig. 8d.

Figure 8. Survey photos of selected location. The location on the map is shown in Fig. 7a: (a) location of Survey-1 in Pengambengan Port, (b) location of Survey-2 in Cupel beach, (c) location of Survey-3 in Baluk beach, and (d) location of Survey-4 in Pebuahan beach.

5. Conclusions

The new simple approach to detect coastline changes due to the construction of man-made structures is presented in this study. It uses low-pass filter to reduce the speckles errors and Otsu thresholding method the create binary image. A closing morphological operation was conducted to improve the quality of binary image. The canny edge detection is used for coastline delineation. The coastlines changes analysis was done by using Digital Shoreline Analysis System (DSAS). The processing steps of this new approach is a semi-automatic process, it does not require manual digitization to extract the coastline. This method was tested using two independent synthetic aperture radar (SAR) satellite images, i.e., ALOS-1 and Sentinel-1 SAR data.

The results show that this new approach produced better results than other well-known methods. This new approach detects clearly the man-made structure and its impact to the coastal zone, i.e., abrasion and accretion. The study area is located in Pengambengan Port, Jembrana Regency, Bali, Indonesia. The accretion occurs in the area of Pengambengan port (south-east part). The erosion areas are mostly located in north-west part of Pengambengan port, i.e., Pebuahan beach. Those results were confirmed by ground surveys at selected areas. The result of this study can be used as evaluation parameters for future development of Pengambengan port or re-design of present structures.

References

[1] Anfuso G, Pranzini E, Vitale G 2011 An integrated approach to coastal erosion problems in northern Tuscany (Italy): Littoral morphological evolution and cell distribution," *Geomorphology* **129**-204-21, doi: 10.1016/j.geomorph.2011.01.023.

[2] Gens R 2010 Remote sensing of coastlines: Detection, extraction and monitoring *International Journal of Remote Sensing* **31** 1819–1836, doi: 10.1080/01431160902926673.

[3] Alesheikh A A, Ghorbanali A, Nouri N 2007 Coastline change detection using remote sensing *International Journal of Environmental Science and Technology* **4** 61–66, doi:10.1007/BF03325962.

[4] Parwata I N S, Shimizu N, Grujić B, Zekan S, Čeliković R, Imamović E, Vrkljan I 2020 Monitoring the Subsidence Induced by Salt Mining in Tuzla, Bosnia and Herzegovina by SBAS-DInSAR Method *Rock Mechanics and Rock Engineering* **53** 5155–5175, doi:10.1007/s00603-020-02212-1.

[5] Spinosa A, Ziemba A, Saponieri A, Navarro-Sanchez D, Damiani L, Serafy G E 2018 Automatic Extraction of Shoreline from Satellite Images: a new approach *IEEE International Workshop on Metrology for the Sea; Learning to Measure Sea Health Parameters (MetroSea)*

[6] Pujianiki N N, Parwata I N S, Osawa T 2021 A New Simple Procedure for Extracting Coastline from SAR Image Based on Low Pass Filter and Edge Detection Algorithm *Lontar Komputer : Jurnal Ilmiah Teknologi Informasi* **12** 175, doi: 10.24843/lkjiti.2021.v12.i03.p05.

IOP Conf. Series: Earth and Environmental Science 1117 (2022) 012042 doi:10.1088/1755-1315/1117/1/012042

Domestic Wastewater Treatment with Multi Soil Layering Method with Variations of Hydraulic Loading Rate (HLT) And Organic Materials at Anaerobic Conditions

A Zubair[1], R Ibrahim[1], N A P Mangarengi[1], T D B Yuswanda[1] and N O Abdullah[1]

[1]Department of Environmental Engineering, Faculty of Engineering, Universitas Hasanuddin, Jl. Poros Malino KM. 6, Bontomarannu, Gowa, South Sulawesi, Indonesia 92172

achmad.zubair@gmail.com

Abstract. Domestic wastewater is the largest producer of wastewater in Indonesia which comes from the remnants of household activities. Hazardous substances such as Biological Oxygen Demand (BOD), Chemical Oxygen Demand (COD), Total Suspended Solid (TSS), Oil and Fat contained in wastewater can cause health problems for humans. In this study, domestic waste treatment was carried out using the Multi Soil Layering (MSL) method. The MSL system is a wastewater treatment method that utilizes the ability of soil as the main medium and several other media such as zeolite, gravel, charcoal, and sawdust which are formed in a multilevel brick pattern. The independent variables are the variation of soil mixture with sawdust (M1) and soil mixture with charcoal (M2) and also variations of Hydraulic Loading Rate namely HLR 500 L/m2.day (H1), 750 L/m^2.day (H2), 1000 L/m^2.day (H3). The characteristics of domestic wastewater include pH 6, BOD 236 mg/L, COD 514 mg/L, TSS 114 mg/L, Oil and Fat 8200 mg/L. The results showed that the best variation in the removal of pollutant levels was a mixture of soil with charcoal (M2) and HLR 500 L/m^2.day (H1), with the average efficiency in reducing BOD, COD, TSS, Oil & Fat levels, respectively. 86.39%, 78.29%, 86.56% and 79.60%.

1. Introduction

The rapid population growth in Indonesia, especially in urban areas, is due to the increasing flow of urban urbanization. The growing number of residents makes the need for clean water also increase. Still, every year the availability of clean water decreases due to the number of polluted rivers and the disposal of wastewater into rivers without prior treatment.

The largest producer of liquid waste in Indonesia comes from the remnants of household activities known as domestic wastewater. Domestic wastewater is the primary source of pollution in rivers because the results of the waste from household activities are directly channeled into drainage channels which then flow into the river. Based on data from the Central Statistics Agency in 2017, there was an increase in the number of moderately polluted rivers from 9 to 11 and the number of heavily polluted rivers from 14 to 19. Sources of domestic waste pollution are black water in the form of toilet water and gray water in the form of bath water, water used for washing clothes, and water used for activities in the kitchen. Hazardous substances such as Biological Oxygen Demand (BOD), Chemical Oxygen Demand (COD), Total Suspended Solid (TSS), Fat Oil, Fecal Coli, and Total Coli contained in wastewater can cause health problems for humans who consume them and can also cause health problems. Life in the water will be disturbed like aquatic plants and animals that live in it.

IOP Conf. Series: Earth and Environmental Science 1117 (2022) 012042 doi:10.1088/1755-1315/1117/1/012042

Various wastewater treatment techniques have been developed. However, each of these wastewater treatment techniques has advantages and disadvantages. One alternative to wastewater treatment that has the advantage of being low-cost is natural sewage treatment using soil. However, applying soil utilization as a conventional method of treating wastewater has limitations, namely clogging and uneven and ineffective distribution of liquid waste in the nitrification/denitrification process. It requires a large land area compared to modern systems [1]. Domestic wastewater treatment using the Multi Soil Layering system (MSL) is a choice that maximizes the function of the soil in the water treatment process. The Multi Soil Layering (MSL) system is a liquid waste treatment method that utilizes the ground's ability as the primary medium in removing pollutant parameters and increasing the capacity of the soil by improving its structure [1]. Multi Soil Layering (MSL) is also known to be simple, easy in terms of operation and control, and environmentally friendly because it uses natural materials and is easy to obtain. This MSL has been tested in several countries. The purpose of this research is as follows: Knowing the characteristics of domestic wastewater samples used in the study; Analyzing the comparison of the effectiveness of organic matter mixed with charcoal and soil mixture with sawdust in the removal of pH, BOD, COD, TSS, Oil, and Fat; Analyzing the effect of Hydraulic Loading Rate (HLR) on the removal efficiency of pH, BOD, COD, TSS, Oil, and Fat.

2. Material and Methods

2.1. Research Design

This type of research is an experimental/pilot quantitatively by testing the effect of the kind of organic material layer and the Hydraulic Loading Rate (HLR) in removing pollutants in domestic wastewater. This study uses research variables: The independent variables affect or cause changes or the emergence of the dependent variable. In contrast, the independent variables are variations in organic material and Hydraulic Loading Rate. The variations in the treatment in this study are as follows.

Table 1. Variations of Treatment

Organic Material (M)	***Hydraulic Loading Rate* (H)**		
	500 L/m2.day	750 L/m2.day	1000 L/m2.day
Soil and sawdust mix	M1H1	M1H2	M1H3
Mixture Soil and charcoal	M2H1	M2H2	M2H3

The variables are measured to determine the magnitude of the effect or the influence of the independent variables, namely the impact of variations in organic material and the importance of variations in the use of Hydraulic Loading Rate (HLR) on the removal of parameters pH, BOD, COD, TSS, Oil, and Fat. The selection of these parameters is based on the Decree of the State Minister of Environment and Forestry of the Republic of Indonesia No. 68 of 2016 concerning Domestic Wastewater Quality Standards.

2.2. Reactor Design

The reactor consisted of 6 units made of glass with dimensions following the design criteria, namely 50 x 15 x 50 cm, and equipped with an influent tub that functions as a waste reservoir to drain into the MSL reactor, which is flowed by gravity, and equipped with an effluent tub to be used as a reservoir for treated

IOP Conf. Series: Earth and Environmental Science 1117 (2022) 012042 doi:10.1088/1755-1315/1117/1/012042

water from the reactor which is placed lower than the MSL reactor so that the reactor outlet is easier to drain into the reactor effluent tub.

a) Rock Layer

The rock layers used for the two MSL reactors are zeolite and gravel. The gravel used is 10 cm in size, while the zeolite used results from a 3-5 mm sieve. Using zeolite and gravel of the same size aims to avoid clogging.

b) Layers of Organic Material

The soil used in this study is a type of volcanic soil formed due to volcanism. The Andosol soil was chosen because it has high nutrients, has a broader surface area for adsorption, and is fine-grained to prevent clogging. The function of charcoal is to purify water. It is because charcoal contains activated carbon which can work by absorption or absorption. In addition to helping absorb organic matter, charcoal also plays a vital role in color adsorption.

The sawdust used is the result of wood construction work. Sawdust is rich in carbon sources that can adsorb contaminants. A mixture of soil and charcoal, as well as a mixture of soil with sawdust, is needed to construct the two reactors. It is estimated that the weight ratio of the layer composition of the organic material used is:

Soil : charcoal = 1 : 1 Soil : sawdust = 1 : 1

This mixture of soil with charcoal or sawdust is then formed into a brick shape, with dimensions of 10 cm long, 15 cm wide, and 5 cm high. The bottom of the tub is filled with gravel with a height of 5 cm, then the entire surface of the gravel is covered with a net. The next second layer is filled with zeolite with a height of 5 cm. The third layer is filled with a mixture of soil with charcoal or sawdust with a distance between bricks of 2 cm and the distance between the second layer 5 cm, and the next layer is filled with zeolite as high as 5 cm. Then the other layers are served the same way to form three layers of soil blocks. For more details, the reactor image can be seen in Figure 1.

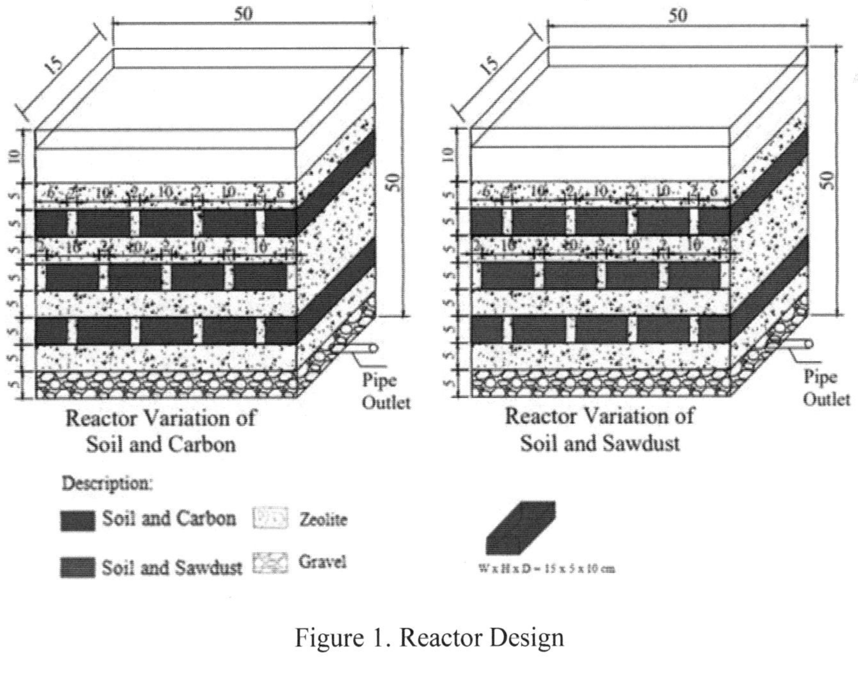

Figure 1. Reactor Design

2.3. Research Implementation

In this study, experiments were carried out continuously. Wastewater drainage is carried out at HLR 500 L/m2.day, 750 L/m2.day, 1000 L/m2.day. HLR is used as such because it refers to previous research as was done in the research of [2]. HLR is the rate of hydraulic loading of wastewater to a surface area in a certain time unit. To be able to drain wastewater into MSL, it is necessary to know the flow rate. The relationship between HLR and discharge mathematically can be seen in the following equation:

$$QHLR = HLR \text{ X Surface Area} \quad (1)$$

From the above equation, the required discharge for HLR 500 L/m^2.day is 26 ml/minute, for 750 L/m^2.day is 39 ml/minute and for HLR 1000 L/m^2.day is 52 ml/minute. The experimental phase begins with testing the pollutant parameters before and after being put into the reactor. The experiment for each HLR was carried out twice.

2.4. Data Analysis

This research was conducted by analyzing laboratory data, namely the results of testing the parameters pH, BOD, COD, TSS, Oil & Fat in domestic wastewater before and after treatment. Data obtained from the laboratory were compared with the required quality standards to determine the feasibility of processing in this study. Furthermore, the test data were analyzed using Two Way Anova on the IBM SPSS statistic 26 application to determine the effect of variations in organic material and hydraulic loading rate in removing pollutants from domestic wastewater. Furthermore, the data is also graphed to find the best take in the study. The data were analyzed using the following formula. The percentage of Removal Efficiency can be seen below:

$$\% \, Removal = (C0 - C1)/Co \text{ x } 100\% \quad (2)$$

Where C0 is the initial concentration and C1 is the final concentration.

3. Results and Discussions

3.1. Characteristics of Domestic Wastewater

The wastewater used in this study is domestic black water and gray water obtained from Pao-Pao Permai Housing, where black water comes from latrines. In contrast, gray water comes from the kitchen, bathroom, and washing activities. Data from laboratory tests on water samples that describe the characteristics of Domestic wastewater data from the research results are presented in Table 2 below.

Table 2. Characteristics of Domestic Water

Parameter	Wastewater (mg/L)	Standard (mg/L)	Criteria
pH	6	6-9	Meet Quality Standard
BOD	236	30	Does not Meet Quality Standard
COD	514	100	Does not Meet Quality Standard
TSS	114	30	Does not Meet Quality Standard
Oil & Grease	8200	10	Does not Meet Quality Standard

Based on the table, it can be seen that the parameters do not meet the required quality standards based on the Regulation of the Minister of Environment and Forestry of the Republic of Indonesia Number 68 of 2016 concerning Domestic Waste Quality Standards. The BOD/COD ratio is 0.45, so it can be treated biologically because it is biodegradable. The ratio of BOD/COD between 0.2 and 0.5 can be processed by biological processes [2].

3.2. Effect of Organic Material Variations

3.2.1. pH
The test data showed that using the Multi Soil Layering system in each treatment variation increased the pH level so that the pH of the water moved towards neutral and even reached a value of 8 with a pH range of 7.23-8.14. The increase in the pH value of domestic wastewater after passing through Multi Soil Layering is influenced by the use of Andosol soil. Andosol soils have a high buffering capacity against changes in chemical and physical conditions due to the influence of microorganism activity and physical reactions caused by the occurrence of wastewater treatment mechanisms in the MSL system, such as changes in pH [3].

3.2.2. Biological Oxygen Demand (BOD)
A two-way ANOVA statistical analysis was carried out to determine the effect of variations in organic matter. The results can be seen in Table 3 below.

Table 3. Analysis of Influence of Organic Material Parameters BOD

Mean Square	F Count	F Table	Description
263.964	0.510	5.11736	H0 rejected

In Table 3 can be seen that the variation of organic material influences the decrease in the BOD value, where the calculated F value < F Table is 0.510 < 5.11736 so that H0 is rejected and H1 is accepted as an effect. To find out the best variety of Organic Materials for removing BOD contaminant levels can be seen in Figure 2 below.

The relationship between the efficiency of removing BOD contaminants and the variation of organic matter depicted in the diagram above shows that the interpretation of organic matter mixed with soil and charcoal (M2) has a higher ability to reduce BOD levels than the variation of organic matter mixed with sawdust (M1).). The average percentage of removal efficiency for the interpretation of soil mixed with charcoal (M2) is 73.62%-86.39%, while for the variation of soil mixture with sawdust (M1), it is 69.31%-77%. It is because charcoal contains 85% to 98% carbon while sawdust only contains 52.25% carbon content [4]. One essential nutrient needed for microorganisms is carbon [5].

3.2.3. Chemical Oxygen Demand (COD)
A two-way ANOVA statistical analysis was carried out to determine the effect of variations in organic matter. The results can be seen in Table 4 below.

Table 4. Analysis of Influence of Organic Material Parameters COD

Mean Square	F Count	F Table	Description
563.584	1.218	5.11736	H0 rejected

In Table 4 can be seen that the variation of organic material influences the decrease in the COD value, where the calculated F value < F Table is 1.218 < 5.11736 so that H0 is rejected and H1 is accepted, which means there is an effect. the best variation of Organic Material is in the removal of COD contaminants, which can be seen in Figure 3 below

The relationship between the efficiency of removal of COD contaminants and the variation of organic matter depicted in the diagram above shows that the interpretation of organic matter in a mixture of soil and charcoal (M2) has a higher ability to reduce COD levels than the variation of organic matter in a combination of sawdust and soil (M1). The average percentage of removal efficiency for the interpretation of soil mixed with charcoal (M2) can be seen from 70.14%-78.29%, while the variation of soil mixture with sawdust (M1) is 58.72%-67.49%. Because charcoal has more delicate pores than sawdust, more adsorption fields will bind ions in wastewater [6].

3.2.4. Total Suspended Solid (TSS)
A two-way ANOVA statistical analysis was carried out to determine the effect of variations in organic matter. The results can be seen in Table 5 below.

Tabel 5. Analysis of Influence of Organic Material Parameters TSS

Mean Square	F Count	F Table	Description
90.182	0.712	5.11736	H0 rejected

In table 5 can be seen that the variation of organic material influences the decrease in the TSS value, where the calculated F value < F Table is 0.712 < 5.11736, so that H0 is rejected and H1 accepted means there is influence. Find out the best variation of Organic Material in removing TSS contaminants. It can be seen in Figure 4 below.

The relationship between the efficiency of removing TSS contaminants and the variation of organic matter depicted in the diagram above shows that the interpretation of organic matter in a mixture of soil and charcoal (M2) has a higher ability to reduce TSS levels than the variation of organic matter in a combination of sawdust and soil (M1). It can be seen from the average percentage of removal efficiency for variations in the mixture of soil and charcoal (M2) of 80.97%-86.56%, while the interpretation of soil mixture with sawdust (M1) is 76.79%-82.19%. It is due to the adsorption process taking place better in a mixture of soil and charcoal than in a mixture of soil and clay sawdust because the pores in the charcoal are more than the surface area of the adsorbent is larger. So that the larger the surface area, the more effective it will be to absorb and reduce the concentration of TSS [7].

3.2.5. Oil and Grease
A two-way ANOVA statistical analysis was carried out to determine the effect of variations in organic matter. The results can be seen in Table 6 below.

Table 6. Analysis of Influence of Organic Material Parameters Oil & Grease

Mean Square	F Count	F Table	Description
177.913	0.449	5.11736	H0 rejected

In Table 6, the variation of organic material has an influence on the decrease in the value of Oil & Fat, where the calculated F value < F Table is 0.449 < 5.11736 so that H0 is rejected and H1 is accepted, which means there is an effect. The best variation of Organic Material is in removing Oil & Fat contaminants, which can be seen in Figure 5 below.

The relationship between the efficiency of removing oil and grease contaminants and the variation of organic matter depicted in the diagram above shows that the interpretation of organic matter in a mixture of soil with charcoal (M2) has a higher ability to reduce oil and fat content compared to the variation of organic material in a mix of soil with charcoal (M2). Sawdust (M1). It can be seen from the average percentage of removal efficiency for the variation of soil mixed with charcoal (M2) of 70.73% - 79.60%, while the interpretation of soil mixture with sawdust (M1) is 64.67%-73.64%. It is caused by charcoal as an adsorbent in oils & fats, which can absorb peroxide values. The process of absorption of peroxide value goes through three stages; the amount of peroxide is adsorbed on the outside, then moves towards the pores, and is adsorbed onto the charcoal wall [8].

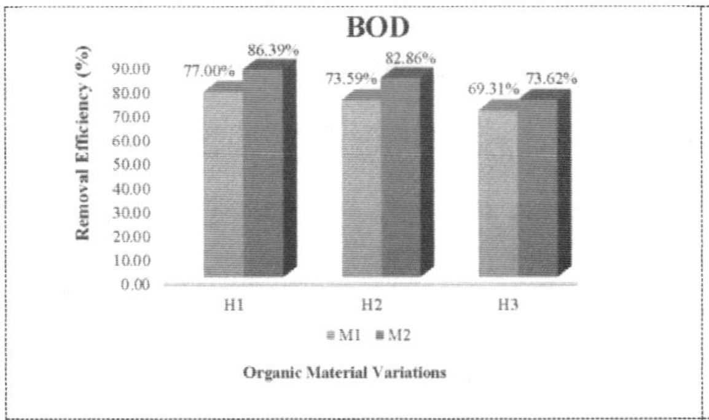

Figure 2. Comparison of BOD removal efficiency between treatment variations of organic material mixed with soil and sawdust and organic material mixed with soil and charcoal

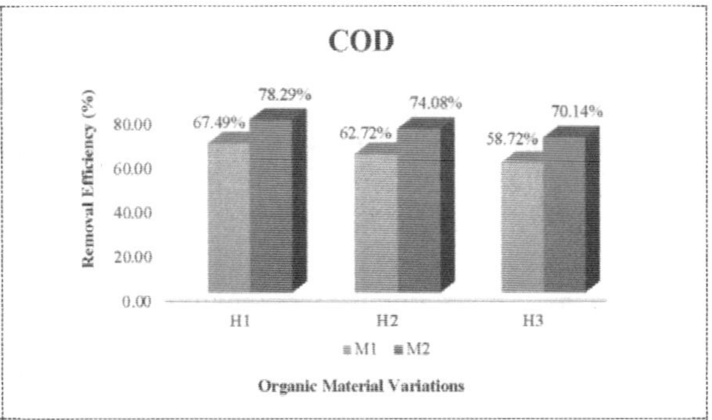

Figure 3. Comparison of COD removal efficiency between treatment variations of organic material mixed with soil and sawdust and organic material mixed with soil and charcoal

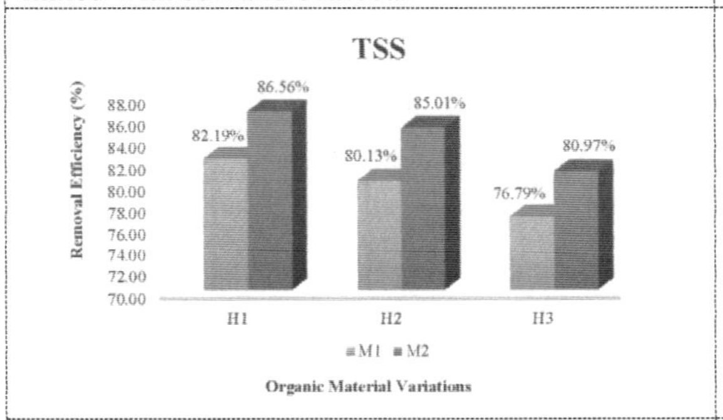

Figure 4. Comparison of TSS removal efficiency between treatment variations of organic material mixed with soil and sawdust and organic material mixed with soil and charcoal

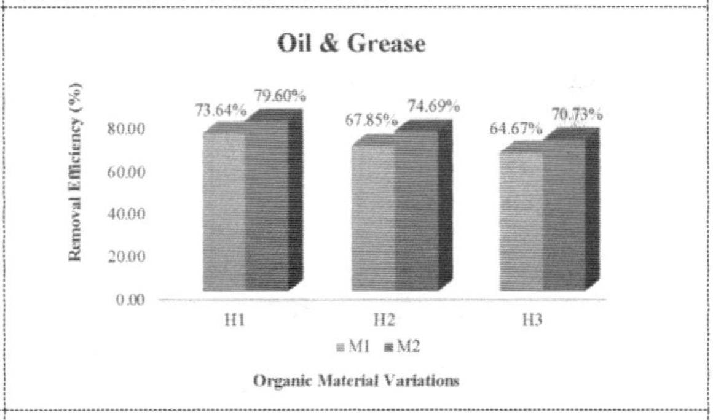

Figure 5. Comparison of Oil & Grease removal efficiency between treatment variations of organic material mixed with soil and sawdust and organic material mixed with soil and charcoal

3.3. Effect of Hydraulic Loading Rate (HLR) Variations

3.3.1. pH

The test data showed that using the Multi Soil Layering system in each treatment variation increased the pH so that the pH of the water moved towards neutral with a pH level of 7.23-8.14. Soil contains acidic and basic cations, so the soil causes changes in pH [9]. Earth can neutralize changes in physical and chemical conditions due to the activity of microorganisms and physical reactions that arise when the wastewater treatment mechanism occurs in Multi Soil Layering.

3.3.2. Biological Oxygen Demand (BOD)

A two-way ANOVA statistical analysis was carried out to determine the effect of variations in the Hydraulic Loading Rate. The results can be seen in Table 7 below.

Table 7. Analysis of the Influence of Hydraulic Loading Rate Parameters BOD

Mean Square	F Count	F Table	Description
162.152	0.314	4.25649	H0 rejected

In Table 7 can be seen that the variation of the Hydraulic Loading Rate influences the decrease in the BOD value, where the calculated F value < F Table is 0.314 < 4.25649 so that H0 is rejected and H1 is accepted, which means there is an effect. The best variation of the Hydraulic Loading Rate in the removal of BOD contaminants can be seen in Figure 6 following.

The relationship between the efficiency of removing BOD contaminants and the variation of the Hydraulic Loading Rate depicted in Figure 15 above shows the variation in HLR 500 L/m^2. Day (H1) has a higher ability to reduce BOD levels than the variation of HLR 750 L/m^2. Day (H2). Meanwhile, HLR 750 L/m^2. Day (H2) has a higher capacity than HLR 1000 L/m2. Day (H3). It can be seen from the average percentage of removal efficiency for H1 of 77%-86.39%, H2 of 73.59%-82.86%, and H3 of 69.31%-73.62%. It means that the lower the speed of Multi Soil Layering, the higher the processing level [10].

3.3.3. *Chemical Oxygen Demand (COD)*

A two-way ANOVA statistical analysis was carried out to determine the effect of variations in the Hydraulic Loading Rate. The results can be seen in Table 8 following.

Table 8. Analysis of the Influence of Hydraulic Loading Rate Parameters BOD

Mean Square	F Count	F Table	Description
107.615	0.233	4.25649	H0 rejected

In table 8 can be seen that the variation of the Hydraulic Loading Rate affects decreasing the COD value, where the calculated F value < F Table is 0.233 < 4.25649, so H0 is rejected and H1 is accepted, which means there is an effect. Find out the best variation of the Hydraulic Loading Rate in removing COD contaminants which can be seen in Figure 7 below.

The relationship between COD contaminant removal efficiency and the variation of the Hydraulic Loading Rate depicted in Figure 16 above shows the interpretation of HLR 500 L/m2. Day (H1) can reduce COD levels more than the HLR variation of 750 L/m^2. Day (H2). Meanwhile, HLR 750 L/m^2. Day (H2) has a higher capacity than HLR 1000 L/m^2. Day (H3). It can be seen from the average percentage efficiency for H1 of 67.49%-78.29%, H2 of 62.72%-74.08%, and H3 of 58.72%-70.14%. The amount of COD removal in the HLR 500 L/m2.day is due to the slower flow rate in the HLR, which makes the ion exchange process that occurs in zeolite and organic material in MSL work more optimally in reducing COD levels [9].

3.3.4. *Total Suspended Solid (TSS)*

A two-way ANOVA statistical analysis was carried out to determine the effect of variations in the Hydraulic Loading Rate. The results can be seen in Table 9 below.

Tabel 9. Analysis of the Influence of Hydraulic Loading Rate Parameters TSS

Mean Square	F Count	F Table	Description
47.118	0.372	4.25649	H0 rejected

In Table 9 can be seen that the variation of the Hydraulic Loading Rate influences the decrease in the TSS value, where the calculated F value < F Table is 0.372 < 4.25649 so that H0 is rejected and H1 is accepted, which means there is an effect. Find out the best variation of the Hydraulic Loading Rate in the removal of TSS contaminants can be seen in Figure 8 below.

The relationship between the efficiency of removing TSS contaminants with variations in the Hydraulic Loading Rate is depicted in Figure 17 below above shows the variation of HLR 500 L/m^2. Day (H1) has a higher ability to reduce TSS levels than the variation of HLR 750 L/m2. Day (H2). Meanwhile, HLR 750 L/m^2. Day (H2) has a higher capacity than HLR 1000 L/m2. Day (H3). It can be seen from the average percentage of removal efficiency for H1 of 82.19%-86.56%, H2 of 80.13%-85.01%, and H3 of 76.79%-80.79%. At the HLR 500 L/m^2.day, the flow rate is slower, so the adsorption process that occurs in MSL works more optimally in the adsorbate adsorption process [11].

3.3.5. *Oil and Grease*

A two-way ANOVA statistical analysis was carried out to determine the effect of variations in the Hydraulic Loading Rate. The relationship can be seen in table 10.

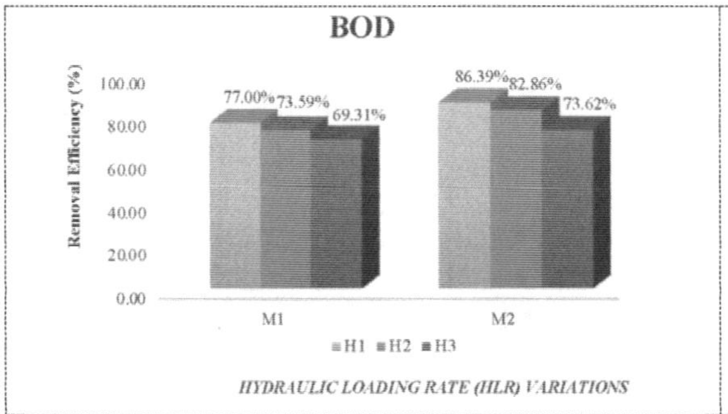

Figure 6. The relationship between the efficiency of removing BOD contaminants and variations in the Hydraulic Loading Rate

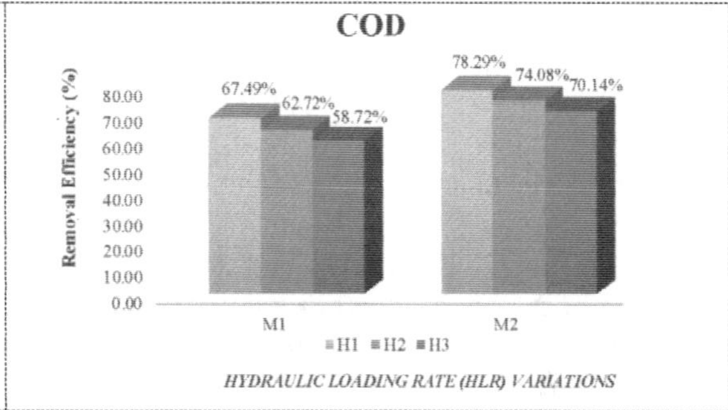

Figure 7. The relationship between the efficiency of removing COD contaminants and variations in the Hydraulic Loading Rate

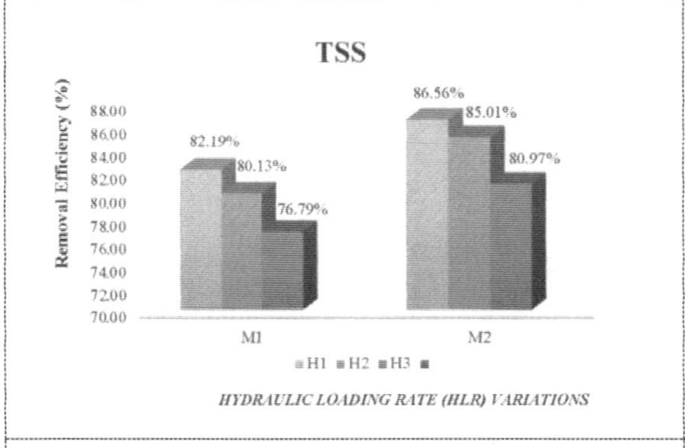

Figure 8. The relationship between the efficiency of removing TSS contaminants and variations in the Hydraulic Loading Rate

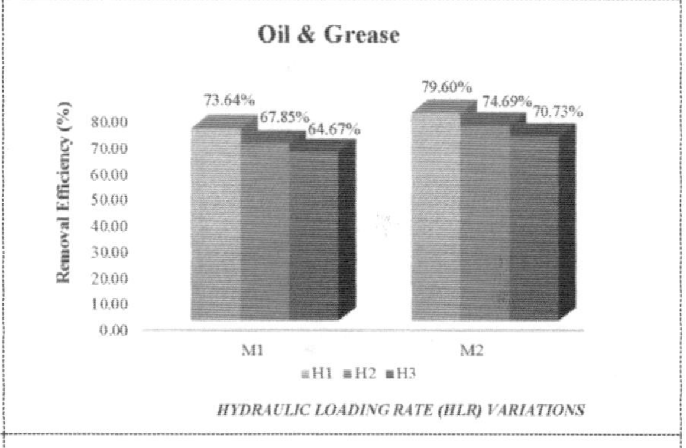

Figure 9. The relationship between the efficiency of removing Oil & Grease contaminants and variations in the Hydraulic Loading Rate

Table 10. Analysis of the Influence of Hydraulic Loading Rate Parameters on Oil & Grease

Mean Square	F Count	F Table	Description
120.886	0.305	4.25649	H0 rejected

In Table 10 can be seen that the variation of thc Hydraulic Loading Rate influences the decrease in the value of oil & grease, where the F value is calculated <F Table, i.e., 0.305 < 4.25649, so H0 is rejected, and H1 is accepted, which means there is an effect. To find out the best variation of the Hydraulic Loading Rate in the removal of oil & grease contaminants can be seen in Figure 9 below.

The relationship between the efficiency of removing oil & grease contaminants and the variation of the Hydraulic Loading Rate depicted in Figure 18 above shows the variation of HLR 500 L/m^2. Day

(H1) has a higher ability to reduce oil & grease content compared to the variation of HLR 750. L/m². Day (H2). Meanwhile, HLR 750 L/m². Day (H2) has a higher capacity than HLR 1000 L/m2. Day (H3). It can be seen from the average percentage of removal efficiency for H1 of 73.64%-79.60%, H2 of 67.85%-74.69%, and H3 of 64.67%-70.73%. Variation of HLR 500 L/m². Day (H1) has a higher ability to reduce levels of Oil & Fat than other HLR variations because HLR 500 L/m².day has a longer contact time. Contact time is an influential factor in adsorption [12,13,14].

4. Conclusion

The characteristics of domestic wastewater have a pH value of 6, BOD of 236 mg/L, COD 514 mg/L, TSS 114 mg/L, and Oils and Fats 8200 mg/L. The levels of these contaminants exceed the required quality standards based on the Regulation of the Minister of Environment and Forestry of the Republic of Indonesia Number 68 of 2016 concerning Domestic Waste Quality Standards. The use of organic material can remove contaminant levels from all test parameters. In organic material, a mixture of soil and charcoal (M2) can remove higher contaminants than organic material mixed with sawdust soil (M1). Using the Hydraulic Loading Rate affects the removal of contaminant levels for all test parameters, where the lower the HLR, the higher the efficiency of removing contaminants. It is because the smaller the HLR, the longer the contact time with MSL

References

[1] Salmariza, Sy. and Sofia. 2011. Application of MSL (Multi Soil Layering) Method to Treat Edible Oil Industrial Wastewater. Journal of Industrial Research, 3:227-238.

[2] Taufiq Ihsan., Shinta Indah. and Denny Helard. 2013. Elimination of Potassium from Liquid Waste of Rice Fields with Multi Soil Lavering (MSL) Method. Journal of Environmental Engineering UNPAD. 10(2): 133-141.

[3] Masunaga, T, et al, 2007. Characteristics Of Wastewater Treatment Using A Multisoil Layering System In Relation To Wastewater Contaminant Levels and Hydraulic Loading Rate. Soil Science and Plant Nutrition.

[4] Adrianto Ahmad. Yelmida. Frisk Irmawanti. 2011. Provision for Fatty Oils Contained in Palm Oil Industry Liquid Waste With Hybrid Bioreactor Anaerobes with Palm Shell Media. Department of Chemical Engineering, University of Riau.

[5] Affandi, F. and Hendri, H., 2011. Effect of Natural Zeolite Activation Method as a Temperature-Reducing Material for Warm Asphalt Mixture. Journal of the Research and Development Center for Roads and Bridges. 28(1):1-8.

[6] Ahmat Nur. 2009. Decrease in COD (Chemical Demand), TSS (Total Suspended Solid), and Color in Industrial Liquid Waste Aero Carbon filter reactor. Yogyakarta: Indonesian Islamic University.

[7] Ain Khaer. 2016. Applied Technology Utilization of Vehicle Washing Car Waste With Coagulation Method and Multimedia Biofilter. Makassar Ministry of Health Polytechnic Vol.11 No.2.

[8] Utari Windy. Hasan Wirsal. Surya Dharma. 2019. The Effectiveness of Activated Carbon in Reducing Peroxide Numbers and Color Clarification in Oil Used Fried. Medan: University of North Sumatra.

[9] Attanandana, T., B. Saitthiti, S. Thongpae, S. Kri Kritapirom and T. Wakatsuki. 2000. Wastewater Treatment Study using the MultiSoil Layering system, Paper Submitted on "Managing Water and Waste in the New Millenium", Johannesburg

[10] Mahmoud, N., Zeeman, G., Gijzen, H., & Lettinga, G. 2004. Anaerobic sewage treatment in a one-stage UASB reactor and a combined UASB- Digester system Anaerobic sewage treatment in a one-stage UASB reactor and a combined Digester System. Water UASB Research, (March). 307-322. https://doi.org/10.1016/j.watres.2004.01.041.

[11] Herman, W., Darmawan, D., & Gusnidar, G. 2017. Utilization of Volcanic Soil in Multiple Soil Layering (MSL) System Against Polluted Irrigation Water Purification. BiBieT Journal, 2(2), 49-59.

[12] Ardina Kusuma. 2007. Decrease in TSS Levels (Total Suspended Solid. And Oil Liquid Waste Fat In VCO (Virgin Control Oil) Processing With Filtration Using Activated Carbon Bed and Kapok. Yogyakarta: Indonesian Islamic University.

[13] Abdullah N, Pandebesie, E. 2018. The infuence of stiring and cow manure added on biogas production from vegetable waste using anaerobic digester. IOP Conf. Series: Environmental Science. 135:1–7, Bristol, United Kingdom: IOP Publishing Ltd. DOI: https://doi.org/10.1088/1755-1315/135/1/012005.

[14] Zubair, A., Abdullah, N.O., İbrahim, R. & Rachma, A.R.D. 2020. Effectivity of constructed wetland using Typha angustifolia in analyzing the decrease of heavy metal (Fe) in acid mine drainage. IOP Conference Series: Earth and Environmental Science, 419, 012160.

IOP Conf. Series: Earth and Environmental Science 1117 (2022) 012043 doi:10.1088/1755-1315/1117/1/012043

Analysis of noise level characteristics on Jalan Metro in Makassar city

M Hustim[1], N Bungawali[1], S A Adisasmitha[2], H Alimuddin[1], A Zakaria[3]

[1] Department of Environmental Engineering, Universitas Hasanuddin, South Sulawesi 90245 Indonesia
[2] Department of Civil Engineering, Universitas Hasanuddin, South Sulawesi 90245 Indonesia
[3] Department of Civil Engineering, Politeknik Negeri Ujung Pandang, South Sulawesi 90245 Indonesia

Email: muraliahustim@ft.unhas.ac.id

Abstract: One of the main collector routes in Makassar that connecting Makassar City and Gowa Regency, is Jalan Metro. The development of transportation in Makassar City caused higher level of noise which coming from transportation. Noise from vehicles is one of the environmental problems on this area. The disruption of public comfort and communication problems are both impacted by the noise from vehicles. The purpose of this study is to evaluate the noise characteristics and level on Jalan Metro.
Quantitative research was used in this study. There were 14 sampling points which were the data was collected for 12 hours from 07.00 - 18.00 WITA in every 10 minutes to represent hourly measurements. The data was analyzed by LAeq calculation analysis and Paired Sample T-Test by using SPSS program.
Noise level in this study were found to be fluctuated in all sampling points. This result might be attributed by vehicles volume, vehicle speed, and amount of horns. The results of the analysis show LAeq Day noise level value on Jalan Metro section has exceeded noise level quality standard required by KepMenLH No. 48 Tahun 1996. The maximum LAeq Day value is 78.23 dB at R2 and the minimum LAeq Day is 72.16 dB at R12. In addition, a comparison of the measurement results with Pedoman PU No. 13/2003 concerning the maximum and minimum limits of Leq10 and LAeq was also carried out.

1. Introduction

The development of transportation in cities in Indonesia causes higher of level noise which coming from transportation. The number of motorized vehicles in Indonesia is increasing rapidly with 133,617,012 units. The motorized vehicle population is divided by category, such as motorcycles, passenger cars, buses, and freight vehicles[1]. The amount of pollution that happens in Indonesia, such as noise, vibration, and air pollution, is rising along with the number of transport modes [2].

Urban noise pollution has been directly related to an increase in ground transportation due to city growth and everyday population demands in terms of increased traffic noise levels[3]. Noise was characterized as undesirable sound due to their impact to the environment and human health. Many previous studies have demonstrated its effect on irritability, stress, and subsequent cardiovascular diseases, sleep disturbances, and further impairments [4-5]. In 2014, the European Environment Agency (EEA) classified urban noise as one of the biggest environmental health hazards in all of Europe [6].

Sources of road traffic noise was vehicles, includes heavy vehicles, light vehicles, and motorcycles, with sources of noise including horn, mechanical friction between tires, the sound of the vehicle exhaust, the volume of the vehicle, and the speed of the vehicle [7]. In metropolitan regions, noise-free periods in urban areas are rare during the day and tend to diminish during night-time [8].

Heavy vehicles such as trucks, buses, etc., and passenger cars are the main sources of noise on the highway. In generally, the noise control strategy is divided into three elements, namely control of noise sources, control of noise paths, and control of noise receivers [9]. The purpose of this study is to analyze the characteristics of the noise level on Jalan Metro Makassar and to analyze the level of noise generated by vehicles on Jalan Metro Makassar.

2. Material and methods

2.1 Study Area

Jalan Metro is one of the roads in Makassar which is included in the type of primary collector road, where this road connects Makassar City with Barombong District, Gowa. The land uses that can be found on Jalan Metro Makassar, such as offices and trade, hospitals, trade and services, housing and settlements, green open spaces, as well as government and public facilities. From the land use mentioned above, the noise caused by motorized vehicles crossing Jalan Metro Makassar can have an impact on communication disturbances and disturbances in the comfort of the people living in that location.

Figure 1. Sampling Point

2.2 Data Collection

In this study, noise level, volume of vehicle, speed of vehicle, number of horns from vehicle, and sampling point characteristics (i.e road width, median width, shoulder width, and other things which related to the sampling point). Noise data was collected by using Sound Level Meter (*10176566 Sound Level Meter Datalogger*). Data was collected for 12 hours from 07.00 - 18.00 WITA in every 10 minutes to represent hourly measurements in 14 sampling points (Figure 1). USB cable, Laptop, Tripod were supporting tools that used for noise field measurement. Moreover, *camera or handphone, speed gun, Counter, Walking Distance Meter* was used to calculate the volume of vehicles, the speed of vehicles, the numbers of horns, and the road characteristics, respectively.

2.3 Data Analysis

- Frequency distribution/frequency table of noise level

The noise calculation can be analyzed by making a frequency distribution/frequency table and analyzing the noise level in indicator numbers. Frequency distribution or frequency table is grouping data into several classes and then counting the number of observations that fall into each class. The frequency distribution is created by calculating the number of class intervals, interval values, class marks/middle values, and frequency.

a. Range or range is the difference between the largest value and the smallest value.

$$\text{Range} = \text{Max data} - \text{min data} \quad [1]$$

Where: Max data = largest value data; Min data = smallest value data

b. Number of classes (k)

$$k = 1 + 3{,}3 \log(n) \quad [2]$$

Where: k = number of classes; and n = number of data

c. Interval is data obtained by measurement, where the distance between two scale points is known.

$$I = (\max - \min)k = rk \quad [3]$$

Where:
I = Interval; max = maximum data value; min = minimum value of data; r = range; k = number of class intervals

d. The class mark is the midpoint of the class interval.

$$\text{Middle} = (BB\text{-}BA)2 \quad [4]$$

Where: BB = lower limit of a class interval; BA = upper limit of a class interval

- Noise Indicator Level

Indicator number system was also called the equivalent index (Leq)). The equivalent indicator number was change (fluctuating) where noise level was measured over a certain time and the amount of which is equivalent to a steady noise level measured at the same time interval.

$$LAeq = Leq_{50} + 0.43\,(Leq_1 - Leq_{50}) \quad [5]$$

where: Leq= Equivalent noise level; Leq50 = 50% noise indicator number; Leq1 = 1% noise indicator

- Normality Test

Normality test is a test in the regression model, the dependent variable, independent, or both have a normal distribution or close to normal. Normality test to determine whether the studied residues are normally distributed or not. The method used to test for normality is to use the Kolmogorov-Smirnov test. If the significant value of the Kolmogorov-Smirnov test result is e" 0.05, then it is normally distributed and otherwise is not normally distributed [10].

- Paired Sample T-Test

Paired-Sample T-Test is an analysis involving two measurements on the same subject to a certain effect. If a treatment has no effect, then the average difference is zero [11]. The decision-making guidelines in the Paired Sample T-Test are based on significant values: (1) If the significant value (Sig.) < 0.05, then the conclusion is that there is a significant difference, which means the effect; and (2) If the value is Significant (Sig.) > 0.05, then the conclusion is that there is no difference, which means that there is no influence [12].

2.4 Quality Assurance

Weather conditions are also one of the things that must be considered when measuring noise levels. If it rains, the measurement must be stopped and postponed until the rain stops or measurements are taken the following week on the same day and at the same time with the same area conditions. In addition, the sound of sirens or long horns, the noise measurement is temporarily stopped. In order to make the noise level data more accurate, the Sound Level Meter is set so that the device can record data every second so that in 10 minutes 600 data will be recorded. Noise level measurements were carried out for 10 minutes representing 1 hour starting from 07.00-18.00 WITA.

3. Result and Discussion

Average equivalent noise level value (LAeq) which represents the 10 minutes measurement was calculated. The stage of calculating the value of the average equivalent noise level (LAeq) begins by making a frequency distribution in the form of range, number of classes, class intervals, and the midpoint of the class interval.

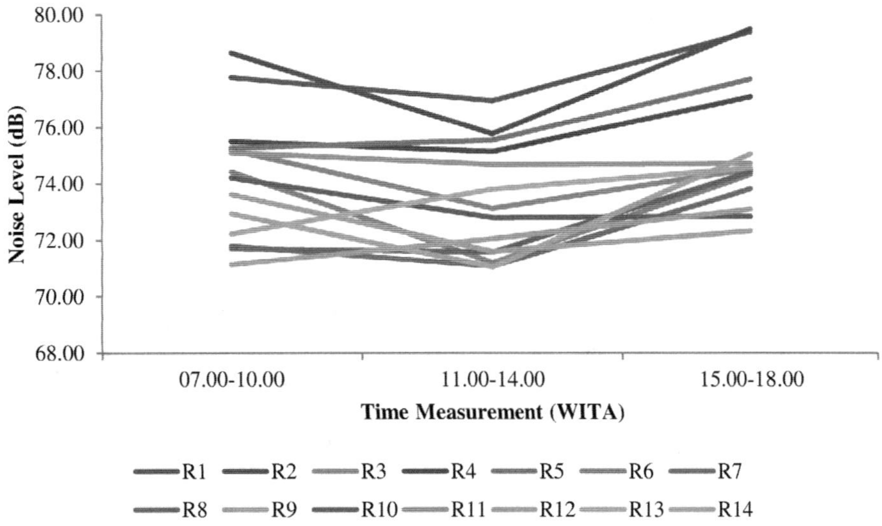

Figure 2. LAeq Value

It can be seen in Figure 2, in the morning the maximum LAeq value is 78.64 dB at the observation point R2 and the minimum LAeq value is 71.13 dB at the observation point R12. During the day, the maximum LAeq value is 76.94 dB at the observation point R1 and the minimum LAeq value is 71.05 at the observation point R14. For the evening, the maximum LAeq value is 79.48 dB at the observation point R2 and the minimum LAeq value is 72.31 dB at the observation point R9. These results could be affected by volume of vehicles, speed of vehicles, number of horns, and the characteristic of the road.

Based on Table 1, it can be seen that the SPSS output results from the significance test of the difference in two measurements were Noise Level Data Relationship between Morning- Afternoon. The significance is 0.025 > 0.05. The result showed that there's no significant difference between the morning and afternoon noise level data statistically. However, between Afternoon-Evening and

Morning-Evening there's a significant different. It can be shown at the mean difference which has minus results.

Tabel 1. Paired Sample Test Data Noise Level

Time Measurement		Paired Differences					
		Mean Difference	***Std. Deviation***	***Std. Error Mean***	**t**	**df**	***Sig. (2-tailed)***
Pair 1	Morning - Afternoon	1	1.38840	.37107	2.535	13	.025
Pair 2	Afternoon – Evening	-2	1.29245	.34542	-5.565	13	.000
Pair 3	Morning - Evening	-1	1.45520	.38892	-2.523	13	.025

One of the factors that affect the high level of noise, the similarities and differences in the data between the noise levels in the morning, afternoon, and evening on the Jalan Metro Makassar was the volume of vehicles. The volume of vehicles was calculated at the same time as the measurement of traffic noise on the Metro Road Section Makassar uses a counter. The types of vehicles that are calculated are vehicles that cross the observation point, motorcycles (MC), light vehicles (LV), and heavy vehicles (HV). According to Figure 3, it can be seen that the highest number of vehicles was found in evening. Volume of vehicles in morning and afternoon was found to be more fluctuated in each sampling points.

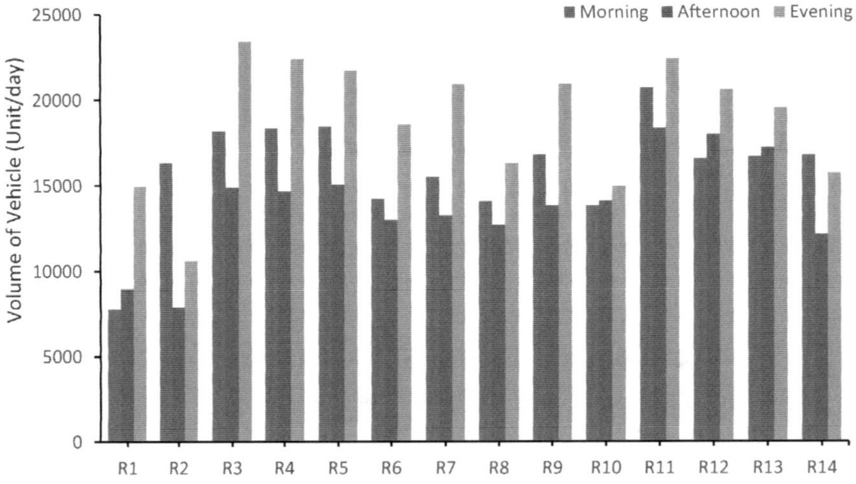

Figure 3. Volume of Vehicles

Based on Table 2, it was established that there was a significant difference between the morning and afternoon traffic volume data using the Paired Samples T-Test difference test, t(13) = 3.057; p > 0.05. The average for the morning data is higher than the average for the afternoon data (Mean = 13854; Std. Deviation = 3007.52975) (Mean = 16013; Std. Deviation = 3041,18183). This indicates that, in terms of numbers, morning traffic is heavier than daytime traffic.

For traffic volume data relationship between morning and evening was calculated by using the Paired Samples T-Test difference test (Table 2). The result was proven that there was a significant difference in the data volume of traffic during the day and evening, t(13) = -7,620; p < 0.05. Afternoon data (Mean = 18777; Std. Deviation = 3737.16073) had a larger average than daytime data (Mean =

13854; Std. Deviation = 3007.52975). This means that, numerically, the traffic volume in the evening was higher than during the day.

According to Table 2, the significance of 0.006 < 0.05 means that there was a significant difference between the morning and evening traffic volume data, which means that they were not statistically the same. By looking at the difference in the mean/average of the morning and evening data, it was -2764. The minus number indicates that the morning mean/average was smaller. To find out the magnitude of the difference can be seen from the mean / average.

Tabel 2. Paired Sample Test Data Volume of Vehicle

Time Measurement		*Paired Differences*					
		Mean Difference	***Std. Deviation***	***Std. Error Mean***	**t**	**df**	***Sig. (2-tailed)***
Pair 1	Morning - Afternoon	2160	2643.45243	706.49238	3.507	13	.009
Pair 2	Afternoon – Evening	-4923	2417.61770	646.13551	-7.620	13	.000
Pair 3	Morning - Evening	-2764	3178.56073	849.50609	-3.253	13	.006

Another factor that affect the noise level was the speed of vehicle. The highest average vehicle speed is at the observation point R14. This is because the observation point R14 is an observation point that can be said to be quite close to the Barombong bridge, where this bridge is an exit or entry access to Jalan Metro or to Barombong. In addition, the high average speed of vehicles at the observation point R14 is caused by the width of the road lane so that it provides a lot of space for vehicles to make it easier for drivers to increase their vehicle speed.

Based on the Figure 4, the speed of vehicles in morning was found to be higher in some sampling points. It might be attributed by the peak time of people to start their activity in the morning. However, the speed of vehicles was found to be more fluctuated in all sampling points. The characteristics of roads might affect the speed of vehicles.

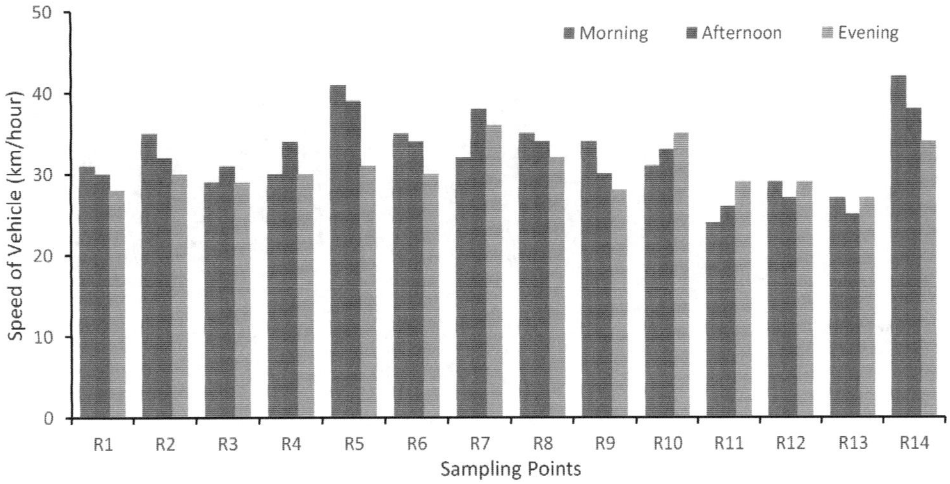

Figure 4. Speed of Vehicles

The morning and afternoon traffic volume statistics did not significantly differ from one another, according to the Paired Samples T-Test difference test. Based on Table 3, it can be concluded, the early traffic volume was greater than the daytime traffic in terms of numbers. Moreover, the relation between afternoon and evening of traffic volume from paired sample T-Test indicated that the volume of traffic was greater during the afternoon than in the evening, in terms of numbers. This result was found to be similar with traffic volume data relationship between morning and afternoon means that numerically the traffic volume in the morning was higher than in the afternoon.

Table 3. Paired Sample Test Data Speed of Vehicle

Time Measurement		*Paired Differences*					
		Mean Difference	***Std. Deviation***	***Std. Error Mean***	**t**	**df**	***Sig. (2-tailed)***
Pair 1	Morning - Afternoon	0	3.024	.808	.354	13	.729
Pair 2	Afternoon – Evening	2	3.028	.809	2.030	13	.063
Pair 3	Morning - Evening	2	4.582	1.225	1.575	13	.139

The horn usually was used to communicate between drivers as a signal to other drivers to avoid possible dangers that will occur. In general, the horn was sounded when the driver is about to overtake another driver, asking for road space, or even when there is a long queue of vehicles, such as at the fork in Jalan G. Bromo-Jalan Metro and U-Turn around the observation point R5. Furthermore, data on the number of horn sounds in the morning, afternoon, and evening can be seen in Figure 5.

According to Figure 5, it can be seen that horn was frequently used in evening. In morning and afternoon time, the numbers of horn were found more fluctuated in each sampling points. This numbers of horn might affect the noise level in that sampling points [13].

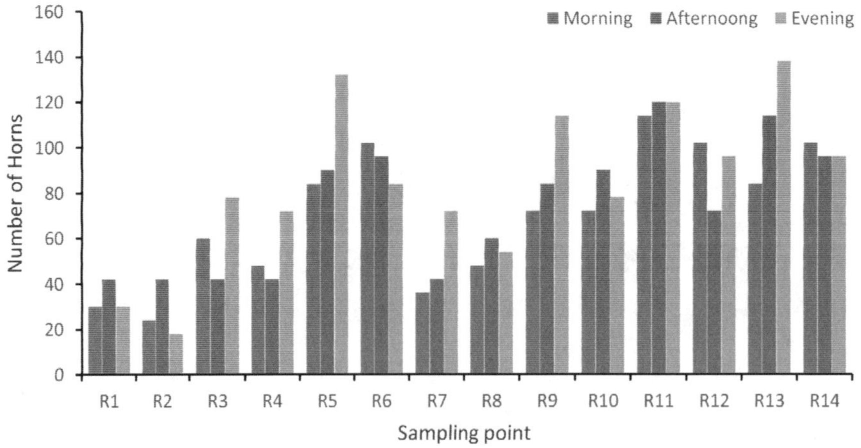

Figure 5. Numbers of Horn

Significant values were 0.000 < 0.05. This study came to the conclusion that the statistics on the quantity of horn sounds in the morning and during the day show a substantial difference, indicating that they are not statistically equivalent. By comparing the mean or average of the morning and afternoon's statistics, we see that the difference is 176. Similar with morning-afternoon statistic, the morning-evening and afternoon-evening were found not statistically equivalent by compared the mean difference in each time measurement with the difference was found 85 and 176, respectively.

Table 4. Paired Sample Test Data Numbers of Horns

Time Measurement		Paired Differences					
		Mean Difference	*Std. Deviation*	*Std. Error Mean*	**t**	**df**	*Sig. (2-tailed)*
Pair 1	Morning - Afternoon	91	42.547	11.371	7.990	13	.000
Pair 2	Afternoon - Evening	85	47.294	12.640	6.713	13	.000
Pair 3	Morning - Evening	176	82.786	22.125	7.942	13	.000

For the LAeq Day value at each observation point on Jalan Metro was calculate by using data from the LAeq value at the time of observation at 07.00-18.00 WITA. Figure 6 shows that the maximum LAeq Day value was found at the observation point R2 with 78.23 dB and the minimum LAeq Day value was found at the R12 sampling point with 72.16 dB. The noise level at point R2 was found to be high even though it has a relatively small number of vehicles when compared to other sampling points. This might be attributed by the activities around sampling point area, not only from transportation. Based on its characteristics, R2 is located close to the floating mosque parking lot, one of the tourist attractions in the city of Makassar. Meanwhile, noise level in sampling point R4 and R5 was found to be increased from R3 with 75,99 dB and 76,31 dB, respectively. This result might be attributed by activites around observation point and characteristic of the road.

Based on the results of the noise level analysis, it shows that the daily equivalent noise level value (LAeq Day) on Jalan Metro Makassar for 14 observation points has exceeded the noise level quality standard required by KepMenLH No. 48 Tahun 1996 for green open space areas, government and public facilities, hospitals, offices and trade, trade and services, as well as housing and settlements, which is 50 to 70 dB. The maximum LAeq day value is at the R2 observation point of 78.23 dB and the minimum LAeq day value is at the R12 observation point of 72.16 dB. In addition, a comparison of the measurement results with the technical limits of road environmental capacity as stated in the PU guidelines No. 13 of 2003 regarding the maximum and minimum limits of Leq10 and LAeq values with the category of commercial local roads for observation points R1, R2, R3, R4, R5, R6, R7, R8, R9, R10, R11, R12 and local residential roads for observation points R13 and R14.

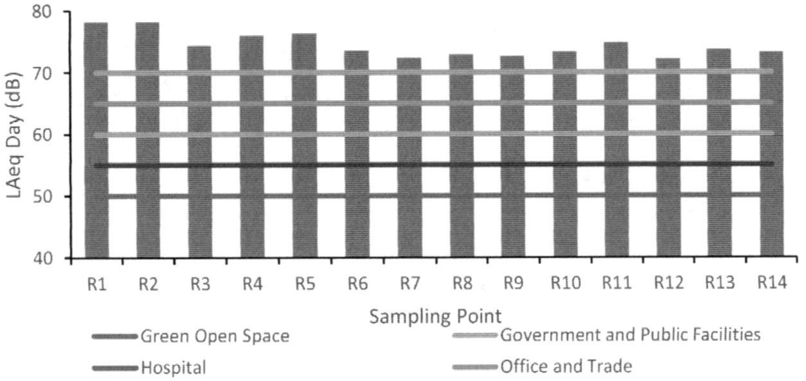

Figure 6. LAeq Day

4. Conclusion

The characteristics of the noise level at all sampling points on Jalan Metro Makassar were different, where each point has a different traffic volume, vehicle speed, number of horns. This is due to differences in road function and land use at each observation point. The characteristics of this noise level were tested using the SPSS program. The results of the Paired Samples T-Test showed the relationship between noise level data in the morning and afternoon, afternoon and evening, as well as in the morning and evening. For the time of measurement in the morning and afternoon, afternoon and evening, morning and evening, each has a significance value of 0.025 < 0.05; 0.000 < 0.05; 0.025 < 0.05, respectively. From this result, it can be concluded that there is a difference significant noise level measurement data at all times statistically the same observations. The LAeqday was describe about the noise level in each sampling points per day. This result was compared between the noise standard from government regulations. Based on the results of the noise level analysis, it shows that the daily equivalent noise level value (LAeq Day) on Jalan Metro Makassar for 14 observation points has exceeded the noise level quality standard required by the KepMenLH No. 48 Tahun 1996 for green open space areas, government and public facilities, hospitals, offices and trade, trade and services, as well as housing and settlements, which is 50 to 70 dB.

Reference

[1] Badan Pusat Statistik, "Perkembangan Jumlah Kendaraan Bermotor Menurut Jenis," *Badan Pusat Statistik*. pp. 1–1, 2018. [Online]. Available: https://www.bps.go.id/linkTableDinamis/view/id/1133

[2] I. Lakawa, L. Samang, M. Selintung, and M. Hustim, "Noise level study based on traffic characteristics, physical and environmental aspects of road," *Int. J. Civ. Eng. Technol.*, vol. 7, no. 1, pp. 188–198, 2016.

[3] M. E. Abdullah, M. K. Shamsudin, N. Karim, I. A. Bahrudin, and S. M. R. Shah, "A Study on Noise Level Produced By Road Traffic in Putrajaya Using Soundplan Road Traffic Noise Software," *Proc. Malaysian Tech. Univ. Conf. Eng. Technol.*, pp. 1–4, 2009.

[4] E. E. M. M. van Kempen, H. Kruize, H. C. Boshuizen, C. B. Ameling, B. A. M. Statsen, and A. E. M. de Hollander, "The association between noise exposure and blood pressure and ischemic heart disease: A meta-analysis," *Environ. Health Perspect.*, vol. 110, no. 3, pp. 307–317, 2002, doi: 10.1289/ehp.02110307.

[5] S. A. Stansfeld and M. P. Matheson, "Noise pollution: non-auditory effects on health," *Br. Med. Bull.*, vol. 68, pp. 243–257, 2003, doi: 10.1093/bmb/ldg033.

[6] European Environment Agency, *EEA Report No 10/2014 - Noise in Europe 2014*, no. 10. 2014. [Online]. Available: https://www.eea.europa.eu/publications/noise-in-europe-2014

[7] A. R. Anindya, W. P. Maryunani, and M. Amin, "Analisis Pengaruh Kecepatan Dan Volume Kendaraan Terhadap Kebisingan Di Suatu Kawasan," *J. Rekayasa Infrastruktur Sipil*, vol. 2, no. 1, p. 1, 2021, doi: 10.31002/.v1i1.2221.

[8] D. Ouis, "ANNOYANCE FROM ROAD TRAFFIC NOISE: A REVIEW," *J. Environ. Psychol.*, vol. 21, no. 1, pp. 101–120, Mar. 2001, doi: 10.1006/JEVP.2000.0187.

[9] S. Djalante, "Analisis Tingkat Kebisingan di Jalan Raya yang Menggunakan Alat pemberi Isyarat Lalu Lintas (APIL) (Studi kasus: Simpang Ade Swalayan)," *J. SMARTek*, vol. 8, no. 4, pp. 280–300, 2010.

[10] A. Zulfatul, "Pengaruh Peran Kepemimpinan Terhadap Kinerja Karyawan Pada Ajb Bumiputera 1912 Kantor Cabang Batu," p. 115, 2016.

[11] Faradiba, *PENGGUNAAN APLIKASI SPSS UNTUK ANALISIS STATISTIKA*. 2020.

[12] F. Wahyuny, "STUDI TINGKAT KEBISINGAN PADA KAWASAN DI SEKITAR JALUR JALAN (FRONTAGE) TOL MAKASSAR," 2021.

[13] M. Hustim, "Road Traffic Noise under Heterogeneous Traffic Condition," vol. 4, no. 1, pp. 109–118, 2012.

[14] T. Watanabe and S. Yamada, "Sound attenuation through absorption by vegetation," *J. Acoust. Soc. Japan (English Transl. Nippon Onkyo Gakkaishi)*, vol. 17, no. 4, pp. 175–182, 1996, doi: 10.1250/ast.17.175.

IOP Conf. Series: Earth and Environmental Science 1117 (2022) 012044 doi:10.1088/1755-1315/1117/1/012044

Analysis of hydrochemical parameters and dissolved zinc interaction by using PHREEQC simulation

H Alimuddin[1], P Andarani[2], K Yokota[3], T Inoue[3], M N Nguyen[3]

[1] Department of Environmental Engineering, Universitas Hasanuddin, South Sulawesi 90245 Indonesia
[2] Department of Environmental Engineering, Diponegoro University, Central Java 50275 Indonesia
[3] Department of Architecture and Civil Engineering, Toyohashi University of Technology, Aichi 441-8580 Japan

Email: hardianti@unhas.ac.id

Abstract: Zinc is frequently reacting with inorganic species in water to form zinc species. Thus, the chemical speciation of Zinc in the aquatic environment has become a significant concern nowadays due to its adverse effect on humans and its potential toxicity in the water system. In this study, to investigate the interaction between dissolved zinc and hydro-chemical factors and to estimate the Zn speciation form, samples of river water were collected spatially from the Umeda River mainstream and tributaries in different seasons for one year. The hydrochemistry and dissolved zinc pollution characteristics of the samples were analyzed. Zn speciation was assessed by PHREEQC simulation. The result show that the main form of Zn speciation was Zinc free ion (Zn^{2+}) in all seasons. However, in summer and spring season, the $ZnCO_3$ and $ZnOH^+$ concentration was higher than the other season. These speciation form may be attributed to the activities around the river in different season according to the similarities in downstream and upstream areas by the hierarchical cluster analysis result.

1. Introduction

The concentration of Zinc in rivers is essential for the aquatic organism. However, the excess zinc concentration can also cause toxicity to the aquatic environment. This concentration may originate from natural sources or human activities [1]. Based on the toxicity in aquatic environment, the Ministry of the Environment Government of Japan established the Environmental Quality Standards (EQS) relating to water pollution for total zinc concentration should be below 0.03 mg/L in 2003.

Previous studies state that the increase of human activities in catchment area of the river can change the water quality in the river including zinc concentration and hydrochemical parameters [2,3].The behaviour of zinc in river water were found to be insoluble but it could reacts with hydrochemical parameters to form zinc species. According to the previous study, many studies concerned about the distribution of chemical metal speciation in different phases and summarized a tendency for metal speciation to migrate into new compartments according to the interaction between phases [4]. Measuring the concentration of zinc speciation can be clarified by using computational modeling software such as PHREEQC, MINTEQA2, and WHAM [5]. In this study, PHREEQC simulation was used to determine the zinc species.

The ability to classify dissolved metal concentration into groups of species is important contribution to the study of water chemistry, geochemistry, and trace metal toxicity. The development of methods for zinc species is especially important because zinc is one of the essential trace metals and its bioavailability almost certain on its chemical forms [6]. In natural waterways, zinc can be found in a variety of chemical forms, including hydrated ions, metal-inorganic complexes, and metal-organic complexes. Furthermore, statistical analysis such as hierarchical cluster analysis was used to analyze the similarity between dissolved zinc concentration and hydrochemical parameters.

The purpose of this study is to: (a) to determine the spatial distribution and seasonal variation of dissolved Zn concentration of Umeda River; (b) to estimate chemical speciation by using the PHREEQC simulation; and (c) to classify the similarity between the form of zinc species based on sampling point by using multivariate analysis and hierarchy analysis cluster.

2. Material and methods

2.1 Study Area

Umeda river stream is one of agricultural river that located in Aichi, Japan. This river flows into Mikawa bay with total length and the area of the river basin is 14.032 km and 86.6 km^2, respectively. According to Fig.1, most of the Umeda River basin is located in Toyohashi in Aichi Prefecture. A small area of the river basin is located in Kosai city in Shizuoka Prefecture, Japan.

There were nine sampling points were taken in this study which were 5 points taken from the Umeda River stream from upstream to downstream, and 4 points (21, 22, 23, and 31) from the tributaries river. Water samples were collected in different seasons from 2019 until 2020.

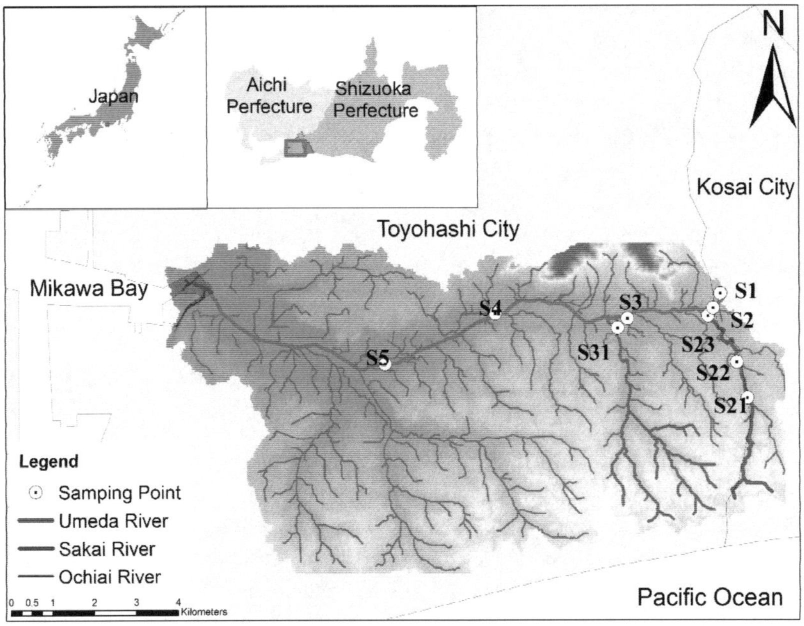

Figure. 1 Study area and sampling points

2.2 Sample collection, preparation, and analysis

There were 36 total water samples from the Umeda River stream and tributaries in summer, autumn, winter, and spring seasons. Water temperature, pH, and electrical conductivity were measured on-site using pH Meter WM-32EP and ECMeter CM-3IP. Surface water was collected and immediately filtered three times (150 ml for each time) in a laboratory using glass microfiber filters (GF/F with pore size 0.7 μm). The filtrate was used to measure the concentration of anions and cations. The concentration of anion (Cl^-, NO_3^- and SO_4^{2-}) and cation (Na^+, K^+, Ca^{2+}, and Mg^{2+}) of the water sample were determined by Ion Chromatography, while HCO_3^- was determined by alkalinity titration method.

The water sample was filtered using Membrane Filters Cellulose Acetate with a pore size of 0.2 μm. Thereafter add 1.0 ml ultrapure analytical reagent HNO_3 68% for 100 ml to the filtrate. For digestion of dissolved metals, 20 ml of sample was heated up on the hotplate at 205°C for 20 minutes.

Dissolved zinc concentrations were measured three times for each sample using Flame and Graphite Furnace Atomic Absorption Spectrophotometer (AAS) with four calibration standards. The detection limits of Zn in water and sediment were 0.0005 mg/L and 2.5 mg/kg, respectively.

2.3 *Estimation of Zn Speciation model*

PHREEQC is a computer modeling program for simulating chemical reactions and transport processes in natural or polluted water [7]. In this study, PHREEQC Version 3 with database (PHREEQC.DAT) was adopted. The model program was used as a speciation program to calculate saturation indices and the distribution of metal speciation. Moreover, PHREEQC was a computer modeling program for simulating chemical reactions and transport processes in natural or polluted water.

2.4 *Quality Assurance*

For analysis laboratory and sample collection, all the equipment and bottle to analyze were alkali-soaked with 5% NaOH for at least 8 hours, then double rinsed with distilled water. Thenceforth, all the equipment was acid-soaked with 10% HCl for at least 24 hours and rinsed three times with distilled water. Before the water sample was collected, rinse the bottle with the water sample 3 times before putting it in the bottle to minimize contamination and interferences. During AAS metal analysis, a quality control sample was routinely included for every six samples.

The Zn concentrations in several sampling points were below the detection limit. Therefore, the undetectable Zn concentration in AAS was assumed to have a concentration of 0.0005 mg/L. It can be presumed as the worst-case.

The charge balance reveals an exceed of anions in some samples ranging from 0.84% to -41.03%. Comparison between the charge imbalance and anion concentration (Cl^-, NO_3^- and SO_4^{2-}) in the water sample showed an R^2 of 0.16, 0.65, and 0.01, respectively. It means that the charge imbalance was due to a positive incline with nitrate concentration. In the present study, nitrate concentrations were adjusted in PHREEQC to achieve a charge balance between anions and cations. [8].

3. Result and Discussion

3.1 *Hydrochemical parameter in Umeda River stream and tributaries*

Hydrochemical characteristics in the Umeda River stream and tributaries in different seasons are shown in Table 1. These parameters were used for zinc speciation form in PHREEQC simulation. All hydrochemical parameters were identified through in-situ measurements or laboratory analysis. The temperature in the river is based on the season. In winter, the temperature in the Umeda river is low (range: 7.50 °C -10.60 °C); meanwhile, the temperature of summer is high (range: 27.70°C-32.00 °C).

The highest range of pH in the river water is in spring, with 6.75-9.95. The pH of 9.95 is weakly alkaline for natural water. It may be attributed to the industrial effluent, farming, or domestic sewage that contains a high concentration of bicarbonate or carbonate [5]. The lowest range of pH in autumn, with 6.77-7.37, expresses the pH of water generally. The highest average electrical conductivity (EC) value is in the winter season with 176.08. The anion and cation concentrations are impacted by the value of EC. An increasing of EC value is accompanied by an increase in anion and cation concentration [9].

The concentrations of all the anions and cations were found to be lower in sample 1 than in sampling point 5. All the concentrations of anions and cations was affected by the activities around the downstream was found to be higher than in upstream. Ion chloride was influenced by domestic effluents, roadways, and industries, meanwhile fertilizer and domestic wastewater containing sulfuric salts are two possible sources of SO_4^{2-} that might increase the quantity of sulphate ions. [10]. NO_3^- ion may have come from anthropogenic sources such as domestic sewage, agricultural runoff, and other waste effluents containing nitrogenous chemicals [11]. Moreover, the concentration of anions in the Umeda River for HCO_3^- increasing upstream to downstream in the Umeda River stream in all seasons. HCO_3^- ion could be released to surface and groundwaters by natural processes such as the dissolution of carbonate minerals and the dissolution of air and soil CO_2 gas [10].

According to Table 1, all the cations concentration reported in the Umeda river stream has a similar trend from upstream to downstream in all seasons. Mg^{2+} and Ca^{2+} often came from carbonate minerals in surface and groundwaters. The anthropogenic sources of these cations (Na^+, Mg^{2+} and Ca^{2+}) were domestic wastewater, housing, and industrial spillages [10]. For K^+ concentration, might be explained by the possibility that the plants' mineral elements cycle annually, with relatively high spring to early summer levels and winter minima. The increased moisture availability in the species growing on wet rocks appeared to be correlated with the higher K^+ concentrations [12].

In contrast, in tributaries showed a more fluctuated trend. The increasing concentration from the upstream area (S1) to downstream (S5) offers some concentration in mid-stream caused by pollution sources in particular areas. Some previous studies stated that ion chemistry could be changed due to pollutant sources, geochemical weathering, and precipitation [3,11,12]

Table 1. The hydrochemical parameter of Umeda River stream and tributaries (n=36).

Variable	Unit	Hydrochemical parameters (sampling point 1, sampling point 5)			
		Summer	Autumn	Winter	Spring
Temp	°C	29.1, 27.7	12.9, 12.3	9.7, 9.4	23.1, 21.9
pH	-	8.26, 7.47	7.37, 7.01	8.41, 6.99	7.70, 7.11
EC	µS/cm	38.2, 176.4	48.5, 198.7	67.2, 203.0	34.7, 155.3
SS	mg/L	4.53, 5.82	14.63, 10.02	2.31, 3.64	2.97, 5.51
DO	mg/L	7.62, 7.78	10.22, 10.36	11.0, 11.08	8.37, 8.55
Cl^-	mg/L	3.52, 45.52	9.15, 46.43	7.35, 35.97	2.29, 39.72
NO_3^-	mg/L	2.27, 15.91	2.58, 28.03	2.49, 30.68	2.27, 16.69
SO_4^{2-}	mg/L	3.77, 39.08	4.11, 50.52	5.29, 60.58	3.62, 30.66
HCO_3^-	mg/L	27.15, 67.96	25.38, 63.59	44.85, 60.34	27.37, 63.10
Na^+	mg/L	5.55, 24.30	7.82, 26.55	7.73, 21.81	4.55, 23.63
K^+	mg/L	0.61, 16.37	0.70, 20.58	1.05, 26.44	0.77, 9.80
Ca^{2+}	mg/L	7.74, 22.25	7.74, 25.96	13.62, 22.44	6.83, 23.01
Mg^{2+}	mg/L	1.18, 6.10	1.61, 8.44	2.21, 7.58	1.23, 7.19

3.2 Dissolved zinc concentration in Umeda River stream and tributaries

Dissolved Zn concentration in water samples seasonally are shown in Fig. 2. The highest Zn concentration in the water sample happens in the winter and autumn season for almost all sampling points.

Upstream area in Umeda river in all different seasons has the lowest concentration, and it increases to the midstream and downstream. This may be attributed to anthropogenic, industrial, and agricultural activities around the midstream. The Zn concentration from the tributary river also contributes to the Zn concentration in the river stream. Moreover, some study conclude that the differences in zinc concentrations in different sampling location suggested that both natural and man-made phenomena, such as transport and dispersion, erosion, urban runoff, redox and dissolution processes, weathering, and rainfall events, were involved [13].

According to Fig. 2, there's no specific trend in dissolved fractions of Zn. The dissolved phase in the autumn and winter seasons is higher than the particulate phase in the downstream area. Some previous study state that if several sampling points revealed that zinc had the largest dissolved to particle fraction ratio, it described that zinc has a high level of bioavailability. [5].

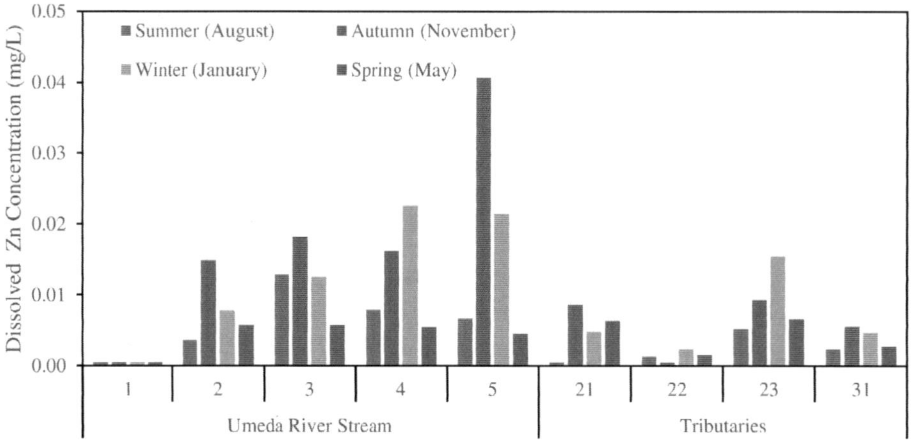

Figure. 2 Concentration of dissolved zinc concentration of Umeda River in summer, autumn, winter and spring seasons

3.3 Zn speciation in Umeda River

The form of Zn speciation in PHREEQC are determined by hydrochemical factors like temperature, pH, DO, anion, and cation concentration. From this simulation model, the concentration of Zn speciation can be estimated. The distribution of Zn speciation can be seen in Fig. 3. Free metal ion for Zn (Zn^{2+}) is the highest compound that formed from PHREEQC in all seasons, followed by $ZnCO_3$, $ZnHCO_3$, and $ZnSO_4$ in river stream which more fluctuated in each season.

In summer season, the main form of zinc species in summer seasons was metal free ion of zinc (Zn^{2+}) in almost all sampling points followed by $ZnCO_3$, $Zn(OH)_2$, $ZnOH^+$, and $ZnSO_4$ (Figure 3a). $ZnCO_3$ was found to be higher in upstream and all tributaries sampling points. In this study, ZnCO3 was found to be higher in those sampling points due to the low concentration of dissolved oxygen in this season that used by aquatic organism and it release CO_2 so it will react to Zn and form $ZnCO_3$.

In addition, $ZnOH^+$ and $Zn(OH)_2$ were found to be higher in sampling point 1 and sampling point 22 in summer season. This concentration might be attributed by pH value, which were the value of pH principally affects the combination of OH^- and Zn. Zinc ion (Zn^{2+}) was the primary form of zinc species, but it will decreased due to the Zn(OH)2 increased. The ratio of Zn^{2+} ions to $Zn(OH)_2$, which was pH-dependent, was reversed at pH 8.5 in favor [14].

Zn^{2+} was the main form of autumn season, followed by $ZnHCO_3^+$ (Figure 3b). $ZnHCO_3^+$ was observed to be high in sampling point 4 and sampling point 5. These concentrations might be influenced by the concentration of HCO_3^- and the EC value [9]. Moreover, in autumn season, $ZnSO^4$ was found to be higher than the other season. Based on the form of $ZnSO_4$ in those sampling points in Umeda River stream and tributary, sampling point 3, sampling point 23, and sampling point 5 have the high concentration of anion SO_4^{2-}.

IOP Conf. Series: Earth and Environmental Science 1117 (2022) 012044 doi:10.1088/1755-1315/1117/1/012044

The main form of Zinc species in winter season was Zn^{2+} in all sampling points. However, in Figure 3c show a small difference between Zn^{2+}, $ZnCO_3$, and Zn(OH)2 in sampling 1. Moreover, the concentration of $ZnCO_3$ sampling 1 was found to be higher than the other sampling point. This might be affected by the high concentration anions concentration of HCO_3^- in sampling point 1 in winter season. As the HCO_3^- concentration increases, the equilibrium of the Zn species in solution moves toward HCO_3^- and CO_3^- species [15].

In practically all sampling points (with the exception of sampling point 22) in the spring season, metal-free zinc ions (Zn^{2+}) were found to be the dominant form of zinc species in the Umeda River stream. According to Figure 3d, the highest form of Zn^{2+} was found in sampling point 4 and the lowest form was found in sampling point 22. The species form of zinc carbonate was found to be higher in tributaries in sampling point 23 and sampling point 21. However, in the Umeda River Stream, $ZnCO_3$ was identified in the highest concentration at sampling Point 1 and the lowest concentration at Sampling Point 4. Dissolved Zinc concentration and HCO_3^- concentration both had an impact on $ZnCO_3$ form.

Figure 3d showed that Zinc hydroxide or Zn(OH)2 was found to be the main form of zinc species in sampling point 22. This might be attributed by the pH value with 9.95 in sampling point 22 in spring season. The high form of $Zn(OH)_2$ in spring season was followed by sampling point 23 and sampling point 21. When the pH is above 7, the $Zn(OH)_2$ speciation form will increase and the Zn^{2+} will decrease [14].

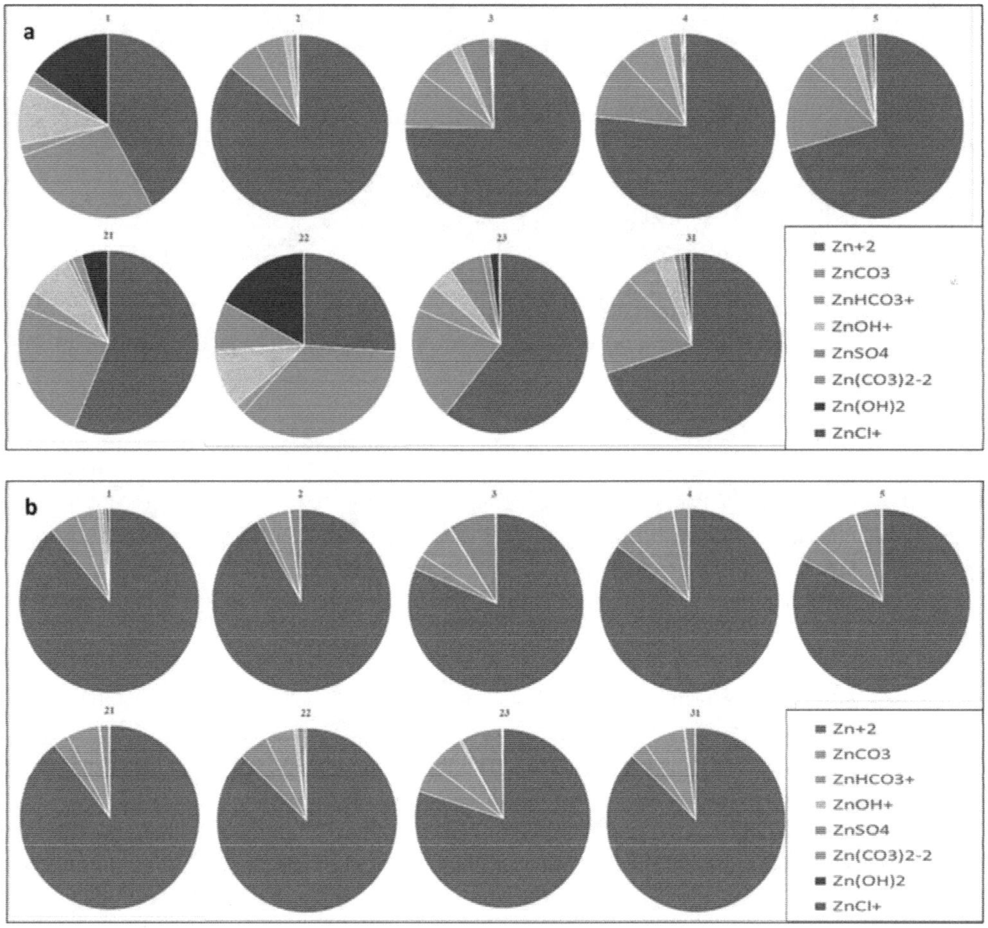

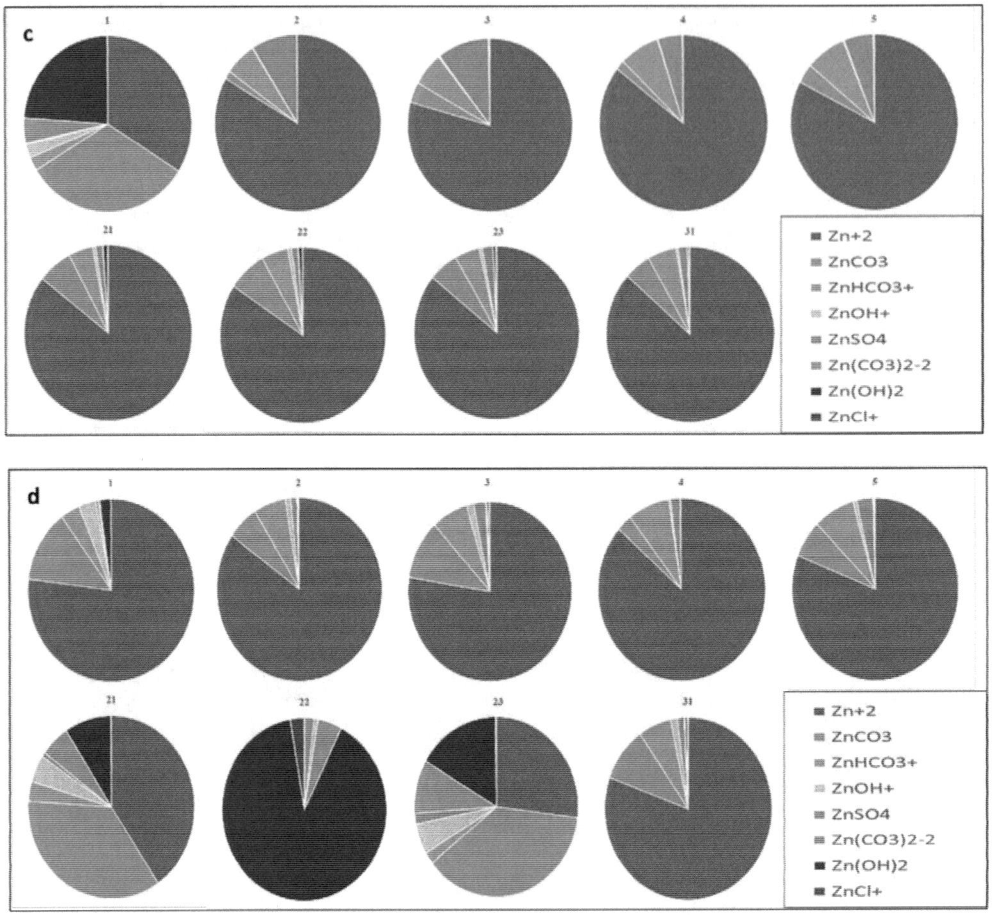

Figure. 3 Zinc Speciation points in (a) Summer; (b) Autumn; (c) Winter; (d) Spring season

3.4 Hierarchical cluster analysis

Many factors could be affected the concentration of dissolved zinc and hydrochemical parameters in river stream. Previous study in Umeda River revealed that the principal sources were point sources, whereas the extremely high concentrations of zinc in the Umeda River during the stormflow were caused by diffuse (non-point) sources [16].

Hierarchical cluster analysis was created to find the similarities between water quality based on the sampling point. Based on Fig. 4, HCA results grouping all the 9 sampling points of the river into two statistically significant clusters in all seasons. Sampling point 4, and 5 are on the same cluster for all season, indicate that downstream area was polluted by same source pollution which were it might be influenced by the activities around the catchment area of these sampling points. In the upstream area (sampling point 1 and 2) are on the same cluster according to the water quality parameter in summer and spring season, illustrate the affect from identical pollution source.

In tributaries river, sampling point 23 has the same cluster with sampling point 4 and 5 in river stream, mean the same major of pollutant source. Sampling points 31 and 23 are in the same cluster in all season, except in spring season due to the high concentration of Zn in spring season for points 23. Industrial effluent source predicted as the major pollutant source in tributaries sampling point area.

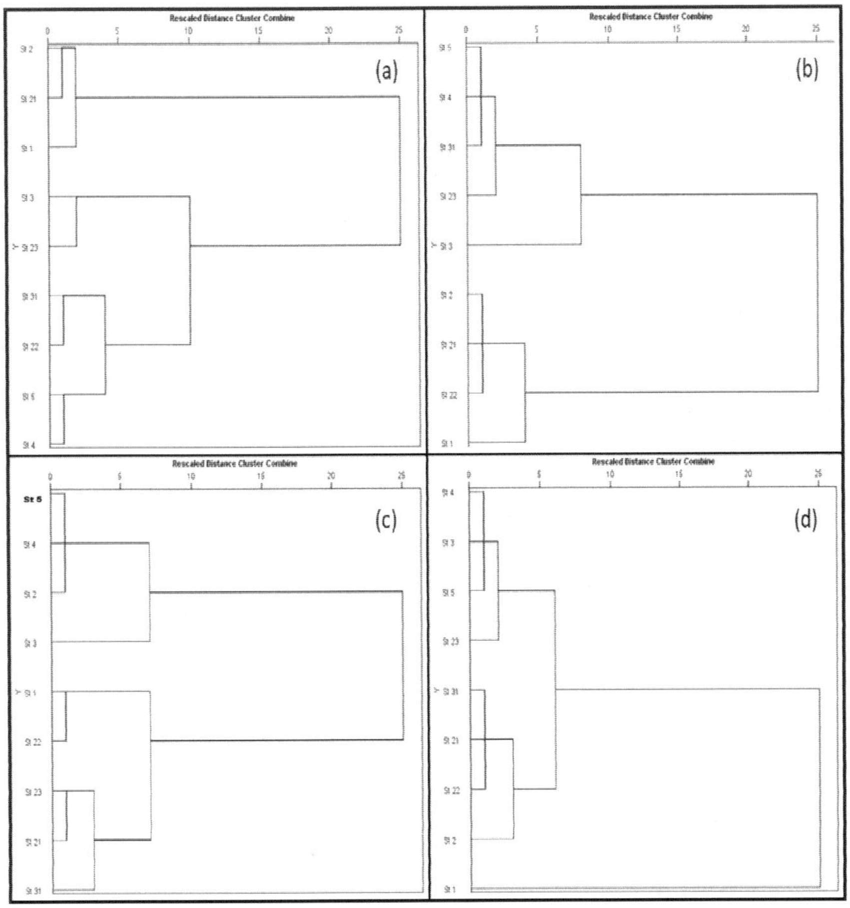

Figure. 4 Dendogram from Hierarchical cluster analysis based on sampling points in (a) Summer; (b) Autumn; (c) Winter; (d) Spring season

4. Conclusion

In this present study, the distribution of Zn species was investigated due to their toxicity in river stream. The influences of hydrochemical parameters and zinc concentration to form zinc speciation was simulated by the PHREEQC model. Zn^{2+}, $ZnCO_3$, $ZnOH^+$, $Zn(OH)_2$, $ZnSO_4$, $ZnCl_2$, and other species were formed fluctuated seasonally. However, free metal ion of Zn (Zn^{2+}) was found to be the highest speciation form by PHREEQC model followed by $ZnCO_3$, $ZnHCO_3$, $Zn(OH)_2$ and $ZnSO_4$ with different levels of concentration for each season. However, in the spring season in sampling point 22, $Zn(OH)_2$ was found to be the highest form of zinc species due to the high pH in this sampling point. The value of pH can contribute the increased form of Zn(OH)2 and decrease the form of Zn2+. In addition, the form of $ZnCO_3$ dan $ZnHCO_3$ indicate that Zn form more complexes with HCO_3^- in comparison between the other cation and anion. The forms of Zn species were affected by pH, EC-value, dissolved oxygen, anion-cation concentration, and zinc concentration. This hydrochemical value and dissolved zinc concentration might be attributed by the land use or activities in the catchment area of this sampling points. Hierarchical cluster analysis revealed the cluster of all sampling points which indicated and identical pollution source.

References

[1] S. Shil and U. K. Singh, "Health risk assessment and spatial variations of dissolved heavy metals and metalloids in a tropical river basin system," *Ecological Indicators,* p. 106, 2019.

[2] R. Nicolau, Y. Lucas, P. Merdy and M. Raynaud, "Base flow and stormwater net fluxes of carbon and trace metals to the Mediterranean sea by an urbanized small river," *Water Res,* vol. 46, p. 6625–6637, 2012.

[3] P. Andarani, K. Yokota, M. Saga, T. Inoue and Y. Matsumoto, "Study of zinc pollution in river water: Average mass balance based on irrigation schedule," *River Research and Applications,* pp. 1286-1295, 2020.

[4] J. M. Parnis and D. Mackay, Multimedia Environmental Models : The Fugacity Approach (Third Edition), Oxon: CRC Press, 2021.

[5] A. Gagoi, K. Taki and M. Kumar, "Seasonal dynamics of metal phase distributions in the perennial tropical (Brahmaputra) river: Environmental fate and transport perspective," *Environmental Research,* vol. 183, 2020.

[6] J. O. Nriagu, ZINC IN THE ENVIRONMENT Part I : Ecological Cycling, United States of America: John Wiley & Sons, Inc., 1980.

[7] D. L. Parkhurst and C. Appelo, Description of input and examples for PHREEQC version 3—A computer program for speciation, batch-reaction, one-dimensional transport, and inverse geochemical calculations: U.S. Geological Survey Techniques and Methods, book 6, chap. A43, 497 p., http://pubs.usgs.gov/tm/06/a43., 2013.

[8] R. T. Kay, G. E. Groschen, G. Cygan and D. H. Dupre, "Diel cycles in dissolved barium, lead, iron, vanadium, and nitrite in a stream draining a former zinc smelter site near Hegeler, Illinois," *Chemical Geology,* 2011.

[9] H. Jing, Z. Chengbin, L. Yunchuan, Y. Qin, Z. Shaoqi, W. Pan, J. Juan, Z. Wei and H. Xuejun, "Interaction analysis of hidrochemical factors and dissolved heavy metals in the karst Caohai Wetland based on PHREEQC, cooccurrence network and redundancy analyses," *Science of the Total Environment,* 2021.

[10] M. Bahar, H. Ohmori and M. Yamamuro, "Relationship between river water quality and land use in a small river basin running through the urbanizing area of Central Japan," *Limnology,* vol. 9, pp. 19-26, 2008.

[11] V. Sheykhi and F. Moore, "Geochemical Characterization of Kor River Water Quality," *Water Qual Expo,* vol. 4, pp. 25-38, 2012.

[12] R. Simth, "Summer and Winter Concentration of Sodium, Potassium, and Calcium in Some Maritime Antarctic Cryptogams," *Journal Of Ecology,* vol. 66, pp. 891-909, 1978.

[13] V. K. Gaur, . S. K. Gupta, S. Pandey, K. Gopal and V. Misra, "DISTRIBUTION OF HEAVY METALS IN SEDIMENT AND WATER OF RIVER GOMTI," *Environmental Monitoring and Assessment,* vol. 102, no. 419–433, 2005.

[14] D. Rabadjieva, A. Kovacheva, S. Tepavitcharova, R. Ilieva, R. Gergulova, I. Vladov and S. Karavoltsos, "Modelling of chemical species of Al, Mn, Zn, and Pb in river body waters of

industrial areas of West Rhodope Mountain, Bulgaria," *Environmental Monitoring Assessment,* vol. 193, p. 430, 2021.

[15] Y. Dang, K. Tiller, R. Dalal and D. Edwards, "Zinc speciation in soil solutions of Vertisols," *Australian Journal of Soil Research,* vol. 34, no. 3, pp. 369-383, 1996.

[16] P. Andarani , H. Alimuddin , K. Yokota, T. Inoue, S. Obaid and M. N. Nguyen, "Baseflow and Stormflow Zinc Loads in a Small Agricultural River Catchment Influenced by an Industrial Area," *Water,* vol. 13, no. 15, p. 2113, 2021.

[17] A. Gogoi, C. G. G. Tushara, A. K. An, D. D. Snow, Y. Li and M. Kumar, "Influence of ligands on metal speciation, transport and toxicity in a tropical river during wet (monsoon) period," *Chemosphere,* 2016.

[18] E. H. Teramoto , H. Gemeiner, M. B. Zanatta, A. A. Menegário and H. K. Chang, "Metal speciation of the Paraopeba river after the Brumadinho dam failure," *Science of the Total Environment, 757,* 2021.

[19] J. Briffa, E. Sinagra and R. Blundell, "Heavy metal pollution in the environment and their toxicological effects on humans," *Heliyon,* vol. 6, no. 9, 2020.

[20] X. Wang, L. Zhao, H. Xu and X. Zhang, "Spatial and seasonal characteristics of dissolved heavy metals in the surface seawater of the Yellow River Estuary, China," *Marine Pollution Bulletin,* pp. 465-473, 2018.

[21] S. Bouragba, K. Komai and K. Nakayama, "Emipircal Approach for Modelling of Partition Coefficient on Lead Concentration in Riverine Sediment," *International Journal of Environment Science and Development,* pp. 352-357, 2020.

[22] V. Hatje, E. Bidone and J. Maddock, "Estimation of the Natural and Anthropogenic Components of Heavy Metal Fluxes in Fresh Water Sinos River, Rio Grande Do Sul State, South Brazil," *Environmental Technology,* vol. 19, no. 5, pp. 483-487, 2010.

[23] S. Krishnawami and S. K. Singh, "Chemical weathering in the river basins of the Himalaya, India," *Current Science,* vol. 89, no. 5, pp. 841-849, 2005.

The Compressive Strengh Of Cohesive Soil Stabilized With Microbial Induced Cementation

Alkadri[1] AR Djamaluddin[2] T Harianto[2] A Arsyad [2]

[1]Doctoral Student of Civil Engineering, Hasanuddin University, Gowa, South Sulawesi, Indonesia, 92171
[2]Departement of Civil Engineering, Hasanuddin University, Gowa, South Sulawesi, Indonesia, 92171

E-mail: alkadri19d@student.unhas.ac.id

Abstract. This research aims to evaluate the physical and mechanical characteristics of soil stablized with microbial induced cementation method. The variation of microbial induced by 4%, 6%, and 8%, while the age variations of bacterial culture of 2, 4, and, 6 days and a curing period of 3,7,14 and 28 days. The properties of the soil studied in this research are as optimum moisture content of 32.19%, dry volume weight (γd) 1.36 gr/cm^3, soil density 2.63 gr/cm^3, atterberg limit test liquid limit 46.89% plastic limit 31 ,94%, plasticity index 14.96% and shrinkage limit 7.49%. The compressive strength of the soil which was stabilized with 6% solution of Bacillus subtilis bacteria for 4 days found to be optimum 19.96 kg/cm^2 at 28 days or an increase of 6 times from the soil was not stabilized with microbial induced 3.02 kg/cm^2

Keywords: Soil, biocementation, Unconfined Compressive test

1. Introduction

Soil in one location has different characteristics from soil in other locations, this is what makes the bearing capacity of the soil as part of the highway construction system different from one location to another, basically not all types of soil can be used in road construction. or structural construction. Soil properties and behavior will be taken into consideration in the planning and implementation of a work, soil improvement has become a very important part and must be considered because of the scarcity of suitable soil is increasingly difficult to obtain and the rate of human population growth continues to increase. In terms of the need for road construction where material replacement and the use of chemicals for subgrade layers is inefficient, it is necessary to use other methods so that the existing soil quality is better for road construction purposes without having to replace the existing soil and damage the environment. There needs to be a new effort to utilize biological processes to modify soil properties and microbes capable of depositing calcium carbonate ($CaCO_3$) compounds called microbial induced calcite precipitation (MICP) by utilizing biological processes to modify soil properties [1] The use of micro-organisms is considered as an alternative in the midst of the proliferation of chemicals and synthetics in civil construction [2] Effective formation of CaCO3 will increase the strength and stiffness of the subgrade and embankment layers while maintaining the permeability characteristics [3] Bio-cementation is a method of soil improvement by utilizing the ability of bacteria that do live in the soil to produce the enzyme urease. The ex-mining land materialafter treated with MICP using Bacillus subtilis, it can be reused and qualifies as a road construction material [4]. The addition of bacteria to stabilize sandy loam using the bio-grouting method succeeded in increasing the carrying capacity of the soil [5] Tropical organic soil along with the addition of sand using a pre-mixing treatment method with a curing period of 3 days. The value of unconfined compressive strength (UCS) increased [6] According to the California Bearing Ratio (CBR) value range criteria (20% ≤CBR<40%) for subbase layer, the use of 3.5 to 6% bacteria can technically be functionally justified as a road material with the modulus of stabilized clay soil reaction value of 68 to 110 kN/m2/mm. The results of this study have technical. Implications for the significance of the use of Bacillus subtilis bacteria as the biotechnology stabilization material to increase the bearing capacity and modulus of soil reaction with high plasticity clay [7] Transformation in sand soil structure after bacterial stabilization for 14 days by varying the injection of bacteria [8]. The gap between this study and previous research is, this study tries to variation the percentages, 4%, 6% and 8% bacterial solution, which have the most optimum compressive strength values and the age of bacterial culture is 2, 4, and 6 days with a curing period of 3, 7, 14 and 28 days.

2. Materials and Testing Methods

2.1. Preparation of material

The material used in the research came from Makassar City, South Sulawesi province. Sampling by conventional excavation using a hoe and a shovel as shown in Figure 1, then the sample is placed in a sample sack and wrapped in plastic to maintain the condition of the moisture content, and labelled with the initials according to the location . The second is the preparation of a preliminary test of the sample that has been taken to determine the physical properties of the soil which includes testing the water content, testing the specific gravity of the soil, testing the limits of Atterberg, sieve analysis test, compaction test and mechanical test of unconfined compression test. The bacteria used in this research is Bacillus subtilis as shown in Figure 2. The reason for choosing Bacillus subtilis is it lives in the soil, forms endospores, which is a period of survival in extreme environmental conditions of temperature and produce calcite.

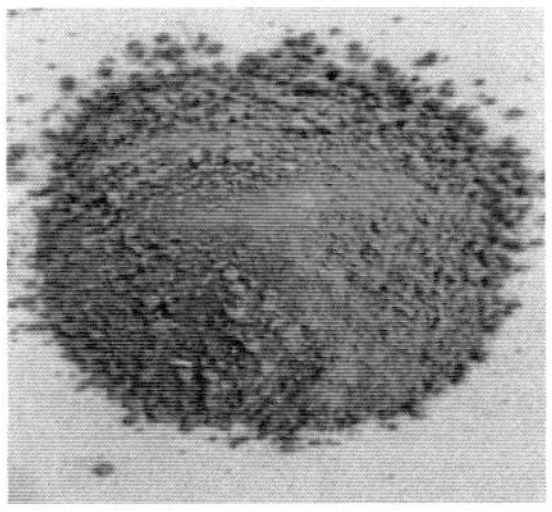

Figure 1. Sample of soil

Figure 2. Bacterical salution

2.2. Test of bacterial growth

The growth test was carried out to determine the growth characteristics of a type of bacteria through a growth curve marked by turbidity in liquid media with the help of a shaker and incubator. Microbiological culture is a method of multiplying microbes on culture media by controlled laboratory culture. Bacterial growth curves were made to determine the growth phases of bacteria, namely the lag, exponential, stationary, and death phases. Lag phase is the adjustment phase of bacteria to the new environment, the exponential or logarithmic phase is characterized by a period of fairly rapid growth. Each cell in the population divides into two, the stationary phase occurs when the rate of bacterial growth occurs, and the death phase is the phase where the death rate is greater as shown in Figure 3. Bacterial culturing process with a composition of 20 g Urea, 3 g Nutrient Brouth, 2.12 g $NaHCO_3$, 4.14 g CaCl, 10 g NH_4Cl then mixed with activated bacteria Bacillus subtilis.

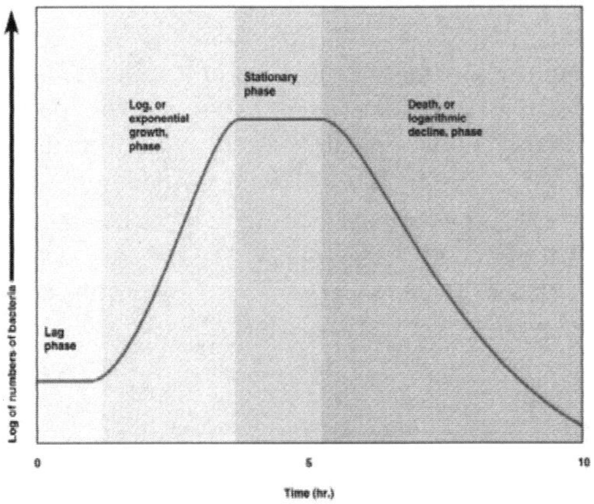

Figure 3. Curve of bacteria growth

2.3. Testing of physical and mechanical properties

Physical properties of the soil are properties related to the elements of the composition of the soil mass, while the mechanical properties of the soil are the behavioral properties of the soil mass structure when subjected to a force or pressure. Laboratory tests to determine physical properties include water content, Atterberg limit, sieve analysis, and specific gravity (specific gravity), while mechanical properties tests include compaction test, and Unconfined Compression Test (UCT) based ASTM D2166-06. Before material testing first measures and weighs the sample to be tested as shown in Figure 4, the surface of the sample looks whitish after curing for a few days as shown in Figure 5, Unconfined Compressive Strength test as shown in Figure 6. The tools that will be used in this research were previously checked for their conditions and capabilities and calibrated first. The procedure and the workings of the tool must be studied carefully, the ability, accuracy, and capacity of the tool must be well understood so that errors do not occur during the implementation of the testing. The Unconfined Compression Test (UCT) testing begins with mixing soil with a bacterial solution with a variation of 4, 6, and 8% solution, soil with bacterial culture 2, 4 and 6 days, UCT samples were made with a diameter of 5.5 cm and a height of 11 cm and then cured for 3, 7, 14 and 28 days, this curing aims to improve the chemical process between the soil and a solution of Bacillus subtilis which is expected to improve the compressive strength of the soil, Then the UCT test is carried out to get the value of axial strain ε (1), average cross-sectional area A (2) and value of compressive strength qu (3)

$$\varepsilon = \frac{\Delta L}{Lo} \text{ x } 100 \quad (1)$$

$$A = \frac{Ao}{1-\varepsilon} \quad (2)$$

$$qu = \frac{K \text{ x } R}{A} \quad (3)$$

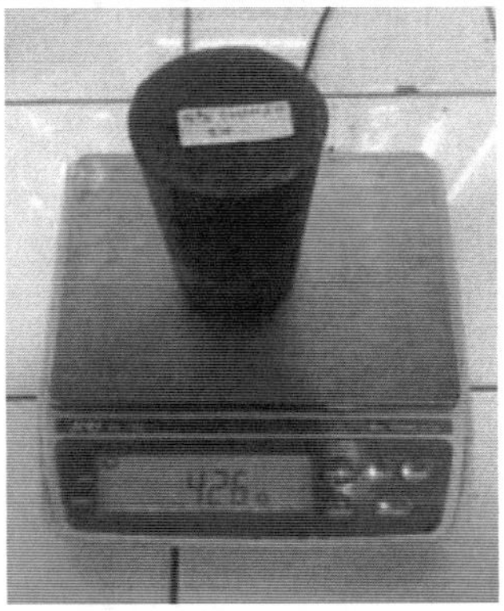

Figure 4 . Samples before UCT test

Figure 5. Surface of the sample is slightly whitish in color

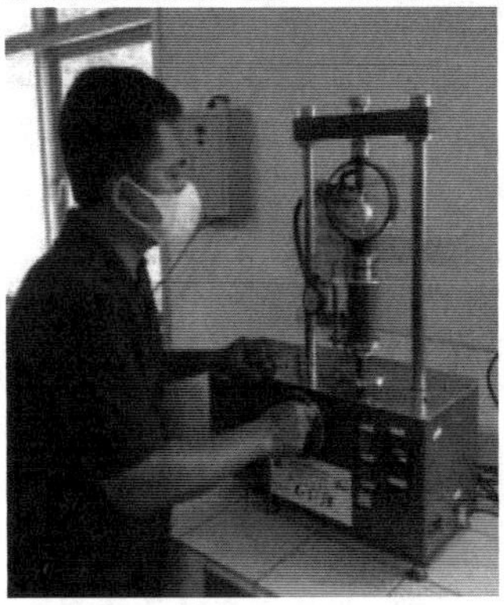

Figure 6. Unconfined compression test

3. Result and Discussion

The results of the physical tests in the laboratory are shown in Table 1. The initial moisture content is 12.12%. The results of the standard Proctor compaction test obtained the optimum moisture content of 32.19% with dry volume weight (γd) 1.36 gr/cm3, soil density 2.63 gr/cm^3, Atterberg limit test liquid limit 46.89% plastic limit 31 ,94%, plasticity index 14.96% and shrinkage limit 7.49%, while the results of the analysis of sand filters are 21.60%, silt 38.32% and clay (clay) 36.88%. The results of the Unified Soil Classification System (USCS) are classified as fine-grained soils with the ML classification, namely silt with low plasticity, while based on AASHTO the soil samples classified into group A-7-5 are clay soils with moderate to poor ratings

Table 1. Result of Soil Properties Testing

Testing	Result
Water (W) % Content	12.92
Specific Gravity (Gs)	2.63
Liquid Limit (LL) %	46.89
Plastic Limit (PL) %	31.94
Plasticity Index (PI) %	14.96
Shrinkage Limit (SL) %	7.49
AASHTO	A-7-5
USCS	ML
Gravel (%)	3.2
Sand (%)	21.6
Silt (%)	38.32
Clay (%)	36.88
Optimum moisture content (Wc) %	32.19
Maximum dry density kg/cm^3	1.36
Sliding angle (ø) kg/cm^2 (°)	33
Cohesion (c) %	0.65

IOP Conf. Series: Earth and Environmental Science 1117 (2022) 012045 doi:10.1088/1755-1315/1117/1/012045

USCS plasticity diagram to obtain a soil classification. Based on the liquid limit value, soils can be classified as high plasticity and low plasticity soils. Soils with high plasticity are characterized by the liquid limit value above 50%, while low plasticity if the liquid limit value is below 50 as shown in Figure 7.

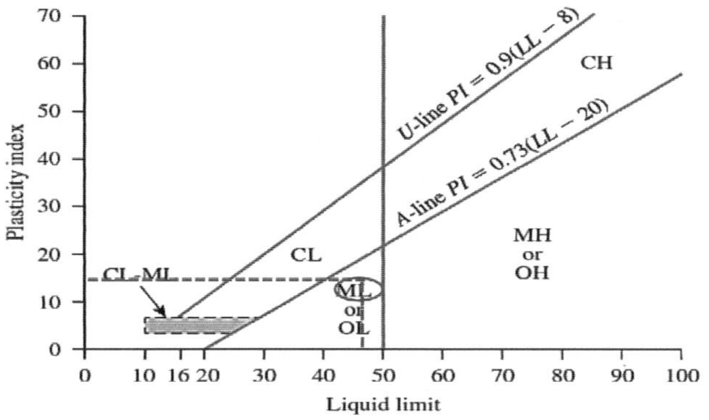

Figure 7. Unified soil classification system plasticity diagram

3.1. Result of Unconfined compression test

The results of the compressive test of stabilized soil with the addition of 4% solution of bacillus subtilis bacteria and treatment periods of 3, 7, 14, and 28. Figure 8, for 2-day culture shows an increase at 28 days of age with a compressive strength value of 17.54 kg/cm2, for 4 days culture at 28 days old showed a compressive strength value of 18.46 kg/cm2, 6-day bacterial culture at 28 days showed a decrease to 12.23 kg g/cm2, the compressive strength value of variation 6% of bacteria culture 4 days at 28 days showed the most optimum compressive strength value 19.96, for a variation of 8% bacteria culture 4 days aged 28 showed a UCT value of 17.44 kg/cm2 as shown in Figure 8,9,10. From the test results showed that the soil was not stabilized with bacteria has a compressive strength value of 3.02 kg/cm2. The age of 4 days of bacterial culture is the highest phase of bacteria because in this phase the bacteria are in a stationary phase, which means the level of calcite production increases compared to 6 days of culture which has decreased due to this phase the bacteria enter the death phase, the longer the curing time, the addition of a concentrated solution. The optimum bacteria is with the value of 6% and culture 4 days mixed into the soil, the greater the value of soil strength or an increase of 6 times the strength of the original soil

IOP Conf. Series: Earth and Environmental Science 1117 (2022) 012045 doi:10.1088/1755-1315/1117/1/012045

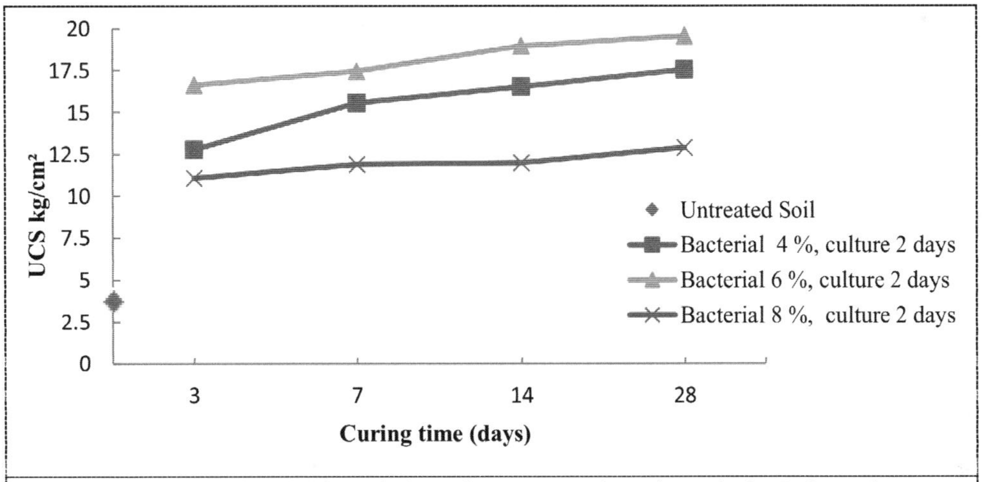

Figure 8. Variations in the solution of 4%, 6%, and 8% bacteria for 2-day culture

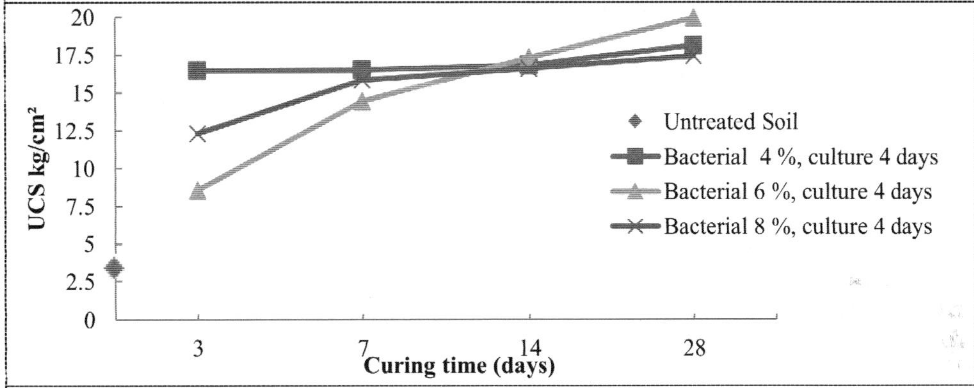

Figure 9. Variations in the solution of 4%, 6%, and 8% bacteria for 4-day culture

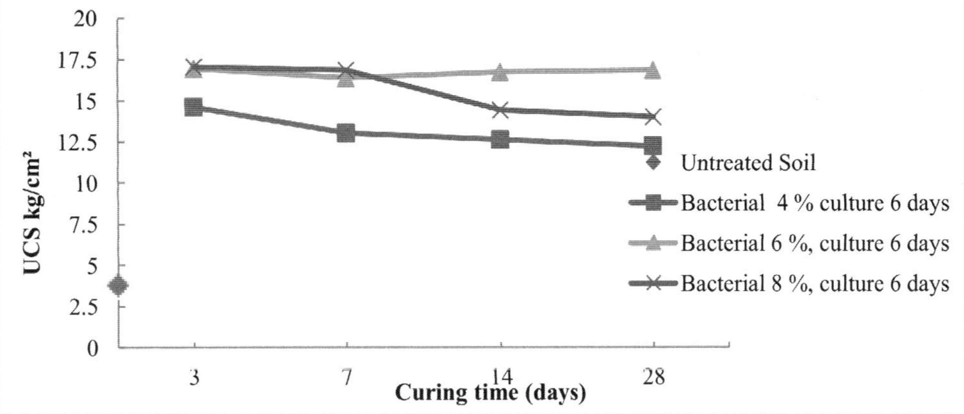

Figure 10. Variations in the solution of 4%, 6%, and 8% bacteria for 6-day culture

4. Conclusion

The compressive strength of the soil stabilized with the addition of 6% solution of bacillus subtilis bacteria and 4 days of culture had the optimum compressive strength value at 28 days with acompressive strength value of 19.96 kg/cm^2, or an increase of 6 times from soil that was not stabilized with bacteria only has a compressive strength value of 3.02 kg/cm^{2,} The research was conducted on a laboratory scale, further development needs to be done by conducting field testing (full scale) to see how the behaviour of bacteria when the conditions are dynamic, such as in dry/hot and wet/rainy conditions

References

[1] DeJong JT, Mortensen BM, Martinez BC, Nelson DC. *Bio-mediated soil improvement.* Ecol Eng. 2010;36(2):197–210

[2] Oliveira PJV, Freitas LD, Carmona JPSF. *Effect of soil type on the enzymatic calcium carbonate precipitation process used for soil improvement.* J Mater Civ Eng. 2017;29(4):1–7.

[3] Mujah D, Cheng L, Shahin MA. *Microstructural and geomechanical study on biocemented sand for optimization of MICP process.* J Mater Civ Eng. 2019;31(4):1–10.

[4] Indriani AM, Harianto T, Djamaluddin AR, Arsyad A. *Bioremediation of Coal Contaminated Soil As the Road Foundations Layer.* Int J GEOMATE. 2021;21(84):76–84.

[5] Harianto T, Hamzah S, Nur SH, Abdurrahman MA, Latief RU, Fadliah I, et al. *Biogrouting stabilization on marine sandy clay soil.* Proc 7th Int Conf Asian Pacific Coasts, APAC 2013. 2020;(Apac):848–52.

[6] Phang IRK, Wong KS, Chan YS, Lau SY. *Effect of microbial-induced calcite precipitation towards tropical organic soil.* AIP Conf Proc. 2018;2020(2018):1–6.

[7] Hasriana, Samang L, Djide MN, Harianto T. *A study on clay soil improvement with Bacillus subtilis bacteria as the road subbase layer.* Int J GEOMATE. 2018;15(52):114–20.

[8] Mwandira W, Nakashima K, Kawasaki S. *Bioremediation of lead-contaminated mine waste by Pararhodobacter sp. based on the microbially induced calcium carbonate precipitation technique and its effects on strength of coarse and fine grained sand.* Ecol Eng [Internet]. 2017;109(September):57–64. Available from: http://dx.doi.org/10.1016/j.ecoleng.2017.09.011

Application of Cation Resin Regeneration for Ferrous (Fe^{2+}) and Manganese (Mn^{2+}) Removal from Shallow Groundwater using Packed-Bed Column with Thomas Model

N A P Mangarengi[1], N O Abdullah[1], and A F Fisu[1]

[1]Environmental Engineering Department, Engineering Faculty, Hasanuddin University, Street of Poros Malino KM. 6, Gowa District, South of Sulawesi, Indonesia 92172

nurannisa.mangarengi@unhas.ac.id

Abstract. One of the factors influencing freshwater contamination is caused by pollutants and organic matter from anthropogenic activities. The appearance of ferrous and manganese metal ions in water systems, especially in shallow groundwater, has become a significant problem. When reacted in the air, an extreme amount of metal ferrous may cause turbidness and severe odor, particularly carrying a dark-brown color to fabrics. Water distribution can also easily be obstructed by iron bacteria in the pipes. It is similar to iron, and manganese has a detrimental effect on the natural environment, human hygiene, and sanitation. The higher influent ferrous and manganese concentration led to the adsorption process getting saturated quicker and a steeper breakthrough curve. The saturation time (ttotal) of the adsorption for ferrous concentration decreased from 430 minute (2 mg/l) to 420 minute (4 mg/l) to 380 minute (6 mg/l) with the elevation of Fe2+ influent concentration. In comparison, the saturation time for manganese concentration dropped from 463 minute (2 mg/l) to 460 minute (4 mg/l) to 420 minute (6 mg/l). Ferrous adsorption capacity (qeq) grew between 0.0046 mg/g and 0.0838 mg/g, followed by a growth in influent concentration. It is similar to manganese. The adsorption performace raised between 0.0216 mg/g and 0.109 mg/g. The Thomas model's R Square (R2) indicates a better fitting (0.788 – 0.8922) with field data.

1. Introduction

Water plays a pivotal role in meeting basic human needs. Water is used daily for bathing, washing, and consumption. The phenomenon of population growth leads to the need for more excellent clean water. The volume of water usage is also affected by each demand, climate conditions, and geographical range. One of the factors influencing freshwater contamination is caused by pollutants and organic matter from anthropogenic activities [1].

The appearance of ferrous and manganese metal ions in water systems, especially in shallow groundwater, has become a significant problem. Because the groundwater is frequently used for domestic and industrial water consumption, a high concentration of heavy metals in groundwater contributes to serious environmental issues and health concerns [2,3]. When reacted in the air, an extreme amount of metal ferrous may cause turbidness and severe odor, particularly carrying a dark-brown color to fabrics. Water distribution can also easily be obstructed by iron bacteria that grow in the pipes [2,4,5]. It is similar to iron, and manganese has a detrimental effect on the natural environment, human hygiene, and sanitation. An excessive of manganese in the water body severely impacts health when water consisting of manganese, iron, and magnesium that exceeds the quality standard is consumed as drinking water continuously as it will possibly accumulate in the human body. A low concentration of manganese does not cause health problems. Still, a large amount can result in a chronic state, such as disturbing the nervous system, kidney and heart related-diseases, and showing symptoms of the disease such as Parkinson's [6,7,8]. Therefore, 0.3 ppm and 0.05 ppm, respectively, are the lowest standards advised by the EPA for ferrous and manganese concentrations in the water [6,9,10].

Due to the serious effects, it is essential to apply certain procedures to get rid of ion heavy metals so that products fulfill quality requirements and don't harm the environment. Reverse osmosis, oxidation, electro-coagulant treatment, and membrane filtering are the most common treatment techniques for heavy metal reduction [11,12,13]. However, most of these methods are expensive and pricey, especially maintenance and operation, and they also consume high energy and produce immense sludge [5,11-13]. Therefore, alternative treatment, such as ion exchange, is needed. Ion-exchange resins are polymerized hydrocarbon compounds containing cross-linking and functional groups of interchangeable ions. As an ion-exchange agent, the resin has beneficial features in chemical analysis, including selling ability, measurement capacity, and exchange selectivity. The resin will release other ions in equivalent equivalence [14]. The regeneration stage is by altering the adsorbed ion with the original ion in the resin matrix and receiving the capacity to the desired level. The regeneration solution must generate the highest point in terms of replacing the regeneration time and the solution usage. Restoring the initial exchange capability can be maximal when the parallel of ions exchange must be the same as the ions removed during the experimental stage. Theoretically, the amount of regeneration solution should be equivalent to the amount of ion eliminated. Therefore, regeneration is only conducted to deliver the proportion of the original exchangeability [15,16]. The effectiveness of cation resin regenerated with NaCl is about 20-50%, and the use of regeneration is much higher than the theoretical necessity [15,16].

Several researchers have conducted ferrous and manganese removal. For instance, chlorine and KMnO4 as oxidants to remove ferrous and manganese from drinking water [7]. Another study from [3] investigated using nanometer-sized polypyrrole coated on rice hush ash to remove Ferrous and Manganese. Furthermore, Vaaramaa K & Letho J (2013), Abo-Farha et al. (2009), and Riveros P A (2004) have analyzed the removal of ferrous and manganese using resin as adsorbent [17, 18, 19]. However, some studies do not use resin cations to remove the ferrous and manganese ion as adsorbents. Most earlier studies used a batch column that was considered less effective when applied on a large scale because most industries use column operation. Because studies about iron (Fe^{2+}) and manganese (Mn^{2+}) elimination from shallow groundwater using the regeneration of cation resin in packed-bed columns are still deficient, it is essential to identify such a study.

Therefore, to prevent the negative impact of the massive level of Fe^{2+} and Mn^{2+}element in shallow groundwater, this study's novelty is to examine the length of exhaustion time required in adsorbing ferrous (Fe) and manganese (Mn) using cation resin regeneration as an adsorbent. This experiment also aims to evaluate the Thomas model's ability to fit field data and the capacity of adsorption in treating ferrous and manganese ions employing cation resin regeneration as adsorbents.

2. Material and Methods

It is vital to establish the optimum conditions of various process parameters through packed-bed experiments.

2.1. Preparation of packed-bed column tools and materials to examine the operation of cation resin adsorbents in the aqueous solution of both Fe and Mn

The necessary equipment and materials must be ready before building the adsorption reactor. The cited equipment and supplies are as follows: As the major media to be filled with adsorbent, two pieces of PVC (polyvinyl chloride) pipe with a 3-inch diameter of line and a 60 cm pipe length were used. With a 3-inch diameter, there are four pieces of PVC pipe covers, two of which are utilized on top and the other two on the bottom. To restrict the flow of water, nine pieces of 3/4" stop taps are used. In addition, two straight joints are used to link parallel pipes, and ten pieces of 3/4" PVC knee are used to connect the other pipes. Five more pieces of 3/4" PVC T joints are also employed in this manner. Five pieces of

3/4" PVC fitting are used to keep some pipes connected, and four pieces of 3/4" external fittings are also used to maintain some lines' connections. Five pieces of a water bolt, 5 meters of 3/4" PVC pipe, PVC adhesive, seal tape, gauze, and a ruler were among the equipment's components. Additionally, two water tanks are used to keep untreated water samples and water samples from tests. Pipe saws, gravel with a diameter of 4.75 to 9.4 mm and a weight of 1.6 kg, and 1.31 kg of cation resin as an adsorbent media are additional ingredients.

2.2. Reactor Design

This adsorption reactor is constructed with a column reactor that has a continuous downward flow from column 1 to column 2 (continuous fixed-bed column). There are two identically sized adsorption columns. The following is the design process for this reactor:

On the same side, one hole is first constructed. Starting from the pipe's lower end, the distance between the centers of each hole is 10 cm. To install the fitting for the bottom treated water pipe, the diameter of the hole is installed for approximately 1 inch. Second, to strengthen the water pressure, each 3-inch PVC pipe cover (cap) on the tube is perforated in the center with a 3/4" diameter hole. Both the outer and inner fittings are then linked, held together with seal tape, and sealed with PVC glue. The filter's top is where the cap is attached. The lid also serves as the top strainer and filter.

The 3-inch pipe's bottom filter is built after the clean-out (CO) installation is complete, and construction then moves on to the fastening of the pipe cover's bottom caps. To help the filter stand straight, flat bottom caps were chosen. PVC glue is used to attach bottom caps to surfaces. The top caps are then put in place, and then the filter and strainer. The taps are installed to control the backwash's input, outflow, and faucets once the top and bottom caps' structures are finished. See Figure 1 below for more information.

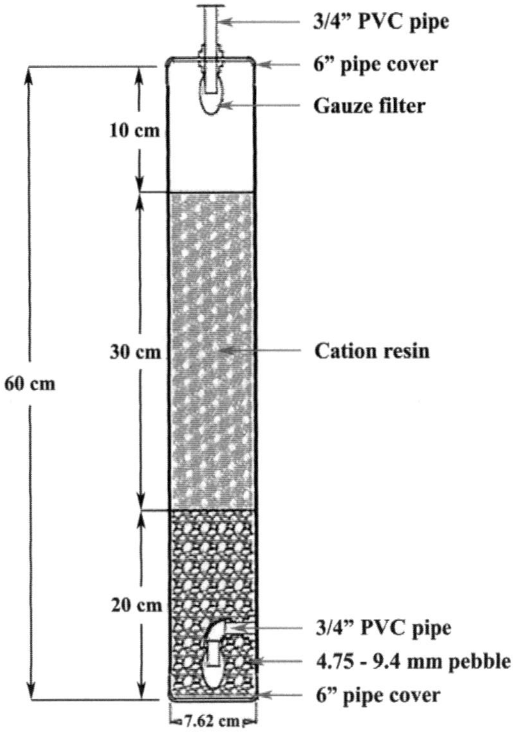

(a)

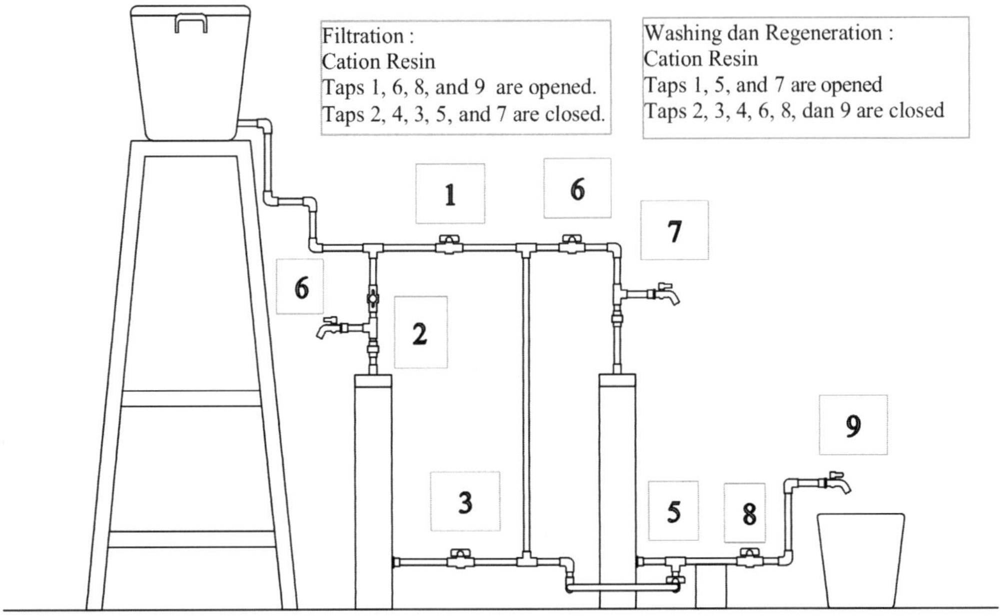

(b)

Figure 1. The Installation of Continous downflow column. Detailed fixed-bed column content (a). A complete reactor installation (b).

Based on Figure 1, the flow of the adsorption processes has two: the adsorption/filtration process using cation resin and the washing and regeneration processes of cation resin. The reactor's pressure is maintained by figuring out the water level in the reservoir, so water continues to flow to fill the column. Then, the process should be waited for the first 15 minutes and stored 500 ml of water should be processed. The process will continue with several contact times, which are 30, 45, 60, 90, 120, 180, 240, 300, 360, 420, and 480 minutes.

In the first process, cation resin adsorption is conducted using the second column by opening taps 1, 6, 8, and 9 and closing taps 2, 4, 3, 5, and 7. The washing and regeneration process uses two columns. In the down flow, the cation resin is the same as the adsorption process, and for the up-flow, taps 1, 5, and 7 are opened, and taps 2, 3, 4, 6, 8, 9 are closed. The flow rate chosen is 110 ml/minute, measured before the adsorption process at the initial experimental stage of the column reactor. The contact time is 15 minutes because the discharge that flows out of the effluent is 15 minutes.

The filter media in the adsorption column reactor use regeneration of cation resin. Meanwhile, gravel with a 4.75-9.4 mm diameter is applied as a physical barrier to prevent the adsorbent layer from falling. The media filter is inserted through the hole of the upper caps in the filter tube. Ensuring that the adsorbent filling is evenly distributed, and better yet, before it is laid into the gravel column and the adsorbent is washed first and given gauze or cloth to prevent the adsorbent not mixing with the gravel. For filling the filter media, the arrangement can be seen in Figure 1.

2.3. Preparation of Solution Samples

Artificial samples of ferrous and manganese were at the Water Quality Laboratory, Environmental Engineering Department, Hasanuddin University.

2.3.1. Preparation of Iron (Fe) and Manganese (Mn) standard solution

Iron (Fe) standard solution is made using the Ferro Sulfate (Fe) standard. The soluble water was generated by gently weighting FeSO4.7H20 as much as 0.6800 grams of iron sulfate. Then the solution was dissolved and combined in a beaker glass. The initial solution was then transported assessably into a 250 ml volumetric flask. It was diluted using distilled water without exceeding the standard solution, so the Artificial sample solution started from 2 ppm, four ppm, and six ppm, which is obtained by pipetting 40 ml, 80 ml, and 120 ml standard solution of 1000 ppm Fe, and diluted with distilled water for 20 liters each. In addition, the manganese (II) sulfate (Mn) standard solution is made. This solution was prepared by gently weighing 0.730 grams of MnSO4 manganese (II) sulfate. Then the solution was dissolved and combined in a beaker glass to acquire a 1000 ppm manganese (II) sulfate solution. The solution was diluted using distilled water without exceeding the standard solution, so the artificial sample solution started from 2 ppm, 4 ppm, and 6 ppm, which is obtained by pipetting 40 ml, 80 ml, and 120 ml standard solutions of manganese (II) sulfate, and diluted with distilled water for 20 liters, respectively.

2.3.2. Regeneration of Cation Resin
According to [16, 20], cation resin regeneration is conducted in the following ways. First, the operation's temperature limit is 120^0C, and the appropriate temperature is 28^0C. The degree of acidity of free pH is 1-14, and the pH used is 7.8-8. Then, the discharge is adjusted. If it is too slow, the resin cannot move, and if it is too fast, the regeneration is ineffective. The discharge used is 100 ml/minute-500 ml/minute. Before regeneration, the cation resin is rinsed with clean and distilled water for about 3-6 times the volume of the cation resin (5L) in the up-flow method. Regeneration is carried out with 50 grams of NaCl for 1 Kg of resin in 1 L of water, while the waiting time is 30 minutes on an up-flow basis. The second rinse water requirement for regeneration is about 3-6 times the volume of the cation resin. The volume of water used is 5 L, and the flow is down.

2.4. Data Analysis

2.4.1. Breakthrough Curve Parameter
Breakthrough curve analysis reveals the saturation time of cation resin to adsorb ferrous metal and manganese metal. The parameter of contact time length (minutes) in the form of the X-axis versus the proportion of the final concentration (Ct) and first concentration (Y-axis) can be thoroughly elucidated through the breakthrough curve. According to the breakthrough curve analysis, the adsorbent circumstance has achieved its equilibrium and exhaustion time when the fraction of the final concentration adsorbed and the first concentration (Ct/Co) approaches the rate of 1, or the rate reaches 90% [21, 22]. The nonlinear correlation between concentration and time is shown by equations such as the following to determine the exhaustion time data with the breakthrough curve.

- The equation for determining the absorbed concentration for each time is calculated from:

$$Cadsorption = Co - Ct \quad (1)$$

$$\frac{Co}{Ct} = \frac{1}{1 + \exp\left(\frac{k_{Th}.qoW}{Q} - k_{ThCot}\right)} \quad (2)$$

$$A = \frac{1}{2}.number\ of\ equivalent\ sides.height \quad (3)$$

$$qtotal = \frac{Q.Co.A}{1000} \quad (4)$$

$$mtotal = \frac{Co.Q.ttotal}{1000} \quad (5)$$

$$\% R = \frac{qtotal}{mtotal}.100 \quad (6)$$

$$qeq = \frac{qtotal}{X} \quad (7)$$

Details: $Cadsorption$ denotes the amount of adsorbed adsorbate (ppm); Co is the adsorbate beginning concentration (ppm), and Ct is the adsorbate ultimate concentration (ppm); K_{Th} is the constant rate (L/min.mg); qo is the capacity of adsorbent (mg/g); W is adsorbent masst; t determines the total flow duration (min); A is the surface area under the curve; the number of parallel sides results from 1- (Ct / Co); height is the time interval (minutes); Q is the water discharge (mL/min); $qtotal$ is the total mass of metal adsorbed (in mg); $mtotal$ is the total mass in mg of the metal ions that have passed the column; $ttotal$ is total saturation time is expressed as $ttotal$ (minutes), the water discharge as Q (mL/min), and the adsorbate beginning concentration as Co (mg/L); $\% R$ is the percentage of metal ions removal (%); qeq is the amount of metal ion removed as a percentage (%), $qtotal$ is the amount of metal adsorbed overall (mg), and X is the adsorbent mass (g).

2.4.2. Model of Thomas

To comprehend the dynamics of the adsorption in packed-bed columns with continuous flow, the Thomas model is frequently applied. The flow equilibrium in this model is predicated on pseudo-first-order kinetics and the Langmuir isotherm [23,24]. Based on a modification of the Bohart-Adams formula, the Thomas model is utilized to depict the kinetics processes. The formula for it is given below:

$$\ln(\frac{Co}{Ct} - 1) = \frac{K_{TH}.qo.m}{Q} - K_{TH}.Co.t \quad (8)$$

where qo denotes the maximum adsorbent capacity (mg/gr), Q denotes the water discharge (ml/min), W denotes the mass of the adsorbent (g), and K_{Th} is the rate constant of the Thomas model (l/mg.min). Thomas model, which is used to describe the value of the constants K_{Th}, may be derived by plotting the linear function of ln $\ln(\frac{Co}{Ct} - 1)$ against time. The qo can be calculated using the intercept and slope of the plot.

3. Result and Discussion

3.1. The effect of the influent concentration of the tested solution and the exhaustion time analysis

Figures 2 and 3 illustrate the effect of influent ferrous (Fe2+) and manganese (Mn2+) concentration on the implementation of breakthrough curves analysis. The parameter of contact time length (minutes) in the form of the X-axis versus the proportion of the final concentration (Ct) and first concentration (Y-axis) can be thoroughly elucidated through the breakthrough curve. Based on the breakthrough curve analysis, the adsorbent circumstance has achieved its equilibrium and exhaustion time when the fraction of the final concentration adsorbed and the first concentration (Ct/Co) approaches the rate of 1, or the rate reaches 90% [25-30].

In this experiment, the weight of adsorbent, flow rate, and bed depth was preserved constantly at 1310 g, 110 mL/min, and 0.6 m, respectively. The pH utilized is 7-8, and the temperature ranges from 25 to 28 degrees Celsius. While the influent concentrations of manganese are added from 2 mg/L to 6 mg/L, the influent concentrations of ferrous solution are added with the same concentration as manganese. As it is shown in Figures 2 and 3, the higher influent ferrous and manganese concentration led to the adsorption process getting saturated quicker and a steeper breakthrough curve [29, 30]. The breakthrough curve declined along with the higher concentration of ferrous and manganese inlet solution. As the initial concentration increases, it also enhances the mass transfer and diffusion coefficient [29, 30, 31, 32, 33]. When the initial solution is low, the breakthrough curve occurs delayed, influenced by the slow transport, transfer zone, and intraparticle diffusion process of ferrous and manganese from the multilayer to the adsorbent surface [34, 35]. Therefore, a faster adsorption rate will result in a higher adsorption capacity (see Table 1). The exhaustion time (ttotal) of initial ferrous concentration decreased from 430 minute (2 mg/l) to 420 minute (4 mg/l) to 380 minute (6 mg/l) with a steady increase in Fe2+ initial concentration, while the exhaustion time of initial manganese concentration dropped from 463 minute (2 mg/l) to 460 minute (4 mg/l) to 420 minute (6 mg/l) (Table 1). The result reveals that alteration in the concentration of a solution affects the capacity exchange, which means that the diffusion process was concentration-dependent. It assumes that cations more easily adsorb the incoming ion's solvation layer in a more concentrated solution compared to ions present in the resin phase [4, 31, 32, 33, 34, 35].

Table 1. The results of adsorption rate (all parameters) of both Fe^{2+} and Mn^{2+} adsorbate in the packed-bed column using cation resin regeneration

Type of Metals	Concentration (C) (ppm)	ttotal (minute)	A Value	qtotal (mg)	m_{total} (mg)	%R	q_{eq} (mg/g)
Ferrous (Fe^{2+})	2	430	135.61	30.88	109.30	28.25	0.0236
	4	420	208.02	98.39	227.04	43.34	0.0751
	6	380	220.71	155.38	337.92	45.98	0.1186
Manganese (Mn^{2+})	2	463	134.98	28.36	100.85	28.12	0.0216
	4	460	195.51	81.72	200.64	40.73	0.0624
	6	420	220.01	142.79	311.52	45.84	0.1090

3.2. Breakthrough Curve Analysis

Using breakthrough curve analysis, the adsorption rates of ferrous and manganese adsorbate were determined. As demonstrated in Table 1, all the adsorption rates (mtotal, qtotal, %R, and qeq) experienced significant growth alongside the increase in influent concentration. For instance, ferrous adsorption capacity (qeq) inclined between 0.0236 mg/g and 0.1186 mg/g, followed by a rise in influent concentration (2, 4, 6 ppm). The amount of qeq (the adsorption capacity), which is 0.0216 mg/g and 0.1090 mg/g, respectively, has gradually increased when compared to manganese. The photon behavior, which is more effective than manganese due to a metal group, affects this occurrence. Unlike manganese metal (Mn), which has an atomic number of 25, iron metal (Fe) has an atomic number of 26 in group 8b. The oxidation process also affects this trend in which manganese's oxidation process is more complex than ferrous, and the ferrous oxidation rate is higher than manganese oxidation [36, 37]. A study conducted by Ye (2011) showed that removal using modified zeolite is better for Fe removal than Mn^{2+} reduction [10, 38].

3.3. Thomas Model Implementation

Figures 4 and 6, as well as Table 2, display the results of the Thomas Model, which were obtained using linear regression and plotted using ln(Co/Ct-1) vs time in accordance with Eq. 8. The Thomas Model's R Square (R2) shows that it fits field data the best (0.788 – 0.8922). Table 2 demonstrates that while influent concentration increased as KTH values decreased, the value of qo increased as KTH values decreased. Manganese's KTh value dropped from 2.45 to 1.08 ml/min.mg and ferrous' from 3.05 to 1.15 ml/min.mg, while both elements' influent concentrations enhanced from 2 to 6 ppm. The ferrous adsorption capacity (qe), which improved from 0.0046 to 0.0838 mg/gr, was also increased. Similar to the manganese adsorption capacity, the trend saw the qe value rise from 0.0081 to 0.0955 mg/gr while the KTh value fell. Additionally, the Thomas Model implementation's adsorption capacity (qo = 0.0046 to 0.0838 mg/g) is lower than the ferrous in column experiment's (qeq = 0.0236 - 0.1186) adsorption capacity. Additionally, the manganese's adsorption capacity in the packed-bed continuous column experiment (qeq= 0.0216 - 0.109) is greater than the adsorption capacity anticipated by the Thomas Model.

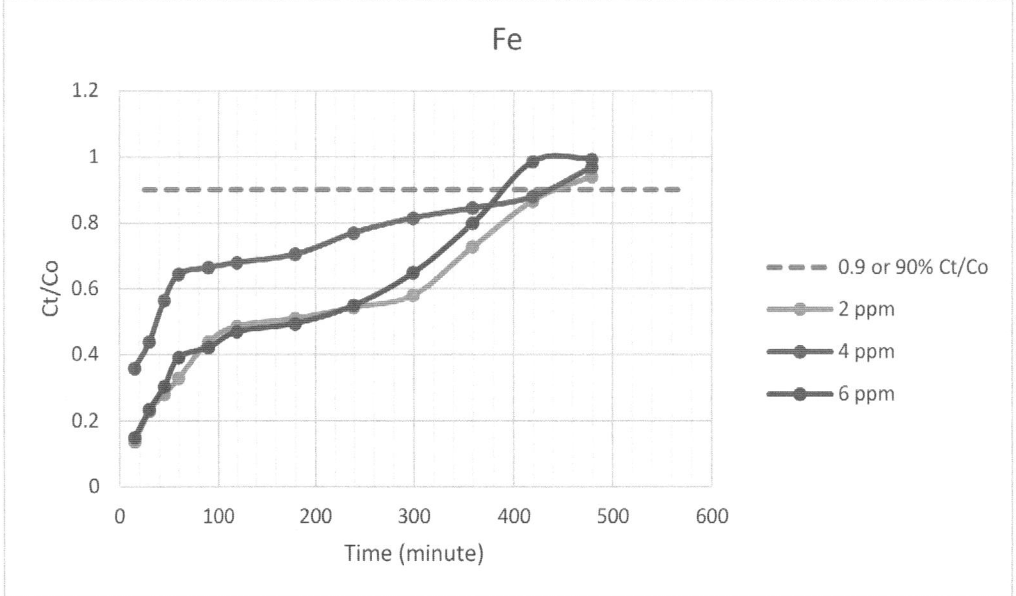

Figure 2. Influence of influent ferrous (Fe^{2+}) concentration on ferrous adsorption breakthrough curve in packed-bed column utilizing cation resin regeneration. Specifications: bed height is 0.6 m; flow rate is 110 ml/min; pH is normal; influent ferrous concentrations are 2, 4, and 6 ppm, respectively.

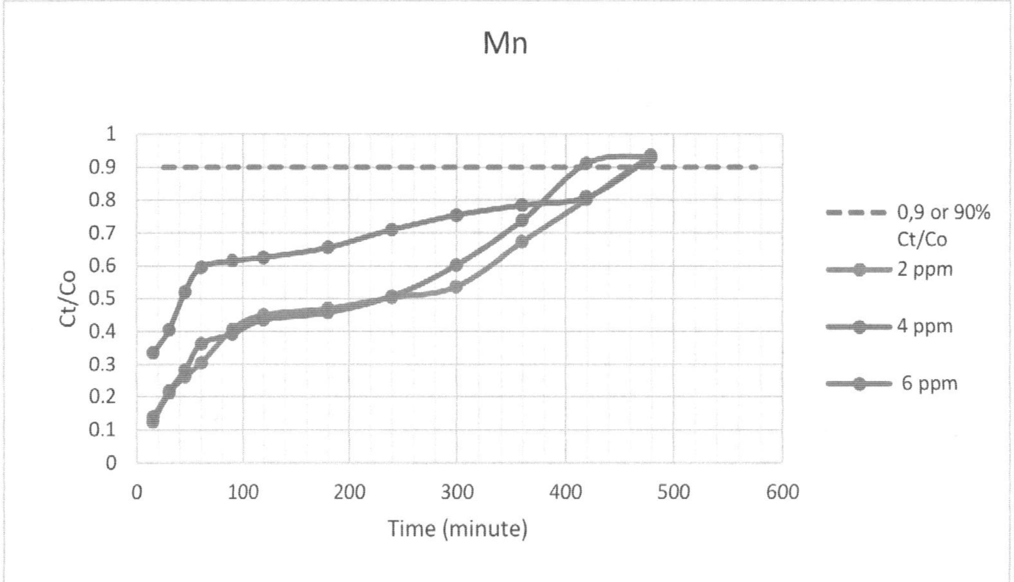

Figure 3. Influence of influent manganese (Mn^{2+}) concentration on manganese adsorption breakthrough curve in packed-bed column utilizing cation resin regeneration. Specifications: bed height is 0.6 m, flow rate is 110 ml/min, pH is normal, and influent manganese concentrations are 2, 4, and 6 ppm, respectively.

Table 2. The results of the Thomas Model on ferrous (Fe^{2+}) and manganese (Mn^{2+}) removal using cation resin regeneration

Variation of Metal	CO (ppm)	Z (m)	X (gr)	*KTh* (ml/min.mg)	qo *Thomas* (mg/g)	R Square
Fe	2	0.6	1310	3.05	0.0046	0.8922
	4	0.6	1310	2.80	0.0043	0.8014
	6	0.6	1310	1.15	0.0838	0.8291
Mn	2	0.6	1310	2.45	0.0081	0.8803
	4	0.6	1310	1.80	0.0568	0.8374
	6	0.6	1310	1.08	0.0955	0.788

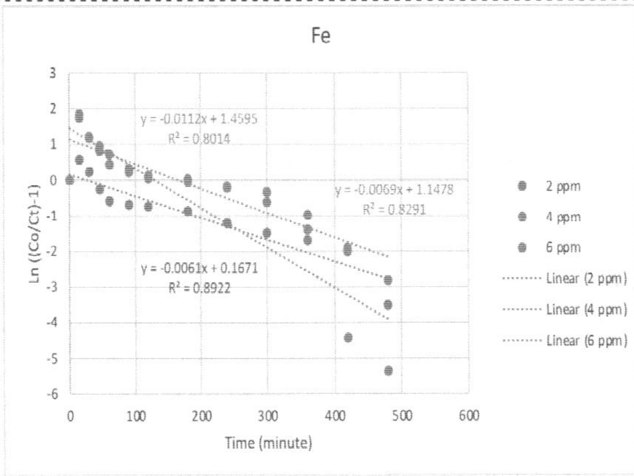

Figure 4. The outcome of applying the Thomas model to regeneration of cation resin in removal of (Fe2+). Details: the water discharge is 110 ml/min, pH is normal; bed height is 0.6 m.

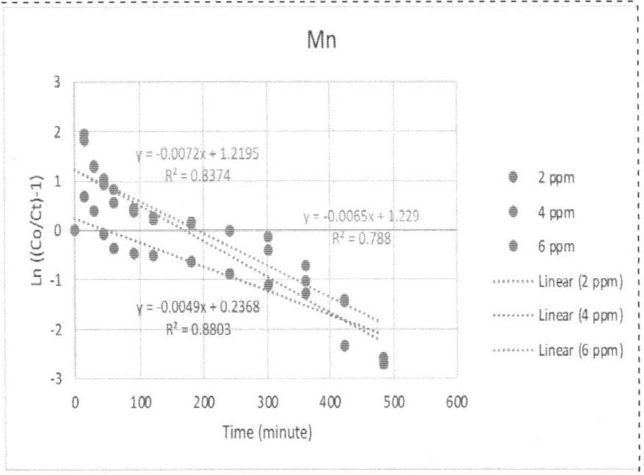

Figure 5. The outcome of applying the Thomas model to the regeneration of cation resin in the removal of manganese (Mn2+). Details: flow rate is 110 ml/min; pH is normal; bed height is 0.6 m.

4. Conclusion

By adjusting the Fe2+ and Mn2+ concentrations, the removal of ferrous (Fe2+) and manganese (Mn2+) utilizing cation resin regeneration in a packed-bed column with continuous flow was examined. Experimental results indicated that fewer Fe^{2+} and Mn^{2+} ions were removed when the breakthrough curve reached attainment. The exhaustion time (t_{total}) of the adsorption for ferrous concentration decreased from 430 minute (2 mg/l) to 420 minute (4 mg/l) to 380 minute (6 mg/l) with the elevation of Fe^{2+} influent concentration. In comparison, the saturation time for manganese concentration dropped from 463 minutes (2 mg/l) to 460 minutes (4 mg/l) to 420 minutes (6 mg/l). Ferrous adsorption capacity (qeq) grew between 0.0046 mg/g and 0.0838 mg/g, followed by a growth in influent concentration. It is similar to manganese. The adsorption capacity raised between 0.0216 mg/g and 0.109 mg/g. The Thomas model's R Square (R2) indicates a better fitting (0.788 – 0.8922) with field data.

Acknowledgments
This work was financially assisted by Hasanuddin University under research grant from LPPM (Lembaga Penelitian dan Pengabdian Masyarakat) Unhas.

References

[1] Asano T, Burton F & Leverenz H 2007 *Water Reuse: Issues, Technologies, And Applications* (McGraw-Hill Education)

[2] Allahdin O, Dehou S C, Wartel M, Recourt P, Trentesaux M, Mabingui J & Boughriet A 2013 Performance of FeOOH-brick-based composite for Fe (II) removal from water in fixed bed column and mechanistic aspects *Chemical Engineering Research and Design* **91** (12) 2732-2742

[3] Ghorbani M & Eisazadeh H 2012 Fixed bed column study for Zn, Cu, Fe, and Mn removal from wastewater using nanometer size polypyrrole coated on rice husk ash *Synthetic metals* **162** (15-16) 1429-1433

[4] Ostroski I C, Barros M A, Silva E A, Dantas J H, Arroyo P A & Lima O C 2009 A comparative study for the ion exchange of Fe (III) and Zn (II) on zeolite NaY *Journal of Hazardous Materials* **161** (2-3) 1404-1412

[5] Onganer Y & Temur Ç 1998 Adsorption dynamics of Fe (III) from aqueous solutions onto activated carbon *Journal of Colloid and Interface Science* **205** (2) 241-244

[6] Lee S M, Tiwari D, Choi K M, Yang J K, Chang Y Y & Lee H D 2009 Removal of Mn (II) from aqueous solutions using manganese-coated sand samples *Journal of Chemical & Engineering Data* **54** (6) 1823-1828

[7] Khadse G K, Patni P M & Labhasetwar P K 2015 Removal of iron and manganese from drinking water supply *Sustainable Water Resources Management* **1** (2) 157-165

[8] Al-Anber M & Al-Anber Z A 2008 Utilization of natural zeolite as ion-exchange and sorbent material in the removal of iron *Desalination* **225** (1-3) 70-81

[9] Ören A H & Kaya A 2006 Factors affecting adsorption characteristics of Zn2+ on two natural zeolites *Journal of hazardous materials* **131** (1-3) 59-65

[10] Ye C, Yang H, Lin J, Zeng H & Yu F 2011 Study on ion exchange property of removing Mn2+ and Fe2+ in groundwater by modified zeolite *Desalination and Water Treatment* **30** (1-3) 114-121

[11] Baran A, Bıçak E, Baysal Ş H & Önal S 2007 Comparative studies on the adsorption of Cr (VI) ions on to various sorbents *Bioresource technology* **98** (3) 661-665

[12] Ahalya N, Ramachandra T V & Kanamadi R D 2003 Biosorption of heavy metals *Res. J. Chem. Environ* **7** (4) 71-79

[13] Erdem E, Karapinar N & Donat R 2004 The removal of heavy metal cations by natural zeolites *Journal of colloid and interface science* **280** (2) 309-314

[14] Silva R A, Hawboldt K & Zhang Y 2018 Application of resins with functional groups in the separation of metal ions/species–a review. *Mineral Processing and Extractive Metallurgy Review* **39** (6) 395-413

[15] Malovanyy A, Sakalova H, Yatchyshyn Y, Plaza E & Malovanyy M 2013 Concentration of ammonium from municipal wastewater using ion exchange process *Desalination* **329** 93-102

[16] Diananto D 2016 Penggunaan Natrium Klorida (Nacl) Sebagai Regeneran Resin Amberlite Ir 120 Na Dalam Menurunkan Kesadahan Total Air Sumur *Jurnal Rekayasa Lingkungan* **16** (2)

[17] Vaaramaa K & Lehto J 2003 Removal of metals and anions from drinking water by ion exchange *Desalination* **155** (2) 157-170

[18] Abo-Farha S A, Abdel-Aal A Y, Ashour I A & Garamon S E 2009 Removal of some heavy metal

cations by synthetic resin purolite C100 *Journal of hazardous materials* **169** (1-3) 190-194
[19] Riveros P A 2004 The extraction of Fe (III) using cation-exchange carboxylic resins *Hydrometallurgy* **72** (3-4) 279-290
[20] Kikuchi Y, Kamiya O & Kobayashi H 1999 Mechanical characteristics of high nitrogen contained weld metal based on 316L stainless steel *Materials science forum* **318** 621-628
[21] Kumar U & Bandyopadhyay M 2006 Fixed bed column study for Cd (II) removal from wastewater using treated rice husk *Journal of hazardous materials* **129** (1-3) 253-259
[22] Han R, Zhang J, Zou W, Xiao H, Shi J & Liu H 2006 Biosorption of copper (II) and lead (II) from aqueous solution by chaff in a fixed-bed column *Journal of Hazardous materials* **133** (1-3) 262-268
[23] Thomas H C 1948 Chromatography: a problem in kinetics *Annals of the New York Academy of Sciences* **49** (2) 161-182
[24] Kumar P A & Chakraborty S 2009 Fixed-bed column study for hexavalent chromium removal and recovery by short-chain polyaniline synthesized on jute fiber *Journal of hazardous materials* **162** (2-3) 1086-1098
[25] Kundu S, Kavalakatt S S, Pal A, Ghosh S K, Mandal M & Pal T 2004 Removal of arsenic using hardened paste of Portland cement: batch adsorption and column study *Water Research* **38** (17) 3780-3790
[26] Han R, Wang Y, Zhao X, Wang Y, Xie F, Cheng J & Tang M 2009 Adsorption of methylene blue by phoenix tree leaf powder in a fixed-bed column: experiments and prediction of breakthrough curves *Desalination* **245** (1-3) 284-297
[27] Han R, Zou L, Zhao X, Xu Y, Xu F, Li Y & Wang Y 2009 Characterization and properties of iron oxide-coated zeolite as adsorbent for removal of copper (II) from solution in fixed bed column *Chemical Engineering Journal* **149** (1-3) 123-131
[28] Aksu Z & Gönen F 2004 Biosorption of phenol by immobilized activated sludge in a continuous packed bed: prediction of breakthrough curves *Process biochemistry* **39** (5) 599-613
[29] Uddin M T, Rukanuzzaman M, Khan M M R & Islam M A 2009 Adsorption of methylene blue from aqueous solution by jackfruit (Artocarpus heteropyllus) leaf powder: a fixed-bed column study *Journal of environmental management* **90** (11) 3443-3450
[30] Chen S, Yue Q, Gao B, Li Q, Xu X & Fu K 2012 Adsorption of hexavalent chromium from aqueous solution by modified corn stalk: a fixed-bed column study *Bioresource Technology* **113** 114-120
[31] Jung K W, Jeong T U, Choi J W, Ahn K H & Lee S H 2017 Adsorption of phosphate from aqueous solution using electrochemically modified biochar calcium-alginate beads: Batch and fixed-bed column performance *Bioresource Technology* **244** 23-32
[32] Luo X, Yuan J, Liu Y, Liu C, Zhu X, Dai X et al 2017 Improved solid-phase synthesis of phosphorylated cellulose microsphere adsorbents for highly effective Pb2+ re- moval from water: batch and fixed-bed column performance and adsorption mechanism *ACS Sustain Chem. Eng.* **5** 5108–5117
[33] Sotelo J L, Ovejero G, Rodríguez A, Álvarez S, García J 2013 Analysis and modeling of fixed bed column operations on flumequine removal onto activated carbon: pH influ- ence and desorption studies *Chem. Eng. J* **228** 102–113
[34] Malkoc E, Nuhoglu Y & Dundar M 2006 Adsorption of chromium (VI) on pomace—an olive oil industry waste: batch and column studies *Journal of Hazardous Materials* **138** (1) 142-151
[35] Baral S S, Das N, Ramulu T S, Sahoo S K, Das S N & Chaudhury G R 2009 Removal of Cr (VI) by thermally activated weed Salvinia cucullata in a fixed-bed column *Journal of Hazardous Materials* **161** (2-3) 1427-1435
[36] Katsoyiannis I A & Zouboulis A I 2004 Biological treatment of Mn (II) and Fe (II) containing groundwater: kinetic considerations and product characterization *Water research* **38** (7) 1922-1932

[37] Ellis D, Bouchard C & Lantagne G 2000 removal of iron and manganese from groundwater by oxidation and microfiltration *Desalination* **130** (3) 255-264
[38] Selintung M, Mangarengi N A P & Zubair A 2020 The reduction of heavy metals Cd and Cr levels in wastewater using bagasse charcoal as an adsorbent *IOP Conference Series: Earth and Environmental Science* **419** (1) 012171

A Low Cost Open Source Remote Monitoring System for Hospital Wastewater Management

Agus Putu Abiyasa[1], I Ketut Nuraga[2], Aryangga Marta Satriadi[3]

[1]Department of Electrical Engineering, Universitas Pendidikan Nasional, Indonesia
[2]Department of Civil Engineering, Universitas Pendidikan Nasional, Indonesia
[3]Department of Engineering, Bali Mandara Hospital, Indonesia

abiyasa@undiknas.ac.id

Abstract. Wastewater can be dangerous if not handled properly, e.g. it spilled out from the tank. Hospital wastewater hazard was even more damaging to environments and humans due to its chemical contents. Therefore, it is desirable to have monitoring system to prevent such incidents. In this research, remote monitoring of wastewater using low cost and open source IOT technology was implemented at Bali Mandara Hospital waste management facility. The system used ESP8266 microcontrollers and HCSR04 ultrasonic sensors to determine wastewater volume. Blynk IOT cloud server was used as the platform to collect and visualize the data online. The implemented IOT system showed robust performance as a monitoring system for hospital wastewater management.

1. Introduction

Hospital wastewaters are dangerous for environment because it contains hazardous substances, such as pharmaceutical residues, chemical hazardous substances, pathogens, and radioisotopes. Hospital wastewaters are considered as major contributor of emerging contaminants and required rigorous treatment before it is released to environment [1],[2]. In order to protect public and environmental health, sophisticated treatment technologies have been developed and strict regulations have also been enforced by regulators [3],[4]. The use of information technology such as Internet of Things (IOT) has become necessary to strengthen hospital wastewaters management process.

IOT technology has been rapidly developed for the past couple of years [5]. Among many IOT technologies, WIFI based connectivity has been rising in implementation due to it ease in integration with existing communication networks [6]. Low cost off the shelf controllers such as ESP8266 with WIFI communication capability makes development of IOT systems much simpler. Many such applications have been developed recently [7],[8],[9],[10] with IOT cloud service has been extensively offered by many providers [11]. IOT has also been implemented for monitoring process on environmental biodegradability and risk analysis [12].

In this work, we implemented a new technological solution to hospital waste management based on IOT technology. Our approach is to design a wastewater monitoring system using low cost and open source IOT technology components available in the market. We then integrate them into a working system and are suitable for field testing. For all device's communication, a WIFI connection is used as data transfer protocol between sensors and server as well as access by users. Our research is focused on testing our proposed IOT solution for wastewater monitoring at a hospital waste management facility in Bali Mandara Hospital located on Bali Province, Indonesia. The result of our research could give useful information to determine whether similar low cost IOT systems can be implemented elsewhere and for different applications.

2. Methods

The research implements IOT technology for wastewater monitoring system in a waste management facility for Bali Mandara Hospital. To conduct the research, the monitoring system was designed making use of HCSR04 sensor, LDR sensor and NODEMCU ESP8266 microcontroller as the main hardware components. The microcontroller software was then programmed to read data from sensors, processed the data and sent the information to cloud Blynk IOT server [11].

The diagram of the system is shown in Figure 1 as part of the waste treatment process. It showed the first stage of the process where all wastewater will be stored into a containment tank. As shown in Figure 1, all the wastewater coming from laundry, kitchen, restroom and even drugs chemicals were collected in the wastewater tank via inlet pipe. After some time, the wastewater will be sent to the next treatment tank by submersible pumps via outlet pipe. The monitoring system was installed to monitor the water level inside the wastewater tank. As part of the monitoring system, an ultrasonic HCSR04 sensor was installed to detect the distance of the water level from the top. LDR sensors were also installed at the pump control panel and used to detect the pump ON/OFF states. These sensors were then connected to NODEMCU ESP8266 microcontroller by wires to get electrical power and also to send the sensor data. NODEMCU ESP8266 will process the sensor data and send the data to the internet using WIFI. These data were then collected by Blynk Server and stored into a database. Using Blynk App, the information can be viewed in a smartphone. Blynk App can also send notification and email to designated users. LCD matrix display was also installed to display information on site.

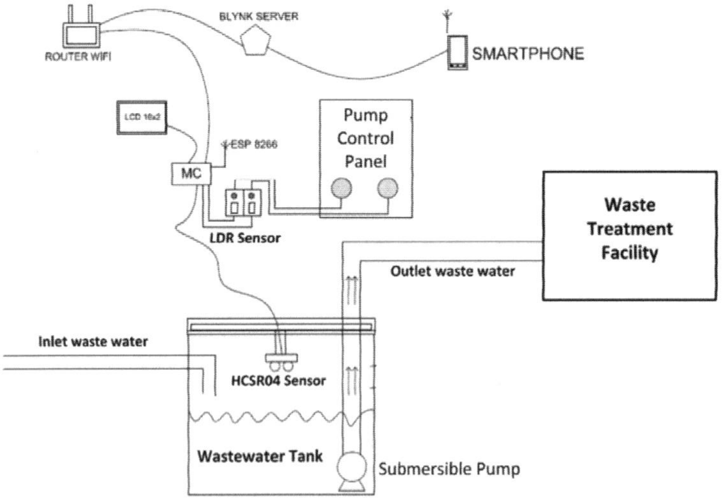

Figure 1. Block diagram of wastewater monitoring system

Figure 2 showed the wiring diagram of the electronic components of the system. HCSR04 and LDR sensors were connected to NODEMCU ESP8266 microcontroller via wires. These wires were the Vcc, GND and digital pin data. LCD matrix display was also connected to NODEMCU ESP8266 and the power for operation was supplied using a USB adapter to the microcontroller. The sensor data collection process was implemented using non-intrusive technique. An HCSR04 sensor emits ultrasonic waves that will be reflected by the water surface hence the distance can be determined. The HCSR04 sensor used two digital pins as the TRIGGER and ECHO pins. The TRIGGER pin is an output digital signal from NODEMCU ESP8266 to HCSR04 and the ECHO is an input digital pin which detects signals from HCSR04 to NODEMCU ESP8266. As for the LDR sensor, it detected the light from the pump indicator lamp so that it can determine the ON/OFF states of the pump. When the indicator lamp is ON, LDR will detect the light as digital input high from the LDR to NODEMCU ESP8266 and vice versa. All these digital data will be read and processed by the software in NODEMCU ESP8266 microcontroller.

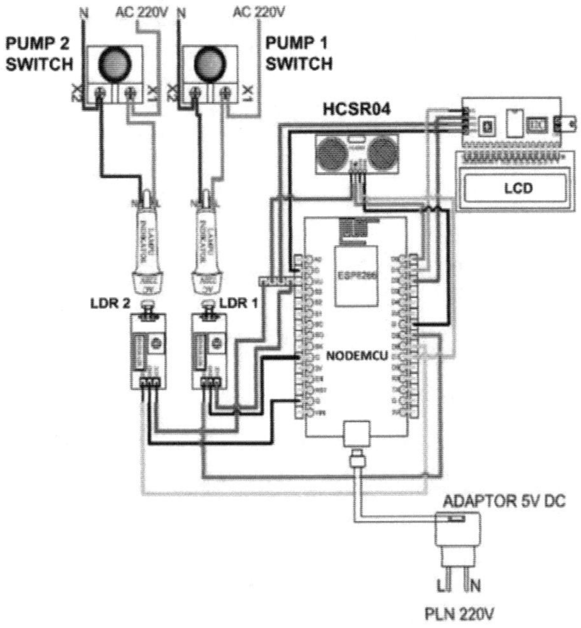

Figure 2. Wiring diagram of NODEMCU ESP8266 microcontroller with sensors and LCD display

The software for NODEMCU ESP8266 microcontroller was designed to work according to specifications of the monitoring system. The program is written in C language using Arduino IDE. Arduino IDE is a programming platform that supports many development boards such as Arduino UNO, NODEMCU ESP8266 and many more. It is also available in Windows, LINUX and MacOS. Arduino IDE can easily transfer the software code into the NODEMCU ESP8266 board using a USB port.

Figure 3 shows the flowchart for the program. The program started by looking into a WIFI connection. This can be done using a hardcoded procedure or by using a user input method such as Wifi Manager. When the WIFI connection has been established, the program continues to check the sensor data. WIFI with internet connection will further establish connection to the online Blynk server. First, HCSR04 sensor data is read and also sent to the Blynk server. The program will check the distance from water level. If the distance is smaller than 15 cm, the program will send request notification to Blynk server. Afterward, warning notification will pop up at user smartphone via Blynk App. If the distance is larger than 15 cm, the program will repeat checking the sensor continuously. Second, the program will also read the LDR sensor data and send the pump states to Blynk server. If the LDR sensor detects light, the program will send data to Blynk server and the pump state will be ON at the users' smartphone via Blynk App. If the LDR sensor did not detect light, the pump state will be OFF at Blynk App. Similar to notification, an email can also be sent to the user email address. Lastly, if a WARNING notification has happened, the user will notify the technician to check the wastewater tank since that is a dangerous situation of wastewater overflow.

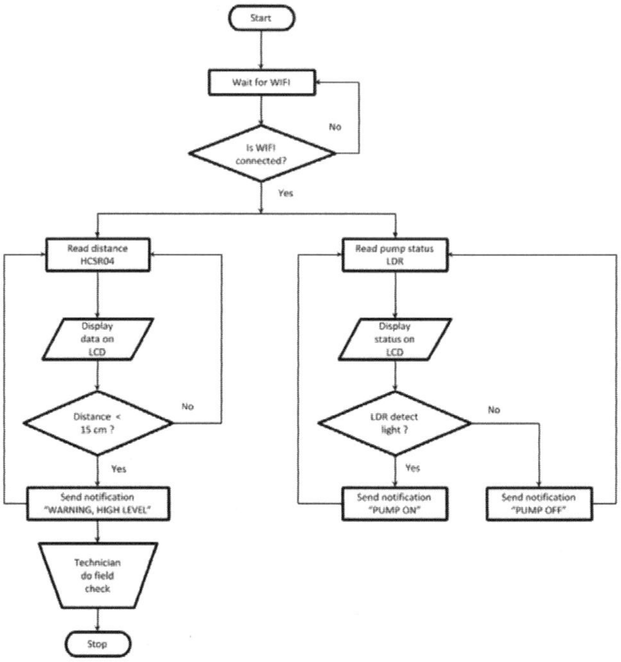

Figure 3. Flowchart diagram of software for NODEMCU ESP8266 microcontroller

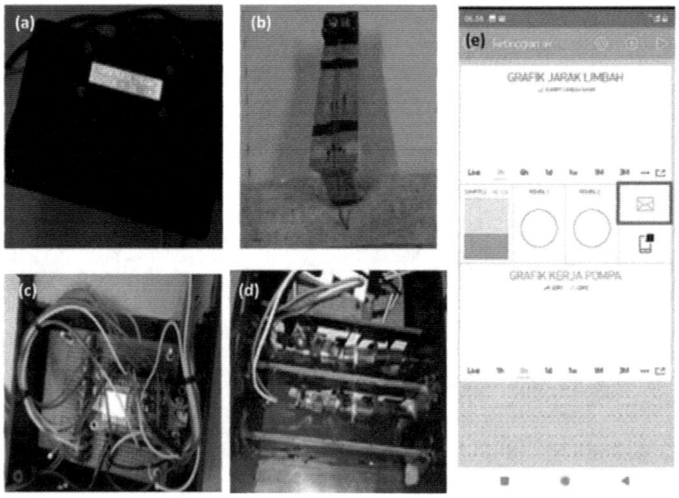

Figure 4. (a) LCD front panel, (b) HCSR04 sensor attached on a piece of wood, (c) NODEMCU ESP8266 microcontroller and the wiring of components, (d) LDR sensors attached to pump switch lights, (e) Blynk App user interface design on smartphone

To test the system, a prototype has been fabricated and installed at the designated location at Bali Mandara Hospital as shown in Figure 4. Figure 4(a) – (d) show the hardware after finished wiring and integrated. Figure 4(e) shows the user interface design on smartphones using Blynk IOT platform. Online datalogging was done at Blynk IOT server and we can download the data later on from the online database. The process of hardware installation can be done with an official permission from the hospital authority. Data about the volume of the wastewater as well as the pump states will be collected every minute and stored online on the Blynk server database. The use of the data was also under official permission of the hospital authority.

3. Results and Discussions

The use of IOT technology has been discussed recently to improve sewage treatment process by Zhu *etal* [13]. The study in their paper mentioned about IOT system needed and its main advantage compare to conventional system is the realization of efficient and cheaper real time monitoring of the sewage treatment plant. The conventional technology used expensive Programmable Logic Controller (PLC), custom software and local server that can cost tens of thousands of dollars to build. However, in their project they used STM32 microcontroller module, WeChat App and cloud services which only cost few thousands of dollars. The approach was similar to our study but we used different low cost and open source components such as ESP8266 microcontroller, Blynk App and cloud services. We also used off the self-ultrasonic sensors as components for our monitoring system. Therefore, to justify our work we need to analyze the sensors used in our system and their validity to be used for field applications.

In this section, the data collected by the sensors will be discussed and analyzed respectively. The result discussions will be divided into several parts, i.e. 1) HCSR04 measurement analysis, 2) LDR measurement analysis, 3) Daily wastewater analysis.

3.1. HCSR04 Measurement Analysis

Figure 5(a) shows the schematic diagram of the waste water tank. We can determine the volume of the wastewater as follows. HCSR04 sensor will measure the distance between the sensor location and the surface of the wastewater. Since the height of the tank offset the sensor location is known, we can measure the height of the wastewater. Hence the waste water volume can be calculated as

$$V = (H - h)\; x\; L\; x\; W \qquad (1)$$

where,

V = wastewater volume
H = height of the tank offset sensor location
h = distance of sensor and water surface
L = length of tank
W = width of tank

In Figure 5(b), the actual installation of the HCSR04 sensor and a measuring rod is shown. A measuring rod was placed to measure the actual height of the wastewater for calibration. The HCSR04 sensor was attached below a piece of wood and placed on top of the tank as shown. The wastewater tank specifications gave the parameters' value of *H*, *L* and *W* as 2.255 m, 3 m and 2.45 m, respectively.

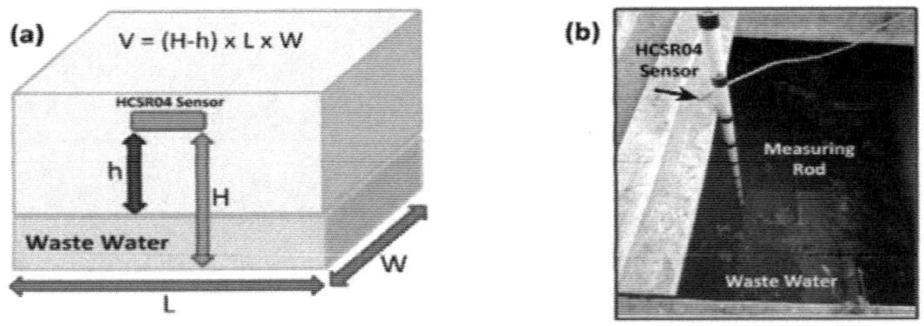

Figure 5. (a) Diagram of wastewater tank with specification parameters, (b) HCSR04 sensor installation at the wastewater tank

Table 1. The Result of Water Level Distance Measurement at Wastewater Tank

Measured by HCSR04 Sensor (cm)	Measured by Measuring Rod (cm)	Calibrated Data (cm)	Calculated Wastewater (Liter)
50	53.5	53.86	12579
60	64.5	64.32	11810
70	74.5	74.78	11041
80	85.5	85.24	10273
90	95.5	95.70	9504
100	106.5	106.16	8735

In order to calibrate the HCSR04 data, test measurements were done according to two test conditions. First, HCSR04 data was calibrated at normal air condition outside the tank where it was found the distance measured by HCSR04 sensor and a measuring tape gave exactly the same values. This was expected as HCSR04 operated using ultrasonic which assumed sound wave travel in air. Second, HCSR04 data is calibrated at special air conditions inside the tank. In this particular measurement, the distance measured by HCSR04 is compared to the measuring rod. Table 1 shows the measurement result of this special condition. It was found that the value measured by the HCSR04 sensor was smaller compared to the value measured by the measuring rod. The data was then calibrated by a linear regression using least square fitting. This resulted in a line equation of $y = 1.53 + 1.05x$, where x is the HCSR04 sensor reading and y is the calibrated data.

The plot of this calibration is shown in Figure 6. It can be seen that this calibration was quite accurate and suitable for the measurement inside the wastewater tank. The measured distance by HCSR04 sensor inside the wastewater tank was affected by the air condition inside the tank due to the chemical vapor by the wastewater. With vaporizing air condition, the speed of ultrasonic waves inside the tank is different from normal air hence affecting HCSR04 measurement result. However, the HCSR04 sensor measurement accuracy has been adjusted using the calibration formula and the calibrated distance was correct when compared to the measuring rod. Furthermore, the calculated wastewater volume has also been shown in Table 1. The value was calculated using Equation 1 with all the known parameters of the tank as given in the specifications. We can then correlate the value of HCSR04 data, e.g. of 50 cm will be equal to 12579 liters of wastewater and so on.

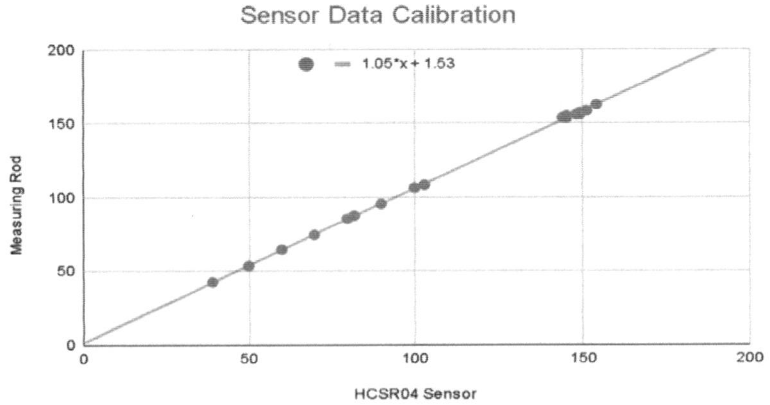

Figure 6. Data calibration of HCSR04 measurement result at the wastewater tank

3.2. LDR Measurement Analysis

The LDR sensors were used to detect the ON/OFF states of wastewater pumps. There were two pumps available as shown in the diagram of Figure 2. These pumps worked alternately to pump out the wastewater into the next stage through the outlet pipe as shown in Figure 1. It is interesting to note the correlation of pumps state and the volume of the wastewater as given in Figure 7. The volume of wastewater at certain times in a day was plotted as a blue bar chart and the state of pump 1 as determined from LDR sensor reading was plotted as a brown line. Please be noted that ON state was measured as LOW and OFF state was measured as HIGH by the LDR sensor. We can see that initially the pump was OFF and when the time reached 11:26 the pump started. From there the volume started to decrease from maximum around 6943 liters. This continued until the time reached 11:35 when the pump stopped and the volume of wastewater stabilized around 5311 liters. This showed that the measurement of volume was in agreement with the pump state. It can be estimated the amount of wastewater pumped out during this particular time was around 1632 liters.

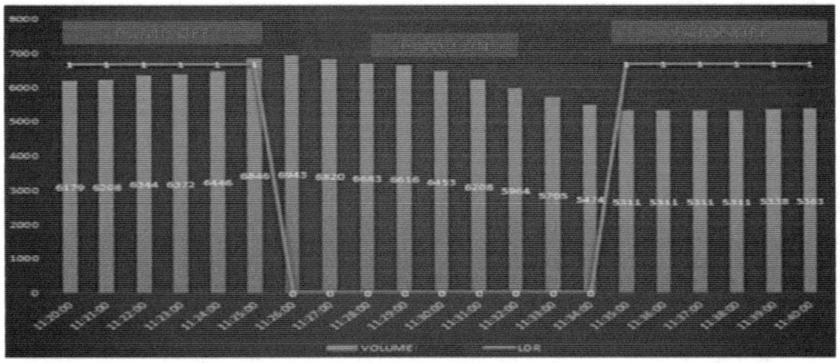

Figure 7. Pump 1 ON/OFF states and its correlation to waste water volume change in the tank

3.3. *A Daily Wastewater Analysis*

After we confirmed our sensors measurement results were accurate, we proceeded to monitor daily wastewater production that has been pumped out for the treatment process. From the observations, pump 1 and pump 2 will operate about two to three times a day. When the pump starts, it will transfer the wastewater to the next stage through the outlet pipe. By calculating the accumulated change of wastewater volume during pumping operation period we can determine how much wastewater has been transferred for treatment every day. The accumulated volume data every day is shown in Figure 8. The average volume of wastewater transferred every day was 6557 liters. If we compare to the total volume of the tank, this was still below 50% of the total tank capacity. At normal operation, the pumps ensure the wastewater tank was safe from possible overflowing.

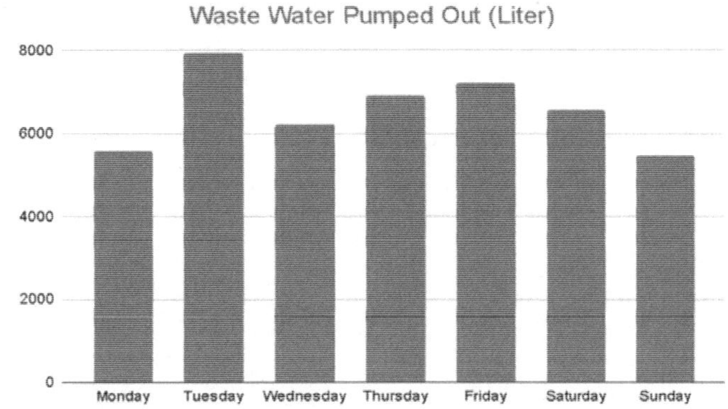

Figure 8. Wastewater daily production volume that will be pumped out for treatment.

4. Conclusions

In conclusion, the monitoring system based on IOT technology has worked for hospital wastewater management implementation at Bali Mandara Hospital. The system to monitor the wastewater volume was fabricated with low cost and open source hardware of NODEMCU ESP8266 microcontroller and integrated with HCSR04 sensor as well as LDR sensors. Due to chemical vapor inside the tank, HCSR04 data has been calibrated to measure the volume of the wastewater. In addition, LDR sensors have also detected the ON/OFF states of the pumps. From the volume calculation, it was found that the average daily wastewater volume pumped out for treatment was around 6557 liters. For further improvement, the monitoring systems should also be implemented at all stages of the waste treatment facilities of Bali Mandara Hospital.

References

[1] M. T. Khan *et al.*, "Hospital wastewater as a source of environmental contamination: An overview of management practices, environmental risks, and treatment processes," *J. Water Process Eng.*, vol. 41, no. February, p. 101990, 2021, doi: 10.1016/j.jwpe.2021.101990.

[2] T. Mackull'ak *et al.*, "Hospital wastewater—source of specific micropollutants, antibiotic-resistant microorganisms, viruses, and their elimination," *Antibiotics*, vol. 10, no. 9, pp. 1–14, 2021, doi: 10.3390/antibiotics10091070.

[3] E. Carraro, S. Bonetta, and S. Bonetta, "Hospital wastewater: Existing regulations and current trends in management," *Handb. Environ. Chem.*, vol. 60, pp. 1–16, 2018, doi: 10.1007/698_2017_10.

[4] P. Verlicchi, "Trends, new insights and perspectives in the treatment of hospital effluents," *Curr. Opin. Environ. Sci. Heal.*, vol. 19, p. 100217, 2021, doi: 10.1016/j.coesh.2020.10.005.

[5] S. Kumar, P. Tiwari, and M. Zymbler, "Internet of Things is a revolutionary approach for future technology enhancement: a review," *J. Big Data*, vol. 6, no. 1, 2019, doi: 10.1186/s40537-019-0268-2.

[6] A. Siswandi, F. S. Srijati, and A. Y. Permana, "Internet of Things Wemos D1 Microcontroller Implementation for Monitoring Water Ph in Real Time on the WWTP (Waste Water Treatment Plant) Based on the Internet of Things," *J. Mantik*, vol. 5, no. 36, pp. 1988–1994, 2021.

[7] T. A. Salih and N. K. Younis, "Designing an Intelligent Real-Time Public Transportation Monitoring System Based on IoT," *Open Access Libr. J.*, vol. 08, no. 10, pp. 1–14, 2021, doi: 10.4236/oalib.1107985.

[8] M. S. Shamsudin and A. S. M. Zain, "Low-Cost IoT Based Carbon Monoxide (CO) Monitoring System," in *Proceedings of Innovation and Technology Competition (INOTEK) 2021*, 2021, pp. 179–180, doi: 10.1016/j.comnet.2010.05.010.

[9] B. N. Mohapatra, R. V. Jadhav, and K. S. Kharat, "A Prototype of Smart Agriculture System Using Internet of Thing Based on Blynk Application Platform," *J. Electron. Electromed. Eng. Med. Informatics*, vol. 4, no. 1, pp. 24–28, 2022, doi: 10.35882/jeeemi.v4i1.2.

[10] T. Malche, S. Tharewal, and D. Pallavi Bhatt, "A Portable Water Pollution Monitoring Device for Smart City based on Internet of Things (IoT)," *IOP Conf. Ser. Earth Environ. Sci.*, vol. 795, no. 1, 2021, doi: 10.1088/1755-1315/795/1/012014.

[11] J. Dedy Irawan and R. Primaswara Prasetya, "IoT Data Logger Using Blynk Framework," *Int. J. Latest Eng. Manag. Res.*, vol. 06, no. 01, pp. 17–23, 2021, [Online]. Available: www.ijlemr.com.

[12] S. Gopikumar, J. R. Banu, Y. H. Robinson, V. Shanmuganathan, S. Kadry, and S. Rho, "Novel framework of GIS based automated monitoring process on environmental biodegradability and risk analysis using Internet of Things," *Environ. Res.*, vol. 194, no. November 2020, p. 110621, 2021, doi: 10.1016/j.envres.2020.110621.

[13] W. Zhu, Z. Wang, and Z. Zhang, "Renovation of automation system based on industrial internet of things: A case study of a sewage treatment plant," *Sensors (Switzerland)*, vol. 20, no. 8, 2020, doi: 10.3390/s20082175.

Acknowledgements

We would like to acknowledge DRTPM Ministry of Education, Culture, Research Technology and Higher Education who has granted Basic Research Grant year 2021 for supporting this research.

IOP Conf. Series: Earth and Environmental Science 1117 (2022) 012048 doi:10.1088/1755-1315/1117/1/012048

Utilization of Solid Waste PT Semen Bosowa as Zeolite for Industrial Waste Water in South Sulawesi

N O Abdullah[1], Safruddim[2], M Tumpu[3]

[1,4] Environmental Engineering Department, Universitas Hasanuddin
[2]Geology Engineering Department, Universitas Hasanuddin
[3]The Graduate School, Universitas Hasanuddin

Email : nurjannahoktorina48@gmail.com.

Abstract. One of the major industries in Eastern Indonesia engaged in manufacturing is PT Semen Bosowa. In the process of processing cement raw materials there are several stages, one of the stages in the process is the mixing process that occurs in the Raw Grinding Mill section. In addition to grinding or smoothing the cement raw material, it also serves as a material dryer. Some residual material is wasted during the process as solid waste and cannot be processed again so that it will have an impact on the environment. so it needs further handling so as not to cause environmental problems. One of the efforts that can be made is to utilize solid waste material as a zeolite replacement adsorbent. The material has a composition similar to zeolite. This study aims to determine the effect of PT Semen Bosowa solid waste as a zeolite substitute on reducing the TDS, BOD, and COD values of industrial wastewater. The method used in the research, namely the wastewater sample is put into the reservoir (tub I), then flowed continuously into tub II which is filled with solid waste material as zeolite. The sampling technique is carried out after the wastewater flows from tub II then collected in a storage container and ready to be analyzed for TDS, BOD, and COD parameters of wastewater. The amount of TDS, BOD and COD reduction from PT Semen Bosowa laboratory wastewater after going through zeolite adsorbent was 3.79%, 58.33%, and 58.35%, respectively.

1. Introduction

Progress in various industrial sectors, including the textile, manufacturing, and food industries, in addition to having a positive impact, also has a negative impact. In industrial activities, during the production process, various materials are wasted or fail that contain metal as residue as well as organic substances/contaminants produced from residual raw materials, residual solvents, washing and rinsing of equipment, such as boiler boilers and cooling water systems and sanitary waste is one of the negative impacts that can pollute the environment and in the end will disrupt the balance of the environmental ecosystem if it is not handled properly.

One of the big industries in Eastern Indonesia which is engaged in manufacturing, namely PT. Semen Bosowa. In this industry, the products produced are expected to have good quality and be able to compete. In the cement raw material processing there are several stages, one of the stages in the process is the mixing process that occurs in the Raw Grinding Mill section. Part of the cement production process equipment has a fairly important role. Besides functioning to grind or refine cement raw materials, it is also a material dryer. However, there is a leak in the Raw Grinding Mill process unit, so there are a number of products or materials that are wasted as solid waste and cannot be processed anymore so that it will have an impact on the environment. So that further treatment is needed so as not to cause environmental problems. One of the efforts that can be done is to utilize the solid waste material as an adsorbent to replace zeolite. Zeolite is a porous material with the main constituent components are silica (SiO_2), iron oxide (Fe_2O_3), alumium oxide (Al_2O_3), calcium oxide, and magnesium oxide. absorption of 99.725% [1]. Natural zeolite can reduce COD, BOD, TSS, pH, and heavy metals in textile wastewater by 88% [2].

Natural zeolites are hydrated alumino silicate compounds, with the main elements consisting of alkaline and alkaline earth cations. This compound has a three-dimensional structure and has pores that can be filled by water molecules. The nature of zeolite as an adsorbent and molecular filter is possible because of the hollow structure of the zeolite, so that the zeolite is able to absorb a large number of molecules that are smaller in size or according to the size of the cavity.

This is what drives research by utilizing solid waste materials from PT. Semen Bosowa as a substitute for natural zeolite to adsorb organic materials, especially the parameters of TDS, BOD, and COD of industrial and domestic wastewater, does not always run smoothly. There are several dominant inhibiting factors and reduced public attention to the environment that affect the availability of water [3]. One of the problems with the availability of clean water that is often encountered is the low quality of water. Water quality is defined as the quality of water, both raw water and treated water that is ready to be distributed. Clean water must comply with existing standards, both in terms of chemical, physical, biological and radioactive parameters. Based on Government Regulation of the Republic of Indonesia Number 22 of 2021 concerning Implementation of Environmental Protection and Management. Drinking water and raw water are classified as first class water. First-class water quality is water that is fit for consumption. It aims to improve public health.

2. Materials and Methods

2.1. Location and type of research

This type of research used in this study is an experimental study and literature review. This research was conducted at the Water Quality Research Laboratory of the Department of Environmental Engineering, Universitas Hasanuddin. The research time was carried out for approximately 2 months.

2.2. Research design

This research is a type of advanced experimental research. The results of the treatment were then analyzed in the laboratory sample to determine the ability of the zeolite substitute waste material as an adsorbent to reduce TDS, BOD, COD, and pH of industrial wastewater.

The materials used in this study were rice husk ash, industrial wastewater samples, aquadest, concentrated sulfuric acid, 2% starch indicator, alkali-iodide-azide solution, manganese sulfate solution, sodium thiosulfate solution ($Na_2S_2O_3.5H_2O$) 0.025N, ferrous ammonium sulfate ($Fe(NH_4)_2(SO_4)_2.6H_2O$) 0.025 N solution, ferroin indicator, and mixed solution of potassium dichromate-mercury sulfate.

The tools used in this research are wastewater treatment tub, mesh sieve, electric oven, burette, funnel, desiccator, erlenmeyer, volume pipette, bucket, cergen, Winkler bottle, COD bottle, measuring flask, incubator, filter paper, analytical balance, and spectrophotometer instruments, namely XRF and pH meters.

2.3. Reactor design

The reactor in this study consisted of 3 units (laboratory scale) made of glass with dimensions in accordance with the design criteria. Unit 1 as a laboratory wastewater container, unit 2 as a treatment reactor filled with solid waste material, and unit 3 as an indicator tub filled with fish.

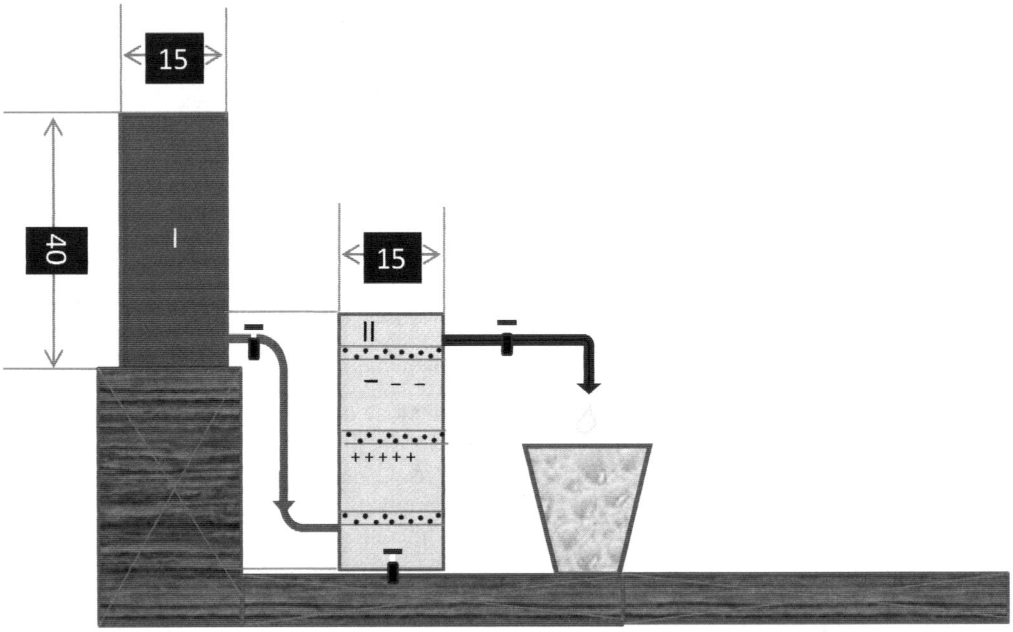

Figure 1. Reactor design.

2.4. Research implementation

In this study, experiments were carried out continuously. Tub 1 contains wastewater with a volume of 20 liters as a reservoir of wastewater before the wastewater is flowed into the treatment tub, tub 2 as a treatment tub filled with solid waste material.

2.5. Data analysis

This research was conducted by analyzing laboratory data, namely the results of testing the parameters pH, BOD, COD, TDS, in laboratory wastewater before and after treatment. Data obtained from the laboratory were compared with the required quality standards to determine the feasibility of processing in this study. Furthermore, the data is also graphed to find the best take in the study. The data were analyzed using the following formula. The percentage of Removal Efficiency can be seen below:

$$\%\,Removal = (C0 - C1)/Co \times 100\% \quad (1)$$

Where C0 is the initial concentration and C1 is the final concentration.

3. Result and Discussions

The wastewater used in this study is laboratory wastewater from PT. Semen Bosowa. Data from laboratory tests on water samples that describe the characteristics of laboratory wastewater data from the research results are presented in Table 1 below. Based on the results of laboratory analysis of the adsorption ability of solid waste in the form of reject material from the Raw Grinding Mill process to organic liquid waste, it can be seen that there are differences in the quality of industrial organic liquid waste (color, pH, TDS, BOD, and COD) before adsorption and after adsorption. There are even some heavy metals that can be adsorbed with the reject material. This is because the reject material solid waste in the Raw Grinding Mill process unit is the remaining material that is wasted during the process. This solid waste contains several components including limestone ($CaCO_3$), clay (Al_2O_3), iron sand, and silica sand (SiO_2), as can be seen in Table 2. These components resemble the components in zeolite ($Al_2O_3.SiO_2$) thus the reject material can also function as an adsorbent. The waste material has voids which, when heated, are able to adsorb the liquid waste passed through them.

Table 1. Characteristics of laboratory water PT.semen bosowa.

Analysis Result				Standard	Removal Efficiency (%)
Parameter	Units	Inlet	Outlet		
TDS	mg/L	2110	2030	4000	3.79
Colour	-	pink	clear		-
pH	mg/L	1.68	6.25	6.0-7.0	-
BOD	mg/L	40	16.67	50	58.33
COD	mg/L	60	24.99	100	58.35
Pb	mg/L	0.0161	0.0029	0.1	81.99
Hg	mg/L	0.263	0.0092	0.02	96.5
Cr	mg/L	0.03	0.01	0.5	66.67
Zn	mg/L	0.263	0.0704	4	73.23

3.1. Material rejected of PT.Semen Bosowa

Material rejected in the Raw Grinding Mill process unit is a solid waste that has a composition similar to zeolite. Reject material has a composition of limestone, clay, silica sand, and iron sand. During the production process, several reject materials are produced and have not been optimally utilized by the industry [4]. Zeolite is a porous material with the main constituent components are silica (SiO_2), iron oxide (Fe_2O_3), aluminum oxide (Al_2O_3), calcium oxide, and magnesium oxide, in their research reported that natural zeolite can reduce metal chromium (Cr) wastewater with an absorption value of 99.725% [5].

3.2. Zeolite mineral

Zeolite minerals have been known since 1756 by Cronstedt when he discovered Stilbit which when heated the mineral will be like a boiling stone due to dehydration of the water molecules it contains. In 1954 zeolites were classified as a separate mineral class, which was then known as molecular sieve materials. Thus, zeolites are minerals consisting of hydrated alumino silicate crystals containing alkali or alkaline earth cations in a three-dimensional framework. The metal ions can be replaced by other cations without damaging the zeolite structure and can absorb water reversibly.

In the zeolite structure, Si and Al atoms form the zeolite structure with oxygen atoms. These atoms make up the zeolite (alumino silicate) framework which is the exchange matrix. Si and Al atoms are both tetrahedral, each atom is mediated by an oxygen atom, causing a negative charge on the Al atom which is a fixed ion. Meanwhile, alkali or alkaline earth ions are opponent ions that can be exchanged easily with other cations in aqueous solutions without affecting the zeolite framework [6]. Zeolites cannot be identified based on chemical composition analysis alone, but must be analyzed for structure. The crystal structure of zeolites where all Si and Al atoms are in the form of tetrahedra (TO_4) is called the Primary Building Unit, zeolites can only be identified based on the Secondary Building Unit (UBS) as shown in Figure 2.

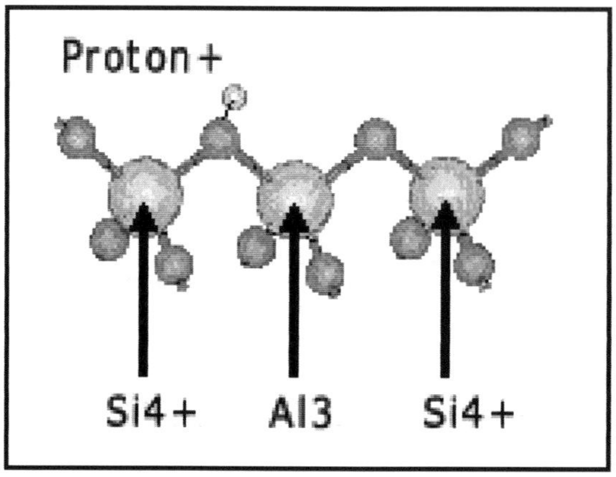

Figure 2. Structure atomic zeolite.

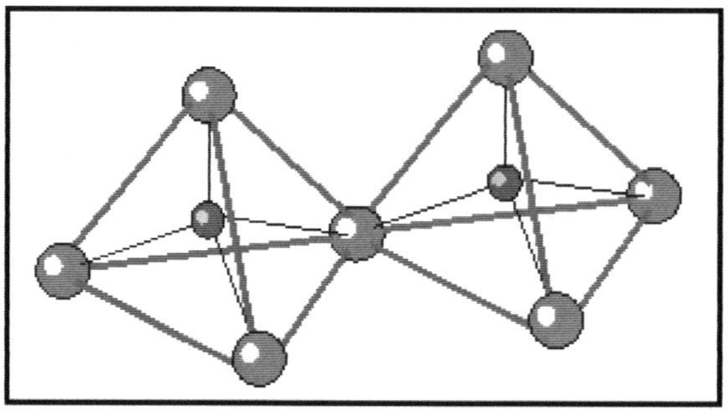

Figure 3. Alumina and silica (TO_4) tetrahedra in zeolite structure.

Due to the unique physical and chemical properties of zeolites, in this decade, zeolites have been used by researchers as versatile minerals. These unique properties include dehydration, adsorbents and molecular filters, catalysts and ion exchangers [7]. Zeolites have dehydration properties (releasing H_2O molecules) when heated. In general, the structure of the zeolite skeleton will shrink. But the basic framework has not changed significantly. Here the H_2O molecule seems to have a specific position and can be removed reversibly. In addition, zeolite crystals that have been dehydrated are selective adsorbents and have high adsorption effectiveness.

The nature of zeolite as an adsorbent and molecular filter is possible because of the hollow structure of zeolite, so that zeolite is able to absorb a large number of molecules that are smaller or according to the size of the cavity. Smaller molecules can pass through while those larger than the vacuum will be retained or rejected [8].

The ability of zeolites as catalysts is related to the availability of active centers in the channels between zeolites. These active centers are formed due to the presence of Bronsted and Lewis type acid functional groups. The ratio of these two types of acids depends on the zeolite activation process and reaction conditions. These acidic active centers can then bind base molecules chemically [9].

While the nature of zeolite as an ion exchanger due to the presence of alkali and alkaline earth metal cations. These cations can move freely in the cavity and can be exchanged with other metal cations in the same amount. Due to the hollow zeolite structure, anions or molecules smaller or equal in size to the cavity can enter and be trapped [9].

Table 2. Wasted material component analysis results PT. semen bosowa.

Component	Concentration (%)
SiO_2	7,26
Al_2O_3	0,62
Fe_2O_3	0,43
$CaCO_3$	48,57
MgO	0,88

As reported by many researchers, including foreign researchers, that zeolites have properties as molecular filters. In addition, zeolites have a three-dimensional cage structure (framework). Zeolites also have cavities and channels, which are occupied by alkali and alkaline earth metal ions such as Na, K, Mg and Ca and water molecules. The nature of zeolites as adsorbents and molecular filters, is possible because of the hollow structure of zeolites, so that zeolites are able to absorb a large number of molecules. smaller or according to the size of the cavity. Smaller molecules can pass through while those larger than the vacuum will be retained or rejected (Jamuddin, 2003). Research conducted by Luna, et al (2007) reported that zeolites made from fly ash material can reduce COD by 43% and TSS by 82%. From some of these studies it can be used as a basis that, because the structure and components in zeolite resemble components in the reject material, the reject material is able to absorb a large number of molecules that are smaller or in accordance with the size of its cavity. From the results of laboratory analysis, it can also be seen that in addition to organic matter TSS, BOD, COD, pH and color that can be reduced also some heavy metals such as Pb, Hg, Cr, and Ni in wastewater can be adsorbed as well.

3.3. Adsorption process

Adsorption is the absorption of a substance on the surface of another substance or the tendency of molecules or ions in solution to aggregate on the surface of a solid. Molecules on the surface of a solid or liquid have an attractive force towards the inside. Adsorption occurs by the attractive force between the adsorbed molecule (adsorbate) and the adsorbing material (adsorbent). A molecule can be adsorbed if the adhesion force between the adsorbate molecule and the adsorbent molecule is greater than the cohesion force in each molecule [10].

Adsorption theory explains that the solution attached to the surface of the adsorbent is caused by the Van Der Wall bond / force. The process of attaching the solution to the adsorbent is more physical than chemical, causing many layers on the surface of the adsorbent. Chemical adsorption (chemisorption) occurs where particles or substances adhere to a surface by forming chemical bonds (usually covalent bonds). Solids can adsorb ions from a solution by way of ions in the solution replacing the exact ions present in the solid. Whereas physical adsorption is where the solid material with a very large inner surface area. The large surface is formed due to the presence of fine pores in the solid. In addition to surface area and pore diameter, distribution density and particle size are the most important characteristics of an adsorbent [10].

The speed of adsorption is affected by concentration differences, adsorbent surface area, particle size, porosity of the adsorbent, molecular size of the material to be adsorbed and viscosity of the mixture to be separated. The maximum flow velocity of the solution to penetrate the exchange layer (the required residence time of the solution in the exchange layer) is determined by the time for the exchange process on each exchange particle. If the speed is too high (too short residence time), not all ions to be separated can be exchanged so that the desired liquid purity requirements are not achieved [11].

3.4. Industrial wastewater

Industrial wastewater is waste in liquid form produced by industrial activities that are discharged into the environment and are thought to reduce the quality of the environment. The amount of wastewater from industry varies greatly depending on the type and size of the industry, supervision of the industrial process, degree of water usage, degree of existing water treatment [12]. To estimate the amount of wastewater produced by industry, it can be considered that 85-95% of the amount of water used is in the form of wastewater if the industry does not reuse the wastewater released. If the industry reuses its wastewater, the amount of wastewater will be even smaller. Industrial wastewater is wastewater that comes from household businesses / industries and or residential activities, restaurants, offices, commerce, apartments, and dormitories [13]. Wastewater is a combination or mixture of water and polluting materials carried by water, both in a dissolved and suspended state that is discharged from domestic sources of offices, housing, trade, industrial sources, and at certain times mixed with groundwater, surface water or rainwater [14].

Regulation of the Minister of Environment and Forestry Number 68 of 2016 concerning Domestic Wastewater Quality Standards, that wastewater is the residue of a business and or activity in the form of liquid. Meanwhile, domestic wastewater is derived from human activities, including businesses/activities, from housing or settlements, offices, shopping/trading centers, restaurants/ eateries and recreational facilities.

4. Conclusion

Based on the results of the research that has been carried out, the data shows that solid waste in the form of material from the Raw Grinding Mill process as an adsorbent can reduce the organic matter of PT Semen Bosowa laboratory wastewater. The decrease in organic parameter tests carried out is TDS before adsorption of 2110 mg/L and after adsorption of 2030 mg/L. BOD before adsorption 40 mg/L and after adsorption 16.67 mg/L. COD before adsorption 60 mg/L after adsorption 24.99 mg/L. and pH becomes neutral (6.25) as well as the color from red to clear. The data is based on the Decree of the State Minister of Environment and Forestry of the Republic of Indonesia No. 68 of 2016 concerning Domestic Wastewater Quality Standards. Due to significant adsorptive properties they can be used for the removal of different pollutants such as heavy metals, oil and organic contaminates. Solid waste of PT. Semen Bosowa as zeolites are low cost adsorbents with excellent selectivity for various cations and good ion exchange capacity.

5. Acknowledgments

This study was supported by the PDPU (Penelitian Dosen Pemula Unhas) 2022 grant funding. We, all the authors, would like to thank Hasanuddin University, especially to LP2M (Lembaga Penelitian dan Pengabdian kepada Masyarakat) Universitas Hasanuddin, who provided the PDPU 2022 grant.

6. References

[1] Sayuti, Adi Iqbal. 2020. Pembuatan Adsorben Berbasis Zeolit Alam Untuk Penyisihan CO_2 Dalam Biogas Melalui Aktivasi Kimia-Fisika. Medan: Universitas Sumatera Utara.

[2] Fitriyanti, Reno, dan Muhrisnyah Fatimura. 2019. Aplikasi Produksi Bersih PadaIndustri Semen. Palembang: Universitas PGRI Palembang.

[3] Jiyah, dkk. 2017. Studi Distribusi Total Suspended Solid (TSS) Di Perairan Pantai Kabupaten Demak Menggunakan Citra Landsat. Jurnal Geodesi Undip.

[4] Ginting, Simparmin Br., dkk. 2020. Sintesis dan karakterisasi zeolite LTA dari coal bottom ash teraktivasi dengan aging system. Jurnal Teknologi dan Inovasi Industri Vol. 01, No.01, Hlm.015- 027.

[5] Ngapa, Yulius Dala, dan Yasinta Embu Ika. 2020. Potensi Zeolit Alam Ende Sebagai Media Adsorben Kompetitif PewarnaBiru Metilena Dan Metil Oranye.Cakra Kimia (Indonesian E-Journal of Applied Chemistry) Volume 8 Nomor 2.

[6] Nurjanah, Siti, dkk. 2017. Penyisihan BOD Dan COD Limbah Cair Industri Karet Dengan Sistem Biofilter Aerob Dan Plasma Dielectric Barrier Dischare (DBD). Jurnal Teknik

Lingkungan, Vol. 6, No. 1.
[7] Muammar, dan Muh. Rais, Patang. 2019. Pengaruh Limbah Industri Terhadap Tingkat Pencemaran Timbal di Perairan Sungai Tallo. Makassar: UNM.
[8] Antaria dan Sukmasari. 2015. Studi Penggunaan Zeolit Alam Sulawesi Selatan Sebagai Bahan Filtrasi Air Baku. Makassar: Universitas Hasanuddin.
[9] Peraturan Pemerintah Republik Indonesia Nomor 68 Tahun 2016 Tentang Baku Mutu Air Limbah Domestik. Jakarta: Kementerian Lingkungan Hidup.
[10] Putri, Suariati Eka, dan Sumiati Side. 2020. Analisis Kandungan Oksida Logam Zeolit Alam Sulawesi Selatan Teraktivasi Asam Klorida. Jurnal Sainsmat, September 2020, Halaman 159-163.
[11] Rais, Wahmin, dkk. 2020. Optimalisasi Sistem Perawatan dan Perbaikan Mesin Vertical Mill di PT. Semen Padang. Jurnal Teknik Mesin Vol. 13 No.1.
[12] Hakiki, Muhammad, dkk. 2021. Pengaruh Lokasi Zeolit Alam Bayah terhadap Adsorpsi Amonium: Studi Kinetika dan Kesetimbangan. Jurnal TeknologiLingkungan.
[13] Abdullah N, Pandebesie, E. 2018. The infuence of stiring and cow manure added on biogas production from vegetable waste using anaerobic digester. IOP Conf. Series: Environmental Science. 135:1–7, Bristol, United Kingdom: IOP Publishing Ltd. DOI: https://doi.org/10.1088/1755-1315/135/1/012005.
[14] Zubair, A., Abdullah, N.O., İbrahim, R. & Rachma, A.R.D. 2020. Effectivity of constructed wetland using Typha angustifolia in analyzing the decrease of heavy metal (Fe) in acid mine drainage. IOP Conference Series: Earth and Environmental Science, 419, 012160.

IOP Conf. Series: Earth and Environmental Science 1117 (2022) 012049 doi:10.1088/1755-1315/1117/1/012049

Strength Behaviour of Sugarcane Bagasse Ash Treated Sewage Sludge-Soil Mixture

S Kartika[1,2], L A Asiyanthi [1], R R Irwan[1], A Hidayat[1]

[1] Environmental Engineering Department, Engineering Faculty Hasanuddin University, Jl. Malino No.8 F, Romang Lompoa, Kec. Bontomarannu, Kabupaten Gowa, Sulawesi Selatan 92171, Indonesia

[2] Correspondence author: kartikasari@unhas.ac.id

Abstract. Laboratory tests were conducted on sugarcane bagasse ash treated sewage sludge-soil mixture with 5%, 10% and 30% bagasse ash to assess its strength behaviour. Unconfined compressive strength (UCS) was performed to observed influence of bagasse ash content (BAC), curing time, water content on value of UCS and axial strain. Soil sample were prepared using standard proctor at 60% of optimum water content. Results recorded value of UCS reached peaked at 20% BAC and then dropped by 46% at 30% BAC. Axial strain increased for every additional BAC. Result shown curing period influence UCS, as curing time longer the value of UCS also rose. Other findings of this study were BAC affected maximum dry density (MDD), 5% BAC and 10% BAC improved MDD but 20% BAC and 30% BAC would reduce value UCS. Finally, it could be concluded 20% BAC was the optimum content to get the optimal UCS improvement of sewage sludge-soil.

1. Introduction

Sugarcane (*Saccharum Officinarum*) is perennial grass, family of Poaceae, primarily planting for its juice as material for sugar manufacture in Indonesia. Production of Indonesia sugarcane is estimated as 2.364.321 ton in 2021 according to Indonesia Directorate General of Estate. It is harvesting every 6 months, then chopped into several pieces and transported to processing plant for extracting the juice [4]. Bagasse, waste of sugarcane after juice extracting, usually burning in incinerator of sugar mills as fuel to produce steam and electricity. Then it becomes ash and fibre in black colour. A large number of bagasse disposal remained as stock pile in open space, and it became environmental issues. Bagasse ash can be fermented and decays when left in the open air, it results in respiratory disease when inhaled in the large doses [5]. It is necessary to utilized bagasse ash to reduce amount of waste in the environment. Several studies have been reported related to utilization of bagasse ash especially in construction, not only as alternative material but also to solve environmental problem. Previous study [1] used bagasse ash to treat lateritic soil as a potential barrier material in waste containment application. The result study concluded that lateritic soil treated with up to 12% bagasse ash can effectively be used as a hydraulic barrier in waste containment systems (liner and covers).

In 2017 [6] study strength behaviour of fly ash and sugarcane bagasse ash treated soil. They reported that optimum moisture content increased and maximum dry density decreased as soil was treated by fly ash and bagasse ash. Furthermore [2] investigated influence of bagasse ash and hydrated lime on strength and mechanical behaviour of expansive soil. The findings indicated increasing of strength and stiffness and mechanical properties with increased bagasse ash content and curing period.

2. Background

Abundant of industrial waste remained untreated and unutilized properly, while production continues every day. As mention above if the waste disposed at open space without covering, it can endanger human health and environment in long time. This study aims to utilized sewage sludge to replace 75% of soil with intention to reduce amount of industrial waste in order to help solving environmental problem. He J et al 2015 has confirmed that using sewage-sludge combine with soil can be used as alternative material for landfill cover. The study reported that unconfined compressive strength increases as curing time increasing, moreover hydraulic conductivity reduced under synthetic leachate.

3. Materials and their properties

3.1 Soil

Image of material tested was shown in Fig.1. Soil tested was classified as lean silt (ML) based on Unified Soil Classified Soil (USCS). Table 1 shown basic properties tested.

Table 1. Basic properties of soil tested

Parameter	Content (mg/kg)
Moisture content (%)	51.94
Spesific Gravity	2.639
PL (%)	31.38
LL (%)	41.32
SL (%)	25.46
IP (%)	9.94
Permeability coefficient (cm/s)	0.094

3.2. Sewage sludge

The sewage sludge used in this study was taken from industrial PT. KIMA Makassar in dry condition. It has water content of 73.78 %, density of 0,97 g/cm3, 10.40 % - passed sieve BS No. 200 and permeability coefficient of 0,0790. The sewage sludge was mixed with soil and Bagasse ash. Table 2 shown chemical content of the sewage sludge tested.

Table 2. Chemical composition of sewage sludge.

Parameter	Content (mg/l)	Quality Standards of Indonesian (mg/l)
Chromium (Cr)	0.0075	< 0.5
Cadmium (Cd)	0.00005	<0.002
Arsen (As)	0.00021	< 0.1
Timbal (Pb)	0.00215	<0.01

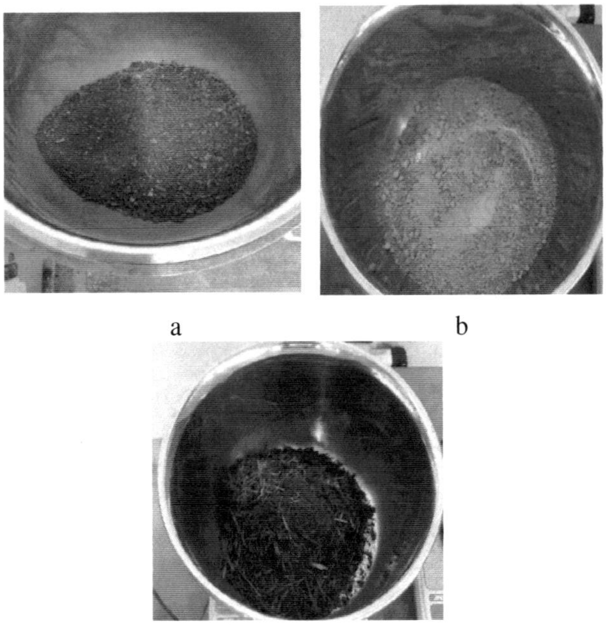

a b c

Figure 1. a. Sewage sludge, b. Soil and c. Bagasse ash

3.3. Bagasse Ash

The bagasse ash tested was collected from bagasse that burnt as fuel in kettle of PT. Perkebunan Nusantara IV Unit Usaha Pabrik Gula Takalar. This material content of 9.45, 0.53, 1.32, 0.50% of Carbon, Nitrogen, Fosfor and Kalium respectively. It was burnt for 6 hours at temperature of 500°C. Some chemical content of bagasse ash tested shown in table 3.

Table 3. Chemical content of bagasse ash.

	Unit	Content (mg/kg)
Chromium (Cr),	mg/kg	7.32
Cadmium (Cd)	mg/kg	0.01
Arsen (As)	mg/kg	0.26
Timbal (Pb)	mg/kg	0.77

4. Preparation specimen and testing procedure

Air dried sewage sludge mixed with air dried soil at ratio of 4:1. Bagasse ash was added into the mixture by variation of 5%, 10%, 20% and 30% of air-dried soil weight, mixed thoroughly. Then conducted compaction test to determined optimum moisture content (OMC) and maximum dry density for every variation of bagasse ash content (BAC) based on ASTM 698-07. After that specimen were prepared for unconfined compressive strength test. Liquid was added into the mixture to reach 60% of optimum moisture content (OMC). Then it was compacted at moulding. The sample was extruded from mould and trimmed into cylindrical specimen 7 cm-high and 3 cm-diameter then sealed with wrapping plastic and kept them for period of 0, 7, 14 and 28 days to allow uniform moisture distributed and curing in the room temperature. After the samples curing, they were set up in a load frame machine drive strain controlled at 0.10% per minute until failure occurred. Unconfined compression strength test was carried based on ASTM D2166-06.

5. Result and Discussion

5.1. Compaction characteristics

Influence of bagasse ash content on Maximum dry density (MDD) and Optimum moisture content (OMC) of sewage sludge-soil are shown in Fig. 2. MDD increased for 5% and 10% of Bagasse ash content (BAC) but dropped for 20 % BAC and 30% BAC. The increase of MDD for 5% BAC and 10% BAC, probably due to the empty space between grain and fine particle of the mixture filled by bagasse ash led to reducing void between particle of specimen. However, if BAC added up to 20% and 30%, the flocculation and agglomeration of clay particle size occurred and mainly affected density of mixture, lead to reducing MDD. The OMC slightly increased as the higher bagasse ash. This was due to the increase of fine content with larger surface area that need more water to react and could be more water required for hydration of bagasse ash.

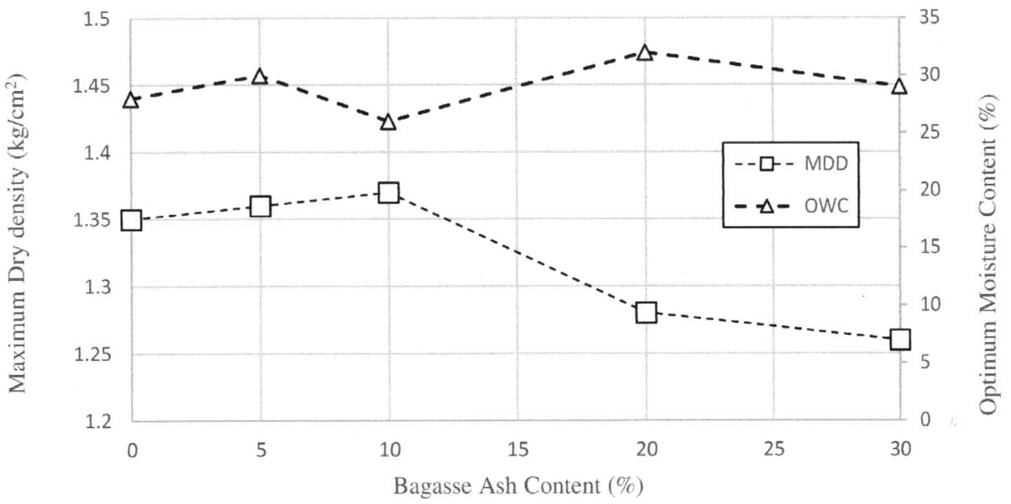

Fig. 2 Maximum dry density and optimum moisture content with bagasse ash content

5.2. Influence of bagasse ash addition on the unconfined compressive strength of sewage sludge-soil

The stress-strain curves attained from unconfined compressive strength tests of bagasse ash treated sewage sludge-soil are depicted in Fig.3. As can be seen from the graph, the stress-strain behaviour of untreated sewage sludge-soil are compared with the bagasse ash up to 30% for 7 days curing time. The untreated sludge-soil reached a peak stress of 2.7 kg/cm2 at strain of 2.1%. The treated sludge-soil mixture reached 4.1 kg/cm2 at axial strain 4%. However, additional 30% bagasse ash, the strength value dropped to 2.2 kg/cm2 at axial strain 5.45%. It indicated that 20% BAC could be the optimum bagasse ash content. It can be noted that 20% BAC can improve 50% of compressive strength untreated sludge-soil and strain. This result was close to result which reported by [2]. They reported that 18% of bagasse ash content in expensive soil was optimum dosage increasing compressive strength of expansive soil tested. This mean when bagasse ash added into sewage sludge-soil, the treated material become more ductile.

Another result tests were shown in Figure.4. The untreated mixture reached 2.1 kg/cm2 at strain 2.7%, by adding 20% bagasse ash and 21 days of curing time, compressive strength of sludge-soil mixture could reach 10.4 kg/cm2 at strain 4%. It means compressive strength improved 3.9 times and increased 48% for strain compared to untreated mixture.

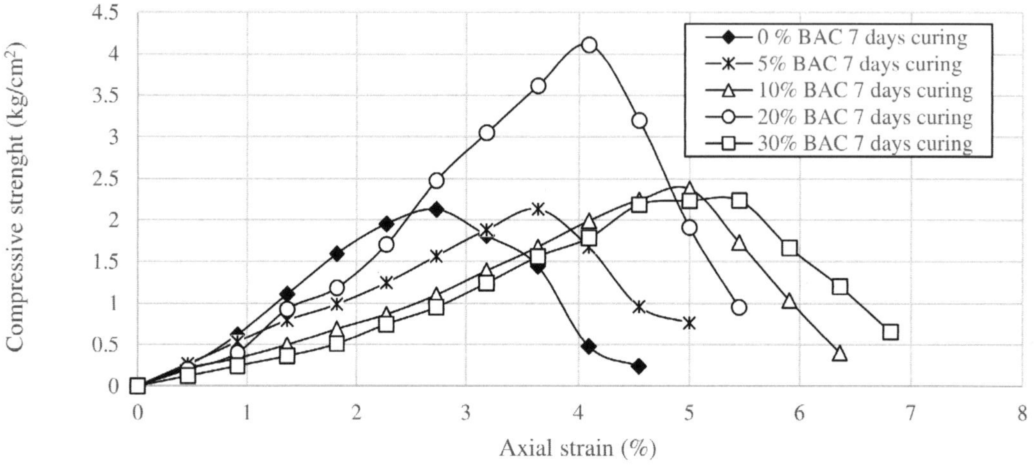

Fig. 3 Compressive strength-axial strain of bagasse ash treated sewage sludge-soil 7 days curing

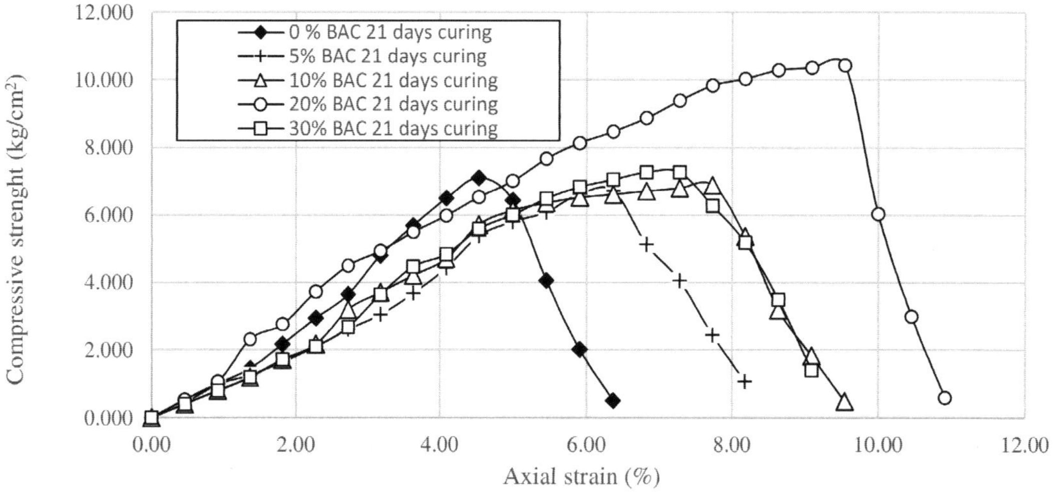

Fig. 4 Compressive Strength-Axial Strain of bagasse ash treated sewage sludge-soil 21 days curing

5.3. Influence of curing time on unconfined compressive strength values of bagasse ash treated sewage sludge-soil

Figure 3 exhibit the variation of unconfined compressive strength of sewage sludge-soil with different curing period, that is 0, 7, 14 and 21 days. The 20 % BAC specimen observed at 21 days of curing time increased compressive strength 6.4 times than without curing period. The graph also indicated that the less curing period the less improvement of unconfined compressive strength value. Comparing 7 and 21 days of curing time, compressive strength of 20% BAC increased 1.4 times when the sample still cured for 14 days more. The increasing of UCS value might be due to the cementitious bond between aggregate of sewage sludge and soil which promoted by pozzolanic in bagasse ash become stronger as the curing period longer.

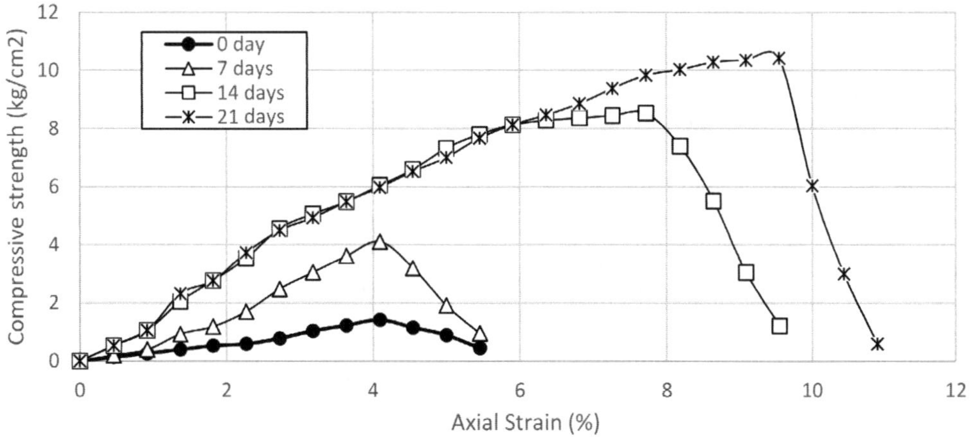

Fig. 5 Compressive Strength versus Axial Strain with various curing time for 20% BAC

5.4. Influence of water content on Unconfined Compressive Strength

The variation of unconfined compressive strength (UCS) with moulding water content is depicted in Fig. 6. Unconfined compressive strength generally decreased with higher moisture content for all cases of bagasse ash treatment. The more water content existed in the mixture, the more deflocculated in sludge-soil fabric and reducing shear resistance. It probably because of cementation between particle of soil and sewage sludge becoming looser and leading reducing value of unconfined compressive strength.

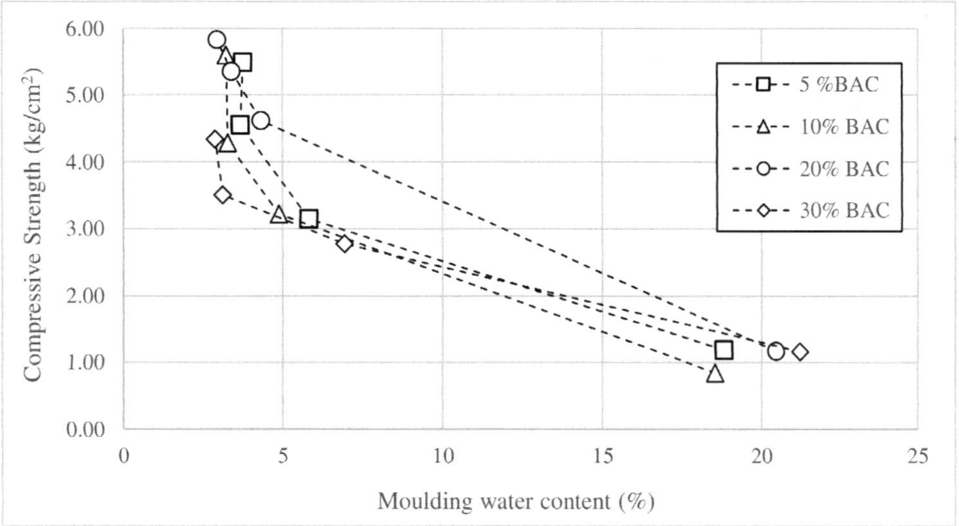

Fig. 6 Compressive shear strength versus moisture content

6. Conclusion

Compaction and unconfined compressive strength have conducted to observe strength behaviour of sugar cane bagasse ash treated sewage sludge-soil mixture. The specimens were added by 5 %, 10 %, 20 % and 30 % of bagasse ash and cured for 7, 14 and 21 days at the constant humid and room temperature. Result showed that MDD increased at 5% BAC and 10% BAC then dropped at 20% BAC and 30% BAC, OWT tend to increase as BAC increased.

The UCS values almost unchanged to 5% and 10% BAC, but urged significantly to 20% BAC. Then followed by drop in strength to 30% BAC, which indicated 20% BAC might be the optimum content for treatment. Axial strain indicated higher as the bagasse ash content increased, nevertheless 20% BAC reached highest among various of BAC tested, this result confirmed 20% BAC could be the optimum content.

Other findings of this study were curing period influence the value of UCS. The longer curing period tested, the higher UCS value. In case of axial strain, there was no significant improvement for 7 days curing time. However, when the specimens were cured up to 14 and 21 days, the UCS value and axial strain increased significantly. In general, it can be concluded that adding sugar cane bagasse ash in sewage sludge-soil mixture could improve UCS value especially 20% BAC.

References

[1] Eberemu A O 2013 Evaluation of bagasse ash treated lateric soil as potential barrier material in waste containment application, *Acta Geotechnica* **9** 407-421

[2] Dang L C, Hassan H, Fatahi B, and Khabbaz H 2015 Influence of Bagasse Ash and Hydrated Lime on Strength and Mechanical Behaviour of Stabilised Expansive Soil Proc. *GeoQuebec 2015 Canadian Geotechnical Society*

[3] He J, Li F, Li Y and Cui X 2015 Modified sewage sludge as temporary landfill cover material *Water Science and Engineering* 8 **3** 257-262

[4] Iryani D A, Hirajima T, Kumagai S 2012 Overview of Indonesiasn Sugarcane Industry and Utilization of Its Solid Waste Proc *Annual Fall Meeting of The Mining and Material Processing Institute of Japan* 2012 *Akita Japan*

[5] Laurianne SA 2004 Farmer Lungs. www.emedicine.com/med/topic771.htm/15/2005

[6] Mahapatra S R and Sahoo 2017 Strength behaviour of soil stabilized with fly ash and sugarcane bagasse ash *Research Journal of Engineering Science* 6 **9** 1-5

The Patommo-Sidrap Landfill Lifespan Estimation

Asiyanthi Tabran Lando[1], Abdul Rahman Djamaluddin[1], Abdul Nasser Arifin [2] and Imam Aprianto[1]

[1] Environmental Engineering Department, Universitas Hasanuddin, Makassar, Indonesia
[2] Agro-Industrial Manufacture Engineering Department, ATI Makassar Polytechnic, Makassar, Indonesia

asiyanthit.lando@unhas.ac.id; asiyanthi@gmail.com

Abstract. Sidenreng Rappang (Sidrap) Regency has a landfill called the Patommo landfill, which has implemented a sanitary landfill model. Patommo landfill is located in Patommo Hamlet, Arawa Village, Watang Pulu District, Sidrap Regency, approximately ± 3 Km from Lawawoi, the capital of Watang Pulu District, and ± 10 Km from Pangkajene, the capital of this regency. This landfill has adequate facilities and infrastructure as a sanitary landfill. The waste transported to the landfill is managed in five stages: weighing, demolition, distribution, solidification, and covering the waste with soil. This study focuses on the lifespan of the Patommo landfill in the Sidrap district. The lifespan is obtained by calculating the amount of total waste per year, waste density, waste volume, and the prediction of the amount of waste in the future based on population growth. This research takes reference data from 2016, which will be compared with current conditions. The results of analyzing data obtained the average value of the waste density of Sidrap District is 1379.51 kg/m^3. Therefore, the volume of waste entering the Patommo landfill is 47,395.95 m^3. The total volume of landfill capacity is 1,708,737.64 m^3, with a volume for soil cover of 624,300.64 m^3, and the waste volume is 1,084,447 m^3. Estimating the lifespan of the Patommo landfill without reduction efforts is 17 years or will be full in 2033, while the estimation of lifespan with reduction efforts is 32 years or will be full in 2048.

1. Introduction

Waste, in general, is a material that is wasted or disposed of from a source as a result of human activities or natural processes that have no economic value. It can even have a negative economic value because it requires considerable costs to handle and disposing or cleaning it [1]. In Indonesia Law (Undang-Undang) No. 18 [2] concerning Waste Management, it is stated that the definition of waste is the rest of daily human activities and/or from natural processes in solid form. A landfill is where waste reaches the final stage of management, starting from the source, collection, transfer/transportation, processing, and disposal. The landfill is where waste is isolated safely so as not to cause disturbance to the surrounding environment. Therefore, it is necessary to provide facilities and proper treatment so security can be adequately achieved.

Landfill types are broadly divided into three types. First is Open Dumping, a waste disposal system carried out openly. It will be a problem if the waste produced is organic waste that decomposes because it causes odour and aesthetic disturbances and becomes a source of disease transmission [3]. The second is Controlled Landfill, an improved open dumping system which is a diversion system for open dumping and sanitary landfill. The closure with a soil cover after the landfill has been fully compacted n accordance with the SNI 19-2454-2002 [4]. The third is the Sanitary Landfill, the destruction of waste by making a hole in the ground, then the waste is inserted and filled with soil as a cover layer and then compacted. This method requires a large area, available land to stockpile it, and large equipment [3].

The problem of landfill waste that often arises is during its operational activities. Several landfills in Indonesia are designed with a sanitary landfill type, but they did not adopt this model at the implementation. As a result, the landfill is abandoned, and its function is decreasing. And it can even cause fires in landfills when methane gas emissions are not controlled, where methane is a greenhouse gas with the most significant potential for global warming. As an illustration, the results showed that the concentration and emission of methane in the Tamangapa landfill were 12 - 425 ppm and 2.44 - 18 Gg per year, respectively [5,6,7].

Waste has become a severe problem in Indonesia and in all over the world. Each people produces waste daily, where the rate of population growth is continuous with the rate of waste production. Waste problems arise from the imbalance between waste production, processing, and management. Some landfills in Indonesia are still using an open dumping system. In this system, waste is collected and dumped in a field hole. In addition, several landfills in Indonesia have over capacity or have reached the maximum limit in accommodating waste. However, the landfill is still active and is still being used. Many impacts will be caused by an over-capacity landfill, such as the expansion of air pollution due to emissions from landfill operational activities and the expansion of water and soil pollution areas around the landfill. It can also damage the aesthetics of the local area in terms of visuals and odors. In addition, the over-capacity landfill is a breeding ground for disease vectors that have the potential to spread disease in the area around the landfill.

In Sidrap Regency (South Sulawesi), Watang Pulu District, Arawa Village, there is one landfill named The Patommo landfill. The waste management system at the landfill has adopted the Sanitary Landfill type, which in outline management is: periodic dredging-bump-compacting-filling (backfill). The landfill area is about 4 Ha, and the land used was about 3 Ha when it was still using the Open Dumping type. In 2015, this landfill turned into a Sanitary Landfill type to improve waste management performance in Sidrap Regency. However, establishing a landfill with the model of a sanitary landfill is not the only solution to improve the quality of waste management in Sidrap Regency. Another solution that can be implemented is to reduce waste so that the landfill's lifespan will be longer. The estimated landfill lifespan can be used as input and consideration for the Sidrap Regency Government, especially the Sidrap Regency Public Works, Spatial Planning and Transportation Office, to find alternatives to extend the lifespan of the Patommo landfill. Several previous studies have estimated the age of landfills, such as in Bantargebang (Jakarta-Indonesia) [8], Jeram (Kuala Lumpur-Malaysia) [9], ASMOC (Brazil) [10], Moree (Australia) [11], Ghana [12], and Illinois landfill (US) [13].

Based on the description above, this study will estimate the landfill lifespan of the Patommo landfill. The objectives of this study are: (1) to find out how large the volume of waste that goes to the Patommo landfill after the waste reduction efforts are implemented, (2) to ascertain the lifespan of the Patommo landfill without implementing waste reduction efforts, and (3) to analyze and assess the lifespan of the Patommo landfill after the implementation of waste reduction efforts. The scope of this study relates to the analysis of capacity development, evaluation, and planning for additional facilities and infrastructure to support the landfill. Moreover, the limitations of the problems discussed in the study are: (1) the research location is the Sidenreng Rappang (Sidrap) Regency, (2) calculating the predicted population growth in Sidrap Regency, (3) collecting data on the weight of waste at the Patommo Village, Watang Pulu District, (4) the data taken is waste entry data starting from 2016 and will be compared with current conditions, (5) calculate the predicted amount of waste generation for the coming years, and (6) estimating the lifespan of the Patommo landfill.

2. Methods

The method used in this study is a quantitative descriptive method. This method describes the condition of the Patommo landfill related to the waste handling process at the landfill, the volume that goes to the landfill, the area of land still open for landfilling, projecting population growth in Sidrap Regency, and associated with the growing quantity of future waste generation. The results of this method can be used as a reference for estimating the landfill's lifespan.

2.1 Study site

Patommo landfill is located in Patommo Hamlet, Arawa Village, Watang Pulu District, Sidenreng Rappang Regency, approximately ± 3 Km from Lawawoi, the capital of Watang Pulu District, and ± 10 Km from Pangkajene, the capital of Sidenreng Rappang Regency. The geographical location of the Patommo landfill is as follows: 3°43'- 4°09' South Latitude (SL) and 119°41' - 120°10' East Longitude (EL), with a land area of 4 Ha. Patommo began operating in February 2015. The Sidrap Regency Government currently owns the land for Patommo landfill with the status of certificate of use rights no. 13 issued by the National Land Agency of the Republic of Indonesia.

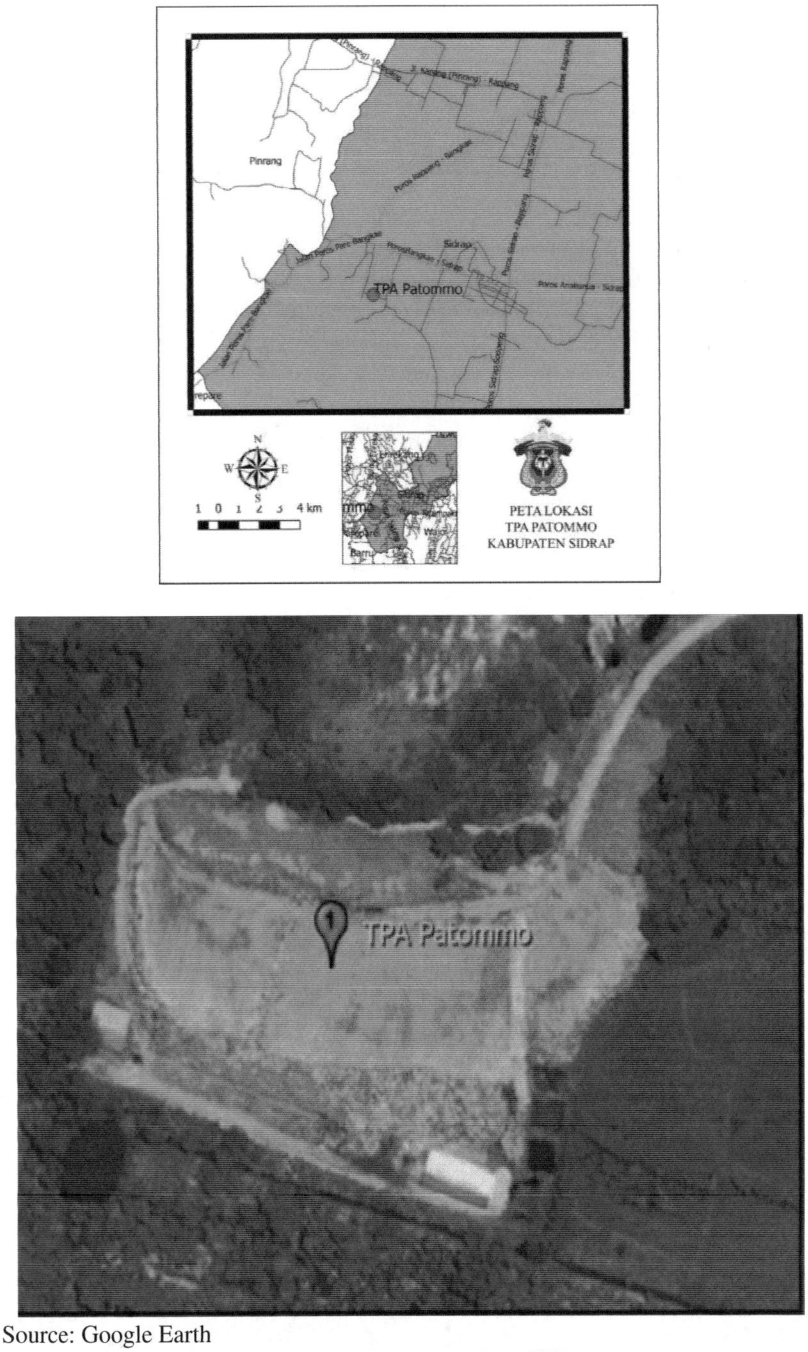

Source: Google Earth

Figure 1. Patommo Landfill

2.2. Study Methods

The object of this study refers to the generation of waste that enters the landfill and measures the capacity of the remaining land to accommodate waste.

The primary data in this study are:

1) The volume of waste that goes to landfill every day can be seen from the weighbridge data found in the Patommo landfill complex. The volume of waste can also be known from the data held by the relevant agencies.
2) The land area is still open for landfilling refers to the number of landfill cells that have been used and also refers to the design drawing of the reservoir to obtain information on the tank's dimensions.
3) Photos of landfill conditions are helpful for study documentation and can be used as additional information.

Moreover, the secondary data in this study are:

1) The study area's data on physical condition, facilities, and infrastructure was obtained from field observations and the Department of Public Works, Spatial Planning, and Transportation of Sidrap Regency archives.
2) Relevant local regulations on waste management were obtained at the Sidrap Regency Public Works, Spatial Planning, and Transportation Office.
3) Methods of handling waste can be seen in the landfill technical planning report (Archives of the Public Works, Spatial Planning, and Transportation Office of Sidrap Regency).
4) The previous year's incorporation of waste into the landfill was obtained from the Sidrap Regency Public Works, Spatial Planning and Transportation Office.
5) The composition of waste in the Patommo landfill was obtained from the Office of Public Works, Spatial Planning, and Transportation of Sidrap Regency.
6) Previous research or related references are helpful for the author to conduct research and process the data obtained.
7) The landfill contour map was obtained from the SRTM (Shuttle Radar Topography Mission) - USGS (United State Geological Survey).
8) Population data of Sidrap Regency was obtained from the Central Statistics Agency (BPS) of Sidrap Regency. Population data from year to year helps get the percentage value of population growth which can then be used to project the population in the coming year.
9) Dimension and weight of the vehicle are needed to calculate the waste weight that goes to the landfill
10) The site plan of the Patommo landfill is needed to determine the reservoir dimensions. Therefore, the volume capacity of the landfill reservoir can be known.

Data analysis was carried out after obtaining the necessary data, namely data on the volume of waste that enters the landfill daily and the land area still open for landfilling. This analysis includes the projection of the amount of waste generated from direct observation of the amount of waste that enters the Patommo landfill daily, measuring the capacity of the remaining land to accommodate waste and the capacity of the landfill. Based on the results of the analysis, then an estimation of the landfill lifespan was carried out.

2.3. Average Amount of Waste Generation per Day

The amount of waste generation that goes to the landfill daily is calculated using the Average Formula (Mean).

$$\overline{X} = \frac{X1+X2+X3+\cdots+Xn}{n} \quad (1)$$

where

$\overline{X}$	= Average value of waste generation per day
$X_1 + X_2 + X_3 + \cdots + X_n$	= The amount of daily waste generation
N	= The number of days of survey time

2.4. Waste Density

Waste density is the weight of waste measured in kilograms compared to the measured volume of waste (kg m^{-3}). Waste density is significant in determining the amount of waste generated. Waste density can be calculated using the following formula:

$$D = \frac{W2 - W1}{\text{Truck Container Volume}} \tag{2}$$

where

D = Density of waste in the truck (kg m^{-3})
W_1 = Weight of the truck before it is filled with waste (kg)
W_2 = Weight of the truck after it is filled with waste (kg)

2.5. Landfill Capacity

The volume of waste per year can be seen by comparing the total amount of waste (kg) in the previous year with the value of the waste density obtained.

$$\text{Waste Volume} = \frac{\text{The total amount of waste generation}}{\text{Waste Density}} \tag{3}$$

$$\text{Daily Waste Volume} = \frac{\text{The total amount of waste generated in a year}}{\text{365 days}} \tag{4}$$

2.6. Prediction of the Total Waste and Population of Sidrap Regency

Prediction of the amount of waste generation is estimating the amount of waste generation and population in the coming year by referring to the increase in the amount of waste generation and population in the previous year using the geometric equation method [14].

$$Pn = Pa\,(1 + r)^{x} \tag{5}$$

where

Pn = Volume of waste per total population in n year projection
Pa = Waste volume per population in the initial year of the projection
R = Average population growth per year (%)
X = Projection time interval (years)

3. Result and Discussions

The Indonesian Department of Public Works [15] has set criteria for the amount of waste generated based on the waste's source and the city's characteristics. It can be used as a guideline for classifying Sidrap Regency based on the amount of waste generated. Sidrap Regency's population in 2016 amounted to 292,985 people, which is classified as a medium-sized city with a population of 100,000 - 500,000 people. Therefore, based on the Waste Generation Specification Standard for Small and Medium Cities in Indonesia [15], the volume of waste generated by residents in one day is around 2.75 - 3.25 litres, or if converted in weight, it becomes 0.70 - 0.80 kg.

3.1. Average Amount of Waste Generation per Day

The average waste generation of Sidrap Regency residents in 2016 was 759,51 liter per day or 759 m3 per day or 179,13 kg per day. Then the total waste generation in 2016 is as follows:

759.51 liter per day x 365 days = 277,222,975 liter per day = 277,221.15 m3 per day
179,132 kg per day x 365 days = 65,383,180 kg per day

3.2. Waste Density

Waste density can be obtained from the ratio of the waste's weight to the garbage truck's volume. The weight of the waste is obtained by measuring the net and gross weight of the waste truck. The trucks used to transport waste at the Patommo landfill are all the same type, namely the Mitsubishi Colt Diesel FE74 HD 125 PS, with the dimensions of the carrying container 2.0 m x 2.0 m x 4.3 m. Therefore, the volume of the dump truck of this type can be found to be 17.2 m^3. From the data analysis, the average density of waste that enters the Patommo landfill for 12 days is 1379.51 kg m^{-3}. So that the volume of waste that entered the Patommo landfill in 2016 was 47,395.95 m^3.

3.3. Landfill Capacity

Based on the Patommo landfill site plan, it can be seen that the Patommo landfill is a landfill with a Sanitary Landfill type. Table 1 shows the recapitulation of the Patommo landfill capacity.

Tabel 1. The Patommo Landfill Capacity

Cell in Patommo landfill		Area (m)	Height (m)	Volume (m^3)	Accumulation (m^3)
LANDFILL A	Lift 1	71187.5	5	355937.5	823105.5
	Lift 2	53390.63	5	266953.15	
	Lift 3	40042.97	5	200214.85	
LANDFILL B	Lift 1	28861.85	5	144309.25	333715.05
	Lift 2	21646.38	5	108231.9	
	Lift 3	16234.78	5	81173.9	
LANDFILL C	Lift 1	18559.25	5	92796.25	214591.35
	Lift 2	13919.44	5	69597.2	
	Lift 3	10439.58	5	52197.9	
LANDFILL D	Lift 1	10425.65	5	52128.25	120546.55
	Lift 2	7819.23	5	39096.15	
	Lift 3	5864.43	5	29322.15	
LANDFILL E	Lift 1	18748.47	5	93742.35	216779.2
	Lift 2	14061.35	5	70306.75	
	Lift 3	10546.02	5	52730.1	
The Volume of Landfill Capacity					**1708737.65**

3.4. Prediction of the Population of Sidrap Regency

The data analysis of population growth in Sidrap Regency shows that the value of population growth is 1.98 % per year. Therefore, this percentage can be predicted for the population of Sidrap Regency in the coming years using the geometric formula. From this prediction, the population in Sidrap Regency in 2019 and 2020 are 319.995.99 and 329.542.11, respectively. Furthermore, the Central Statistics Agency (BPS) [16] of Sidrap Regency in 2021 shows the population in 2019 and 2020 were 301,972 and 319,900. Therefore, the predictions that have been made are acceptable.

3.5. Prediction of Total Waste Generation in Sidrap Regency

In predicting the amount of waste generation in Sidrap Regency in the coming year, the results from the calculation of the solid waste density and the prediction results of the population are needed. In 2016, the volume of waste generation in Sidrap Regency was 47,395.95 m^3, so the daily volume was 129.85 m^3. From the daily volume, it can be seen the volume of waste generated per person per day, namely from the division of the daily volume by the total population of Sidrap Regency in 2016, which is 0.000432 m^3. Moreover, the prediction of the volume of waste in Sidrap Regency in the coming years can be seen in Figure 1.

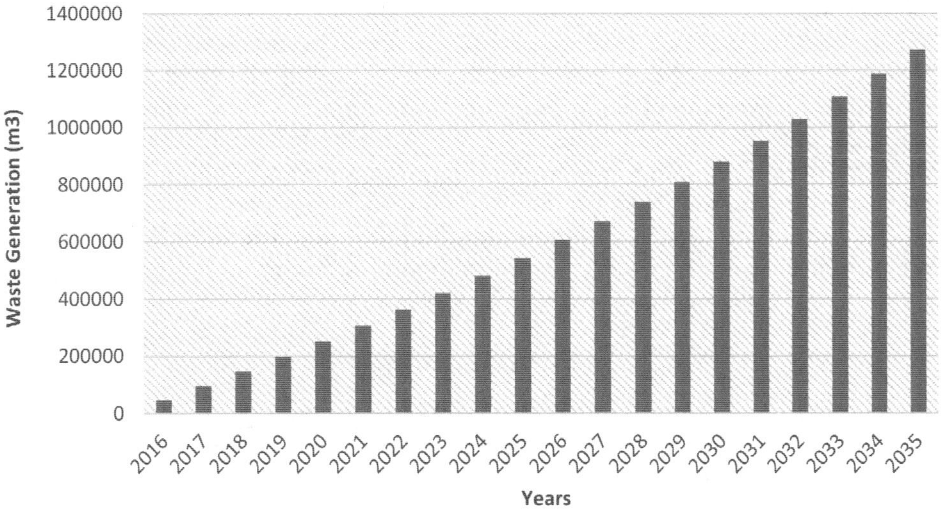

Figure 2. Prediction of Waste Generation in Sidrap Regency

3.6. Estimation of the Landfill Lifespan without Reduction Effort

The total volume capacity of the Patommo landfill is 1,708,737.64 m^3. This total volume includes the volume of landfill and covered soil. The volume of the cover soil heap is 624,300.64 m^3, and the volume of the waste heap is 1,084,437 m^3. Estimation of the landfill lifespan without reduction effort from 2016 can be seen in Table 2.

Table 2. Estimation of the Landfill Lifespan without Reduction Effort

Year	The Volume of Waste Generation (m^3)	The Volume of Landfill Capacity (m^3)
2016	47395.95	1037041.05
2017	96205.45	988231.55
2018	146471.04	937965.96
2019	198236.15	886200.85
2020	251545.52	832891.48
2021	306445.21	777991.79
2022	362982.67	721454.33
2023	421206.76	663230.24
2024	481167.78	603269.21
2025	542917.57	541519.43
2026	606509.47	477927.52
2027	671998.45	412438.55
2028	739441.09	344995.90
2029	808895.69	275541.31
2030	880422.25	204014.75
2031	954082.60	130354.40
2032	1029940.38	54496.62
The Patommo Landfill will be full in 2033		

3.7. Estimation of the Landfill Lifespan with Reduction Effort

The lifespan of a landfill can be extended by carrying out reduction efforts. One reduction effort is sorting waste through 3R (Reuse, Reduce, Recycle) efforts, which are carried out starting from the source of waste, namely domestic waste from households, and are also carried out at the Integrated

Waste Treatment Plant (TPS3R) through composter and incinerator facilities [17,18]. Under the auspices of the Bina Marga Office of Sidrap Regency, the Environment Agency conducted a waste composition study in 2016. Therefore, each volume of generation from each waste composition can be known. The percentage of each waste composition is 56% of food waste; 9.6% of wood, twigs, and leaves waste; 6.2% of paper waste; 6.6% of plastic waste; 4.2% of glass; 4.2% of rubber and leather; 4% of metal; 3.6% of cloth and textiles; and 10.3% of other waste. It is known that the volume of waste generated in Sidrap Regency in 2016 was 47,395.95 m^3.

Based on the theory of waste reduction by Tchobanouglous et al.,1993 [19], several types of waste can be reduced, with the maximum percentage of potential reduction or recovery factor, which is 80% of organic waste, 50% of plastic waste, and 40% of paper waste. Furthermore, the results of estimating the volume of waste generated after reduction efforts were made in 2016 and subsequent years can be seen in Table 3.

Table 3. Estimation of the Landfill Lifespan with Reduction Effort

Year	The Volume of Waste Generation Without Reduction Efforts (m^3)	The Volume of Reduced Waste (m^3)	Accumulation of Reduced Waste Volume (m^3)
2016	47395.60	19934.59	19934.59
2017	48809.50	20529.28	40463.87
2018	50265.59	21066.19	61530.05
2019	51765.11	21772.41	83302.46
2020	53309.37	22421.92	105724.38
2021	54899.69	23090.81	128815.19
2022	56537.46	23779.66	152594.84
2023	58224.09	24489.05	177083.90
2024	59961.03	25219.61	202303.50
2025	61749.78	25971.96	228275.46
2026	63591.90	26746.75	255022.22
2027	65488.98	27544.66	282566.88
2028	67442.64	28366.38	310933.26
2029	69454.59	29212.60	340145.86
2030	71526.56	30084.07	370229.93
2031	73660.34	30981.54	401211.47
2032	75857.78	31905.78	433117.26
2033	78120.77	32857.59	465974.85
2034	80451.27	33837.80	499812.65
2035	82851.29	34847.25	534659.91
2036	85322.91	35886.81	570546.72
2037	87868.26	36957.39	607504.11
2038	90489.55	38059.90	645564.02
2039	93189.03	39195.30	684759.32
2040	95969.05	40364.58	725123.90
2041	98831.99	41568.73	766692.64
2042	101780.35	42808.82	809501.46
2043	104816.66	44085.89	853587.35
2044	107943.5542	45401.06	898988.41
2045	111163.73	46755.46	945743.87
2046	114479.96	48150.27	993894.14
2047	117895.13	49586.69	1043480.83
The Patommo Landfill will be full in 2048			

3.8. Estimation of the Landfill Lifespan With and Without Reduction Efforts

From the analysis of the landfill lifespan, it can be estimated that without the implementation of reduction efforts, starting from 2016, the Patommo landfill will be complete in 2033. While referring to the analysis of the landfill lifespan with reduction efforts, the Patommo landfill will be full in 2048. Both estimations of landfill lifespan can be seen in Figure 2.

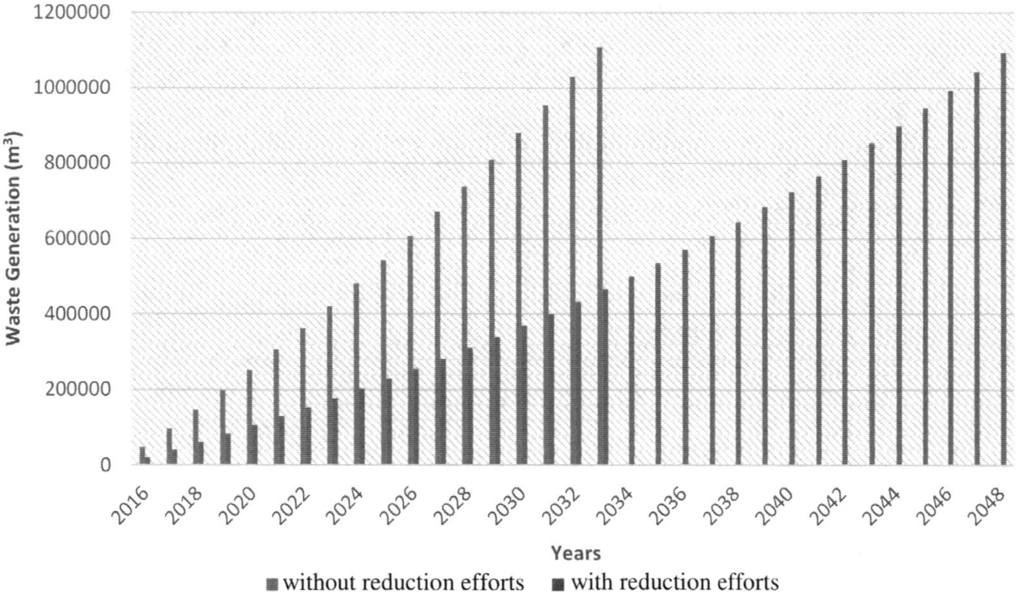

Figure 3. Estimation of the Landfill Lifespan With and Without Reduction Efforts

4. Conclusions

Based on the study on estimation of Patommo landfill lifespan, the following conclusions can be obtained:

After analyzing the data obtained, the following conclusions can be drawn:

1) The average value of the solid waste density of Sidrap Regency is 1379.51 kg m^{-3}. From this density value, the volume of waste generation in Sidrap Regency in 2016 was 47,395.95 m^3.
2) Based on an analysis of the estimation of the Patommo landfill lifespan without reduction efforts, this landfill will be full in 2033.
3) Based on an analysis of the estimation of the Patommo landfill lifespan with reduction efforts, this landfill will be full in 2048.

5. References

[1] Gunawan T 2007 Landscape Ecosystem Approach as a Basis for Environmentally Based Regional Development Yogyakarta, UGM.
[2] Indonesian Law No. 18 2008 *Waste Management*
[3] Iqbal W and Chayatin N 2009 *Public Health Sciences: Theory and Application* Jakarta, Salemba Medika
[4] SNI 19-2454-2004 2004 *Urban Waste Management Operational Techniques* Direktoral Jenderal Cipta Karya, Departemen Pekerjaan Umum, Jakarta
[5] Lando A T, Nakayama H, and Shimaoka T 2015 *Development of Laser Scanning Method for Estimation of Spatial Distribution of Methane Emission in A Landfill* Memoirs of The Faculty of Engineering Kyushu University Vol. 75 (1) pp 29-46
[6] Lando A T, Arifin A N, and Nurhayati 2018 *Effect of Bio-activator Addition and Stirring on Biogas Production from Rice Husks and Cow Dung* E3S Web of Conference 73 07015

[7] Lando A T, Rahim I R, Sari K, Djamaluddin I, Arifin A N, and Sari A M 2021 *Estimation of Methane Emissions from Municipal Solid Waste Landfill in Makassar City based on IPCC Waste Model* IOP Conf.Series: Earth and Environmental Science 841 012002
[8] Warlina L and Listyarini Sri 2022 *The Study of Estimation of Landfill Capacity through Dynamic System Approach* Hindawi Scientifica Article ID:1068111 https://doi.org/10.1155/2022/1068111
[9] Azizan N S, Tea M R T M A, and Senin S F 2022 *An Estimation of Jeram Sanitary Landfill Lifespan using Artificial Neural Network (ANN) Modelling Analysis* ESTEEM Academic Journal Vol. 18, March 2022, 71-80
[10] Filho F O H C and Junior F H D C 2019 *Estimating the Lifespan of The Expansion of A Landfill in Brazil by Means of A Model Equation based on Socio-Economic Indicators* International Jornal of Environment and Waste Management Vol.24 No.22 pp. 161-177
[11] Proterra Group 2019 *Landfill Lifespan Assessment Report – Moree Waste Management Facility – Moree Plains Shire Council.*
[12] Akyen T, Boye C B, and Ziggah Y Y 2017 *Landfill Lifespan Estimation: A Case Study*. Ghana Mining Journal Vol. 17 No.2 pp. 1-5
[13] Illinois Environmental Protection Agency 2020 *Illinois Landfill Disposal Capacity Report.*
[14] Sudjana 1996 *Regression and Correlation Analysis Techniques* Bandung, Tarsito
[15] The Indonesian Department of Public Works 1993 Waste Generation Specification Standard for Small and Medium Cities in Indonesia Bandung, LPMB
[16] Central Statistics Agency (BPS) of Sidrap Regency 2021
[17] Lando A T, Rahim I R, Sari K, Djamaluddin I, Arifin A N, and Nurhidayanti 2021 The Effectiveness of Composter at the Integrated Waste Treatment Plant in the Campus of Engineering Faculty Hasanuddin University IOP Conf.Series: Earth and Environmental Science 841 012001
[18] Lando A T, Djamaluddin I, Arifin A N, Oktorina N, Danah R, and Sulistyowati MF 2020 The Effectiveness of Incinerator at the Integrated Waste Treatment Plant in the Campus of Engineering Faculty- Hasanuddin University IOP Conf.Series: Materials Science and Engineering 875 012017
[19] Tchobanoglous G, Thiesen H, and Eliasen R 1993 *Integrated Solid Waste Management* New York, McGraw Hill Publishing Company

Performance of Covid-19 Waste Management at Health Service Facilities in North Toraja Regency

I R Rahim[1], M Selintung[1], I Jamaluddin[1], M Ihsan[2], E Y Tiranda[1]

[1]Department of Environmental Engineering, Hasanuddin University, Gowa, Indonesia
[2]Department of Civil Engineering, Sekolah Tinggi Teknik Baramuli, Pinrang, Indonesia

irwanrr@eng.unhas.ac.id

Abstract. Community health centers are the frontliners of healthcare services which in its activities produces medical waste and non-medical waste both in solid and liquid form. This waste needs to be managed in accordance with existing regulations where environmental management must be carried out systematically and sustainably. Regarding the management of Covid-19 medical waste in the health facilities of North Toraja Regency, this study aims to determine the management process, management constraints and compare the management of wastewater and hazardous and toxic solid medical waste generated from handling Covid-19 patients at hospitals and health centers in Toraja Regency. North. The research method used is descriptive research method with case study design. Data collection techniques are carried out by observations, interviews, and documentation of secondary data. The results show that the management of hazardous and toxic solid medical waste water and waste for Covid-19 patients in private hospitals leaves the processing to third parties and government hospitals destroy solid hazardous medical waste with incinerators, while at community health center the wastewater is directly discharged into water bodies and medical hazardous waste. handed over to a third party. The process of treating waste water and hazardous and toxic solid medical waste for Covid-19 patients at the health facilities of North Toraja Regency has not been in accordance with the guidelines issued by the Indonesian Ministry of Health. The main obstacle in processing solid medical waste water and hazardous waste for Covid-19 patients at the health facilities of North Toraja Regency is the lacking of operational costs and delays in disbursing operational costs. Government-owned hospitals were found to be slow to comply the Covid-19 Health Facilities Waste Management Guidelines issued by the Ministry of Health of the Republic of Indonesia.

Keywords: *medical waste, liquid waste, hazardous and toxic solid medical waste, covid-19, hospital, health centes*

1. Introduction

The COVID-19 pandemic is continuing to give an impact on the healthcare system and waste management system in all countries including Indonesia. In Indonesia, the number of cases infected with Covid-19 continues to reach million cases as per July 2022. This is what encourages the public to be obliged to implement health protocols to reduce the number of cases of the spread of Covid-19, by using masks, washing hands or using hand sanitizer in daily activities, and the mandatory use of complete personal protective equipments (PPEs) for medical personnel in carrying out all activities both in hospitals and in other health care facilities.

In line with the current state of the Covid-19 pandemic, the focus of this review is on the implementation of handling medical waste with an infectious character. Covid-19 medical waste in Indonesia reached 18,460 tons as of 27 July 2021 [1]. Referring to Ministrial Regulation Permen LHK No.P.56/MENLHK-SETJEN/2015, infectious waste is the waste which is contaminated with pathogenic organisms that are not commonly present in the environment and these organisms are in sufficient quantity and virulence to transmit disease to susceptible humans. In general, this infectious medical waste is categorized as hazardous and toxic waste [2].

North Toraja Regency is one of the regencies in South Sulawesi where the spread of Covid-19 based on data from the Covid-19 Task Force of North Toraja Regency, all patients confirmed positive for COVID-19 as of June 3, 2020, totaling 233 cases and 12 fatalities. Therefore, it is very important to pay attention to the management of Covid-19 medical waste at the Health Facilities in North Toraja Regency whether it is in accordance with applicable regulations as was done during the pandemic. In connection with the importance of managing waste generated from handling Covid-19 patients, which is a problem because there are no previous regulations made by the government, before the Indonesian Ministry of Health has issued Guidelines for Waste Management for Referral Hospitals, Emergency Hospitals, and Health Centers that Handle Covid-19 Patients. to be a reference for managing waste generated from handling Covid-19 patients. Because the location of the health facilities in North Toraja Regency is close to settlements, offices, and trade places, it is necessary to supervise and monitor the management of medical waste in accordance with the Ministry of Health Guidelines which regulates Wastewater Management, Domestic Solid Waste Management, and hazardous and toxic solid medical Waste Management. Considering the waste generated during the pandemic has increased from that generated before the pandemic and the dangers of medical waste for Covid-19 patients which are classified as infections, they must be treated properly because the waste can be a medium for transmission of the Covid-19 virus.

2. Literature review

Increased volumes of discarded materials from health activities, including medical diagnosis, treatment and scientific research are also a huge burden to the sanitation sector. Safe and efficient disposal of the particular medical waste is a crucial element of a comprehensive response effort [3]–[9]. Medical waste is waste from service activities that are not used or not useful. Medical waste tends to be infectious and toxic and can affect human health and worsen environmental sustainability if not managed properly. Health center medical waste is all waste generated from Community Health Center activities in solid and liquid form. According to the Decree of the Minister of Health of the Republic of Indonesia No. 1204 of 2004, solid non-medical waste is waste generated from activities in hospitals outside of medical origin from kitchens, offices, parks and yards that can be reused if there is technology. The occurrence of the Covid-19 pandemic at the end of 2019 to 2020 resulted in a significant increase in the generation of medical waste in health facilities around the world.

3. Method

The research was carried out in 2 (two) hospitals, namely one of the government hospital of North Toraja Regency and one of the private hospitals which belonged to a church foundation in North Toraja Regency, and 2 (two) Community Health Center. The research was carried out from June 2020 to October 2020. Research data collection was carried out during hospitals and Community Health Center working hours, at 07.00 – 16.00 local time.

Secondary data collection is conducted by studying literature and data from various agencies related to the research conducted. The data is in the form of Covid-19 hazardous and toxic solid medical waste generated by particular health facilities in North Toraja Regency. The data collected is then reduced, by selecting, focusing and simplifying the data collected, to provide data that are relevant to the problems and research objectives. After the data is reduced, then the data is presented in written form, presenting information that allows it to be used as the basis for drawing research conclusions..

4. Results and Discussion

4.1. Analysis of Factors Affecting the Feasibility of Management of Wastewater and Hazardous and Toxic Solid Medical Waste

Waste originating from handling Covid-19 patients is categorized as hazardous waste, because it has the potential to contain microorganisms, especially the SARS-CoV-2 virus which is highly contagious, both in the form of liquid waste and solid waste. Based on the guidelines issued by the Indonesian Ministry of Health, it is known that for Covid-19 case wastewater that must be treated is all waste water including faeces, originating from activities handling Covid-19 patients which may contain microorganisms, especially the Corona virus, toxic chemicals, blood and body fluids. other fluids, as well as fluids used during patient isolation activities, including fluids from the mouth and/or nose or patient's mouthwash and washing water for work equipment, patient eating and drinking utensils and/or laundry linen, which are hazardous to health, sourced from the activities of the isolated patient with Covid-19, treatment rooms, examination rooms, laboratory rooms, equipment and linen washing rooms.

From the results of interviews and filling out questionnaires with the Head of Health from Private Hospitals, there are still some criteria that are not in accordance with the Waste Management Guidelines for Referral Hospitals, Emergency Hospitals and Health Centers that Handle Covid-19 Patients issued by the Indonesian Ministry of Health, namely in point 4 , 5, 8, and 9. In points 4 and 5 in terms of checking the sewerage which should be done every day, the inspection from the hospital is only done when there is a problem. Whereas in point 8 regarding the management of the sludge produced by the wastewater treatment plant (WWTP), it should not be used as compost but submitted to a third party or carried out burial in accordance with the Regulation of the Minister of Environment and Forestry number P.56 of 2015, as for point 9, measurements made by the the hospitals is only limited to measuring pH.

4.2. Hazardous and Toxic Solid Medical Waste Management

Based on information from resource persons and filled out questionnaires, regarding the processing of hazardous and toxic solid medical waste resulting from the handling of Covid-19 patients, from the Health Facilities Waste Management Guidelines, there are some criteria which have not yet been implemented, due to the damage of hazardous and toxic solid medical waste processing equipment (incinerator) owned by the hospital, so that some points were not implemented. However, the hospital handed it over to a third party for further processing and transportation of solid hazardous and toxic medical waste every 2 months. While waiting for hazardous and toxic medical waste to be collected, it is then handed over to a third party.

For the government-owned hospital there are still several points that have not been implemented based on the guidelines issued by the Indonesian Ministry of Health, namely at points 4, 5, 8, and 9, regarding the guidelines issued by the Indonesian Ministry of Health, for instance, flow checks are still carried out once a week according to the schedule that is owned by the hospital. For point 8, the sludge resulting from wastewater treatment is dried in a mud dryer, and the dried mud is immediately buried. From the results of the burial interviews conducted by the health facilities, it was not in accordance with the Regulation of the Minister of Environment and Forestry Number 56 of 2015. For point 9, the health facilities only carried out pH measurements at the outlets of the WWTP.

From the results of questionnaires and interviews with resource persons related to the management of solid hazardous and toxic medical waste resulting from handling Covid-19 patients, at this Government Hospital there are several points that have been implemented but are still not optimal, as in point 5 where all solid hazardous and toxic medical waste produced should be recorded. in the logbook but there are several that are not recorded, the 15 points of the hospital do not yet have a freezer/cold-storage at the waste disposal site, the hospital also admits that once disposal of solid hazardous and toxic medical waste requires a large amount of money, processing is only carried out 2-3 times a month because they are waiting for medical waste to be collected, and for point 26 there has been no report to the ministry of environment or local government regarding the waste being managed.

From the table above, it is known that the management of wastewater at Community Health Center A is still not in accordance with the guidelines of the Ministry of Health of the Republic of Indonesia, namely at points 4 to 9. The Community Health Center has wastewater treatment equipment but it has not been operated because the electrical power needed is quite large and the electric power at the Community Health Center is not able to operate the equipment, in addition to increasing the electric power the Community Health Center is still constrained by costs, the Community Health Center has also asked for help from the government concerned but has not received a response, because the WWTP at the Community Health Center is not operating, therefore points 7, 8, and 9 are not implemented.

The results of interviews with informants said that the Community Health Center did not yet have a waste disposal site so that the waste produced was only in the place behind the Health Center which was placed in a container, the tools used were not sufficient. Such as the transportation is done manually, the provision of PPE is still very minimal and there is a lack of awareness from cleaning service officers to use PPE

From the results of interviews with resource persons, Community Health Center B has not implemented it because it does not yet have an WWTP, so the waste water that is produced is directly discharged into water bodies.

From the results of questionnaires and interviews with sources at Puskesma B, the results are relatively the same as for Community Health Center A, there are still many provisions of the Covid- 19 Health Facilities Waste Management Guidelines that have not been implemented.

4.3. Analysis of Measurement of Health Facilities Compliance with Guidelines

To find out the description of the compliance of each health facility with the Guidelines for Waste Management for Referral Hospitals, Emergency Hospitals and Health Centers that Handle Covid-19 Patients, a descriptive analysis was used based on the responses to the questions in the questionnaire. Question items in the guidelines are described in the form of descriptive tables. The weight for each question is 2, a value of 2 can be obtained

if the respondent answers "Yes" and can provide evidence in the form of documentation, a value of 1 if the health facility can only answer "Yes" and not accompanied by documentary evidence and a value of 0 if it answers "No". For the categories, namely: From the calculations in the table above, a score is obtained for each health facility, so that if it is included in the category of health facility compliance with the Waste Management Guidelines for Referral Hospitals, Emergency Hospitals and Health Centers that Handle Covid-19 Patients.
It is known that the category for Private Hospital is "Good" with a score of 12, Government Hospital is also in the "Good" category with a score of 14, for Community Health Center A "Quite Good" with a score of 5 and Community Health Center B "Less" with a score of 4. "Very Good" due to the fact that many provisions in the Guidelines for Waste Management for Referral Hospitals, Emergency Hospitals and Health Centers that Handle Covid-19 Patients issued by the Indonesian Ministry of Health are not carried out or are not complied with by health facilities.

4.4. Analysis of Hazardous and Toxic Solid Medical Waste Management Compliance Measurement

The assessment is the same as the wastewater management assessment, namely the weight for each question is 2, a value of 2 can be obtained if the respondent answers "Yes" and can provide evidence in the form of documentation, a value of 1 if the health facility can only answer "Yes" and is not accompanied by documentary evidence. and a value of 0 if you answer "No". It can be concluded that private hospitals achieved a score of 40 so that it can be said "Very Good" in processing hazardous and toxic solid medical waste, Government Hospitals can be said "Good" because they get a score of 33, while for Community Health Center A "Quite Good" with a score of 16 and Community Health Center B only gets a score of 10 out of 50 so it can be categorized as "Very Poor" in terms of hazardous and toxic solid medical waste management.

From the score above, it can be seen that the management of hazardous and toxic solid medical waste in the two health centers still has many points from the guidelines that have not been implemented. From the results of interviews, the obstacles between the two health centers are relatively the same in the management of medical waste, namely the costs that are still borne by each health center and as is known the operational costs of the Community Health Center itself are quite large and there is no assistance from the local government.

Conclusions

For waste water management, namely all waste water including feces, originating from activities handling Covid-19 patients must be disposed of through sanitary equipment (eg sinks) or put in the provided collection container, using closed channels with adequate pipe diameters, WWTP at least consisting of from the initial sedimentation process, biological processes (anaerobic and/or aerobic), final sedimentation, sludge treatment, and sedimentation by chlorination, and sludge produced by WWTP, can be treated using a sludge dryer or press machine, burned in an incinerator or sent to a processing service company. hazardous and toxic waste. If both cannot be done, burial can be carried out according to theRegulation of the Minister of Environment and Forestry number P.56 of 2015

For the management of hazardous and toxic solid medical waste, namely goods or materials left over from activities that are not reused that have the potential to be contaminated by infectious substances or contact with patients and/or officers at Health Facilities who handle Covid- 19 patients, then the container uses yellow plastic bags which are with the symbol “biohazard”, only hazardous and toxic solid medical waste can be entered, collection is carried out every 24 hours, recorded and stored at waste disposal site, all waste entering waste disposal site must be disinfected as well as empty containers/bins, transportation using special transportation, then for management can use incinerators/autoclaves/microwaves or hazardous and toxic waste management service providers. All officers must use complete PPE and after serving must immediately clean themselves with running water, any waste that is processed by the health facilities or handed over to a third party must have a report to the local government.

From the 4 health facilities, namely 2 hospitals and 2 Community Health Center, from the score for wastewater management which consists of 9 questions and hazardous and toxic solid medical waste consisting of 25 questions, the results for the management of wastewater private hospitals get a score of 12 which is equivalent to "Good". Government Hospital with a score of 14 is equivalent to "Good", Community Health Center A with a score of 5 "Poor" and Community Health Center B with a score of 4 "Poor" with a score of 4. For hazardous and toxic solid medical waste management, Private Hospitals with a score of 40 are equivalent to " Very Good" with a score of 40, Government Hospital with a score of 33 equals "Good", for Community Health Center A with a score of 16 it is equivalent to "Enough Good" and Community Health Center B with a score of 10 is equivalent to "Less Good".

References

[1] I. N. Amanah, "Environmental Cost Accounting: Behind the Major Project for the Construction of a Covid-19 Hazardous Waste Management Facility," *Jurnal Perencanaan Pembangunan: The Indonesian Journal of Development Planning*, vol. 6, no. 2, pp. 186–201, 2022.

[2] Ministry of Forestry and Environment, "Minister's Regulation No. 56, 2015, Tata Cara dan Persyaratan Teknis Pengelolaan Limbah Bahan Berbahaya dan Beracun dari Fasilitas Pelayanan Kesehatan (Procedures and Technical Requirements for the Management of Hazardous and Toxic Waste from Health Care Facilities)." Ministry of Forestry and Environment, The Republic of Indonesia, 2015.

[3] A. Kodir, A. Tanjung, M. Rosyendra, and M. Saputra, "Challenges of Covid-19 Medical Waste Management in Indonesia: A Multi-stakeholder Perspective," 2021, vol. 2, pp. 52–59.

[4] I. S. Wulandari, S. Soemarno, and K. Koderi, "An Analysis on Household waste Management during Covid-19 Pandemic Era (Study at Suzuki Residents, North Minahasa)," *Indonesian Journal of Environment and Sustainable Development*, vol. 12, no. 1, 2021.

[5] T. Tahmar, E. A. Jayadipraja, and R. A. Surya, "Hazardous and Toxic Waste Management in The Covid-19 Referral Hospital in Kendari City," *IJHSRD*, vol. 3, no. 3, pp. 104–108, 2021.

[6] N. T. Nurwahyuni, L. Fitria, O. Umboh, and D. Katiandagho, "Pengolahan Limbah Medis COVID-19 Pada Rumah Sakit (COVID-19 Medical Waste Treatment in Hospitals)," *Jurnal Kesehatan Lingkungan*, vol. 10, no. 2, pp. 52–59, 2020.

[7] A. Ganing, H. Ahmad, R. Adiningsih, S. Wahyuni, and R. H. Saeni, "Description of Waste Management of COVID-19 Patients without Symptoms in Self-Isolation Period Mamuju District," *International Journal of Research and Review*, vol. 9, no. 2, pp. 328–331, 2022.

[8] B. K. Firmalasari and R. Rasyidah, "The Impact of the Covid-19 Pandemic on the Environment: Environmental Diplomacy on Handling Covid-19 Medical Waste in Indonesia," *WIMAYA*, vol. 1, no. 02, pp. 19–30, 2020.

[9] T. Prasetiawan, "Permasalahan Limbah HAZARDOUS AND TOXIC medis (padat) Covid-19 di Indonesia (The problem of Covid-19 (solid) medical hazardous waste in Indonesia)," *Pusat Penelitian Badan Keahlian DPR RI.*, vol. 12, no. 9, May 2020.

IOP Conf. Series: Earth and Environmental Science 1117 (2022) 012052 doi:10.1088/1755-1315/1117/1/012052

Analysis of NO_2 gas concentration from the transportation sector through direct measurement and the Caline 4 dispersion program on the Makassar City Toll Road

R Zakaria[1], M I Ramli[2], M Hustim[1], H Alimuddin[1], Anisah Pratiwi[1]

[1]Department of Environmental Engineering, Universitas Hasanuddin, South Sulawesi 90245 Indonesia

[2]Department of Civil Engineering, Universitas Hasanuddin, South Sulawesi 90245 Indonesia

rasdianazakaria@unhas.ac.id

Abstract. Makassar as a developing city is experiencing development from the transportation sector. This resulted in the increasing number of vehicles. The increase in the number of vehicles produces emissions that cause air pollution, including NO_2 emissions from motor vehicles. Currently NO_2 concentration can be determined by direct measurement using the Gries Saltman method and pollutant prediction using the Caline program. This research was conducted on the Makassar city toll road which is divided into nine sample points. This study aims to determine the level of NO_2 concentration from the transportation sector on the Makassar city toll road. Concentration results from direct measurements and predictions of caline are compared to show the level of concentration present. The results of the NO_2 concentration are still below the threshold value that has been set in the latest applicable regulations, which is 200 g/Nm^3 with a measurement time of 1 hour. The high or low value of NO_2 concentration is influenced by the vehicle volume factor, where the NO_2 concentration increases along with the increase in vehicle volume. In addition, it is also influenced by meteorological factors. Based on the results of the Caline-4 software output, the results obtained from the distribution of NO_2 concentrations at each receptor found on the Toll Road in Makassar City, it can be stated that it is still below the threshold value set in the latest applicable regulations, which is 200 g/Nm^3 with time measurement for one hour.

1. Introduction

Air quality is very important for the environment and human health. At both the local and global levels, air pollution is one of the most obvious contributors to negative externalities. At both the local and global levels, air pollution is one of the most obvious contributors to negative externalities. On a global scale, greenhouse gas emissions are a factor in climate change, meanwhile in local scale revealed that the current levels of air pollution seen in many cities are linked to health concerns [1]–[3].

Population growth, technological advances applied to combustion, industrial and agricultural processes, and the indiscriminate use of territory are causing the continuous evolution of atmospheric pollution characteristics, which need to be understood to intervene effectively [4]. Additionally, the growth of transportation in developing countries is contributing to the decreased of air quality [5].

Makassar is one of the big cities in Indonesia. A previous study stated that a dense population was found to be correlated to the use of vehicles [6]. Based on data obtained from the Makassar City Statistics Agency, the use of motorized vehicles in Makassar City has a ratio of 27.4% increase in the number of motorized vehicles, from 2011 as many as 976.004 units to 1,338,738 units in 2015 and in October 2018, vehicles reached 1,563,608 units [7].

Some previous study stated that the potential for air pollution was found to be higher due to the increasing use of vehicle fuels which could produce emissions from these vehicles. Vehicles' diesel exhaust is a major source of PM2.5, air hazardous pollutants, and NOx emissions (one of the precursors

IOP Conf. Series: Earth and Environmental Science 1117 (2022) 012052 doi:10.1088/1755-1315/1117/1/012052

to ozone), all of which have the potential to be harmful to human health [8]. The concentration of each pollutant is also very dependent on the meteorological conditions of an area such as wind speed [9].

According to the background, the aim of this study is to know the distribution pattern of Nitrogen Dioxide (NO_2) and the estimation of the impact of the pollutant distribution of Nitrogen Dioxide (NO_2) received in the Makassar City Toll Road area by applying the Caline-4 program.

2. Material and method

2.1. Study Area

Makassar toll road was located in Makassar city which connects several important places such as Soekarno-Hatta Port, the city center, Panakkukang commercial center, Makassar Industrial Area, and Sultan Hasanuddin International Airport with a total length of 21.92 km. This toll road was divided into 2 area i.e. Jl. Toll Reformasi and Jl. Toll Insinyur Sutami.

There were 9 (nine) points of sampling point within Makassar City Toll road area, which were 3 points taken from Jl. Toll Reformasi and 6 points from Jl. Toll Insinyur Sutami. These sampling points were chosen to find out if there is a change in air quality at that point due to changes in vehicle volume every time there are openings on the toll road.

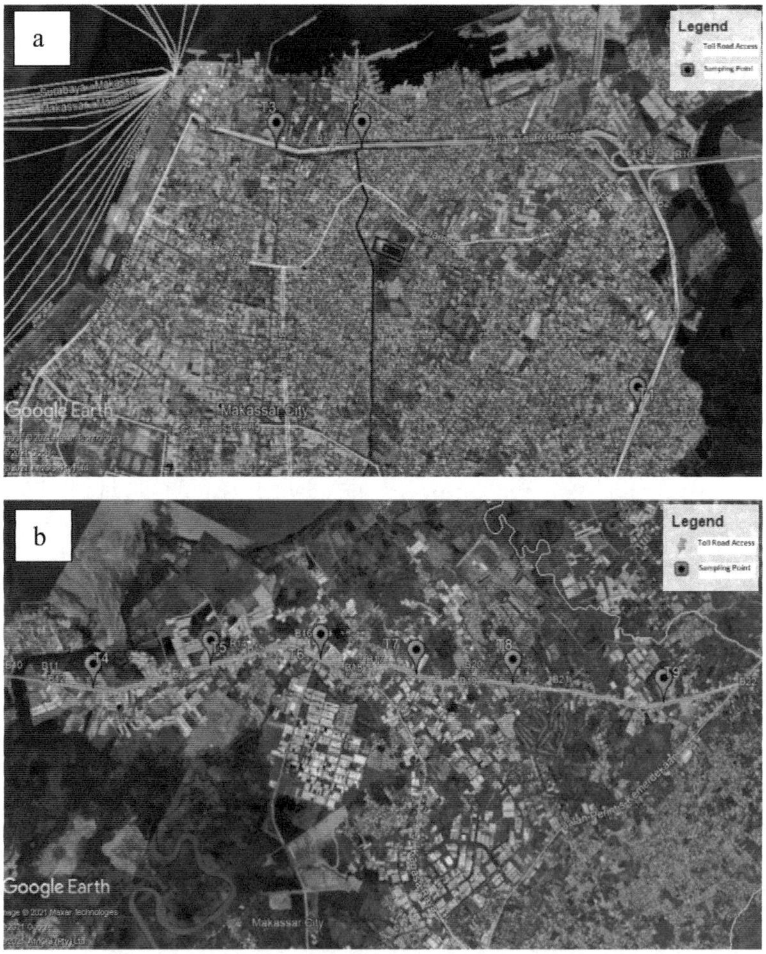

Figure 1. Study area and sampling Point (a) Jl. Toll Reformasi; (b) Jl. Insinur Sutami

2.2. *Sample collection, sample preparation, and analysis*

Data was collected in 9 days from Tuesday, 27 April 2021 to Wednesday, 5 May 2021. Volume of vehicles and air sample for calculate the concentration of nitrogen dioxide (NO_2) emissions from vehicles were taken at each sampling points. Measurements were carried out for 1 (one) day with a measurement duration of 1 (one) hour by taking a representative of the peak hour at the afternoon interval between 15.00-18.00 WITA on the Makassar City Toll Road where it was assumed that the activity density of vehicles passing through the area was high.

Concentration of NO_2 was measured by took air sample on-site. However, according to SNI 19-7119.2-2005, that describes about the procedure for measuring the concentrations of nitrogen dioxide (NO_2) in ambient air, materials need to be prepared before data collection in the field such as adsorption solutions and calibration curves [10]. The adsorbent solution was prepared by dissolving 70 g of glacial acetic acid and 3 g of sulfanilic acid. Add 800 ml of distilled water after stirring. Lastly, added 20 ml of NEDA solution, 10 ml of acetone, and distilled water, then homogenized. Moreover, calibration curves was prepared by making standard solutions and measure their respective absorptions with a spectrophotometer. In this study, we used 0 mL, 0.1 mL, 0.2 mL, 0.4 mL, 0.6 mL, 0.8 mL, dan 1.0 mL as a standard solution for NO_2 concentration.

Impinger was used as a data collection tools. Moreover, the impinger migdet adsorbent bottle and silica bottle were used to store the adsorbent solution during the measurement process and to store silica which is connected to the adsorbent bottle to prevent moisture from entering the impinger, respectively. Air sample was taken and directly analyzed in laboratory.

2.3. *Data analysis*

Gasoline and diesel are two different types of fuel released by engines found in vehicles and have almost the same pollutant composition. In addition, the value of the emission factor was directly proportional to the amount of pollutants contained by the combustion residues released by vehicles on the road.

In this study, the emission factor value for the metropolitan area was assumed from the exhaust gas emission factor value based on the type of vehicle and the type of fuel. The calculation of the emission factor value has been regulated and stipulated by the government in PerMen LH No. 12 of 2010 which regulates the Implementation of Air Pollution Control in the Regions [11]. The quantity of emissions produced by each type of vehicle was calculated by using equation 1.

$$q = \frac{\sum_{i=1}^{n} (\mathrm{EFi \times Vi})}{\mathrm{T}} \qquad \text{(Equation 1)}$$

Where: q = Emission Quantity (gram/km); EF = Vehicle Emission Factor (grams/km); V = Vehicle Volume (vehicles/hour); I = Vehicle Type/Type; and T = Total Vehicle

2.4. *Simulation the dispersion of NO_2*

In this study, first the data will be analyzed using WRPLOT software by inputting time data and meteorological data when conducting research. This software serves to determine the condition of the wind direction and the dominant wind speed [12]. Afterwards, data will be processed by applying Caline-4 software. CALINE-4 is a program that can simulate air quality from line sources to estimate the dispersion of a pollutant [13]. The application of this software was carried out to get the results of the estimation of the amount of NO_2 pollutant around the sampling point which was used as a receptor point. The files that were inputted in this software were Job Parameters, Run Conditions, Link

Geometry, Link Activity, Receptor Position, and 1 output, namely result. The last step was by using software Golden surfer 11. This application was used to determine the distribution pattern of NO_2 pollutants on each road segment at the research site.

3. Result and discussion

3.1. Traffic volume studies

One of the major sources of air pollution in this study area is vehicular emissions. The numbers of vehicle number is determined by carried out for 1 (hour) data collection from each point. The measurement is carried out by taking representatives of peak hours in this case occurring at the afternoon interval, which is between 15.00-18.00. Motor vehicles are divided into motorcycles (MC), gasoline cars (LV Gasoline), diesel cars (LV Solar), trucks, and buses. This grouping by type is done to make it easier when calculating the volume of motorized vehicles expressed in vehicles per hour. The results of data collection of traffic volume can be seen in the following figure 2.

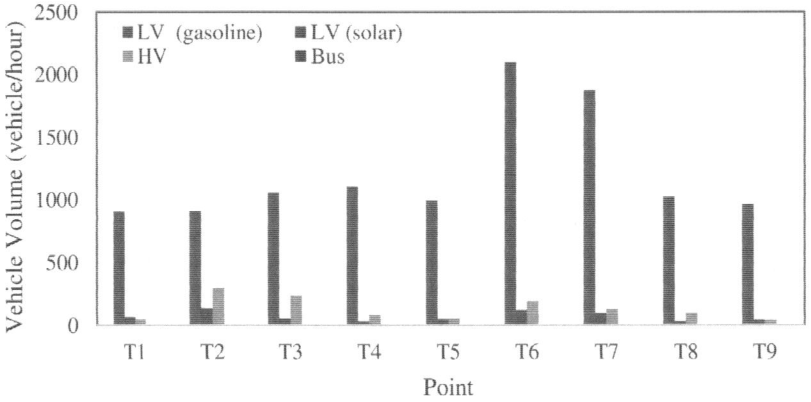

Figure 2. The number of vehicle volume

Based on the vehicle volume figure above, it can be seen that the maximum LV Gasoline vehicle volume is at the T6 observation point as much as 2099 units/hour with a percentage of 19.21%. For the minimum LV Gasoline volume at the observation point T1 as much as 907 units/hour with a percentage of 8.30%. The maximum LV Solar volume is at the T2 observation point as much as 138 units/hour with a percentage of 21.73%. The minimum LV Solar volume is at the observation point T8 as much as 30 units/hour with a percentage of 4.72%. The maximum HV volume is at the observation point T2 as much as 303 units/hour with a percentage of 25.36%. The minimum HV volume is at the observation point T9 as much as 43 units/hour with a percentage of 3.60%. Meanwhile, the bus volume is only at points T2 and T6 as much as 1 unit/hour with a percentage of 50%. The number of vehicle volume and dthe fuel consumption must be determined because traffic emissions are directly related to the fuel consumption of each vehicle. It caused by incomplete combustion process in the engine system [14].

3.2. Vehicular emissions

The value of the amount of emission for each type of vehicle is obtained from the calculation results of measuring the volume of vehicles on the toll road that passes at each measurement point. Calculation of the value of the amount of emission for each type of vehicle using the emission factor. The results of emission from vehicle can be seen in the following figure 3 and figure 4.

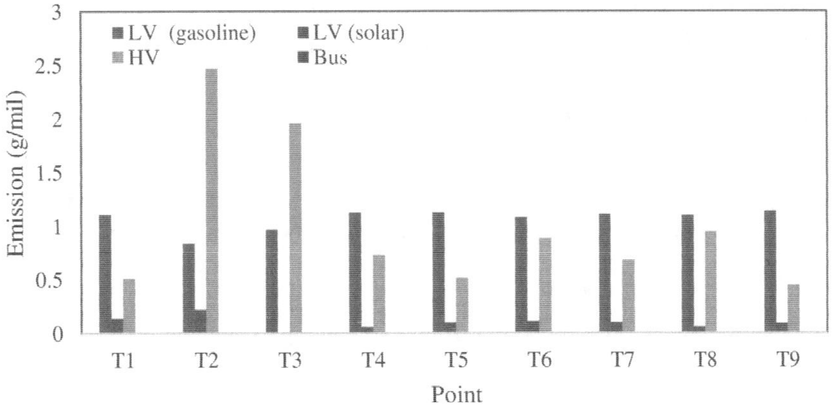

Figure 3. NO_2 emissions by type of vehicle

Based on Figure 3, it can be seen that the highest amount of NO_2 emission is 3.53 gr/mile at T2, while the lowest emission 1.68 gr/mile. Based on the value of the amount of emission, it can be seen that the value of the amount of emission based on the value of the emission factor is influenced by the volume of vehicles of each different type of motorized vehicle.

Gasoline-fueled vehicles (LVb) are the most dominant types of vehicles that contribute to NO_2 emissions, which is 48.62%. Truck type vehicle (HV) is the type of vehicle that contributes the second highest NO_2 emission at 46.47%. Diesel-fueled vehicles (LVs) contributed 4.87% of NO_2 emissions while Buses (BS) contributed 0.04% of NO_2 emissions.

Each type of vehicle that contributes different amounts of emissions is influenced by differences in vehicle volume and emission factors. The high volume of LVb vehicles makes the value of the emission load contributed by LVb is also large. This shows that on the Toll Road in Makassar City, the dominant vehicle passing is the LVb type vehicle, which is 48.62%. the emission load of each type of vehicle at each point is then averaged. The average value of the emission load at each point can be seen in the following figure 4.

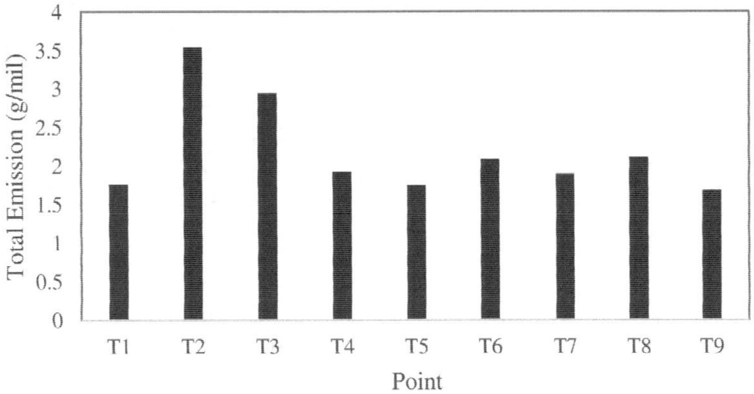

Figure 4. Emission load per point

From figure 4, the highest average concentration was obtained for 1 hour of measurement, which was 97.1 g/Nm^3 at T3, and the lowest average concentration was 30.3 g/Nm^3 at T1. The value of NO_2

concentration based on the output of Caline-4 software is influenced by the input variables, namely vehicle volume, wind direction and speed, background concentration, photolysis rate, and NO_2 ratio.

3.3. NO_2 concentration by direct measurement

Measurement of NO_2 concentration in ambient air is carried out using an impinger with one impinger measurement point or sampling point on each road. Test method for nitrogen dioxide (NO_2) levels using the Griess Saltzman method using a spectrophotometer. Concentration results from direct measurements at each point can be seen in Figure 5.

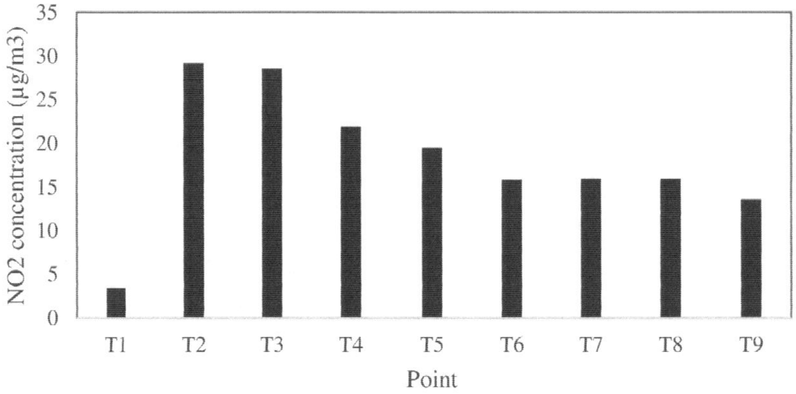

Figure 5. Concentration of direct measurement results at each point

3.4. NO_2 concentration prediction

Prediction of the spread of NO_2 pollutants that experience dispersion into free air with pollutants contained in each receptor on the Toll Road area in Makassar City using a software called Caline-4. The data used to operate the Caline-4 is different for each parameter. For the NO_2 parameter, data on background concentrations of O_3, NO, and NO, the ratio of NO_2/NOx, and the appropriate rate of NO_2 photolysis are required. The concentration results of Caline-4 are in ppm which will then be converted to g/Nm^3 for comparison with ambient quality standards. NO_2 Concentration Prediction results can be seen in Figure 6.

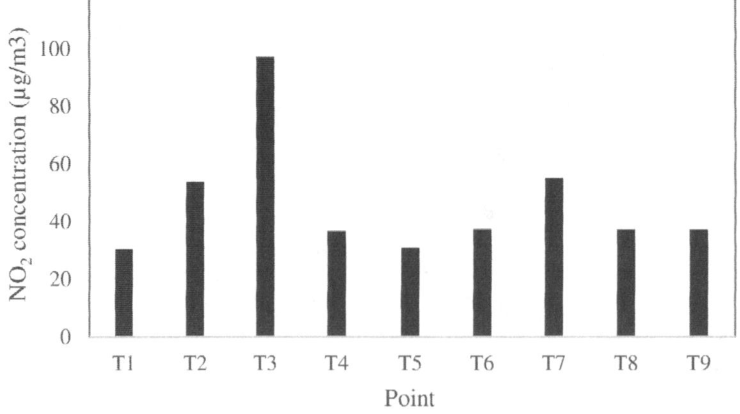

Figure 6. NO_2 concentration prediction

3.5. Comparison of NO_2 concentration from measurement results and caline dispersion

A recapitulation of the pollutant concentration of nitrogen dioxide (NO_2) direct measurement using impinger and Caline-4 can be seen in Figure 7

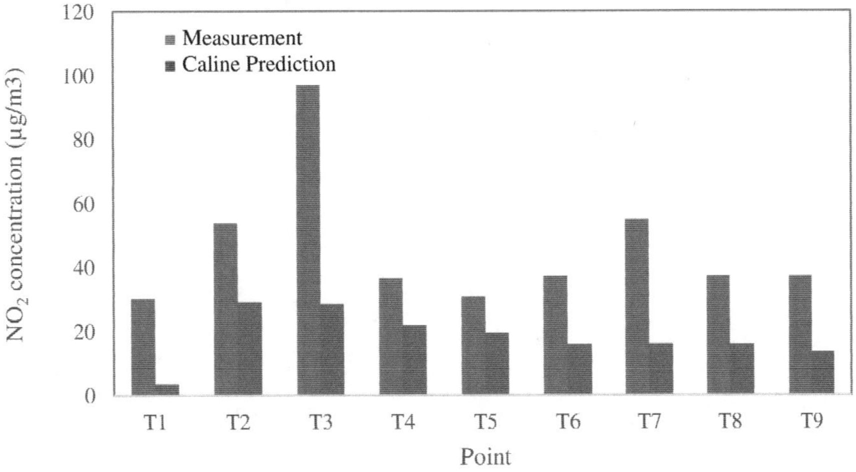

Figure 7. Comparison of pollutant concentrations of nitrogen dioxide (NO_2)

The concentration of Nitrogen Dioxide (NO_2) from caline 4 was higher than direct measurement using impinger. Caline-4 tends to be influenced by factors from transportation and the most important is the use of emission factors that affect the calculation of average emissions and the existing traffic volume.

4. Conclusion

In accordance with the results of measurements that have been carried out directly on the Toll Road in Makassar City using the impinger tool, the results of the NO_2 concentration are still below the threshold value that has been set in the latest applicable regulations, which is 200 g/Nm^3 with a measurement time of 1 hour. The high or low value of NO_2 concentration is influenced by the vehicle volume factor, where the NO_2 concentration increases along with the increase in vehicle volume. In addition, it is also influenced by meteorological factors. Based on the results of the Caline-4 software output, the results obtained from the distribution of NO_2 concentrations at each receptor found on the Toll Road in Makassar City, it can be stated that it is still below the threshold value set in the latest applicable regulations, which is 200 g/Nm^3 with time measurement for 1 hour.

References

[1] Jonathan Samet and D. Krewski, "HEALTH EFFECTS ASSOCIATED WITH EXPOSURE TO AMBIENT AIR POLLUTION," *J. Toxicol. Environ. Heal. Part A*, vol. 70, pp. 227–242, 2007.

[2] P. Taylor *et al.*, "Integrating Health on Air Quality Assessment — Review Report on Health

Risks of Two Maj or European Outdoor Air Pollutants : PM and NO_2," pp. 37–41, 2014, doi: 10.1080/10937404.2014.946164.

[3] M. Ravina and D. Panepinto, "Air Quality Planning and the Minimization of Negative Externalities," 2019, doi: 10.3390/resources8010015.

[4] A. Manafia, E. A. Jayadipraja, and R. A. Surya, "RISK OF ENVIRONMENTAL HEALTH RISK PARAMETERS NO_2 AND COMPLAINTS OF RESPIRATION ON PUBLIC TRANSPORTATION DRIVERS (NEW CAMPUS-CITY) IN KENDARI CITY," vol. 3, no. 1, pp. 97–107, 2021, doi: 10.36566/ijhsrd/Vol3.Iss1/61.

[5] WHO, *Health effects of transport-related to air pollution*. WHO Regional Office for Europe, 2005.

[6] G. D. Ramady, R. G. Wowiling, P. Studi, T. Elektro, S. Tinggi, and T. Mandala, "ANALISA PREDIKSI LAJU KENDARAAN MENGGUNAKAN METODE LINEAR REGRESION SEBAGAI INDIKATOR TINGKAT KEMACETAN," vol. 12, no. 2, pp. 22–28, 2017.

[7] Badan Pusat Statistik, "Perkembangan Jumlah Kendaraan Bermotor Menurut Jenis (Unit), 2018-2020," 2018. bps.go.id/indicator/17/57/1/perkembangan-jumlah-kendaraan-bermotor-menurut-jenis.html

[8] The Federal Highway Administration, "Freight and Air Quality Handbook," 2010.

[9] S. Noftri, M. Faizal, and R. Mohadi, "Indonesian Journal of Fundamental and Applied Chemistry," no. 2, pp. 58–61, 1999, doi: 10.24845/ijfac.v2.i4.58.

[10] S. N. Indonesia and B. S. Nasional, "Udara ambien – Bagian 2: Cara uji kadar nitrogen dioksida (NO_2) dengan metoda Griess Saltzman menggunakan spektrofotometer," 2005.

[11] Kementrian Lingkungan Hidup, "Peraturan Menteri Negara Lingkungan Hidup Nomor 12 Tahun 2010 Tentang Pelaksanaan Pengendalian Pencemaran Udara Di Daerah," *Pelaksaan Pengedalian Pencemaran Udar. Di Drh. Menteri Negara Lingkung. Hidup*, pp. 1–199, 2010.

[12] M. A. Johnson, "WRPLOT View Wind and Rain Rose Plots for Meteorological Data," 2016.

[13] P. Benson, *CALINE4 Dispersion Model for Predicting Air Pollutant Concentrations Near Roadway*. Sacramento, CA: California Department of Transportation, 1898.

[14] S. H. Mohd Shafie and M. Mahmud, "Urban air pollutant from motor vehicle emissions in kuala lumpur, malaysia," *Aerosol Air Qual. Res.*, vol. 20, no. 12, pp. 2793–2804, 2020, doi: 10.4209/aaqr.2020.02.0074.

IOP Conf. Series: Earth and Environmental Science 1117 (2022) 012053 doi:10.1088/1755-1315/1117/1/012053

Identification of Green Lane in its ability to absorb emissions on Main Arterial Street in Makassar City

N M R Harusi[1], S H Aly[1], Y Mitani[2], M Junaid[3]

[1]Department of Environmental Engineering, Faculty of Engineering, Makassar City, University of Hasanuddin, South Sulawesi, Indonesia 90245
[2] Department of Civil Engineering, Faculty of Engineering, University of Kyushu, Fukoka 819-0395, Japan
[3]Department of Agrotechnology, Faculty of Agriculture, University of Hasanuddin, Makassar City, South Sulawesi , Indonesia 90245
email: nurulmasyiah@unhas.ac.id (correspondence author).

Abstract. This study aims to know the number and variation of plant species and to analyze the potential for CO_2 absorption from each type of plant in the green lane. The focused green lane is located on the main arterial road with length of 4.2 km, St. Urip Sumoharjo which is known as the longest and most congested road in Makassar City. This study was carried out by surveying and observation the tree species on the road and used image processing to validate the suspect of tree species through App. Meanwhile, the calculation of CO_2 absorption is analyzed using the per plant species method by multiplying between the number of existing trees with the absorption capacity per plant species. As a result of the inventory, it was found that there were thirty-eight (38) tree species with a total of 2,587 trees. The dominant tree species were *Bougainvillea glabra* with 693 trees (26.8%), and *Tabebuia chrysantha* with 339 trees (13.1%) located in the median side. With the total trees, it has the capacity to absorb CO_2 of about 2,478,996 gr/hour. Additionally, Segment 5 and 6 have a highest CO_2 absorption with 1,278,274 gr/hour and 786,879 gr/hour respectively. It was caused by having the dominant typical plant species which are capable of absorbing the highest CO_2 concentration, such as *Bougainvillea glabra* and *Samanea saman* at 1,643.83 gr/hour.plant and 3,252.1 gr/hour.plant.

1. Introduction

The city, as the central area of various human activities, has been developed continually. The increase in development has become one major problem related to the decline in the number of Green Open Space (GOS). Makassar City, as the Gateway of Eastern Indonesia, has experienced a population growth rate of 1.29% with a total population of 1,508,154 people [1]. In addition, this city has a surface area of 199.26 km2 which covers 15 districts. Harusi (2019) revealed that urban expansion has increased in Makassar City by about 4,935 ha from 1997 to 2015, and affected to decline of 43% of green fields such as forests, mangrove forests, swamp fields, paddy fields, garden fields, and shrub field [2]. Based on the monitoring of the Paotere Maritime Meteorological Station from 1993 to 2012[3], Makassar has experienced the global warming phenomenon which indicated the increase of average temperature in the atmosphere by 0.5°C over 19 years. According to this phenomenon, many micro and macro disasters have happened in recent years such as floods, high-temperature surfaces, unhealthy air, food security, and climate change.

In Law Number 26 of 2007 about Spatial Planning in Indonesia, it is stated that GOS provision is at least 30% of the total urban area [4]. The proportion of 30% is a minimum measure to ensure the stability of the city's ecosystem which are the hydrological system and microclimate system, as well as other ecological systems. This stability influences the availability of clean air needed by the community and increases the aesthetic value of the city. However, based on the Regional Regulation of Makassar City No. 4 of 2015 about the Makassar City Regional Spatial Plan for 2015-2034, it states that the area of GOS in Makassar City is only 8.31% of the total area of Makassar City [5]. Meanwhile, a green belt area of 71.21 Ha or 0.4% of the total area. This condition become a major issue due to the percentage being still low compared with the regulation.

In order to overcome this condition, Green Open Space (GOS) has a simple idea to balance the function of using space in urban areas. One type of GOS for urban areas is green lanes on roads which have the ability to reduce air pollution from motor vehicle emissions such as carbon dioxide (CO_2) [6] and are referred to as urban lungs. Trees in the green lane that grow on medians and roadside are known to contribute absorbed the carbon dioxide (CO_2) emissions for photosynthesis. Photosynthesis is the formation of carbohydrates (C6H12O6) from CO_2 gas in the atmosphere and water molecules (H2O) from the soil with the help of sunlight and chlorophyll [7].

The ability to absorb CO2 in ambient air is certainly different for each plant. According to the results of Kusminingrum (2008), the vegetation in the green lane is able to reduce CO and CO2 by about 23.47% to 92.22% depending on the types of plants [8]. Hence, this study was conducted to know the number and variation of plant species and to analyze the potential for CO2 absorption from each type of plant in the green lane. The focused green line is located on the main arterial road in Makassar with 4.2 km, St. Urip Sumoharjo which is known as the longest and most congested road. Hopefully, the result can become information in terms of monitoring, especially for the City Planting Office, Regional Government, and South Sulawesi Regency, to reduce air pollution and prevent the global warming phenomenon.

2. Methodology

2.1. Study Area

This research was conducted on the main arterial road, namely St. Urip Sumoharjo (Figure 1) with a road length of 4.2 km. Then, this road will be divided into 6 segments due to the different geometric characteristics and dimensions of the road, as shown in Table 1.

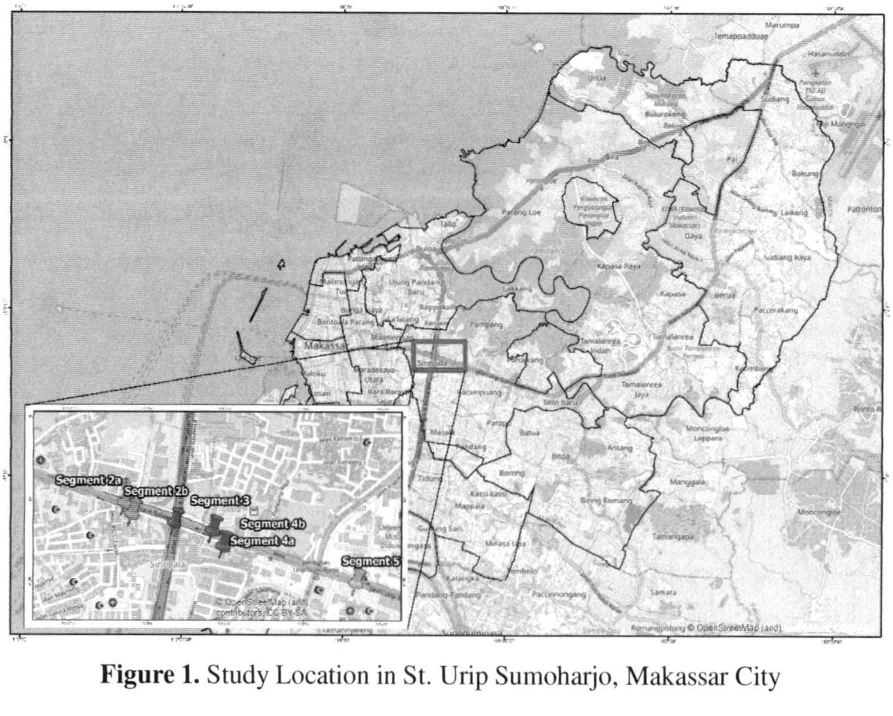

Figure 1. Study Location in St. Urip Sumoharjo, Makassar City

Table 1. Location of Measurement Points and Boundaries in Each Segment

Segments (Seg.)	Length (km)	Coordinate		Area of Boundaries	
		X	Y	West	East
1	1,015	5°8'5,14" S	119°25'48,77"E	Start from St. Bawakareng until St. Masjid Raya	West edge of St. Flyover
2	0,325	5°8'10,36 "S and 5°8'11,26 "S	119°26'18,02"E and 119°26'17,79"E	West edge of St. Flyover	Signaled Intersection on St. A.P. Pettarani, St. Urip Sumoharjo, and St. Barawaja
3	0,627	5°8'13,19 "S	119°26'25,44"E	West edge of St. *Flyover*	East edge of St. *Flyover*
4	0,302	5°8'14,20 "S and 5°8'15,17 "S	119°26'30,12"E and 119°26'29,84"E	Signaled Intersection on St. A.P. Pettarani, St. Urip Sumoharjo, and St. Barawaja	East edge of St. *Flyover*
5	1,181	5°8'18,55 "S	119°26'41,58"E.	East edge of St. *Flyover*	St. Prof. Abdur Rahman Basalamah in front of the Governor's Office
6	1,992	5°8'33,89 "S	119°27'29,90"E	St. Prof. Abdur Rahman Basalamah in front of the Governor's Office	T-junction of St. Urip Sumoharjo, St. Perintis Kemerdekaan, and St. Dr. Leimena

(a) *Soil ground*

(b) *Concrete slab*

Figure 2 Median Types

This difference will affect the number of trees and the ability of CO_2 absorption in each of the existing trees on every single side of the green corridor. This study also classifies the geometric location of the road for plant placement based on Minister of Public Works Regulation No.5 (2008) [9]. St. Urip Sumoharjo has a variety of structures in green lanes, divided into the side of the road, the median of the road, and the traffic island. The green lane on the side of the road is divided into 2 lanes, namely the North side (the direction from St. Bawakaraeng to St. Perintis Kemerdekaan), and the South side (direction from St. Perintis Kemerdekaan to St. Bawakaraeng). For the time of collection data on the road, it was carried out from January 29th - February 5th, 2022. In addition, the green line has different dimensions and the plant placement system, such as the base cover is divided into two, namely soil ground (ground cover) and concrete slab (a cement pavement cover), as shown in Figure 2.

2.2. Data Collection Techniques

Determination of plant species is done by searching for literature on plant species that are generally planted on green lanes. Additionally, this phase utilizes the Agrio App to get more pictures and plant characteristics. From the picture, it will be adjusted to the type of plant at the research location. In the end, to improve accuracy, image processing is carried out using the Google Lens App and PictureThis–Plant Identifier App. By taking directly pictures of trees using the Apps, the system automatically identifies plant species and shows the type and name of plant.

2.3. Analysis of Carbon Dioxide (CO_2) Absorption on Vegetation on Green Lane Road

To estimate the CO2 absorption ability of the vegetation on green lane road, the method used is per plant species. This calculation is based on Equation (1) by multiplying between the number of existing trees with the absorption capacity per plant species [10,11,12]. The plant absorption value of each plant was obtained from previous research through measurement and analysis of plant leaves in the laboratory, as followed in Tabel 2[10,11,12,13,14,15,16]. Meanwhile, to calculate the total CO_2 absorption of the green lane road in a certain segment, calculations are carried out according to Equation (2).

$$C_{sink} = CO_2 \; absorption \; per \; 1 \; plant \; species \times number \; of \; plants \quad (1)$$

$$C_{sink\ total} = \sum_{i=1}^{n} C_{sink\ ke-i} \quad (2)$$

where:

$C_{sink\ total}$: The total of CO_2 absorption of vegetation on each structure of green lane
n : Number of plant species on each structure of green lane
i : The plant species

3. Result and Discussion

3.1. Identification of vegetation types on the green lane of Urip Sumoharjo Road

3.1.1. Structure in Green Lane

Segments 1, 5, and 6 have two structures in the green lane: edge and median. For the placement of green lane trees in Segment 1, it tends to follow the pattern of line patterns that have not tightened. Meanwhile, the placement corresponds to a line pattern while the median road tends to have a clustered pattern with soil ground cover. In Segment 6, the placement of trees on the edge of the road tends to follow a lined pattern while at the median road it tends to cluster. However, in some parts of the green lane on the edge of Segment 6 road, it is converted into a sidewalk and parking area, namely in the Panaikkang Market area. This condition can reduce the area of vegetation planting.

Segments 2 and 4 have various types, of median and road islands, but segment 4 only has one road island. On the edge and median of the road, the placement of trees tends to follow a lined pattern, while on the island the road tends to follow a group pattern. In addition, Segment 4 is included in the minor lane at a non-diamond-shaped intersection on Jalan Urip Sumoharjo Makassar City. Both segments have a large median. Meanwhile, segment 3 is quite different because it does not have a green lane on the edge or traffic island. The green line in this segment is only on the median of the road with flower pot plant containers located on shaped lids. This segment is a flyover or major road at a non-diamond-shaped crossroads on St.Urip Sumoharjo.

3.1.2. Vegetation Types

From the results of the inventory, it was found that there were thirty-eight (38) tree species with a total of 2,587 trees on Jalan Urip Sumoharjo, as shown in Table 2. The dominant tree species were *Bougainvillea glabra* with 693 trees (26.8%), *Tabebuia chrysantha* with 339 trees (13.1%), *Polyalthia longifolia* with 267 (10.3%), *Tabebuia rosea* with 261 trees (10.1%), and *Samanea saman* with 248 trees (9.6%). Those dominant plant species have a canopy shape of umbrellas and pagodas such as Bougainvillea glabra and Polyalthia longifolia [17]. In addition, *Tabebuia chrysantha* and *Tabebuia rosea*, and *Samanea saman* are trees that can beautify the urban environment and are resistant to hot climates.

Based on the segment, the highest number of trees is located in segments 5 and 6 with 927 trees and 923 trees. Then, it was followed by Segments 1, 2, and 4 with 271 trees, 232 trees, and 176 trees, respectively.

Meanwhile, the lowest value is found in Segment 3 with a value of 58 trees. However, segments 1 dan 6 have various types of trees: 22 and 27 tree species. Moreover, the highest plant species for six segments were *Tabebuia chrysantha* (73 trees), *Polyalthia longifolia* (62 trees), *Bougainvillea glabra* (58 trees), *Terminalia mantaly* (32 trees), *Bougainvillea glabra* (511 trees), and *Tabebuia chrysantha* (170 trees) respectively.

In Figure 3. the dominant value of plant located on the median side with 1,361 trees. It was continuously followed by the South Side (489 trees), North Side (421 trees), Median in North Side (130 trees), Traffic Island in North Side (118 trees), Traffic Island in South Side (41 trees) Median in South Side (27 trees). Due to the characteristics of the St. Urip Sumoharjo road which are divided into two-way four-lane roads (4/2 D), the dominant vegetation planted in the median side.

Tabel 2. Type and amount of vegetation in each segment on St. Urip Sumoharjo

No.	Scientific Name of Tree Type	Ability of CO_2 absorption (gr/hour.trees)	Segments						Total	Percentage (%)
			1	2	3	4	5	6		
1	Acacia auriculiformis	165[10,11,12]	1	-	-	-	3	1	5	0.2%
2	Pterocarpus indicus	310,52[10,11,12]	4	26	-	1	83	88	202	7.8%
3	Tamarindus indica	0,17[14]	-	-	-	-	-	8	8	0.3%
4	Pithecellobium dulce	165[10,11,12]	1	-	-	-	-	-	1	0.0%
5	Bambusa multiplex	0,39[10,11,12]	3	1	-	-	-	-	4	0.2%
6	Ficus banjamina	1.146,51[10,11,12]	3	2	-	1	3	6	15	0.6%
7	Cerbera manghas	96,9[10,11,12]	-	-	-	8	2	-	10	0.4%
8	Vitex cofassus	702,17[13]	3	-	-	-	-	3	6	0.2%
9	Bougainvillea glabra	1.643,83[15]	19	27	58	1	511	77	693	26.8%
10	Casuarina equisetifolia	45[10,11,12]	-	1	-	-	-	-	1	0.0%
11	Polyalthia longifolia	719,74[10,11,12]	36	62	-	28	15	126	267	10.3%
12	Syzgium malaccense	44,59[10,11,12]	-	-	-	-	-	2	2	0.1%
13	Tectona grandis	12,41[10,11,12]	-	-	-	-	6	11	17	0.7%
14	Plumeria acuminate	25,114[16]	-	-	-	-	-	2	2	0.1%
15	Lannea coromandelica	45[10,11,12]	-	-	-	-	2	-	2	0.1%
16	Diospyros celebica	589,73[13]	-	-	-	-	-	6	6	0.2%
17	Barringtonia asiatica	165[10,11,12]	3	1	-	-	1	3	8	0.3%
18	Spathodea campanulata	24,16[10,11,12]	3	6	-	12	-	-	21	0.8%
19	Muntingia calabura	0,6[10,11,12]	1	-	-	-	2		3	0.1%

IOP Conf. Series: Earth and Environmental Science 1117 (2022) 012053 doi:10.1088/1755-1315/1117/1/012053

No.	Scientific Name of Tree Type	Ability of CO_2 absorption (gr/hour.trees)	Segments						Total	Percentage (%)
			1	2	3	4	5	6		
20	Terminalia mantaly	24,16[10,11,12]	4	8	-	32	2	14	60	2.3%
21	Leucaena leucocephala	165[10,11,12]	-	-	-	-	4	11	15	0.6%
22	Swietenia macrophylla	3.112,43[10,11,12]	4	3	-	-	31	24	62	2.4%
23	Mangifera indica	51,96[10,11,12]	2	2	-	1	-	1	6	0.2%
24	Arthocarpus heterophyllus	22[10,11,12]	-	-	-	-	-	1	1	0.0%
25	Wodyetia bifurcata	0,39[10,11,12]	4	-	-	-	1	17	22	0.9%
26	Phoenix sylvestris	0,39[10,11,12]	-	-	-	3	-	-	3	0.1%
27	Dypsis lutescens	0,39[10,11,12]	33	-	-	-	-	5	38	1.5%
28	Veitchia merrillii	32,6[10,11,12]	-	-	-	6	1	49	56	2.2%
29	Dracaena draco	0,39[10,11,12]	-	-	-	1	-	3	4	0.2%
30	Oleina syzygium	155,58[10,11,12]	33	-	-	23	70	39	165	6.4%
31	Alstonia scholaris	1.319,35[10,11,12]	4	-	-	1	-	1	6	0.2%
32	Annona muricata	8,9[13]	-	3	-	-	-	-	3	0.1%
33	Artocarpus altilis	22[10,11,12]	-	1	-	1	-	-	2	0.1%
34	Tabebuia chrysantha	24,2[10,11,12]	73	14	-	20	62	170	339	13.1%
35	Tabebuia rosea	24,2[10,11,12]	23	59	-	23	37	119	261	10.1%
36	Mimusops elengi	67,58[10,11,12]	5	1	-	12		2	20	0.8%
37	Samanea saman	3.252,1[10,11,12]	9	15	-	2	88	134	248	9.6%
38	Hibiscus tiliaceus	183,58[14]	-	-	-	-	3	-	3	0.1%
Total			**271**	**232**	**58**	**176**	**927**	**923**	**2587**	**100%**

126

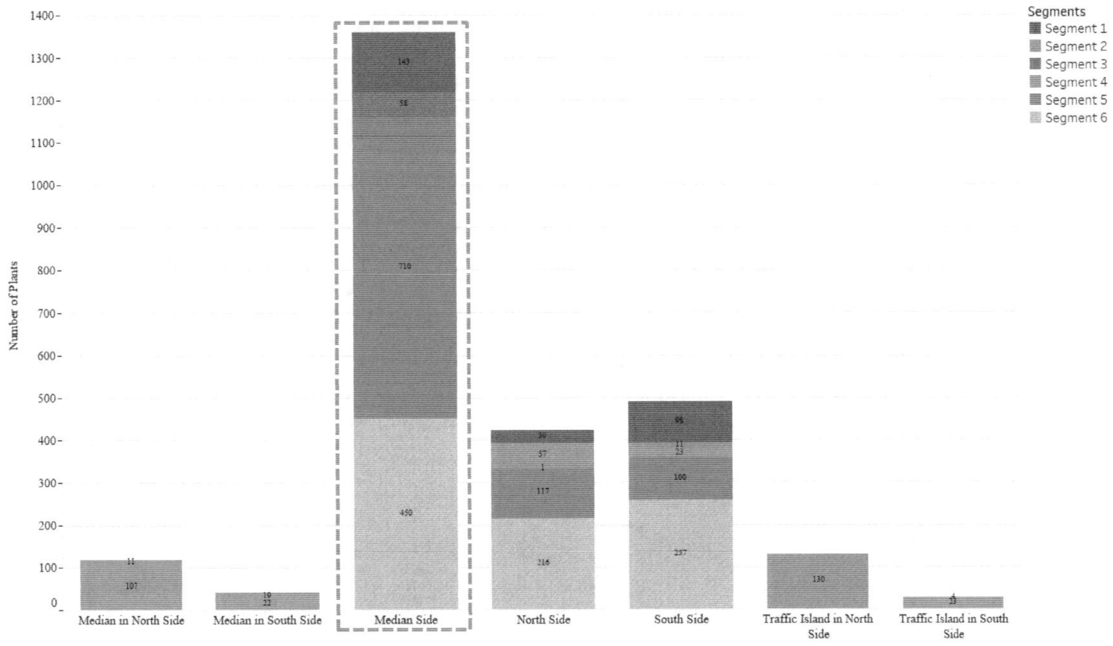

Figure 3. Number of Plant based on structure in green lane

3.2.Ability of CO2 absorption on plant species in each segment of St. Urip Sumoharjo

According to calculations using the plant species method, the ability to absorb CO_2 on St. Urip Sumoharjo is 2,478,996 gr/hour with total 2,587 trees. Figure 4 shows that the highest values are found in Segments 5 and 6 of 1,278,274 gr/hour and 786,879 gr/hour. While the lowest value is in Segment 3 (95.342 gr/hour) and Segment 4 (38.616 gr/hour). It is because Segments 5 and 6 have a large number of trees compared to other segments, 927 and 923 trees and have the longest roads. Meanwhile, In Table 4, the 5 dominant tree species located in these two segments are *Bougainvillea glabra* (588 trees), *Samanea saman* (222 trees), *Pterocarpus indicus* (83 trees), *Oleina syzygium* (70 trees), and *Tabebuia chrysantha* (232 trees), *Polyalthia longifolia* (126 trees), *Tabebuia rosea* (119 trees).

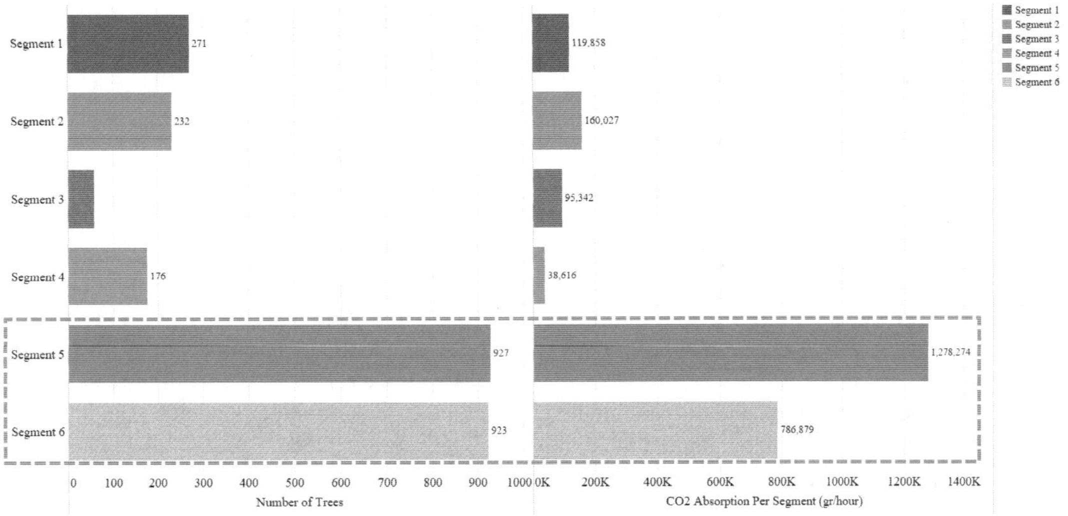

Figure 4. The comparison between number of plant and CO_2 Absorption per segments

Figure 5 clearly shows that the dominant trees in segments 5 and 6 are typical plant species which are capable of absorbing the highest CO_2 concentration, such as *Bougainvillea glabra* and *Samanea saman* at 1,643.83 gr/hour.plant and 3,252.1 gr/hour.plant. The high absorption capacity can be proven based on the anatomy and physiology types of trees. Both species have different stomata sizes as their leaves ranges 15.65-24.28 m length and 12.41-19.48 m width [18]. The finding suggests that these species are capable of receiving more photosynthesis. The higher of stomata density, the higher the potential to absorb pollutants and air particles [19] Additionally, *Samanea saman* is potential spesies as well to absorb carbon dioxide and air pollutants in the atmosphere [20,21]. In addition, trees that can live in an environment with low light intensity will have larger, thinner leaves, larger stomata size, fewer leaves, and larger intercellular spaces. Conversely, in an environment with high light intensity, the leaves will be smaller, thicker, stomata small and numerous, as well as the number of leaves that are shadier [22,23]. Therefore, each type of plant has a different absorption capacity, this is influenced by many factors including leaf area, relative leaf thickness, number of stomata, plant age and environmental factors where the plant grows [24]. The absorption of CO_2 per unit time of each plant is different, depending on the type of plant itself, especially on the morphology of the leaves.

In segment 5, the CO_2 absorption capacity of green line vegetation is influenced by 511 *Bougainvillea glabra* which have the ability to absorb CO_2 of 839.997 gr/hour. Then, it was followed by *Samanea saman* with a total of 88 trees and a total CO_2 absorption capacity of 286,184 gr/hour. While in segment 6, the two types of trees are still relatively small compared to segment 5 with value of the ability of CO_2 absorption: *Samanea saman*: 435,781.4 gr/hour and *Bougainvillea glabra*: 126,574.91 gr/hour respectively.

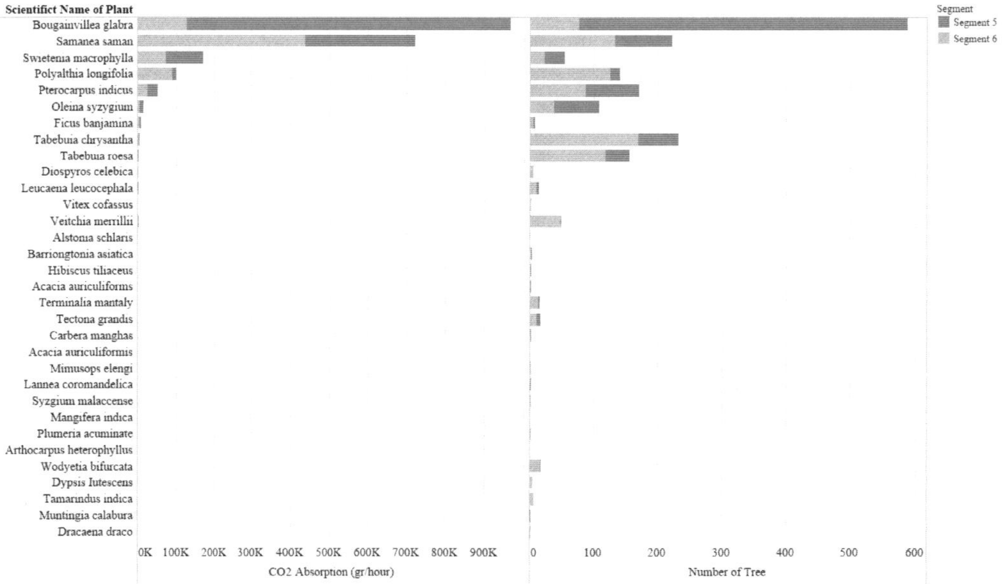

Figure 5. The value between number of plant and CO_2 Absorption in the segment 5 and 6

According to the comparison between the total number of trees and the value of absorption ability, there is not liner value between those variables, as shown in Figure 3. The number of trees in segments 5 and 6 has not significantly the difference about 4 trees. However, the difference in CO_2 absorption values in both the segment reached 491,394 gr/hour (23%). It was caused by the different distribution value of plant species in every segment, which can be seen in Figure 4. Moreover, the location of the trees on the green lane and the dominant direction of the wind also greatly affects the tree coverage area to absorb CO_2 on Jalan Urip. In these two segments, the distribution of dominant trees is on the Median Side and South Side with the ability to absorb CO_2 of 882,681 gr/hour and 159.545 gr/hour respectively, as shown in Table 4.

Table 3. Dominant Plant and CO2 Absorption in Segment 5 and Segment 6

Ranking	Segment 5			Segment 6		
	Scientific Name	Total of Trees	Total of CO_2 Absorption (gr/hour.trees)	Scientific Name	Total of Trees	Total of CO_2 Absorption (gr/hour.trees)
1	*Bougainvillea glabra*	511	839,997.13	*Tabebuia chrysantha*	170	4,114
2	*Samanea saman*	88	286,184.8	*Samanea saman*	134	435,781.4
3	*Pterocarpus indicus*	83	25,773.16	*Polyalthia longifolia*	126	90,687.24
4	*Oleina syzygium*	70	10,890.6	*Tabebuia rosea*	119	2,879.8
5	*Tabebuia chrysantha*	62	1,500.4	*Bougainvillea glabra*	77	126,574.91

*Gray line = have high ability to absorb CO_2

Table 4. The number of Trees and CO_2 Absorption per structure of green lane

Segment	Structure of green lane	Number of Trees per structure	CO_2 Absorption per structure (gr/hour.total)
5	North Side	117	274440.05
	Median Side	710	882681.18
	South Side	100	121152.73
6	North Side	216	221238.72
	Median Side	450	159545.878
	South Side	257	406094.54

4. Conclusion

The CO_2 absorption was calculated using the per plant species method. The result founded that there were thirty-eight (38) tree species with a total of 2,587 trees on St. Urip Sumoharjo. The dominant tree species were *Bougainvillea glabra* with 693 trees (26.8%), and *Tabebuia chrysantha* (13.1%). Meanwhile, the capacity total trees for CO_2 absorption are 2,478,996 gr/hour. The highest CO_2 absorption values were in Segments 5 and 6. Those segments have the dominant species trees which had a high ability to absorb CO_2, namely *Bougainvillea glabra* and *Samanea saman*. These species have higher of stomata density which are capable of receiving more photosynthesis related to decline air pollutants in the atmosphere. However, this method has disadvantage which does not use the actual size of plant morphology in the green lane. It means that can reduce the accuracy of result CO_2 absorption. Therefore, the further research requires to compare the analysis methods of potential CO_2 absorption.

5. References

[1] Statistic of Central Bureau. (2017). Makassar in Figures. Makassar City.
[2] Harusi, N.M.R (2019). Spatial-Temporal Analysis of Landscape Ecological Connectivity Changes in Makassar City. *International Symposium on Earth Science and Technology, page 78-84.* Japan.
[3] Taylor, J. (2013). *Climate Change Vulnerability Assessment in Makassar City.* Jakarta: UNDP Indonesia.
[4] Presiden Republik Indonesia. Undang-Undang Republik Indonesia No. 26 of 2007 about Spatial Planning. Jakarta: Indonesia

[5] Government of Makassar City. (2015). Makassar City Regional Regulation Number 4 of 2015 concerning Spatial Planning of Makassar City. Makassar: Regional Development Planning Agency.
[6] Mercy, A., Hamzah, B., Rahim, D. (2018). An Analysis of The Adequacy of Green Lane Vegetation in Absorbing the Carbon Monoxide (CO) of Transportation Activities. International Journal of Engineering Science Invention (IJESI): pp 59-64.
[7] Ingen-Housz, J. (1779). *Experiments upon Vegetables.* London: Printed for P. ELMSLY and H. PAYNE.
[8] Kusminingrum, N. (2008). Potential of Plants in Absorbing CO2 and CO to Reduce Global Warming Impact. *Jurnal Permukiman*, hal. 96-105.
[9] Ministry of Public Works of the Republic of Indonesia. (2008). Ministerial Regulation No.5/PRT/M about Guidelines for Provision and Utilization of Green Open Space in Urban Areas, Jakarta, Indonesia.
[10] Roshintha, R. R., and Mangkoediharjo, S. (2016). Analysis of the Adequacy of Green Open Space as an Absorber of Carbon Dioxide (CO_2) Gas Emission in the ITS Sukolilo Campus Area, Surabaya. *JURNAL TEKNIK ITS*, hal. D123-D137.
[11] Gratimah, RD. G. (2009). Analysis of Urban Forest needs as an Anthropogenic CO_2 absorber in the center of Medan City. Thesis. University of North Sumatra, Medan.
[12] Purwaningsih, S., (2007). The ability of Carbon Dioxide (CO_2) Uptake in Urban Forest Plants in the Bogor Botanical Gardens. Thesis. University of Bogor Agriculture, Indonesia.
[13] Sa'iedah, A. (2018). Correlation Between Green Open Space With Concentration Of Carbon Dioxide (CO_2) and Oxygen (O_2) in UIN Sunan Ampel Surabaya Campus. Surabaya: Environmental Engineering Study Program, Faculty of Science and Technology, State Islamic University of Sunan Ampel Surabaya, Indonesia.
[14] Environmental Agency in Surabaya City. (2017). Report of Environmental Carrying Capacity Study for City Parks in Surabaya. Surabaya: City Environmental Service in Surbaya City.
[15] Dubal, K., Ghorpade, P., Dongare, M., & Patil, S. (2013). Carbon Sequestration in the Standing Trees at Campus of Shivaji University, Kolhapur. *Natural Environmental and Pollution Technology an International Quarterly Scientific Journal*, 725-726.
[16] Suryaningsih, L., Haji, A. T., & Wirosoedarmo, R. (2019). Spatial Analysis of Deficiency of Green Open Space (RTH) in Mojokerto City. *Jurnal Sumberdaya Alam dan Lingkungan*, 1-10.
[17] Mansur, M., and Pratama, A.B. (2014). Potential Absorption of Carbon Dioxide in Wayside Trees. Jurnal Biologi Indonesia 10 (2): 149-159.
[18] Mustika, S. (2018). Analysis of Plant Stomata Size and Type in Pontianak City. Thesis. University of Tanjungpura, Indonesia.
[19] Muntadhiroh, C. (2015). Anatomical Characteristics and Potential of Trembesi Leaves (Albizia Saman) in Malang City Street as an Accumulator of Heavy Metal Lead (Pb). Thesis. Maulana Malik Ibrahim State Islamic, University of Malang, Indonesia.
[20] Ristyawati, Y.D. (2021). Analysis of the Density and Stomata Size of Trembesi Leaves (Samanea saman (Jacq.) Merr) based on Differences in Altitude in Mojokerto Regency as a Biology Study Source. University of Muhammadiyah Malang, Indonesia.
[21] Leopold, A. & P. Kriedemann. 1975. *Plant Growth and Development.* New York: McGraw Hill Book Co.
[22] Marisha, S. (2018). Analysis of the Ability of Trees to Absorb CO2 and Store Carbon on a Green road in the Sub-region of Tegalega City, Bandung City. Bandung Institute of Technology, Indonesia.
[23] Sukmawati, Tria. Fitrihidajati, H. Indah, N.,K. (2015). The Carbon Dioxide Absoption of Plant of the Urban Forest in Surabaya. LenteraBio Vol. 4 No.1: 108-111.

Potential Marine Plastic Debris Detection using Sentinel-2 Multi-Spectral Instrument (MSI)

A D Damayanti[1], Z A Hanami[1], K Hirose[2], S Ainun[3], R D Putra[4]

[1]Department of Environmental Engineering, Hasanuddin University, South Sulawesi 90245 Indonesia
[2]Japan Space System, Japan
[3]Department of Environmental Engineering, Institut Teknologi Nasional, West Java 40124 Indonesia
[4]Departement of Environmental Engineering, Bandung Institute of Technology, West Java 40116 Indonesia

annisadwidam@gmail.com

Abstract: Plastic debris has a long-term and quite severe impact on the marine ecosystem. Population growth in the developing area, Makassar coastal zone, will contribute significantly to the land's waste. Consequently, floating plastic debris comes directly from the land's plastic leaks to rapidly increase marine plastic debris. Due to being a major environmental issue in the coastal zone and reducing options for removal are very limited, this study becomes to detect the potential marine plastic debris in Makassar by Sentinel-2 Multi-Spectral Instrument (MSI) using Kernel Normalization Vegetation Index (KNDVI) and Floating Debris Index (FDI) for satellite image processing. The plastic debris was corrected atmospheric effect by Sen2Cor and processed by Sentinel Application Platform (SNAP). The resolution of images was set to 10 x 10 meters. The target area was calculated as 173.47 Ha. The results show that KNDVI and FDI combined are a developed and powerful approach to detecting plastic debris. KNDVI has more representative indices (SE 0.38) than FDI indices (SE 0.47). Through the discriminant analysis were verified pixels of KNDVI (3506 pixels) and FDI indices (701 pixels). Meanwhile, the Sensitivity Analysis Value (SAV) was well performed to detect the potential of marine plastic debris by KNDVI with SAV 7.5 than FDI indices with SAV 2.9.

Keywords: Plastic Debris, Sentinel-2 MSI, NIR, KNDVI, FDI, Sensitivity Analysis Value

1. Introduction

Marine debris has become an environmental issue that is a significant concern globally. The United Nations expert states that 45 to 70% of all floating marine debris is made up of plastic and potentially hazardous pollutants [1]. The persistent plastic debris can accumulate in high concentrations in the ocean and break down into very small pieces and never biodegrade because of the lack of UV radiation and oxygen in the marine environment [2, 3, 4, 5, 6, 7].

Plastic debris is a direct result of plastic leakage on land. Several studies have been looking for a comprehensive study on this topic. For example, not only Gallo F et al. (2018) research shows 8% million tons of plastic debris entering the ocean [8] but also Guffogg J A et al. (2020) and Borelle S B et al. (2020) showed 0.8–2.7 million of ocean plastics [9,10]. Jambeck et al. (2015) estimated that 4.8 to 12.7 million metric tons of plastic from land-based activities entered the world's oceans [11]. This is equivalent to approximately between 4,762,000,000 – 12,700,000,000 kg [11]. Approximately plastic waste that ends up in the coastal water is 14 million tons every year [12]. Indonesia is the second largest marine plastic debris producer after China [3]. At least, 0.49 to 0.86 million tons/year of plastic debris was detected in Indonesia [3].

Population growth and the amount of waste will rapidly increase marine plastic waste [13, 14] and adversely affect aquatic life and potentially human health [12, 7]. Based on Central Statistics Agency (2021), the population growth rate of Makassar City is 0.6% [16]. According to the National Waste Management Information System (SIPSN), the plastic waste composition in Makassar is 21.51% [15].

Plastic debris has a long-term and severe impact on coastal and marine ecosystems. The result of plastic pollution has a collective effect on ecological and economic aspects. Marine debris is effortless to be carried away by water flows scattered in the waters and can damage an ecosystem's habitat [16, 6]. Pollutants enter the open ocean and a chain reaction can occur when one group of organisms is affected [12]. For example, fish and shellfish absorb plastic microfibers and other particles directly from their environment. Although they may only absorb small amounts at a time, they can remain in the fish body for months or even longer, and humans will be consumers [12, 17]. The assessment of ecosystem impact analysis shows significant risk and economic cost [18]. The direct ingestion of plastic can be defined as increasing ecological risks of plastic contamination [19]. The annual cost to reduce marine natural capital in the ocean is between $3300 to $33,000 [20]. According to the Asia Pacific Economic Cooperation report, marine debris will cost individuals, communities, and the economy $200 billion by 2050 [20].

For several decades plastic has become a significant environmental effect, particularly in the open ocean, and reducing options for removal are very limited [6, 14]. Therefore, several programs have been started. UN's 2030 Agenda for oceans, seas, and marine resources contribute to plastic debris in water bodies through Goal 6, Goal 12, and Goal 14 [7, 12, 21]. Indonesia's National Plan of Action on Combating Marine Plastic Debris aims to reduce marine debris [1].

Despite the issue of marine debris, several fundamental studies explore the potential to detect plastic pollution with surveys and observations in laboratories [8, 12]. However, assessing marine plastic debris was time-consuming and costly. Therefore, they provide satellite data and programs with a temporal resolution to detect and monitor marine debris, as well as remote sensing data from Sentinel-2 images were currently identifying plastic spatial in the ocean [2, 5, 7, 21, 22, 23, 24, 25, 29].

To this end, this study aims to detect the potential marine plastic debris in Makassar by Sentinel-2 Multi-Spectral Instrument (MSI), which is calculated by indices processing with SNAP (Sentinel Application Program) and atmospheric correction using Sen2Cor.

2. Methods and Material

The objective of this study is to detect the potential of marine plastic debris using remote sensing in the coastal area of Makassar. Remote sensing is a based-ideal method for detecting the basic estimated amount of plastics entering the marine environment [8]. The steps of the methodology are outlined in detail below, as shown in Figure 1.

2.1. Data Collection and Study sites

The study utilized Sentinel-2 MSI Top of Reflectance (TOA) reflectance Level 1-C, which was collected from Copernicus Open Access Hub [26]. Sentinel-2 MSI is a high-resolution multi-spectral satellite image for monitoring the study of land, coastal areas, etc. [27]. This satellite image was divided into 100 km tiles in UTM/WGS84 projection. Sentinel-2 MSI Level 1-C data take from ground geometry [28] and it has been converted to data Sentinel-2 MSI Level 2-A (Table 1).

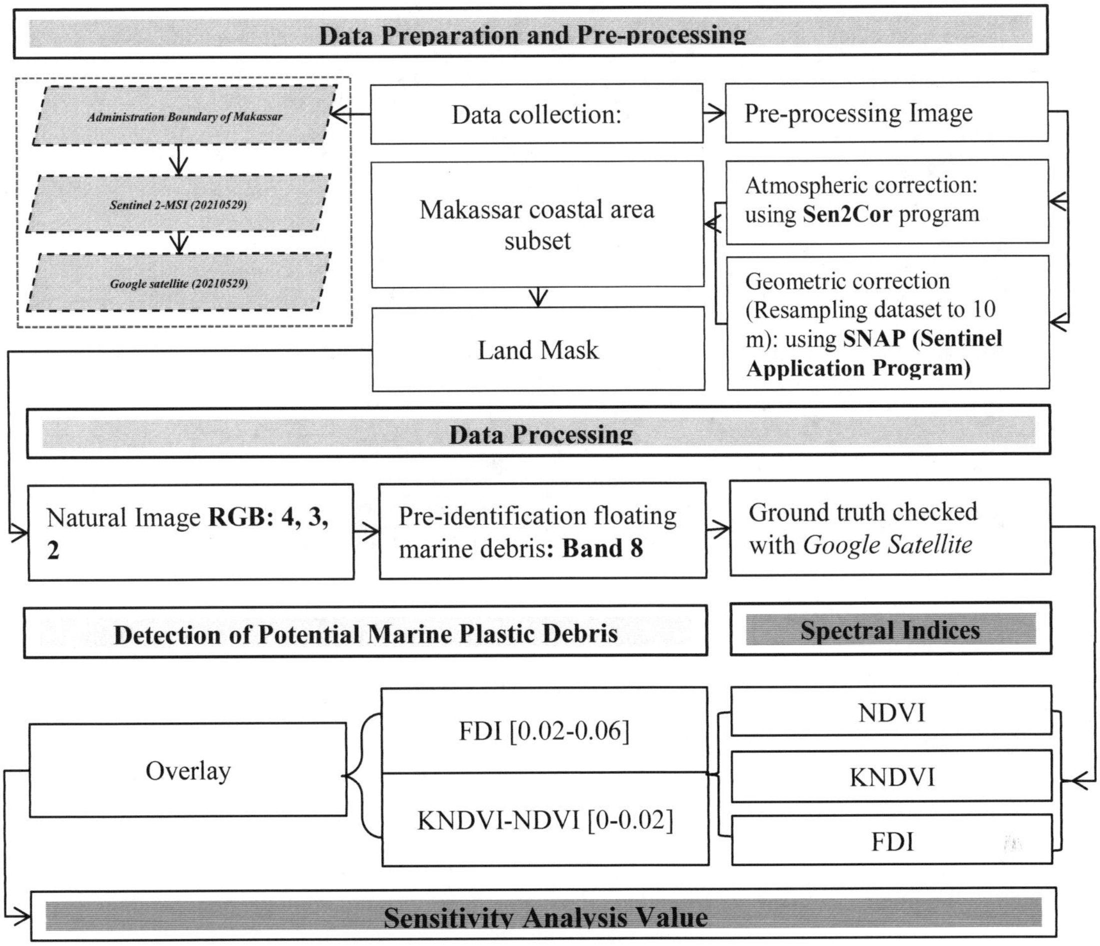

Figure 1. The flowchart of workflow to detect potential plastic debris

Table 1. List of Sentinel-2 images captured on 25 May 2021 over Coastal Zone of Makassar, Indonesia

Image Filename
S2A_MSIL1C_20210529T022331_N0300_R103_T50MQV_20210529T055350.SAFE
S2A_MSIL2A_20210529T022331_N9999_R103_T50MQV_20220612T133853.SAFE

There are 13 spectral bands in Sentinel-2 MSI, which can be seen in Table 2. Each spectral band of Sentinel-2 depends on the spatial resolution of satellite data. Data collected with a range of cloud cover 0-1%. However, 8 bands have been used in this study i.e. visible bands (2, 3, 4), Red Edge bands (5, 6, 7), NIR band (band 8), and SWIR band (band 11).

This study was conducted around the coastal zone in Makassar, Indonesia (Figure 2). This site was selected because the amount of waste has increased in the last two years by 6.51 % [15]. Plastic waste in Makassar is the second largest composition of waste after organic/food waste, which is 21.51 % of the total amount of waste [15].

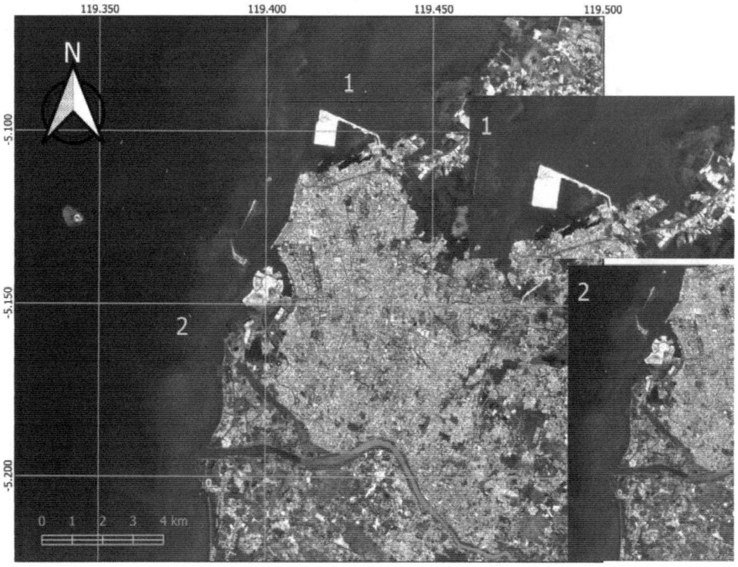

Figure 2. Sentinel-2 scenes over the coastal area of Makassar on the 29th of May 2021 with visible band combination (RGB: 2, 3, 4)

2.2. Pre-processing Satellite Image

Pre-processing images related to the satellite data acquisition process, noise filtering, and image quality enhancement [29]. The pre-processing image has applied the following operations consecutively atmospheric correction and geometric correction. In the marine environment, atmospheric correction is necessary to develop high-resolution data on coastal water [29]. Atmospheric correction is best suited for the identification of materials using satellite data [29]. According to Sannigrahi et al. (2021), atmospheric correction is needed to be used if there is no in-situ data [25]. Therefore, the atmospheric correction (AC) that has been used in this study is Sen2Cor.

Sen2Cor is a European Space Agency (ESA) algorithm for atmospheric correction on Sentinel-2 images, which corrects the atmospheric effect from Sentinel-2 Level 1-C Top of Atmosphere (TOA) products to a Level 2-A Bottom of Atmosphere (BOA) reflectance product [29]. Themistocleous K et al. (2020) and Topouzelis K et al. (2018) study found that the target marine plastic debris was conducted using Sen2Cor which generated a high reflectance value [7, 25]. The central and closest wavelengths in the hyperspectral dataset with different bands after atmospheric correction are provided in **Table 2.**

Table 2. Sentinel-2 band numbers and their hyper-spectral wavelengths

Sentinel-2	Band Name	Central λ (nm)*	R_{min}** λ (nm)	R_{max}** λ (nm)	R ** λ (nm)
B1	Aerosol	443	412	456	442.7
B2	Blue	490	456	533	492.7
B3	Green	560	538	583	559.8
B4	Red	665	646	684	664.6
B5	Red Edge 2	705	695	714	704.1
B6	Red Edge 2	740	731	749	740.5
B7	Red Edge 3	783	769	797	782.8
B8	NIR	842	760	907	832.8
B8A	VNIR	865	837	881	864.7

IOP Conf. Series: Earth and Environmental Science 1117 (2022) 012054 doi:10.1088/1755-1315/1117/1/012054

Sentinel-2	Band Name	Central λ (nm)*	R_{min**} λ (nm)	R_{max**} λ (nm)	R ** λ (nm)
B9	Water Vapour	940	932	958	945.1
B10	SWIR Cirrus	1375	1337	1412	1373.5
B11	SWIR 1	1610	1539	1682	1613.7
B12	SWIR 2	2190	2078	2320	2202.4

**Spectral resolution bands [36]*
*** Remote sensing reflectance after correction of atmospheric with Sen2Cor*

Furthermore, the next pre-processing image is geometric correction, which was done also in SNAP Software. The geometric correction was used to resample the images to a similar grid at 10 meters for every band to provide high resolutions and then the small features in the coastal area could be detected. Not only provides the high resolutions but also the extracted spectral pixels consist of reflectance spatial resolution [2] to minimize the error effect in detecting potential marine plastic debris.

The next image processing was subsets and masks of the Makassar land area. The image subsets were processed by cropping thus reducing the size of the image and speeding up the performance. Masking is proposed in this study to track small materials from the coastal surface.

2.3. Detection of Marine Plastic Debris

The plastic patches have been extracted using spectral indices which were applied in the study area. To detect potential marine plastic debris, there are spectral indices that have been used in this study, i.e. Floating Debris Index (FDI), Normalized Difference Vegetation Index (NDVI) [23,7], Kernel Normalized Difference Vegetation Index (KNDVI) [26].

Earlier studies [14], have found that the Floating Debris Index (FDI) generates high performance to detect potential plastic debris depending on the amount of material in a given pixel. FDI has given spectral signature value of floating materials and water results [26]. The FDI was expressed by equations (1) and (2). The threshold of FDI extracted to 0.02 – 0.06 as represented in Biermann L et al. (2021) [30].

$$FDI = R_{rs}NIR - R'_{rs}NIR \quad (1)$$

$$R'_{rs}NIR = R_{rs}RE2 + (R_{rs}SWIR1 - R_{rs}RE2) * \frac{(\lambda B8 - \lambda B4)}{(\lambda B11 - \lambda B4)} * 10 \quad (2)$$

Where R_{rs} is remote sensing reflectance intensity of NIR (B8), Red Edge-2 (B6), and SWIR-1 (B11). The index also used reflectance which was very close to the central wavelength of B8 (λB8, 842 nm), B4 (λB4, 665 nm), and B11 (λB11, 1610 nm) by the satellite.

Sannigrahi et al. (2021) proposed that Normalized Difference Vegetation Index (NDVI) quantified vegetation greenness. NDVI measured the ratio between the red and near-infrared bands [7, 22, and 30]. The vegetation was absorbed in amounts of red light and reflected in high amounts in NIR light, and that is characteristic to identify the presence of vegetation. NDVI values tend to have close to zero values for marine plastic debris [30, 22]. The equation of NDVI was developed to identify spectral signatures expressed as equations (3).

$$NDVI = \frac{R_{rs}NIR - R_{rs}RED}{R_{rs}NIR + R_{rs}RED} \quad (3)$$

Furthermore, Kernel Normalized Difference Vegetation Index (NDVI) also calculated the pixel of reflectance in near-infrared and red spectra. KNDVI using based on NDVI to approach the high dimensional of indices. KNDVI was simplified to measure the similarity of two bands, the near-infrared and red bands [26]. The simplified of KNDVI is calculated as follows:

$$KNDVI = \tanh\ (NDVI * NDVI) \quad (4)$$

2.4. Performance Evaluation

To measure the level of accuracy of the indices that have been applied to detect marine plastic debris in coastal water bodies, the KNDVI and FDI distribution needed to calibrate and validate. In cases, where no training data are available, the validation method was obtained from ground-truth checks from *Google Earth Pro* on the same date as the Sentinel-2 data (May $29^{th,}$ 2021). Moreover, to evaluate indices, we used to extract statistical method, Sensitivity Analysis Value (SAV) [7], to validate which index can detect potential marine plastic debris using the KNDVI and FDI index. The equation is expressed in equation 5.

$$SAV = \frac{DV*NPD}{Dmax-Dmin} \quad (5)$$

where, DV is the discriminative value of two indexes, NPD is the number of pixels detected using each index, D-min and D-max are the minimum and maximum values of detected pixels. Discriminative value (DV) can describe as the possibility of a particular pixel of each index that can discriminate between the indicated plastic and non-plastic debris.

3. Results and Discussion

The results of plastic debris detected in this study were divided into three parts; preliminary plastic debris identification, spectral indices of plastic debris detection, and performance evaluation of the indices.

3.1 A Preliminary Plastic Debris Identification

Preliminary identification was performed by displaying the multispectral images with a natural image (RGB: 4, 3, 2). Natural images have been set to realize the materials which present on the coastal surface (Figure 3a), i.e. vegetation, ships/boats, and fishing net cages. On the other hand, band 8 (NIR) was used to detect floating debris on the coastal surface. In a previous study, band 8 is the unique spectral signature for floating debris detection, which is the key role of detection [24]. Near-infrared wavelength was efficient to identify clear water [30]. Pixel was a key role in absorbing the intensity of the reflectance signal. Band-8 images showed that the offshore area was a darker color than the land area (Figures 3b and 3c). Regarding Ciappa A C (2020), the darker color of coastal surfaces depends on the water depth. A near-infrared band is easily used to detect floating debris which has a bright signature in NIR [31]. As in Figures 3b and 3c, not only plastic materials could be detected but also other floating materials such as boats, ships, wakes, and other materials with natural and NIR images. It would be brighter if band 8 was extracted with a reflectance threshold of 0.02 [2]. The study also mentioned that reflectance thresholds have proven good results on dark surfaces, like water.

One of the important passive remote sensing methods is masking. The number of pixels after land mask processing was captured in 17347 pixels. It was usually done between the values for the land and water area in marine remote sensing studies. The land mask was created with band 8 because it has the best contrast in reflectance for discriminating land from water [32]. The infrared wavelength when exposed to water and reflected by the sensor will produce a dark color. The dark color is due to the very small reflectance value produced in waters. The low reflectance value in waters was affected by the loss of energy at near-infrared wavelengths which is completely absorbed by water molecules. Therefore when infrared waves hit objects on land, infrared waves are still left and due have a brighter color on the land. The land mask was used to interpret the value of the land and water area respectively zero and one. The results represented can be seen in Figure 4a.

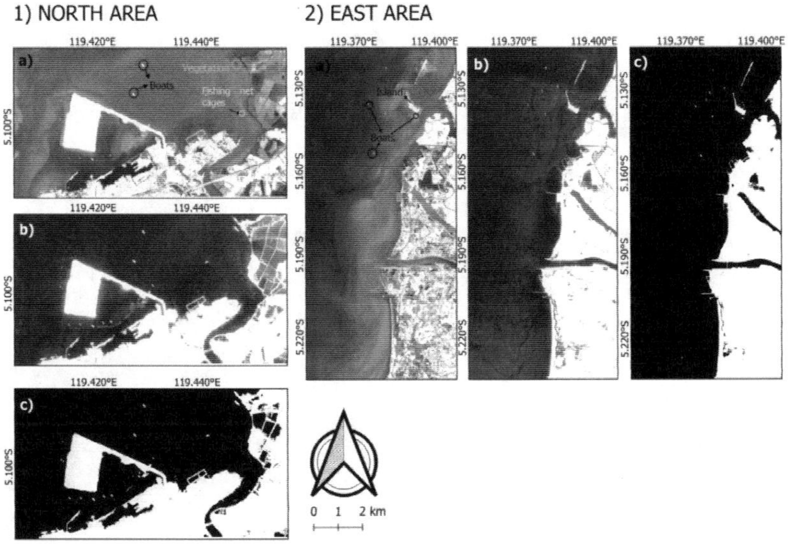

Figure 3. Multi-spectral images after atmospheric correction
a. Natural Image (RGB: 4, 3, 2)
b. Band 8 image
c. Band 8 image with reflectance threshold of 0.02

Figure 4a shows that the land area with no color has been overlaid with the ground-truth image. The coastal area showed a different color which means invalid ocean patches color, which may be contaminated [31]. If a masking image was extracted with a natural image, the presence of floating materials with a faint red color would be presented in the common area (Figure 4b). However, it is not enough to consider there was plastic debris in that coastal surface area because the color domination could be changed depending on different floating materials [22]. As Moshtaghi M et al. (2021) mentioned that one of the important factors to discriminate the color of plastic signals from water in the spectrum was the color of plastic objects [5].

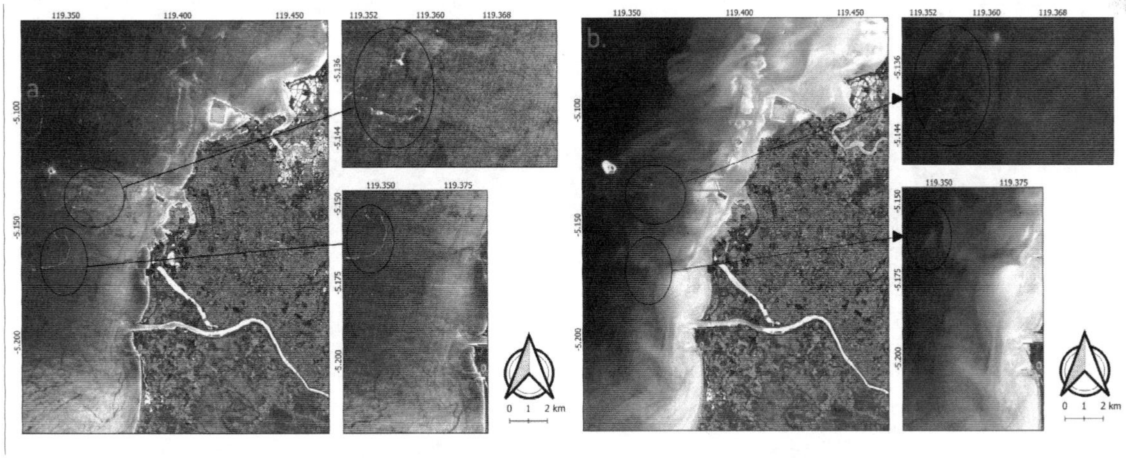

Figure 4. a) Masked out the land area using NIR band and overlaid with the ground-truth image; b) Natural image (RGB 4, 3, 2) with indicated floating materials and transect line (Red line)

Spectral views acquired from horizontal transect line images are presented in Figure 5. The coastal surface spectral in each pixel showed almost null. The results are generally reflected with a flat signal spectrum in natural color [22, 25], which means water has a low reflectance [25]. The highest reflectance in several pixels showed that the transect line can be seen at a 495-pixel number.

According to Mikelsons K et al. (2021), the natural spectrum on the land is typically much higher than in water [32]. On the other increased reflectance bands indicated that transect lines through several floating materials, i. e ships, and unknown materials which indicated plastic.

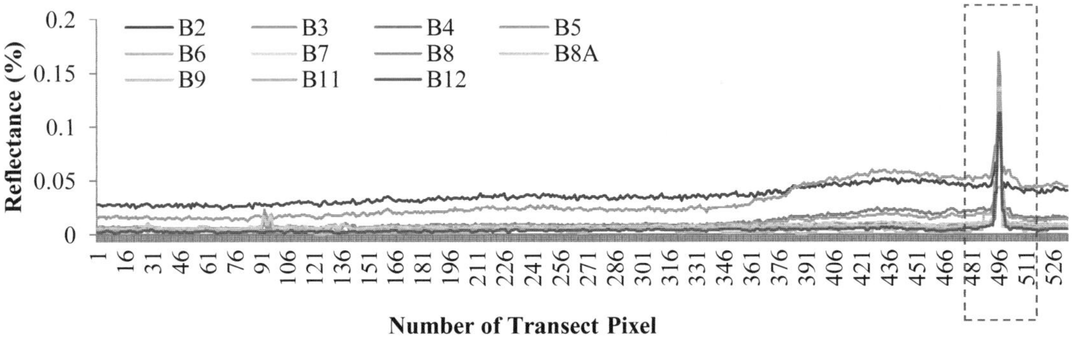

Figure 5. Spectral signature of Sentinel 2-MSI after atmospheric correction of coastal area in Makassar

3.2 Plastic Debris Spectral Indices

Finding plastic debris algorithms from an earlier study [26,30], which are Kernel Normalized Difference Vegetation Index (KNDVI) and the Floating Debris Index (FDI). KNDVI and FDI combined are a developed and powerful approach to detecting plastic debris. The distribution of each index was performed in Figure 6. KNDVI has high reflectance values which are detected with white and black color shifting to the land area as seen in Figure 7. Aquatic vegetation has lower NDVI values, ranging from 0.4 to 0.7; however, it is higher than NDVI values to detect plastic debris [22]. On the other hand, Tasseron P et al. (2021) study showed that the FDI index can also detect plastic types and vegetation [22]. According to a previous study, 86 % accuracy of KNDVI range from 0 to 0.2, and FDI range from 0.02 to 0.06 [5]. Therefore, this study tried to use a reflectance threshold based on a previous study [5]. The overlaid result of KNDVI and FDI can be seen in Figure 8a. Regarding KNDVI and FDI pixel distribution results, we have noted that using these indices could detect potential marine plastic debris.

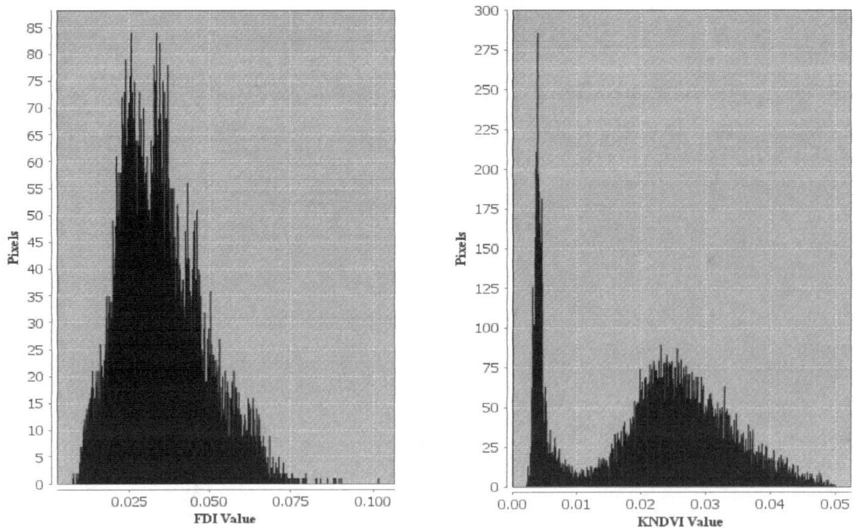

Figure 6. Pixel Distribution Value of FDI and KNDVI

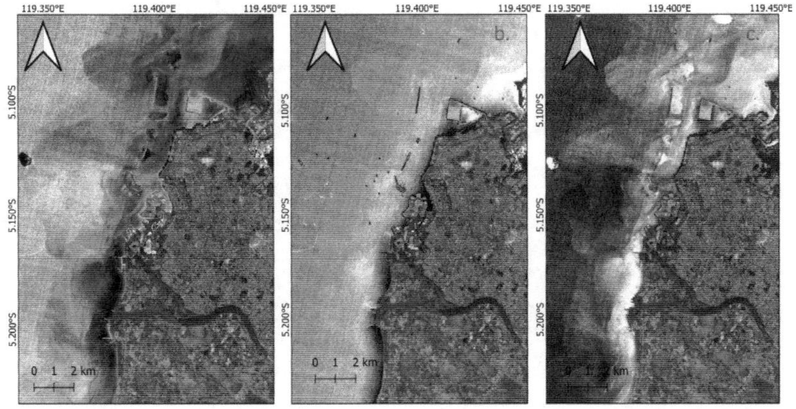

Figure 7. Plastic Debris Spectral Indices: a) NDVI, b) FDI, c) KNDVI

The spectrum illustrated the spectral signatures of floating materials derived with Sentinel-2 MSI (Figure 8b). The spectrum expressed that plastic has a higher reflectance than water. The same result was also mentioned in Moshtaghi M., et al (2021) due to the absorption effect increasing in near-infrared [5]. The intensity of reflectance appeared to be primarily dependent on marine plastic debris within pixels. Furthermore, the spectral signatures of plastic debris and vegetation have a higher peak in the near-infrared on the ocean, although vegetation leads to a significantly higher in the near-infrared. The fact means that Sentinel-2 is capable of detecting marine plastic debris.

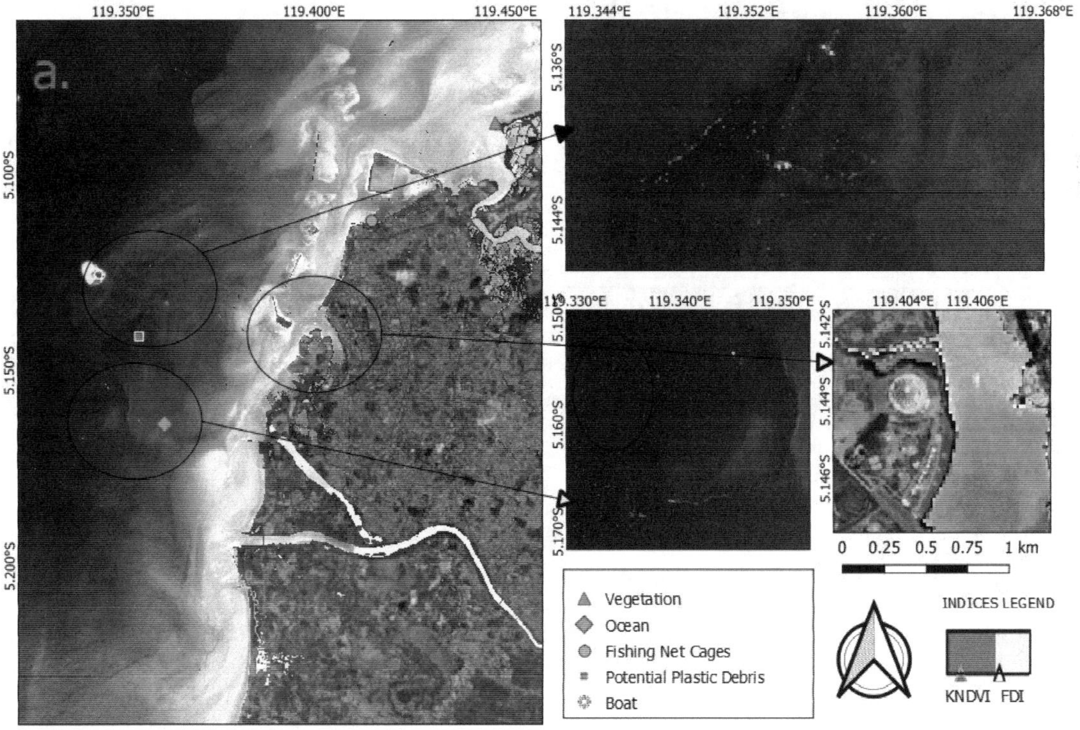

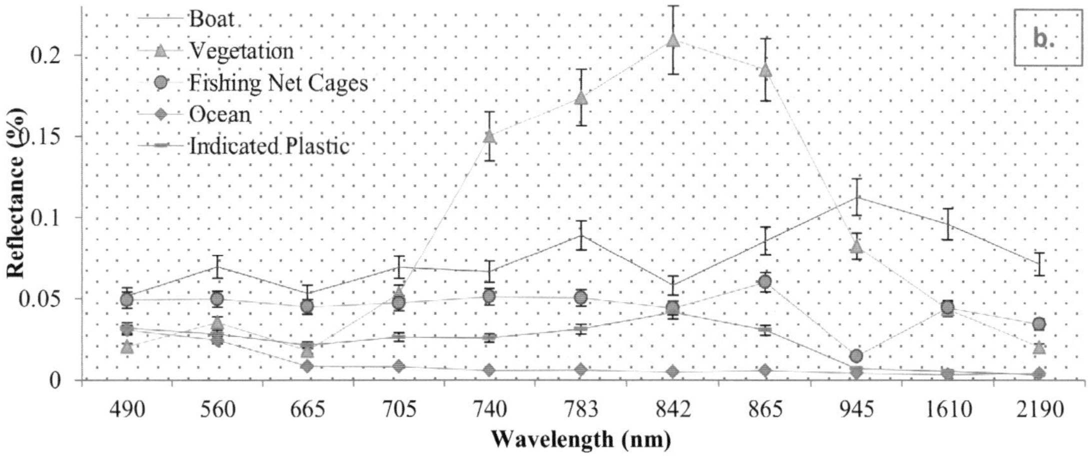

Figure 8. a) Indices Overlay with FDI threshold from 0.02 to 0.06 and KNDVI from 0 to 0.2; b) Spectrum View of Plastic Debris and other Coastal Patches.

3.3 Performance Evaluation Indices with Statistic Data

Performance evaluation for FDI and KNDVI were assessed using statistical data. The statistical data proposed to validate the performance of each index has the best capability to detect plastic debris in coastal waters. The sensitivity analysis value has been performed as the most representative index to reveal potential marine plastic debris as can be seen in Table 3.

Table 3. Statistic Summary of Indices

Indices	Number of considered pixels	Index Min Value (D-min)	Index Max Value (D-max)	Standard error	Number of Plastic Detected-NPD (pixels)	Discriminant value (DV)	SAV
KNDVI	17347	0.0078	0.1015	0.38	3506	0.2	7.5
FDI	17347	0.0022	0.0498	0.47	701	0.2	2.9

The statistical data evaluation reveals that the KNDVI was more representative indices (SE 0.38) than FDI indices (SE 0.47). It was influenced by the number of plastic detected for KNDVI indices (3506 pixels) is more than FDI indices (701 pixels). It also supported by the sensitivity analysis value (SAV) was performed that KNDVI has the best indices (SAV 7.5) than FDI indices (SAV 2.9). Meanwhile, the low accuracy of FDI may be caused by the possibility of submerged plastic in the water. Therefore FDI value is close to zero [22]. This is likely due to increased absorption of water in NIR to SWIR which turns out in the loss of spectral of submerged plastics. A scatterplot of discriminating values between potential plastic and non-plastic was illustrated in Figure 9.

4. Conclusions

Marine plastic debris detection depends on the capability of satellite images and their optical bands. Sentinel-2 MSI Level 2A was found in 17347 pixels in Makassar coastal zone on May 29^{th,} 2022. The target area was calculated by 173.47 Ha with a 10-meter resolution image. Through preliminary and indices identification, plastic has a higher reflectance of water. It can be seen in the near Infrared image-based land masking process which identified the floating patches, including plastic debris.

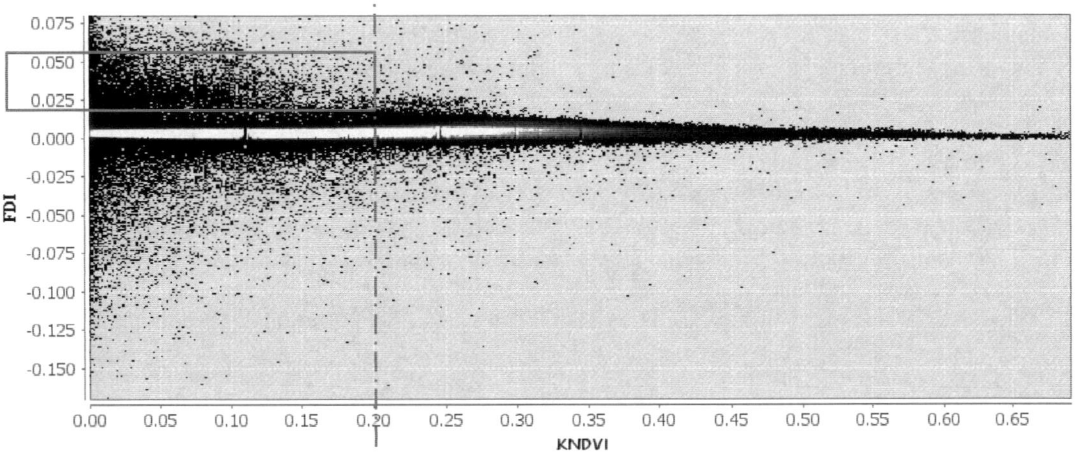

Figure 9. KNDVI indices (0 to 0.2) and FDI indices (0.02 to 0.06)

This study highlighted using Sentinel-2 MSI through two indices, the Kernel Normalized Difference Vegetation Index (KNDVI) and the Floating Debris Index (FDI). The KNDVI and FDI pixels, numerically, calculated were 3506 and 701 pixels (with SE 0.47 and 0.38). The KNDVI indices were more effective to detect potential plastic debris than FDI indices, which has validated with a Sensitivity Analysis Value (7.5). The challenge from this result is the number of plastic detected with Sentinel-2 appeared to depend on how large the plastic patches of varying shapes and sizes. Further research needs a plastic training dataset and mask-only training sets of remote sensing grids. In addition, further analysis needs to discriminate plastic debris based on its composition and other materials using machine learning algorithms to generate the model accuracy.

References

[1] The Government of Republic Indonesia 2017 Indonesia's Plan of Action on Marine Plastic Debris 2017-2025. Accessed on June 18th 2022. https://maritim.go.id/portfolio/indonesias-plan-action-marine-plastic-debris-2017-2025/

[2] Ciappa A C 2020 Marine Litter Detection by Sentinel-2: A Case Study Adriatic (Summer 2020) *Remote Sens* **14** 2409

[3] D A A Sari, Suryanto, A S Sudarwanto, S Nugraha, R Utomowati 2021 Reduce Marine Debris Policy in Indonesia *IOP Conf. Ser.: Earth Environ. Sci.* **724** 012118

[4] Ocean.si.edu. 2018 Marine Plastics. Accessed on June 18th 2022. https://ocean.si.edu/conservation/pollution/marine-plastics

[5] Mostaghi M, Knaeps E, Sterckx S et al 2021 Spectral Reflectance of Marine Macroplastics in the VNIR and SWIR Measured in a Controlled Environment *Sci Rep* **11,** 5436.

[6] Boris W, Heike K L, Isabelle J, Chris W, Jenna J 2017. Plastic as a Persistent Marine Pollutant *Annual Review of Environment and Resources* **42**:1 1-26

[7] Themistocleous K, Papoutsa C, Michaelides S, Hadjimitsis D 2020 Investigating Detection of Floating Plastic Litter from Space Using Sentinel-2 Imagery *Remote Sens* **12** 2648

[8] Guffogg J A, Blades S M, Soto B M, Bellman C J, Skidmore A K, Jones S D 2021 Quantifying Marine Plastic Debris in a Beach Environment using Spectral Analysis *Remote Sens* **13** 4548

[9] Gallo F, Fossi C, Weber R, Santillo D, Sousa J, Ingram I, Nadal A, Romano D 2018 Marine Litter Plastics and Microplastics their Toxic Chemicals Components: The Need for Urgent Preventive Measures *Environ. Sci. Eur.* **30** 13

[10] Borrelle S B, Ringma J, Law K L, Monnahan C C, Lebreton L, McGivern A, Murphy E, Jambeck J, Leonard G H, Hilleary M A 2020 Predicted Growth in Plastic Waste Exceeds

Efforts to Mitigate Plastic Pollution *Science* **369** *1515–1518.*

[11] Jambeck J R, Geyer R, Wilcox C, Siegler T R, Perryman M, Andrady A, Narayan R, Law K L 2015 Marine pollution. Plastic Waste Inputs from Land into The Ocean. *Science.* **13** 347(6223) pp 768-771.

[12] Portals.iucn.org 2020 The Marine Plastic Footprint. Accessed on June 19th, 2022. https://portals.iucn.org

[13] Damayanti A D, Suprihanto N 2020 Metode G-ALDIT dan G-ALDITLcR untuk Evaluasi Kerentanan Air Tanah Dangkal Akibat Pengaruh Intrusi Air Laut (Studi Kasus: Air Tanah Dangkal Kawasan Pesisir Bagian Utara dan Selatan Kota Makassar) *Jurnal Lingkungan dan Bencana Geologi* **12(2)** pp 107-123

[14] Wilcox C, Van S E, Hardesty B D 2015 Threat of Plastic Pollution to Seabirds is Global, Pervasive, and Increasing *Proceedings of The National Academy of Sciences* **112(38)** pp 11899-11904

[15] The Central Statistics Agency Makassar in 2021. Accessed on June 22th 2022. https://makassarkota.bps.go.id

[16] Ministry of Environment and Forestry. (2020). National Waste Management Information System. Accessed on June 24th 2022. https://sipsn.menlhk.go.id/sipsn/

[17] Mathalon A, Hill P. (2014. Microplastic Fibers In The Intertidal Ecosystem Surrounding Halifax Harbor, Nova Scotia. Mar. Pollut. Bull. 81: 69–79

[18] Nicola J. Beaumont, Margrethe Aanesen, Melanie C. Austen, Tobias Börger, James R. Clark, Matthew Cole, Tara Hooper, Penelope K. Lindeque, Christine Pascoe, Kayleigh J. Wyles. (2019). Global ecological, social and economic impacts of marine plastic. *Marine Pollution Bulletin, Volume 142, Pages 189-195.*

[19] Miller, M. E., Hamann, M., & Kroon, F. J. (2020). Bioaccumulation and biomagnification of microplastics in marine organisms: A review and meta-analysis of current data. *PloS one, 15(10), e0240792.*

[20] University of Wollongong. (2020). Marine debris costs Asia-Pacific economies US$10.8B annually: report - University of Wollongong – UOW. Accessed on June 25th 2022.

[21] Welden, Natalie. (2020). The environmental impacts of plastic pollution. *10.1016/B978-0-12-817880-5.00008-6.*

[22] Tasseron P, Van E T, Peller J, Schreyers L, Biermann L 2021 Advancing Floating Macroplastic Detection from Space using Experimental Hyperspectral Imagery *Remote Sens* **13** 2335

[23] Kikaki A, Konstantinos K, Caroline A P, and Dionysios E R 2020. Remotely Sensing the Source and Transport of Marine Plastic Debris in Bay Islands of Honduras (Caribbean Sea). *Remote Sensing* **12 (11)** 1727

[24] Garaba S, Aitken J, Slat B, Dierssem H, Lebreton L, Zielinski O, Reisser J 2018 Sensing Ocean Plastics with An Airborne Hyperspectral Shortwave Infrared Imager. *Environ. Sci. Technol.* **52** pp 11699–11707

[25] Topouzelis K, Apostolos P, Shungudzemwoyo P G 2018 Detection of Floating Plastics From Satellite and Unmanned Aerial Systems (Plastic Litter Project 2018) *International Journal of Applied Earth Observation and Geoinformation* **79** pp 175-183

[26] Sannigrahi S, Basu B, Arunima S B, Francesco P 2021 Detection of marine floating plastic using Sentinel-2 imagery and machine learning models *Marine Pollution Bulletin* **178** 113527

[27] Martínez V V, Clark J R, Corradi P, Aliani S, Arias M, Bochow M, Bonnery G, Cole M, Cózar A, Donnelly R, et al 2019. Measuring Marine Plastic Debris from Space: Initial Assessment of Observation Requirements *Remote Sens.* **11** 2443

[28] Main-Knorn M, Pflug B, Louis J, Debaecker V, Müller-Wilm U, Gascon F 2017. Sen2Cor for Sentinel-2 *SPIE Proceedings* **10427**

[29] Sentinel Copernicus. Overview. Accessed June 29th 2022. https://sentinels.copernicus.eu/web/sentinel/user-guides/sentinel-2-msi/overview

[30] Biermann L, Clewley D, Martinez-Vicente V. et al 2020. Finding Plastic Patches in Coastal Waters using Optical Satellite Data. *Sci Rep* **10** 5364
[31] Ciappa A C 2021 Marine Plastic Litter Detection Offshore Hawai'i by Sentinel-2 *Marine Pollution Bulletin* **168** 112457
[32] Mikelsons K, Menghua W, Xiao-Long W, Lide J 2021 Global Land Mask for Satellite Ocean Color Remote Sensing *Remote Sensing of Environment* **257** 112356

IOP Conf. Series: Earth and Environmental Science 1117 (2022) 012055 doi:10.1088/1755-1315/1117/1/012055

Estimation of Makassar's Landfill Surface Temperature and Its Surroundings Using Remote Sensing

ZA Hanami[1*], AD Damayanti[1], T Takeda[2], H Alimuddin[1]

[1]Department of Environmental Engineering, Universitas Hasanuddin, Makassar, Indonesia
[2]Japan Space Systems, Japan

Email: zaraharwienyhanami@unhas.ac.id*

Abstract. Remote sensing is one of the cost and time-effective methods that can be used to assess change in the land, especially the land surface temperature and land cover. This study aims to estimate Makassar's landfill surface temperature as a vital waste management monitoring parameter and the surrounding area using remote sensing. The Landsat 5 and Landsat 8 data in the years 2006, 2009, 2013, 2015, 2018, and 2020 were used in this study to see the change in the temperature and land cover. The study confirmed a significant increase in landfill surface temperature, with the highest value of 67°C found in 2018 and the minimum is 30°C in the year 2006. It also showed that compared to the surrounding area, the landfill has a higher temperature, followed by the built-up area of land cover that expanded from time to time. The accuracy of supervised classification of the land cover in 2006-2020 ranged between 58%-98%, and the Kappa coefficient ranged from 0.42 to 0.97.

1. Introduction

Makassar is one of the big cities in Indonesia, with a population of approximately 1,5 million people in 2019. In 2021, Makassar generated 1,023.71 tons/day, increasing by around 6% from the previous year. However, this increase in waste volume every day is not facilitated by good waste management, especially the final processing in the landfill. Most of Makassar citizens' waste was disposed of in the Tamangapa landfill located in the eastern part of Makassar, which still uses the open dumping method in the landfill operation [1]. The waste characteristics disposed of in Makassar city were dominated by organic waste sources from food scraps and leaves. Based on data from the Indonesia National Waste Management Information System [2], in 2020, the composition of organic waste from food waste in Makassar was 58.42%, the highest compared to other types of waste, followed by plastic. This high composition of organic waste in landfill has the potential to produce a high rate of greenhouse gas (GHG) emissions and heat [3,4], especially for landfill which poorly managed and has non-functioned facilities, as in the Tamangapa landfill [5,6].

Land surface temperature, typically described as the surface temperature of the Earth, one of the many land surface characteristics, is crucial in regulating the majority of the Earth system's physical, chemical, and biological processes [7]. A landfill site is one of the vital areas in which the surface temperature needs to be monitored to evaluate the landfill's effects on the environment and prevent fires in landfill [4]. However, obtaining Landfill surface temperature data through ground measurements at landfill sites requires a high cost for devices and the expert. Direct monitoring of landfill also has difficulties in determining the measurement point, instrument position, and analysis [8]. Therefore, using another method, such as remote sensing data generated from satellite images, is a feasible and effective alternative to monitoring the landfill surface temperature [4].

Several studies on mapping and monitoring a landfill's surface temperature with the surrounding area have already been reported. For instance, the Al-Qurain in Kuwait reported that the surface of the landfill site had a higher temperature than the nearby desert using Landsat Thematic Mapper (TM) [4]. Also, a study was conducted in two Vietnam landfills (Nam Son and Da Phuoc landfill) using Landsat 8 data [9]. These studies show that landfill sites may be identified and mapped using satellite-based remote sensing technologies that employ time series analysis and suitable image processing methods based on variations between baseline surface temperatures and their surroundings [10]. However, there is no study found to investigate the land surface temperature in Makassar landfill.

Furthermore, the land surface temperature can also be compared with land cover since the research on land cover classification methods based on Landsat images has been an important topic over decades, especially with the current effects of climate change [11]. Regarding all the reasons above, this study aimed to map the landfill surface temperature and the surrounding area, then correlate it to the land cover change.

2. Methodology

2.1. Study location

This study is located in Makassar, Tamangapa Landfill, and the area surrounding with a radius <2.9 km to cover different types of land cover. The landfill is situated at coordinates 5,17520 S; 119,49350 E. This landfill is open dumping and has been in service since 1993 with an area of 14.3 hectares. However, there is an extension of land for numerous zones as a result of the rising amount of waste in Makassar. As a result, the Tamangapa landfill area expanded to 16.8 hectares in 2015 [5,6]. There is no daily cover dirt, and the waste reported has reached a height of 20 meters [6].

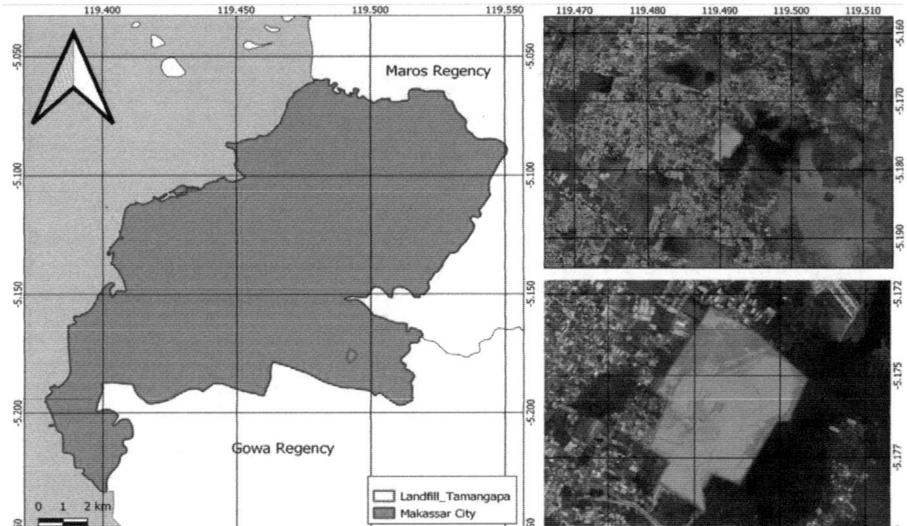

Figure 1. Location and an aerial view of Tamangapa Landfill and its surrounding in this study

2.2. *Data collection and image processing*

The data collected in this study were satellite data and solid waste amount in landfill collected from Makassar Environmental Agency. The Landsat Collection 2 Level 2 satellite images of the study location (GeoTIFF format) were downloaded from the U.S. Geological Survey (USGS) Earth Explorer online archive (https://earthexplorer.usgs.gov). The Landsat satellite data were used to detect the land surface temperature of the landfill and the area surrounding it and for land cover analysis. All relevant data was generated in the dry season (April – September) between 2006 – 2009 (Landsat 5) and 2013 – 2020 (Landsat 8) with less than 10% overall cloud coverage. In addition, the SRTM DEM data was also collected to map the elevation contour. Detailed information about data acquired in this study is provided in Table 1.

Table 1. Data collection information

Data	Data acquired	Spatial Resolution (m)	Cloud cover (%)
Landsat 5 TM C2 L2	July 13th, 2006	30 m	< 10%
Landsat 5 TM C2 L2	July 5th, 2009	30 m	< 10%
Landsat 8 OLI/TIRS C2 L2	April 27th, 2013	30 m	< 10%
Landsat 8 OLI/TIRS C2 L2	Sept 8th, 2015	30 m	< 10%
Landsat 8 OLI/TIRS C2 L2	Sept 16th, 2018	30 m	< 10%
Landsat 8 OLI/TIRS C2 L2	Aug20th, 2020	30 m	< 10%
SRTM 1 Arc-Second Global	Sept 23rd, 2014		

The data was then processed using QGIS 3.22.6. For the land surface temperature processing, the data in Landsat 5 uses band 6, and Landsat 8 uses band 10. The data is already in Kelvin, which scale factor provided in metadata has to be applied by pre-processing step using equation 1 [12] in raster calculation.

$$ST = 0.00341802\ (DN) + 149.0 \tag{1}$$

DN is the digital number, 0.00341892 is the scaling factor for the Surface Temperature Data in the Landsat image, and 149 is the offsite value [12]. After applying the scale factor, the data in the Kelvin unit will be converted to Celcius by minus 273.15.

The methods of land cover classification to be applied to Landsat images were visual analyses starting from 2006 to 2020, followed by supervised pixel-based classification methods using maximum likelihood in the Semi-Automatic Classification Plugin of QGIS. Supervised classification is the method of taking a sample of pixels from the image and utilizing them to set thresholds to distinguish different types of ground cover. Hence, it requires prior knowledge of land cover types before classification [11]. To meet the requirement, this study uses Google Earth Pro historical images to help classify the land cover type and the technical guidelines for interpreting medium-resolution satellite images provided by the Ministry of Environment and Forestry Indonesia [13]. The classification class used in this study were Building and housing, vegetation, paddy field (wet), agriculture (dry), water bodies, landfill, and open area. The assessment for the classification accuracy in the land cover analysis was also performed in this study using the accuracy tool in Semi-Automatic Classification in QGIS. The flowchart of this study methodology showed in Figure 2.

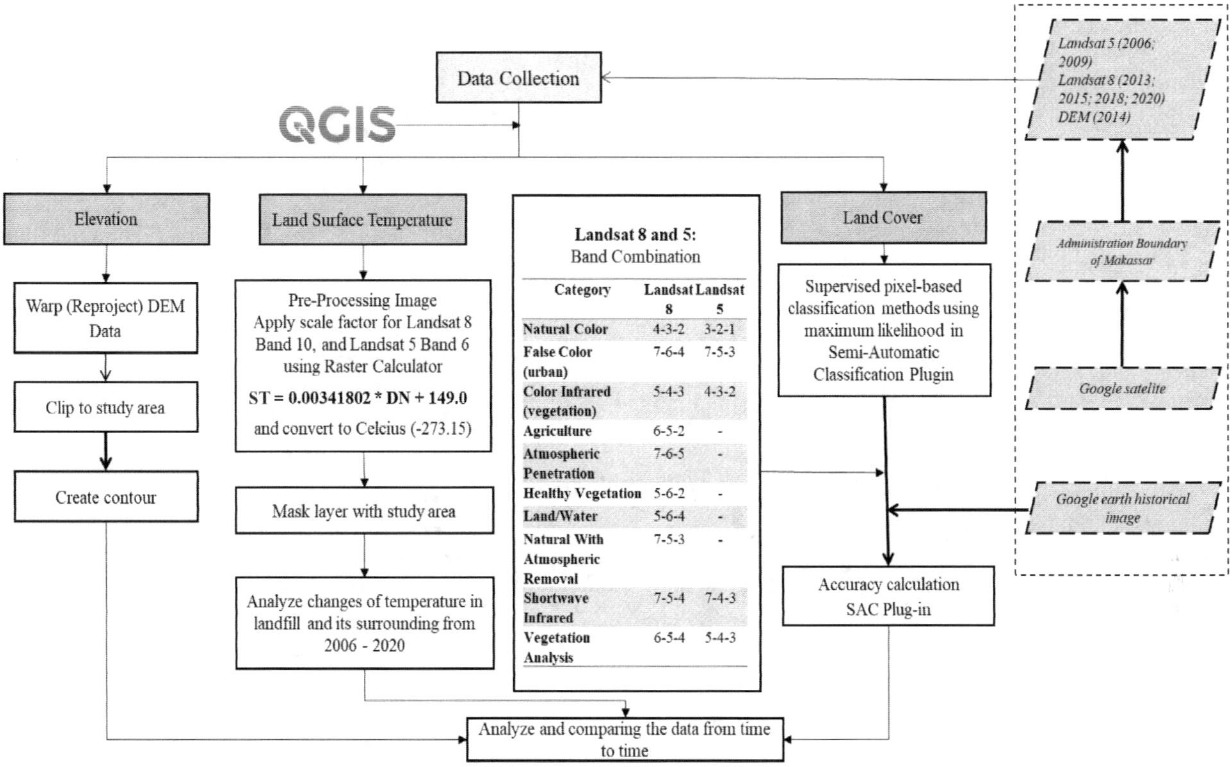

Figure 2. Data collection and image processing methodology flowchart

3. Results and Discussion

3.1. Waste data and landfill change

The waste data in the Tamangapa landfill was acquired directly from the Makassar Environmental Agency in the year used for this study, shown in Figure 3.

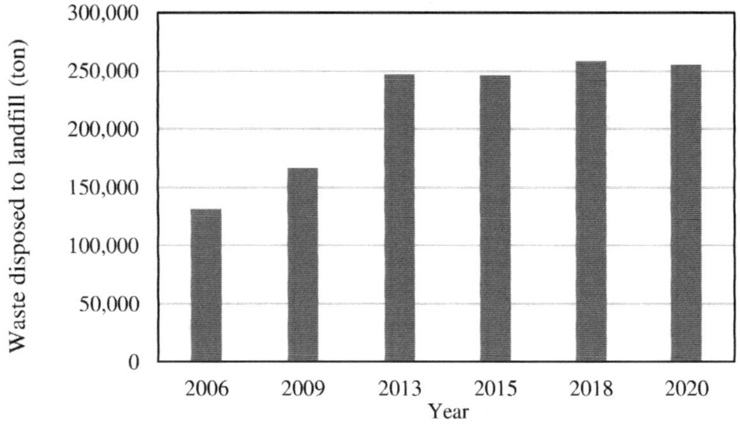

Figure 3. The annual amount of waste generated in landfill from 2006-2020

The waste disposed to the Tamangapa landfill in Makassar has an increasing trend from 2006 to 2020. As can be seen, the highest waste amount generated is in the year 2018. The amount generated in the landfill was increased then decreased but not too significant until 2020. Figure 4 illustrates the visual landscape change of landfill depicted on Google Earth.

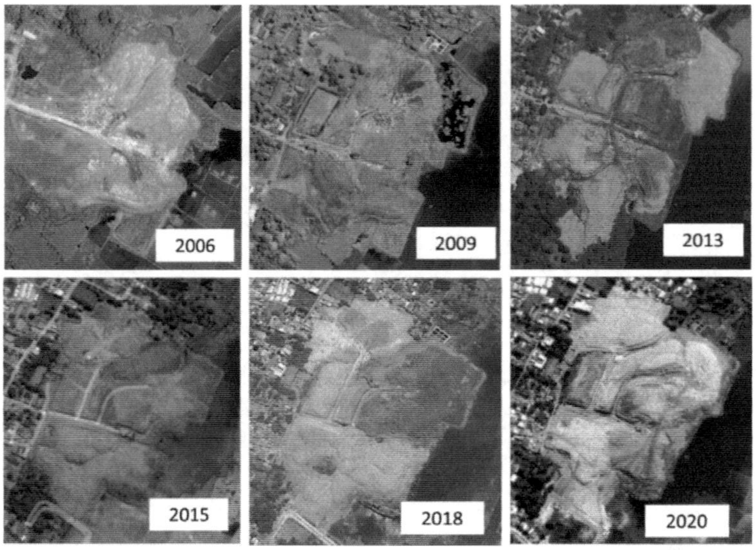

Figure 4. The landscape of Tamangapa Landfill Makassar 2006-2020

In Figure 4, the Tamangapa landfill has changed due to the amount of disposed waste in the landfill as well as the land cover in the landfill site. The area of the landfill becomes larger than its capacity resulting in the land addition for the landfill site.

3.2. Landfill Surface Temperature

The landfill surface temperature is one of the important parameters in landfill monitoring. The estimation of the landfill surface temperature in Tamangapa Landfill, Makassar, was determined using Landsat 5 and 8. The result is provided in Figure 5. Comparing each year's results showed a significant increase in landfill surface temperature from 2006 to 2018. However, the temperature declined from the year 2018 to 2020. The temperature increases with the highest temperature found in 2018 at 67°C, whereas the lowest was found in 2006 at 30°C.

The landfill surface temperature can be influenced by several factors. According to Qdais [4], the amount of solid waste in landfill has resulted in a good correlation with landfill surface temperature due to the waste generation that keeps increasing from time to time. This is to be expected since when more waste is disposed of in landfills, more organic materials will be subjected to exothermic biodegradation reactions, which are often linked to heat and methane production. The heat generated in landfills is the by-product of the biological decomposition of organic waste along with leachate and gas [3,10]. As in Figure 3, which has already been discussed, the amount of waste disposed of in the Tamangapa landfill generally has an increasing trend from year to year. As reported in Santoso [5], in 2014, the composition of organic waste in the Tamangapa landfill was 70.43%.

Furthermore, Based on Lando et al. [6], in 2015, there was an additional 2 hectares of landfill. Therefore, as stated in Qdais et al. [4], the change in the land cover of landfill (Figure 4) is one of the reasons for a temperature difference, as shown in Figures 5c and 5d. In 2015 the temperature became higher than in previous years since the zone that was previously not used to dump the waste was used. Meanwhile, Figure 5e shows the highest temperature of the landfill surface.

IOP Conf. Series: Earth and Environmental Science 1117 (2022) 012055 doi:10.1088/1755-1315/1117/1/012055

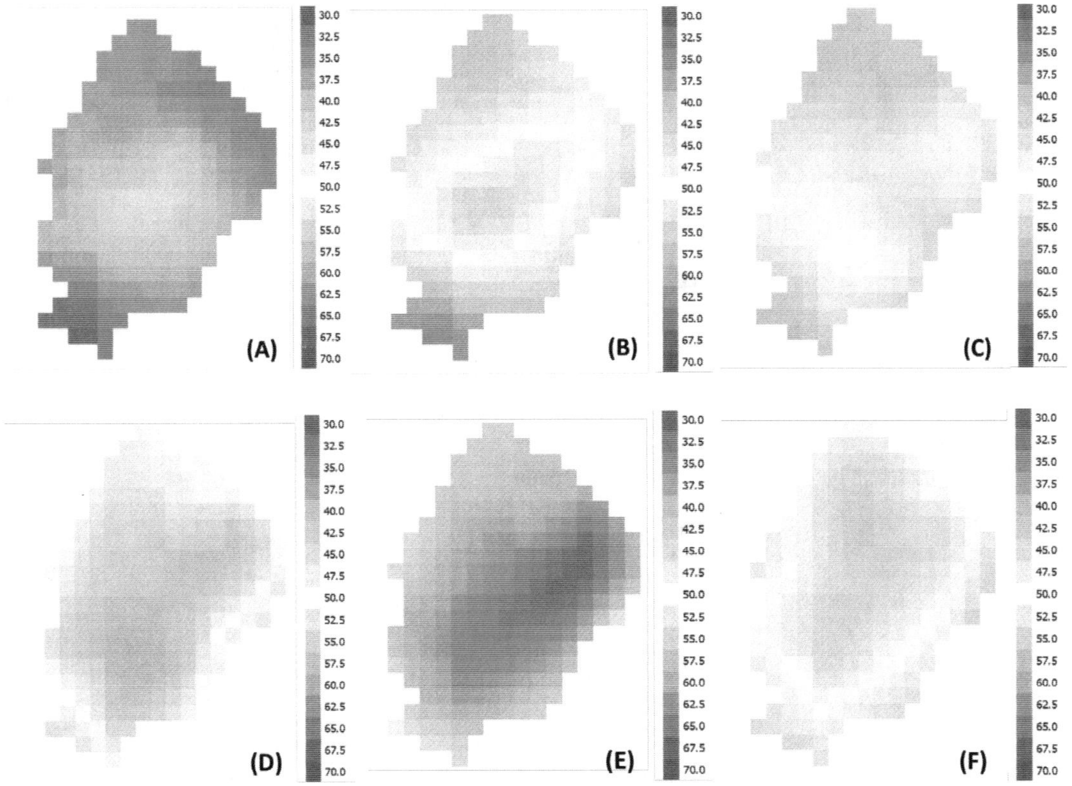

Figure 5. Tamangapa Landfill surface temperature (Celcius) in (a) July 13th 2006 (b) July 5th 2009 (c) April 27th 2013 (d) September 8th 2015 (e) September 16th 2018 (f) August 20th 2020

In 2018, the landfill surface temperature ranged from 51-67°C. It indicated that due to the high amount of waste disposed of and accumulated from the previous years, according to data from Makassar Environmental Agency, in 2017, the waste generated reached 290,222 tons higher than years before. Moreover, surface temperature variations are also influenced by a variety of factors, such as weather conditions, for instance, sun radiation, humidity, and so on [14]. Thus, it is also indicated that the high landfill temperature was probably influenced by the meteorological condition in 2018. The Landsat data used in this study for the year 2018 was also acquired before the landfill fire due to the hot climate in October [15]. However, in this study, meteorological data was not collected. Hence, an in-depth discussion regarding the meteorological aspect and landfill surface temperature needs to be investigated for the next study. It also needs to be highlighted that the estimation using remote sensing data only provides a general estimation for the intensity of the waste decomposition process at a largescale level in the landfill [8]. Direct measurement in the same atmospheric conditions as the landfill itself is also required to validate the result of satellite data [8]. Hence, further research is needed.

3.3. Land Surface Temperature Surrounded Landfill

In order to see the difference in landfill surface temperature to the area surrounding the landfill, the land surface temperature of the area nearby was investigated. The land surface temperature was obtained from Landsat data and layered with the elevation data depicted in Figure 6.

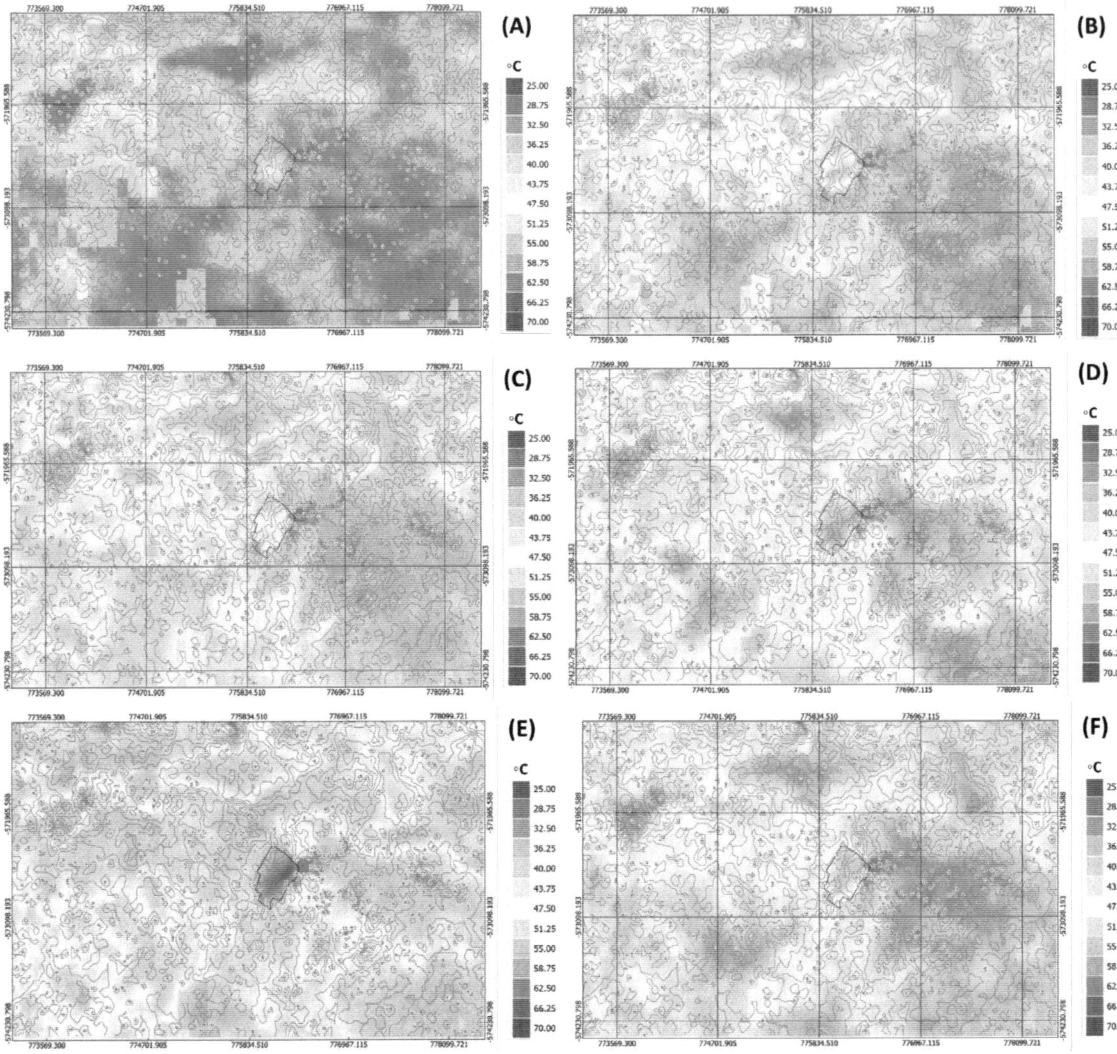

Figure 6. Land surface temperature (Celcius) surrounding landfill in (a) July 13th 2006 (b) July 5th 2009 (c) April 27th 2013 (d) September 8th 2015 (e) September 16th 2018 (f) August 20th 2020

In Figure 6, the surface temperature increased from 2006 to 2018 and then decreased in 2020, similar to the trend of landfill surface temperature data in Figure 5. Based on the result, it shows that the landfill body has a higher temperature than its surroundings ranging from around 1-19°C temperature difference, implying there is an activity in the landfill that can produce heat. The same phenomena also occurred in Nam Son landfill and Da Phuoc landfill in Vietnam, where the landfill surface temperature was so much higher than the surrounding area [9]. It is assumed that it is due to the presence of a biodegradation process inside the landfill body related to gas emissions through the surface of the landfill, as also reported in Qdais et al. [4]. Moreover, as mentioned in Zhang et al. [14], the difference in surface temperature is affected by factors including land cover.

3.4. Land cover change

The result of land cover classification using supervised classification for the area surrounding the landfill from 2006 to 2020 is illustrated in Figure 7.

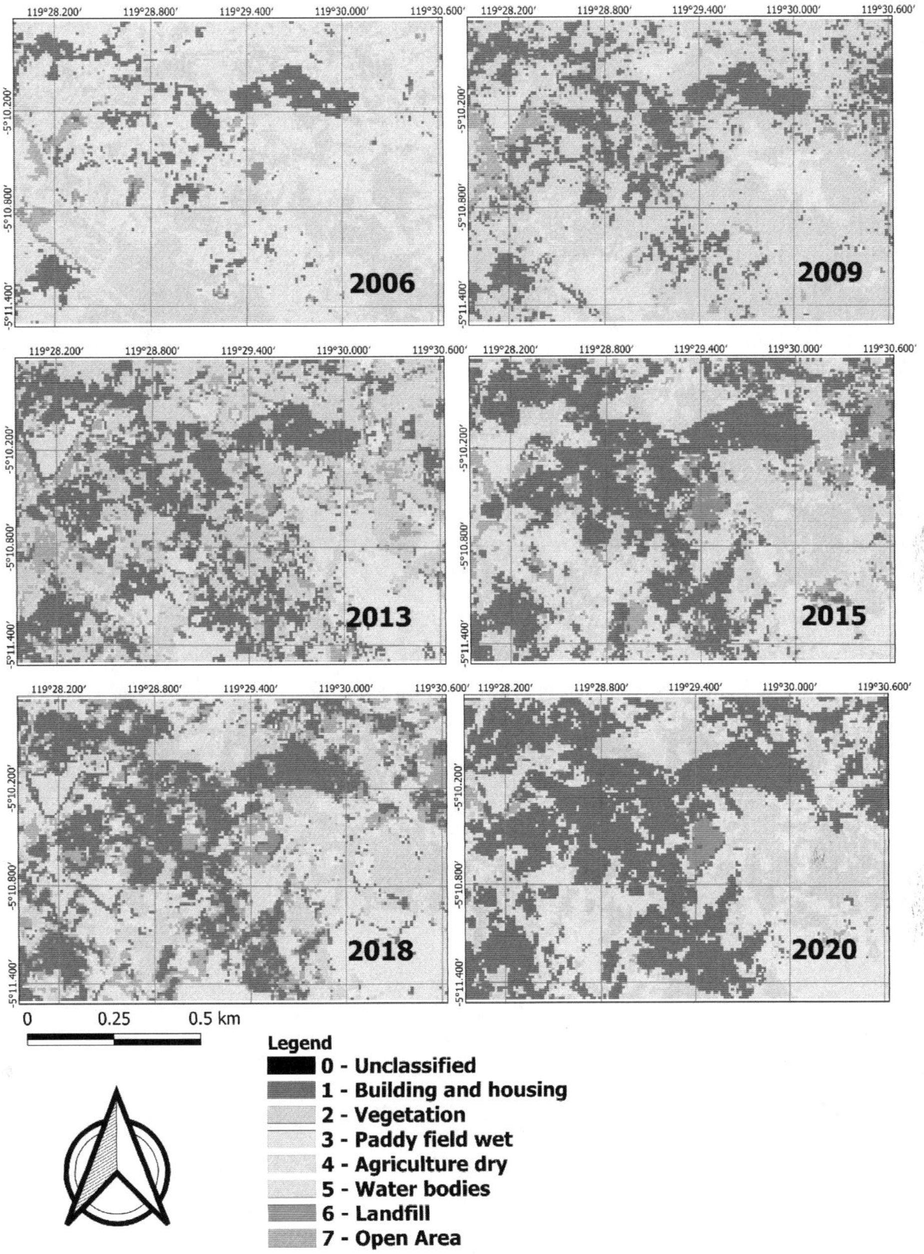

Figure 7. Land cover surrounding landfill in (a) July 13th 2006 (b) July 5th 2009 (c) April 27th 2013 (d) September 8th 2015 (e) September 16th 2018 (f) August 20th 2020

The result in Figure 7 presented that there is a change detected in land cover type for each classification category. The percentage of the land cover area for each category is presented in Table 2. According to Table 2, the majority of the land cover category is the building and housing (built-up) category which tends to expand from 2006 to 2020 due to activity and population increase. Whereas the vegetation area increased until 2009 and then declined. The same result was also found in Saputro et al. [16], who investigated changing land cover areas in Makassar increasing in 1999, 2010, and 2015.

Table 2. The land cover percentage for 2006 - 2020

Category	Percentage (%)					
	2006	**2009**	**2013**	**2015**	**2018**	**2020**
Building and Housing	9.53	23.02	30.02	35.00	31.06	42.93
Vegetation	29.28	60.55	35.42	22.76	17.77	18.23
Paddy field (wet)	3.31	1.38	10.23	8.55	9.84	22.80
Agriculture (dry)	52.99	6.75	9.16	22.84	23.06	12.04
Water bodies	1.64	2.64	4.92	1.12	2.23	1.71
Landfill	1.48	0.63	1.85	2.09	0.56	1.03
Open area	1.76	5.03	8.38	7.63	15.48	1.26

Based on Table 2, the landfill area percentage was having fluctuation and had a different pattern from its visual change in Figure 4. This is because of the limitation of the method used and the need for field data for better results. Compared to the land surface temperature in Figure 6, the higher temperature was in the built-up area and the landfill. As mentioned in do Nascimento [17], the relationship between the land surface temperature and land cover classification is the temperature will be higher in regions with higher densities of people, followed by lower densities area and vegetation cover, then high vegetation densities area and water bodies have lower temperatures which are similar to the result of this study. Furthermore, the classified image in land cover identification needs to be assessed for accuracy to evaluate the accuracy with which the pixels were sampled into the appropriate land cover classes on a quantitative level so that it can be used as input for any applications or further study [18]. The land cover accuracy of this study was calculated directly in the QGIS presented in Table 3.

Table 3. Land cover accuracy for 2006 - 2020

Year	Overall accuracy (%)	Kappa coefficient
2006	98.09	0.970
2009	61.51	0.496
2013	97.3	0.965
2015	90.66	0.877
2018	58.09	0.424
2020	94.08	0.918

The overall accuracy in Table 3 ranged between 58% - 98% and the Kappa coefficient ranged from 0.42 to 0.97. The Kappa coefficient of or near 1 is in perfect agreement, whereas a number near zero indicates that the agreement is poor [18]. Based on the rating criteria of Kappa statistics [18], the land cover classification of 2006, 2013, 2015, and 2020 are in almost perfect agreement. It also has high accuracy with a value above 90%, where overall accuracy of 85% is widely used as a minimum for acceptable results [19]. Hence, the approach of supervised classification is acceptable and well suited for classifying the land covers in Landsat 5 and 8 data. Meanwhile, the 2009 and 2018 data have an overall accuracy percentage lower than 85%, thus, affecting the result of the land cover classification. The low accuracy indicates confusion during the land cover classification process since the method used is a supervised classification that requires prior knowledge and primary field survey data to classify the land cover type. However, the Kappa coefficient is still in the moderate category according to Rwanga and Ndambuki [18].

4. Conclusion and Recommendation

The study confirmed that the land surface temperature and the area surrounding the landfill increased from 2006 until 2018. It declined in 2020 which was influenced by the amount of waste generated and other factors. The minimum landfill surface temperature was found in 2006 at 30°C, and the highest was in 2018 at 67°C. The study also found that there is a correlation between land surface temperature and land cover change, where the highest temperature comes from the landfill and built-up areas. The accuracy of supervised classification of the land cover is ranged between 58% - 98%, and the kappa coefficient is ranged from 0.42 to 0.97. As a recommendation, further validation is needed through primary observation and measurement in the field.

References

[1] Sari WW 2021 Waste Management Services in Makassar City *Undergraduate thesis* Universitas Hasanuddin, Makassar
[2] Ministry of Environment and Forestry of Indonesia 2020 Indonesia National Waste Management Information System. Accessed in https://sipsn.menlhk.go.id/
[3] Grillo RJ 2014 Energy Recycling – Landfill Waste Heat Generation and Recovery *Curr Sustainable Renewable Energy Rep* **1** 150–6
[4] Qdais A, Hani, Shatnawi N 2019 Assessing and Predicting Landfill Surface Temperature Using Remote Sensing and an Artificial Neural Network *International Journal of Remote Sensing* **40** 9556–71
[5] Santoso GD 2017 Study of Tamangapa Landfill Unit Age Assessment in Makassar City *Undergraduate thesis* Universitas Hasanuddin, Makassar
[6] Lando AT, Rahim IR, Sari K, Djamaluddin I, Arifin AN, Sari AM 2020 Estimation of methane emissions from municipal waste landfill in Makassar city based on IPCC waste model *IOP Conference Series* **84** 1:012002
[7] Boori MS, Vozenílek V, Balzter H 2014 Satellite Datasets and there Scaling Factor for Land Surface Temperature *J Geol Geosci* **3** e118.
[8] Lacoboaea C, Petrescu F 2013 Landfill monitoring using remote sensing: a case study of Glina, Romania *Waste Management & Research* **31** 1075-80.
[9] Trinh LH, Zablotskii VR, Vu DT *et al* 2021 Mapping and Assessing Landfills Surface Temperature Using Landsat 8 Satellite Data. A Case Study in Vietnam *Izv. Atmos. Ocean. Phys* **57** 1098–1107
[10] Nazari R, Alfergani H, Haas F, Karimi ME, Fahad MGB, Sabrin S, et al. 2020 Application of Satellite Remote Sensing in Monitoring Elevated Internal Temperatures of Landfills *Applied Sciences* **10** 6801
[11] Phiri D, Morgenroth J 2017 Developments in Landsat Land Cover Classification Methods: A Review *Remote Sensing* **9** 967
[12] U.S. Geological Survey 2022 Landsat 8-9 Collection 2 (C2) *Level 2 Science Product (L2SP) Guide Version 4.0* (South Dakota: USGS)
[13] Ministry of Environment and Forestry of Indonesia 2020 *Technical Guidelines for interpreting medium resolution satellite images for updating national land cover data* (Jakarta: Ministry of Environment and Forestry Indonesia)
[14] Zhang L, Liu C, Li J 2021 An Investigation of influencing factors of land surface temperature based on the iButton and MODIS temperature data *IOP Conf. Ser.: Earth Environ Sci.* **861** 042054
[15] Ikbal B, Kaelan C, Nurdin A 2020 The impact analysis of fire disaster at Tamangapa, Makassar *STRADA Jurnal Ilmiah Kesehatan* **9** 848-60

[16] Saputro EA, Rifani A, Invanni I, Maru R 2017 Study of land surface temperature using remote sensing satellite imagery in Makassar, South Sulawesi *The 9th international graduate students and scholars conference of Indonesia*
[17] do Nascimento, Lima AC, et al 2022 Comparison between Air Temperature and Land Surface Temperature for the City of São Paulo, Brazil *Atmosphere* **13** 491
[18] Rwanga SS, Ndambuki JM 2017. Accuracy assessment of land use/land cover classification using remote sensing and GIS *International Journal of Geoscineces* **8** 611-22
[19] Congalton RG, Green K 2019 *Assessing the Accuracy of Remotely Sensed Data: Principles and Practices. 3rd ed* (Boca Raton: CRC Press)

CBR laboratory investigation on Lightweight Modular Block (LMB) using recycled dredged soil mixed with EPS

N Marfu'ah[1], T Harianto[2], A B Muhiddin[2] and R Irmawaty[2]

[1]Doctoral Student of Civil Engineering, Hasanuddin University, Gowa, South Sulawesi, Indonesia, 92171
[2]Departement of Civil Engineering, Hasanuddin University, Gowa, South Sulawesi, Indonesia, 92171

E-mail: asnm20d@student.unhas.ac.id

Abstract. Changing material into a lightweight geomaterial such as EPSCS is one of the considerable methods to improve the geotechnical performance in the field. This material is already used in numerous construction projects due to its light characteristics. Lightweight Modular Block or LMB is a type of EPSCS composed of dredged Soil, cement, and Expanded Polystyrene (EPS). The amount of cement composition are 3%, 5%,7% and 9%. Meanwhile, Expanded Polystyrene (EPS) are 0.5% and 0.75%. This paper aims to briefly explain the performance of LMB as road foundation materials, such as subbase and base, through CBR laboratory tests. Based on the data, adding EPS can reduce density by 18% - 29% depending on the chosen mixture. Whereas adding more cement tends to increase the CBR value. All the test results will be analyzed and compared to SNI 03- 3438-1994 for its qualification as a road foundation layer.

Keywords: LMB; Dredged Soil; EPS; CBR; Road foundation layer

1. Introduction

Soil is an important material that can not be separated from our life due to its function as a supporting system in all construction projects. However, soil with poor mechanical properties is often found, so efforts are needed to improve the geotechnical performance until specific technical requirements are met. Soil improvement with chemical stabilization by adding cementitious material is the most common method [1]–[4]. The efforts to improve the stability of the soil are carried out depending on the needs. In Japan, changing material into a lightweight geomaterial named Expanded Polystyrene Composite Soil (EPSCS) has been practiced since 1980 [5]. This material has been used in numerous construction projects such as alternative backfill, bridge abutment, and road pavement layer. The advantages of EPSCS as construction materials are: its light characteristic reduces the self-weight of material with higher strength than soil.

This paper uses Lightweight Modular Block or LMB as the name of EPSCS. LMB is an eco-friendly geomaterial composed of EPS, cement, and dredged soil from the Bili-Bili Dam disposal area. During the operational phase of the Bili-Bili Dam, the deposit of dredged soil from sedimentation is very abundant. It continues to increase. Based on echo-sounding observations at the Bili-Bili Dam, the volume of sediment reached 60,959,000 m^3 in 2008 [6] and increased to 110,371,000 m^3 in 2019 [7]. Based on the data, it can be concluded that sedimentation is one problem that must be managed well.

This paper is an overview of the performance of LMB using dredged soil under CBR laboratory test. CBR laboratory test is conducted to identify the capacity of this material as a road pavement layer, such as subbase and base layer.

2. Materials and Method

2.1. Materials

LMB is composed of three main constituents such as dredged-soil (DS), Expanded Polystyrene (EPS) beads, and cement (C) as a binder material. These materials are generally described as follows:

2.1.1. Dredged Soil. The primary material used in this study is dredged soil/dredged sediment obtained through dredging activities at the Bilibili Dam in Gowa Regency, South Sulawesi. This material was dredged and disposed to a specific area at 5 17'05" S and 119 35'06" E coordinates. The sample was taken from the disposal area to the laboratory to be investigated.

a. b.

Figure 1. a) Quarry map (Source: Google Earth), **b.)** dredged soil after drying and sieving.

In Table 1, we can see the summary of the physical and mechanical properties of the dredged soil. Based on the Unified Soil Classification System (USCS), this dredged soil is classified as inorganic silt with low plasticity.

Table 1. Dredged Soil Properties

Properties	Value
Specific Gravity, Gs	2.664
Initial Water Content, (%)	11.26
Consistency Limit	
Plasticity Index, PI (%)	6.50
Liquid Limit, LL (%)	38.00
Plastic Limit, PL (%)	31.50
Shrinkage Limit, SL (%)	27.16
Grain Size Analysis	
Sand (%)	9.20

Silt (%)	83.10
Clay (%)	7.50
USCS Classification	ML
Standard Proctor Test	
MDD (kN/m^3)	14.455
w opt (%)	27.20
CBR (%)	
Unsoaked	15.88
Soaked	6.37

2.1.2. EPS. There are many kinds of EPS beads distributed in the market, but this present study utilized EPS beads with a density ranging from 0.27 kN/m^3 to 0.37 kN/m^3, with an average density of 0.33 kN/m^3. Based on measurements with a vernier caliper, the diameter utilized in this investigation varied between 2 mm and 4 mm.

2.1.3. Cement. In this research, cement is utilized as a binder agent between dredged Soil and EPS, and no particular characteristics were observed. Only PT. Semen Tonasa cement is used to avoid the probability of extra variables from different cement.

2.2. Mix Design

An experimental model with a cylindrical soil sample with a diameter = 15.3 cm and height = 17.5 cm were conducted to determine the capacity of the materials during the CBR laboratory test. The CBR test refers to the ASTM D1883-07 in two environment conditions (unsoaked and soaked). Soaked condition is a simulation test when the material on its worst condition, which is being flooded for four days. Table 2. is attached the mix design of this present study.

Table 2. Mix Design

Composition	Cement/C (%)	EPS (%)	Curing Time (days)	Environment Condition
Dredged Soil (DS)	-	-	-	Unsoaked
Dredged Soil (DS)	-	-	-	Soaked
Dredged Soil (DS)	3,5,7, and 9	0,5 and 0,75	28	Unsoaked
Dredged Soil (DS)	3,5,7, and 9	0,5 and 0,75	28	Soaked

Sample preparation consists of 2 methods. First, the dredged soil specimens were dynamic compacted in three different blows, which are 10 x, 25 x, and 56 x. Second, LMB was compacted with the static compaction method.

3. Result and Discussion

Based on Figure 2. We can see the CBR value of 10, 25, and 56 blows plotted into a CBR design chart. From this chart, we found the trend line equation, which is used to determine the CBR value of the dredged soil on its MDD. From the plotted MDD = 14,455 kN/m^3 obtained from the standard proctor test, the CBR value of dredged soil reached 15.88% and 6.37% in unsoaked and soaked conditions, respectively.

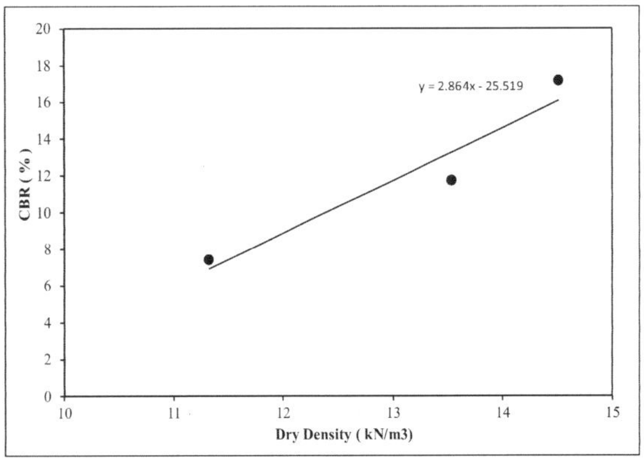

a.

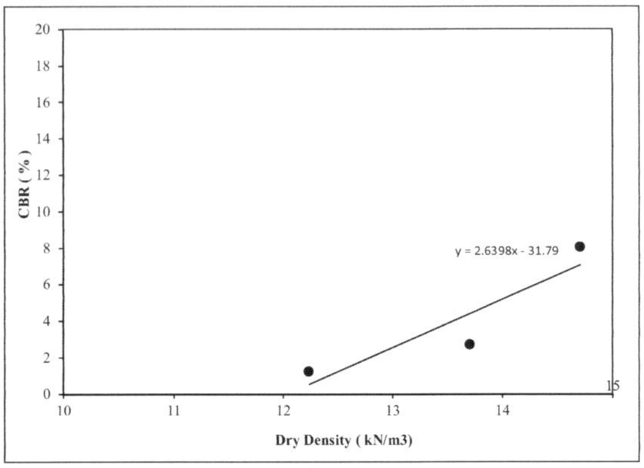

b.

Figure 2. a) CBR unsoaked Dredged Soil b) CBR soaked Dredged Soil

Table 3 shows the data from several CBR tests conducted on four different laterite soils.

Table 3. Comparison of some types of soil

CBR Value	Saing, 2018 [8]			Tangkeallo, 2020 [9]	Present Study
	LS 1 (Sorowako)	LS 2 (Sorowako)	LS 3 (Sorowako)	LS 4 (North Toraja)	DS (Bili Bili Dam)
Unsoaked (%)	11.77	22.99	22.8	18.7	15.88
Soaked (%)	7.67	10.33	8.5	Not Conducted	6.37

The CBR value of dredged soil is between the CBR values of laterite soil. The value of dredged soil meets the requirements for use in subgrade layers with a minimum standard of 6%. This data shows that the dredged soil has good potential if recycled as a construction material.

However, it requires more strength when this material is used on a road pavement layer such as the subbase and base layer. Also, when EPS beads are added, it requires cement as a binding material for the EPS beads and the dredged soil. Based on the Indonesian National Standard requirement or SNI 03-3438-1994 for cement stabilized soil, the minimum CBR value for application as a subbase layer is 20% and 80% for base [10].

In Figure 3 were attached CBR test results for LMB specimens with 0.5% and 0.75% EPS. In the DS + C + EPS mixture under unsoaked conditions, when compared with the value of the CBR test results on untreated dredged soil, it increased by 1.5 x - 4.5 x with the addition of 0.5% EPS and 1.1 x - 3.5 x with the addition of 0.75% EPS. While in the soaked condition, the CBR value decreased from the unsoaked condition. However, compared to the soaked conditions of untreated dredged soil, this value is still much higher, with the bearing ratio increased up to 2.3 x – 9.9 x and 1.9 x – 4.8 x, with EPS 0.5% and EPS 0.75%. The bearing ratio's value depends on the amount of cement used. Increasing the percentage of cement tends to increase the value of the bearing ratio in all types of mixtures. Almost all of the data have the same trend, but some mixtures do not meet the criterion and even have lower performance than dredged soil, such as DS + C5% + EPS 0.75% for soaked conditions, and all of the mixtures with C3%.

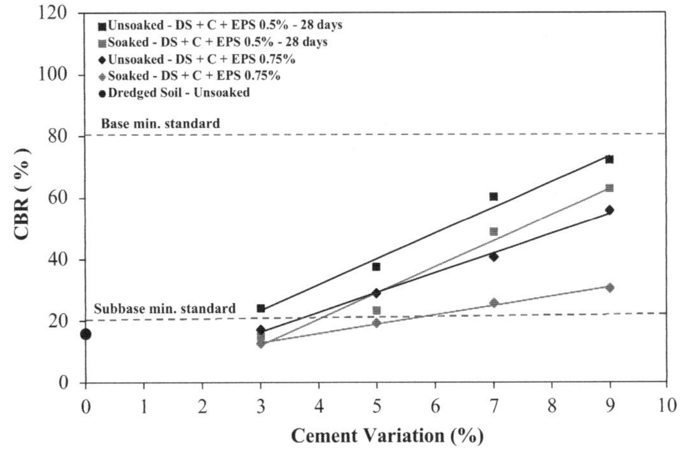

Figure 3. Summary of CBR Value

In addition, based on Figure 4. We can see that adding more EPS decreased the density and also the CBR value. Where for dredged soil is 14.45 kN/m^3. The results show that the higher the ratio of EPS in the mixture, the greater the density reduction occurred. The addition of 0.5% EPS reduced the density of 18% - 19% while the 0.75% EPS decreased it to 28% - 29%.

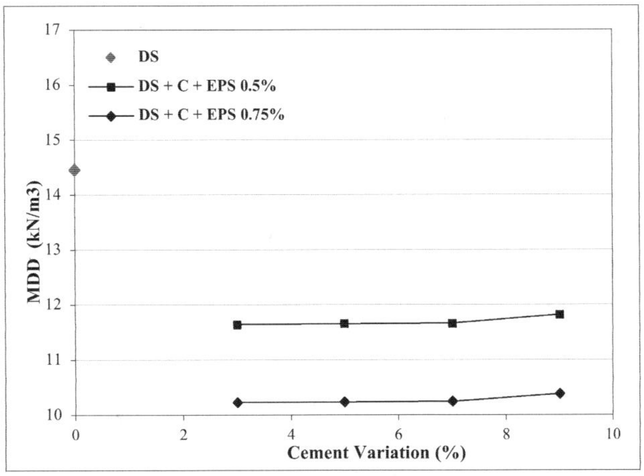

Figure 4. γ dry of the mixture

Conclusions

1. The untreated dredged soil reached the minimum requirement of subgrade layer.

2. Based on the results that have been stated previously, it can be concluded that additional EPS decreases the CBR Value in all conditions but still has a better performance than untreated dredged-soil conditions, except:

- DS + C3% + EPS 0.5%
- DS + C3% + EPS 0.75%
- DS + C5% + EPS 0.75%

Where the value decreased from the untreated dredged soil, this mixture is not recommended for further research related to the application of a lightweight modular block that utilizes the same dredged soil as its main constituents. The addition of 7% and 9% cement increased the performance of LMB with a more stable trend than that with 3% and 5%.

3. Adding EPS reduce density by 18% - 29% depending on the chosen mixture. With a smaller density, most of the mixture shows an increase in the CBR value. All of the data, except the mixture in point 1, meet the CBR minimum requirement for the subbase, but none reached the base's minimum value.

4. References

[1] A. Kumar, B. S. Walia, and A. Bajaj, "Influence of Fly Ash, Lime, and Polyester Fibers on Compaction and Strength Properties of Expansive Soil," *J. Mater. Civ. Eng.*, vol. 19, no. 3, pp. 242–248, 2007, doi: 10.1061/(ASCE)0899-1561(2007)19:3(242).

[2] R. M. Brooks, "Soil Stabilization With Flyash and Rice Husk Ash," *Int. J. Res. Rev. Appl. Sci.*, vol. 1, no. 3, pp. 2076–734, 2009.

[3] T. Harianto, F. Sitepu, and Jasruddin, "Strength improvement of cement stabilized soil by binder mineral additive," *Low. Technol. Int.*, vol. 21, no. 2, pp. 90–97, 2019.

[4] Z. Saing, L. Samang, T. Harianto, and J. Patanduk, "Study on characteristic of laterite soil with lime stabilization as a road foundation," *Int. J. Appl. Eng. Res.*, vol. 12, no. 14, pp. 4687–4693, 2017.

[5] J. Liu, "A New Light-weight Geomaterial – EPS Composite Soil," *Halifax*, pp. 992–996, 2009.

[6] A. R. Asrib, Y. J. Purwanto, and S. Sukandi, "Dampak Longsoran Kaldera terhadap Tingkat Sedimentasi Di Waduk Bili-Bili Provinsi Sulawesi Selatan," *J Hidrolitan*, vol. 2, no. 3, pp. 135–146, 2011.

[7] E. T. Sanjaya, M. Lukman, and B. Tanan, "Analisis Volume Sedimen Berdasarkan Hasil Pengukuran Dengan Echosounder Dalam Waduk Bili-bili Kabupaten Gowa," *Paulus Civ. Eng. J.*, vol. 2, no. 3, pp. 192–196, 2020, doi: 10.52722.

[8] Z. Saing, L. Samang, T. Harianto, and J. Palanduk, "Bearing capacity characteristic of subgrade layer quicklime treated laterite soil," *MATEC Web Conf.*, vol. 181, pp. 1–7, 2018, doi: 10.1051/matecconf/201818111001.

[9] M. M. Tangkeallo, L. Samang, A. Bakri Muhiddin, and A. R. Djamaluddin, "Experimental Study on Bearing Capacity of Laterite Soil Stabilization using Zeolite Activated by Waterglass and Geogrid Reinforcement as Base Layer," *J. Eng. Appl. Sci.*, vol. 15, no. 6, pp. 1496–1501, 2020, doi: 10.36478/jeasci.2020.1496.1501.

[10] SNI03-3438-1994, "Tata cara pembuatan rencana stabilisasi tanah dengan semen portland," *Natl. Stand. Agency Indones.*, pp. 1–15, 1994.

Acknowledgments
The author would like to thank the Directorate General of Higher Education of the Ministry of Education and Culture for providing PMDSU Scholarship.

IOP Conf. Series: Earth and Environmental Science 1117 (2022) 012057 doi:10.1088/1755-1315/1117/1/012057

Shear Strength Behaviour of Sedimentary Soil Stabilized by Fly Ash

S.H Nur[1], Nofriana[1]

1) Department of Civil Engineering, Faculty of Engineering, Hasanuddin University Jl.Poros Malino KM. 06 Bontomarannu, Gowa, 92171, Indonesia

hijraininur@unhas.ac.id, nofriana04@gmail.com

Abstract. Sedimentary soil is made up of deposits that are the consequence of the weathering of rocks that have already been deposited elsewhere by various natural processes, including water, wind, and others. A location with sedimentary soil is the Bili-bili Dam in the South Sulawesi Province's Gowa Regency. Sedimentary soil at this location is quite abundant and has not been used in general, so it can be investigated for its use, namely stability against the influence of external forces on the carrying capacity of the sedimentary soil. Stabilizing the soil is one approach to boost its bearing capacity, particularly for sedimentary soil. The original soil is categorized as group A-4, or silty soil, according to the AASHTO classification system and as ML, or silt with low plasticity, according to the USCS classification system, based on the findings of physical characteristics testing. It has been demonstrated that the soil's cohesiveness and shear angle increase with the addition of fly ash. It has been demonstrated that adding fly ash stabilizing material to the sedimentary soil of the Bili-bili Dam and allowing it to cure increased the cohesion value from the initial soil cohesion value of 0.72 kg/cm2 to 1.70 kg/cm2 after 28 days. With a 20% fly ash mixture and a curing period of 28 days, the shear angle value increased from the initial soil shear angle value of 22 to 31; this is a 40% increase in the shear angle value.

1. Introduction

Sediment is material resulting from the erosion process that settles in waterways, rivers, and reservoirs. The process of depositing broken down material as a result of erosion is known as sedimentation. Sedimentation can have an effect as a result of erosion, such as raising the riverbed so that the water level will likewise rise, perhaps resulting in flooding. In the meantime, heavy sedimentation at the reservoir's bottom will reduce the reservoir's effective volume, which will shorten the reservoir's design life. Because they potentially question the sustainability of the reservoir's function, sedimentation events that take place in the Bili-bili reservoir are becoming more and more concerning. This this issue must be seriously addressed in order to prevent further consequences.

The process of stabilizing soil involves combining it with specific elements to enhance its technical characteristics and make it more suitable for use in building. In general, soil stabilization aims to maintain or maintain the existing soil capacity so that it does not decrease due to environmental influences, both from the outside (external effect) and the inside (internal effect), as well as to increase various types of soil capacity in accordance with the needs of construction engineering.

Fly ash has historically been used in concrete and soil stabilization. Lime and Portland cement are manmade ingredients, whereas fly ash is a by product of burning coal for electricity. The characteristics of fly ash, like those of other by-products, can change greatly depending on the type of coal used and the procedures used throughout the coal burning process[6]. Fly ash, which is spherical, smooth, and pozzolanic in nature, is a by product of burning coal in a steam power plant furnace. Because fly ash contains pozzolanic substances such chemical silica (SiO_2), alumina (Al_2O_3), Ferrum (Fe_2O_3), and Calcium Oxide (CaO) gray, its pozzolanic nature is under doubt. Indonesia produces between 400,000 and 500,000 tons of fly ash annually. Fly ash will keep expanding until it is very

IOP Conf. Series: Earth and Environmental Science 1117 (2022) 012057 doi:10.1088/1755-1315/1117/1/012057

huge, at which point it will become a significant environmental issue that requires immediate attention. These characteristics apply to this substance. pozzolan with features of high silica and alumina content [10] the function of fly ash. Fly ash's ability to alleviate shrinkage and cracking issues that typically occur when using cement as a soil stabilizing agent is due to its pozzolanic qualities. Fly ash can be utilized as a stabilizing substance to improve the soil's bearing capacity[5].

According to AASHTO and ASTM, these by-products can be roughly divided into class C (self-cementing) and class F (non-self cementing) fly ash. CaO, a significant amount of lime, is present in Class C fly ash, but practically all of it is mixed with glassy silicates and aluminates. As a result, when combined with water, a reaction known as hydration occurs that is comparable to the hydration of Portland cement. Similar to Portland cement, this hydration reaction creates free lime, which, depending on how it reacts, can either cause a pozzolanic reaction or a pozzolanic reaction with the silica and/or alumina in the soil. While glassy silica and/or alumina are almost entirely present as pozzolans in Class F ash, it includes very little lime. As a result, additions like Portland cement or lime, which are available sources of free lime, are needed to activate these pozzolans. When fly ash and water are combined, hydration or "cementitious" processes as well as pozzolanic reactions take place, creating the compounds that bind soil grains or agglomerates together to provide strength within the soil matrix. As was previously said, maintaining a high system pH is necessary for fly ash-soil mixes to build strength over time[7.8].

2. Experimental Process

2.1 Location

The process of making specimens, testing the physical and mechanical properties of the soil was carried out at the Soil Mechanics Laboratory, Hasanuddin University in Gowa Regency, South Sulawesi.

2.2 Data Collecting

Data on the soil used in the study for the creation of test objects was gathered by first choosing the soil, then visually identifying it and determining whether it was present in nature, and finally by determining the soil's characteristics to determine whether it was suitable for use with the stabilizing material to be used with the composition. refers to the goals of the research that will be conducted.

2.3 Material

2.3.1. Soil

The soil used in this study, as shown in Figure 1, is a sedimentary soil located in the Bili-Bili Dam, Gowa Regency, South Sulawesi. The color of the soil is gray and fine grained.

Figure 1. Bili-bili dam Sedimentary Soil

2.3.2. Fly Ash

The fly ash used in this research is fly ash from PLTU Barru, the granules that pass the No. sieve. 200 or smaller 0.075 mm.

Figure 2. Fly ash

2.4. Testing Standard

The values of the basic characteristics of the materials used in this study were tested based on the reference to the standards issued by ASTM, as shown in table 1, as follows:

Table 1. Testing Standards

Test Type	Testing Standard
Water Content	D2216 - 98
Speific Gravity	D854 - 14
Atteerg Limit	D4318 -05, D4943 - 08
Sieve Analysis and Hydrometer	D422 - 63
Compaction	D698 - 07
Direct Shear	D3080

2.5. Sample Tes

Sample testing is divided into 2 types of tests, namely testing the physical properties and mechanical properties of the soil with fly Ash material.

- Physical Properties Test

  The physical properties test on the soil aims to determine the index properties needed in the classification and determination of the stabilizing material and the composition of the stabilizing material and its effect on the soil used. In this study, physical properties tests were carried out on the original soil, namely testing water content, specific gravity, sieve and hydrometer analysis, and Atterberg limits.
- Mechanical Properties Test

  Testing the mechanical properties of the soil aims to determine the behavior of the soil or the resistance of a soil that has been given a stabilizing material to the pressure or load given, the soil mechanical property test is a parameter that can be seen to determine the bearing capacity of the soil. The mechanical properties test carried out in this study was a direct shear and compaction test.

3. Results and Discussion

3.1 Characteristics and Physical Properties of Sedimentary Soil

- Specific gravity

  In the specific gravity test, the original soil density value was 2.6.
- Organic Level

  In testing the organic content of the original soil, the organic content value was 1.8%.
- Atterberg's Limit

  In the Atterberg limits test, the values obtained are as follows:
  1. Plastic Limit

     The plastic limit is the lowest water content at which the soil begins to become plastic. The plastic limit value obtained is 31.99%.
  2. Liquid Limit

     The liquid limit is a certain water content at which the behavior of the soil changes from a plastic state to a liquid state. The liquid limit value obtained is 38.53%. The liquid limit graph can be shown in Figure 3, below:

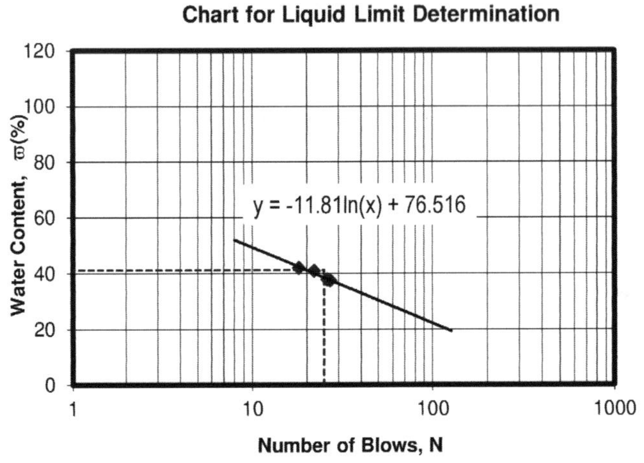

Figure 3. Graph of Liquid Limit Test Results

3. Shrinkage Limit
   The shrinkage limit is the water content limit where the soil with water content below this value does not shrink anymore (no change in volume). The value of the shrinkage limit obtained is 30.11%
4. Plasticity Index
   The value of the plasticity index is obtained from the difference between the search limit and the plastic limit, so that the value of the soil plasticity index is 6.54%.

- Sieve Analysis and Hydrometer
  In the sieve analysis test, it was found that the value of more than 50% of the soil passed the No. sieve. 200, which is 88.20%. The graph of the test results is shown in Figure 4.

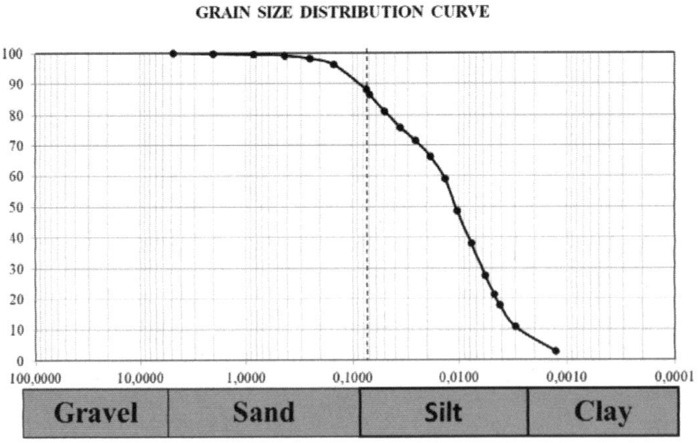

Figure 4. Soil Gradation

3.2 *Natural Soil Classification*

From the results of testing the physical properties of the original soil used in the study, the following are the results of classifying the sedimentary soils of the Bili-bili Dam, Gowa Regency:

- American Association of State Highway and Transportation Officials (AASHTO)
  From the test results of sieve analysis and hydrometer and Atterberg limits, the percentage of native soil grains that passed the No. 200 sieve was 88.20%. The liquid limit value is 38.53%,

IOP Conf. Series: Earth and Environmental Science 1117 (2022) 012057 doi:10.1088/1755-1315/1117/1/012057

and the plasticity index is 6.54%. From the values obtained, the soil type can be determined based on the AASHTO classification system. The determination of the type of soil in question can be seen in Table 2 below:

Table 2. Soil classification for road subgrade (AASHTO Classification System)

Classification Soil		Filter No. 200	Liquid Limit*	Index Plasticity	Type Material	Evaluation As soil base
A-4		Min. 36	Max. 40	Max. 10	Soil Silted	Normal until Bad
A-5		Min. 36	Min. 41	Max. 10	Soil Silted	Normal until Bad
A-6		Min. 36	Max. 40	Min. 11	Soil Clay	Normal until Bad
A-7	A-7-5	Min. 36	Min. 41	Min. 11 And PL≤ LL-30	Soil Clay	Normal until Bad
	A-7-6	Min. 36	Min. 41	Min. 11 And PL>LL-30	Soil Clay	Normal until Bad

Based on the table above, with reference to the data obtained, the original land is included in group A-4, namely the soil is classified as silty soil.

- Unified Soil Classification System (USCS)

  By connecting the plastic limit value and the plasticity index value, the soil type tested is classified as ML (Milt with low plasticity) which means silt soil with low plasticity. . The relationship that shows these values can be seen in Figure 5.

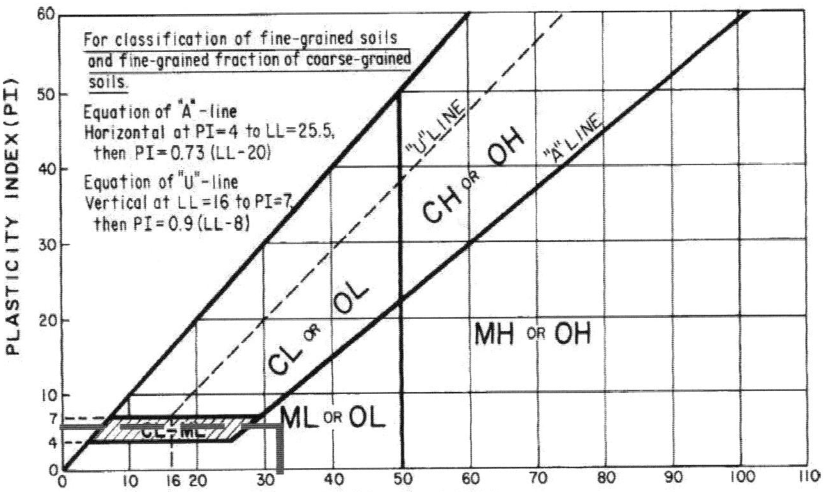

Figure 5. Classification of Unified Soil Classification according to the USCS System

3.3 Characteristics of Mechanical Properties of Soil

a. Compaction Test

   The compaction test is a test to obtain the maximum soil density value with standard energy so that the maximum dry density and optimum moisture content of the research sample can be known.

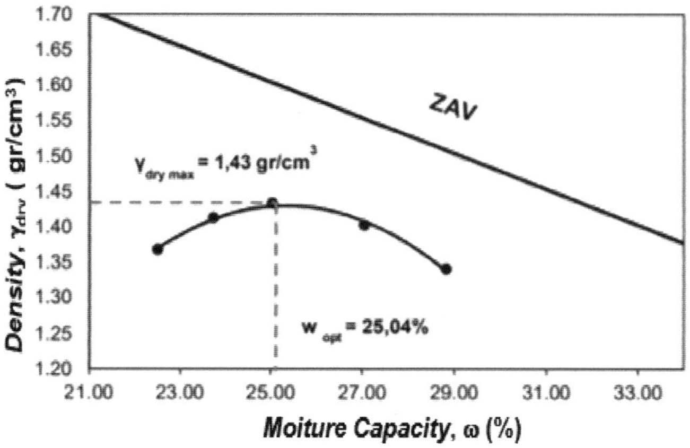

Figure 6. Compaction Results of Natural Soil

Based on the test results obtained the maximum dry weight value (γdry max) = 1.43 gram/cm^3 which is achieved at the optimum moisture content (ωopt) = 25.04%.

b. Direct Shear

Direct shear test is carried out to determine the value of cohesion and shear angle in the soil as a parameter of soil bearing capacity. The test was carried out with 3 loading trials, namely 10, 20, and 40 kg. From the test results, the cohesion value in the original soil is 0.72 kg/cm^2 and the shear angle in the soil is 22°. The following is a graph of the relationship between shear stress and deformation so that the value of the internal shear angle can be seen in Figure 7 below:

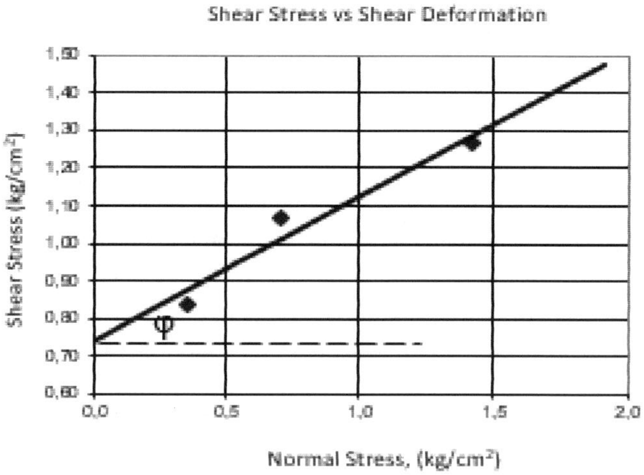

Figure 7. Normal stress and shear stress

3.4 Characteristics of Direct Shear Test Results on the Addition of Fly Ash to Sedimentary Soil

Mixing was carried out on the original soil, namely the sedimentary soil of the Bili-bili Dam, Gowa Regency withfly ash to determine its effect on changes in the value of the bearing capacity of the soil, in this case the changes in the cohesion value and the shear angle of the soil through a direct shear test experiment.

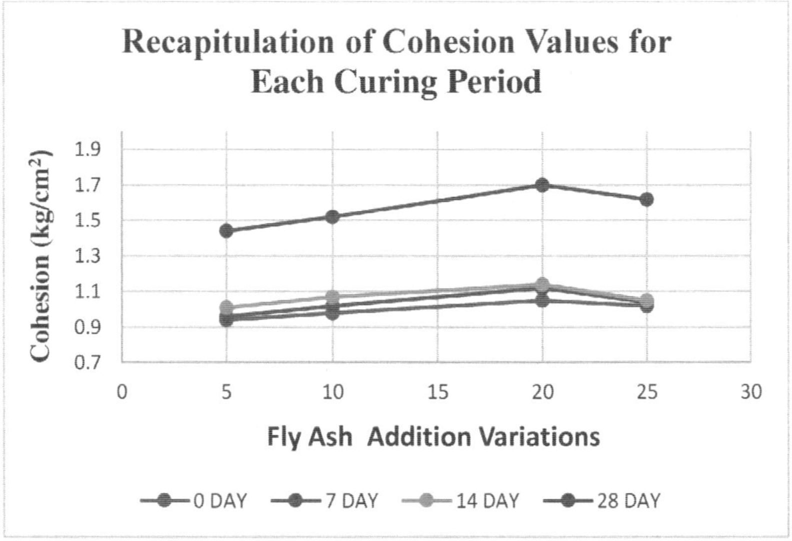

Figure 8. Cohesion values vary with the amount of fly ash used at each curing stage.

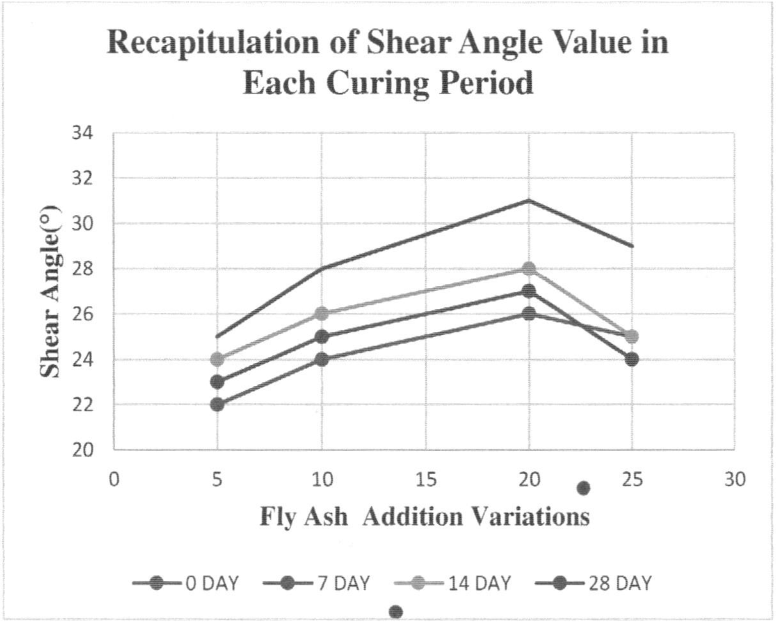

Figure 9. Variations in the addition of fly ash each time of curing cause changes in the shear angle value in the soil

According to the test results, the cohesion value of the mixed sedimentary soil with 20% fly ash had a highest increase of 1.70 kg/cm^2 with a curing period of 28 days. At 0 days of curing, the cohesion value of the soil mixture with 20% fly ash was 1.18 kg/cm^2, and without the addition of fly ash (original soil cohesion) it was 0.84 kg/cm^2. The same is true for the value of the shear angle, which tends to rise as a result of the addition of fly ash during each curing session. When compared to the internal shear angle value of 25° during the 0 day curing period and the preceding shear angle of the original soil of 22°, the highest shear angle value obtained in each variation of the 20% fly ash mixture is 31°.

4. Conclusion

- The sedimentary soil of the Bili-bili Dam is categorized as soil group A-4, or silty soil, according to the AASHTO classification system, and as ML, or silt with low plasticity, according to the USCS classification system, based on the findings of physical characteristics testing
- According to the test results, soil cohesiveness and shear angle increased with curing times of 0, 7, 14, and 28 days. In this instance, the addition of 20% Fly Ash increased the value's maximum cure period to 28 days. After 28 days of curing, the cohesion value increased from the initial soil cohesion value of 0.72 kg/cm^2 to 1.70 kg/cm^2. Compared to the natural soil shear angle value of 22°, which was increased to 31°, there was a 40% rise in the shear angle value.

5. Reference

[1] AASHTO. Standard Specifications for Transport Materials and Methods of Sampling and Testing. 14th Edition, American
[2] Arora, S. and Aydilek, A.H, “Class F Fly-Ash-Amended Soils as Highway Base Material." J. Materials in Civil Eng., 17(6), 2005, 640-649.
[3] Asrib, AR, Erizal, Purwanto, Y., & Sukandi. (2011). Impact of Caldera Avalanche on Sedimentation Levels in Bili-Bili Reservoir, South Sulawesi Province. J. Hydraulics. Vol. 2 : 3 : 135-146, 2011, 135-146
[4] ASTM. Annual Book of ASTM Standards. Section 4, Vol. 4.02, 4.08 and 4.09, West Conshohocken, 2003
[5] Das, BM (1995). Soil Mechanics (Principles of Geotechnical Engineering) Volume I. Jakarta : Erlangga.
[6] FADIAH, Ir. Darmanto, Dipl. HE., M.Sc. (2006). Study of the Effect of Land Erosion on Sedimentation in Bili-bili Reservoir, Gowa Regency, South Sulawesi Province
[7] Hardiyatmo, HC 1992. Soil Mechanics I. Jakarta : PT. Main Library Gramedia.
[8] Magdi M. E. Zumrawi 2015. Scientific Journals: Stabilisation of Pavement Subgrade by Using Fly Ash Activated by Cement, university of Khartoum, Khartoum, Sudan
[9] Scientific journals Denny Boy Pinasang, O.B.A. Sompie, Freddy Jansen, , “Analysis a mixture of Lime-Fly Ash and Lime Gray Rice Husk Against Expansive Clay” Journal of Media Engineering , vol. 6, no. 3,Seoptember.2016
[10] Wahyuni Nanda Dwi et al, 2021.Performance of Fly Ash on Soft Soil Stabilization as Subgrade Improvement Material.

IOP Conf. Series: Earth and Environmental Science 1117 (2022) 012058 doi:10.1088/1755-1315/1117/1/012058

Evaluation of Unconfined Compression Strength Value of Expansive Soil Stabilized Fly Ash

S.H Nur, [1], R Djamaluddin[1], S Nurafifa. Y[1]

1) Department of Civil Engineering, Faculty of Engineering, Hasanuddin University Jl. Poros Malino KM. 06 Bontomarannu, Gowa, 92171, Indonesia

hijraininur@unhas.ac.id nurafifasiti2@gmail.com

Abstract. Expansive soils are referred to as swelling soils as a result of changes or modifications in the water content in the soil. This study aims to determine how the addition of fly ash affects the value of the expansive soil's free compressive strength as well as how the curing period affects the value of the expansive soil's free compressive strength that has been stabilized with fly ash. Variations of the addition of fly ash were made by 5%, 10%, 20%, and 25% with curing times of 0 days, 7 days, 14 days, and 28 days to see how adding fly ash affected the Free Compressive Strength Value. The value of the Free Compressive Strength increases as fly ash addition is increased in percentage and curing time is extended. The variation of the addition of 10% fly ash with a curing period of 28 days of 31,622 kg/cm2 and 25% with a curing period of 0 days, 7 days, 14 days, and 28 days produced the best compressive strength value. The value of the Free Compressive Strength increases as fly ash addition is increased in percentage and curing time is extended. The variation of the addition of 10% fly ash with a curing period of 28 days of 31,622 kg/cm2 and 25% with a curing period of 0 days, 7 days, 14 days, and 28 days produced the best compressive strength value. The value of the Free Compressive Strength increases as fly ash addition is increased in percentage and curing time is extended. The variant of 10% fly ash addition with a 28-day curing period produced the maximum compressive strength value of 31,622 kg/cm2.

1. Introduction

Soil plays a crucial role and can frequently be problematic if it has bad qualities. Poor and undesirable soil characteristics, such as excessive plasticity, poor shear strength, substantial compression or volume fluctuations, and a high propensity for shrinkage, when utilized as the foundation of a building or constructing [1]. When the water content rises above its reference level, clay expands. The swelling, shrinking, and cohesiveness qualities are influenced by the water content. When the water content falls below the shrinkage limit from its reference value, shrinkage occurs. If a clay soil's plasticity index is greater than 20, it is typical to anticipate a significant volume change (expansive). Expansive soil, sometimes referred to as swelling soils, is a significant swelling and contracting phenomenon that occurs in that area as a result of variations in the water content [1][5].

Water has a significant impact on the physical and mechanical behavior of soil with a high shrinkage value [7]. Buildings frequently sustain damage from this expanding soil, including broken walls, higher foundations, uneven roads, etc. [3][6]. The soil stabilization method is one of the methods to produce soil qualities that meet specific technical requirements. [2]. To improve the technical qualities of the soil, soil must be mixed with specific materials. Alternatively, soil stabilization can refer to an effort to alter or improve the technical properties of the soil in order to comply with specified technical standards. [4].

There are two basic categories for classifying soil stabilization techniques: those based on their technical characteristics and those based on their intended application, where different modifications may be employed. Stabilization can be categorized into three categories based on its technical characteristics:

mechanical stabilization, physical stabilization, and chemical stabilization [2]. The cheapest and most efficient way to stabilize expansive soil is to add certain chemicals, which can bind clay minerals to solidify and reduce expansive clay's swelling and shrinkage [2][8]. Fly ash was employed in this study to stabilize the soil. The physical and mechanical qualities of the soil can be enhanced by the addition of fly ash [10]. Fly Ash, a solid byproduct of combustion at the PLTU, is removed by the combustion gas stream and collected using an electrostatic precipitator. Poszolanic qualities can be found in fly ash [4][9].

2. Research methods

2.1. Research sites

The Laboratory of Soil Mechanics, Faculty of Engineering, Hasanuddin University, in Gowa Regency, South Sulawesi, manufactured the specimens and examined the soil's physical and mechanical characteristics.

2.2. Method of collecting data

Data on the soil used in the study for the creation of test objects was gathered by first choosing the soil, then visually identifying it and determining whether it was present in nature, and finally by determining the soil's characteristics to determine whether it was suitable for use with the stabilizing material to be used with the composition. refers to the goals of the research that will be conducted..

2.3. Material

2.3.1 Natural Soil

The soil employed in this study is a sedimentary soil from the Bili-Bili Dam in the Gowa Regency of South Sulawesi, as illustrated in Figure 1. The dirt is fine-grained and gray in tone.

Figure 1. Expansive Soil

2.3.2 Fly Ash

Fly ash from PLTU Barru that is less than 0.075 mm and passes No. 200 sieve is used in this study.

Figure 2. Fly Ash

2.4 Testing Standard

Based on the standards published by ASTM, as indicated in Table 1, the values of the fundamental properties of the materials utilized in this study were evaluated as follows:

Table 1. Testing Standards

Test Type	Testing Standard
Water Content	D2216-98
Specific Gravity	D854-14
Atterberg Limit	D4318-05, D4943-08
Sieve Analysis and Hydrometer	D422-63
Kompaction	D698-07
Direct Shear	D3080

2.5 Sample Test

The two methods of sample testing are examining the soil's mechanical and physical qualities using fly ash as a test material.

2.5.1 Physical Properties Test

The purpose of testing soil physical properties is to identify the index of soil characteristics required for classification and soil characteristics that are used as a guide to identify the proper stabilizing agent and the estimated amount of stabilizing material that can be added to the soil to be stabilized.

2.5.2 Mechanical Properties Test

This test is performed to ascertain the soil's strength and bearing capacity. The compaction test (Standard Proctor Compaction) and the unconfined compression strength were performed as mechanical tests.

3. **Results and Discussion**

3.1. Characteristics of Expansive Soil Physical Properties

3.1.1 Specific gravity

In the specific gravity test, the original soil density value was found to be 2.7

3.1.2 Organic Level

In testing the organic content of the original soil, the organic content value was 1.8%. So the soil is included in the category of inorganic soil.

3.1.3 Atterberg's Limit

The Atterberg limits test is carried out to determine the nature of the soil to changes in the given water content.

3.1.4 Plastic Limit

The plastic limit obtained is 26.77%.

3.1.5 Liquid Limit

The liquid limit is a certain water content where there is a change in soil behavior that changes from a plastic state to a liquid state. The liquid limit value obtained from the test results is 82.75%. The graph of the results of the liquid limit test can be seen in Figure 3 below

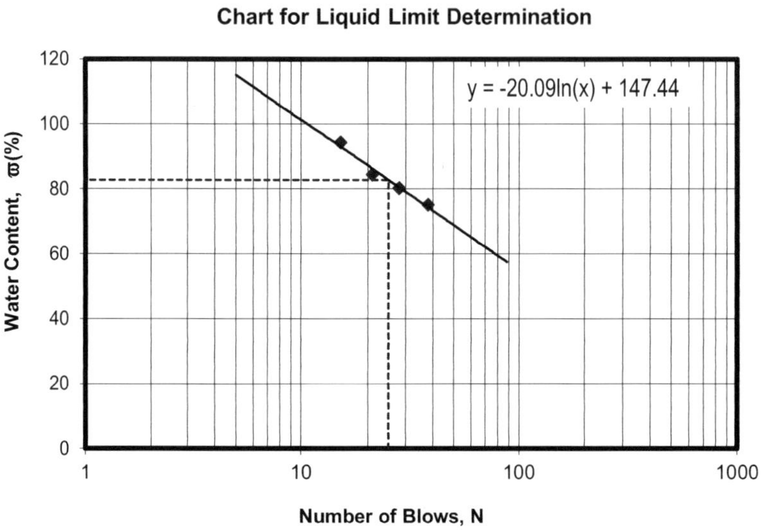

Figure 3. Graph of Liquid Limit Test Results

3.1.6 Shrinkage Limit

The shrinkage limit is the water content limit where the soil with a water content below the obtained water content does not cause a change in volume (shrinkage) again in the soil. The shrinkage limit value obtained is 13.88%.

3.1.7 Plasticity Index

The plasticity index is the value of the difference between the liquid limit and the plastic limit. From the test results, the plasticity index value is 55.98%.

3.1.8 Sieve Analysis and Hydrometer

In the sieve analysis test, it was found that more than 50% of the soil passed the No. sieve. 200, which is 96.6%. The graph of the test results is shown in Figure 4.

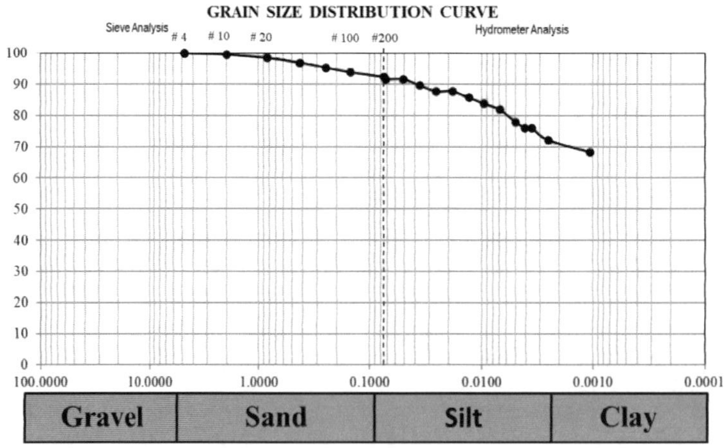

Figure 4. Soil Gradation

3.2. Natural Soil Classification

The classification of the sedimentary soils at the Bili-bili Dam in Gowa Regency was determined by testing the physical characteristics of the original soil utilized in the study

3.2.1 American Association of State Highway and Transportation Officials (AASHTO)

According to hydrometer and sieve analysis test results as well as Atterberg limitations, 92.4% of native soil grains made it through the No. 200 sieve. The plasticity index is 55.98%, while the liquid limit value is 82.75%. Based on the results obtained, the AASHTO classification system can be used to identify the type of soil. Table 2 below shows the determination of the soil type in question:

Table 2. Soil Classification Based on AASHTO

General Classification	*Granular materials* *(35% or less passing No. 200 Sieve (0.075 mm)*							*Silt-clay Materials* *More than 35% passing No. 200 Sieve (0.075 mm)*			
Group Classification	A-1		A-3	A-2				A-4	A-5	A-6	A-7
	A-1-a	A-1-b		A-2-4	A-2-5	A-2-6	A-2-7				A-7-5 A-7-6
(a) Sieve analysis: Percent Passing (i) 2.00 mm (No. 10) (ii) 0.425 mm (No. 40) (iii) 0.075 mm (No. 200)	50 max 30 max 15 max	50 max 25 max	51 min 10 max	35 max	35 max	35 max	35 max	36 min	36 min	36 min	36 min
(b) Characteristics of fraction passing 0.425 mm (No. 40) (i) Liquid limit (ii) Plasticity index	6 max		N.P.	40 max 10 max	41 min 10 max	40 max 11 min	41 max 11 min	40 max 10 max	41 min 10 max	40 max 11 min	41 min 11 min
(c) Usual types of significant Constituent materials	Stone Fragments Gravel and sand		Fine Sand	Silty or Clayey Gravel Sand				Silty Soils		Clayey Soils	
(d) General rating as subgrade	Excellent to Good							Fair to Poor			

**If plasticity index is equal to or less than (liquid Limit-30), the soil is A-7-5 (i.e. PL > 30%)*
If plasticity index is greater than (Liquid Limit-30), the soil is A-7-6 (i.e. PL < 30%)

3.2.2 *Unified Soil Classification System*(USCS)

The soil type is determined to belong to the ML group (Milt with Low Plasticity), which stands for Silty Soil with Low Plasticity, by connecting the Plastic Limit Value and Plasticity Index Value. Figure 5 depicts the relationship between these variables.

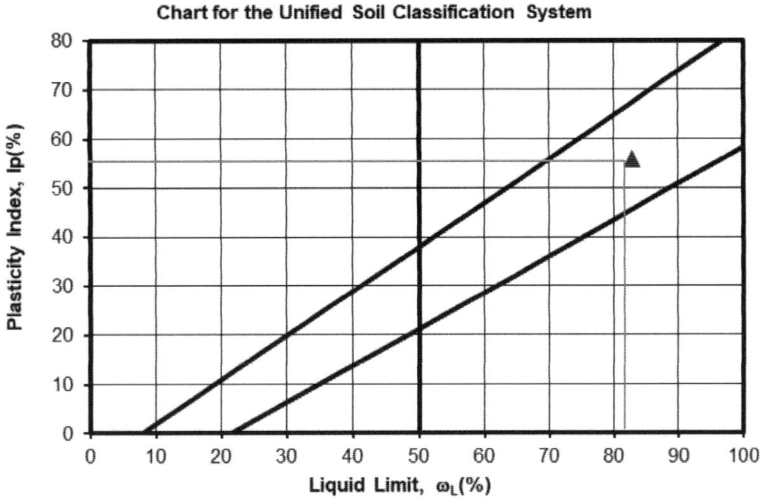

Figure 5. Classification of Nature Soil Classification according to the USCS System

3.3. Mechanical Properties of Expansive Soil

3.3.1 Compaction (Compression Test)

In order to determine the soil density based on the maximum dry density and ideal water content of the study sample, compaction using a compass is an attempt to achieve the maximum soil density with a standard energy.

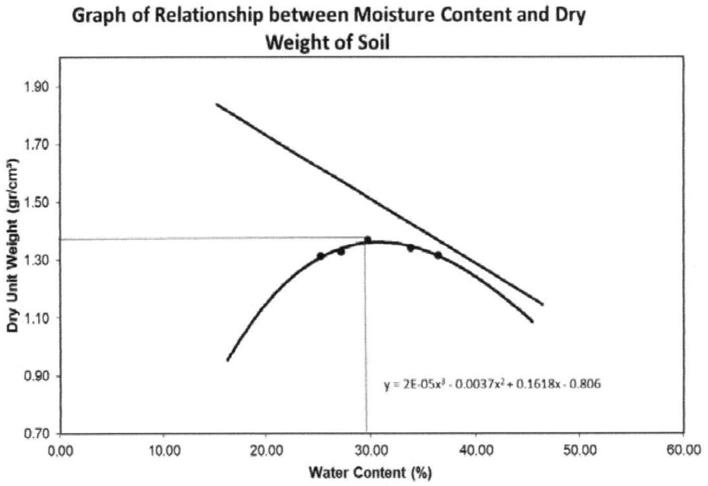

Figure 6. Graph of the Relationship of Moisture Content with Dry Fill Weight of Compaction Results in Original Soil

According to the test results, the ideal moisture content (opt) of 29.7% allowed for the achievement of the maximum dry weight value (drymax) of 1.51 gr/cm3.

3.3.2 UCS Test

To determine the value of the first soil free compressive strength to be tested, UCS testing is done. In this test, two samples were used, resulting in an average compressive strength value of 0.33 kg/cm2.

3.4 Characteristics of Stabled Expansive Soil Free Compressive Strength Test Results Fly Ash

To measure the compressive strength of the original soil, expanding soil with fly ash, mixing was done. The addition of fly ash was varied, adding 5%, 10%, 20%, and 25%, with curing times of 0 days, 7 days, 14 days, and 28 days, respectively.

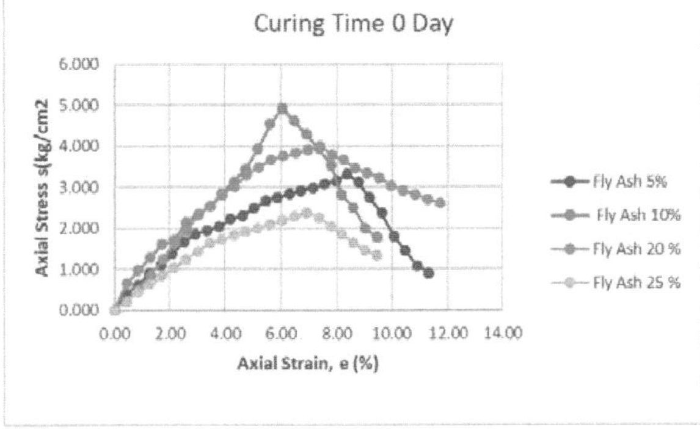

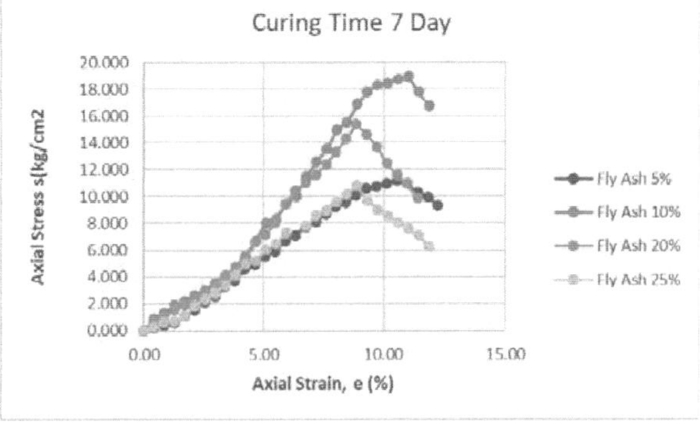

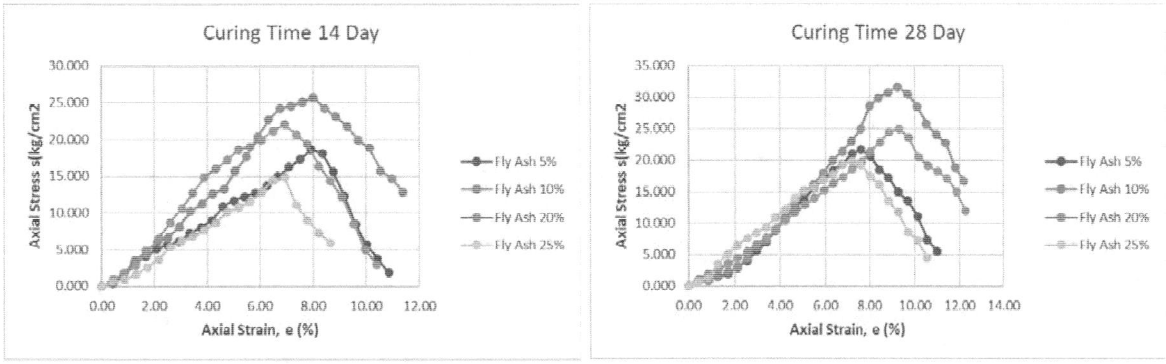

Figure 7. Stress-Strain Relationship of Addition of Fly Ash

The test showed that the amount of fly ash mixture in expansive soil may raise the free compressive strength value throughout each curing phase. With 10% fly ash stabilization and a 28-day curing interval, the expansive soil mixture with the greatest free compressive strength value has a free compressive strength that is 31.22 times more than the original soil's free compressive strength value.

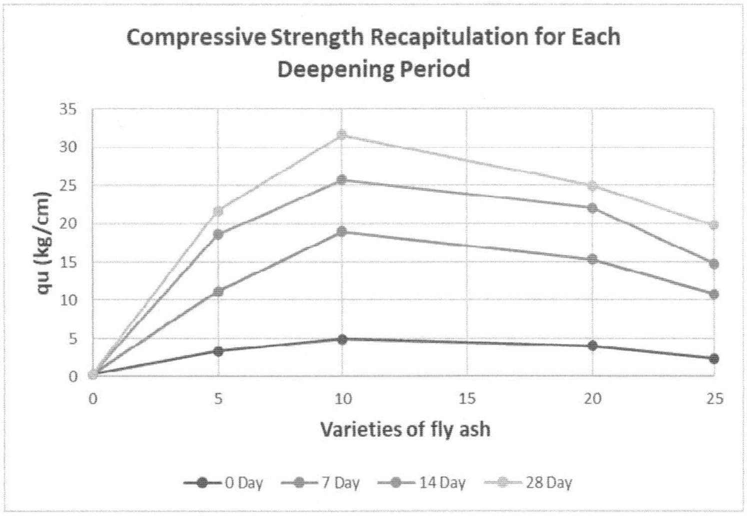

Figure 8. Recapitulation of Compressive Strength Value Changes on Test Objects After Curing Period

4. **Conclusion**

1. The original soil was categorized as ML, or silt with low plasticity, according to the USCS classification system and as A-7-5, or loamy soil, according to the AASHTO classification system, based on the findings of physical characteristics testing.
2. Fly ash can be added to expansive soils to improve specific gravity while lowering the soil's plasticity index.
3. In expansive soils, the addition of a stabilizer, such as fly ash, can lower the value of the ideal moisture content and raise the value of the soil's dry density.
4. The addition of stabilizing material, specifically fly ash, has been shown to increase the soil's bearing capacity as the curing period lengthens; in this case, the value of the Free Compressive Strength. In a mixture of sedimentary soil with 10% fly ash content and a curing time of 28

days, the highest Free Compressive Strength value that was successfully achieved was 16.25 times that of the original soil..

5. **Reference**

[1] Leliana, Arinda. 2015. The Effect of Addition of Fly Ash to the Value of Free Compressive Strength in Expansive Clay Soil in Magetan, East Java. Research Thesis. Surabaya: UNESA.
[2] Ingles, OG, and Metcalf, JB, 1972, Soil Stabilization, Butterworths, Sydney.
[3] Sudjianto, AT, 2006. Study of Development Potential and Pressure on Expansive Clay at Variation of Moisture Content, Research Report on Research Grants from the Ministry of National Education PDM Skim, Faculty of Engineering, Widyagama University Malang (unpublished)
[4] Hardiyatmo, C. Hary, 2010. Soil Stabilization for Road Pavement. Yogyakarta: Gajah Mada University Press
[5] Bowles, Joseph E. 1984. Physical and Geotechnical Properties of Soil (Soil Mechanics). Jakarta: Erlangga.
[6] Hardiyatmo, Christady, Harry. 2006. Soil Mechanics I. Gadjah Mada University. Yogyakarta.
[7] Das, BM, 1995, Soil Mechanics (Principles of Geotechnical Engineering), Jakarta, Erlangga
[8] Fernandez, GJW 2007. Study of Bobonaro Clay Characteristics in East Nusa Tenggara Province. Research and Development Center for Roads and Bridges, Bandung.
[9] Apriyanti, Yayuk. 2014. Peningkatan Nilai CBR Tanah Lempung Dengan Menggunakan Semen Untuk Timbunan Jalan. Tugas Akhir, Jurusan Teknik Sipil. Universitas Bangka Blitung. Bangka Belitung.
[10] Tallama, Djumarno, Astrid. 2010. Utilization of Waste Coal Ash for Subgrade Coating. Master's Thesis Civil Engineering study program. Gadjah Mada University, Yogyakarta

IOP Conf. Series: Earth and Environmental Science 1117 (2022) 012059 doi:10.1088/1755-1315/1117/1/012059

Concrete Pile Defect Identification: Insights from Cross-Hole Sonic Logging and High Strain Dynamic Pile Test

Chungyean Ly[1*], Chandoeun Eng[2], Muoy Yi Heng[3], Phanny Yos[4]

[1,2,3,4] Faculty of Geo-resources and Geotechnical Engineering, Institute of Technology of Cambodia, Russian Federation Blvd., P.O. Box 86, Phnom Penh, Cambodia
[4] Research Innovation Center, Institute of Technology of Cambodia, Materials Science and Structure, Institute of Technology of Cambodia

Corresponding author: chungyean_ly@gsc.itc.edu.kh

Abstract. This paper aims to identify the types of defects, location of defects, size of defects, and capacity of five concrete piles using a High strain dynamic pile test and Cross-hole sonic logging test. In this study, Cross-Hole Sonic Logging method (CSL) is a potent quality-control tool that exclusively detects much smaller flaws with high accuracy with utilizing velocity and first arrival time (FAT) from ultrasonic waves to find out the integrity of the drill shaft. From CSL test, the bearing capacity needs to be calculated manually using a waterfall diagram and soil profile. Moreover, the high strain dynamic pile test is used to identify integrity and bearing capacity of concrete pile via resistance analysis down to the pile and split of the base to shaft resistance when only pile head measurements are taken then using Case Method and CAPWAP analysis. Through this analysis, the bearing capacity can be interpreted; the Belta (damage factor) is applied to estimate location and size of the defect in the concrete pile. Based on the result of CLS test for three concrete piles, P01, P04, and P10, indicated that P01 and P04 are "Good" pile integrity with the bearing capacity of 4583kN and 6600kN, while P10 showed as "Potential defect at depth 9.25m to 10.25m" of pile integrity with bearing capacity of 6600kN. Furthermore, High strain dynamic pile test of two concrete piles, P2E and P2P, showed that the integrity category is "Ok" with bearing capacity 9896kN.

1. Introduction

The construction sector in Phnom Penh City has accelerated in the recent decade. The construction types are high-rise residential buildings, factories, hotels, commercial buildings, institutional structures, as well as bridges. With the high demand for construction, the quality of the foundation is very needed to be inspected, especially deep concrete pile foundations because the defects in form of necking/bulging may have occurred. These defects occur due to failure of soft vertical strata after excavation of soil during lowering of the cage [1]. The defects occupying up to 15% of the cast in situ pile section could remain undetected. [3] recorded that out of 441 piles tested on different projects in South Carolina, more than 70% of the projects had at least one pile containing a defect. The presence of these defects in the concrete pile may affect the capacity and ability to transmit the design loads.

Several methods are currently used in construction companies to ensure the quality of concrete piles, particularly non-destructive testing (NDT). Cross-hole sonic logging test (CSL) and High strain dynamic pile test are the common and most reliable techniques among methods of NDT. CSL is non-invasive and capable of detecting the subsurface across a vast region and deeper, and the high strain dynamic pile test allows to measure the capacity of the pile. Moreover, both methods provide benefits such as being cost-effective and time-saving. Therefore, this paper aims to estimate the pile resistances and pile integrity of the IBC Multi-Purpose Building Project, which is located in street 63, Boeng Kengkorng 1 commune, Boeng Kengkorng district, Phnom Penh, Cambodia (Figure 1) and HSC Building 310 Project, which is located in street 310, Khan Chamkamorn, Phnom Penh, Cambodia (Figure 1).

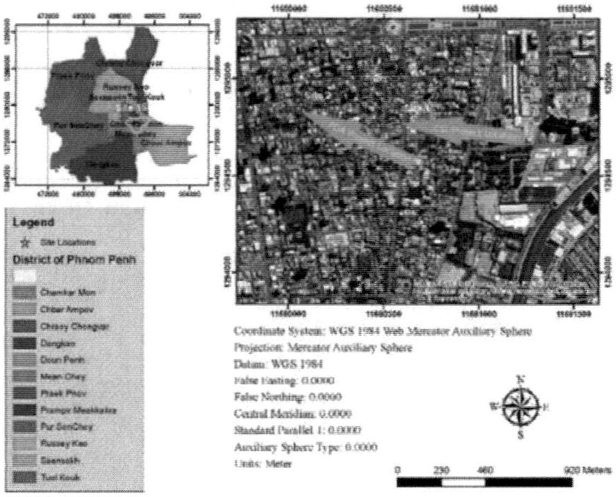

Figure 1. Construction Sites (The arrows indicate IBC Multi-Purpose Building and HSC Building 310 Projects)

2. Methodology

Two tests are applied in this study, such as Cross-hole Sonic Logging Test (CSL) and High Strain Dynamic Pile Test (PDA) to determining the integrity and bearing capacity of concrete piles (Figure 1). CSL is employed at HSC Building 310 Project for three dug piles (P01, P04, P10). P01 has diameter of 800 mm, while P04 and P10 have diameter of 1000 mm. These three concrete piles have same length of 41 m and same age 28 days. In addition, PDA is employed for concrete pile, P2E and P2P, with same diameters of 1200mm and lengths of 40.9m and 37.2m, respectively. The ages of concrete pile P2E and P2P is 28 days.

2.1. Cross-Hole Sonic Logging Test

Cross-hole sonic logging test (CSL) uses for deep foundation investigation, which identifies the pile foundation integrity (structural defects) and bearing capacity of concrete piles. A signal transmitter or emitter is placed at the bottom of one duct, while a receiver is placed at the bottom of another duct. The transducer and receiver are then both increased in tandem at a set pace. Changes in the pile's or its component materials' properties can impact the signal's travel duration or quality when it reaches the receiver (Figure 2) [8].

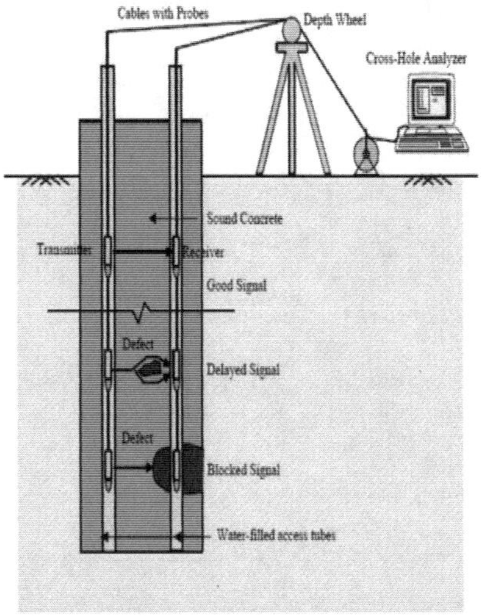

Figure 2. Schematic of cross-hole sonic logging test

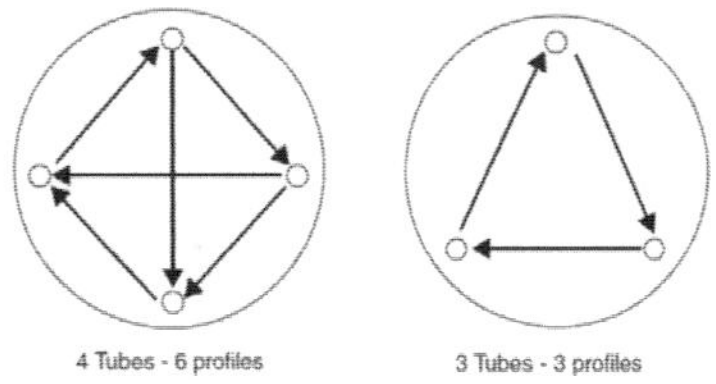

Figure 3. Cross-hole sonic logging access tube layout in plan view

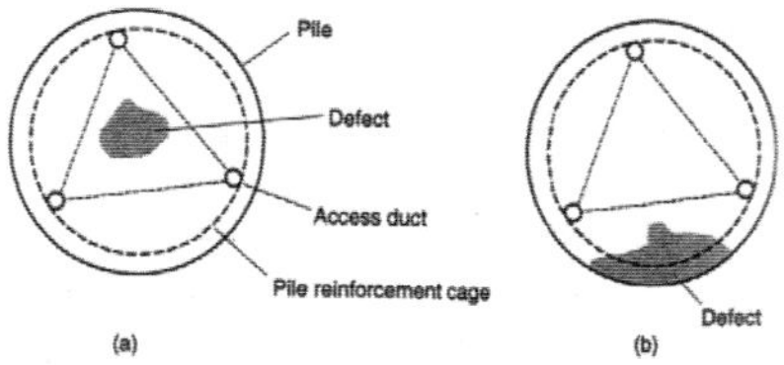

Figure 4. Possible defects off the shortest direct path between probes: (a) defect in center, (b) defect on perimeter

Table 1. An overview of the recommended number of access tubes for different pile diameter

Pile diameter (mm)	Tijou (1984)	Turner (1997)	O'Neil and Reese (1999)	Works Bureau (2000)	Thasnanipan et al. (2000)	JGJ 106-2003 (MOC 2003)
600-750	2	3	2	3 to 4	2	2
750-1000	2 to 3	3 to 4	2 to 3	3 to 4	3	2 to 3
1000-1500	4	4 to 5	4 to 5	3 to 4	4	3
1500-2000	4	4 to 5	5 to 7	3 to 4	6	3
2000-2500	4	4 to 5	7 to 8	3 to 4	6	4
2500-3000	4	4 to 5	8	3 to 4	8	4

2.1.1. Identify defects in concrete piles

The signal velocity can be determined in CSL analysis using:

$$\text{Signal Velocity} = \frac{\text{Distance}}{\text{First Arrival Time}} \qquad (1)$$

To ensure the distance between probes is minimized (probes spacing does not exceed 1.5m to ensure good strength signals), the First arrival time(FAT) to evaluation criteria for the CSL test seen in Table 2 and the average velocities should be compared to the recommended wave velocities in concrete, which are described in Table 3.

Table 2. Evaluation criteria for CSL test

Class	Conclusion	FAT	Energy Reduction
Satisfactory	(G)(Good)	FAT Increase 0 to 10%	< 6db
Anomaly	(Q)(Questionable)	FAT Increase 11% to 20%	< 9db
Flaw	(P/F)(Poor/Flaw)	FAT Increase 21% to 30%	9 to 12 db
Defect	(P/D)(Poor/Defect)	FAT Increase >31%	> 12 db

Table 3. Velocity criterion of concrete quality [1]

Pulse velocity (km/sec)	Grading of concrete quality
> 4.5	Excellent
3.5 to 4.5	Good
3.0 to 3.5	Fair/Medium
< 3.5	Doubtful

2.1.2. Bearing capacity of concrete piles stand on soil extending into rock

Calculate bearing capacity of isolate pile following 4 steps by using waterfall diagram of a sonic test and soil profile along piles

Step1: Calculate end bearing capacity of isolate pile

$$Q_p = \frac{\pi B^2 q_{\max}}{4} \tag{2}$$

Coates (1967): $q_{\max} = 3q_{uc}$ (3)

Rowe and Armitage (1987): $q_{\max} = 2.7q_{uc}$ (4)

Zhange an Eistein (1998): $q_{\max} = 4.83(q_{uc})^{0.51}$ (5)

The unconfined compressive strength (MPa)

$$q_{uc} = \min(f_{ck}, q_{uc}) \tag{6}$$

► After calculation

Choosing Q_p minimum because it is secure

Step2: Calculate the frictional capacity of the isolated pile

$$Q_s = \pi BL\tau_{\max} \tag{7}$$

Carter and Kulhawy (1988): $\tau_{\max} = 0.63P_a\sqrt{\frac{q_{uc}}{P_a}}$ (8)

Kulawy and Phoon (1993): $\tau_{\max} = \frac{q_{uc}}{2}\left(\frac{q_{uc}}{2P_a}\right)^{-0.5}$ (9)

Zhange and Eistein (1998): $\tau_{\max} = 0.4(q_{uc})^{0.5}$ (10)

Where

- L is the length of the pile along the rock surface
- B is the diameter of the pile shaft
- P_a is atmospheric pressure ($\approx 0.1MPa$)

► After calculation

Choosing Q_s minimum because it is secure

Step3: Calculate ultimate load carry capacity

IOP Conf. Series: Earth and Environmental Science 1117 (2022) 012059 doi:10.1088/1755-1315/1117/1/012059

$$Q_u = Q_p + Q_s \tag{11}$$

Step4: Calculate allowable capacity of isolate pile

$$Q_{all} = \frac{Q_u}{FS} \tag{12}$$

FS safety factor (Generally used range from 2 to 3)

2.2. High strain dynamic pile test

PDA of deep foundations or high-strain dynamic testing is the standard method for determining: pile integrity, load bearing capacity, vertical displacement, and pile-soil load transfer connections. In dynamic pile testing, the pile is loaded by a drop weight that applies a very brief duration impact (5 ms) load at the pile's head. A wave will be reflected up the pile if the pile is resisted or there is a change in impedance, such as at the pile tip. It is feasible to analyze the dynamic pile reaction by comparing the waves traveling up and down the pile. It is feasible to infer the comparable static pile load-displacement behavior by further study see Figure 5.

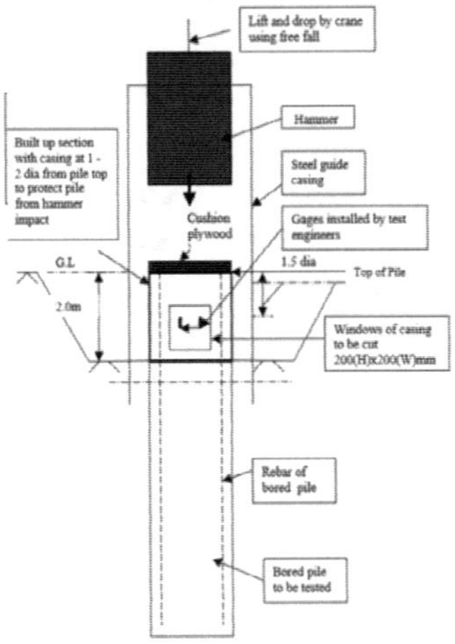

Figure 5. A typical arrangement for high strain dynamic testing of a deep foundation

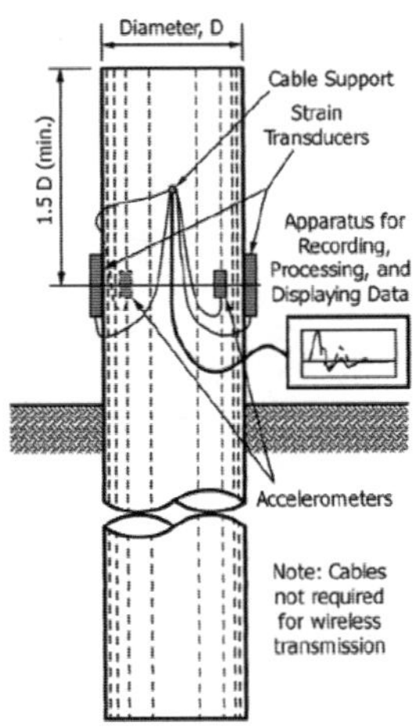

Figure 6. Schematic Diagram of Apparatus for Dynamic Monitoring of Deep Foundations

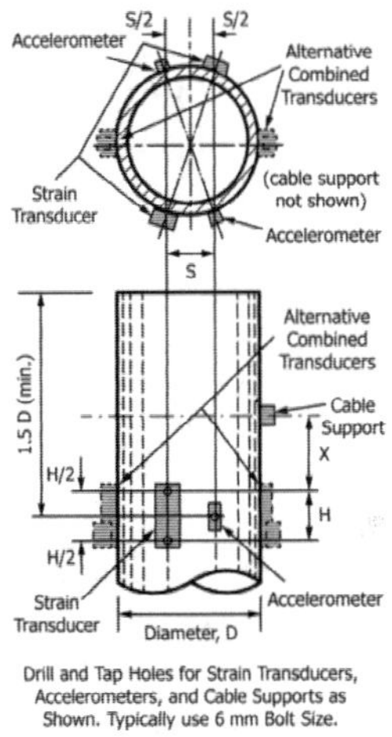

Figure 7. Typical Attaching Transducers to Concrete Piles

2.2.1. Identify defects in concrete piles

Pile damage: BTA, LTD

- Pile damage causes a tension reflection before 2L/c
- The time at which the tension reflection arrives at the gage location indicates the distance to the damage: LTD=t_{damage}(c/2)
- The extent of the damage is quantified with damage factor BTA (β)

$$v = \frac{F}{Z} \Rightarrow Z = \frac{F}{v} \tag{13}$$

Where

- Z is pile impedance
- F is force
- v is pile velocity

$$c = \sqrt{\frac{E_p}{\rho_p}} \tag{14}$$

Where

- c is wave velocity in the pile
- E_p is Young's modulus of the pile
- ρ_p is the density of the pile

$$Z = \frac{E_p A_p}{c} = \rho_p \times c \times A_p \tag{15}$$

Where

- A_p is the pile's cross-sectional area

$$\beta = Z_2 / Z_1 \tag{16}$$

$$\beta = \frac{(F_{d,1} + F_{u,1})}{(F_{d,1} - F_{u,1})} \tag{17}$$

Where

- $F_{u,1}$ is the reflection wave that indicates the extent of the impedance reduction (damage)
- $F_{d,1}$ forces downward
- $F_{u,1}$ forces upward

Table 4. Belta value for evaluating concrete integrity

$\beta = Z_2 / Z_1$	Damage assessment
1.0	Uniform
0.8 – 1.0	Slight damage
0.6 – 0.8	Damage
< 0.6	Pile with a major discontinuity

2.2.2. Bearing capacity of concrete piles

For bearing capacity verify with axial stress receives a signal from the strain transducer. The top of the pile is given a drop weight. To further estimate the final pile capacity, CASE and CAPWAP.
The static load-set graph is based on the static resistance parameters generated by CAPWAP and the pile's elastic compression properties.

Force stress: $FMX = \varepsilon_{av} EA$ (18)
Compressive stress: $CSX = FMX / A$ (19)
Tension stress: $TSX = CTX / A$ (20)

Where

- $CTX = Max\left[\left(W_{d,\max-tension} + W_{u,\min-comp}\right), CTN\right]$ (21)
- $CTN = F_{\uparrow}(t_2) + F_{\downarrow}(t_3)_{\min}$ (22)

- Case Method

Total Resistance = Static + Dynamic Resistance (23)
$RTL = R_{static} + R_D \Rightarrow R_{static} = RTL - R_D$ (24)

Where

- $RTL = F_{u,2} + F_{d,1}$ (25)

- $R_D = J_c.Z.v_{toe}$ Where $v_{toe} = (2F_{d1} - RTL)/Z$ (26)
- $\Rightarrow R_{static} = F_{u,2} + F_{d,1} - J_c(2F_{d,1} - F_{d,1} - F_{u,2})$ (27)
- $R_{static} = (1 - J_c)F_{d,1} + (1 + J_c)F_{u,2}$ (28)

CAPWAP Method

- Set up pile and soil model and assume R_{shaft} and R_{toe}
- Apply measured W_{DM} to pile model at top and calculate complementary W_{UC}
- Compare W_{DM} with measured W_{UC}
- Adjust R_{shaft} and R_{toe}
- If no satisfactory match: Go to step2

3. Results and Discussion

3.1. Cross-hole sonic logging test

To response the objective of research for evaluate concrete pile using cross-hole sonic test, this section will descript about research finding which are identify types, location, size of defects and bearing capacity of isolate bored piles.

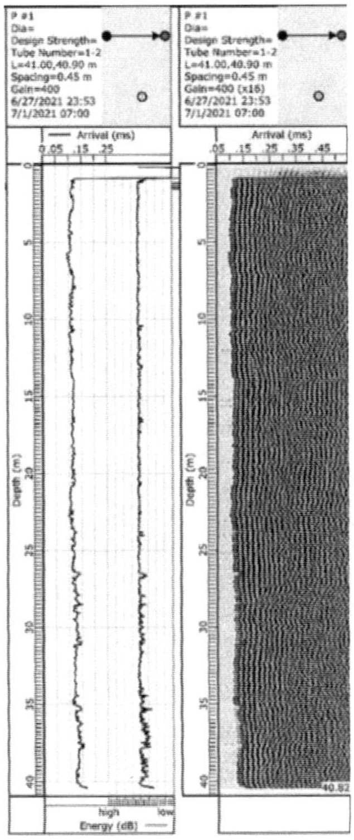

Figure 8. P01(Tube 1-2) of the CSL test

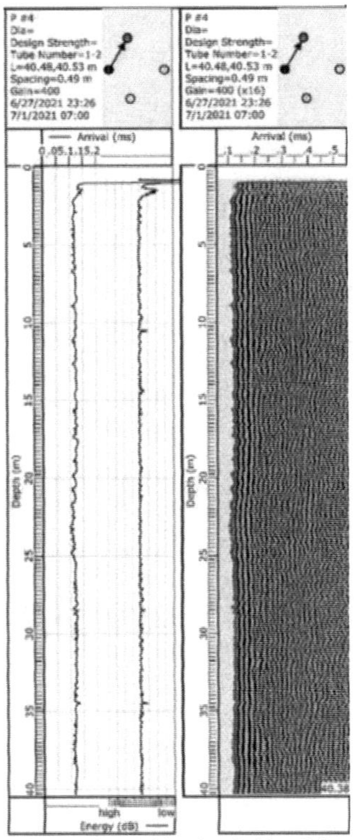

Figure 9. P04(Tube 1-2) of the CSL test

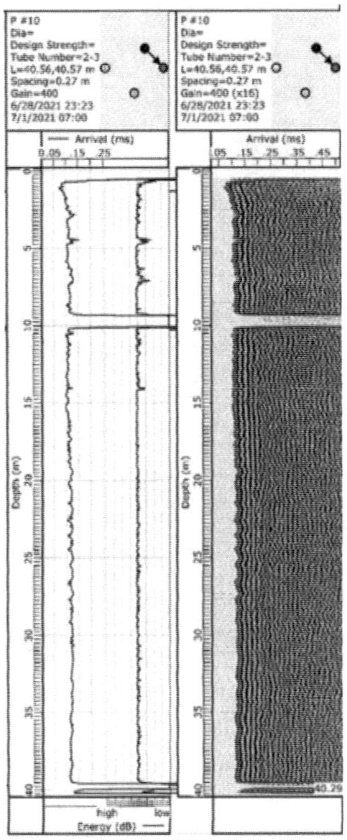

Figure 10. P10(Tube 2-3) of the CSL test

3.1.1. Location and size of defects

The results from CSL tests allow concluding that:

- All piles are good except P10 at depth 9.25m to 10.25m (based on diagram waterfall see in Figures 10)

Table 5. Summary location and size of CSL test

No	Type	Pile Integrity Category	Comments
1	P01	Good	None
2	P04	Good	None
3	P10	Poor/Defect	Defect at depth 9.25m to 10.25m

3.1.2. Identify types of defects

Following analysis of velocity, FAT, and about problematic in the bored pile (P01, P04, P10), Seeing that P1 and P04 have velocity>3500m/s, FAT<20%, and Energy Reduction <9db (data interpretation in waterfall diagram see in Figure 8,9,10) conclude that integrity is "Good". On the other hand, P10 has a velocity<3500m/s, FAT>30%, and Energy Reduction >12db (data interpretation in waterfall diagram see in Figure 8,9,10) so it can summary that integrity is "defects whole 6 profiles". This type of defect tends to occur with tremied piles case under bentonite and soil profile of P10 at the "depth 9.25m to 10.25" is clay layer it is easy to collapse soil and defect on the concrete pile when casting concrete, It is seen that it is probably typical of "Neck" in concrete bored pile cause by failure of aspect of pile forming process, whether during pile construction or thereafter. Seeing the type of defect seen in Table 6

Table 6. Summary types of defects using the CSL test

No	Type	Pile Integrity Category	Type of defect
1	P01	Good	None
2	P04	Good	None
3	P10	Poor/Defect	Potential of Neck

3.1.3. Bearing capacity of concrete piles

Based on a calculation using the waterfall diagram in Figure 8,9,10 and soil profiles we can obtain the bearing of the concrete pile as illustrated in Table 7.

Table 7. Summary isolate pile capacity of CSL test

No	Type	End capacity(kN)	Frictional capacity(kN)	Ultimate Capacity(kN)	Allowable Capacity(kN)	Safety Factor
1	P01	8362	5386	13750	4583	3
2	P04	13065	6735	19800	6600	3
3	P10	13065	6735	19800	6600	3

3.2. High strain dynamic pile test

This section will provide the research's findings, which include the types, locations, sizes of flaws, and bearing capacities of isolation drilled piles, in response to the research's goal of evaluating concrete piles utilizing high strain dynamic pile tests

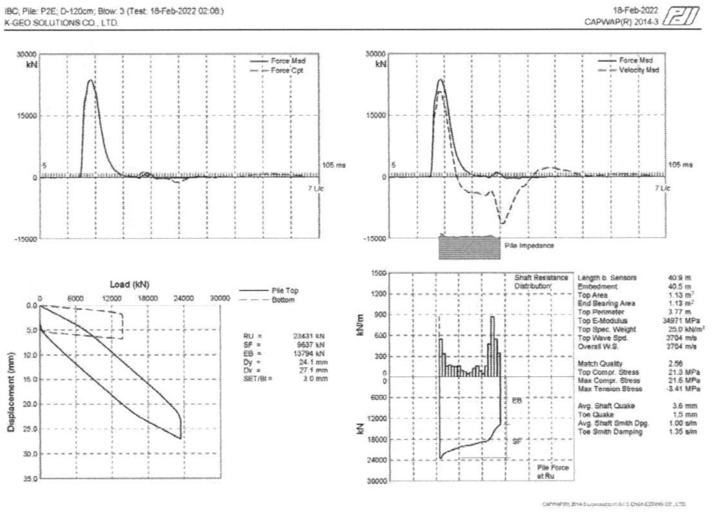

Figure 11. PDA-CAPWAP results (P2E)

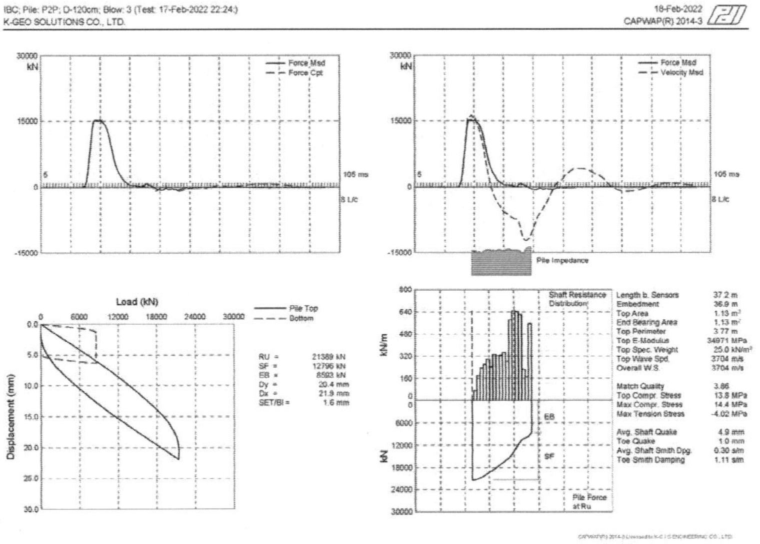

Figure 12. PDA-CAPWAP results (P2P)

3.2.1. Location and size of defects

Based on belta inspection can conclude the location and size of defects in concrete piles See Table 8.

Table 8. Summary location and size of defects using PDA test

No	Type	Pile Integrity Category	Comments
1	P2E	Defect, Slight damage	Potential defect at depth 2m, 38.8m to 40.9m
2	P2P	Defect, Slight damage	Potential defect at depth 6m, 32.2m to 33.2m, 35.2 to 37.2 m

3.2.2. Identify types of defects

The results of the high strain dynamic pile test include P2E has potential defect slight damage of Bulb and step because belta (damage factor) is 1.15, 0.92, 0.93 and 0.89 below normal rank equal 1 and soil profile of P2E at the "depth 2m" is made ground layer it is easy to defect on the concrete pile and at the "depth 38.8m to 40.9m" is Yellow grey heavy weathered Sand Stone layer it is the hard layer that is potential of step when casting concrete. P2P has possibly slight damage and defect of Bite, Waist, and Expansion because belta (damage factor) is 0.90, 0.89, 0.90, and 1.13, 1.15, 1.17 below normal rank equal 1, and soil profile of P2P at the "depth 6m" is Brown medium stiff silt layer is easy to collapse soil and defect, at the "depth 32.2 m to 33.2m" is yellow very hard sandy silt is easy to get the waist of concrete and at the "depth 35.2 m to 37.2m" is Yellow grey heavy weathered SANDSTONE is easy to get expansion when casting concrete. Even though P2E and P2P have potential various kinds of the defect but it is reducing a small area of the pile in the rank of acceptable belta inspection See Table 9 and 10.

Table 9. Summary defects types of P2E

Pile ID	Impedance (kN/m/s)	Inspection β	Type of defect
P2E	12274.8	1.15	Potential of Bulb
	9827.7	0.92	Potential of Step
	9877.9	0.93	
	9510.4	0.89	

Table 10. Summary defect type of P2P

Pile ID	Impedance (kN/m/s)	Inspection β	Type of defect
P2P	9589	0.90	Potential of Bite
	9452.1	0.89	Potential of Waist
	9563.8	0.90	
	12061.7	1.13	Potential of Expansion
	12328.2	1.15	
	12458.9	1.17	

3.2.3. Bearing capacity of concrete piles

The relevance of the PDA tests is shown in a static load-displacement graph. The following results capacity of concrete pile verified with pile design requirement using CAPWAP and CASE methods are summarized in the table below.

Table 11. Pile design requirement

Pile No:	P2E & P2P
Design working load (ton)	1000
Design ultimate load (ton)	2000

Table 12. Summary PDA testing results

Pile ID	Diameter	Capacity			Working Load (kN)	Safety Factor	Integrity	Remark
		Total Capacity (kN)	Skin Friction (kN)	End Bearing (kN)				
P2E	1200	23431	9637	13794	9896	2.37	OK	
P2P	1200	21389	12796	8593	9896	2.16	OK	

4. Conclusion and Perspective

According to the results of Cross-hole sonic logging test and high strain dynamic pile test on five concrete piles, the quality of concrete pile can be concluded as:

- **Cross-hole sonic logging test**
  - Location and size of defects in concrete piles for P01 and P04 "None defects" but for P10 has defects at "Depth 9.25m to 10.25m"
  - Types of defects in concrete for P01 and P04 have "None" but P10 has types of "Neck" in concrete bored piles.
  - Bearing capacity for P01 is 4583kN, P04 is 6600kN and P10 is 6600kN
- **The perspective of the CSL Test:**
  - Propose that relevant engineer(s) use the CSL findings with soil borings, and drill shaft construction records to verify results.
  - Solution 1 for "Neck" defects of P10 is to build a new pile which the same criteria next to the old one which is site construction already installed.
  - Solution 2 for "Neck" defects of P10 can repair the defect and cast concrete back on that defect pile (P10) and foundation loading (PDA test) data to assess foundation acceptability in terms of design criteria.
- **High strain dynamic pile test**
  - Location and size of defects in the concrete pile for P2E has "Potential defect at depth 2m, 38.8m to 40.9m" and P2P has "Potential defect at depth 6m, 32.2m to 33.2m, 35.2 to 37.2m"
  - Types of defects in concrete piles for P2E have "Potential of Bulb and Step" and for P2P has "Potential of Bite, Waist, and Expansion" in concrete bored piles.
  - The bearing capacity for P2E is 9896kN and P2P is 9896kN.
- **The perspective of the PDA Test:**
  - Looking at the detail of defects in the concrete pile, it can cause large damage for a long time with the impact of change in underground soil and water.

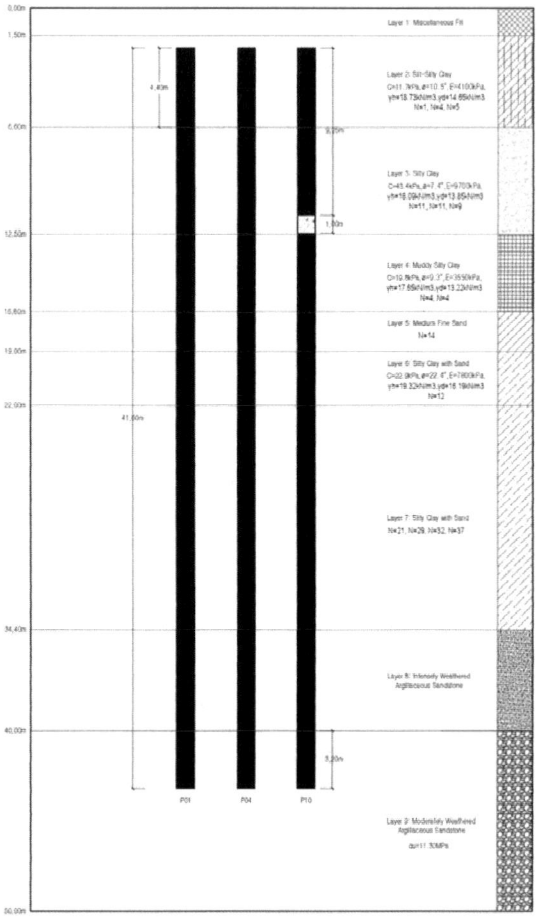

Figure 13. Bored piles on soil extend into rock profile using the CSL test

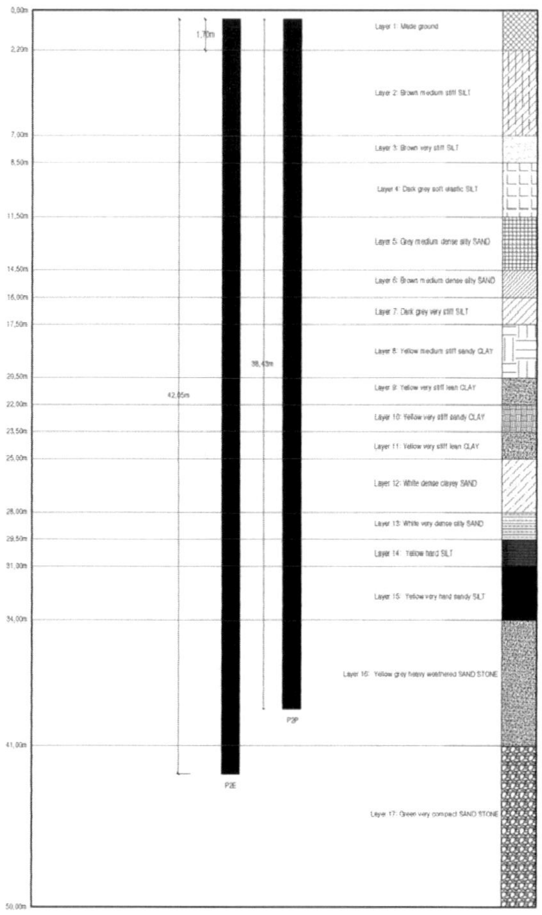

Figure 14. Bored piles on soil extend into the rock profile using the PDA test

Acknowledgments

We would like to thank you to K-GIS Engineering Co., Ltd. for providing an opportunity to conduct this research. Sincere thanks also go to the Cambodia Higher Education Improvement Project (HEIP-ITC-SGA#14) (Credit No. 6221-KH) for providing funding and the opportunity to conduct research at the Institute of Technology of Cambodia.

References

[1] Bagui, S. K., Puri, S. K., & Subbiah, K. (2020). Cross hole sonic test results for the analysis of pile load test. Advances in Bridge Engineering, 1(1), 1-16.

[2] Baker, C. N., Drumwright, E. E, Briaud J.-L., Mensah-Dumwah,F. and Parikh, G. (1993). Drilled Shafts for Bridge Foundations, FHWA Publication No. FHWA-A-RD920004.Washington, DC: FederalHighway Administration.

[3] Camp WM III, Holley DW, Canivan GJ (2007). Cross-hole Sonic Logging (CSL) of South Carolina drilled shafts: a five-year summary. GeoDenver 2007 New Peaks in Geotechnica, Denver (CD Rom).

[4] CIRIA Report 144 (1997) Integrity Testing in Piling Practice.
London: Construction Industry Research and Information Association.

[5] Cameron, G. and Chapman, T. (2004). Quality assurance of bored pile foundations. Ground

Engineering, 37(2), 35–40.
[6] Fleming, K., Weltman, A., Randolph, M., and Elson, K. (2009). Piling Engineering (3rd Edition). London: Taylor & Francis.
[7] Finno, R. J., Gassman, S. L., & Osborn, P. W. (1997). Non-destructive evaluation of a deep foundation test section at the Northwestern University national geotechnical experimentation site. A Report Submitted to the Federal Highway Administration Office, Northwestern University, Evanston, Illinois.
[8] Hertlein, B. and Davis, A. (2006). Nondestructive Testing of Deep Foundations. New York: John Wiley.
[9] Holeyman, A. E. (1992). The technology of pile dynamic testing. In Proceedings of the International Conference on the Application of Stress Wave Theory to Piles (ed Barends, F. B. J.), 21–24 September 1992, The Hague, The Netherlands. Rotterdam: Balkema, pp. 195–215.
[10] Institution of Civil Engineers (2007). ICE Specification for Piling and Embedded Retaining Walls (2nd Edition). London: Thomas Telford
[11] Iskander, M., Roy, D., Earley, C. and Kelley, S. (2001). Class-A Prediction of Construction Defects in Drilled Shafts. Transportation Research Record 1772, Paper No. 01-0308. Washington, DC: Transportation Research Board, pp. 73–83.
[12] Preiss, K. and Shapiro, J. (1979). A statistical estimation of the number of piles to be tested on a project. RILEM Commission on Non-Destructive Testing, Stockholm.
[13] Randolph, M. F. (2003). Science and empiricism in pile foundation design. Géotechnique 53(10), 847–875. DOI: 10.1680/geot.53.10.847.37518.
[14] Rausche, F., Goble, G. G. and Likins, G. E. (1985). Dynamic determination of pile capacity. ASCE Journal of Geotechnical Engineering, 111(3), 367–383.
[15] Stain, R. T. (1992). SIMBAT – A dynamic load test for bored piles. In Piling: European Practice and Worldwide Trends (ed. Sands, M.J.). London: Thomas Telford, pp. 198–205.
[16] Sarhan HA, O'Neill MW, Tabsh S (2003). Response of cast in situ pile with minor flaws to axial and lateral loads. Eng Struct 25(1):47–56.
[17] Turner, M. J. (1997) Integrity Testing in Piling Practice. CIRIA Report 144. London: Construction Industry Research and Information Association.
[18] Williams, H. T., and Stain, R. T. (1987). Pile integrity testing: Horses for courses. In Proceedings of the International Conference on Foundation and Tunnels, 22–26 March. London: Engineering Technics Press, pp. 184–191.
[19] Li, D., Zhang, L., ASCE, M., Tang, W., ASCE, H., (2005). Reliability Evaluation of Cross-Hole Sonic Logging for Bored Pile Integrity. ASCE, 9. DOI: 10.1061/(ASCE)1090-0241(2005)131:9(1130).
[20] Chernauskas, L., Paikowsky, S., (2016). Defect Detection and Examination of Large Drilled Shafts Using a New Cross-Hole Sonic Logging System. ASCE, 18.
[21] Bagui, S., Puri, S., Subbiah, K., (2020). Cross hole sonic test results for the analysis of pile load test. Advances in Bridge Engineering, 16. https://doi.org/10.1186/s43251-020-00017-4.
[22] Sibit, D., Handayani, G., (2016). The Crosshole Sonic Logging (CSL) Measurement System to Measure the Quality of Physical Model of Bored Pile. 6th Asian Physics Symposium, 8. doi:10.1088/1742-6596/739/1/012051.
[23] Das, B., Sivakugan, N., (2019). Drilled-Shaft Foundations-Drilled Shafts Extending into Rock. The United States of America. Cengage Learning, Inc.
[24] Zhang, L., (2005). The axial load capacity of drilled shafts in rock-capacity of drilled shaft related to rock. A.A.Balkema Publishers.
[25] Rowe, R.,(2001). Geotechnical and Geoenvironmental Engineering Handbook. Spring US.

doi:10.1088/1755-1315/1117/1/012060

Study of Yeh Embang Watershed Characteristics for Sustainable Water Management

I Gusti Agung Putu Eryani[1,*], Made Widya Jayantari[2], I Nyoman Nuri Artana[3]

[1]Civil Engineering Department, Warmadewa University, Denpasar, Bali, 80239, Indonesia
[2]Civil Engineering Department, Udayana University, Badung, Bali 80361
[3]Architectural Engineering Department, Warmadewa University, Denpasar, Bali, 80239, Indonesia

*corresponding author: eryaniagung@gmail.com

Abstract. One of Bali's perennial watershed that is experiencing the effects of climate change is the Yeh Embang Watershed. If it is not managed holistically, the sustainability of the current water availability will be impacted. This research will analyse the Yeh Embang Watershed's characteristics in order to evaluate the watershed's condition and decide how to manage the sustainability of the water resources. For this research, topographic information, boundary maps of the Yeh Embang watershed, maps of river network, maps of land use, and maps of soil types were all used. The appropriate sustainable management for Yeh Embang Watershed should be determined based on the characteristic. According to the characteristics of the Yeh Embang Watershed, integrated management from upstream to downstream be necessary for the watershed to be managed sustainably. In the upstream area, it is hoped that the forest will be protected by forest conservation to reduce the risk of erosion that could lead to flash floods, and the downstream portion of the Yeh Embang watershed will be structured to prevent flooding from harming the surrounding area. Infiltration wells can be used as a water-saving measure in residential areas in the Yeh Embang watershed to increase the amount of water that is absorbed and lessen surface runoff.

Keywords: Watershed characteristics, sustainability, integrated management.

1. Introduction

Climate change has been seen as a global issue that may pose a threat to human life [1], [2]. This climate change causes negative impacts on various sectors. One of the sectors affected is water resources. Climate change causes changes in rain patterns and an increase in the frequency of extreme climates which poses a serious challenge for the sustainability of water resources [3], [4]. One of the nations in the world affected by climate change is Indonesia. In Indonesia, the frequency of floods and droughts continues to increase [5].

This also happened in the Yeh Embang watershed, Jembrana Regency, in 2018 and 2020 extreme floods occurred which caused some damage to public facilities such as roads, bridges, and several residential residents [6], [7]. However, in 2019, the Yeh Embang River experienced a long dry season which caused a water crisis for the area around the Yeh Embang River [8], [9]. According to Suriyani [10] in 2019 the condition of the Yeh Embang river was dry, even if there was water it could not be used for clean water, there were only puddles between large rocks. Where in general, based on the research of Suyarto and Kusmayati [11]. Yeh Embang is one of the potential rivers that has a minimum discharge of 210 m^3/second which is larger than some rivers in Jembrana Regency such as the Bilukpoh, Yeh Sumbul, Medewi and Pulukan rivers.

Sustainable watershed management is a step that must be taken to maintain the sustainability of water resources in order to combat the effects of climate change on the Yeh Embang watershed. Analysis of watershed characteristics is one of the important components as the basis for the preparation of watershed management patterns [12]–[15].

Nowaday, to facilitate analysis of watershed characteristics, geographic information systems can be assisted [16]–[18].

As a result, it is necessary to plan the management of the Yeh Embang watershed based on its characteristics, which can be accomplished through an analysis of the Yeh Embang watershed's characteristics using geographic information systems.

The results of this analysis can later be used as a first step to identify, study and better understand the characteristics of the Yeh Embang watershed so that it can be used as a reference in determining sustainable watershed management patterns.

2. Research Methods

2.1. Research Location

The Yeh Embang Watershed is located in Mendoyo District, Jembrana Regency with an area of ± 61,561 Ha [19]. The length of the Yeh Embang river is 23 km [20]. Yeh Embang River is an perennial river with a minimum discharge of 210 m^3/second, with an estimated flow of 150 liters/second [21].The map of Yeh Embang Watershed can be seen in Figure 1.

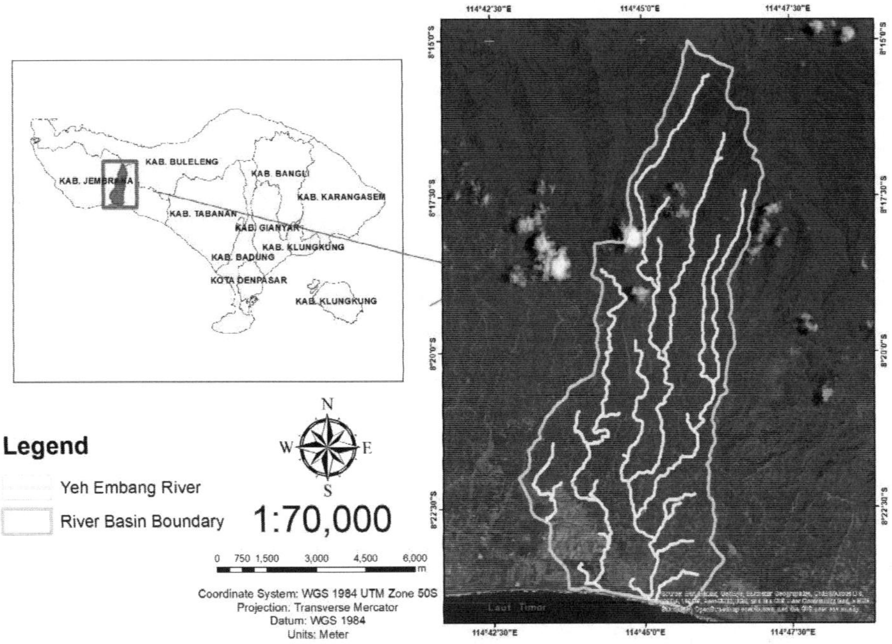

Figure 1. Yeh Embang Watershed

2.2. Research Data

The data used in this study are digital maps of watershed boundaries, land use maps, maps of the Yeh Embang river network obtained from the Bali-Penida River Region Office. In addition, Yeh Embang's digital elevation data (DEM) data was also used which was obtained from DEMNAS data (https://tanahair.indonesia.go.id/demnas/#/demnas). The Yeh Embang watershed's topographic information was obtained using the DEM map, which was categorized using the QGIS 3.1 program. The Center for Environmental Research at Udayana University's soil map was digitized to create the soil type map.

2.3. Data Analysis Methods

This research is a descriptive research conducted by analyzing the characteristics of the Yeh Embang watershed. The characteristics analyzed were morphometric and morphological characteristics.
The morphometric characteristics analyzed were classification of watershed area, river length, watershed shape, bifurcation ratio, and drainage density.

As for the morphological characteristics analyzed in the form of soil type, watershed topography, slope, and land use.

After obtaining the characteristics of the watershed, then a pattern of sustainable water resource management was selected that was in accordance with the characteristics of the Yeh Embang watershed. Flowchart data analysis can be seen in Figure 2.

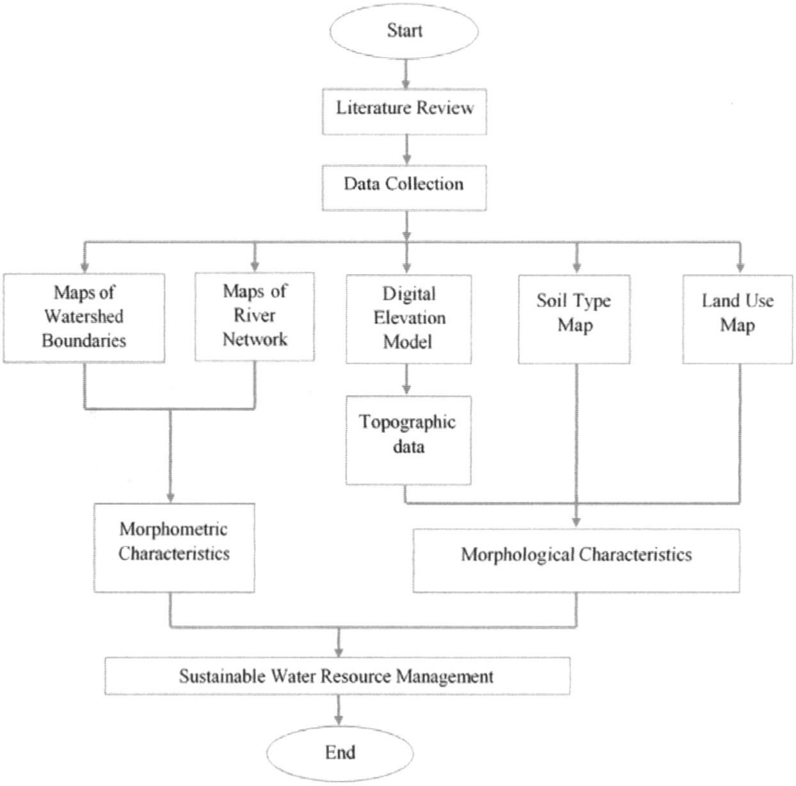

Figure 2. Flowchart Data Analysis

3. Result and Discussion

3.1. Morphometric Characteristics of Yeh Embang Watershed

3.1.1. Length of Yeh Embang River
Based on the Bali-Penida River Region Office [22], the length of the Yeh Embang Watershed is 23.915 km.

3.1.2. Area of Yeh Embang Watershed
Based on the Regulation of the Directorate General of Watershed Management and Social Forestry [23], the area of the watershed can be classified into classes, namely the area of the watershed greater than 1,500,000 Ha which is classified as a very large watershed, the area of the watershed ranging from 500,000 to 1,500,000 ha which is classified as a large watershed. Watershed area ranging from 100,000 to 500,000 are classified as medium watershed, watershed area ranging from 10,000 between 100,000 are classified as small watersheds, and watershed areas of less than 10,000 are classified as very small watersheds.

According to the Bali-Penida River Region Office's watershed boundary map, the Yeh Embang watershed covers an area of 6631.50 ha, or roughly 66.3150 km2. When categorized by area, the Yeh Embang watershed, which has a range of less than 10,000 Ha, is categorized as a very small watershed.

3.1.3. Bifurcation Ratio (Br) of Yeh Embang Watershed
The hydrographic characteristics, watershed flow patterns, and drainage system effectiveness are all impacted by the composition of the river network. The first order is the most upstream river channel that does not branch, the second order is the meeting between the first order, and so on until the main river is marked with the largest serial number [24]. This method can be used to determine the order of a river.

The branching pattern of a river channel that symbolizes the main river pattern of the watershed is known as river order. The bifurcation ratio can be determined using Equation 1.

$$Br = \frac{N_u}{N_{u+1}} \quad [1]$$

Where:

Br	: Bifurcation Ratio
Nu	: Total streams for the order-u
Nu + 1	: Total streams for the order to (u + 1)

According to the results of equation 1, there are three categories for the bifurcation ratio value, with a Br value of less than 3 indicating a rapid rise in floodwater levels in the river channel and a gradual decline. Br values between 3 and 5 denote a moderate increase and decrease in the level of the floodwaters in the river channel. Moreover, a Br value over 5 indicates that flood waters are rising quickly in the river channel, indicating that the decline is occurring quickly.

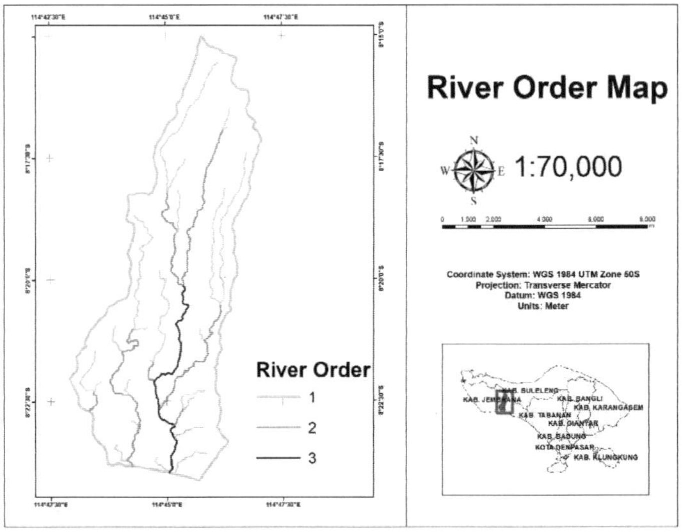

Figure 3. Yeh Embang River Order

Table 1. Bifurcation Ratio

No	Description	Total
1	Streams for the order-1	23
2	Streams for the order-2	13
	Bifurcation Ratio (Br)	1.77

Figure 3 and Table 1 show that the Yeh Embang watershed has the characteristics of a river channel with a rapid rise in flood level and a slow decline if the value of Br (the number of comparisons between first and second order rivers) is 1.77 or less than 3.

3.1.4. Density / Drainage Density (Dd) of Yeh Embang Watershed

Based on the Guidelines for Identification of Watershed Characteristics in Sriyana [24], river flow density is measured as the length of the river per square kilometre of the watershed. The flow density (Dd) value can be calculated by divide the total length of river (Ln) in km and the area of watershed (A) in km^2 as be seen in Equation 2.

$$Dd = \sum Ln/A \quad [2]$$

The higher the value of Dd, the better the drainage system in the area. It also shows that less groundwater is stored in the area and causes small infiltration the more total surface water there is.

According to the Regulation of the Directorate General of Watershed Management and Social Forestry [23], flooding is predicted for the river area if the flow density value is less than 1 mile/mile^2 (0.62 km/km^2), whereas drought is frequently experienced in the watershed if the flow density value is greater than 5 miles/mile^2 (3.10 km/km^2).

The flow density value for the Yeh Embang watershed was calculated based on its area and length, and the results were 0.36 km/km^2, which was categorized as medium flow density and indicates a moderate groundwater reservoir in the watershed. The Yeh Embang watershed will frequently experience flooding because the flow density is less than 1 mile/mile^2 (0.62 km/km^2).

3.1.5. *The shape of Yeh Embang Watershed*

A watershed can be categorized into one of four shapes: complex, radial, parallel, or feather-shaped. Yeh Embang watershed can be categorized as a parallel type of watershed based on the map of the watershed. Two main rivers in the parallel watershed eventually joined upstream. This shape has a significant flood discharge, particularly where the two river branches meet. At the location where the two river branches meet, this frequently causes flooding.

3.2. *Morphological Characteristics of Yeh Embang Watershed*

3.2.1. *Soil Type of Yeh Embang Watershed*

Based on Figure 4 and Table 2, in general the Yeh Embang Watershed is dominated by Brown Latosol and Litosol soil type. Latosol soil or Insepticol soil is soil that has a slum layer. The layers of slums on Latosol soil are often thick, even thick. The topography of this land is not very clear. Soil texture is usually sticky. Soil structure is usually in the form of detritus with a loose consistency. Litosol soil is a type of soil formed from the weathering process of igneous and sedimentary rocks. Litosol soil has a characteristic coarse grain in the form of gravel. This soil is very poor in nutrients so it is not fertile and not good for agriculture. Due to the infertile nature of the soil, this land is only suitable for planting large trees in the forest. Alluvial soil is a sedimentary soil, formed from silt and fine sand that undergoes soil erosion. Many are found in the lowlands, around river mouths, swamps, valleys, and on the right and left of large rivers. This soil contains a lot of sand and clay, not many nutrients. Its characteristics are gray with a slightly loose structure and are sensitive to erosion. Fertility levels are moderate to high depending on the part of the parent and climate. The properties of this alluvial soil are mostly derived from the materials transported and deposited [25].

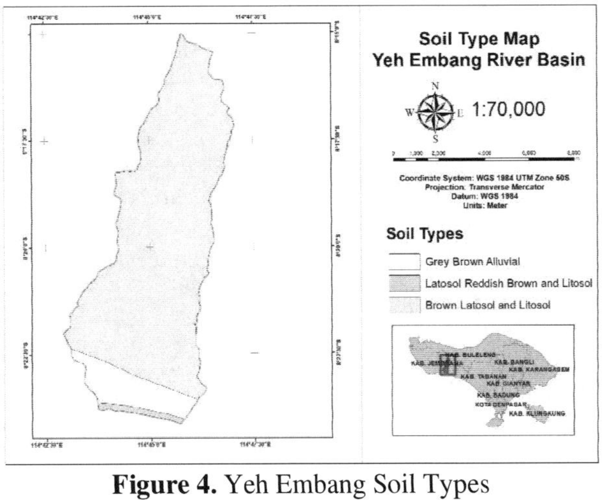

Figure 4. Yeh Embang Soil Types

Table 2. Soil Types of Yeh Embang Watershed

Soil Type	Area (km^2)	Percentage
Brown Latosol and Litosol	58.871	88.43%
Grey brown Alluvial	6.799	10.21%
Latosol Reddish brown and Litosol	0.903	1.36%

3.2.2. *Topography of Yeh Embang Watershed*

According to Figure 5 and Table 3, the Yeh Embang Watershed's height generally ranges from 0 to 1500 m, with the dominant height being less than 300 m at 69.36% of the watershed area. The Yeh Embang watershed has a high risk of flooding due to its area being predominately lowlands [26][27][28].

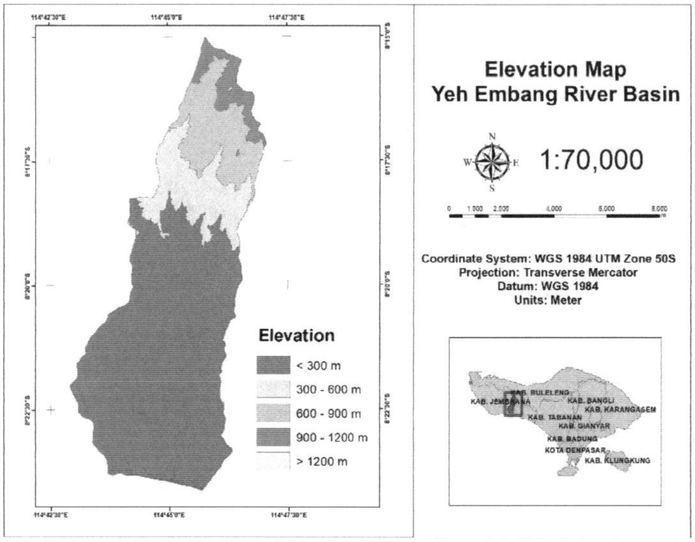

Figure 5. Yeh Embang Elevation Map

Table 3. Topography of Yeh Embang Watershed

Elevation	Area (km^2)	Percentage
< 300 m	45.995	69.36%
301-600 m	8.713	13.14%
601-900 m	8.521	12.85%
901-1200 m	3.074	4.64%
1201-1500 m	0.013	0.02%

3.2.3. Slope

The slope is the ratio of vertical (land height) to horizontal (length of flat land) in percentages; the flatter the slope, the greater the risk of flooding; and vice versa. The steeper the slope, the less vulnerable it is to flooding [29]. Table 4 shows the distribution of slope in the Yeh Embang Watershed and shows that the slope is dominated by a range of 25-45%, or 29.94% of the watershed.

Table 4. Slope of Yeh Embang Watershed

Slope	Area (km^2)	Percentage
0 - 8 %	9.547	14.47%
8 - 15 %	11.109	16.84%
15 - 25 %	13.726	20.80%
25 - 45 %	19.752	29.94%
>45 %	11.847	17.95%

3.2.4. Land Use of Yeh Embang Watershed

According to Figure 6 and Table 5, the land use of the Yeh Embang watershed is dominated by Dryland Farming (51.06%), followed by Dryland Forest (34.98%), and irrigated areas which account for 10.58% of the total area.

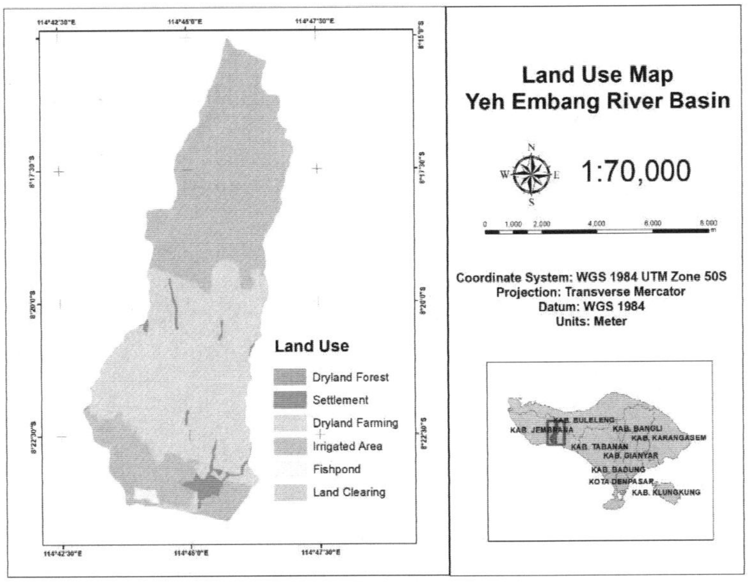

Figure 6. Yeh Embang Elevation Map

Table 5. Land Use of Yeh Embang Watershed

Land Use	Area (km^2)	Percentage
Dryland Forest	26.080	35.98%
Dryland Farming	37.003	51.06%
Settlement	1.358	1.87%
Irrigated Area	7.667	10.58%
Fishpond	0.337	0.47%
Land Clearing	0.030	0.04%

3.3. Yeh Embang Watershed's Sustainable Water Management Policy

Water availability can be sustained by enacting policies in water resource management, particularly in natural resource management systems and water distribution to meet water needs in the Yeh Embang watershed. Based on the morphometric and morphological characteristics, the Yeh Embang watershed has a large potential for flooding, especially in the downstream, besides being a meeting place for 2 rivers, the slope tends to be flat (range 0-8%), alluvial soil types with low infiltration rates have high flood potential [29]. For this reason, proper management for the Yeh Embang watershed is integrated management from upstream to downstream, in the upstream area it is hoped that the forest will be protected by forest conservation so that the potential for erosion that can cause flash floods can be prevented, as well as structuring the downstream part of the Yeh Embang watershed to prevent flooding from damaging the surrounding area. Yeh Embang watershed, infiltration wells can also be used as water conservation in residential areas so that more water that falls is absorbed and reduces surface runoff that occurs.

4. Conclusion

Based on the explanation above, it can be concluded as follows.

1. The Yeh Embang watershed is classified into very small watersheds with an area range of less than 10,000 Ha based on the analysis of morphometric characteristics. The Yeh Embang

watershed is a parallel type of watershed with a Br value of 1.77 or less than 3. According to that Br value, the Yeh Embang watershed has characteristics such as a rapid rise in flood level in the river channel.

2. According to an analysis of morphological characteristics, latosol and alluvial soil types predominate in the Yeh Embang watershed as a whole. The Yeh Embang Watershed's topography ranges from 0 to 1500 meters, with a dominant height of less than 300 meters at 30.75% of the area. The Yeh Embang watershed's land use is dominated by dryland farming (51.06%), followed by dryland forest (34.98%), and irrigated area (10.58%).
3. According to the characteristics of the Yeh Embang Watershed, integrated management from upstream to downstream be necessary for the watershed to be managed sustainably. In the upstream area, it is hoped that the forest will be protected by forest conservation to reduce the risk of erosion that could lead to flash floods, and the downstream portion of the Yeh Embang watershed will be structured to prevent flooding from harming the surrounding area. Infiltration wells can be used as a water-saving measure in residential areas in the Yeh Embang watershed to increase the amount of water that is absorbed and lessen surface runoff.

5. Acknowledgments

Acknowledgments submitted to the Bali-Penida River Region Office for providing secondary data to complete the analysis, as well as those who contributed to the completion of this paper.

References

[1] Kundzewicz Z W Krysanova V Benestad R E Hov Piniewski M and Otto I M, 2018 Uncertainty in climate change impacts on water resources *Environ. Sci. Policy* **79**, October 2017 p. 1–8.

[2] Hussain M *et al.*, 2020 A comprehensive review of climate change impacts, adaptation, and mitigation on environmental and natural calamities in Pakistan *Environ. Monit. Assess.* **192**, 1.

[3] Wasko C and Sharma A, 2017 Global assessment of flood and storm extremes with increased temperatures *Sci. Rep.* **7**, 1 p. 1–8.

[4] Qiu J Shen Z Hou X Xie H and Leng G, 2020 Evaluating the performance of conservation practices under climate change scenarios in the Miyun Reservoir Watershed, China *Ecol. Eng.* **143**, November 2019 p. 105700.

[5] Adi T B and Subrata B A G, 2019 Impact of Climate Change on Social, Economics and Culture in Indonesia *Agricultura* **109**, 2 p. 183–191.

[6] Baliexpress, 2020, Air Sungai Meluap, Warga Yehembang Waspada.

[7] Baliberkarya, 2020, Hujan Deras, Akses Jalan di Yehembang Kauh Putus Ribuan KK Terisolir, Jembrana.

[8] Tribun Bali, 2019, Tiga Bulan Warga Desa Yehembang Jembrana Krisis Air Bersih, 300 KK Hanya Andalkan Kiriman BPBD, *Tribun Bali*, Jembrana.

[9] Balipuspanews, 2019, Giliran Water Canon Turun ke Yehembang, *balipuspanews*, Negara.

[10] Suriyani L D, 2019, Mengingat Banjir Bandang, Menengok Hutan Jembrana, *mongabay*. [Online]. Available: https://www.mongabay.co.id/2019/08/12/mengingat-banjir-bandang-menengok-hutan-jembrana-1/. [Accessed: 21-Nov-2021].

[11] Suyarto R and Kusmawati T, 2016, Conditions and Problems of Water Resources and Agricultural Land in Bali Conditions and Problems of Water Resources and Agricultural Land in Bali.

[12] Pemerintah Republik Indonesia, 2012 *Peraturan Pemerintah Republik Indonesia Nomor 37 Tahun 2012 Tentang Pengelolaan Daerah Aliran Sungai* .

[13] Fulazzaky M A, 2014 Challenges of integrated water resources management in Indonesia *Water (Switzerland)* **6**, 7 p. 2000–2020.

[14] Tejada-Guibert J A, 2015, Integrated Water Resources Management (IWRM) in a Changing World, in *Sustainability of Integrated Water Resources Management: Water Governance, Climate and Ecohydrology*, M. C. Setegn, S.G., and Donoso, Ed. (Switzerland: Springer Nature

Switzerland AG), p. 65–85.

[15] Meshesha, T. W., and Khare D, 2019 Towards integrated water resources management considering hydro-climatological scenarios: an option for sustainable development *Environ. Syst. Res.* **8**, 1.

[16] Rai P K Mishra V N and Mohan K, 2017 A Study of Morphometric Evaluation of The Son Basin, India using Geospatial Approach *Remote Sens. Appl. Soc. Environ.* **7** p. 9–20.

[17] Mohamed S A and El-Raey M E, 2020 Vulnerability assessment for flash floods using GIS spatial modeling and remotely sensed data in El-Arish City, North Sinai, Egypt *Nat. Hazards* **102**, 2 p. 707–728.

[18] Eryani I G A P Manan A M A and Jayantari M W, 2021 Comparative Analysis of Watershed Characteristics in Bali Province for Sustainable Water Resources Management *Indones. J. Urban Environ. Technol.* **4**, 2 p. 210.

[19] Pemerintah Kabupaten Jembrana, 2013 *Buku putih sanitasi Kabupaten Jembrana .*

[20] Badan Pusat Statistik Provinsi Bali, 2018, Nama-Nama Sungai dan Panjangnya Menurut Kabupaten/Kota di Provinsi Bali. [Online]. Available: https://bali.bps.go.id/statictable/2018/04/10/51/nama-nama-sungai-dan-panjangnya-menurut-kabupaten-kota-di-provinsi-bali.html. [Accessed: 21-Nov-2021].

[21] Suyarto R and Kusmawati T, 2016 *Kondisi dan Permasalahan Sumberdaya Air dan Lahan Pertanian Di Bali Kondisi dan Permasalahan Sumberdaya Air dan Lahan Pertanian Di Bali .*

[22] BWS Bali-Penida, 2012, Rencana Pengelolaan Sumber Daya Air untuk Wilayah Sungai Bali - Penida (Tahap I), Denpasar.

[23] Direktorat Jenderal Bina Pengelolaan Daerah Aliran Sungai dan Perhutanan Sosial, 2013 *Pedoman Identifikasi Karakteristik Daerah Aliran Sungai* Indonesia, p. 52.

[24] Sriyana, 2011 Study of the Characteristics of the Tuntang Watershed and the Integrated Watershed Management Model *TEKNIK* **32**, 3.

[25] Iqbal M, 2021, Characteristics of Red Brick By Comparing Clay Soil Malang, Kediri, Tulungagung, Muhammadiyah Malang University.

[26] Miguez M G Veról A P De Sousa M M and Rezende O M, 2015 Urban floods in lowlands-levee systems, unplanned urban growth and river restoration alternative: A case study in Brazil *Sustain.* **7**, 8 p. 11068–11097.

[27] Merten J *et al.*, 2020 Flooding and land use change in Jambi Province, Sumatra: Integrating local knowledge and scientific inquiry *Ecol. Soc.* **25**, 3 p. 1–29.

[28] Ikkala L Ronkanen A K Utriainen O Kløve B and Marttila H, 2021 Peatland subsidence enhances cultivated lowland flood risk *Soil Tillage Res.* **212**.

[29] Eryani I G A P and Jayantari M W, 2021 Mapping of Flood Vulnerability and Mitigation Strategy of Unda River Basin Using Geographical Information System *Int. J. Eng. Sci.* **10**, 10 p. 53–61.

Wave Transformation on Wave Catcher Shore Protection – Dual Slope (WCSP-DS) Zig-Zag Model with Wave Focused Wall

MA Thaha[1], AID Puspita[1], MP Hatta[1], C Paotonan[2], WZ Prihatini[1], DA Suriamiharja[3], and F Patittingi[4]

[1]Department of Civil Engineering, Faculty of Engineering, Universitas Hasanuddin
[2]Department of Ocean Engineering, Faculty of Engineering, Universitas Hasanuddin
[3]Department of Geophysics, Faculty of Sciences, Universitas Hasanuddin
[4]Department of Civil Law, Faculty of Law, Universitas Hasanuddin

Email: arsyad999@gmail.com

Abstract. The Wave Catcher Shore Protection Dual Slope (WCSP-DS) is a coastal structure with a dual function of shore protection and wave power generation. The WCSP-DS structure captures seawater through overtopping into the crest reservoir resulting in a higher reservoir water level than the sea. This elevation difference will be the pressure height (head) and is represented by the freeboard height (Fb) and together with the overtopping discharge unit (q) will be the main parameter for electrical power. This study aims to increase the discharge value (q) at the maximum freeboard height (Fb) through engineering effort in the wave transformation process by adjusting the setting position of the WCSP in a zig-zag manner and adding a wave-focused wall (WFW) to produce a maximum deformed wave height (Hdf) in front of the WCSP-DS. Located in Hasanuddin University's Coastal & Environmental Engineering Laboratory, the study was carried out experimentally using geometrically scaled 1:20 physical model simulations. Three Dimensional (3D) wave basins in the size of 10 m x 15 m x 1 m equipped with an automatic wave generator and data recording instrument were utilized. Two type models namely the M1 (without collector) and the M2 (with collector) were simulated with relative water depth (d/z), freeboard height (Fb), incoming wave height, and period (Hi & T). The results indicated that the deformed wave height (Hdf) increased significantly ranging from 300% to 900% on M1 and from 350% to 1100% on M2. The increase in the value of Hdf also increases the discharge (q) that enters the reservoir. By these results, a combination of Fb and q can be selected which can generate optimal electrical power.

1. Introduction

Indonesia is a maritime country where 2/3 of its territory is the sea and has the second longest coastline in the world, and therefore, Indonesia has a large potential for marine energy such as wave energy, currents, tides, and temperature differences. These potentials have not been fully exploited and Indonesia has so far relied on fossil energy and coal, whose availability is dwindling. In conditions of increasing demand and the availability of abundant non-renewable energy sources, It is time to create new renewable energy sources, and the sea wave is one of those sources. The wave energy is a function of the wave height and period [1]. These two parameters thus assume a crucial role. In particular, for port cities and remote islands in Indonesia, the characteristics of wave energy sources are very suitable to be developed to meet energy needs. Due to the energy transfer process from the wind in the generating area, sea waves are one of the hydrodynamic activities that occur in the sea and on beaches [2]. The waves that travel along the coast from deep waters to shallow waters have a lot of energy, sometimes even enough that it is known that the wave has a lot of destructive power. Large and expensive breakwaters may occasionally be required to be built to destroy the wave energy in order to protect coastal areas and certain port areas. This paradigm appears to be out of date and needs to be modified to reflect the circumstances of both the present and the future. It is essential to create a multifunctional coastal protection system that serves as both a wave power plant and a barrier against wave attack [3].

Except for Indonesian waters bordering the Indian Ocean, which have not been developed, research and exploration efforts on the use of wave energy have been made in many regions of the world, particularly in the Pacific Ocean ring area, Atlantic Ocean, and the Indian Ocean. According to information, there are places with a sufficient wave energy potential ranging from 10 to 20 kW per wave meter along the South Coast of Java to Nusa Tenggara. According to the findings of several studies, wave energy at some locations in Indonesia can reach 70 kW/m. Additionally, the southern coast of Java Island's western portion and the west coast of Sumatra Island both have potential wave energy of about 40 kW/m [4].

The generation of intermittent electric current poses a number of challenges for the development of wave power plants, necessitating the use of adequate infrastructure and technology to ensure stability. Another barrier is that it costs a lot of money to build a wave power plant, which makes electricity more expensive and makes it difficult for it to compete with on-land power plants [5]. In order to lower the investment cost, it is therefore important to consider multiple functions for the wave energy conversion rather than just one.

This study is a component of work on the Zig-Zag Model Wave Catcher Shore Protection, also known as the Overtopping Wave Energy Converter (OWEC) (WCSP-DS). WCSP-DS Zig-Zag model is a model of a coastal protection structure in the form of a caisson box with dual slopes on the front wall, namely 45° slope of the upper wall and 90° or upright at the sub wall [6]. The combination slopes of this wall are intended to accommodate wave runups that can produce large volumes of overtopping to be accommodated in the crest reservoir. This research is the development of a series of OWEC breakwater studies studied by Thaha et al at the Research Center for Coastal Engineering, Faculty of Engineering, Hasanuddin University, Indonesia [7],[8],[9], and [10]. The two problems mentioned above are the focus of the entire research series. Because wave potential energy is used, the OWEC type can overcome the instability of the electric current produced. While waiting, a multipurpose coastal structure can be built to solve the issue of financial inadequacy. Engineering efforts to produce a wave transformation that generates a large wave height (H) in front of the model structure are one of the significant aspects that are being studied. It is anticipated that the maximum wave height will generate a significant overtopping discharge, increasing the amount of wave energy that can be captured and used to turn the turbine.

1.1. Problem Formulation & Research Objectives

The new paradigm that underlies this series of research is to change the working mechanism of the breakwater from destroying energy to capturing wave energy in its function of protecting the coast. The captured potential energy is then flowed to turn a turbine to produce electrical power. The amount of energy that can be utilized depends on the amount of overtopping discharge that can be caught. The magnitude of the overtopping discharge is determined by the height of the breaking wave on the WCSP-DS wall. The height of the deformed waves that arrive at the structure model that travels from the deep sea determines the height of this breaking wave. The zig-zag setting shape of WCSP-DS and the installation of a wave-focused wall under shoaling conditions are the two main factors that will be examined in this study to determine the extent to which the deformed wave height (Hdf) will increase.

2. Literature Review

2.1. Wave Energy Converter

The wave-activated body, point absorber, oscillated water column, and overtopping are the four concepts that have been successfully applied to convert wave energy. According to Lopez in Puspita et al., the point absorber type of WEC is the most widely used and developed, whereas the overtopping WEC is the less developed type. The Overtopping Wave Energy Converter (OWEC) converter works on the principle of accommodating the overtopped water into the reservoir, where potential energy is then transformed into mechanical energy by a turbine from the reservoir at the top of the converter with an elevation higher than sea level (Figure 1). The first overtopping-related invention was Tapered Chanel technology, or simply Tapchan (Figure 1). The first WEC overtopping sink of this kind was created in Norway in 1980. The principle used in Tapchan is similar to that of traditional hydroelectric power, in which ocean waves are concentrated into the reservoir through an overtopping mechanism and released back into the sea via a turbine house from the reservoir, which is located above sea level.

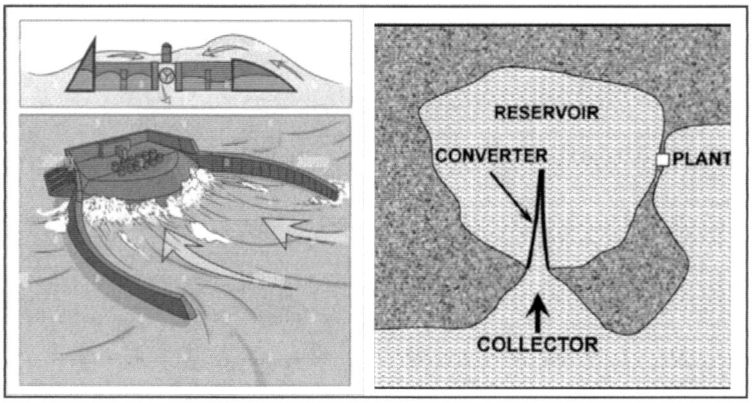

Figure 1. Wave converter Dragon type (left side) and Tapchan type (right side) [11]

2.2. Wave and Wave Transformation

Based on Laplace's equation for irrotational flow and boundary conditions on the seabed and the water's surface, the theory of airy waves is developed. Some of the wave parameters that propagate in the *x-axis* direction which are generally used in calculations as shown in Figure 2 are water depth (*d*), fluctuations in water level to still water level (x, t), which can be calculated by the equation $= \ a\ cos\ (kx.t)$, where, a: amplitude, wave height $(H\ =\ 2a)$, wavelength (L), Wave period (T), Wave propagation speed, $C\ =\ L/T$; wave number $(k\ =\ 2\pi/L)$; wave frequency $(\sigma =\ 2\pi\ /T)$, acceleration due to gravity $(g\ =\ 9.81\ m/s2)$. The relationship of wave propagation speed with wavelength and depth is:

$$C = \frac{gT}{2\pi} tanh \frac{2\pi d}{L} \tag{1}$$

And the relationship wavelength (L) in the function of water depth (d) is:

$$L = \frac{gT^2}{2\pi} tanh \frac{2\pi d}{L} \tag{2}$$

where,

$$n = \frac{1}{2}\left(1 + \frac{2kd}{Sinh2kd}\right) \tag{3}$$

If a series of waves propagate towards the coast, then the wave will undergo a change in shape or transformation including the process of refraction, shoaling, diffraction, reflection, and breaking waves. The idea of equivalent deep-sea waves, or the height of deep-sea waves if refraction does not occur, is frequently used in wave transformation analysis. The use of the transformation wave aims to determine the height of the wave undergoing refraction, diffraction, and other transformations so that it is easier to estimate the transformation and deformation of the wave. The form provides the equivalent deep sea wave height.

$$H_o' = K'.\ K_r.\ H_o \tag{4}$$

where Ho' is the equivalent height of deep-sea waves, Ho is the height of deep-sea waves, K' is the diffraction coefficient, and Kr is the refraction coefficient.

A coastal structure will reflect some or all of the energy of an incoming wave. Waves will completely reflect off of the smooth, impermeable walls of vertical structures. The reflection coefficient (Kr), which is the product of the reflection wave height Hr and the incident wave height Hi, determines the extent to which a structure can reflect waves:

$$Kr = \frac{Hr}{Hi} \tag{5}$$

The modification brought on by shoaling is another wave of deformation. The Shoaling Coefficient, which is the proportion of deep-water wavelengths to beach wavelengths, can also be used to measure shoaling.

$$Ks = \sqrt{\frac{n_0 L_0}{nL}} \tag{6}$$

where the value of n changes to 1 in shallow waters and $n0$ = the depth factor in deep sea waters.

2.3. *Wave Energy and Wave Power*

Kinetic energy and potential energy of waves are the two types of wave energy. The orbital velocity of particles caused by wave motion generates kinetic energy. While the potential energy is caused by a wave's motion shifting the water's level. The components of potential energy and kinetic energy are the same for the small-amplitude wave theory if the wave energy is set relative to a stationary water surface and all waves propagate in the same direction. Consequently, one wavelength's total energy per unit wavelength is [12]:

$$Et = Ek + Ep = \frac{\rho g H^2 L}{8} \tag{7}$$

While the average energy per unit area is:

$$E = \frac{Et}{L} = \frac{\rho g H^2}{8} \tag{8}$$

Wave power (P) is the wave energy per unit time in the direction of wave propagation, where Ek is the kinetic energy per unit wavelength ($joules/m$), Ep is the potential energy per unit wavelength ($joules/m$), Et is the total energy per unit wavelength ($joules/m$), E is the average wave energy per unit area ($joules/m2$), H is the wave height (m), is the density of water ($kg/m3$), and g is the acceleration due to gravity (m).

$$P = \frac{nE}{T} \tag{9}$$

The available power in the reservoir is [13]:

$$D = Qh\rho g \tag{10}$$

Where h is the difference height between the water level in the reservoir and the sea surface.

2.4. *Wave Energy Conservation*

A wave that propagates enters a narrowed area, then its energy will be conserved at the 2 points under consideration. Figure 2 presents the conservation of wave energy in a narrowed channel.

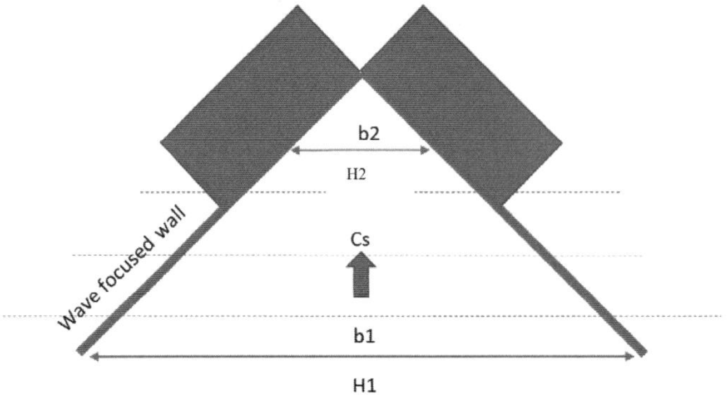

Figure 2. Conservation of wave energy on Focusing Wall

$$(EnC)_1 b_1 = (EnC)_2 b_2 \tag{11}$$

Then, the wave height $H2$ can be solved:

$$H_2 = H_1 K_s \sqrt{b_1 / b_2} \tag{12}$$

Where, $C = L/T$ = Wave propagation velocity (m/s); Ks = Shoaling coefficient; $H1$ = wave height at point 1, while $H2$ = wave height at the point 2; $b1$ = width of the focused wall at the point 1, while $b2$ = width of the focused wall at the point 2.

3. Research Method

The study was carried out at the Faculty of Engineering, Hasanuddin University, Indonesia's Coastal & Environmental Engineering Laboratory. In a wave basin measuring 10 m by 15 m and having a depth of 1 m, experimental studies on 3D physical models are being conducted. Three different wave heights and periods (H & T), five different freeboard heights (Fb), and two different WCSP-DS models with and without wave energy-focusing collector walls (WFW) are simulated at three different elevations of the water level in relation to the vertical wall ($d/z = 0.857$; $d/z = 1$ & $d/z = 1.13$) as shown in Figure 3. The figure presents a layout physical simulation in a 3D wave basin 4.

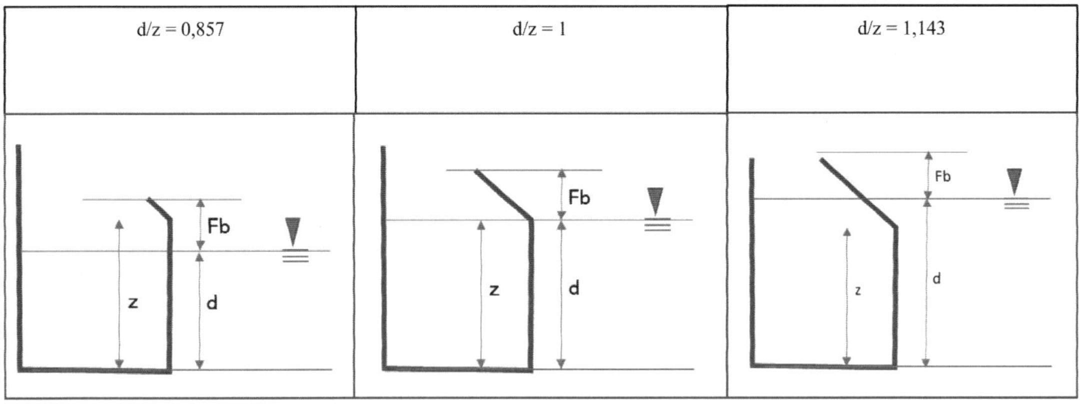

Figure 3. Three different types of water level elevation profiles in relation to a vertical wall

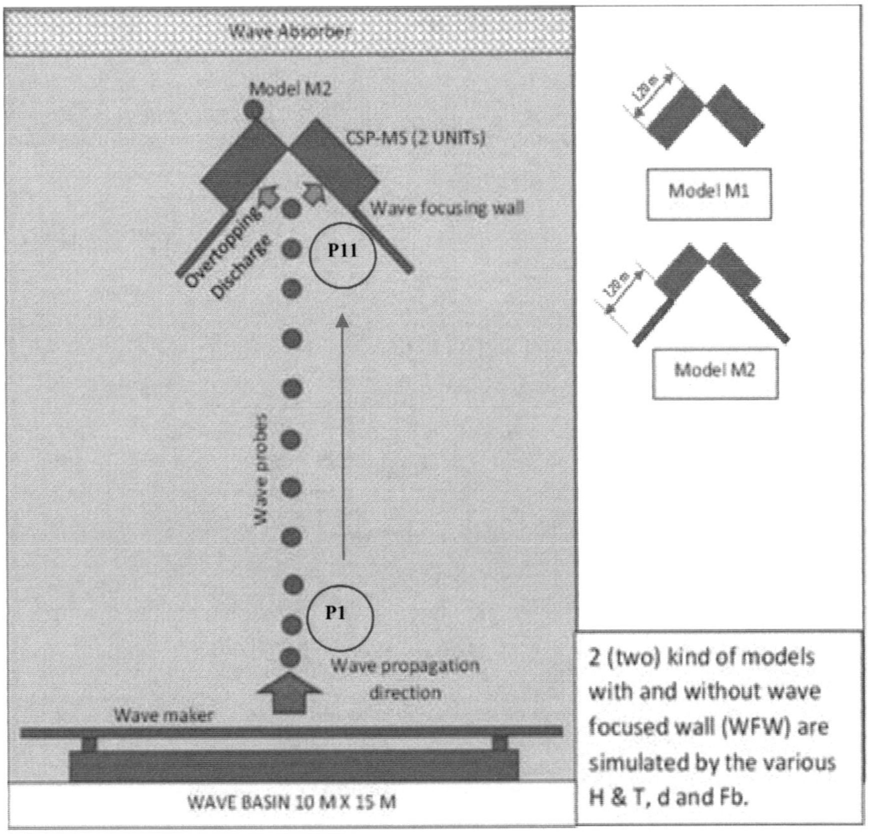

Figure 4. The layout of 3D physical simulation

From the wave generator to the WCSP-DS model, the positions of the probes P1 to P11 are listed in order. The wave generator device is set up in accordance with the simulation plan after the model is constructed and placed in the wave basin in accordance with the plan shown in Figure 4. Thereafter, the 11-piece calibration of the generator and wave height measuring probes is completed. The following step involves wave height simulation and data collection for all probes. Other variations are tried out by repeating the process.

4. Result and Discussion

In order to find data trends, relationships, or effects between parameters, the water level fluctuation recording data from the simulation results are then converted to wave height (H) on all probes (P1 - P11). A sample of the simulation parameter data and the recorded deformation wave height for the zigzag model without collector (M1) and with the collector (M2) are shown in Table 1.

Table 1. Simulation parameter data for d/z =1.143 and T = 2 s in the M1 and M2 experiments

Model	Running Cyrcles	Water Depth	Freeboard		Height of Vertical Wall	Water Depth Relative	Period	Wave Length in Deep Water	Wave Length	Wave Amplitude	Running Time
	Code	d (m)	Rc Lab	Fb (m)	z (m)	d/z	T (dt)	Lo (m)	L (m)	a (cm)	Sekon (s)
M1	V11	0.4	40	0.35	0.35	1.142857	2.0	6.2400	3.8837	4	180
	V12	0.4	40	0.35	0.35	1.142857	2.0	6.2400	3.8837	5	180
	V13	0.4	40	0.35	0.35	1.142857	2.0	6.2400	3.8837	6	180
	U11	0.4	35	0.3	0.35	1.142857	2.0	6.2400	3.8837	4	180
	U12	0.4	35	0.3	0.35	1.142857	2.0	6.2400	3.8837	5	180
	U13	0.4	35	0.3	0.35	1.142857	2.0	6.2400	3.8837	6	180
	etc	....		....	....	....	....	....	....	....	....
M2	X11	0.4	40	0.35	0.35	1.142857	2.0	6.2400	3.8837	4	180
	X12	0.4	40	0.35	0.35	1.142857	2.0	6.2400	3.8837	5	180
	X13	0.4	40	0.35	0.35	1.142857	2.0	6.2400	3.8837	6	180
	W11	0.4	35	0.3	0.35	1.142857	2.0	6.2400	3.8837	4	180
	W12	0.4	35	0.3	0.35	1.142857	2.0	6.2400	3.8837	5	180
	W13	0.4	35	0.3	0.35	1.142857	2.0	6.2400	3.8837	6	180
	etc	....		....	....	....	....	....	....	....	....

Table 2. A part of recorded deformation wave height (Hdf) and overtopping unit discharge (q) for each simulation cycle at relative depth (d/z) = 1.143 and T = 2s

Model	Running Cyrcles	Wave Height Record at 11 Probes (m)											Overtopping Discharge Unit
	Code	P1	P2	P3	P4	P5	P6	P7	P8	P9	P10	P11	q/1000 (m3)
M1	V11	0.073	0.073	0.068	0.114	0.123	0.112	0.117	0.105	0.096	0.051	0.152	0.000
	V12	0.094	0.094	0.095	0.156	0.145	0.141	0.134	0.141	0.127	0.060	0.183	0.000
	V13	0.113	0.113	0.121	0.188	0.178	0.179	0.155	0.168	0.159	0.080	0.194	0.000
	U11	0.073	0.073	0.077	0.122	0.134	0.119	0.135	0.120	0.111	0.057	0.161	0.000
	U12	0.097	0.097	0.106	0.169	0.161	0.157	0.146	0.151	0.140	0.065	0.187	0.000
	U13	0.128	0.128	0.128	0.207	0.201	0.192	0.168	0.189	0.181	0.088	0.215	0.026
	etc	etc	....		....	....	....	....	....	....	....	....	....
M2	X11	0.095	0.084	0.175	0.202	0.086	0.107	0.159	0.163	0.062	0.226	0.430	0.009
	X12	0.121	0.088	0.243	0.230	0.107	0.145	0.213	0.195	0.061	0.284	0.473	0.415
	X13	0.143	0.102	0.262	0.282	0.127	0.148	0.202	0.219	0.075	0.301	0.540	1.161
	W11	0.095	0.086	0.172	0.213	0.088	0.105	0.178	0.166	0.070	0.231	0.433	0.054
	W12	0.121	0.095	0.226	0.241	0.106	0.126	0.204	0.189	0.062	0.284	0.484	0.893
	W13	0.114	0.104	0.187	0.214	0.105	0.133	0.177	0.157	0.072	0.258	0.462	1.662
	etc	etc	....		....	....	....	....	....	....	....	....	....

The relationship between several simulation parameters, such as relative depth (d/z), Freeboard height (Fb), overtopping unit discharge (q), and deformation wave height (Hdf), is then examined using all acquisition data, such as Table 2 above. From this recorded data, it can be seen that all simulation cycles resulted in an increase in the height of the deformation wave with the maximum height occurring at Probe 11 (at front of the model) (at front of the model).

4.1. *Effect of Relative Depth* (d/z)

The first analysis that needs to be performed compares the impact of the water level's vertical position on the WCSP's vertical wall height. Three vertical positions are examined for their effects: $d/z = 1.143$, which indicates that the water level is on an inclined upper wall; $d/z = 1$, which indicates that the water level is at the intersection of an inclined upper wall and an upright sub wall; and $d/z = 0.857$, which indicates that the water level is at a sub wall (upright wall). In Figure 5, the deformation wave profiles are compared for the three different water level positions in the model without and with WFW under the conditions of freeboard height (Fb), amplitude (a), both at wave period (T) $= 2.0\ s$, and at $T = 2.3\ s$.

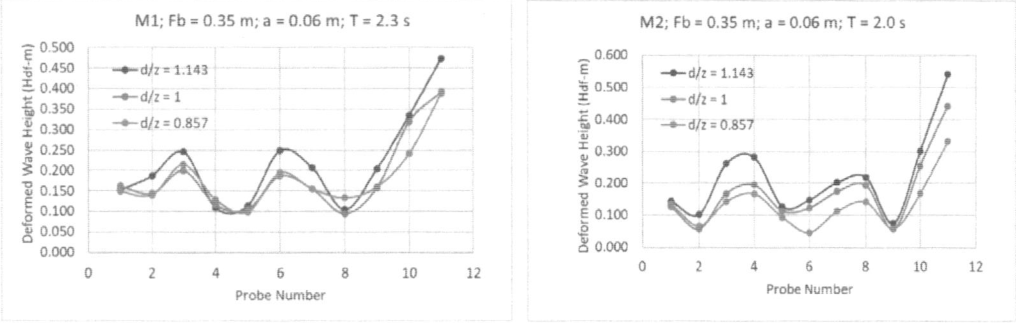

Figure 5. Comparison of three different face elevation positions' deformation wave profiles

Figure 5 shows that from Probe 1 to Probe 11, the wave height significantly increased for all deformation wave profile curves that occur in the M1 model, the model without WFW, as well as the M2 model with WFW. The three curves in both models demonstrate that $d/z = 1.143$ is the relative face elevation position that has the greatest impact on an increase in wave height. This indicates that the water level elevation is on the upper wall (inclined 45° wall) or above a sub-wall (upright wall). Therefore, the d/z value will be discussed further.

4.2. *Relationship of Freeboard Height (Fb), Overtopping Discharge (q), and Deformed Wave Height (Hdf)*

The freeboard height is one of the key factors in the OWEC study (Fb). The overtopping discharge unit (q) that enters the reservoir is significantly influenced by the freeboard height, and q itself is significantly influenced by the deformation wave height (Hdf) formed. The deformation wave height (Hdf) is a significant variable in this study that affects the magnitude of the difference in water levels between the reservoir and the sea (h), which affects how much power the reservoir can produce. To increase the Hdf, engineering is carried out by placing the WCSP model box in a zig-zag manner and providing an additional structure that changes the collector wall to concentrate the wave energy. There is a triangular relationship that influences each other between the freeboard height (Fb), the overtopping unit discharge (q), and the deformation wave height (Hdf). Figure 6 and figure 7 present the relationship between the three parameters and their comparison in the M1 and M2 models.

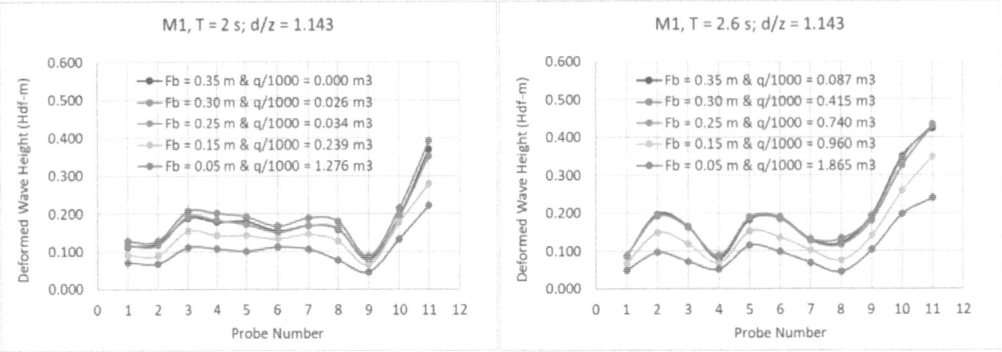

Figure 6. The relationship between Fb, q and Hdf at $d/z = 1.143$ and for $T = 2\ s$ & $T = 2.6\ s$ of M1 model (zigzag model without WFW)

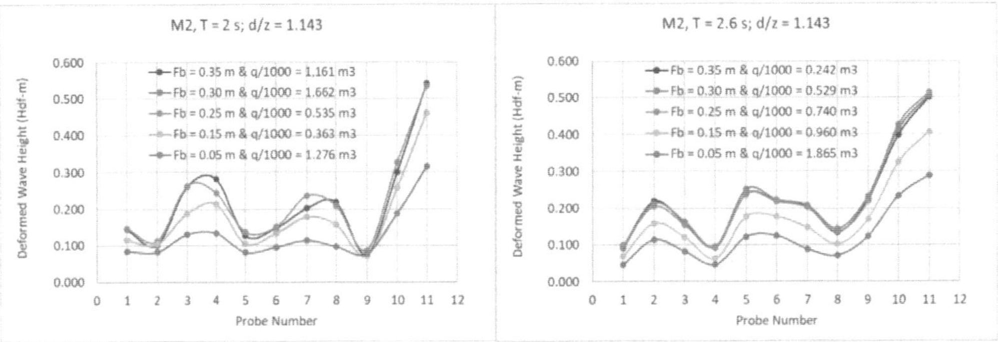

Figure 7. The relationship between Fb, q and Hdf at $d/z = 1.143$ and for $T = 2\ s$ & $T = 2.6\ s$ of M2 model (zigzag model with WFW)

Figure 6 and figure 7 show the wave transformation curve that occurs during propagation from point 1 (in the deep sea) to point 11 which is located in front of the WCSP-DS Model. The effect of the wave period (T) within the limits of the studied values has no significant effect on Hdf for both model conditions. The presence of an additional energy-focusing collector structure has a significant effect on the Hdf height, especially in front of the model (P11), where the Hdf achievement on M2 is greater than that of M1 for all Fb conditions. It can be seen that the larger the Fb, the larger the recorded Hdf, and the smaller the overtopping discharge (q) that is caught. This condition can also be explained that the greater the discharge (q) that flows into the reservoir due to the lower Fb, the smaller the Hdf that occurs. This can be understood in accordance with the theory that the Hdf height is formed by 3 mechanisms, namely energy concentration with the application of the law of energy conservation, the formation of standing waves due to reflection by walls with a height of Fb, and the shoaling process due to the slope of the seabed (Ks). This finding is also consistent with the research of Hatta M. P. et al. and Puspita A. I. D. et al. The value of Hdf decreases with increasing overtopping discharge as shown in the bottom 2 curves in Figure 6 and Figure 7. The top three curves tend to be high and close together because there is no or almost no overtopping discharge into the reservoir. These results indicate that there is an ideal combination of the height of Fb and the value of q which will give the best value to produce optimal electrical power.

4.3. Increase in Deformed Wave Height

Waves will change in height as they move from the deep sea to the shallow water along the coast. The amount of energy captured increases with increasing wave height. Figure 8 shows the wave steepness scale's (Hi/L) percentage increase in wave height from point 1 in the transition ocean to point 11 in front of the model.

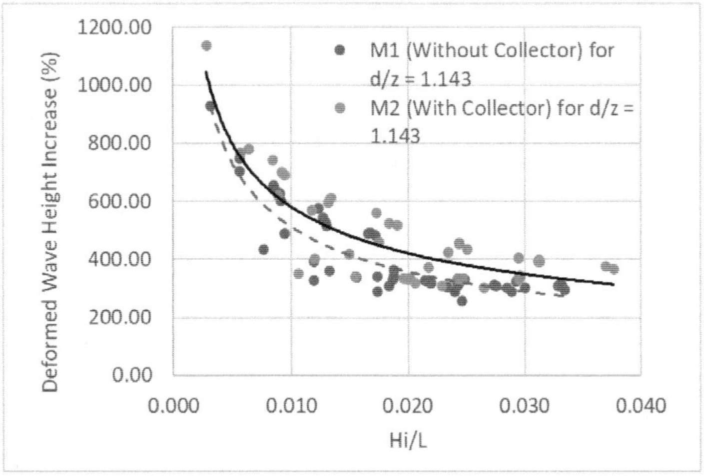

Figure 8. Relationship of increase in Hdf (%) with wave steepness (Hi/L) for Model M1 (without collector) and M2 (with collector) at relative depth $(d/z) = 1.143$

From Figure 8 it is clear that the wave steepness (Hi/L) is a significant factor in the rise in the deformed wave height (Hdf). This is because the greater the value of Hi/L, the greater the chance of the waves breaking on the walls of the structure causing the reflection of the waves to be small so that standing waves are not formed. This is also in accordance with what was also found by Puspita A. I. D et al. It can be seen that the higher the steepness of the Hi/L wave, the less the deformation wave height increases by percentage (Hdf). It is also seen that the WCSP model with a wave energy focusing collector (M2) gives a higher percentage value than M1 (without a collector). The Hdf height increase in the ranges of 300% to 900% on zigzag WCSP models without additional collectors, while the Hdf increase in the ranges of 350% to 110% for the models with additional collectors (M2) with a length equal to the length of the WCSP model. Based on the small difference between both models, then it can be recommended for the M1 type for suitable implementation in the field.

5. Conclusion

Following is a list of conclusions that can be drawn from the analysis's findings.

1. From the analysis of the relationship between relative water depth (d/z) and Hdf, the largest value is obtained at $d/z = 1.143$, or the position of the water level elevation is on a sloping wall above the WCSP-DS upright wall for both conditions of the model studied.
2. Obtained a triangular relationship that influences each other between the freeboard height (Fb), the overtopping unit discharge (q), and the deformation wave height (Hdf).
3. At a high Fb value, the overtopping discharge value becomes small, causing the Hdf value to also be high due to the formation of standing waves due to high reflection energy and vice versa.
4. The results show that there are ideal conditions for the combination of the height of Fb and the value of q which will give the best value to produce optimal electrical power according to Equation (10).
5. The percentage increase in Hdf height ranges from 300% to 900% on zigzag WCSP models without additional collectors and ranges from 350% to 1100% increases in models with additional collectors (M2) with a length equal to the length of the WCSP model.
6. With the small difference between both models, it can be considered to choose the M1 model for implementation in the field.

Acknowledgment
This study is part of the work within the program RUNHAS LPPM Universitas Hasanuddin.

References

[1] Dean, R. G. A. Dalrymple, 1984, Water Waves Mechanics for Engineer and Scientist, Prentice Hall, Inc., New Jersey; Englewood Cliffs.
[2] Triatmodjo B., 1999, Teknik Pantai. Beta Offset. Yogyakarta
[3] Thaha M A, Maricar F, Aboe A F, Puspita A I D 2015 The breakwater, from wave breaker to wave catcher *Procedia Engineering Elsevier*, 6e 91-698.
[4] Mustapa M 2017 Wave energy device and breakwater integration: A review *Renewable and Sustainable Energy Reviews*, 43-58.
[5] Thaha M. A., Puspita A. I. D., Suriamiharja D. A., Patittingi F., Paotonan C., Mustari A. S., 2021, Estimation of The Potential Utilization of Wave Power Surrounding Bali Beach using CSP-MS., *IOP Conference Series: Earth & Environmental Science*, doi:10.1088/1755-1315/841/1/012022.
[6] Puspita A. I. D., Thaha M. A., Pallu M. S., Maricar F., 2020, Effect of Wave Steepness to Relative Wave Runup on OWEC Breakwater, IOP Conference Series: Earth & Environmental Science, doi: 10.1088/1755-1315/419/1/012117.
[7] Hatta M.P., Puspital A. I. D., Thaha M. A., Karamma R., Pongmanda s., Mustari A. S., and Ibrahim M., 2020, Experimental Study of Wave Reflection in Breakwater Overtopping Catcher Model, IOP Conference Series: Materials Science and Engineering, doi:10.1088/1757-899X/875/1/012026.
[8] Puspita A. I. D., Pallu M. S., Thaha M. A., Maricar F., 2020, The Effect of Wave Reflection Coefficient to The Breaker Parameter on OWEC Breakwater, IOP Conference Series: Earth & Environmental Science, doi:10.1088/1755-1315/419/1/012135.
[9] Puspita A. I. D., Thaha M. A., 2021, Experimental Investigation of Wave Reflection at a Wave Energy Converter Breakwater, IOP Conference Series: Earth & Environmental Science, doi:10.1088/1755-1315/841/1/012029.
[10] Puspita, A. I. D., Pallu M. S., Thaha M. A., Maricar F., 2020, The Effect of Wave Deformation on Overtopping Discharge in Wave Energy Converter (OWEC)-breakwater: Journal of Engineering and Applied Sciences, Page No.: 2058-2064, 2020.
[11] Alamy, 2018, Wave Dragon, Wave Energy Converter, Illustrationhttps://www.alamy.com.
[12] CERC., 1984, "Shore Protection Manual", Department of The Army Waterway Experiment Station, Corp of Engineering Research Centre Fourth Edition, US Government Printing Office, Washington.
[13] Triatmodjo B., 1996, Hidrolika II. Beta Offset. Yogyakarta.

The Effectiveness of Tidal Power Plant Model

A A Rusvan[1], F Maricar[2], M A Thaha[2], C Paotonan[3]

[1] Doctor Program of Civil Engineering, Hasanuddin University, Makassar, South Sulawesi, Malino Street, INA
[2] Department of Civil Engineering, Hasanuddin University, Makassar, South Sulawesi, Malino Street, INA
[3] Department of Ocean Engineering, Hasanuddin University, Makassar, South Sulawesi, Malino Street, INA

email: alfa_fangky@yahoo.co.id (correspondence author).

Abstract. Since tidal energy is renewable and the Indonesian archipelago has a beach length of roughly 80,000 km, the country has a very large potential for tidal energy. The possibility for utilizing tidal energy is highly open because there are several places in Indonesia where the high tide and low tide are quite different from one another. When calculating the quantity of tidal energy based on the greatest and lowest sea level readings, tides are a crucial factor. The moon and sun's gravitational pull are what bring about the tides. The structural characteristics of pristine beaches, coastal waterways, such as bays and narrow straits, and the depth near the shore are additional local elements that affect.. The results of this study use the assumption that the tidal height difference that occurs is 5 m for 1 tidal cycle (12 hours) or with a single daily tidal type model (Diurnal Tide) or in one day there is one high tide and one low tide accompanied by Reservoir length 150 m with a reservoir width variation of 60 m produces a maximum discharge of 12.50 m^3/s with a maximum energy potential of 439.92 kWh, a variation in the width of a pool of 80 m produces a maximum discharge of 16.67 m^3/s with a maximum energy potential of 586.56 kWh and a variation of the width of the reservoir 100m produces a maximum discharge. 20.83 m^3/s with a maximum energy potential of 733.20 kWh. The results above make it abundantly evident that the potential energy produced increases with the size of the pond formed, the magnitude of the tidal height difference that occurs, and the duration of the production processes. In order to maximize tidal energy as a potential alternative energy source, it may therefore be inferred that tidal energy consumption can be used as a reference in future development planning. This article is a part of ongoing research on single-pond dual cycle systems for tidal energy empowerment.

1. Introduction

Tidal power, solar thermal power, geothermal power, wind power, and hydrodynamic magnetic fields are examples of renewable energy sources that could be used. In the world of industry and the economy, the amount of energy produced by these primary sources is still quite little.

Being an archipelago, Indonesia is encircled by two sizable oceans, the Indian Ocean and the Pacific Ocean, giving Indonesia a significant tidal potential. The energy produced by this power plant is pretty enormous, reaching 240 MW, and it is exactly located in Kuala Rance, which is between Saint Maro and Dinard, Brittany, France. France constructed a tidal power from sea water power plant in 1968. [1].

The majority of Indonesia continues to use traditional fuels. Petroleum and other conventional energy sources are nonrenewable (unrenewable) and will eventually run out even when used sustainably.

Continuous energy demands will result in a decline in the crucial energy requirements for human survival. Because conventional energy is not renewable, its use will have an effect on the environmental situation. By harnessing the potential energy held in each flow in the reservoir, tidal energy can be employed as a substitute energy source. An alternate source that can currently be explored is tidal energy. [1].

As a renewable source of energy that doesn't contribute to pollution, tidal energy in Indonesia has a very high potential to promote sustainable economic development (pollution free) [3]. Due to Indonesia's equatorial location, which produces relatively big tides, winds, waves, and ocean currents, as well as the Indonesian archipelago's about 80,000 km of coastline, there are sometimes quite huge and variable tidal differences, particularly in Indonesian waters. This article is a part of a series of studies being done on two-way systems for capturing tidal energy.

2. Methodology

2.1. Tidal

According to equilibrium theory, the rotation of the earth on its axis, the revolution of the moon with respect to the sun, and the revolution of the earth with respect to the sun are the factors that influence tides. Meanwhile, based on dynamic theory, it is the depth and breadth of waters, the influence of the earth's rotation (Coriolis force), and basic friction. The bathymetry of the seabed, the width of the strait, the shape of the bay, and other local factors can also have an impact on the tides in a body of water, resulting in different tidal characteristics in different places. [5]. The tides in Indonesia are divided into 4, namely: single daily tides (Diurnal Tide), double daily tides (Semi Diurnal Tide), single mixed tides (Mixed Tide, Prevailing Diurnal), and mixed tides mixed tide mixed daily double skew (Mixed Tide, Prevailing Semi Diurnal) [5].

For this study, because it uses the assumption of 1 tidal cycle for 12 hours, the closest assumption is a single daily tide (Diurnal Tide) or in one day there is one high tide and one low tide with a tidal height difference of 5m in 1 day. tidal cycle (12 hours).

2.2. Tidal Energy in dual cycle system

The potential of tidal energy is obtained by using a single pond system simulation and a double cycle system. This single pond system is a combination of a single tide and tide cycle [4].

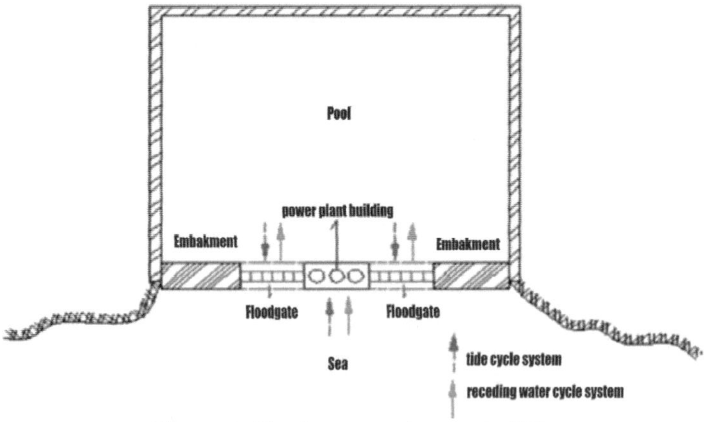

Figure 1. Single reservoir system [1]

This system is a blend of a single tide and tide cycle system. This method is very useful since it is possible to generate electricity at high tide and at low tide. The water gates are opened when the T1-T2 of the sea is experiencing a low tide so that the water in the reservoir can drain out along with the retreating sea water, causing it to reach its lowest point. When the storage reservoir's level is equal to that of the sea, or T2, the floodgates are closed, leaving the reservoir empty. When the minimum height difference is reached, the water starts to be fed into the reservoir through the turbine at T3, which will cause electricity to be produced between T3 and T4. The waiting period between T2 and T3 is necessary so that the water has a minimum height difference (Hmin) that is considered capable of moving the turbine. The floodgates were shut during T4 because the water level couldn't turn. The period of time between T4 and T5 is needed to allow the reservoir's water level to match that of the sea, which is receding, The floodgates are opened at T5 to release water into the retreating sea from the reservoir. The creation of electricity in the ensuing cycle will be the same as it was in the one before, as shown in the figure below.

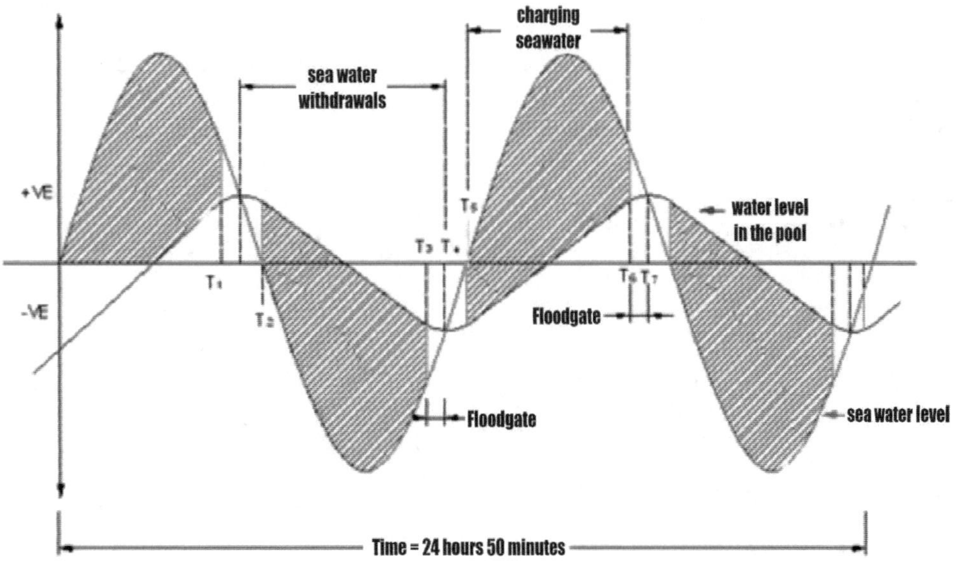

Figure 2. Double cycle system in single reservoir [1]

When the reservoir is filled (at high tide) or drained (during low tide), electrical energy can be produced (during low tide). The amount of electrical energy produced with this technique depends on how long it is used for production. Estimates of predicted electrical energy can be determined using the reservoir's area, its function, and the difference between the tide's height and the discharge that results, namely the volume of the reservoir inflow, V (m^3);

$$V = A \times \Delta h \qquad (1)$$

water discharge is average; Q (m^3/sec);

$$Q = \frac{V}{t} \qquad (2)$$

Theoretical discharge is average; Qt (m^3/sec);

$$Q_t = \frac{1}{4}\pi \times D^2 \times \sqrt{2gH} \qquad (3)$$

Discharge coefficient is average; Cd;

$$C_d = \frac{Q}{Q_t} \qquad (4)$$

The amount of energy that can be produced, P (Watt);

$$P = 0{,}7 \times \rho \times Q \times g \times H \qquad (5)$$

IOP Conf. Series: Earth and Environmental Science 1117 (2022) 012062 doi:10.1088/1755-1315/1117/1/012062

Energy is average; E (kWh);

$$E = \frac{1}{2} \times (\eta \times \rho \times g \times A \times H^2)$$

3. Result and Discussion

In the calculation method for the estimation of electric power potential, 3 methods are used [1], in the three methods the first assumption is that the water flow is steady (*steady flow*). This runs counter to the fact that, at a given site, the amount of pressure and the speed of the flow of sea water into a reservoir or reservoir water into the sea fluctuate with time (unsteady flow). The second supposition is that only tidal production is included in the analysis of energy availability. The third assumption is that the size of the reservoir used is 150m long with variations in width, namely 60m, 80m and 100m. The assumption of the type of tidal used is diurnal tide or In one day there is one high tide and one low tide with a tidal height difference of 5m in 1 tidal cycle (12 hours). The estimated electrical energy is calculated while taking into account the fact that the value obtained for estimating tidal power potential only serves to roughly define the availability of potential energy. This information on energy availability can be utilized to map the amount of energy that is present in the review region for more in-depth uses.

Another thing that affects the potential for tidal power generation is the amount of water discharge which is strongly influenced by the width of the submerged hole (1.77m) and also the greater the water discharge prouced, it will be directly proportional to the increase in the value of the discharge coefficient.

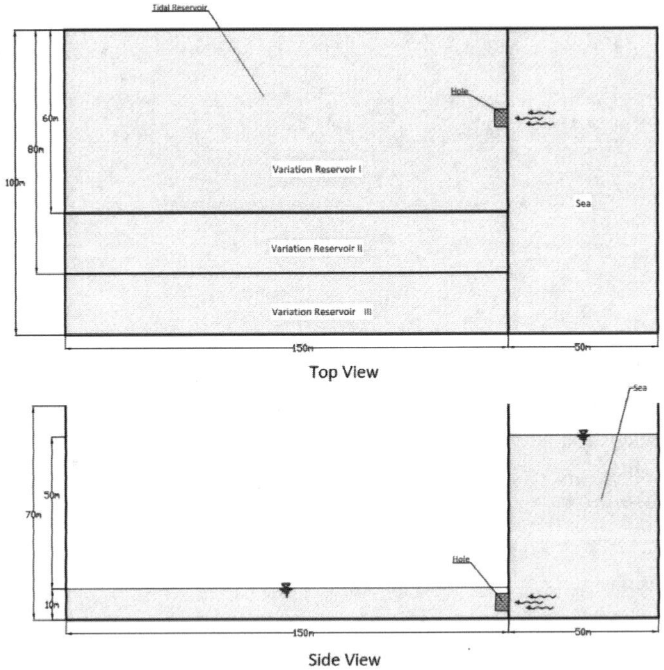

Figure 3. Tidal power plant reservoir variation model design

Table 1. Calculation of discharge and reservoir hole area for 1 tidal cycle (12 hours)

Tidal height difference (m)	Water discharge (m^3/s) width 60m	Water discharge (m^3/s) width 80m	Water discharge (m^3/s) width 100m	Discharge coefficient (Cd) width 60m	Discharge coefficient (Cd) width 80m	Discharge coefficient (Cd) width 100m	Hole width (a) (m)
5.00	12.50	16.67	20.83	0.71	0.95	1.19	1.77
4.58	11.46	15.28	19.10	0.68	0.91	1.14	1.77
4.17	10.42	13.89	17.36	0.65	0.87	1.09	1.77
3.75	9.38	12.50	15.63	0.62	0.82	1.03	1.77
3.33	8.33	11.11	13.89	0.58	0.78	0.97	1.77
2.92	7.29	9.72	12.15	0.55	0.73	0.91	1.77
2.50	6.25	8.33	10.42	0.50	0.67	0.84	1.77
2.08	5.21	6.94	8.68	0.46	0.61	0.77	1.77
1.67	4.17	5.56	6.94	0.41	0.55	0.69	1.77
1.25	3.13	4.17	5.21	0.36	0.48	0.60	1.77
0.83	2.08	2.78	3.47	0.29	0.39	0.49	1.77
0.42	1.04	1.39	1.74	0.21	0.27	0.34	1.77

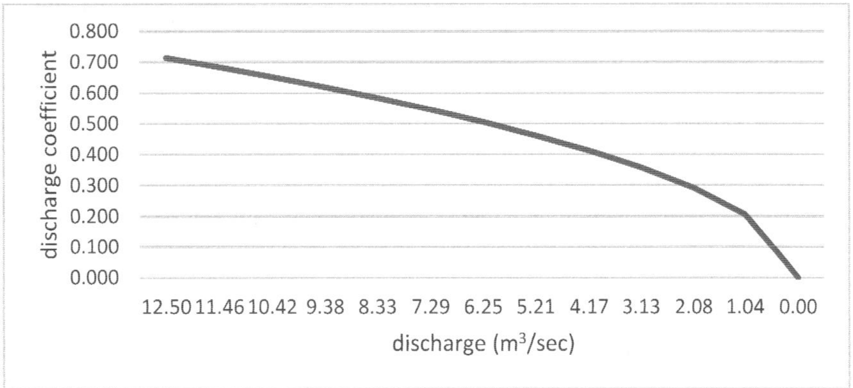

Figure 4. Ratio of discharge and discharge coefficient with a variation of the width of the reservoir 60m

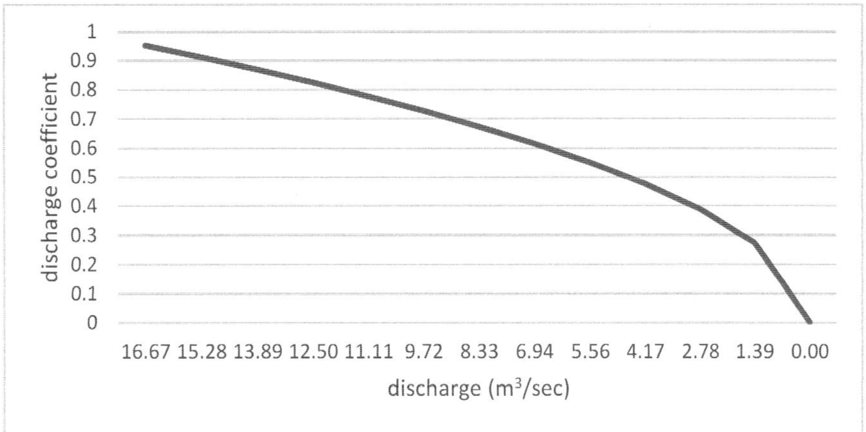

Figure 5. Ratio of discharge and discharge coefficient with a variation of the width of the reservoir 80m

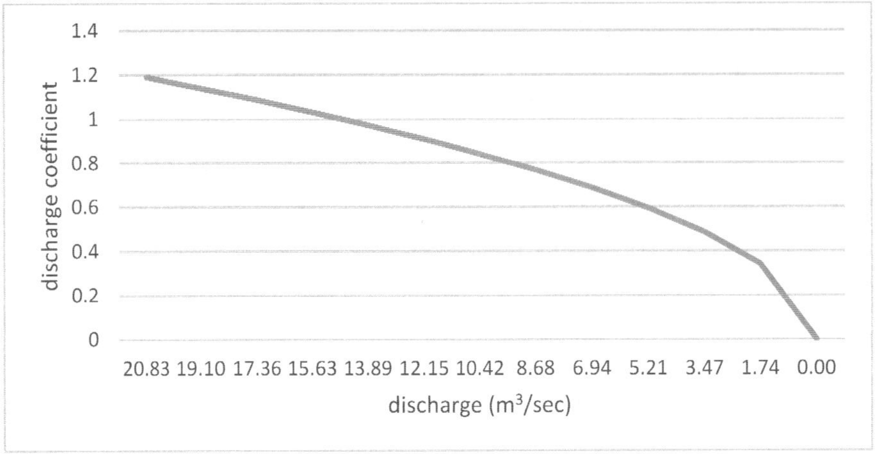

Figure 6. Ratio of discharge and discharge coefficient with a variation of the width of the reservoir 100m

Table 2. Tidal energy potential for 1 tidal cycle (12 hours)

Reservoir length (m)	Reservoir width (m)	Time (hour)	Water discharge (m^3/s) width 60m	Water discharge (m^3/s) width 80m	Water discharge (m^3/s) width 100m	Energy (kWh) width 60m	Energy (kWh) width 80m	Energy (kWh) width 100m
150	60, 80,100	12	12.50	16.67	20.83	439.92	586.56	733.20
150	60, 80,100	11	11.46	15.28	19.10	369.65	492.87	616.09
150	60, 80,100	10	10.42	13.89	17.36	305.50	407.33	509.16
150	60, 80,100	9	9.38	12.50	15.63	247.45	329.94	412.42
150	60, 80,100	8	8.33	11.11	13.89	195.52	260.69	325.86
150	60, 80,100	7	7.29	9.72	12.15	149.69	199.59	249.49
150	60, 80,100	6	6.25	8.33	10.42	109.98	146.64	183.30
150	60, 80,100	5	5.21	6.94	8.68	76.37	101.83	127.29
150	60, 80,100	4	4.17	5.56	6.94	48.88	65.17	81.47
150	60, 80,100	3	3.13	4.17	5.21	27.49	36.66	45.82
150	60, 80,100	2	2.08	2.78	3.47	12.22	16.29	20.37
150	60, 80,100	1	1.04	1.39	1.74	3.05	4.07	5.09

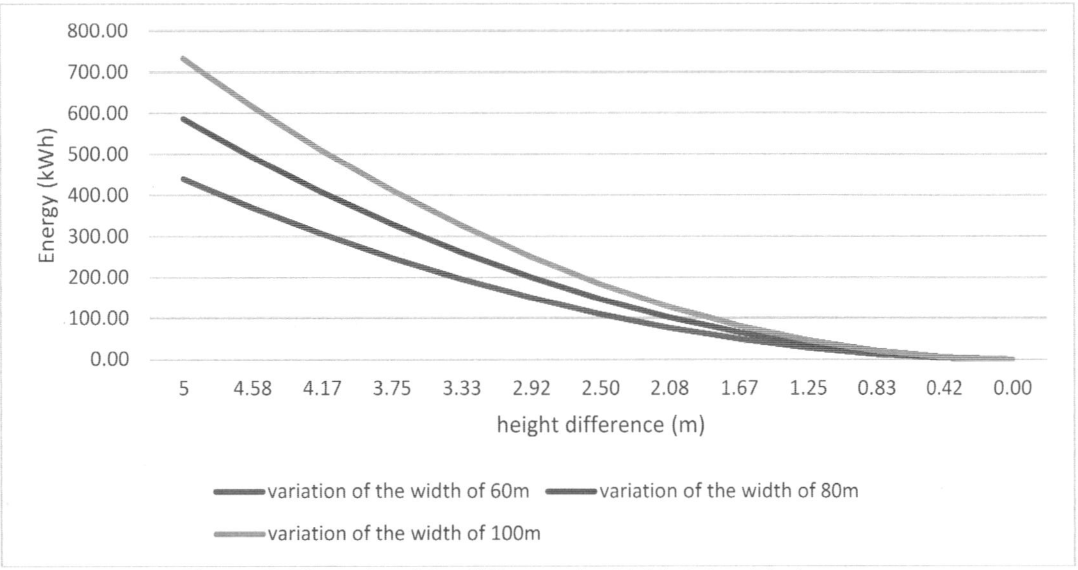

Figure 7. The ratio of the difference in tidal height with the energy

Tides are a phenomenon of the movement of rising and falling sea levels accompanied by horizontal movements of the mass of sea water periodically rhythmically due to the attraction of the moon and sun [6]. With the existence of these tides, it can be a source of renewable energy to support the existing electricity sector.

From the table and graph above, it can be seen that from a reservoir length of 150 m with a variation of reservoir width of 60 m resulting in a maximum discharge of 12.50 m^3/s with a maximum energy potential of 439.92 kWh, a variation of reservoir width of 80 m produces a maximum discharge of 16.67 m^3/s with a maximum energy potential of 586.56 kWh and the variation of the reservoir width of 100m produces a maximum discharge of 20.83 m^3/s with a maximum energy potential of 733.20 kWh.

The potential energy produced increases with the size of the reservoir formed, the magnitude of the observed difference in tidal height, and the duration of the production operations. This allows for the use of known energy as a guide when planning future developments to achieve energy maximum tide.

4. Conclusion

Based on the results of this study, it can be concluded that the larger the tidal water reservoir, the greater the potential for tidal energy to be generated. From the variation of the cross-sectional width of the reservoir 60m, 80m and 100m with the assumption that the tidal height difference that occurs is 5m for 1 tidal cycle (12 hours), then the largest and maximum tidal energy potential occurs in order of 100m variation, which is 733.20 kWh, variation 80m is 586.56 kWh and 60m variation is 439.92 kWh.

5. References

[1] Dandekar, M.M dan Sharma, K.N. (1991). Pembangkit Listrik Tenaga Air. Universitas Indonesia : Jakarta.
[2] Manuaba, IBG dan Kadek. (2008). Potensi Energi Pasang Surut (Tidal Energy) Untuk Alternatif
[3] Sumotarto, U. (2003). Pemanfaatan Energi Pasang Surut. Jurnal Sains Indonesia, 05(5):85-93.
[4] Tantrawati, E. (2007). Potensi Pembangkit Listrik Tenaga Pasang Surut Sistem Daur Ganda Dengan Kolam Tunggal. UII,Yogyakarta, 04(1): 64-76.
[5] Wyrtki, K. (1961). Phyical Oceanography of the South East Asian Waters. NagaReport Vol. 2 Scripps, Institute Oceanography, California.
[6] Nontji, A. (1987). Laut Nusantara. Djambatan. Jakarta

Sengkang City Runoff Management in Primary Drainage Channels with One Dimensional SWMM model

M idris[1], R T Lopa[2], R Karamma[2] dan F Maricar[2]

[1]Magister Course, Department of Civil Engineering, Faculty of Engineering, Hasanuddin University, Indonesia.

[2]Department of Civil Engineering, Faculty of Engineering, Hasanuddin University, Indonesia.

E-mail: muhammad.idris.st@gmail.com

Abstract. The main problem in Sengkang City is the occurrence of flood with a height between 0,5 to 1 m in the city center which is an economic area, these is visually caused by the insufficient capacity of the drainage channel to drain runoff water. Topographically, the economic center of Sengkang city is a basin area which is the downstream part of 4 primary channels, namely channels A, B, C, and D. The purpose of this study was to analyze the existing conditions of inundation and volumes and to make modification plans to reduce the area and volume of flooding in Sengkang City. In this study, topographic measurements, data on channel cross-sectional conditions and elevations were carried out as well as data collection on rainfall and land use to be used as input data for the rainfall-runoff SWMM simulation that was appropriate and under existing conditions. Simulations were carried out using rainfall intensity 5-year and 10-year return periods and modifications were made to an additional storage capacity of 2.000 m3 in primary channel A. From the simulation results, the flood volume for 5-year return periods (R5) of 96.375m3 (11 inundation areas), 10-year return periods (R10) of 148.901 m3 (12 inundation areas), and modification of 10-year return periods (R10) of 43.990 m3 (9 inundation areas) with a percentage decrease of 70,46% and reduce of 3 inundation areas.

1. Introduction

Based on data from the BNPB 2022, within a period of 15 years for the South Sulawesi Province, natural disasters were dominated by floods and tornadoes with 463 and 469 incidents, respectively. Specifically, the incidence of flooding has increased in the last 10 years [1]

Flood is the process of inundation of residential land due to overflowing rivers, caused by heavy rains or flooding sent from other areas that are at higher altitudes. Indonesia has high rainfall, which ranges from 2000-3000 mm/year, so flooding is easy during the rainy season which is between October to January [2].

The phenomenon of flood in urban areas is a problem faced by almost all parts of Indonesia. This is due to the development of urban areas that are built in flood-prone plain areas close to rivers or beaches. The location is considered more strategic because it is close to transportation facilities and community economic routes.

The increasingly rapid development of urban areas, which is marked by the development of extensive urban facilities and infrastructure, attracts residents to move to urban areas and occupy settlements in water catchment areas. Changes in land use change which is quite drastic which is not accompanied by regional arrangement will increase the risk of flooding. This is due to increased surface runoff in urban areas.

In this study, Sengkang City, Tempe sub-district, Wajo district was used as the object of research. The population growth rate of the tempe sub-district reached an average of 0,91% per year, exceeding the district average of 0,33%. This has an impact on the level of population density in the sub-district and directly affects changes in land use in the Sengkang city area. The effect experienced by the population is an increase in the incidence of flooding 5 to 6 times a year in the central market area and Andi Paggaru road with inundation heights between 0,5 to 1 m.

Topographically, Sengkang City, Tempe District, Wajo Regency is characterized by a hilly area with valleys, with 4 primary channels that cross the city and end at the Walannae River. The river route crosses densely populated areas with inlets from secondary channels in residential areas.

The modeling of rainfall and runoff (RR) that occurs in the research area is used to analyze the condition of existing events in the primary canal. This model is used in various hydrological modeling applications to estimate the amount of runoff that occurs because of land-use changes. One application that uses a rainfall and runoff modeling system is the Storm Water Management Model (SWMM). SWMM is a software capable of analyzing water quantity and quality problems related to urban runoff. Storm Water Management was developed by the EPA (Environmental Protection Agency – US) since 1971 [3].

The use of SWMM has been used in various urban areas in hydrological and runoff models. Warwick and Tadepalli have calibrated and validated the SWMM to model a ± 10.000 km^2 urban watershed in Dallas State of Texas [4].

The purpose of this study was to obtain 5- and 10-year return flood volumes with SWMM modeling with primary channel modeling and alternative drainage network system modeling in reducing the impact of flood volume in the Sengkang City area.

2. Material and Methods

This research was conducted in Sengkang City. Sengkang City is the capital of the Wajo Regency. It is a small town located in South Sulawesi Province and is located between 30°39'01" – 40°16'01" South Latitude and 119°053'02" – 120°027'02" East Longitude. The total area of Sengkang city is 38,27 km^2, including one sub-district, Tempe District, which consists of 16 villages. The location of the research is shown in Figure 1.

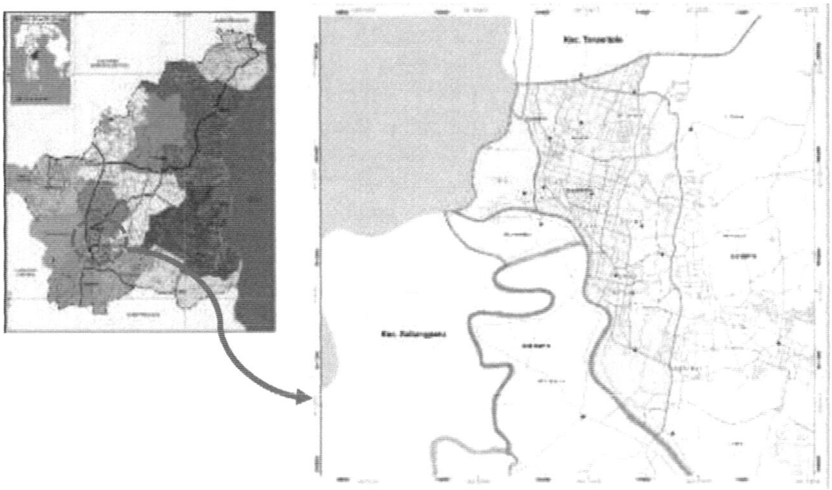

Figure 1 Sengkang city

The data used in this study include:

a. Rainfall data in the form of TRMM (1998 to 2021) from the National Institute of Aeronautics and Space (LAPAN).
b. Sengkang city channel elevation measurement data and primary channel cross-section with geodetic and manual measurements.
c. Map of the type of soil in Sengkang City from the Balitbangda government office Wajo City.
d. Sengkang city elevation data is in the form of the National Digital Elevation Model (DEMNAS) from the Geospatial Information Agency and a map of the layout Sengkang City.

Rainfall data from TRMM were analyzed to obtain rainfall intensity with 5-year and 10-year return periods. Mononobe rainfall intensity formula is as follows:

$$R_t = \frac{R_{24}}{t}\left(\frac{t}{T}\right)^{2/3} \quad (1)$$

For Indonesia t = time of rain concentration used 6 hours

The curve number data used uses references from Urban Hydrology for Small Watersheds (TR-55) [4], with the following table:

Table 1 Curve Number

Cover description		Curve numbers for hydrologic soil group			
Cover type and hydrologic condition	Average percent impervious area 2/	A	B	C	D
Fully developed urban areas (vegetation established)					
Open space (lawns, parks, golf courses, cemeteries, etc.) 3/:					
Poor condition (grass cover < 50%)		68	79	86	89
Fair condition (grass cover 50% to 75%)		49	69	79	84
Good condition (grass cover > 75%)		39	61	74	80
Impervious areas:					
Paved parking lots, roofs, driveways, etc. (excluding right-of-way)		98	98	98	98
Streets and roads:					
Paved; curbs and storm sewers (excluding right-of-way)		98	98	98	98
Paved; open ditches (including right-of-way)		83	89	92	93
Gravel (including right-of-way)		76	85	89	91
Dirt (including right-of-way)		72	82	87	89
Western desert urban areas:					
Natural desert landscaping (previous areas only) 4/		63	77	85	88
Artificial desert landscaping (impervious weed barrier, desert shrub with 1- to 2-inch sand or gravel mulch and basin borders)		96	96	96	96
Urban districts:					
Commercial and business	85	89	92	94	95
Industrial	72	81	88	91	93
Residential districts by average lot size:					
1/8 acre or less (townhouses)	65	77	85	90	92
1/4 acre	38	61	75	83	87
1/3 acre	30	57	72	81	86
1/2 acre	25	54	70	80	85
1 acre	20	51	68	79	84
2 acres	12	46	65	77	82
Developing urban areas					
Newly graded areas (Previous areas only, no vegetation) 5/		77	86	91	94

The Soil group as follows:

Group A soils have low runoff potential and high infiltration rates even when thoroughly wetted. They consist chiefly of deep, well to excessively drained sand or gravel and have a high rate of water transmission (greater than 0,30 in/hr).

Group B soils have moderate infiltration rates when thoroughly wetted and consist chiefly of moderately deep to deep, moderately well to well drained soils with moderately fine to moderately coarse textures. These soils have a moderate rate of water transmission (0,15- 0,30 in/hr).

Group C soils have low infiltration rates when thoroughly wetted and consist chiefly of soils with a layer that impedes downward movement of water and soils with moderately fine to fine texture. These soils have a low rate of water transmission (0,05-0,15 in/hr).

Group D soils have high runoff potential. They have very low infiltration rates when thoroughly wetted and consist chiefly of clay soils with a high swelling potential, soils with a permanent high-water table, soils with a claypan or clay layer at or near the surface, and shallow soils over nearly impervious material. These soils have a very low rate of water transmission (0-0,05 in/hr).
These formulations for the rainfall-runoff simulation uses the SWMM program:

a. The height of inundation or rain runoff D (mm) in each sub-watershed is as follows:
$$D_1 = D_t + R_t \quad (2)$$
b. The outflow discharge from sub-watershed runoff is calculated by the Manning equation:
$$v = \frac{1}{n} D_2{}^{2/3} S^{1/2} \quad (3)$$
$$Q = vBD_2 \quad (4)$$
c. The water level of the sub-watershed from rain, infiltration and outflow is obtained through the following equation:
$$D_{t+\Delta t} = D_2 - (\frac{Q}{A})\Delta_t \quad (5)$$
d. Process number 1 to 4 is repeated until all sub-das calculations are complete.
e. The discharge entering the channel is calculated by adding the discharge from the land (Q_{oi}) with the discharge from the upstream channel (Q_{gi}).
$$Q_{in} = \sum Q_{oi} + \sum Q_{gi} \quad (6)$$
f. Changes in water level due to increased discharge in a channel are:
$$Y_1 = Y_t + (Q_{in}/A_g)\ \Delta_t \quad (7)$$
g. Manning's equation is used to calculate the channel outflow discharge.
$$v = \frac{1}{n} R^{2/3} S^{1/2} \quad (8)$$
$$Q_g = vA_c \quad (9)$$
h. The results of the water depth in the channel from the inflow and outflow are calculated by the continuity equation as follows:
$$Y_t + \Delta_t = Y_1 + (Q_{in} - Q_g)\Delta_t/A_g \quad (10)$$
i. Steps 6 to 9 are repeated until all channels have been counted. [5]

3. Result and discussion

The land use conditions Sengkang City are based on the results of LAPAN information processing with the results of the mapping in Figure 2

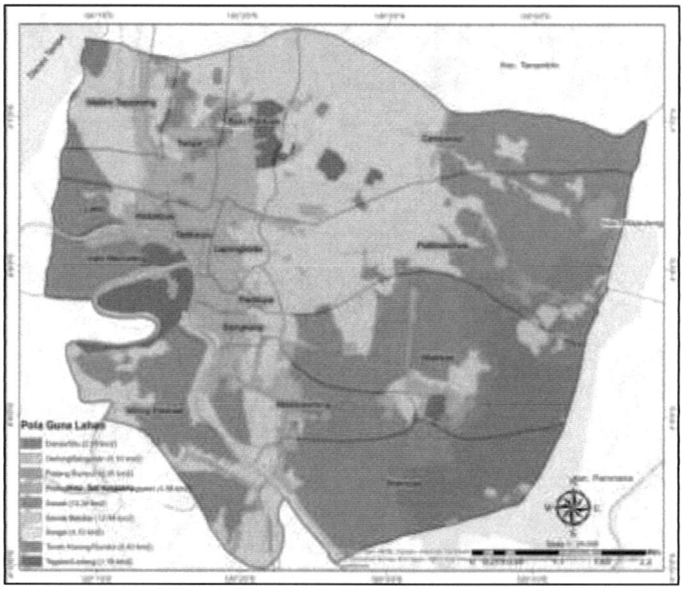

Figure 2 Land map use of Tempe city

Digitizing the sub-catchment area of each catchment that affects the junction of the primary canal, there are 47 sub-catchments accompanied by the area, soil type, and average slope of the sub-catchment shown in figure 3

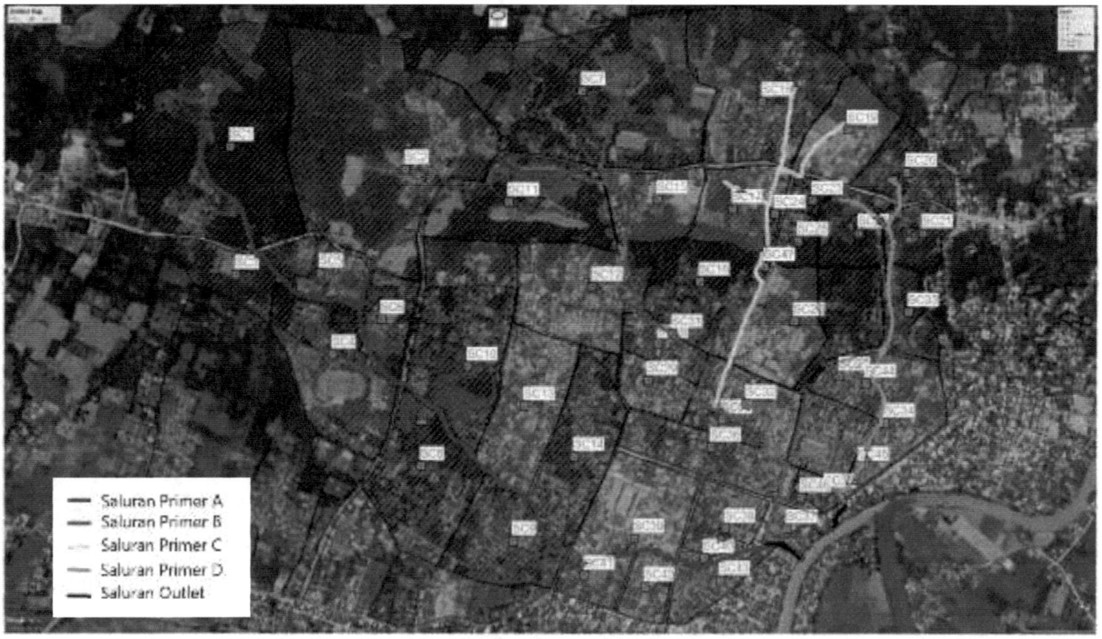

Figure 3 sub-catchment Map

IOP Conf. Series: Earth and Environmental Science 1117 (2022) 012063 doi:10.1088/1755-1315/1117/1/012063

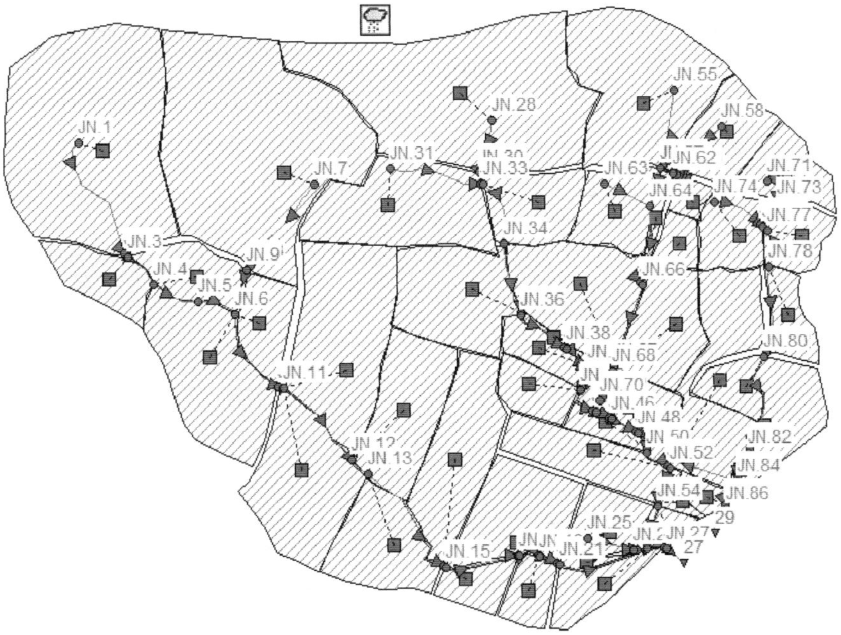

Figure 4 SWMM Modelling Map

Mononobe rainfall intensity for Sengkang city

Table 2 Mononobe rainfall intensity

t	Rt	Rainfall net (Rn, mm) with Repeat Time (Year)	
		R5	R10
		116,115	139,038
(Hour)	(%)	Rainfall Net = Rn x Rt	
1	55,032%	63,900	76,515
2	14,304%	16,609	19,888
3	10,034%	11,651	13,951
4	7,988%	9,275	11,106
5	6,746%	7,833	9,379
6	5,896%	6,847	8,198

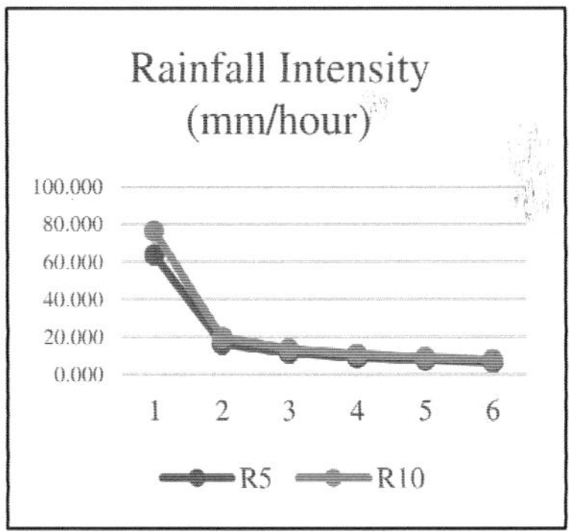

Figure 5 rainfall intensity

3.1. Calibration Model

Model calibration is to ensure that the results of the field survey are the same as the results of the SWMM modeling, by comparing the water level. The water level used from field survey data was taken on June 10, 2022.

Table 3 SWMM model calibration recapitulation

Junction	Water Level Observation (cm)	Model Water Level (cm)	% Error
JN 18	10	9	0,11
JN 22	11	10	0,10
JN 65	15	12	0,25

The calibration value is acceptable if the resulting error value is less than 10% [3]. Based on the table above, the % error obtained is less than 10% so that the model can be accepted.

3.2. Running modelling with R5

Running modelling SWMM one dimension with R5 shown in table 4 and figure 6

Table 4 Junction that is flooding R5

Node	Maximum Rate CMS	Total Flood Volume 10^6 ltr	Total Flood Volume m3
JN.12	0,319	0,01	5,00
JN.14	20,605	55,77	55.768,00
JN.15	1,832	0,02	18,00
JN.16	4,194	18,28	18.277,00
JN.45	1,949	2,11	2.108,00
JN.50	0,804	1,04	1.042,00
JN.55	2,33	5,54	5.538,00
JN.58	0,661	1,39	1.392,00
JN.66	1,72	3,46	3.457,00
JN.67	0,957	0,84	840,00
JN.68	2,231	7,93	7.930,00
	Total		96.375,00

Running model R5 there is an excess of channel capacity on the primary channel path, resulting in flood events with a total flood volume of 96.375 m^3 (11 inundation areas).

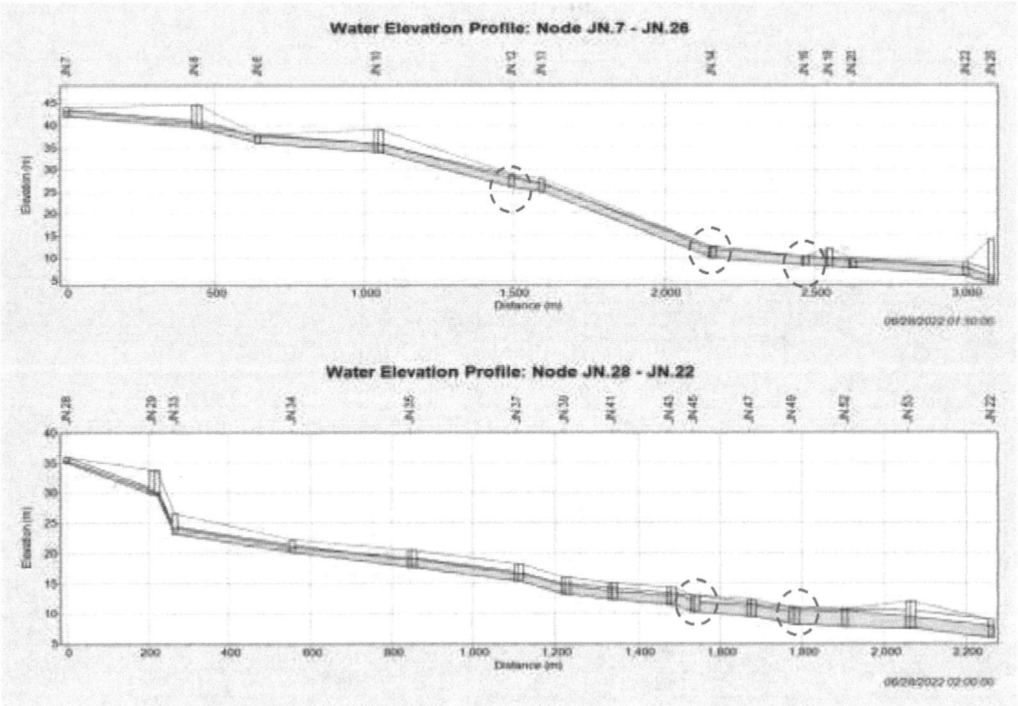

Figure 6 Primary channel water evaluation A,dan B with R5

IOP Conf. Series: Earth and Environmental Science 1117 (2022) 012063 doi:10.1088/1755-1315/1117/1/012063

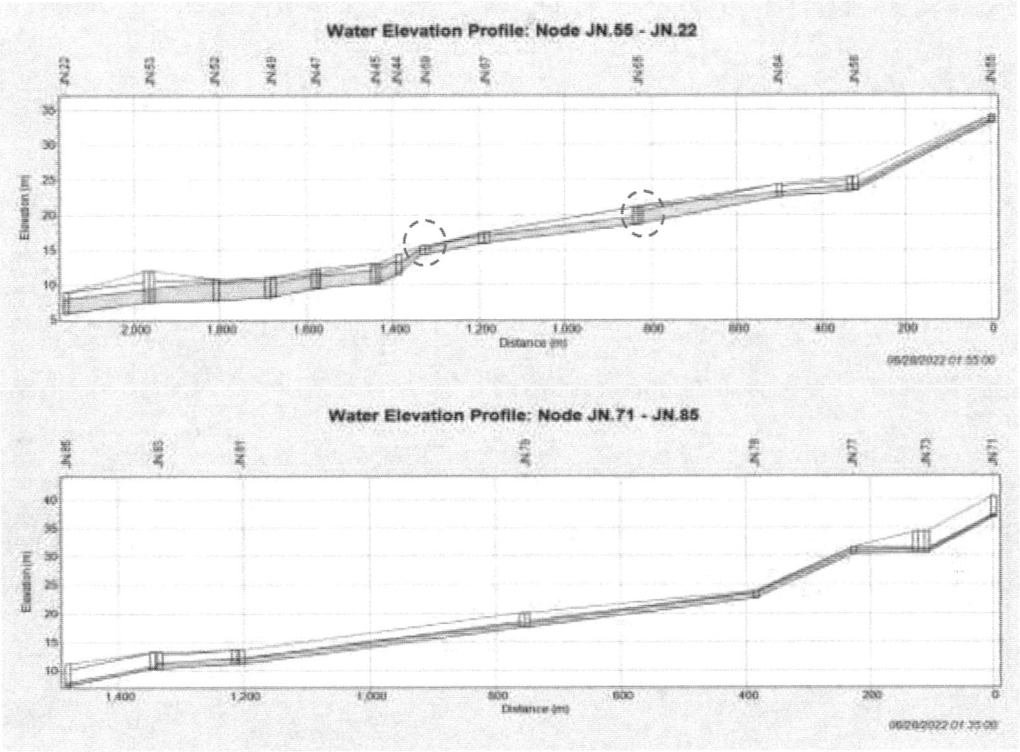

Figure 7 Primary channel water evaluation C dan D with R5

3.3. Running modelling with R10

Running modelling SWMM one dimension with R10 shown in table 5 and figure 7

Table 5 Junction that is flooding R10

Node	Maximum Rate CMS	Total Flood Volume 10^6 ltr	Total Flood Volume m3
JN.12	4,22	5,617	5.617,00
JN.14	22,203	77,511	77.511,00
JN.15	2,282	0,027	27,00
JN.16	4,814	23,883	23.883,00
JN.45	4,146	7,812	7.812,00
JN.50	1,658	3,538	3.538,00
JN.55	3,298	8,692	8.692,00
JN.58	0,991	2,339	2.339,00
JN.66	2,312	5,453	5.453,00
JN.67	1,977	2,773	2.773,00
JN.68	2,243	10,852	10.852,00
JN.78	0,51	0,404	404,00
	Total		148.901,00

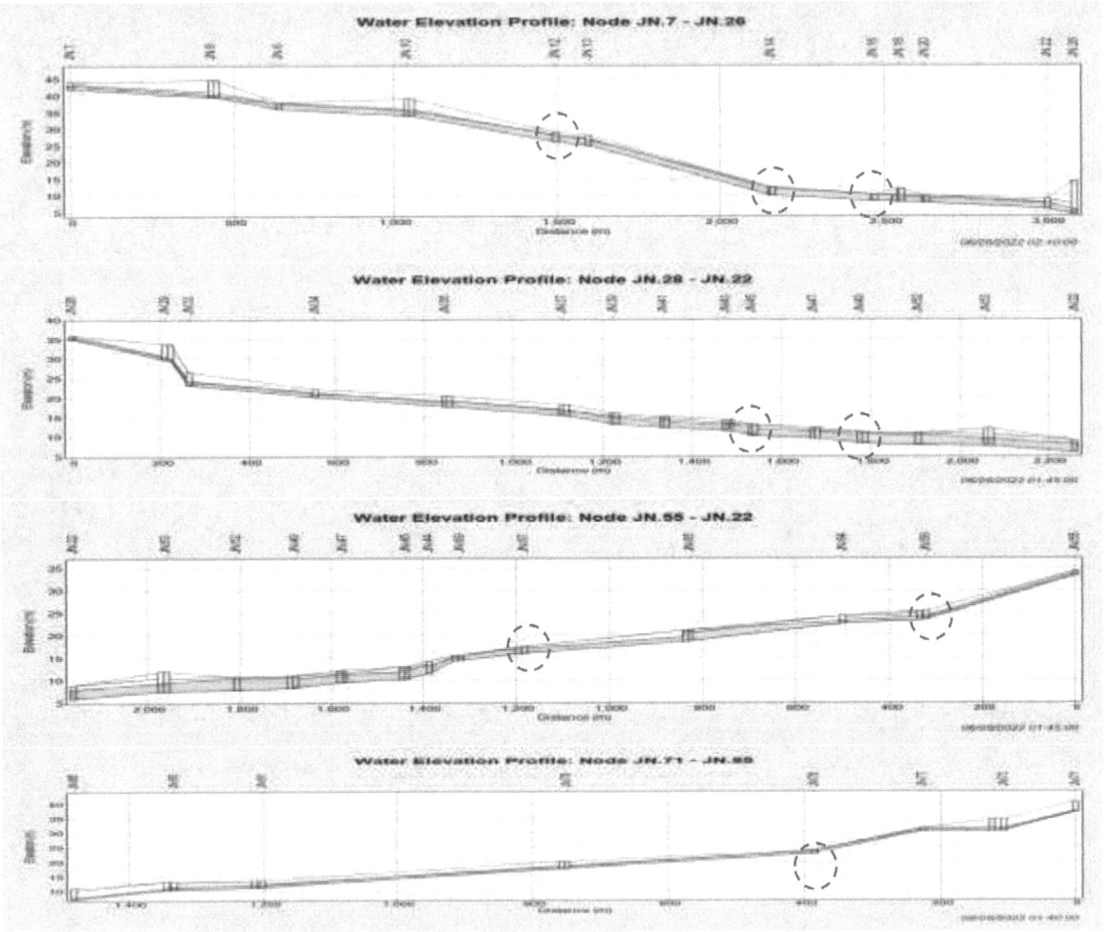

Figure 8 Primary channel water evaluation A, B, C dan D with R10

In the R10 run model, there is excess channel capacity in the primary channel path, which results in flooding events in the table with a total flood volume of 148.901 m^3 (12 inundation areas) in Figure 7

If a comparison is between R5 and R10, there is an increase in the total flood volume by 54,5%. based on field analysis of flood events with R10 approaching the incidence of flood conditions at the research location. The results of the calculation of the Coeff runoff are in the interval of 0,73 to 0,93 (figure 8).

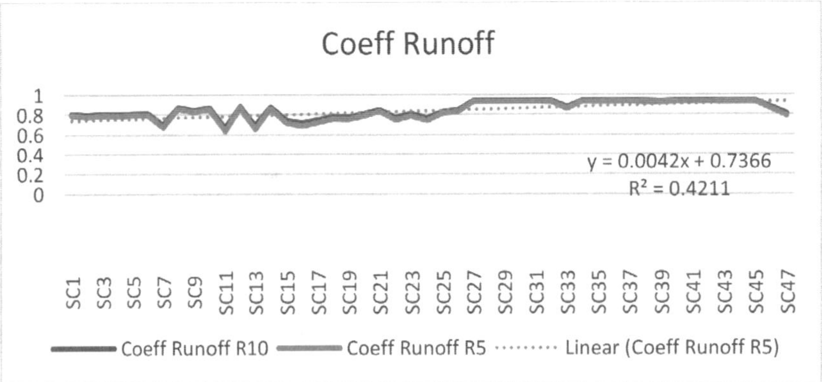

Figure 9 Coeff Runoff

3.4. Running model R10 Modification

Flood disaster mitigation efforts were carried out by modifying the primary channel network model with the addition of 2.000 m^3 storage capacity at the Rusa Street. This was because the largest total volume of water flowed in primary channel A.

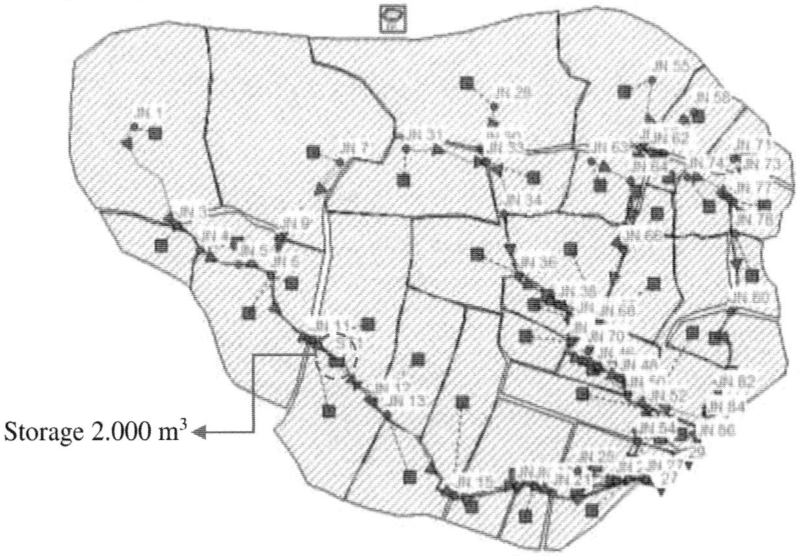

Figure 10 Storage Site Plan

Running modelling SWMM one dimension with R10 modification shown in table 6 and figure 10

Table 6 Junction that is flooding R10 modification

Node	Maximum Rate CMS	Total Flood Volume 10^6 ltr	Total Flood Volume m3
JN.16	2,111	2,135	2.135,00
JN.45	4,141	7,808	7.808,00
JN.50	1,659	3,532	3.532,00
JN.55	3,298	8,692	8.692,00
JN.58	0,991	2,339	2.339,00
JN.66	2,297	5,454	5.454,00
JN.67	1,976	2,773	2.773,00
JN.68	2,243	10,852	10.852,00
JN.78	0,51	0,405	405,00
	Total		43.990,00

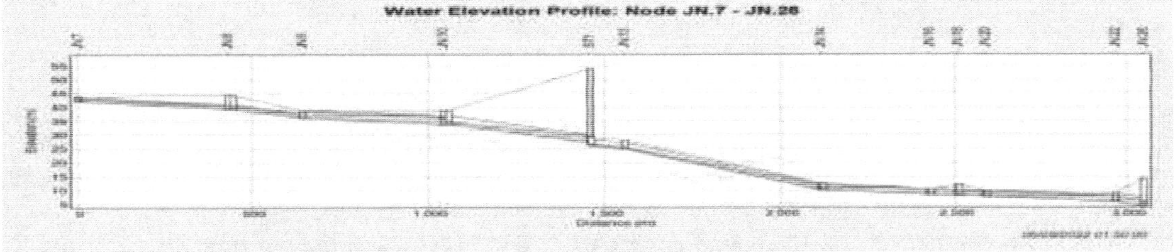

Figure 11 Primary channel water evaluation A with R10 modification

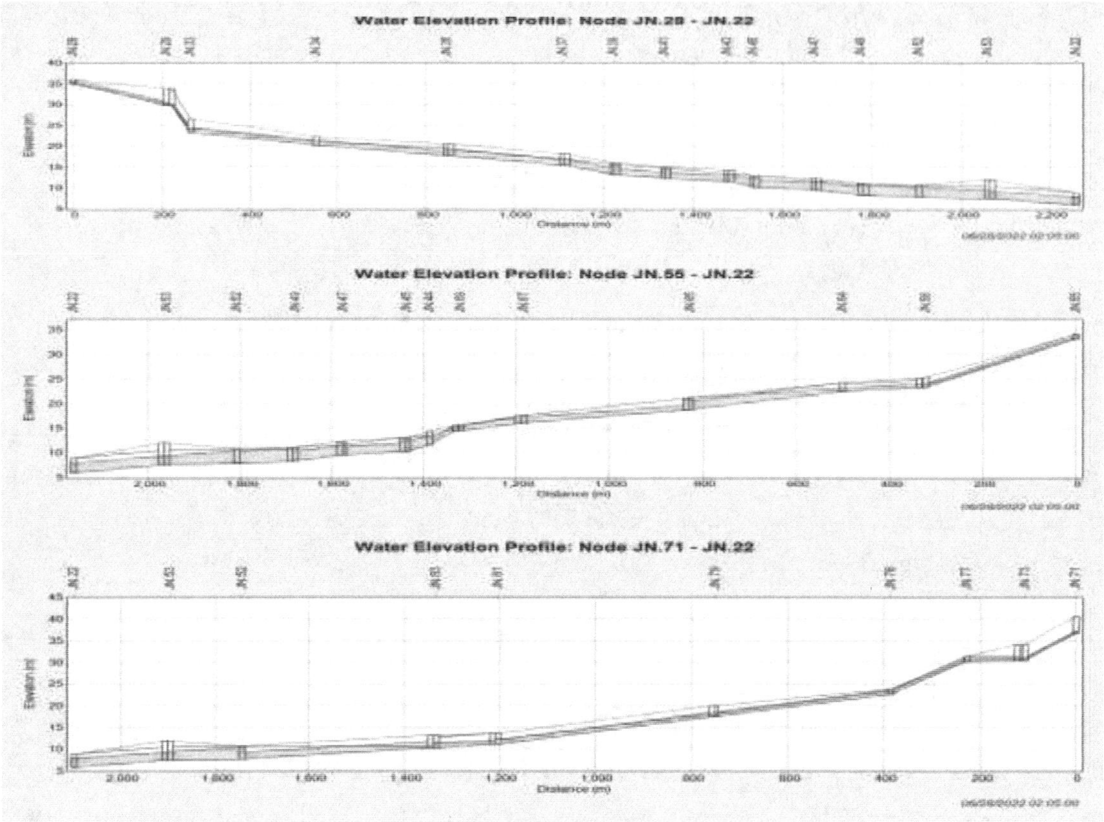

Figure 12 Primary channel water evaluation B, C dan D with R10 modification

Based on the results of running the modified R10 model, the total flood volume is 43.990 m^3 (9 inundation areas).

The evaluation results show that disaster mitigation by modifying primary channel A with the addition of a storage capacity of 2.000 m^3 can reduce the flood potential of R10 by 70,46% and reduce the inundation area by 3 areas (Junction: JN12, JN14 and JN15).

4. Conclusions

- The flood event at the research location was confirmed according to the flood location using the SWMM rainfall-runoff modelling method, total flood volume of 5-year return periods (R5) 96.375 m^3 (11 inundation areas) and 10-year return periods (R10) 148.798 m^3 (12 inundation areas) with Coeff runoff 0,73 – 0,93.
- Modification of the primary channel by installing a storage capacity of 2.000 m^3 reduces the flood potential of 10-year return periods (R10) by 70,46% and reduces 3 inundation areas (JN12, JN14 and JN15).

References

[1] Bnpb 2022 *https://dibi.bnpb.go.id/*
[2] A Findayani 2015 *Community Preparedness in Flood Management in Semarang City* Journal Geography Unnes Semarang
[3] L A Rossman 2015 *Storm Water Management Model User's Manual Version 5.1* EPA-600/R-14/413b www2.epa.gov/water-research
[4] J J Warwick and member 1991 *Efficacy of SWMM Application* Journal of Water Resources Planning and Management Vol.117 Issue 3 ASCE Library

[5] R H McCuen and members 1986 *Urban Hydrology for Small Watersheds-TR-55* United States Department of Agriculture (USDA)
[6] Anwar N dan members 2010 *Strom Water Management Model (SWMM) Application for the Deluwang Situbondo Watershed, East Java* Surabaya Institute of technology
[7] R T Lopa 2012 *Accuracy of Index Flood Method Applied in Two Different Regions in Indonesia* Civil Engineering Forum. Volume XXI/1-January 2012
[8] R Karamma 2018 *Comparison of Model Hydrograph Synthetic Units (HSS) with the Model of Hydrograph Observations on DAS Jeneberang Gowa Regency, Indonesia* International Journal of Innovative Science and Research Technology ISSN No: -2456 - 2165

IOP Conf. Series: Earth and Environmental Science 1117 (2022) 012064 doi:10.1088/1755-1315/1117/1/012064

Study of Sediment Characteristics in the Jeneberang River's Upstream

A L Wildany[1]*, R T Lopa[2], F Maricar[3]

Master of Civil Engineering, Universitas Hasanuddin, Gowa, South Sulawesi
Civil Engineering Department, Universitas Hasanuddin, Gowa, South Sulawesi
Civil Engineering Department, Universitas Hasanuddin, Gowa, South Sulawesi

Email : azkalayyina@yahoo.co.id

Abstract. This study aims to determine the characteristics of sediments found in the Jeneberang River specifically in the upstream part. The data required includes the results of sieve analysis, grain diameter of the sediment and the density of the sediment. The study was conducted by taking samples from the segments around Sabo 7-1. The data obtained indicate that the gradation of sediment particles consists of fine sand, medium sand, coarse sand, fine gravel and coarse gravel with a median diameter (D50) of 0.5-13.10 mm and specific gravity values ranging from 2.659-2.682 gr/ cm.

Key words : Sediment characteristics, sediment disaster, Jeneberang River

1. Introduction

Jeneberang River is one of the rivers that disgorge on Mount Bawakaraeng. The collapse of Mount Bawakaraeng on March 26, 2004, resulted in the deaths of 32 people, 635 cows missing, the destruction of several houses and an elementary school building, and the buried area of 1,500 hectares of agricultural land. The volume of debris is estimated at 200 - 300 million m3. Mountain material flows in the Jeneberang river and results in very large sedimentation of the Bili-Bili Reservoir. Subsequent disasters due to debris flow in January 2006 caused the Daraha Bridge to be washed away.

The collapse of the caldera wall of Mount Bawakaraeng in March 2004, followed by flash floods that killed and injured dozens of people, and the destruction of dozens of houses have created a continuing threat of disaster for local residents. This was proven when flash floods hit the same area again in February 2007 which caused the isolation of thousands of residents.

The biggest threat is the deposit of debris from the landslide in March 2004 which disrupts the preservation of the function of the Bili-Bili reservoir downstream of Mount Bawakaraeng as a source of water for local residents. This problem is getting worse when this loose and unstable debris material with a very large volume can at any time move into a flash flood/landslide disaster if heavy rains occur. This then becomes a secondary sediment disaster or what is also often called cold lava.

The movement of sediment mass can be grouped into several types, including landslides, debris flow, transitional flow between debris flow and individual flow (bedload), which is called hyper concentration flow. [1]

Debris flow is a mass flow in the form of a mixture of water and sediment with a very high concentration. Once this flow starts (because the static equilibrium between the shear forces generated is greater than the shear forces that resist), then the amount of mass that flows, its height, and its velocity will always increase (has an acceleration). At a certain level, due to local boundary conditions, for example, changes in slope that become gentle, reduced water mass, changes in sediment character and so on. This debris flow process will experience a slowdown, the amount of mass flowing is reduced, then a certain amount of mass will be deposited. The initial criteria for the occurrence of moving debris flow due to the surface at the bottom of the groove formed by loose-grained deposits modeled by Takahashi.

In general, debris flow is caused by high rainfall intensity and lasts a long time in the upstream area of the watershed. The beginning of the occurrence of debris flow can be divided into three types including, first, landslides on cliffs that turn into debris flows, second is the collapse of natural dams due to sediment buildup upstream, and the third is sediment flows that move sequentially following a steep channel bottom [2].

According to Zaini (2005), in his understanding, Sabo is a technical terminology derived from the Japanese "Sa" and "Bo" which in a broad sense means erosion and sediment control works. The term Sabo means sand and gravel management which is essentially an effort to prevent mountainous land from being damaged by erosion, protecting people and downstream infrastructure against the threat of disasters due to erosion and sediment. [3]

The target of the sabo work is all sand, gravel and stones of various sizes that are upstream, both those in riverbeds, river cliffs and hills that are expected to landslide or collapse or are called sediment production source areas (Cahyono, 2010). 2000). [4] Sediment production areas are areas located on upstream slopes with a slope of > 6". Countermeasures for cold lava flooding in sediment production can be anticipated by constructing sediment retaining structures, consolidation dams and flow control dams.

To identify and determine the river fluctuation and advice the best position of proposed sabo facilities along the main river course especially at the upstream, it is necessary to study the characteristics of sediments found in the Jeneberang River. These sediment characteristics include sediment density, sediment diameter size and sediment grain size analysis

2. Research Method

2.1. Research Design

This research was conducted in the upstream Jeneberang River and this type of research used a descriptive approach to determine the characteristics of sediments that could be used as a basis for determining the type and location of the right sabo facility as a form of sediment disaster mitigation.

2.2. Research Site

The research location was determined at the upper reaches of the Jeneberang River (Figure 1). This upstream part was chosen because the existing sabo facility suffered a lot of damage, so it was deemed necessary to review the design by reviewing the characteristics of the sediment first.

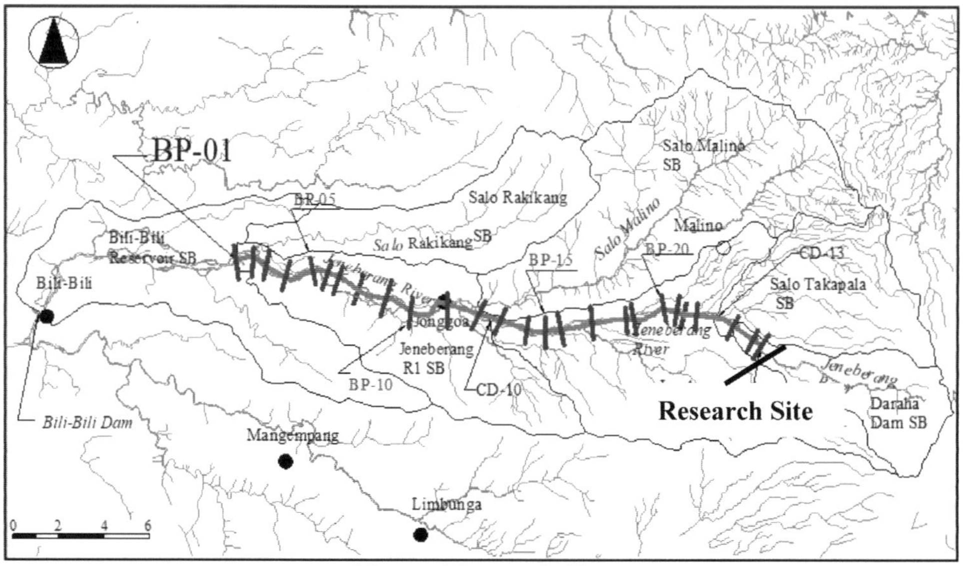

Figure 1 Research Site

2.3. *Field Activities*

The data collection method was carried out by taking samples in the upperstream of the Jeneberang River for the segments before Sabo 7-1 and after Sabo 7-1. In each segment, at least 2 different types of samples are taken. This segment difference is taken as a comparison and supporting data to see the performance of the existing Sabo Series 7 facilities. The location of the segment in question is shown in Figure 2.

Figure 2. Location of Segments

2.4. *Data Analysis*

The data presented has been clearly and systematically arranged, for further analysis. The data obtained from laboratory and field results are sieve analysis and specific gravity testing. The entire data is presented in an easy-to-understand form, namely in the form of test results tables and graphs. The data obtained in this study is primary data.

3. Result and Interpretation

3.1. Sediment Specific Gravity

From the results of the examination and calculations, the values of sediment density (Gs) are obtained which are presented in Table 1:

Table 1. Specific Gravity Test Results

Sample Depth & Inclination	m	2		E1		3		E2		1a		1b		4	
Number of Volumetric Flask	-	A	B	A	B	A	B	A	B	A	B	A	B	A	B
Weight of Vol. Flask + Soil (W2)	Gram	45.8	53.0	45.7	47.2	46.7	54.1	49.6	53.3	82.78	82.26	81.58	80.86	81.82	83.52
Weight of Vol. Flask (W1)	Gram	20.75	28.02	20.65	22.21	21.72	29.12	24.58	28.32	32.78	32.26	31.58	30.86	31.82	33.52
Weight of Dry Soil (Ws=W2-W1)	Gram	25.00	25.00	25.00	25.00	25.00	25.00	25.00	25.00	50.00	50.00	50.00	50.00	50.00	50.00
Temperature, T (oC)	Degree	28	28	28	28	28	28	28	28	25	25	25	25	25	25
Weight of Vol. Flask+Water at T (W4)	Gram	68.90	77.63	68.56	76.53	68.79	76.83	68.91	77.65	129.40	131.32	128.43	134.13	127.94	135.26
Weight of Vol. Flask+Water+Soil (W3)	Gram	84.58	93.22	84.22	92.15	84.38	92.50	84.51	93.28	160.64	162.61	159.59	165.41	159.24	166.71
Unit Weight of Water at T, g_T	Gram/Cm^3	0.9963	0.9963	0.9963	0.9963	0.9963	0.9963	0.9963	0.9963	0.9971	0.9971	0.9971	0.9971	0.9971	0.9971
Temp. Corr. Coefficient, $a=g_T/g_{20}{}^0C$	-	0.9980	0.9980	0.9980	0.9980	0.9980	0.9980	0.9980	0.9980	0.9988	0.9988	0.9988	0.9988	0.9988	0.9988
Weight of Soil (Wu=(Ws+W4-W3))	Gram	9.3	9.4	9.3	9.4	9.4	9.3	9.4	9.4	18.76	18.71	18.84	18.72	18.70	18.55
Specific Gravity of Soil (Gs=a*Ws/Wu)	-	2.677	2.652	2.671	2.660	2.652	2.674	2.654	2.663	2.662	2.669	2.651	2.668	2.671	2.692
Average of Gs	-	2.664		2.666		2.663		2.659		2.666		2.659		2.682	
Remarks:				Unit Weight of Water,		$g_{w,20}{}^0C=$ 0.99823									

The density of sediment is generally around 2.65 g/cm^3 except in the form of heavy metals (Hambali et al., 2016). The density of the bottom sediment in the Jeneberang River ranged from 2.659-2.682 g/cm^3 with an average of 2.665 g/cm^3 (see Table 2). The distribution pattern of sediment density tends to be regular.

Table 2 Recapitulation of Riverbed Sediment Density (g/cm^3)

No	Sample Code	Sediment Density
1	2	2,664
2	E1	2,666
3	3	2,663
4	E2	2,659
5	1A	2,666
6	1B	2,659
7	4	2,682

3.2. Sediment Diameter Test Results

Determination of sediment diameter in this case is through sieve analysis experiments conducted in the laboratory, so that from the results of these experiments can be obtained a uniform grain diameter value or d50 of the sediment. The value of the diameter of the sediment grains (d50) obtained is presented in table 3 below.

Table 3 Value of d_{50}

No	Sample Code	d_{50} (mm)
1	2	5,21
2	E1	6,06
3	3	9,03
4	E2	13,10
5	1A	10,92
6	1B	3,15
7	4	0,50

3.3. Results of Sediment Grain Size Analysis

The results of the recapitulation of sieve analysis for each sample are presented in the table of sieve analysis test results in Table 4. Figure 3 shows a graph of the recapitulation of the gradation of sediment grains in each sample tested.

Table 4 Recapitulation of Percent Results Passing the Sieve Analysis Test

Sieve Number	**Sieve Diameter**	**Percent Passing (%)**						
	Sample Code	**2**	**E1**	**3**	**E2**	**1A**	**1B**	**4**
3"	75.00	99.2	99	98.68	98.44	99.48	100	100
2"	50.00	99.2	99	89.6	98.04	98.04	100	100
11/2"	37.50	97.48	94.6	84.16	97.44	92.8	98.36	100
1"	25.00	93.56	87.08	77.36	88.24	81.8	94.56	100
3/4"	19.00	90.08	80.4	72.08	81.8	61.88	86.64	100
3/8"	9.500	78.68	66.64	51.04	38.32	47.92	77.08	99.6
4	4.750	46.44	43.6	28.96	22.8	31.12	59.08	98.8
10	2.000	25.36	24	12.2	12.2	21.88	43.44	79.56
20	0.840	12.6	12.2	5.12	5.88	14.64	30.2	61.92
40	0.425	6.24	6.24	2.68	2.68	8.68	19.04	47.16
60	0.250	2.12	4.28	2.12	1.8	5.8	7.8	28.2
100	0.150	0.52	2.48	1.6	0.72	2.16	3.36	4.16
200	0.075	0.24	0.72	0.72	0.16	0.6	1.4	1.8
Pan	-	0	0	0	0	0	0	0

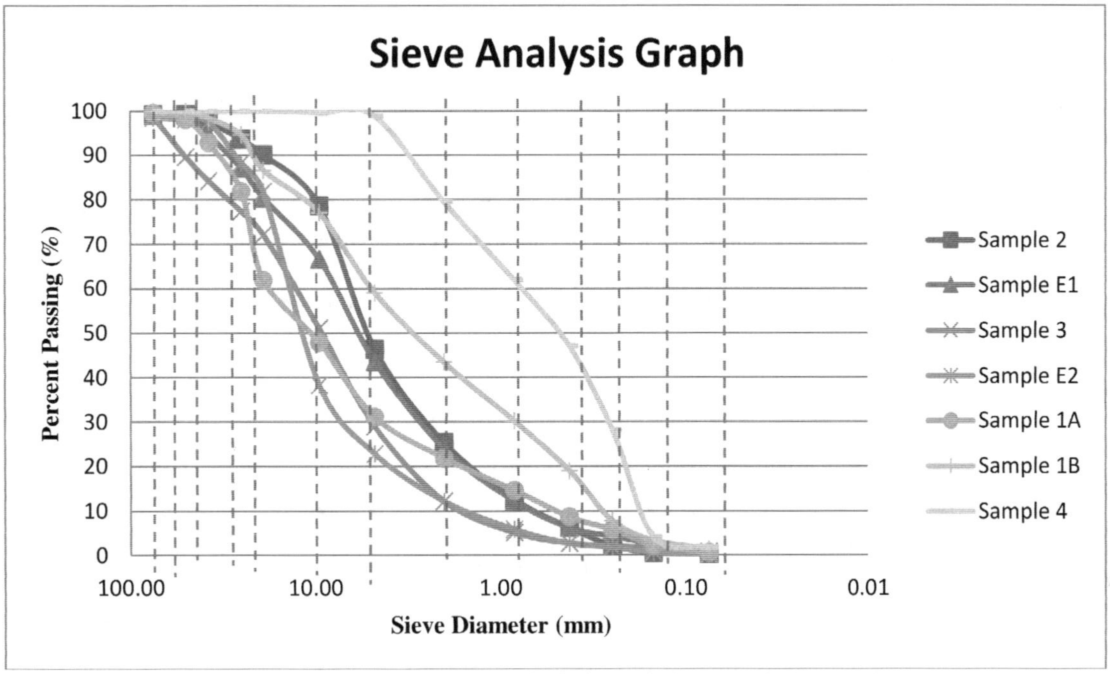

Figure 3 Graph of the Recapitulation of Sieve Analysis Results

4. Conclusion

Based on the data and discussion of the research, it can be concluded that the gradation of sediment particles consists of fine sand, medium sand, coarse sand, fine gravel and coarse gravel with a median diameter (D50) of 0.5-13.10 mm and the specific gravity values range from ranged from 2.659-2.682 g/cm. These data could be use to determine the best sabo plan for upstream area; include for the right structure, the right position etc.

References

[1] Subarkah 2005 *Aliran Debris. Materi Kuliah Program Pascasarjana Magister Pengelolaan Bencana Alam* (Yogyakarta : Universitas Negeri Gadjah Mada)

[2] Maricar F, Lopa RT *Studi Perilaku Bangunan Pengendali Sedimen yang Berwawasan Lingkungan* (Makassar : Universitas Hasanuddin)

[3] Zaini, M.F 2005 *Effectiveness of Sabo Dam BO-07 in Boyong River, Master Thesis, Faculty of Enginering* (Yogyakarta : Universitas Gadjah Mada)

[4] Cahyono, Joko 2000 *Pengantar Teknologi Sabo* (Yayasan Teknologi Sabo)

The Effect of Climate Changes on Sediment Concentration in the Saddang Irrigation Area

G H Syamsuddin[1], M Saleh Pallu[2], F Maricar[3], B. Bakri[3]

[1]Doctoral program student of Civil Engineering Department, Faculty of Engineering, Hasanuddin University, South Sulawesi, Indonesia
[2]Professor of Civil Engineering Department, Faculty of Engineering, Hasanuddin University, South Sulawesi, Indonesia
[3]Associate Professor of Civil Engineering Department, Faculty of Engineering, Hasanuddin University, South Sulawesi, Indonesia

Abstract. Saddang irrigation area does not get water evenly this is due to, among other things, the condition of the network experiencing a decrease in capacity. The decrease in capacity was caused by sedimentation at the bottom of the channel and the condition of the channel lining was damaged, resulting in seepage/leakage and the building and sluice gates were not functioning properly. This study aims to analyze seasonal changes to sediment concentrations in the main channel Rappang Pinrang Regency. This research was conducted in the Rappang main channel, Pinrang Regency with the coordinates of 3°41'44.93"S 119°40'43.12"E. Data collection was divided into 3 research locations, namely the upstream, middle and downstream of the Rappang main channel with a distance of ± 33 km from upstream to downstream. Sediment sample testing was carried out at the Civil Engineering Soil Mechanics Laboratory, Hasanuddin University, Makassar. Data collection in the form of flow velocity, suspended load and flow rate was carried out in the upstream, middle and downstream areas of the Rappang main channel. The expected result of this study is to find the value of the effect of seasonal changes on sediment concentration in the Rappang Pinrang Regency.

Keywords: Rappang Main Channel, Sediment Concentration, Pinrang Regency.

1. Introduction

In the saddang river, it can be seen that there have been fluctuations in floating sediment in recent years which can be seen that the sediment transport activity in the middle is higher than the upstream and downstream rivers, this can be caused by erosion in the upstream river which causes degradation so that it carries large amounts of sediment transport. to the middle of the river then in the middle area of the river there is deposition and aggradation occurs so that for several years it can be seen when it reaches the downstream the sediment transport activity decreases and by using the empirical formula of Meyer Peter Muller the results of the analysis of the last few years in the upstream and downstream of the Saddang river there is a sediment transport process different drift rates, namely floating sediment discharge (Qs) upstream of 104,204 kg/day or 0.104 tons/day, in the middle of 1347,473 kg/day or 1,347 tons/day and downstream of 2128,906 kg/day or 2,128 tons/day [1] Saddang River in the rainy season with river water the water turns cloudy, and along the downstream of the Saddang river there are sediment deposits that cause siltation so that the ability of the river to drain water is getting smaller. sediment concentration (Cs), the estimated floating sediment is 93,389,256 tons/year and bottom sediment is 18,677,851 tons/year (Nurhidayah Basri, 2019) The decrease in capacity is caused by sedimentation at the bottom of the channel and the condition of the channel lining is damaged, so that seepage/leakage occurs and buildings and sluice gates are not functioning properly.

2. Method and Material

2.1 Research Location and time

This research is an experimental in the field and was carried out in November 2021 and May 2022. This research was carried out on the Upstream to Downstream Rappang Main Channel. The distance between the Upstream to Downstream of the Rappang Main Channel is +- 33 Km. The distance between the upstream to the middle of the research point is +- 18 Km, while the distance between the middle to the downstream channel is +- 15 Km by examining the flow velocity and sediment drift in the Rappang primary channel. The research sample was tested at the Soil Mechanics Laboratory, Department of Civil Engineering, Faculty of Engineering, Hasanuddin University, Gowa.

Figure 1. Research Location

2.2 Data Collection Technique

To obtain data as the main material in this study, two methods of data collection were used, namely:

1. Literature study, to obtain secondary data by reading a number of books, scientific articles as a theoretical basis for the perfection of this research.
2. Secondary data collection consisting of:
   a. Rainfall data collection at Benteng Rain Station and Mallaga Rain Station
   b. Discharge data from the Rappang DI Saddang main channel in 2010 - 2021.
   c. Rappang main line Existing Data

3. Primary data collection by:
   Primary data collection was carried out through field surveys to observe field conditions in detail about the condition of the Rappang main canal. In addition, primary data collection was also carried out by measuring the flow velocity in the channel and taking *suspended load* samples from the primary channel in Saddang Irrigation Area.

   a. Measurement of flow velocity in the Rappang Mains channel using a *Current Meter Model Propeller A. OTT. measuring instrument*

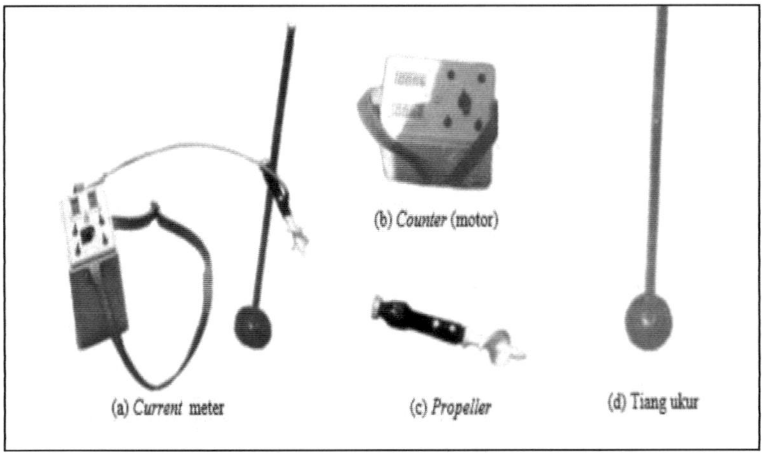

Figure 2 . *Current Meter Propeller A. OTT*

The measurement procedure using a current meter is as follows:
1. Measure the depth of the channel with the measuring rod of the current meter tool
2. Choose a propeller that is suitable for the channel depth, so that it can be used for several vertical points, namely (0.2h, 0.6h, and 0.8h) where h is the channel depth.
3. The current meter is mounted on a measuring pole (static) with a depth of 0.2h, 0.6h, and 0.8h, then the measuring pole is immersed in the water until the measuring pole is located at the bottom of the channel with the propeller facing the direction of flow.
4. The number of revolutions per unit time, which occurs at each water depth is calculated.

b. Sampling of floating sediment in the Rappang main channel using the *Van Dorn Bottle Samplers tool.*

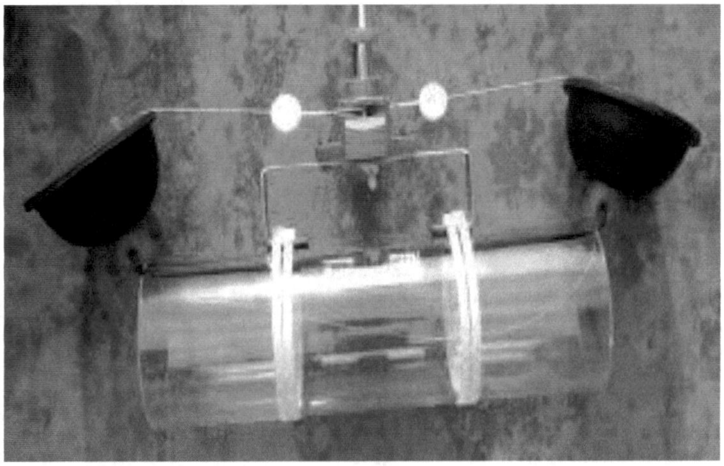

Figure 3 . *Van Dorn Bottle Samplers*

The procedure for taking floating sediment samples using Van Dorn Bottle Samplers is :

1. The fishing line of the two tube covers is attached to the hook and clamped with the hook clamp.
2. After the tube cover is open, the tool is lowered to the desired depth by using a tube hanging rope with an additional iron pole as a support.
3. When Van Dorn has been in the desired position, the metal clamp opening is dropped to open the clamp so that the rubber connecting the cover can function to close the tube that already contains the desired water sample.
4. After that Van Dorn was ready to be picked up again.

2.3 Research Data Collection

Table 1 shows a matrix of research variables in the Rappang trunk. Each research location is divided into 5 segments of the channel width by taking data on flow velocity, floating sediment and bottom sediment at each location.

Table 1 . Research data retrieval matrix.

No	Research sites	Examined variables	
		Velocity	Suspended Load
1	Upstream		
	Segment 1	0.2H, 0.6H & 0.8H	0.6 of Depth
	Segment 2	0.2H, 0.6H & 0.8H	0.6 of Depth
	Segment 3	0.2H, 0.6H & 0.8H	0.6 of Depth
	Segment 4	0.2H, 0.6H & 0.8H	0.6 of Depth
	Segment 5	0.2H, 0.6H & 0.8H	0.6 of Depth
2	Middle		
	Segment 1	0.2H, 0.6H & 0.8H	0.6 of Depth
	Segment 2	0.2H, 0.6H & 0.8H	0.6 of Depth
	Segment 3	0.2H, 0.6H & 0.8H	0.6 of Depth
	Segment 4	0.2H, 0.6H & 0.8H	0.6 of Depth
	Segment 5	0.2H, 0.6H & 0.8H	0.6 of Depth
3	Downstream		
	Segment 1	0.2H, 0.6H & 0.8H	0.6 of Depth
	Segment 2	0.2H, 0.6H & 0.8H	0.6 of Depth
	Segment 3	0.2H, 0.6H & 0.8H	0.6 of Depth
	Segment 4	0.2H, 0.6H & 0.8H	0.6 of Depth
	Segment 5	0.2H, 0.6H & 0.8H	0.6 of Depth

2.4 Sediment Sample Test

Sediment sample testing was carried out in the soil mechanics laboratory by testing suspended load sediment samples in November 2021 and May 2022 from upstream, middle and downstream of the Rappang Kab. Pinrang. The data that will be obtained in testing the suspended sediment sample is the sediment concentration.

3. Result and Discussion

Table 2. Flow Velocity and Sediment Concentration at Upstream Rappang Main Channel in November 2021

Upstream	Depth (H)	Vertical Point	Depth (m)	Velocity V(m/s)	V Average	Cs (mg/l)
T1	2.2	0.2	1.76	0.143	0.130	320
		0.6	0.88	0.127		
		0.8	0.44	0.120		
T2	2.2	0.2	1.76	0.145	0.138	310
		0.6	0.88	0.141		
		0.8	0.44	0.129		
T3	2.5	0.2	2.00	0.161	0.150	300
		0.6	1.00	0.154		
		0.8	0.50	0.135		
T4	2	0.2	1.60	0.147	0.136	330
		0.6	0.80	0.139		
		0.8	0.40	0.123		
T5	1.85	0.2	1.48	0.143	0.128	340
		0.6	0.74	0.125		
		0.8	0.37	0.116		

In Table 2 the highest average velocity upstream of the Rappang main channel in November 2021 is located at T3 at a depth of 2.5 m with an average velocity of 0.150 m2 / s with a sediment concentration value of 300 mg/l and the smallest velocity at T5 at a depth of 1.85 m with velocity of 0.128 m2 / s with a sediment concentration value of 340 mg/l.

Table 3. Flow Rate and Sediment Concentration in the Middle of the Rappang Main Channel in November 2021

Middle	Depth (H)	Vertical Point	Depth (m)	Velocity V(m/s)	V Average	Cs (mg/l)
T1	1.58	0.2	1,264	0.138	0.126	200
		0.6	0.632	0.123		
		0.8	0.316	0.117		
T2	1.9	0.2	1.52	0.142	0.135	200
		0.6	0.76	0.136		
		0.8	0.38	0.126		
T3	2.05	0.2	1.56	0.157	0.145	180
		0.6	0.78	0.151		
		0.8	0.39	0.127		
T4	1.95	0.2	1.64	0.135	0.130	230
		0.6	0.82	0.133		
		0.8	0.41	0.122		
T5	1.4	0.2	1.12	0.135	0.123	240
		0.6	0.56	0.119		
		0.8	0.28	0.116		

IOP Conf. Series: Earth and Environmental Science 1117 (2022) 012065 doi:10.1088/1755-1315/1117/1/012065

In Table 3 the highest average velocity in the middle of the rappang main channel in November 2021 is located at T3 at a depth of 1.9 m with an average velocity of 0.145 m2 / s with a sediment concentration value of 180 mg/l and the smallest velocity at T5 at a depth of 1.4 m at a speed of 0.123 m2 / s with a sediment concentration value of 240 mg/l.

Table 4. Flow Rate and Sediment Concentration at the Rappang Main Channel Downstream in November 2021

Downsteam	Depth (H)	Vertical Point	Depth (m)	Velocity V(m/s)	V Average	Cs (mg/l)
T1	1.7	0.2	1.36	0.133	0.121	100
		0.6	0.68	0.116		
		0.8	0.34	0.113		
T2	2	0.2	1.60	0.138	0.130	90
		0.6	0.80	0.133		
		0.8	0.40	0.119		
T3	2.2	0.2	1.76	0.149	0.139	80
		0.6	0.88	0.143		
		0.8	0.44	0.123		
T4	2	0.2	1.60	0.129	0.124	110
		0.6	0.80	0.126		
		0.8	0.40	0.116		
T5	1.3	0.2	1.04	0.127	0.117	120
		0.6	0.52	0.114		
		0.8	0.26	0.110		

In Table 4 the highest average velocity downstream of the Rappang main channel in November 2021 is located at T3 at a depth of 2.2 m with an average velocity of 0.139 m^2 / s with a sediment concentration value of 80 mg/l and the smallest velocity at T5 at a depth of 1.3 m with a velocity 0.117 m^2/s with a sediment concentration value of 120 mg/l.

Table 5. Flow Rate and Sediment Concentration in Upstream Rappang Main Channel in May 2022

upstream	Depth H	Vertical Point	Depth (m)	Speed V(m/s)	V Average	Cs (mg/l)
T1	2.3	0.2	1.84	0.152	0.143	7704
		0.6	0.92	0.149		
		0.8	0.46	0.127		
T2	2.5	0.2	2.00	0.171	0.156	7744
		0.6	1.00	0.152		
		0.8	0.50	0.145		
T3	3.3	0.2	2.64	0.182	0.163	7267
		0.6	1.32	0.158		
		0.8	0.66	0.148		
T4	2.1	0.2	1.68	0.152	0.141	7814
		0.6	0.84	0.147		
		0.8	0.42	0.123		
T5	1.8	0.2	1.44	0.120	0.109	9120
		0.6	0.72	0.111		
		0.8	0.36	0.097		

In Table 5 the highest average velocity upstream of the Rappang main channel in May 2022 is located at T3 at a depth of 3.3 m with an average velocity of 0.163 m^2/ s with a sediment concentration value of 7267 mg/l and the smallest velocity is at T5 at a depth of 1.8 m with velocity 0.109 m^2/s with a sediment concentration value of 9120 mg/l.

Table 6. Flow Rate and Sediment Concentration in the Middle of the Rappang Main Channel in May 2022

Middle	Depth H	Vertical Point	Depth (m)	Speed V(m/s)	V Average	Cs (mg/l)
T1	1.3	0.2	1.04	0.142	0.130	4824
		0.6	0.52	0.132		
		0.8	0.26	0.116		
T2	1.7	0.2	1.36	0.152	0.146	4946
		0.6	0.68	0.149		
		0.8	0.34	0.136		
T3	1.8	0.2	1.44	0.160	0.152	4482
		0.6	0.72	0.155		
		0.8	0.36	0.141		
T4	1.7	0.2	1.36	0.151	0.143	5283
		0.6	0.68	0.143		
		0.8	0.34	0.133		
T5	1.2	0.2	0.96	0.138	0.130	5901
		0.6	0.48	0.129		
		0.8	0.24	0.123		

In Table 6 the highest average velocity in the middle of the rappang main channel in May 2022 is located at T3 at a depth of 1.8 m with an average velocity of 0.152 m2 / s with a sediment concentration value of 4482 mg/l and the smallest velocity at T5 at a depth of 1.2 m at a speed of 0.130 m^2/s with a sediment concentration value of 5901 mg/l.

Table 7. Flow Rate and Sediment Concentration at Rappang Main Channel Downstream in May 2022

Downstream	Depth H	Vertical Point	Depth (m)	Velocity V(m/s)	V Average	Cs (mg/l)
T1	1.3	0.2	1.04	0.113	0.100	3487
		0.6	0.52	0.103		
		0.8	0.26	0.084		
T2	1.7	0.2	1.36	0.114	0.096	3341
		0.6	0.68	0.090		
		0.8	0.34	0.084		
T3	2.1	0.2	1.68	0.141	0.127	2644
		0.6	0.84	0.123		
		0.8	0.42	0.116		
T4	1.5	0.2	1.2	0.120	0.109	3522
		0.6	0.6	0.111		
		0.8	0.3	0.097		
T5	1.5	0.2	1.2	0.116	0.106	4707
		0.6	0.6	0.110		
		0.8	0.3	0.092		

IOP Conf. Series: Earth and Environmental Science 1117 (2022) 012065 doi:10.1088/1755-1315/1117/1/012065

In Table 7 the highest average velocity downstream of the Rappang main channel in May 2022 is located at T3 at a depth of 2.1 m with an average velocity of 0.127 m2 / s with a sediment concentration value of 2644 mg/l and the smallest velocity is at T5 at a depth of 1.5 m with velocity 0.106 m2 / s with a sediment concentration value of 4707 mg/l.

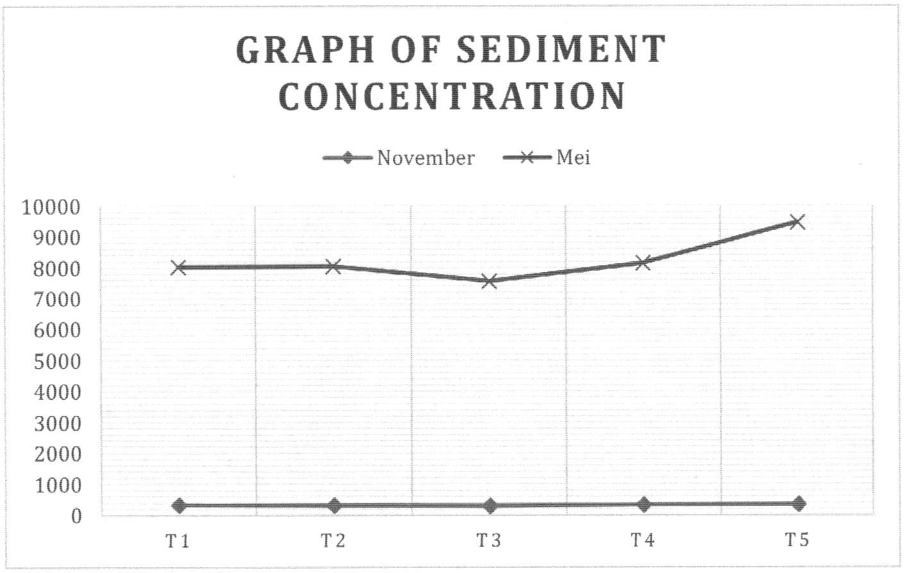

Figure 4. Graph of upstream sediment concentration rappang main channel

In Figure 4 explains that in the upstream of the Rappang main channel, the largest concentration is in May with an average sediment concentration value of 7930 mg/l and in November average sediment concentration value of 320 mg /l. This can prove that seasonal changes have resulted in very significant changes in sediment concentration.

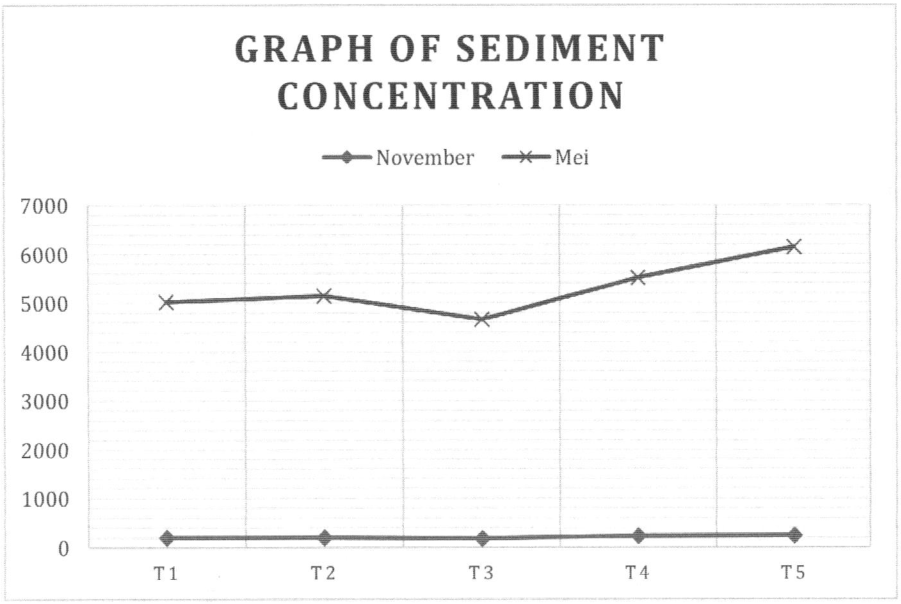

Figure 5. Graph of middlestream sediment concentration rappang main channel

In Figure 5 explains that in the middlestream of the Rappang main channel, the largest concentration is in May with an average sediment concentration value of 5087 mg/l and in November average sediment concentration value of 210 mg/l. This can prove that seasonal changes have resulted in very significant changes in sediment concentration

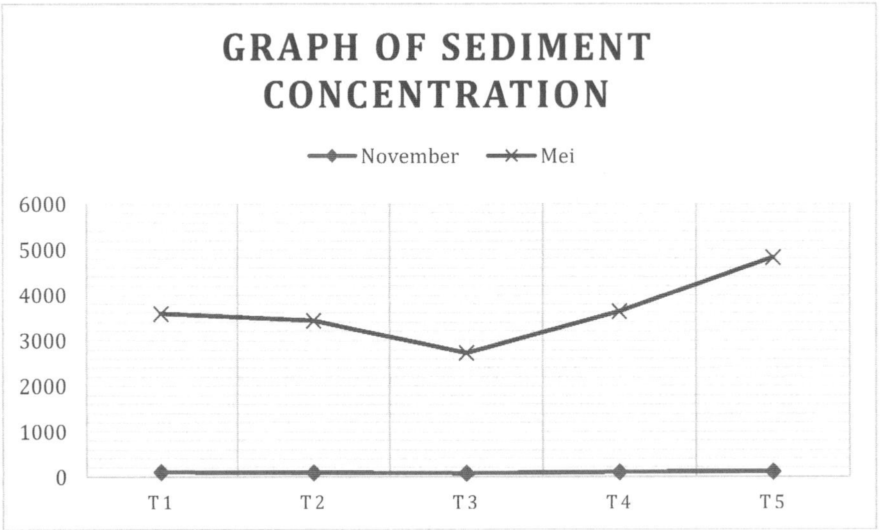

Figure 6. Graph of downstream sediment concentration rappang main channel

In Figure 6 explains that in the downstream of the Rappang main channel, the largest concentration is in May with an average sediment concentration value of 3540 mg/l and in November average sediment concentration value of 100 mg/l. This can prove that seasonal changes have resulted in very significant changes in sediment concentration

4. Conclusion

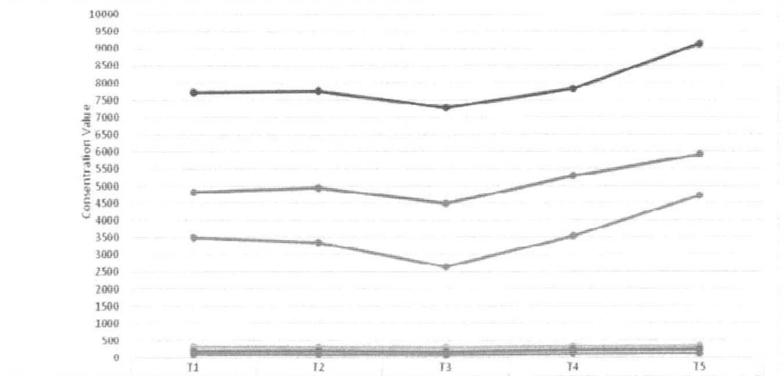

	T1	T2	T3	T4	T5
May Upstream	7704	7744	7267	7814	9120
May midstream	4824	4946	4482	5283	5901
May Downstream	3487	3341	2644	3522	4707
Nov Upstream	320	310	300	330	340
Nov Midstream	200	200	180	230	240
Nov Downstream	100	90	80	110	120

Figure 7. Sediment concentration value during season

In this Research knows that the sediment concentration greatly influences of seasonal changes In this study, it can be seen that the highest flow velocity and sediment concentration are located in May 2022. the largest concentration is in May with an average sediment concentration value of 7930 mg/l and in November average sediment concentration value of 320 mg /l. This can prove that seasonal changes have resulted in very significant changes in sediment concentration.

5. References

(1) XX Lu, et al. 2013." Sediment loads response to climate change: a preliminary study of eight large Chinese rivers." International Journal of Sediment Research
(2) Jan Pietro´n and Jerker Jarsjo. 2014." Sediment transport patterns and climate change: the downstream Tuul River case study, Northern Mongolia." Geophysical Research
(3) Putra, IBG, Saadi, Y., and Suroso, A., 2014, Analysis of Sediment Grain in the Jangkok and Tembiras Downstream Rivers, Proceedings of the National Conference on Civil Engineering (Context) 8Volume 2: Transportation- Geotechnical- Materials and Water Resources, Bandung
(4) AR Solomon. 2015." Analysis of Sedimentation Rate in Irrigation Channels in the Sanrego Irrigation Area, Kahu District, Bone Regency, South Sulawesi Province."
(5) Arief Fahmi, et al. 2015." Analysis of Sedimentation in the Jetty Building Muara Sungai Ketahun, Kab. North Bengkulu."
(6) Indonesia, National Standards and National Standardization Agency. 2015. "Measurement of River Flow Discharge and Open Channels Using Current Measuring Instruments and Buoys."
(7) Inan Najiah Arifin, et al. 2016." Analysis of Floating Sediment Profiles and Estimation of Sedimentation Rates in the Lonrong Secondary Canal, Bissua Irrigation Area, Kab. Gowa." Journal of Agritechno
(8) Suci Handayani, et al. 2017. "Identification of Types of River Bend Base Sediment Materials in Sungai Duri Village, Bengkayang District, Bengkayang Regency."
(9) Ching-Nuo Chen, et al. 2018." Climate Change Impacts on Soil Erosion and Sediment Yield in a Watershed
(10) Joanna Sziło, and Robert Józef Bialik. 2018." Grain Size Distribution of Bedload Transport in a Glaciated Catchment (Baranowski Glacier, King George Island, Western Antarctica)
(11) Zulfikar Ali Ahmad, et al. 2019." Correlation Between Flow Discharge and Suspended Load in the Data River 'Pinrang Regency.
(12) M. Iqbal Pratama, et al. 2019." Analysis of Sediment Transport and the Effect of Mining Activities on the Sombe River, Palu City, Central Sulawesi." Journal of Water Resources Engineering
(13) MP Hatta and SS Yuni. 2019." Changes in rainfall and discharge in Saddang Watershed, South Sulawesi, Indonesia." IOP Conference Series: Earth and Environmental Science.
(14) Putra, Indra Pratama, Kartini, and Nurhayati. 2019. "Analysis of Sediment Transport in Muara Parit Berkat." Journal of Civil Engineering Students UNTAN.

Analysis Of Sediment Concentration With Flow Speed In The Primary Channel In Saddang

Nur Asrida[1] and M S Pallu[2]

[1]Magister Student Department of Civil Engineering, Faculty of Engineering, Hasanuddin University, Makassar, Indonesia
[2]Lecturer Department of Civil Engineering, Faculty of Engineering, Hasanuddin University, Makassar, Indonesia

E-mail: nurasrida2@gmail.com

Abstract. The decrease in channel capacity can cause sedimentation at the bottom of the channel and the condition of the channel lining is damaged so that seepage / leaks occur and buildings and sluice gates do not function properly. This study aims to analyze the concentration of sediment on the flow velocity. Measurements were carried out in the field to observe field conditions in detail related to sediment velocity and concentration. This research was carried out on the Rappang Main Channel from Upstream to Downstream of the Rappang Main Channel. The distance between upstream to downstream of the Rappang Main Channel is +- 30 Km. The distance between the upstream to the center of the research point is +- 18 Km, while the distance between the middle to the downstream channel is +- 12 Km. Velocity measurement using A.OTT propeller current meter model and sediment collection in the channel using grab sampler and van dorn tools. The results showed the effect of velocity on the value of sediment concentration in the dry season.

1. Introduction

Water is one of the factor determinant in production process of agriculture. To fulfill water needs for various necessity effort farm, channel irrigation becomes very important. The process of distributing water through channel irrigation must be given in amount, time, and right quality. Plant will experience disturbance growth if channel irrigation not managed with good impact on production agriculture, so irrigation becomes very important and strategic in skeleton water supply for agriculture[1].

One of influencing factors that decrease in water channel is the existence of sediment that settles on base channel. Pattern deployment sedimentation depends on topography channel and area Genre sediment. Precipitate sedimentation on base channel g can cause reduced effective water volume and channel will influence age operational channel [2].

Primary channel is a cannal that carries water from building tap going to channel secondary and to plots irrigated tertiary. End of the limit primary channel is on building for the latter. Good primary channel could fulfill water requirement in land to be watered. In technique civil knowledge, good building must have two important requirements, a good quality and economical quality. This thing applied not only in plan building graded or street highway, but also on planning the water line too. The water channel on the question here is water line for necessity irrigation, between the othe primary channel [3].

Sedimentation process on a river includes the process of erosion, transportation, deposition and compression from sedimentation that alone. As known , sedimentation in rivers occur because there is a deposition process concentration sediment on Genre sourced river from results erosion in part upstream river . This also applied on channel irrigation in a weir. Damage area Genre river caused increase transport carried sediment Genre to channel irrigation. If the speed Genre is low so will there is a deposition process in the channel irrigation[4].

Right now, the irrigation in Saddang areas don't get water equally. The reason is because the conditions network experience is having decrease capacity. Decrease capacity could cause a sedimentation at the bottom channel and damaged channel lining cindition so that occur seepage / leak as well as building and floodgate not working properly.

Soewarno (1991) in (Rosmiati et al., 2020) said that the volume of sediment transport mainly depends on the flow velocity due to changes in the rainy and dry seasons and changes in velocity influenced by human activities. The result of changes in the volume of sediment transport is the occurrence of scouring in some places and the occurrence of deposition in other places at the bottom of the irrigation canal, thus the dimensions of the channel will change so that the volume of water carried is also reduced[4].

Afiato (2016) in (Hermawan & Afiato, 2021) studied about distribution speed Genre and concentration sediment on Genre not harmonym on channel irrigation Mataram Yogyakarta. From the result of this study showing that concentration sediment suspension the bigger it approaches to the base channel and concentration sediment vertical average suspension tend the bigger it approahes to the middle area cross section channel[5].

Based on the background and the previous studies that had explained, this study then emphasizing on sedimentation Flying and sedimentation base on three point in the primary channel DI Saddang so that writer tries to study a thesis with title : **" Analysis " Concentration Sediment with Speed Streams in the Primary Channel DI Saddang "** .

2. Research Method

2.1 Research location and time

This research was conducted from November 2021 to May 2022. This research was carried out on the Rappang Main Channel from Upstream to Downstream of the Rappang Main Channel. The distance between upstream to downstream of the Rappang Main Channel is +- 30 Km. The distance between the upstream to the center of the research point is +- 18 Km, while the distance between the middle to the downstream channel is +- 12 Km. The research sample was tested at the Soil Mechanics Laboratory, Department of Civil Engineering, Faculty of Engineering, Hasanuddin University, Gowa.

2.2 Data collection technique

In this study, two methods of data collection were used , namely:

1. Literature study, to obtain the results of previous research, relevant research and scientific articles as a theoretical basis to complete this research.
2. Secondary data collection.
   a. collection of Building Schemes and irrigation network schemes of UPTD Alitta.
   b. Data on discharge of secondary irrigation canals in Saddang District in 2013 - 2020.
   c. Collection of administrative maps of DI Saddang government.
3. Primary data collection by:
   Primary data collection was carried out through field measurements to observe field conditions in detail related to sediment velocity and concentration.
   - Primary data collection consists of two, namely:
     a. Flow speed measurement
     Measurement of flow velocity is carried out by direct measurement of velocity and wet cross section in the channel. Measurement of flow velocity in the secondary channel using a *Current Meter* Propeller A.OTT, measuring instrument.

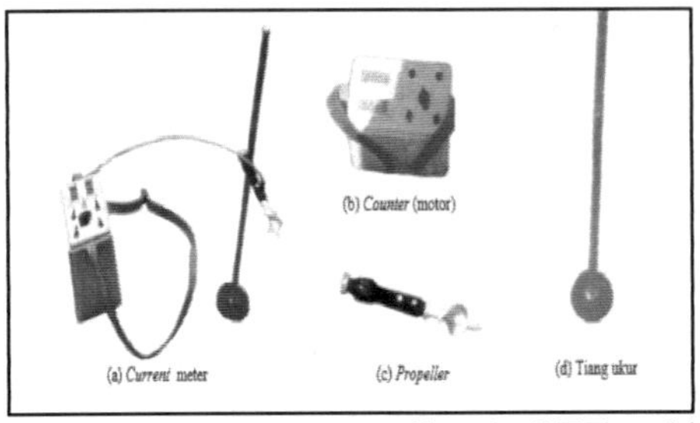

Figure 1. Current meter propeller A. OTT and its part.

The measurement procedure using a current *meter* is as follows:

a) Measure the depth of the channel with the measuring rod of the current meter tool
b) Choose a propeller that is suitable for the channel depth, so that it can be used for several vertical points, namely (0.2h, 0.6h, and 0.8h) where h is the channel depth.
c) The current meter is mounted on a measuring pole (static) with a depth of 0.2h, 0.6h, and 0.8h, then the measuring pole is immersed in the water until the measuring pole is located at the bottom of the channel with the propeller facing the direction of flow.
d) The number of revolutions per unit time, which occurs at each water depth is calculated.

b. Sediment removal in the channel

1. Grab Sampler

Figure 2. Grab Sampler.

a) Prepare a grab complete with rope and messenger.
b) Open the bottom valve of the Ekman grab, then hook the valve wire to the lever at the top of the grab.
c) Make sure the valve is open with the wire properly attached.
d) Enter the grab slowly in the water until it reaches the bottom of the water to be sampled.
e) Then release the messenger, and the grab valve will close.
f) Drain sediment through top valve, not bottom

2. Van Dorn

Figure 3. Van Dorn.

a) The fastening rope of the two tube covers is pulled and then tied to the hook and clamped with the hook clamp.
b) After the tube cover is open, the tool is lowered to the desired depth using a tube hanging rope
c) When the van dorn is in the desired position, the metal clamp opening is dropped to open the hook clamp so that the rubber connecting the cover can function to close the tube which already contains the desired water sample.
d) Van Dorn is ready to be reappointed

Sediment sampling was also carried out for each channel as many as 5 points at the upstream, middle and downstream of the channel.

2.3 Research implementation

The research implementation process is principally divided into several parts, namely data collection, data processing, data calculation and analysis, and discussion.

➢ The procedure for conducting research in the field follows several stages, including:
1. Preparation in this preparatory stage required data on channel cross-sectional images in the field.
2. The implementation of research in this stage is carried out by analyzing, among others:
   a) Measure the width of the channel after that stretch the rope along the width of the channel
   b) Measure the channel Width after that stretch the string along the channel width at each channel input and output.
   c) Performing Flow Velocity measurements with A.OTT Propoler Current Meter
   d) Make a distance interval of every 50 cm on each rope on each Channel width.
   e) Measuring the depth of the channel by paying attention to the sediment on the channelfloor.
   f) After getting the channel depth data, it is entered into the formula 0.2h, 0.6h, and 0.8h.
   g) Retrieve flow rate data at each predetermined distance interval on each channel type.
   h) Conducting Sediment Trap at the Top and Bottom of the Channel.
   i) The flow velocity data for each channel and sediment data in the channel are entered into the observation table and then data analysis is carried out.

Table. 1 Data collection in the field

No	Research Sites	**Research Variable**	
		Speed Data	Sediment Removal
1	Hulu Channel		
	T1	0.2h, 0.6h, 0.8h	0.6 of the channel depth
	T2	0.2h, 0.6h, 0.8h	0.6 of the channel depth
	T3	0.2h, 0.6h, 0.8h	0.6 of the channel depth
	T4	0.2h, 0.6h, 0.8h	0.6 of the channel depth
	T5	0.2h, 0.6h, 0.8h	0.6 of the channel depth
2	Middle Channel		
	T1	0.2h, 0.6h, 0.8h	0.6 of the channel depth
	T2	0.2h, 0.6h, 0.8h	0.6 of the channel depth
	T3	0.2h, 0.6h, 0.8h	0.6 of the channel depth
	T4	0.2h, 0.6h, 0.8h	0.6 of the channel depth
	T5	0.2h, 0.6h, 0.8h	0.6 of the channel depth
3	Downstream Channel		
	T1	0.2h, 0.6h, 0.8h	0.6 of the channel depth
	T2	0.2h, 0.6h, 0.8h	0.6 of the channel depth
	T3	0.2h, 0.6h, 0.8h	0.6 of the channel depth
	T4	0.2h, 0.6h, 0.8h	0.6 of the channel depth
	T5	0.2h, 0.6h, 0.8h	0.6 of the channel depth

- Sediment Sample Test

  Sediment sample testing was carried out at the soil mechanics laboratory at Hasanuddin University Makassar by testing suspended load sediment samples from upstream, middle and downstream in the main channel Rappang Kab. Pinrang. The data that will be obtained in testing the suspended sediment sample is the sediment concentration.

3. Result and Discussion

3.1 Result and discussion of the upstream channel section

Table. 2 Flow velocity at the upstream of the channel

T	H	Vertical point	Depth (m)	V (m/s)	V Average (m2/s)	Qs (mg/l)
T1	2.3	0.2	1.84	0.152	0.143	7704
		0.6	0.92	0.149		
		0.8	0.46	0.127		
T2	2.5	0.2	2	0.171	0.156	7744
		0.6	1	0.152		
		0.8	0.5	0.145		
T3	3.3	0.2	2.64	0.182	0.163	7267
		0.6	1.32	0.158		
		0.8	0.66	0.148		
T4	2.1	0.2	1.68	0.152	0.141	7814
		0.6	0.84	0.147		
		0.8	0.42	0.123		
T5	1.8	0.2	1.44	0.120	0.109	9120
		0.6	0.72	0.111		
		0.8	0.36	0.097		

On the Table. 1 it can be seen that the highest average flow velocity in the upstream part of the channel in Mei is located at point T3 at a depth of 3.1 m with an average velocity of 0.163 m2 $^{/}$s with a sediment concentration value of 7267 mg/l.

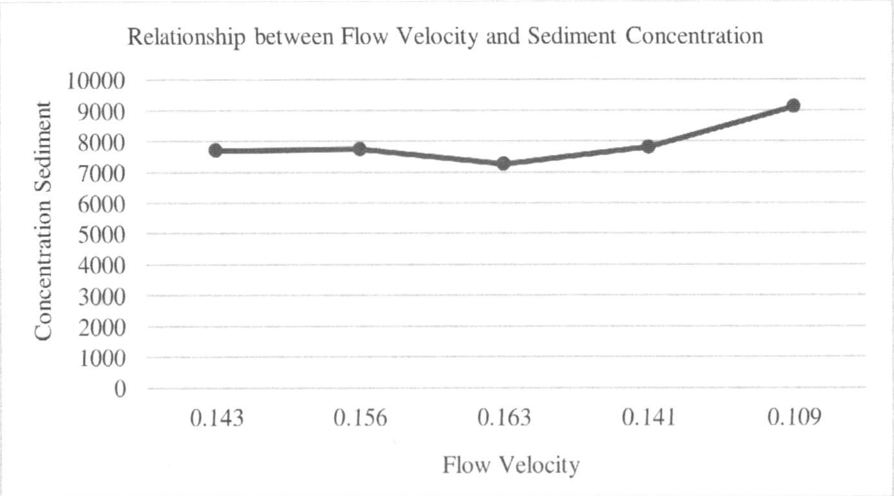

Figure 4. Relationship of flow velocity with concentration of sediment in the upstream of the channel

On Figure. 4 it can be seen that the largest sediment concentration in the upstream is at point T5 with a value of 9120 mg/l where the flow velocity at that point is 0.109 m2 $^{/}$s and the lowest sediment concentration is at point T3 with a value of 7267 mg/l which where the flow velocity is 0.163 m ^{2}/s

3.2 Result and discussion of the middle channel section

Table. 3 Flow velocity at the middle of the channel

T	H	Vertical point	Depth (m)	V (m/s)	V Average (m2/s)	Qs (mg/l)
		0.2	1.04	0.142		
T1	1.3	0.6	0.52	0.132	0.130	4824
		0.8	0.26	0.116		
		0.2	1.36	0.152		
T2	1.7	0.6	0.68	0.149	0.146	4946
		0.8	0.34	0.136		
		0.2	1.44	0.160		
T3	1.8	0.6	0.72	0.155	0.152	4482
		0.8	0.36	0.141		
		0.2	1.36	0.151		
T4	1.7	0.6	0.68	0.143	0.143	5283
		0.8	0.34	0.133		
		0.2	0.96	0.138		
T5	1.2	0.6	0.48	0.129	0.130	5901
		0.8	0.36	0.123		

On the Table. 3, it can be seen that the highest average flow velocity in the center of the channel in Mei is located at point T3 at a depth of 1.8 m with an average velocity of 0.152 m2 / s with a sediment concentration value of 4482 mg/l.

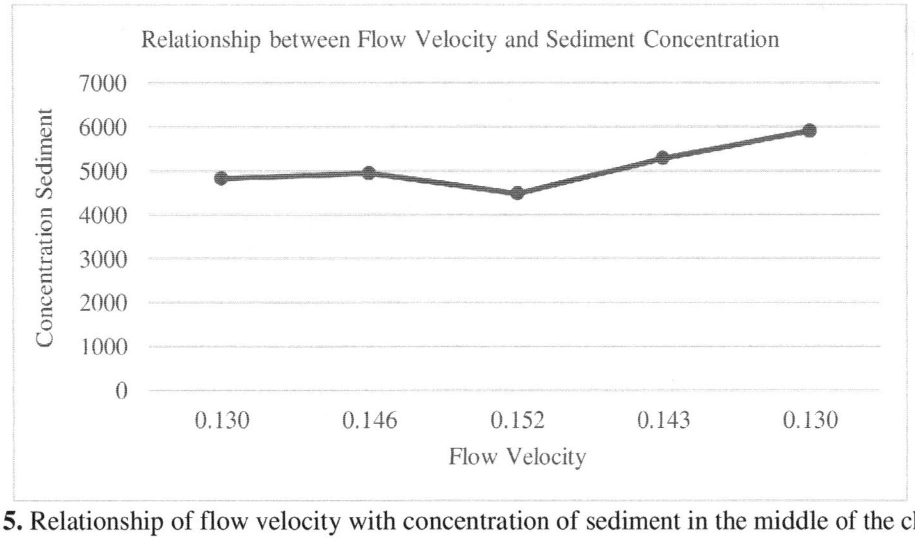

Figure 5. Relationship of flow velocity with concentration of sediment in the middle of the channel

On Figure. 5 it can be seen that the largest sediment concentration in the middle is at point T5 with a value of 5901 ml/g where the flow velocity at that point is 0.130 m2 / s and the lowest sediment concentration is at point T3 with a value of 4482 mg. /l where the flow rate is 0.152 m ^{2}/s

3.3 Result and discussion of the downstream channel section

Table. 4 Flow velocity at the downstream of the channel

T	H	Vertical point	Depth (m)	V (m/s)	V Average (m2/s)	Qs (mg/l)
T1	1.3	0.2	1.04	0.113	0.100	3487
		0.6	0.52	0.103		
		0.8	0.26	0.084		
T2	1.7	0.2	1.36	0.114	0.096	3341
		0.6	0.68	0.090		
		0.8	0.34	0.084		
T3	2.1	0.2	1.68	0.141	0.127	2644
		0.6	0.84	0.123		
		0.8	0.42	0.116		
T4	1.5	0.2	1.2	0.120	0.109	3522
		0.6	0.6	0.111		
		0.8	0.3	0.097		
T5	1.5	0.2	1.2	0.116	0.106	4707
		0.6	0.6	0.110		
		0.8	0.36	0.123		

On the Table. 4 it can be seen that the highest average flow velocity downstream of the channel in Mei is located at point T3 at a depth of 2.1 m with an average velocity of 0.127 m2 $^{/}$s with a sediment concentration value of 2644 mg/l.

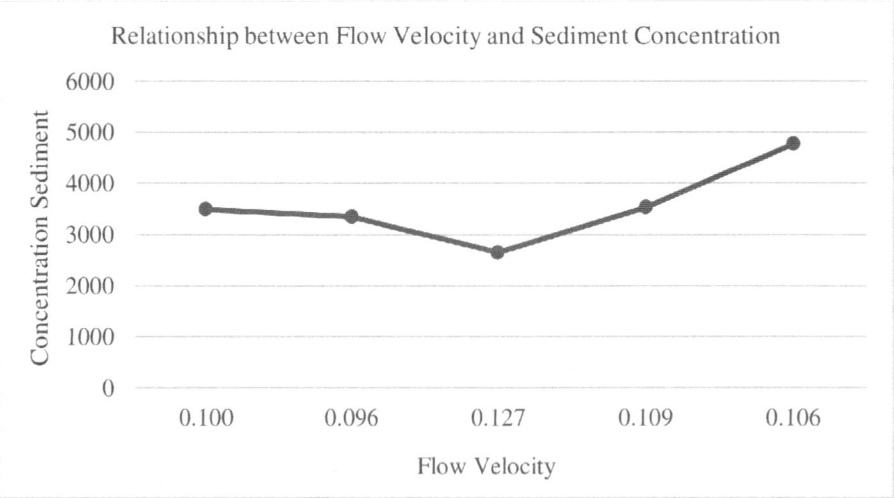

Figure 6. Relationship of flow velocity with concentration of sediment in the downstream of the channel

On Figure. 6 it can be seen that the largest sediment concentration downstream is at point T5 with a value of 4707 ml/g where the flow velocity at that point is 0.106 m2 $^{/}$s and the lowest sediment concentration is at point T3 with a value of 2644 mg. /l where the flow velocity is 0.127 m ^{2}/s.

4. Conclusion

In this study it can be concluded that the concentration of sediment in the channel can affect the flow velocity. The higher the sediment concentration value, the lower the flow velocity in the channel, and vice versa, the lower the sediment concentration value, the lower the flow velocity in the channel.

References

[1] Lusiantorowati, E., Prawitosari, T., & Achmad, M. (2015). *Efficiency of Water Distribution in Pekkabata Main Canal in North Saddang Irrigation Area, North Regency* . *8* (2), 116–122.

[2] Satriawansyah, T. (2018). *Analysis of Sedimentation at the Beringen Sila Weir, Stowe Berang Village, Utan District, Sumbawa Regency* . *2* , 78–87, https://doi.org/10.31227/osf.io/26tmk.

[3] Irrigation, D., & Flowers, DI (2016). Planning the Economic Dimensions of Primary Channels. *Cycle* , *2* (1), 47–57.

[4] Rosmiati, AYU, Studies, P., Agriculture, T., & Agriculture, JT (2020). *Analysis of the sedimentation rate in the irrigation canal of Kekalik Gerisak, Kekalik Gerisak, Mataram City* .

[5] Hermawan, A., & Afiato, E. (2021). Analysis of Basic Sediment Transport (Bed Load) in Yogyakarta Mataram Irrigation Channel. *Technician* , *XXVI* (1). https://doi.org/10.20885/teknisia.vol26.iss1.art3

Numerical Modelling of Shoreline changes of Galesong, Takalar's coast

R Karamma[1], M P Hatta[1], AS Mustari[1], SP Manda[1], A.Ilda [1] M Ihsan[2]

[1]Department of Civil Engineering, Hasanuddin University, Gowa, Indonesia
[2]Department of Civil Engineering, Sekolah Tinggi Teknik Baramuli, Pinrang, Indonesia

E-mail: riswalk@unhas.ac.id

Abstract : Morphological changes of Talakar's coastline covering three administrative subdistrict has been modelled numerically in terms of sedimentation build up. Shoreline changes is one of the dynamic processes as parts of natural processes for coastline to reach their subsequent equilibrium resulting in bathimetrical alteration of coastal bed. Sedimentation modeling simulation has been conducted for the coastlines of North Galesong, Galesong, and South Galesong subdistricts. As the location is physically affected by the oceanoghrapic dynamics of Makassar Strait, Flores Sea as well as the river hydraulics of Jeneberang and Pappa rivers' estuaries. Shoreline accretion and abbrasion are distincted by data analysis. Sedimentation model shows shallowing process (pendangkalan) of 0.0136m to 0.05977m with 0.000006 m/day to 0.013 m/day growth near shorelines of North Galesong, 0.025202m to 0.1194 deepening and 0.02526m to 0.16736m shallowing with 0,0013 m/day to 0,013 m/day growth near shorelines of Galesong, and 0.014148 m to 0.11948 deepening and 0.01351 m - 0.16736 shallowing with 0.01m/day to 0.026 m/day growth near shorelines of South Galesong.

Keywords : shoreline changes, bed morphology, bathymetry

1. Introduction

Understanding variability of shoreline, including the erosion-accretion processs is a crucial issue in the eyes of researcher and stakeholders in the field of coastal science, engineering, and management. As the interface of of water and land masses, the dynamic nature of coastline on the temporal and spatial scale creates an importance of various definitions of shoreline indicators [1]. The coastline is the boundary line of the meeting between land and sea water whith tendencies to displace in effect of tidal conditions, resulting in coastal defined as accreting if there are indications of continuous deposition, and defined as retreating if the abrasion process is still ongoing [2]. Coastal process is a continuous adjustment towards a natural balance to the impacts that occur that affect shoreline changes. Coastline change is a fairly dynamic process which includes changes in the bathymetry of coastal waters [3]. Coastline changes are influenced by sediment movement, longshore current, sea surface waves and land use [4]. Changes in coastlines can occur from time to time on a seasonal or annual scale, depending on the durability of coastal conditions in the form of topography, rocks and their properties with ocean waves, tides and wind [3]. Changes in coastlines are indicated by changes in their position, not only determined by a single factor but by a number of factors and their interactions which are the result of a combination of natural and human processes. Natural factors come from the influence of hydro-oceanographic processes that occur in the sea such as wave crashes, changes in current patterns, tidal variations. In response to the locality, some research have been conducted in near by region [5]–[11]

2. Methodology

The computation modelling of shoreline changes have been conducted in MIKE software by Danish Hydraulics Institute. The stages include the pre-processing of model development. The pre-processing includes secondary data collection, as the basic input in carrying out the modeling; determination of model boundaries, as the boundaries in making the domain consist of open boundaries, friction boundaries, and close-boundaries; development of a grid model based on bathymetric data, the grid used is an irregular grid (flexible-mesh) built in MIKE Zero.

The research location as the domain of the modeling is the coasts of Takalar Regency, Spouth Sulawesi, Indonesia, which includes the coastlines of North Galesong, Galesong, and South Galesong subdistricts, as shown in Fig. 1.

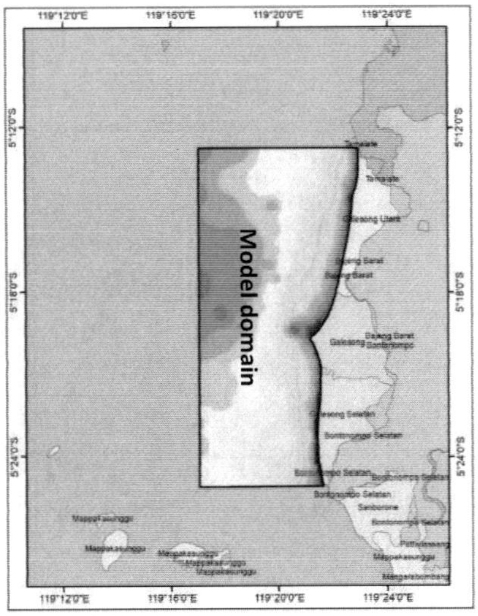

Figure 1. Model domain in coast of Takalar

2.1 Data

Bathymetry Data

The detailed bathymetry data used is sourced from secondary data obtained from the National Bathymetry data scale 1: 50000 6-arcsecond resolution issued by the Indonesian Geospatial Information Agency.

Tidal Data (open boundaries conditions)

The tidal data used is derived from the results of modeling forecasting on Mike 21 tidal analysis. The elevation data given is 1 year with an interval of 1 hour in the period from 2020 to 2021.

Wind data (friction condition)

Wind data is used in the model scenario to provide surface friction against current movement. This will have a direct effect on the transport force exerted by the current as a carrier for the diffusion of sedimentation.

2.2 Hydrodynamic modeling

The hydrodynamic modeling in this paper is carried out using a mathematical equation as the governing equation and the boundary conditions of the model.

- Continuity equation

$$\frac{\partial u}{\partial x} + \frac{\partial v}{\partial y} + \frac{\partial w}{\partial z} = S \quad (1)$$

- Horizontal momentum equation for x direction

$$\frac{\partial u}{\partial t}+\frac{\partial u^2}{\partial x}+\frac{\partial vu}{\partial y}+\frac{\partial wu}{\partial z}=fv-g\frac{\partial \eta}{\partial x}-\frac{1}{\rho_0}\frac{\partial p_a}{\partial x}-\frac{g}{\rho_0}\int_z^{\eta}\frac{\partial \rho}{\partial x}dz-\frac{1}{\rho_0 h}\left(\frac{\partial S_{xx}}{\partial x}+\frac{\partial S_{xy}}{\partial y}\right)+F_u+\frac{\partial}{\partial z}\left(v_t\frac{\partial u}{\partial z}\right)+u_sS \quad (2)$$

- Horizontal momentum equation for y direction

$$\frac{\partial v}{\partial t}+\frac{\partial v^2}{\partial y}+\frac{\partial uv}{\partial x}+\frac{\partial wv}{\partial z}=-fu-g\frac{\partial \eta}{\partial y}-\frac{1}{\rho_0}\frac{\partial p_a}{\partial y}-\frac{g}{\rho_0}\int_z^{\eta}\frac{\partial \rho}{\partial y}dz-\frac{1}{\rho_0 h}\left(\frac{\partial S_{yx}}{\partial x}+\frac{\partial S_{yy}}{\partial y}\right)+F_v+\frac{\partial}{\partial z}\left(v_t\frac{\partial v}{\partial z}\right)+v_sS \quad (3)$$

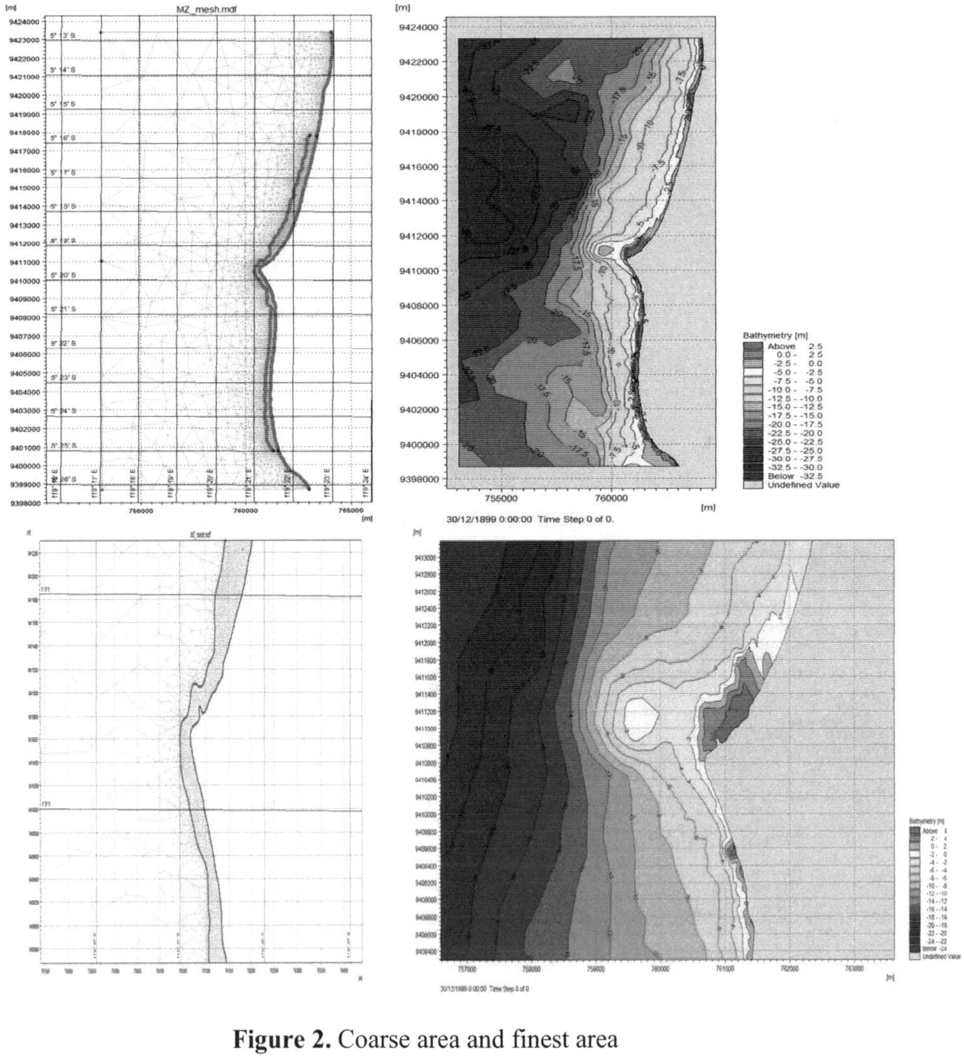

Figure 2. Coarse area and finest area

From the continuity equation, the value of the change in water elevation with time η(t) could be obtained, while from the horizontal equation for the momentum in the x,y direction, the current velocity value will be obtained. The mesh model for the current model is carried out using input data in the form of shoreline and bathymetry data. The converted points from the bathymetric data are entered as the basic morphology in the domain area. The domain area is divided into two parts in one large square, namely a large area (coarse area) and a small area (finest area). Large area (coarse area) is an area that supports the coverage of small areas. Takalar waters with an area of 209.3 km2. The large area has 29,103 elements with 15,239 connected nodes, the maximum element area is 0.0002° with the smallest angle of the triangle element is 30° . Small area (finest area) is the area that becomes the main objective in the implementation of modeling, studied within the scope of sedimentation. This area covers the three sub-districts, namely North Galesong District, Galesong District and South Galesong District as well as in areas close to the coastline. The area close to the shoreline has a maximum area of 8e-008° and the smallest element angle is 30° (Figures 2 and 3). The presentation of the modeling results in the sedimentation model is shown in cross-shore and long-shore sections, to describe changes in the modeling period (Figure 3).

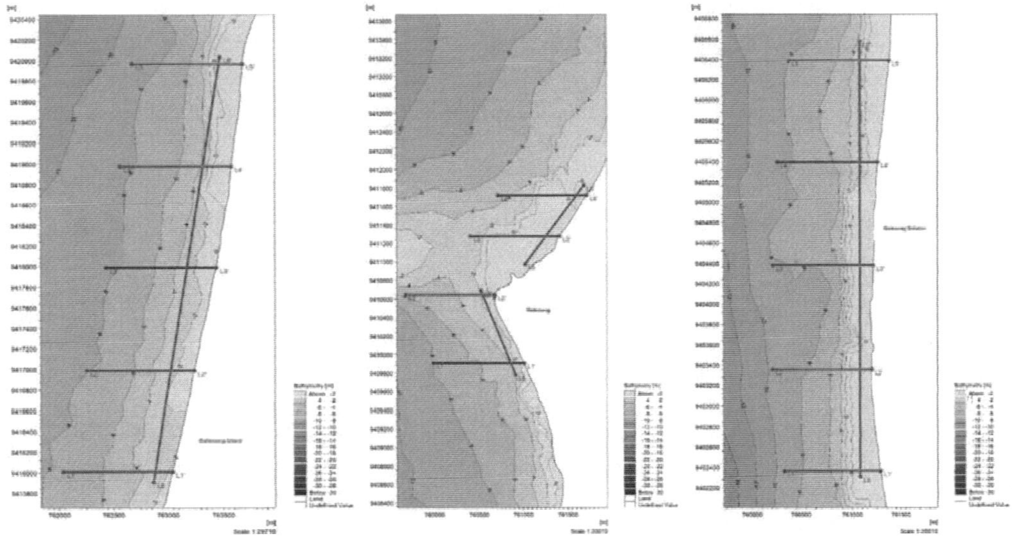

Figure 3. Cross section of the area of North Galesong District, Galesong District and South Galesong District

Open boundary conditions are given at the west, north and south boundaries, each of which is given 1(one) open boundary condition. The closed boundary condition is the land boundary given to the east, the coastline of the western coastal waters of Takalar Regency (Figure 4)

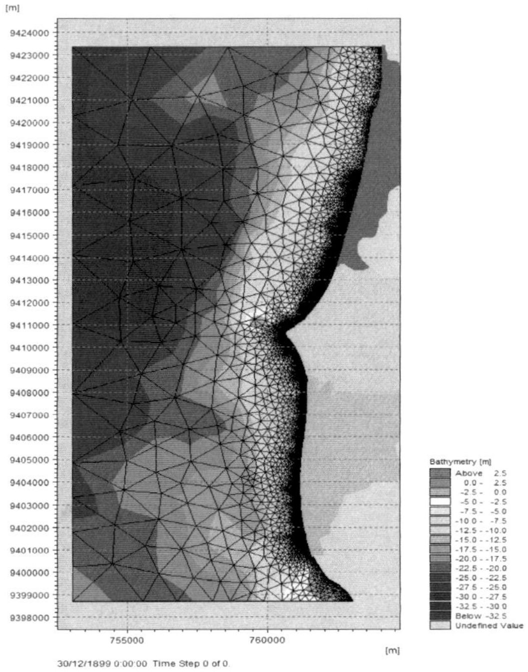

Figure 4. Closed and Open boundary conditions in area of modelling in the coast of Takalar

3. Results and discussion

Sedimentation modeling in coastal waters of Takalar Regency

The results of the sedimentation simulation in the Takalar waters in the coastal area of North Galesong District obtained the results of sediment distribution patterns, changes in basic morphology, and sedimentation as shown in Figures 5, 6, 7.

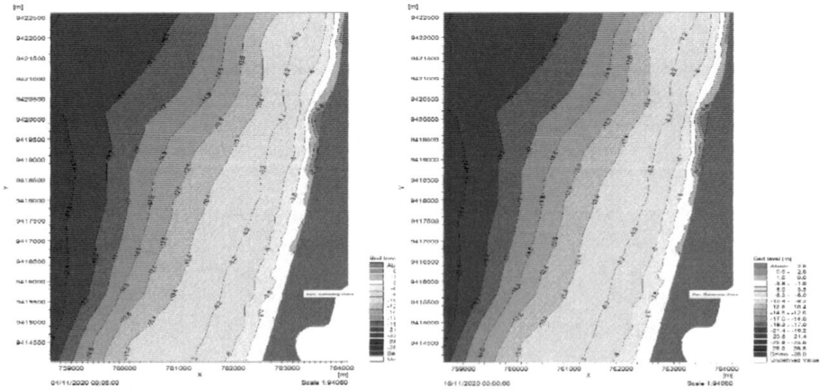

Figure 5. Depth of North Galesong waters before (Δt = 0) and after simulation (Δt = 1)

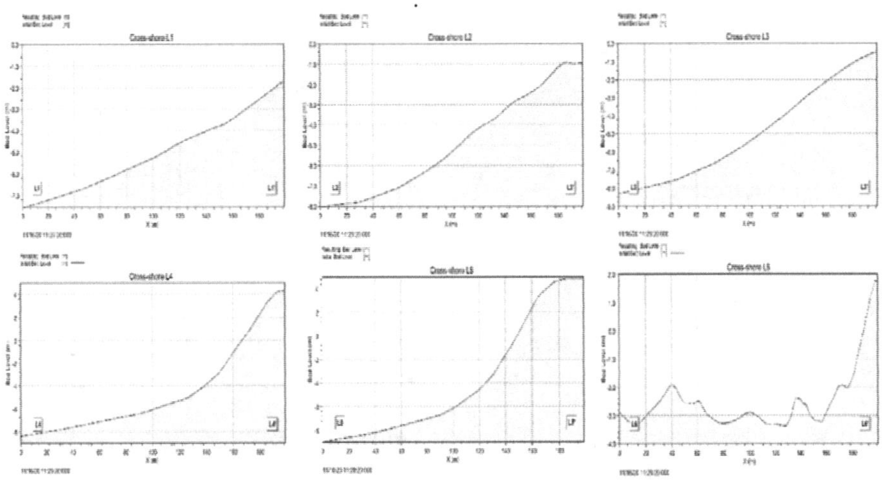

Figure 6. Profile of depth in of North Galesong waters

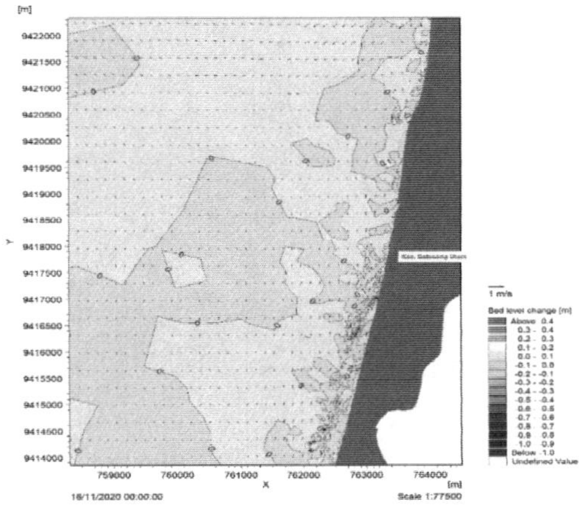

Figure 7 Results of morphological change simulation of North Galesong waters after simulation

The modeling results show the results on a cross-section of the coast in North Galesong District L_1,L_2,L_3,L_4,L_5, and L_6 as shown in Table 1

Table 1. Average values for changes in coast bed morphology in North Galesong District

Morphological change	Lines					
	L1 (m)	L2 (m)	L3 (m)	L4 (m)	L5 (m)	L6 (m)
Deepening	0.06448	0.026827	0.089068	0	0.019119	0.03385
Shallowing	0.02365	0.05977	0.02035	0.0236	0.0136	0.03588

Simulation of sedimentation in the Takalar waters in the coastal area of Galesong District produces sediment distribution patterns, basic morphological changes, and sedimentation as shown in Figures 8, 9, and 10.

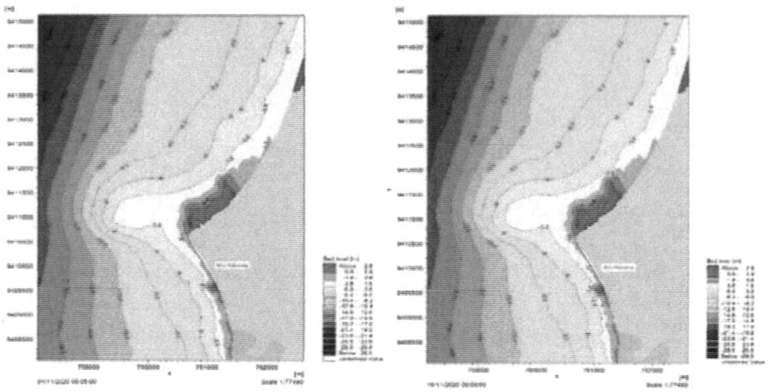

Figure 8. Depth of Galesong waters before (Δt = 0) and after simulation (Δt = 1)

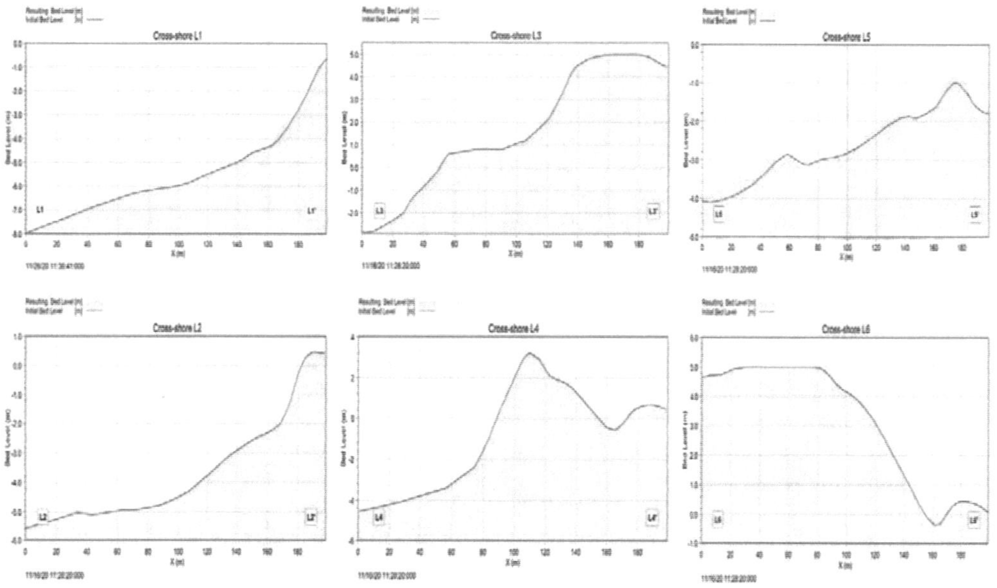

Figure 9. Profile of depth in of Galesong waters

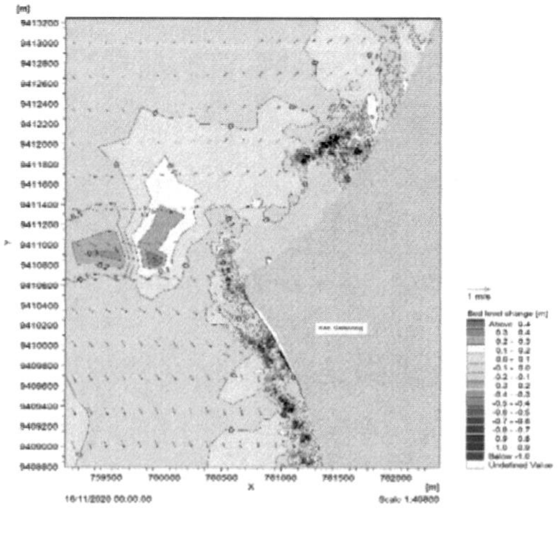

Figure 10. Results of morphological change simulation of Galesong waters after simulation

The modeling results show the results on a cross-section of the coast in Galesong District L_1,L_2,L_3,L_4,L_5, and L_6 as shown in Table 2

Table 2. Average values for changes in coast bed morphology in Galesong District

Morphological change	Lines					
	L1 (m)	L2 (m)	L3 (m)	L4 (m)	L5 (m)	L6 (m)
Deepening	0.025202	0.044968	0.045344	0.026613	0.11948	0.039544
Shallowing	0.1041	0.03524	0.02605	0.02526	0.16736	0.22366

Simulation of sedimentation in the Takalar waters in the coastal area of South Galesong District produces sediment distribution patterns, basic morphological changes, and sedimentation as shown in Figures 11, 12, and 13.

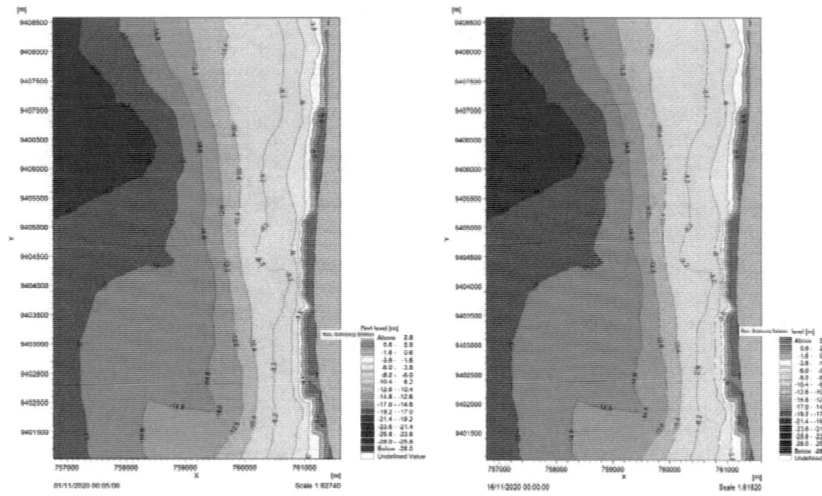

Figure 11. Depth of South Galesong waters before (Δt = 0) and after simulation (Δt = 1)

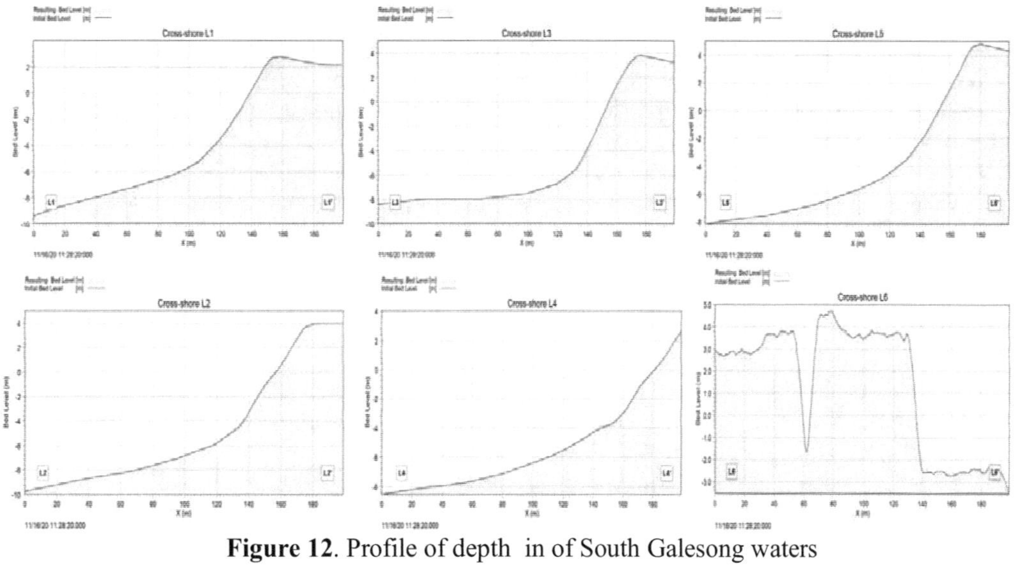

Figure 12. Profile of depth in of South Galesong waters

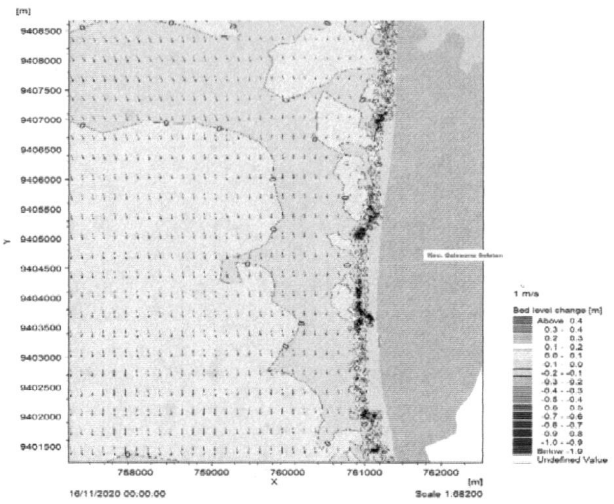

Figure 13. Results of morphological change simulation of Galesong waters after simulation

The modeling results show the results on a cross-section of the coast in South Galesong District L_1,L_2,L_3,L_4,L_5, and L_6 as shown in Table 3

Table 3. Average values for changes in bed morphology of water in South Galesong District

Morphological change	Lines					
	L1 (m)	**L2 (m)**	**L3 (m)**	**L4 (m)**	**L5 (m)**	**L6 (m)**
Deepening	0.027304	0.014148	0.019003	0.017392	0.11948	0.109836
Shallowing	0.03961	0.02205	0.01351	0.08067	0.16736	0.11177

4. Conclusion

1. The results of sedimentation modeling in the area of North Galesong District show that there is a bottom deepening between 0.019119 m - 0.089068 m and a shallow depth of 0.0136 m - 0.05977 m. In the area near the shoreline, the depth changes towards siltation with variations in depth reduction from 0.0001 m to 0.2 m or 0.000006 m/day – 0.013 m/day.
2. The results of sedimentation modeling in the Galesong District area show that there is a bottom deepening between 0.025202 m - 0.11948 m and a shallow depth of 0.02526 m - 0.16736 m. In the area near the shoreline, the depth changes towards siltation with variations in depth reduction from 0.02 m to 0.2 m or 0.0013 m/day – 0.013 m/day.
3. The results of sedimentation modeling in the area of South Galesong District show that the depth of the water base is between 0.014148 m - 0.11948 m and siltation of 0.01351 m - 0.16736 m. In areas near the shoreline, the depth changes towards siltation with variations in depth reduction from 0.15 m to 0.4 m or 0.01 m/day – 0.026 m/day.

5. References

[1] E. H. Boak and I. L. Turner, "Shoreline definition and detection: a review," *Journal of coastal research*, vol. 21, no. 4, pp. 688–703, 2005.
[2] R. Dolan, M. S. Fenster, and S. J. Holme, "Temporal analysis of shoreline recession and accretion," *Journal of coastal research*, pp. 723–744, 1991.
[3] J. P. Mills, S. J. Buckley, H. Mitchell, P. Clarke, and S. Edwards, "A geomatics data integration technique for coastal change monitoring," *Earth Surface Processes and Landforms: The Journal of the British Geomorphological Research Group*, vol. 30, no. 6, pp. 651–664, 2005.

[4] W. J. Emery and R. E. Thomson, *Data analysis methods in physical oceanography*, 2nd and rev. ed ed. Amsterdam ; New York: Elsevier, 2001.

[5] R. Karamma, M. S. Pallu, M. Arsyad Thaha, and M. P. Hatta, "Numerical modeling of water mass structure distribution at the estuary Jeneberang river, Makassar," *International Journal of Advanced Research in Engineering and Technology*, vol. 11, no. 5, pp. 420–431, 2020, doi: 10.34218/IJARET.11.5.2020.044.

[6] Y. Arafat, M. S. Pallu, F. Maricar, and R. T. Lopa, "Hydrodynamics and Morphological Changes Numerical Model of the Jeneberang Estuary," *International Journal of Innovative Research in Advanced Engineering (IJIRAE)*, vol. 3, no. 8, pp. 21–29, 2016.

[7] R. Karamma, M. S. Pallu, M. A. Thaha, and M. P. Hatta, "Stratification model of seawater mass structure at the estuaries of Jeneberang River and Tallo River and the influences to current pattern in Makassar coastal areas," 2020, vol. 419, no. 1. doi: 10.1088/1755-1315/419/1/012132.

[8] R. Karamma, M. S. Pallu, M. A. Thaha, M. P. Hatta, A. S. Mustari, and A. Syarif Sukri, "Analysis of Longshore Sediment Transport at Theestuaries of Jeneberang River and Tallo River Caused by Waves on Coast of Makassar," 2020, vol. 797, no. 1. doi: 10.1088/1757-899X/797/1/012010.

[9] R. Karamma, M. S. Pallu, M. A. Thaha, M. P. Hatta, and M. Ihsan, "Spatial mapping of water mass structure in the estuary of Jeneberang river," *IOP Conf. Ser.: Earth Environ. Sci.*, vol. 841, no. 1, p. 012023, Aug. 2021, doi: 10.1088/1755-1315/841/1/012023.

[10]B. Bakri, A. Sumakin, Y. Widiasari, and M. Ihsan, "Distribution pattern of water salinity analysis in Jeneberang river estuary using ArcGIS," 2020, vol. 419, no. 1. doi: 10.1088/1755-1315/419/1/012116.

[11]Y. Arafat, M. S. Pallu, F. Maricar, and R. Lopa, "Hydrodynamics and Morphological Changes Numerical Model of the Jeneberang Estuary," *International Journal of Innovative Research in Advanced Engineering*, vol. 3, no. 08, p. 9.

IOP Conf. Series: Earth and Environmental Science 1117 (2022) 012068 doi:10.1088/1755-1315/1117/1/012068

2D modeling of current patterns and sediment patterns on Pasir Beach, Kebumen

AID Puspita[1], **MA Thaha**[2], **and B Hidayah**[3]

[1]Lecturer, Department of Civil Engineering, Faculty of Engineering, Universitas Hasanuddin
[2]Professor, Department of Civil Engineering, Faculty of Engineering, Universitas Hasanuddin
[3]Lecturer, Department of Geological Engineering, Faculty of Engineering, Universitas Hasanuddin

Email: ildhaunhas@gmail.com

Abstract. Pasir beach is one of the beaches in the row of the South Coast of Java Island which is located in the Kebumen Regency, Central Java Province. On the sand beach, there is a fish landing port which is in poor condition. Some of the main port facilities are not functioning at all due to sedimentation problems. The purpose of this study is to determine the current pattern and sedimentation pattern on the Pasir beach, especially in the fish landing port area with 2-dimensional modeling. The results of 2-dimensional modeling with RMA2 and SED2D found that fluctuations in current velocity are not significant because the velocity is small. The increase in current velocity fluctuations is more dominantly caused by the influence of the narrow width of the port mouth, the body of water spreads so that the current velocity will also spread. In the sediment simulation, after simulating the distribution of sediment concentration for 10 days, in the location of the Pasir fish landing port, there is a change in the bed elevation due to the pattern of sediment distribution from the river. This basic change is strongly influenced by the concentration of sediment and sedimentation velocity of sediment grains at each location.

1. Introduction

During the period of Regional Autonomy, Regency Regional Governments generally tried to make various efforts to identify and utilize natural resources, both unrenewable and renewable. The utilization of non-renewable natural resources requires careful thought to avoid the ecological damage caused. For regions that do not have reserves of non-renewable natural resources, inevitably they must utilize renewable natural resources, such as agriculture, forestry, livestock, and fisheries, for marine, land, and brackish fisheries.

Kebumen Regency is a district that has both of these natural resources. Therefore, it is necessary to use other natural resources such as fishery resources. Fishery resources are renewable and if managed responsibly will provide many benefits, and can be continuously improved. In addition, the fishery business can absorb a large number of workers, become a source of community income, regional income, and foreign exchange, improve nutrition, and stimulate related businesses (multiplier effect) and regional development. The negative impacts caused by fishing businesses are relatively mild and easy to overcome. Likewise, the investment requirement is relatively smaller.

The Pasir beach is used to land fish by fishermen so the local government has built Pasir Fish Landing Port on the Pasir Beach. The Pasir beach area is considered to have considerable potential as well as its strategic value when compared to other areas for the development of Fisheries City in the future. Pasir Beach with its strategic and potential values including a large enough alternative land, the number of fishermen and fish production, as well as the proximity to the city center of Kebumen, would be the right choice for a regional development model that can be used as a pilot project for other regions. And so that the program can run optimally, it requires careful, effective, efficient, and comprehensive planning. One thing that is required to support the achievement of a fisheries city is the development of the current Fish Landing Port infrastructure.

In addition to the infrastructure that needs to be developed to increase future potential, the existing condition of the Fish Landing Port infrastructure on the Pasir beach during the study period (2004) also experienced many problems. Based on the observations made during the survey, the physical condition of Pasir Fish Landing Port, in general, requires very serious attention, care, and arrangement to function properly and provide optimal results.

The main problem related to the operation of the current Pasir Fish Landing Port is that almost all basic facilities/infrastructure of the fish landing port are not functioning, such as shipping lanes that are covered by sedimentation, short breakwaters, and the structures that have collapsed, as well as other problems.

Based on information from the local community that less than 2 years since Pasir Fish Landing Port was built in 1994, the pond and fish landing dock can no longer function because the pond inlet is shallow and eventually covered by sand sedimentation. On the east side of the inlet channel, there is a 115 m long breakwater but it cannot function to hold sediment from entering the harbor channel. The purpose of this study is to determine the current pattern and sedimentation pattern on the Pasir beach, especially in the fish landing port area with 2-dimensional modeling, so that a solution can be formulated by the parties concerned.

2. Study Area

The location of this study is in the Pasir Beach area which is a row of the South Coast of Java Island which is located in the Kebumen Regency, Central Java Province. Geographically, Pasir Beach is located at 7°27' – 7°50' South Latitude and 109°22' – 109°50' East Longitude. The boundaries of its territory are to the west of Cilacap and Banyumas Regencies, to the east of Purworejo Regency, to the north of Wonosobo and Banjarnegara Regencies, and to the south of the Indonesian Ocean. The total area of Kebumen Regency is 12,2170 km^2 which consists of 82880 km^2 of garden land and 0.1889 km^2 of irrigated rice fields, 0.6597 is used as rainfed rice fields, 1.1419 km^2 in the form of fields/fields, 1.1419 km^2 of bushes and 0.4823 km^2 as settlements. The Pasir beach location is shown in figure 1.

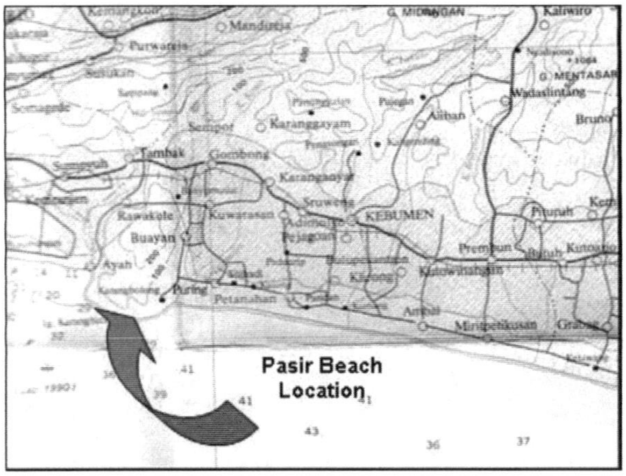

Figure 1. Pasir Beach Location

IOP Conf. Series: Earth and Environmental Science 1117 (2022) 012068 doi:10.1088/1755-1315/1117/1/012068

Topographically Pasir Beach is flanked by two rocky hills, namely on the west and east. On the east side, there is an estuary, namely the estuary of the Pasir River. The current physical condition of the Pasir fish landing port facility is as follows: Total area ± 15 Ha, breakwater Rublemount mound boulder 50-60 cm and concrete tetrapod 700 kg, 115m long and not functioning. Anchored pond with an area of 14 ha that is not functioning because it has been covered by sediment. The shipping channel with a width of 30-40 m and dredged as deep as 2 m is flanked by breakwaters and western cliffs, so far it has not functioned due to sedimentation. The pier with a length of 160 m is currently not functioning because the harbor pool is not functioning. A sloping beach was used for landing fishing boats. There is a fish shop building with a size of 12 x 30 m^2 and the road to the location is paved. The condition of Pasir Fish Landing Port is shown in figure 2.

Figure 2. The condition of Pasir Fish Landing Port

Hydro-Oceanographic conditions on Pasir Beach are as a relatively flat and sandy seabed with contours parallel to the coast and getting deeper towards the south. The contours of the seabed become somewhat steeper after a depth of more than 2.5 meters. For a distance of 0 - 200 meters from the shoreline, it has a depth of between 0.00 - 2,500 meters LWS. Depth – 4.0 located about 350 m to the sea from the shoreline position during the LWS, with a wave height with a 50-year return period of 4.9 m at Tipar from Research and Development Center measurements. The tidal type is daily double, with a tidal range of 325.5 cm, MSL 225 cm (zero palms), HWL 110.5 cm above MSL, and LWL 110 cm below MSL.

Figure 3. The pool of the Pasir Fish Landing Port that experiencing sedimentation

The natural condition of Pasir Fish Landing port which is surrounded by rocky hills with hill terraces directly adjacent to the harbor pond has caused all surface runoff that carries sediment into the pond causing rapid siltation, see figure 3, and the position of Pasir Fish Landing Port respect to the direction of the waves from the Indian Ocean shown in figure 4.

Figure 4. The position of Pasir Fish Landing Port to the direction of the waves from the Indian Ocean

3. Current and Sediment Transport

One of the important aspects of oceanography to determine the hydrodynamics of a beach is the pattern of the current movement. In general, ocean currents that affect the characteristics of waters in Indonesia are ocean currents generated by wind and tides [1]. In addition to waves, currents also determine patterns of sedimentation on the coast. Current data need is known to determine the sedimentation pattern and erosion in the area around the building coast guard after the structure is built [2]. Sand grain sedimentation is affected by the current velocity at which the grains are sediment and silt settles when the velocity current decreases [3]. Some of the previous studies that used RMA2 for tide modeling and current are [4] and [5]. The two studies modeled tides and currents for the port area which often happens sedimentation and cause problems in the ship traffic. Research related to modeling coastal sediments in a certain location with other methods has also been carried out a lot, such as that carried out by

Sediment transport is mobilized by the presence of working currents. Thus the current pattern becomes an important factor in studying sediment transport. On the coast, currents can be generated by breaking waves that form an oblique angle with the shoreline. In addition, currents are also generated by tidal waves. To study more accurately the mechanism of sediment transport, the software can be used. Simulation of hydrodynamics and the process of sea level change using a mathematical model, namely a mathematical model packaged in the Boss SMS software package. The module contained in the Boss SMS software is RMA2 (Resources Management Association Inc.), which is a two-dimensional horizontal finite element hydrodynamic model with mean depth. With this numerical model, it is possible to predict flow patterns, water level elevation, and horizontal velocity components of the flow, both under steady flow and unsteady flow conditions.

3.1. RMA2

The RMA 2 mathematical model used to simulate flow hydrodynamics is based on 2 basic equations, namely the mass conservation equation (the continuity equation) as shown in equation 1, and the momentum equation, as shown in equations 2 & 3.

Equation of continuity

$$\frac{\partial h}{\partial t} + h\left(\frac{\partial u}{\partial x} + \frac{\partial u}{\partial x}\right) + u\frac{\partial u}{\partial x} + v\frac{\partial v}{\partial y} = 0 \tag{1}$$

Momentum Equation

The momentum equation for two-dimensional flow in the x and y directions can be written in the following form:

$$\frac{\partial u}{\partial t} + u\frac{\partial u}{\partial x} + v\frac{\partial u}{\partial y} + g\left(\frac{\partial h}{\partial x} + \frac{\partial a_0}{\partial x}\right) - \frac{\varepsilon_{xx}}{\rho}\frac{\partial^2 u}{\partial x^2} - \frac{\varepsilon_{xy}}{\rho}\frac{\partial^2 u}{\partial y^2} + \frac{gu}{C^2 h}\sqrt{u^2 + v^2} = 0 \tag{2}$$

$$\frac{\partial v}{\partial t} + u\frac{\partial v}{\partial x} + v\frac{\partial v}{\partial y} + g\left(\frac{\partial h}{\partial y} + \frac{\partial a_0}{\partial y}\right) - \frac{\varepsilon_{xx}}{\rho}\frac{\partial^2 v}{\partial x^2} - \frac{\varepsilon_{yy}}{\rho}\frac{\partial^2 v}{\partial y^2} + \frac{gv}{C^2 h}\sqrt{u^2 + v^2} = 0 \tag{3}$$

with u = horizontal velocity of the x −direction flow, v = horizontal velocity in the y −direction, t = function of time, g = acceleration due to gravity, h = depth of water, $a0$ = elevation from the base of the view, ρ= density, ε_{xx} = x −direction normal turbulence exchange coefficient, ε_{xy} = x −direction tangential turbulence exchange coefficient, ε_{yx} = y −direction tangential turbulence exchange coefficient, ε_{yy} = y −direction normal turbulence exchange coefficient C = Chezy roughness coefficient (or Manning coefficient, $n = \frac{1}{c}h^{1/6}$)

3.2. SED2D

The numerical model in the Boss SMS software package which is used for calculating the coastal erosion-sedimentation process is the SED2D model which is based on 2D flow, the simulation result of the RMA2 model. This numerical model only examines one effective grain size for each run, so a separate run model is needed for each effective grain size. The software does not calculate the hydrodynamics of the flow, therefore this data must be calculated externally using the RMA2 hydrodynamic calculation. The explanation of the two models is given below.

The SED2D module is an application for sediment transport with clay or sand as the base material. This SED2D can only work for one-grain size (uniform basic grain gradation). The SED2D module is a continuation of the RMA2 program. The SED2D model is based on the suspended sediment convection-diffusion equation, as shown in equation 4.

$$\frac{\partial C}{\partial t} + \overline{U}\frac{\partial C}{\partial x} + \overline{V}\frac{\partial C}{\partial y} = \frac{\partial}{\partial x}\left(Dx\frac{\partial C}{\partial x}\right) + \frac{\partial}{\partial y}\left(Dy\frac{\partial C}{\partial y}\right) + \alpha_1 C + \alpha_2 \tag{4}$$

with C = concentration, U = average flow velocity in the x direction, V= average flow velocity in the y direction, Dx = effective diffusion coefficient in the x direction, Dy = effective diffusion coefficient in y direction, α_1, α_2= coefficient of the source term.

In the above equation, the turbulent diffusion coefficient in the direction of the solid flow direction is formulated as follows (Elder's equation), $Dx = 5.93H\,U*$, While the turbulent diffusion coefficient perpendicular to the direction of flow is, $Dy = 0.23\,H\,U*$, with sheer speed, is calculated by the equation 5.

$$U* = \frac{\sqrt{g}\bar{U}n}{H^{1/6}} \tag{5}$$

where H = depth of flow, $U*$ = sheer speed

4. Data and Simulation

A coastal hydraulics survey is very important to do to get the data needed in planning or building coastal buildings and facilities. These data include current velocity, both coastal currents and river currents, tidal data, dominant current direction, and wave height. For the wave height, some secondary data is used.

4.1. Bathymetry

In this modeling study, RMA2, SED2D, and CGWAVE were used, where the general data required for simulation is the bathymetry data. Bathymetric measurements are carried out simultaneously with topographic measurements. During these measurements, continuous tidal measurements/observations were also carried out which were observed every hour for 15 days. Tidal measurements/observations were carried out to determine the middle seat and correct the bathymetric measurement/mapping results. GPS binding is done using BM that is near the location. This binding will later be used to unify bathymetric measurements and topographic measurements. Bathymetric measurements were carried out covering a water area of approximately 6 kilometers along the coast and 1.5 km offshore as shown in figure 5.

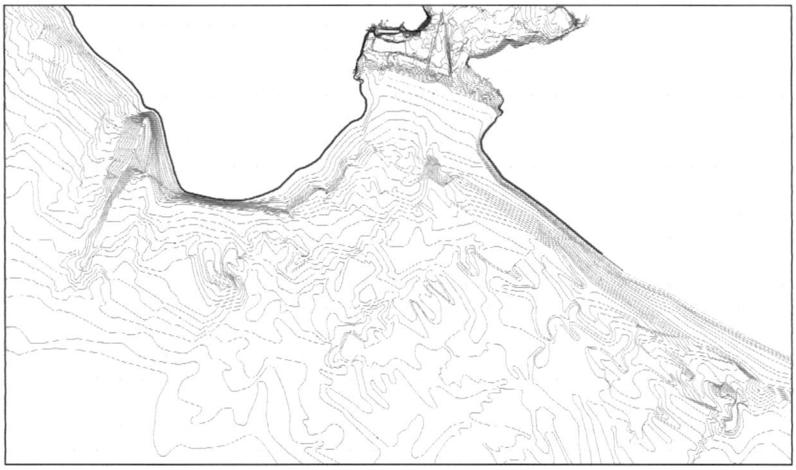

Figure 5. Bathymetry map around Pasir Beach

4.2. Tidal

Tidal measurements at Pasir Beach are carried out for 15 days. The data from the tidal measurements at Pasir Beach are presented in figure 6.

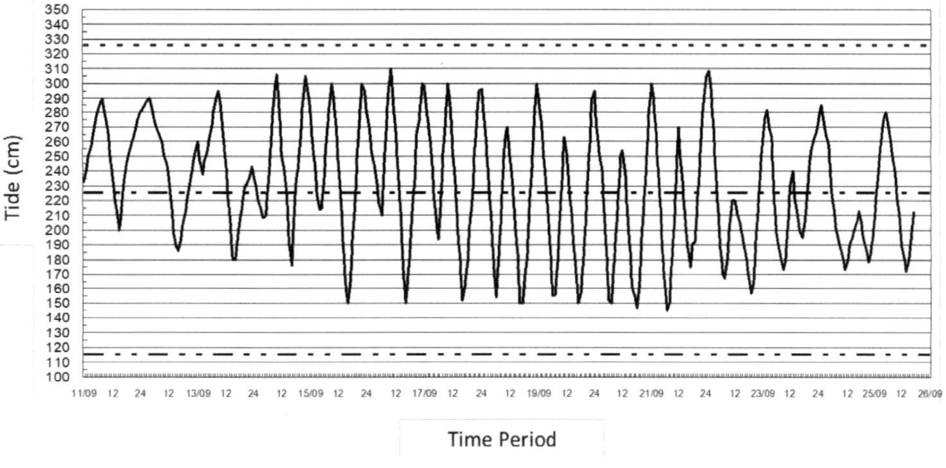

Figure 6. Tides on the Pasir beach

From the results of measuring the tides on the Pasir beach for 15 days, data is obtained that the type of tide is double daily. HHWL elevation : 2.15 m; MHWL: 1.65m; MSL : 1.05 m; MLWL: 0.45 m; and LLWL : 0.0 m. The mean and maximum tidal heights are 1.2 m and 2.15 m, respectively.

4.3. *Waves*

The wave data used is secondary data, due to the unavailability of primary data. The secondary data used is wave data which is determined based on wave height from JICA data, compared with data from the Water Research and Development Center, namely that the waves from the deep sea (Ho) are 4 m.

The waves with a 5-year return period have a wave height of 3.8 m, a wave with a 10-year return period has a wave height of 4.1 m, a wave with a 20-year return period has a wave height of 4.5 m, and a wave with a 50 year return period has a wave height of 4.9 m.

4.4. *Current data*

The intended coastal current data is the tidal current, both its speed and direction as well as the dominant current that will affect the Pasir beach (the Pasir Fish Landing Port). The simplest measurement is with a buoy, namely by measuring the travel time of the buoy to cover a certain distance. With a relatively small coastal current, this measurement method can provide more accurate results than other methods. The results of these measurements are presented in figure 7.

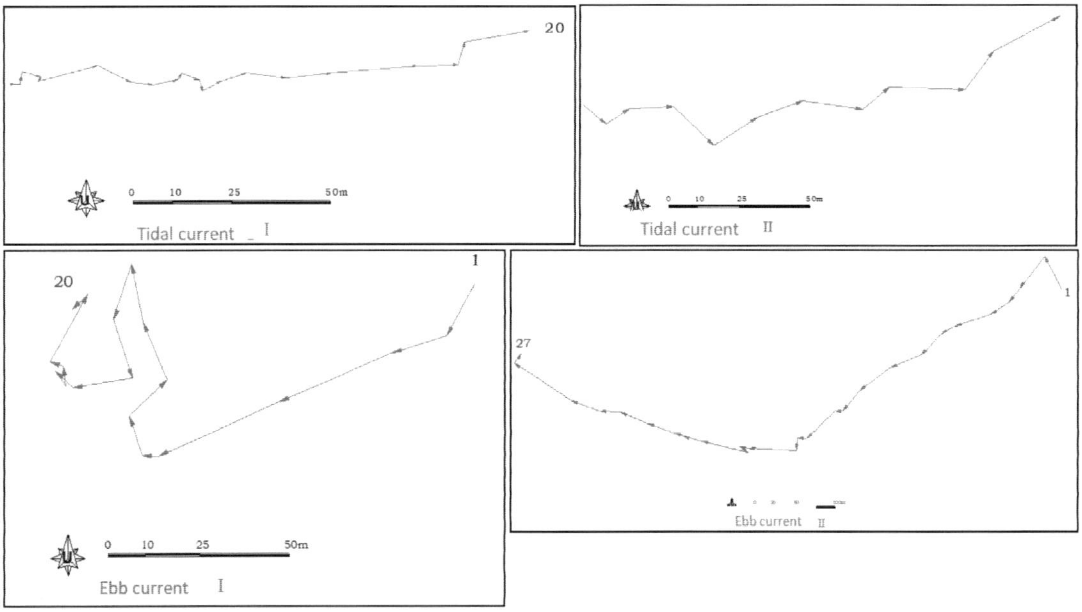

Figure 7. Pasir beach's current measurement results

4.5. *Sediment data*

Sediment data is needed to know the characteristics of the sediment supply carried by the river to the estuary which will later affect the sediment supply in the waters of the Pasir beach location. This sediment sampling was carried out on three rivers that affect the waters around the Pasir beach location, the rivers are the Jintung River, Pecaron, and Pasir River which finally empties into Pasir Beach.

4.6. *Current simulation procedure*

The model domain is the study water area where current patterns and water level fluctuations are simulated. The domain of this model will be represented in a discrete mesh. The element net covers the Pasir Beach waters section along 6826 m from west to east and 2191 m from the shoreline to the offshore area. In the domain model, a mesh is made which is a representation of the physical domain in the mathematical domain. The element net consists of 2137 elements and 5778 count points (nodes). The geometry of the existing element mesh was compiled based on the bathymetric map from the latest measurements.

The data from the bathymetry measurements of Pasir Beach are used as the basis for making hydrodynamic model meshes. Calibration and running the model to obtain the magnitude of the model parameters, including the river bed roughness coefficient (Manning coefficient), and the flow diffusion coefficient. If the value of the model parameter has been obtained, then a simulation of the flow (current pattern) and the distribution of sediment from existing rivers is carried out. In each scenario, the simulation of current patterns and water level fluctuations is carried out by taking into account the position of the tidal motion of seawater, namely the spring tide period, which is when the sun-earth-moon positions are in line, and neap, which is when the sun-earth-moon positions are aligned. The mesh of the domain model around the Pasir Beach water area is shown in figure 8.

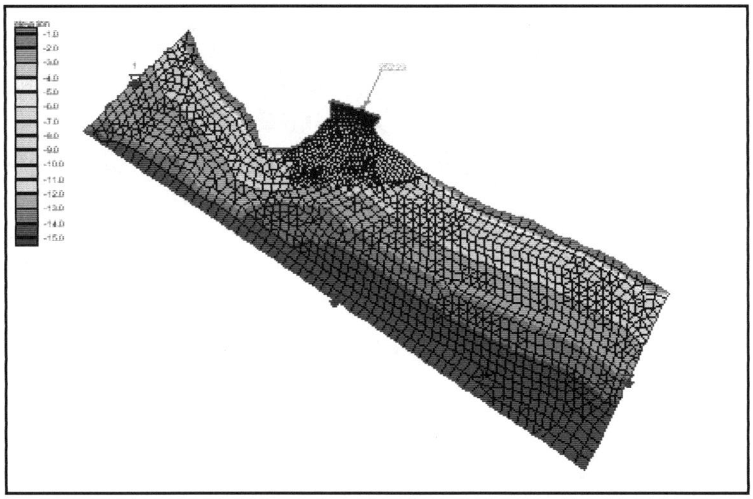

Figure 8. The mesh of the domain model around the Pasir Beach water area

4.7. Sediment simulation data
The data used for simulating sediment distribution is sediment concentration data from rivers, namely from previous studies. The bottom sediments of the coastal waters of the Sandy Coast are classified as sand, with a concentration of sediment distribution ranging from 10-20 $\frac{mg}{l}$.

4.8. Sediment simulation procedure
The model domains and elemental nets for running SED2D are the same as those used for running RMA2. Current data from running RMA2 is used for running simulations of sediment distribution. In the study of sediment distribution, simulations were carried out for 15 days for each tidal period. By doing simulations for 15 days continuously, conditions will be obtained where there is a full moon (spring) and a neap tide. Thus will be observed changes in both the current pattern and the distribution of sediment which results in bed changes due to the influence of tidal conditions.

5. Results and discussion

5.1. Current pattern
The current pattern simulation was evaluated in two tidal events, namely at spring tide (full tide) where the moon, earth, and sun are in a straight line, and at neap tide (near tide) where the moon, earth, and

sun form an angle of 90 degrees. By entering tidal data and river flow discharge, mathematical model simulations for flow hydrodynamics can predict the pattern of current distribution that will occur.

The current pattern simulation at the study site illustrates the flow pattern that occurs, where the Pasir Beach waters area is an open beach directly facing the open sea (incoming waves). The description of the current distribution pattern for the conditions at high and low tide is presented in the form of vectors and velocity contours and can be seen in figures 9 and figure 10.

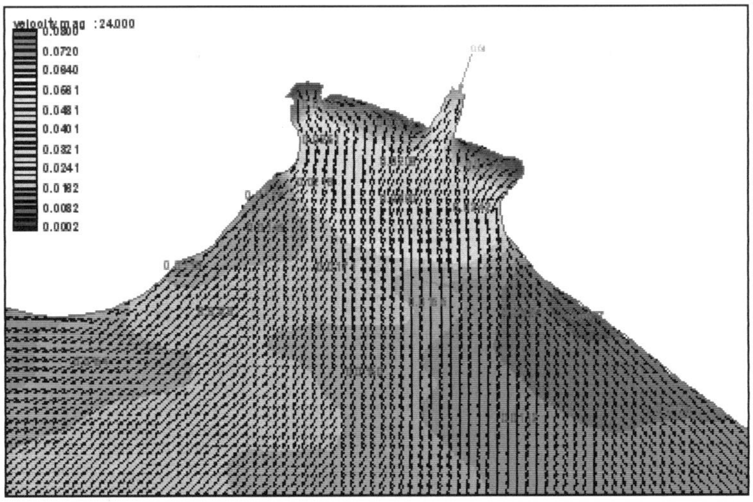

Figure 9. The largest current velocity in the spring condition

Figure 10. The largest current velocity in the neap condition

The current pattern that occurs during spring tide is visually seen around the port location ranging from 0.013 m/s ~ 0.021 m/s, while the velocity in the open sea is around 0.016 m/s. The current pattern with the highest velocity condition flows from the coast towards the open sea.

The current pattern that occurs during neap tide is visually seen around the port location ranging from 0.0084 m/s ~ 0.0118 m/s, while the velocity in the open sea is around 0.088 m/s. The current pattern with the highest velocity condition flows from the open seas towards the coast.

5.2. Sediment distribution

The study area under review is the same as the study area for the simulation of current patterns. Likewise, the meshes used are the same as those used for simulating current patterns. Sediment data used for material $d50 = 0.25\ mm$ assuming sediment concentration = $0.02\ \frac{kg}{m^3}$. Sediment is poured into the waters of Pasir Fish Landing Port, namely from the Cilauteureun River with an annual average discharge of $0.64\frac{m^3}{s}$ continuously for 10 days (240 hours). The hydrodynamic conditions for the sediment simulation used the output of the RMA2 simulation which was simulated for 48 hours for each condition, namely during neap tide and spring tide. The results of the simulation of sediment distribution can be seen in figure 11.

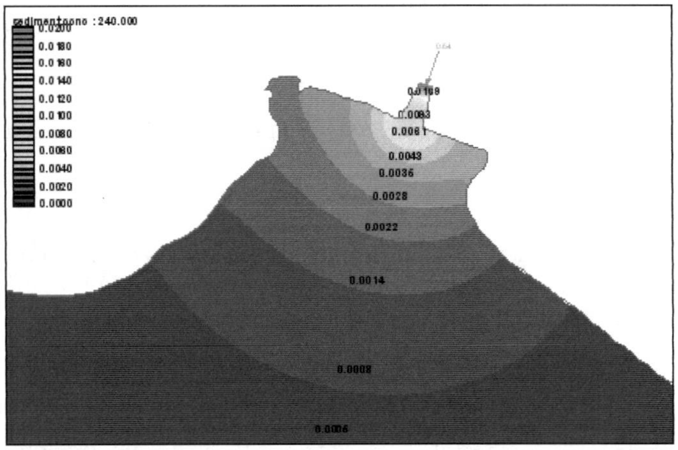

Figure 11. Sediment distribution pattern (for 10 days)

After the simulation was carried out for 10 days the concentration of sediment distribution spread from the estuary with concentrations ranging from 0.01 ~ to $0.02\ \frac{kg}{m^3}$. The concentration is carried by currents and spreads in the waters of the planned Pasir Fish Landing port development in the range of 0.0028 ~ $0.0081\ \frac{kg}{m^3}$.

After simulating the distribution of sediment concentration for 10 days, in the plan of the Pasir Fish Landing Port, there is a change in bottom elevation due to the pattern of sediment distribution from the river, as seen in figure 12. This bed change is strongly influenced by sediment concentration and depositional velocity of sediment grains at each location.

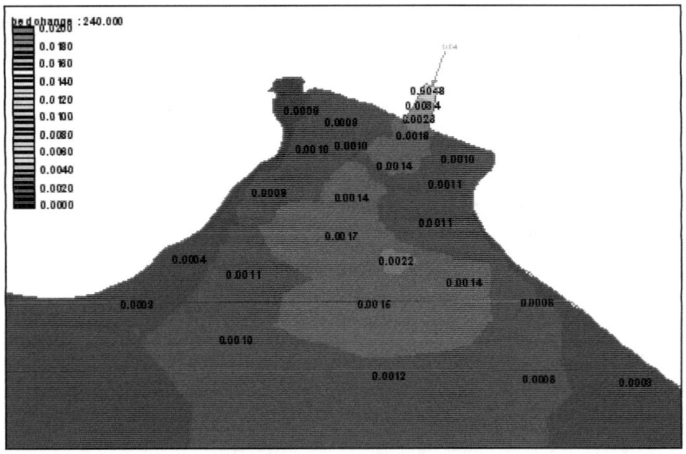

Figure 12. Bed change (for 10 days)

Visually, the bed changes that occur in the Pasir fish landing port pool plan are around 0.0024 m for a month and the outer port pool plan is 0.0030 m a month.

6. Conclusion

From the results in the previous chapter, it can be concluded that fluctuations in current velocity are not significant because the velocity is small. The increase in current velocity fluctuations is more dominantly caused by the influence of the narrow width of the port mouth, the body of water spreads so that the current velocity will also spread. In the sediment simulation, after simulating the distribution of sediment concentration for 10 days, in the plan for the Pasir Fish Landing Base harbor, there is a change in the base elevation due to the pattern of sediment distribution from the river. This bed change is strongly influenced by the concentration of sediment and sedimentation velocity of sediment grains at each location.

Acknowledgment

This study is part of the work within the program "Pembinaan Pengelolaan Pembangunan Jaringan Pengairan Wilayah Tengah (P4JWTg) PUPR.

References

[1] Sugianto D N, Ads A 2012 Pola Sirkulasi Arus Laut di Perairan Pantai Provinsi Sumatera Barat *ILMU KELAUTAN: Indonesian Journal of Marine Sciences* **12** (2) 79-92

[2] Wurjanto A, Ajiwibowo H 2020 Analisis Hidrodinamika di Perairan Lemong, Kabupaten Lampung Barat, Provinsi Lampung Menggunakan Piranti Surface-Water Modeling System *REKAYASA: Journal of Science and Technology* **13** (2) 154-163

[3] Febribusmadian, Puwranto, Hariadi 2012 Studi pola arus dan sedimentasi di dekat breakwater bagian luar kolam labuh Pelabuhan Tanjung Emas Semarang *Journal of Oceanography* **1**(2), 178-185

[4] Mawarda M, Kurniawati N, Ulqodry T Z 2010 Simulasi pemodelan arus pasang surut di kolam Pelabuhan Tanjung Priok Jakarta menggunakan perangkat lunak SMS 8.1 (Surface-water Modeling System 8.1) *Maspari Journal*, **1**(1), 48-52

[5] Supiyati, Suwarsono, Hutami K 2012 Pola Tidal Currents at Pulai Harbor Baai Bengkulu Using Software Surface-water Modeling System (SMS) 8.1 *Journal of Gradients* **8**(2), 792-795

[6] P Yao, M Su, Z Wang, LCV Rijn, C Zhang and MJF Stive 2018 Modelling tidal-induced sediment transport in a sand-silt mixed environment from days to years: Application to the Jiangsu coastal water, China: *Coastal Engineering* **141**, 86-106

Effects of Fluctuation in Discharge and Head on Flow Parameters and Head Losses in Pipeline Networks

B. Bakri[1], S.Pallu[1], Sulhairi[2], S.Pongmanda[1], Y.Arai[3], M. Ihsan[4]

[1]Department of Civil Engineering, Makassar, Hasanuddin University
[2]Ministry of Public Work, Republic of Indonesia
[3]Department of Civil and Environmental, Tokyo Metropolitan University, Tokyo, Japan
[4]Baramuli College of Engineering, Pinrang

Abstract. Discharge, head, flow velocity and Reynolds number are flow parameters in pipe networks. On the other hand, energy loss is one of the factors that affect the capacity of the pipe as a means of distributing water flow. Energy loss is caused by several factors including the roughness of the pipe walls associated with the pipe material. Due to the lack of attention to the interrelationship of these factors, sometimes there are piping networks that do not work optimally or do not function as expected. The purpose of this study was to determine the effect of changes in discharge and head on flow parameters and energy loss in the piping network. This study uses experimental research conducted in the laboratory. The primary data collection is quantitative analysis with the main data collection through measurements using a series of pipes as a test model, and is preceded by the collection of supporting data, namely the analysis of the characteristics of the water sample. The characteristics of the water samples used are density (ρ) 1,000 kg/m3, kinematic viscosity 0.804 x 10-6 m2s, dynamic viscosity (μ) 0.801 x 10-3 Nd/m2, viscosity 8.5x10-7 m2/ sec, and specific gravity (y) 1.00355 gr/cm3, and a temperature of 29.50C. There are 3 discharge variations used, namely Q1 = 0.004 m3/s, Q2 = 0.003m3/s and Q3 = 0.002 m3/s and variations in head H1 = 2 meters, H2 = 1 meter and H3 = 0.5 meters. The results showed that due to changes in discharge and fall height caused an effect on flow parameters in the piping network. The greater the discharge, the greater the energy loss, and the greater the head, the smaller the energy loss. Changes in flow rate have a more significant effect on changes in energy loss when compared to changes in head.

1. Introduction

Fluid flow in a closed channel is important so that an optimal planning process is needed [1,2,3]. Pipes as distribution media have been widely used in industry and public water distribution. The advantages of the pipe as a closed channel as a water distribution include lower water loss, and easier maintenance when a blockage occurs. Another advantage is that it can be more easily used in areas where the source has higher elevation than the distribution area. In addition, the pipeline flow is not directly affected by air but by hydraulic pressure, the required land used is not extensive, less disruption to current activities, and safety because the channel is closed, the risk to humans is less than an open channel during bad weather, the size is larger [4]. A rather smaller size than the open channel gives closed channel more flexiblity in its placement. It proveides a better water quality than the open channel because the water that flows is protected by the surface of the pipe.

In the installation of pipelines there will be flow resistance caused by factors of the pipeline installation itself such as flow velocity, changes in pipe dimensions and the presence of turns, as well as changes in the roughness of the pipe material. The flow resistance will cause a decrease in energy and pressure [5].

The process of designing and planning a closed channel (pipeline network) is a very broad aspect and requires an interrelated analysis of water demand (discharge), velocity, pipe diameter, pressure head fall and so on. Due to the lack of attention to this energy loss, some of the existing piping networks do not work optimally or in other words the network does not function as expected. Energy loss is one of the factors that affect the capacity of the pipe as a means of distributing water flow. One of the important things that affect network performance is energy loss. As a result of errors in calculating this energy loss, it can cause problems, namely sometimes what is planned is not in accordance with what is happening in the field, especially related to capacity and pressure on the piping network.

2. Theoretical background

Flow in closed channel

Closed channels which are usually circular in cross-section are used to flow fluids with a full flow section. The fluid flowing through the pipe can be a liquid or a gas and the pressure can be greater or less than atmospheric pressure. If the liquid in the pipe is not full then the flow is included in open channel flow or because the pressure in the pipe is equal to atmospheric pressure, the flow is included in open flow. The pressure at the surface of the liquid along an open channel is atmospheric pressure. The basic difference between flow in an open channel and flow in a closed channel is the presence of a free surface which is (almost always) air in the open channel. So if the flow pipe is not full so that there is still a cavity filled with air, then the nature and characteristics of the flow are the same as the flow in an open channel.

Reynolds number

Reynolds number (Re) is a number that does not have dimensions which is an important factor in analyzing the type of flow (laminar, turbulent or transition). Re is the ratio of the inertial force to the viscosity force. Reynolds number:

$$Re = \frac{V\rho d}{\mu} \;..... (1)$$

Where

Re = Reynolds Number

v = average velocity of flow in closed channel (m/s)

d = pipe's inner diameter (m)

ρ = density of fluid (kg/m3)

μ = dynamic viscosity of fluid (kg/m.s)

Head losses

Energy loss (Head losses) is a factor that affects the capacity of the pipe as a means of conducting flow of both water and pipes [6,7,8]. Energy losses are divided into major and minor losses. Meanwhile, minor losses are caused by changes in cross-section, joints, bends, valves and other pipe components. Major Headloss (major losses) Major losses are losses in the flow in the pipe caused by friction that occurs along the fluid flow that flows against the pipe wall. In a liquid flowing in the boundary plane, there will be shear stress and velocity gradient throughout the flow field because of the viscosity. The shear stress will cause a loss of energy during flow. The formula for finding the major pressure loss is as follows:

$$hf = f\frac{L}{D}\frac{v^2}{2g} \qquad ......(2)$$

Minor Headloss (minor losses) Losses that occur in the pipe system are caused by bends, elbows, joints, valves. And Others are called minor losses [5].

Viscosity

The viscosity of a fluid is a measure of its resistance to the rate of deformation [6]. Viscosity or the viscosity of a fluid is very important in analyzing fluid conditions and fluid motion. Real liquids are substances that have a viscosity, while ideal liquids have no viscosity. This happens because of the nature of cohesion between the liquid particles. The difference in particle velocity in the flow field occurs because of the viscosity of the liquid. Liquid particles adjacent to the boundary wall will have zero velocity or rest, while those located at a certain distance from the wall will move. The change in velocity is a function of the distance from the boundary wall. The flow of real liquids is called viscous flow. This flow can be divided into 2 (two). If the effect of viscosity (viscosity) is dominant enough so that the particles of the liquid move regularly in a straight line, the flow is said to be laminar. Laminar flow occurs when the viscosity is large and the velocity is small. With the decrease in the effect of viscosity or the flow rate increases, the flow will change from laminar to turbulent [9].

Pitot tube speed

Bernoulli's law statement is the sum of the pressure, kinetic energy per volume, and potential energy per volume at any point along the fluid flow is the same. That is, as the fluid flow increases, the fluid pressure will decrease. Thus, the potential energy of the fluid will also decrease. Conversely, when the fluid flow velocity decreases, the fluid pressure will increase. This law can be applied to various types of fluid flow with several assumptions, as follows:

a) The fluid is incompressible.
b) The fluid has no viscosity (inviscid).
c) Fluid flow does not change with time (steady).
d) Laminar fluid flow (constant, no eddies).
e) There is no loss of energy due to friction between the fluid and the wall. And there is no energy loss due to turbulence.
f) No heat energy is transferred to the fluid either as heat gain or loss. Bernoulli's equation is as follows:

$$P_1 + \frac{1}{2}\rho v_1^2 + \rho g h_1 = P_1 + \frac{1}{2}\rho v_2^2 + \rho g h_2 \ldots(3)$$

where:
P = Pressure (Pascal)
ρ = Density of fluid (kg/m3)
v = Fluid Velocity
g = gravitational force (g= 9.8 m/s 2)
h = height (m)

Bernoulli's law is very useful in everyday life and is used in the following applications:

a) Pitot tube

Pitot tube is an instrument for measuring pressure in fluid flow. The Pitot tube was invented by the French engineer, Henry Pitot [10] and modified by the French scientist, Henry Darcy [11]. Pitot tubes have been widely used to determine the speed of aircraft and measure air and gas velocity in industrial applications [12]. The velocity of the air in the pipe can be calculated using the equation:

$$v = \sqrt{2gh} \ldots(4)$$

b) Venturimeter

In ideal conditions in fluid dynamics the Bernoulli equation and the continuity equation are the basic equations used by the venturimeter pipe, through the translation and substitution of the equations, the formula for determining the flow rate in the venturi meter pipe is obtained. Based on the continuity theory that the inlet and outlet holes of the venturimeter pipe which have different cross-sectional sizes are so regular, the discharge between the holes is the same.

3. Research Method

The research was carried out in the laboratory by making a distribution network pipe model. To obtain research data, the source of data used comes from primary data, namely data obtained directly from simulations of physical models in the laboratory and secondary data obtained from literature and existing research results, both those that have been carried out in the laboratory and carried out elsewhere. related to velocity distribution research. The design of the network test equipment carried out in this study is as shown in the following figure.

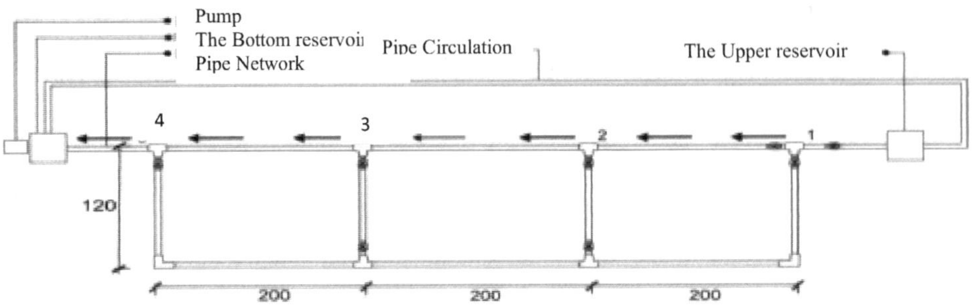

Figure 1. Set up test equipment

Figure 1 shows a model of a rectangular test equipment set up. This network is equipped with a lower water reservoir and an upper water reservoir as a place for water circulation, a valve as a discharge regulator, a water pump machine that regulates the water supply from the lower reservoir to the upper reservoir and a flow velocity measuring device.

This pipe test is intended to observe the flow characteristics that occur from the beginning of the flow (upstream) to the end (downstream) with variations in discharge (Q). The initial stage includes the preparation of tools and materials needed during the research. The calibration of the equipment was carried out on manual pitot tubes and digital pitot tubes to obtain the relationship between the readings of these tools. Manual pitot tube calibration is carried out with a digital pitot tube and installing the pitot tube at the Measurement Point (TP) in the test pipe. And pitot is placed with the same height, which is 2 cm from the bottom of the test pipe. This test is carried out based on changes in discharge and fall height. Variations in flowrate changes are Q1 = 0.004 m3/s, Q2= 0.003 m3/s and Q1 = 0.002 m3/s. While the change in height of fall is H1 = 2 m, H2 = 1 m and H3 = 0.5 m.

4. Results and Discussion

Results

The characteristics of the water samples used in this study are specific gravity = 1000 kg/m3, Density = 995.2 kg/m3, viscosity = 0.0804 x 10^-6 m2/sec and temperature = 29.5 oC. Meanwhile, the results of the high pressure test due to changes in discharge and fall height on a straight pipe network can be seen in the following table.

Table 1. The results of the high pressure test on the Pitot tube

Running	TP	h pitot (cm)		
		Q1	Q2	Q3
HEAD 200 cm	TP1	49.49	37.40	22.40
	TP2	45.71	34.10	18.72
	TP3	41.39	32.84	17.35
	TP4	31.88	19.20	5.00
HEAD 100 cm	TP1	48.08	37.01	18.75
	TP2	44.93	36.52	12.64
	TP3	43.52	32.15	12.36
	TP4	19.09	11.47	5.00
HEAD 50 cm	TP1	31.00	19.93	5.83
	TP2	25.18	17.39	5.50
	TP3	22.55	15.10	5.38
	TP4	16.15	10.80	5.00

Discussion

Effect of changes in discharge and fall height on flow velocity

To calculate the flow velocity on the network, Equation 2 is used, the results of which are shown in the following figure:

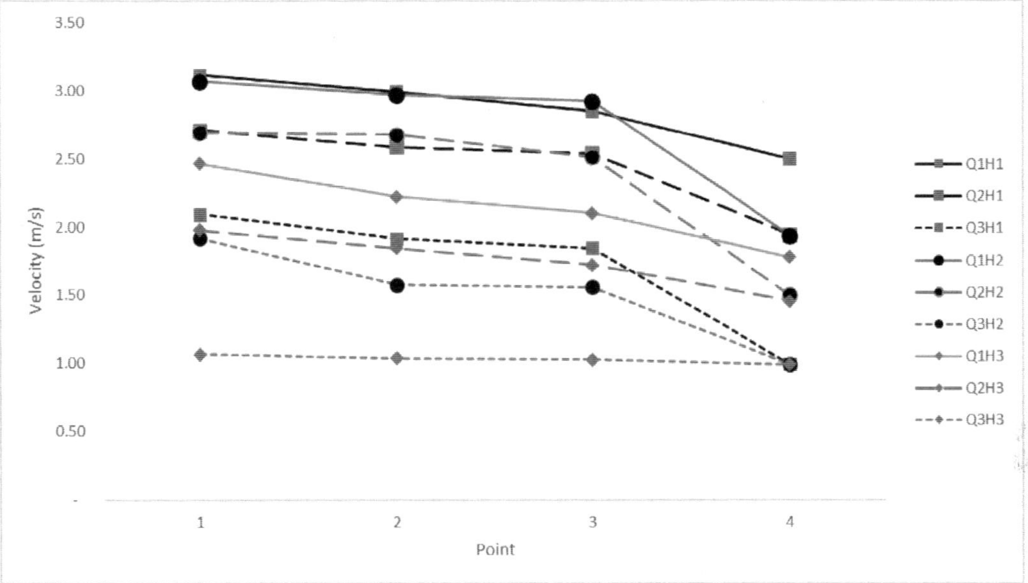

Figure 2. Effect of changes in discharge and fall height on flow velocity

From the picture above shows the effect of changes in flowrate on speed that the greater the discharge, the speed also increases, and vice versa if the discharge decreases, the speed also decreases. In the same condition, due to a decrease in head also results in a decrease in speed.

Effect of changes in discharge and height of fall on the Reynolds Number

The effect of changes in discharge and height of fall on the Reynolds number can be seen in the following figure

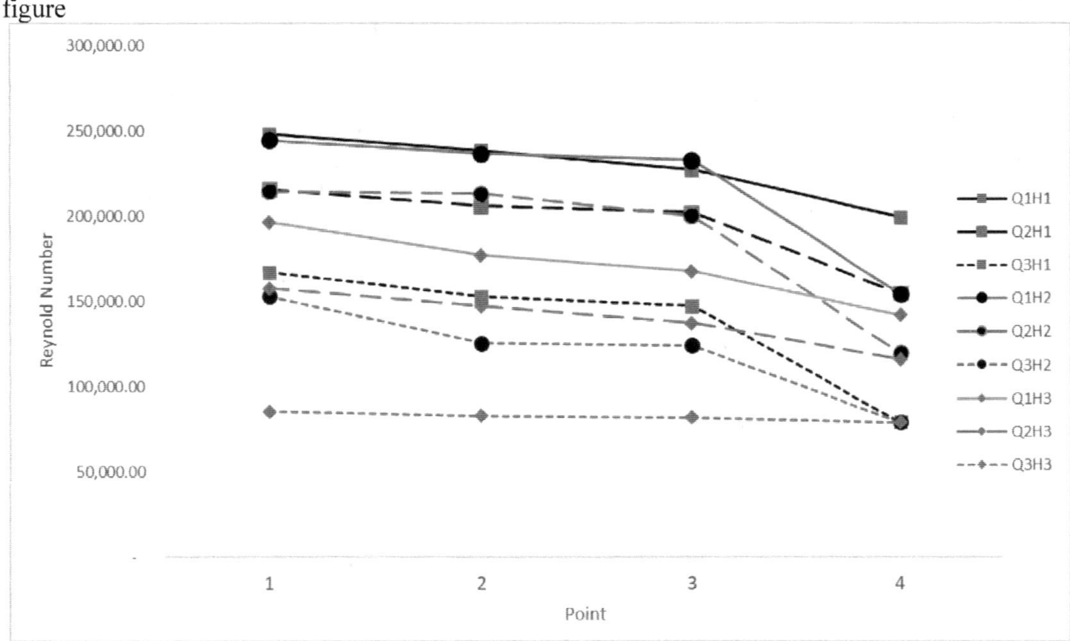

Figure 3. Effect of changes in discharge and height of fall on the Reynolds Number

In the picture above, it can be seen the effect of changes in discharge and height of fall on the Reynolds Number. The smaller the discharge, the Reynolds Number also decreases, and vice versa if the discharge is large, the Reynolds Number will increase. Likewise, the decrease in the value of the Reynolds Number is caused by a decrease in the height of the fall. This is because a decrease in both discharge and fall height causes a decrease in velocity which results in a decrease in the Reynolds number. A large speed will cause an increase in the Reynolds number in the pipeline network.

Effect of changes in discharge and fall height on energy loss

The major losses caused by friction of the liquid with the pipe wall and minor losses due to the connection were taken into account in the study. The effect of changes in discharge and height of fall on the complete energy loss is shown in the following figure

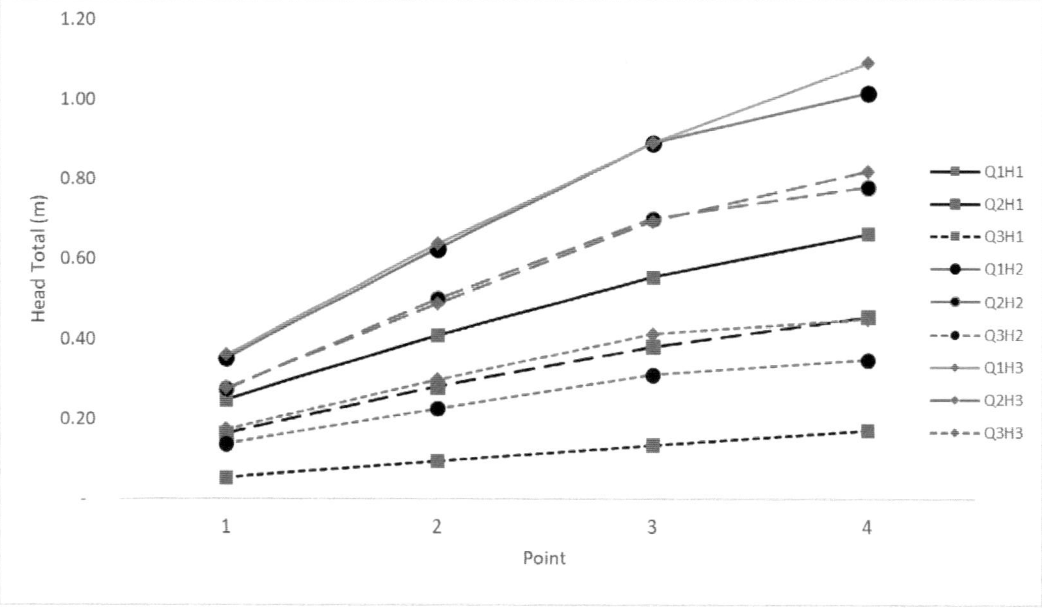

Figure 4. Effect of changes in discharge and fall height on energy loss

The effect of changes in discharge on energy loss (head loss) in straight pipe networks is that the greater the discharge, the more energy loss (head loss) will be, and vice versa if the discharge decreases, the energy loss (head loss) will also decrease. This can be seen at point (TP1) at the H1 fall height (0.2 m), there is a decrease in the level of energy loss (head loss) from Q1 to Q2, namely from 0.25 meters to 0.17 meters or by 33.62%. Likewise, the decrease in energy loss (head loss) from Q2 to Q3 is 0.17 meters to 0.05 meters or by 66.87%. And at the point (TP2) at the H1 fall height, there was a decrease in the level of energy loss (head loss) from Q1 to Q2, namely from 0.41 meters to 0.28 meters or 31.59%. Likewise, the decrease in energy loss (head loss) from Q2 to Q3 is 0.28 meters to 0.1 meters or 65.96%. In TP3 there is a decrease in the level of energy loss (head loss) from Q1 to Q2, from 0.56 meters to 0.38 meters or 31.29%. Likewise, the decrease in energy loss (head loss) from Q2 to Q3 is 0.38 meters to 0.14 meters or by 64.55%. TP4 there is a decrease in the level of energy loss (head loss) from Q1 to Q2, namely from 0.66 meters to 0.46 meters or 31.16% from Q2 to Q3 which is 0.46 meters to 0.17 meters or 62.18%.

On the other hand, the effect of changes in head on energy loss (head loss) is that the greater the height of the fall, the smaller the energy loss (head loss). This can be seen at point (TP1) with Q1 discharge an increase in energy loss (head loss) from H1 to H2, from 0.25 meters to 0.35 meters or 41.98%. Likewise, the increase in energy loss (head loss) from H2 to H3 is 0.35 meters to 0.36 meters or 2.4%. At point (TP2) with Q1 discharge there is an increase in energy loss (head loss) from H1 to H2, namely from 0.41 meters to 0.6 meters or by 52.66%. Likewise, the increase in energy loss (head loss) from H2 to H3 is 0.626 meters to 0.639 meters or 2.05%. At point (TP3) with Q1 discharge there is an increase in energy loss (head loss) from H1 to H2, namely from 0.556 meters to 0.892 meters or 60.25%. Likewise, the increase in energy loss (head loss) from H2 to H3 is 0.892 meters to 0.893 meters or 0.12%. At point (TP4) with Q1 discharge there is an increase in energy loss (head loss) from H1 to H2, namely from 0.664 meters to 1.017 meters or 53.11%. Likewise, the increase in energy loss (head loss) from H2 to H3 is 1,017 meters to 1,093 meters or 7.4%.

Level of Effect of Changes in Discharge and Head Falls on Energy Loss
In this study, the discharge variation carried out is from Q1 = 4 lt/sec to Q2 = 3 lt/sec and Q3 = 2 lt/sec. This means that the decrease in discharge from Q1 to Q2 is 25% and Q1 to Q2 is 50%. While the variation of the drop in height fell from H1 = 2 m to H2 = 1 m and H3 = 0.5 m. So the percentage decrease in discharge from H1 to H2 is 50% and H1 to H3 is 75%. Figure 5 shows the level of influence of changes in discharge and head on energy loss due to a decrease in discharge and fall height. The purple color represents energy loss during Q1H1 which is the starting point for changes in discharge and head.

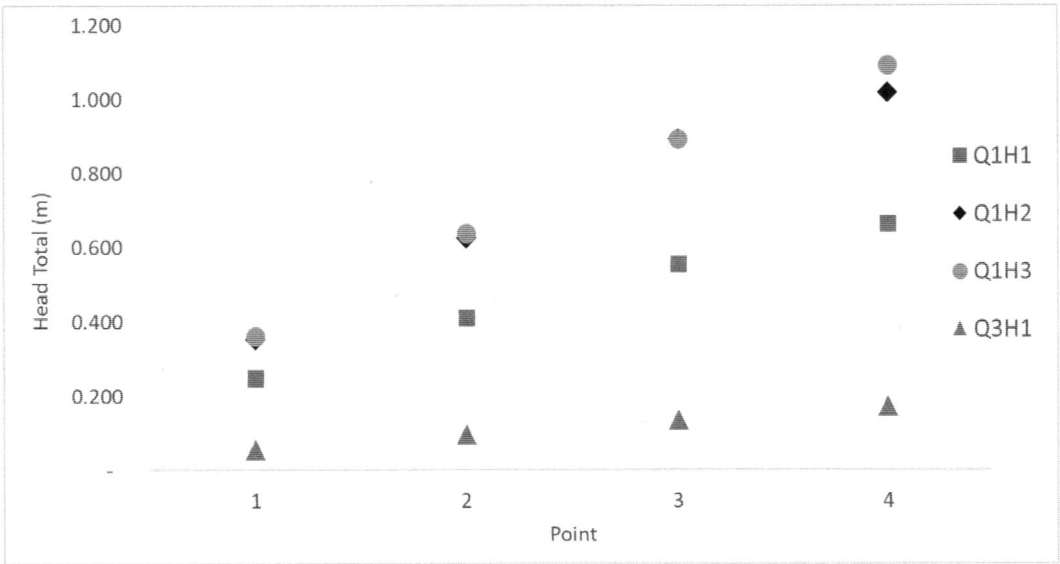

Figure 5 Energy loss in straight pipe due to decrease in discharge and height of fall
Based on the picture above, it can be seen that changes in discharge result in greater energy losses when compared to changes in height falling at the same observation point. For example Observation Point 1, the change from Q1H1 to Q1H2 is from 0.249 to 0.353 m or a change of 41.97% with a 50% decrease in H1 to H2 while the change from Q1H1 to Q3H1 is from 0.249 to 0.055 m or a change of 78% with a decrease in Q1 to Q3 by 50% or equal to the percentage change in height falling from H1 to H2. Likewise at Observation Point 2, the change from Q1H1 to Q1H2 is from 0.410 to 0.626 m or a change of 52.66% while the change from Q1H1 to Q3H1 is from 0.41 to 0.096 m or a change of 76.71%. Observation point 3, the change from Q1H1 to Q1H2 is from 0.556 to 0.892 m or a change of 60.25% while the change from Q1H1 to Q3H1 is from 0.556 to 0.136 m or a change of 75.64%. Observation point 4, the change from Q1H1 to Q1H2 is from 0.664 to 1.017 m or a change of 53.11% while the change from Q1H1 to Q3H1 is from 0.664 to 0.173 m or a change of 73.96%.

5. Conclusion

Based on the purpose of this study, it can be concluded several things from the results of the research that has been done, namely: The effect of changes in discharge on energy loss can be concluded that the higher the discharge, the level of energy loss also increases. Vice versa, the smaller the discharge, the lower the energy loss. This is because the increased discharge causes the mass of the liquid that rubs or contracts with the pipe wall to increase in number. On the other hand, the effect of a change in the height of the fall on the tissue is inversely proportional to the change in energy loss. The greater the height of the fall, the less energy loss (head loss) at the same discharge condition. This is because an increase in head will cause an increase in speed, an increase in speed causes an increase in the Reynolds Number. Changes in discharge are more significant in influencing changes in energy loss when compared to changes in head.

REFERENCES

1. Bakri, B. Arai, Y. Inakazu, T, Koizumi. A, Pallu S., Yoda H, A multi-step genetic algorithms model for ensuring cost-effectiveness and adequate water pressure in a trunk/limb mains pipe system, Journal of Water Supply; Research and Technology – AQUA, 64(2) (2015).
2. Hong, B. Li, Z. Di, G. and et all. An integrated MILP model for optimal planning of multi-period onshore gas field gathering pipeline system. Computers & Industrial Engineering. Volume 146, August 2020. https://doi.org/10.1016/j.cie.2020.106479.
3. Dbouk, H.M. Hayek, H. Ghorayeb, K. Modular approach for optimal pipeline layout. Modular approach for optimal pipeline layout. Volume 197, February 2021. https://doi.org/10.1016/j.petrol.2020.107934
4. Lehtola, M.J. Miettinen, I.T. Lampola, T and et all. Pipeline materials modify the effectiveness of disinfectants in drinking water distribution systems. Water Research. Volume 39, Issue 10, May 2005. https://doi.org/10.1016/j.watres.2005.03.009
5. Sukharev, M.G. Kulaeleva, A.M. Identification of model flow parameters and model coefficients with the help of integrated measurements of pipeline system operation parameters. Energy. Volume 232, 1 October 2021. https://doi.org/10.1016/j.energy.2021.120864.
6. Yang, X.C. Cao, Y.G. Effects of head loss, surface tension, viscosity and density ratio on the Kelvin–Helmholtz instability in different types of pipelines. Physica D: Nonlinear Phenomena. Volume 424, October 2021. https://doi.org/10.1016/j.physd.2021.132950.
7. Annan, M. Gooda, E.A. Effect of minor losses during steady flow in transmission pipelines – Case study “water transmission system upgrade in Northern Saudi Arabia”. Alexandria Engineering Journal. Volume 57, Issue 4, December 2018, Pages 4299-4305. https://doi.org/10.1016/j.aej.2018.12.002
8. Osra, F.A. A laboratory study of solid-water mixture flow head losses through pipelines at different slopes and solid concentrations. South African Journal of Chemical Engineering. Volume 33, July 2020, Pages 29-34. https://doi.org/10.1016/j.sajce.2020.04.001
9. Diaou, Z. Jiang, J. Ni, L. Mebarki, A. Shen, G. Electrification hazard of turbulent pipe flow: Theoretical approach and numerical simulation. Process Safety and Environmental Protection. Volume 159, March 2022. https://doi.org/10.1016/j.psep.2021.12.045
10. Pitot, Henri. 1732. "Description d'une machine pour mesurer la vitesse des eaux courantes et le sillage des vaisseaux" Histoire de l'Académie royale des sciences avec les mémoires de mathématique et de physique tirés des registres de cette Académie
11. Darcy, Henry. 1858. "Note relative à quelques modifications à introduire dans le tube de Pitot" Annales des Ponts et Chaussées
12. Gean koplis, C J. 2003. Transport processes and separation process principles (include unit operations) (4th ed) New Jersey, Prentice Hall.

IOP Conf. Series: Earth and Environmental Science 1117 (2022) 012070 doi:10.1088/1755-1315/1117/1/012070

Analysis The Influence of Hydrometeorological Disaster in Kera River Wajo Regency

M R Pratama[1], F Maricar[2], R Karamma[2] and R T Lopa[2]

[1]Magister Student Department of Civil Engineering, Faculty of Engineering, Hasanuddin University, Makassar, Indonesia
[2]Lecturer Department of Civil Engineering, Faculty of Engineering, Hasanuddin University, Makassar, Indonesia

E-mail: ryanryanpratama1@gmail.com

Abstract. Recently, Indonesia has been dominated by hydrometeorological disasters which include landslides, tornadoes, droughts and floods. Hydrometeorological disasters cannot be separated from the consequences of climate change that occur. The purpose of this study is to examine the influence of climate change on hydrometeorological disasters in the Kera River of Wajo Regency. This research uses quantitative methods. The data used are primary data in the form of field surveys and secondary data in the form of river topographic data and TRMM rainfall. The climate change greatly affects the increase in rainfall in the watershed, and flooding in the Watershed. Changes in rain characteristics in the watershed, in the form of annual rain height and maximum daily rain tend to increase, resulting in flooding around the Kera River area. The highest discharge that occurs at 2 year return period is 147.1 m^3/s, 5 year return period is 212.5 m^3/s, 10 year return period is 251.3 m^3/s, 25 year return period is 296.1 m^3/s. Therefore, it is necessary to carry out flood disaster mitigation with structural methods shown, by normalizing rivers, building flood embankments respectively. The results of this study found that climate change that occurred in the Kera River had an effect on the number of hydrometeorological disaster events, especially floods.

1. Introduction

A disaster is an event or series of events that threatens and disrupts people's lives and livelihoods caused by natural and/or non-natural factors as well as human factors, environmental damage, property losses, and psychological impacts. Flood is the natural disasters that often occur in Indonesia. Climate can encourage the potential for this hydrometeorological disaster to occur is climate. In the era of global warming, Indonesia has experienced quite drastic temperature changes. The timing of the transition of the seasons is increasingly difficult to detect due to the presence of global warming which changes the intensity of rainfall[1]. There are two climatic elements that are commonly used as parameters of climate change, ie air temperature and rainfall[2]. Changes in rain patterns that occur can have an impact on the lives of people around the watershed. A major flood event can significantly change the morphology of a river in the sedimentary channel of a stream. Therefore, previous flood disaster events cause losses, and can be used as a study to determine the potential for disasters that occur in the future[1].

Flood hydrometeorological disasters are the dominating disasters in Indonesia in 2010 – 2020 in addition to landslides, and tornadoes. Hydrometeorological disasters cannot be separated from the current climate change in all regions of the world including Indonesia[3]. Indications of global climate change in the form of rising surface temperatures, changes in precipitation patterns, evaporation, sea level rise and surpluses or deficits in water availability can threaten life anywhere sustainable development policy strategy like land use is needed. Land use change will bring serious problems according to soil and water conservation rules and the land capability classes. Climate change will have an impact on the existence of water use in several watersheds , especially on meteorological, hydrological and land water supplies. From a hydrological aspect, changes in land use will have a direct effect on the characteristics of land cover so that it will affect the water system in each watershed.

Climate change that occurs today, is also thought to cause an increase in hydrometeorological disasters such as floods, landslides and tornadoes. Currently flooding is a problem that is also a major concern in many regions of many countries. This problem is further exacerbated due to global climate change[4].

To determine the type of distribution that will be used in determining the return period (frequency analysis), statistical parameters are sought from regional rainfall data both normally and logarithmically. To estimate the amount of rainfall with various re-periods, a frequency analysis of rainfall data is carried out.

Frequency analysis of hydrological data aims to determine the value of the magnitude of extreme events related to the frequency of occurrence through the application of probability distributions. Frequency analysis using random variables and probability distributions is part of the statistical method. In the statistical analysis of data, there are parameters that can help in determining the right type of distribution. These parameters are divided into 4 (four) large parts of measurement, namely, central tendency measurement, variability measurement, skewness measurement and kurtosis measurement. The types of distributions used are Normal Distribution, Normal Log Distribution, Gumbel Distribution, Pearson III Log Distribution[5].

Flood disaster mitigation is an effort to reduce the risk due to flood disasters According to Yulia S. (2015) "Flood disaster mitigation is an effort made to prevent or reduce the risk of flood disasters.

Mitigation in flood disasters is also divided into 2 types, namely structural mitigation and non-structural mitigation. Structural Mitigation is an effort made to minimize disasters such as by building special canals to prevent flooding and by making technical engineering of disaster-resistant buildings, as well as waterproof building infrastructure. Meanwhile, non-structural mitigation is an effort made in addition to structural mitigation such as with regional planning and &ing insurance. In this non-structural mitigation, we really expect from the development of increasingly advanced technology. The hope is that technology can predict, anticipate & reduce the risk of a disaster occurring[6].

The SCS Synthetic Unit Hydrograph is a hydrograph dimensionless unit, where this method calculates the influence of soil type and land use patterns and can be calculated using HEC-HMS software[7]. HEC-HMS (Hydrologic Engineering Center's Hydrologic Modeling System) is software developed by the U.S Army Corps of Engineering. HEC-HMS is used for hydrological analysis by simulating rainfall processes and run off from a river basin.

HEC-RAS stands for Hydraulic Engineering Center-River Analysis System. The program was created by the Hydrologic Engineering Center (HEC) which is a division within the Institute for Water Resources, under the US Army Corps of Engineers (USAGE). HEC-RAS is a permanent and non-permanent flow one-dimensional model (steady and unsteady one dimensional flow model).

2-D modeling using HEC RAS software for unsteady flow conditions uses input data in the form of longitudinal cross sections (long section) and cross section (cross section), boundary condition data upstream and downstream, channel base slope data, manning coefficient and DEM (Digital Elevation Model) data to describe the topographic conditions of the watershed[8]. HEC-RAS software is a hydrodynamic model used in this study. HEC-RAS 5.03 was used to model river hydraulics so that flood flow profiles could be simulated including their inundation characteristics[8].

2. Research Method

Kera River in Wajo regency, is located in the central part of South Sulawesi Province with a distance of 242 km from Makassar which is the capital of South Sulawesi Province. The area of Wajo Regency is 2,506.19 km2 or 4.01% of the area of South Sulawesi Province.

Floods occurred in Wajo Regency in 2020 affecting 9 (nine) sub-districts and 78 villages/kelurahan, where one of the sub-districts affected by severe floods was Keera District. The inability of the Kera River to accommodate and drain the flood discharge that occurred resulted in runoff which affected settlements, agricultural land and community ponds in the area. Based on information obtained from the community, the worst flooding incident that has ever occurred reached a height of ± 2 m in residential areas and inundation on national roads which also resulted in the interruption of transportation access in the area.

Figure 1 shows the research location is in Kera river, Wajo Regency, South Sulawesi Province. Hydraulic tracing for flood analysis was carried out along 32 km with a watershed area of 172.37 km^2, namely in the upstream area towards the estuary or at coordinates 3°51'2.04" LS; 120°21'46.87 BT to 3°48'58.93" LS; 120°13'24.11 BT.

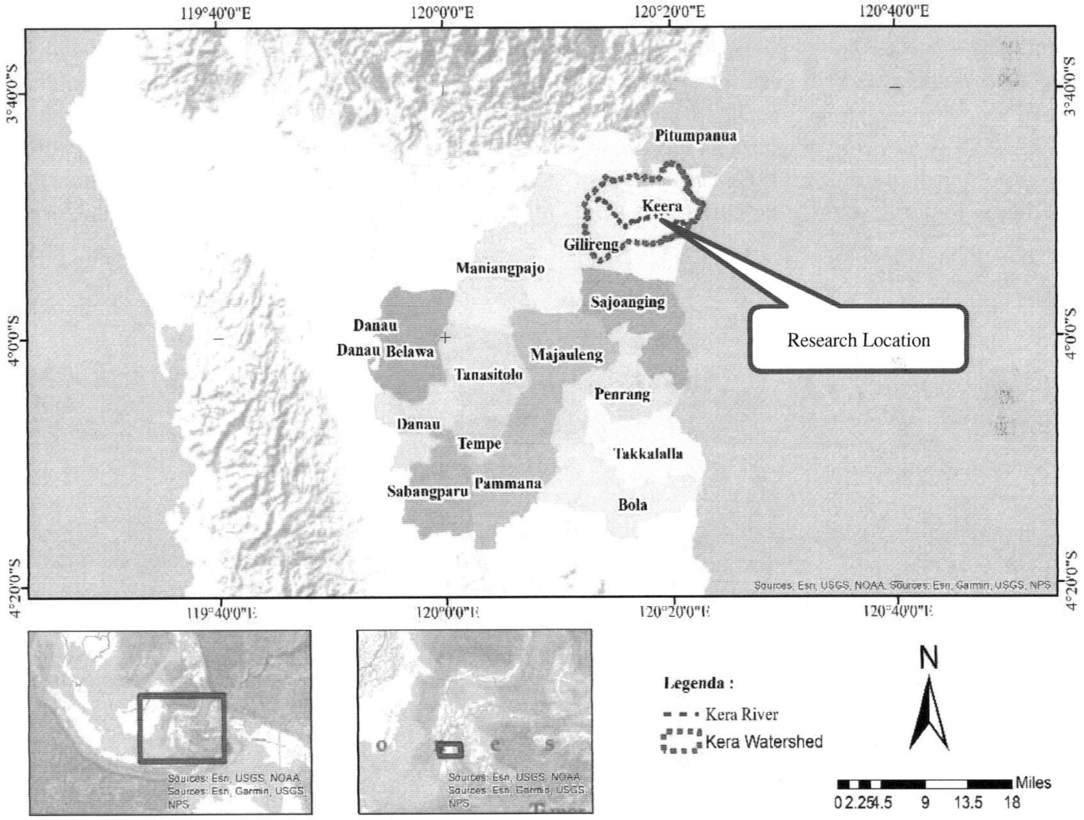

Figure 1. Research Location Map.

Primary data is obtained from survey results to collect information related to flood events while secondary data is obtained from related agencies. Information data retrieval techniques are 1). Prepare everything that will be used for surveys to the field so that the research runs smoothly, 2). Seeing and paying attention to the condition of buildings in the field, 3). Coordinating/communicating to relevant agencies for data collection, 4). Collect data and input into Microsoft Office which will be used to analyze.

The method to determine the flood-prone areas of The Kera River in Wajo Regency is carried out based on the flood discharge plan resulting from hydrological analysis which has previously been verified with discharge data at the time of Q2, Q5, Q10, Q20, Q25 in accordance with the Regulation of the Minister of Public Works and Public Housing of the Republic of Indonesia concerning the Determination of the River Boundary Line and Lake Boundary Line No. 28 / PRT / M / 2015 for the regency / city capital to flow the plan discharge (Q10 - Q20). Furthermore, the flood discharge data of the selected plan and the cross-sectional data of the river are used as input data for 2D modeling using the HEC-RAS application to see the flood distribution area. From this modeling, it can be known the extent of inundation, the speed of flood flow, and the duration of floods which are then overlayed using the Arcgis 10.5 Application to analyze the determination of flood-prone areas based on the Regulation of the Head of the National Disaster Management Agency Number 02 of 2012 concerning General Guidelines for Disaster Risk Assessment and making maps of flood-prone areas. From the results of regional modeling, it is continued by mitigating flood disasters with structural methods to minimize the impact of floods.

3. Result and Discussion

Based on TRMM rainfall data for 1998-2021 shows that the rainfall in the watershed averages more than 90 mm / month. A positive increase in annual rainfall will increase the availability of water resources if they can be managed. Rainfall in the watershed has increased and decreased every year (Figure 2) shows that climate change does affect hydrometeorological disasters , especially floods. However, if there is successive rains with high intensity, the flood event will be even greater. In addition, high-intensity rains will also cause more flood overflow events beyond river embankments.

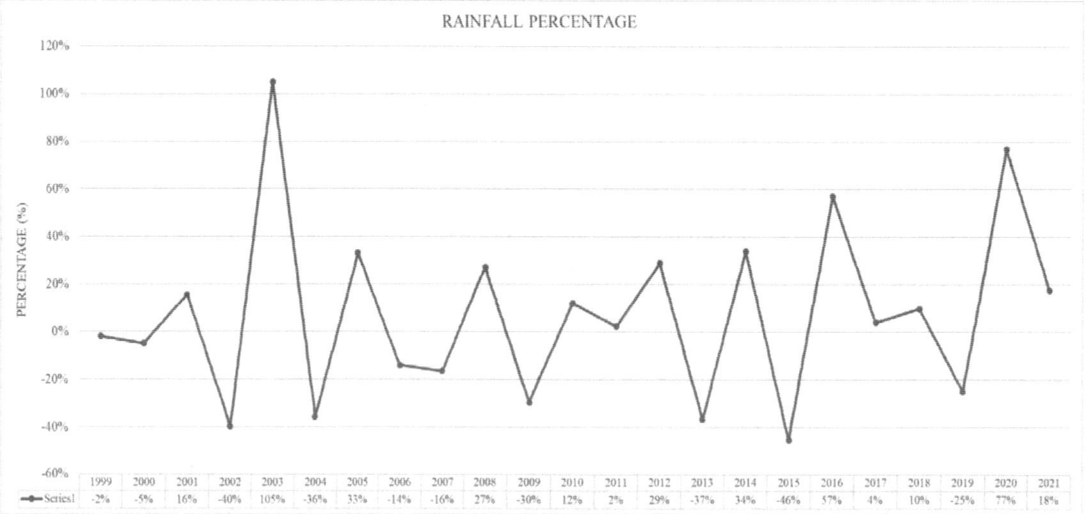

Figure 2. Rainfall Percentage

The trigger for flooding in the Watershed area is due to the high intensity of rainfall. Based on TRMM rainfall data in 1998-2021, the maximum rainfall that occurred in the Kera river was 122 mm/day in June 2001, 140 mm/day in April 2003, 128 mm/day in May 2005, 146 mm/day occurred in December 2021. The results of the rainfall analysis that has been carried out by the rainfall data in the category of very heavy rain in accordance with the threshold of the rain intensity value :

a. 0 mm/day (gray) : Cloudy
b. 0.5 – 20 mm/day (green) : Light rain
c. 20 – 50 mm/day (yellow) : Moderate rain
d. 50 – 100 mm/day (orange): Heavy rain
e. 100 – 150 mm/day (red) : Very heavy rain
f. >150 mm / day (purple) : Extreme rain

Figure 3 shows that the hydrograph of floods due to climate change greatly affects the peak of floods. Discharge with a high re-time, quantitatively experienced a higher increase than the discharge with a lower re-time. The highest discharge that occurs at 2 year return period is 147.1 m^3/s, 5 year the return period is 212.5 m^3/s, 10 year return period is 251.3 m^3/s, 25 year the return period is 296.1 m^3/s.

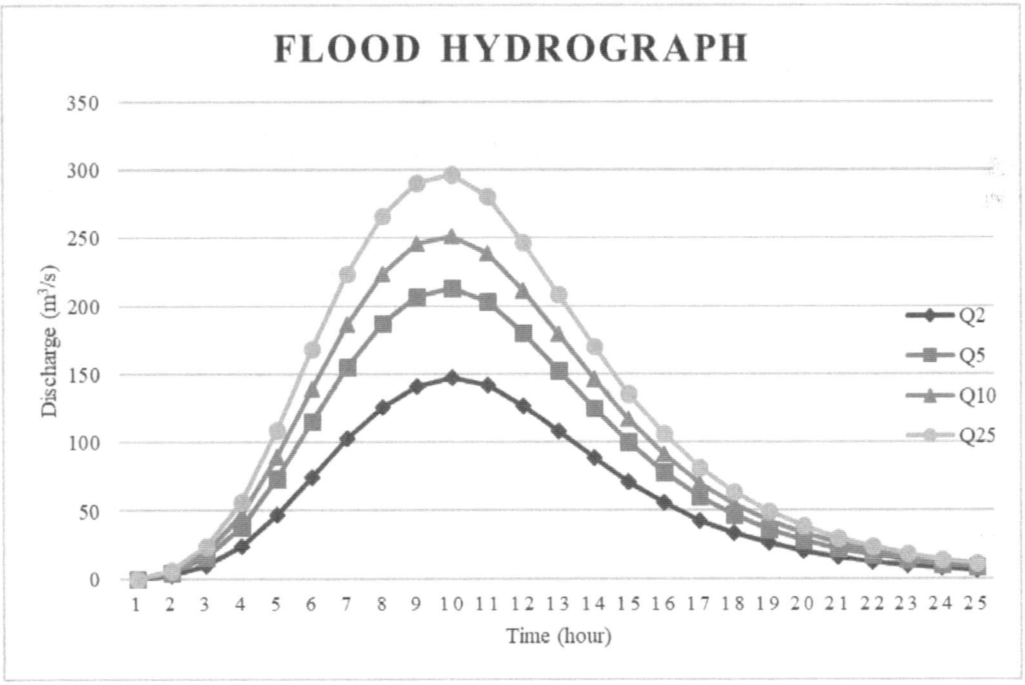

Figure 3. Flood Hydrograph

This study presents a flood-prone map of the River (Figure 3), where typical land characteristics and the potential for flooding can provide information about a flood-prone condition related to these geomorphological and hydrological characteristics.

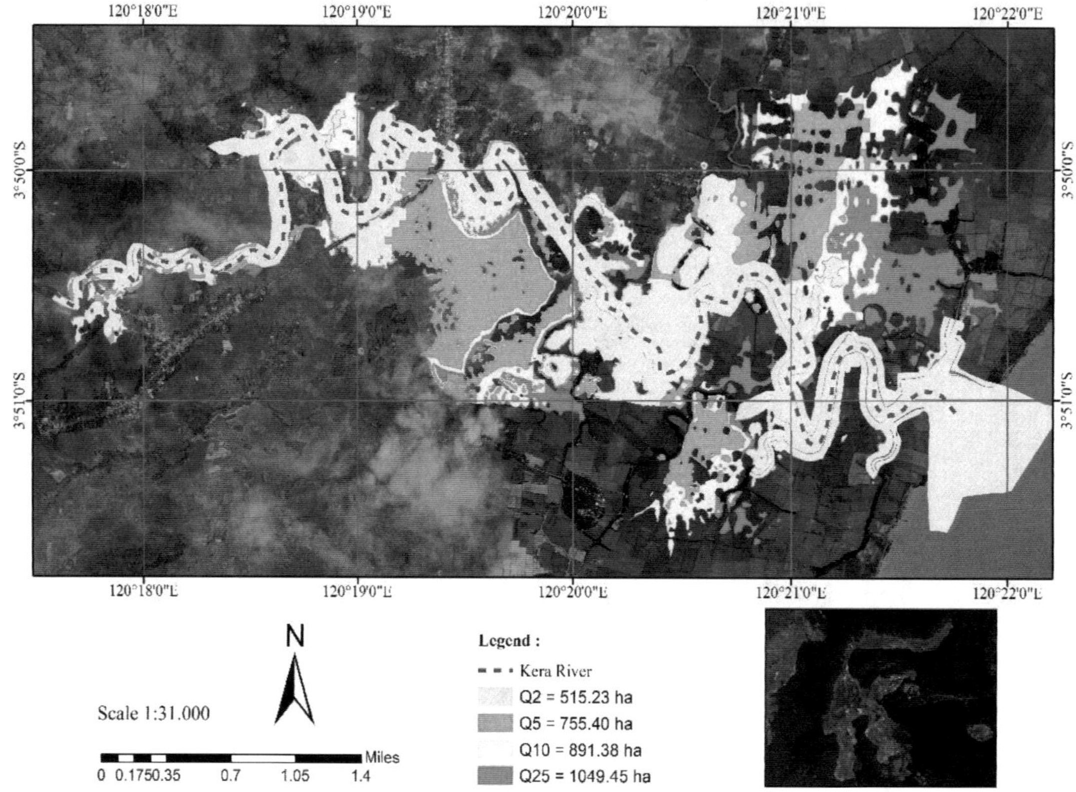

Figure 4. Map prone to Flooding of the Ape River with Various return period

Based on the simulation results, it describes a model of the distribution of flood inundation, so that information on the affected areas is obtained. Efforts to overcome river problems and damage due to flooding that can be applied to the Watershed according to the location and causes can be carried out disaster mitigation, by structural methods. Flood control efforts with the structural method in question are to prevent flood runoff to a certain height with flood embankments.

Based on the flood simulation study above, the structural flood control alternatives for the Watershed are as follows:

3.1 Flood Embankment

The embankment is one of the flood control buildings that serves to prevent the overflow of floodwaters, generally made from landfills, but under certain conditions it can be made of stone or concrete pairs. The soil used to form the embankment body at least meets the requirements, including high imperviousness, high cohesion value, and in a water-saturated state the inner shear angle is high, concentrated and the pore number is low. A mixed soil between sand and loam with a proportion of ± 1/3 part of sand and ± 2/3 part of clay is a fairly adequate embankment material.

In general, flood embankments are planned to follow the existing flow of the river. Except for the parts of the river channel that are hostage, the flood embankment is made straight (it does not always follow the flow of the river so that the alignment of the embankment becomes shorter).

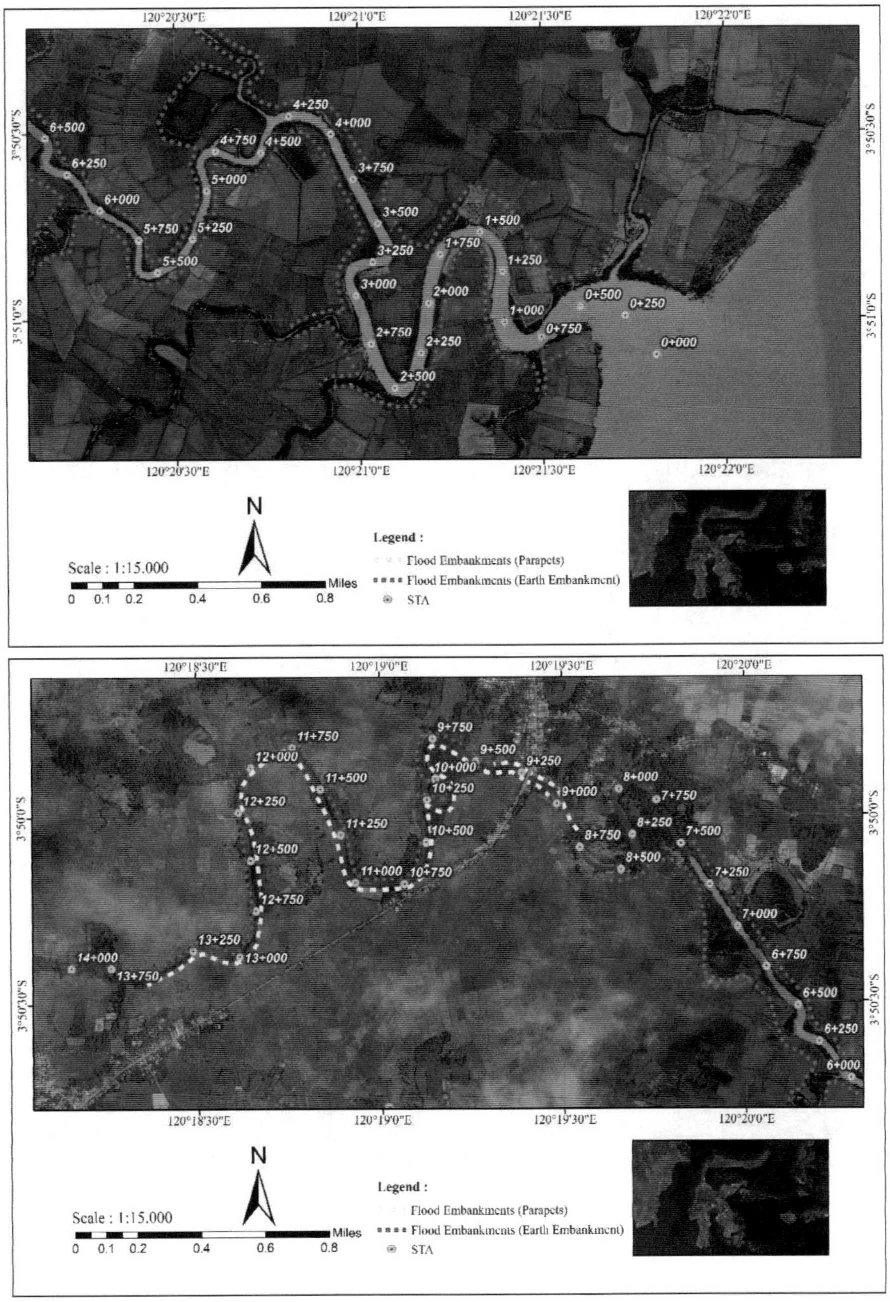

Figure 5. Location Layout of Flood Control and Ape River Safety Buildings

Based on figure 5, it can be seen that the installation of flood embankments (earth embankments) at Sta 0+000 – Sta 13+500, while at Sta 8+750 - Sta 13 +500 , flood embankments (parapets) are installed so as to reduce the impact of floods due to climate change that occurs in the Kera River.

4. Conclusion

Climate change greatly affects the increase in rainfall in the watershed, and flooding in the watershed. Changes in rain characteristics in the watershed, namely the annual rain height and maximum daily rain tend to increase causing inundation, resulting in flooding around the Ape River area. The highest discharge that occurs at 2 year return period is 147.1 m^3/s, 5 year return period is 212.5 m3/s, 10 year return period is 251.3 m3/s, 25 year return period is 296.1 m3/s respectively. The simulation results for 2 year return period have an inundation area is 515.23 ha, 5 year return period is 755.40 ha, the the return period 5 year is 891.38 ha, the the return period 5 year 1049.45 ha. It is necessary to carry out flood disaster mitigation with structural methods, building flood embankments. Therefore, it can be concluded that climate change that occurs in the Ape River affects the number of hydrometeorological disaster events, especially floods

References

[1] M. D. A. Satria, "Climate Change Analysis of the Chances of Hydrometeorological Disaster Events in the Komering Watershed of South Sumatra Analysis of Climate Change on Chances of Hydrometeorological Disaster Events in The Komering Sub-Watershed South Sumatra," in Proceedings of the *World Water Day National Seminar*, 2019, pp. 44–51.

[2] R. Milanda and B. Setiawan, "Proceedings of the World Water Day National Seminar," 2019.

[3] M. Azizah, A. Subiyanto, S. Triutomo, and D. Wahyuni, "The Effect of Climate Change on Hydrometeorological Disasters in Cisarua District - Bogor Regency," *PENDIPA J. Sci. Educ.* , vol. **6**, no. 2, pp. 541–546, 2022, doi: 10.33369/pendipa.6.2.541-546.

[4] A. Rosyida, R. Nurmasari, and Suprapto, "Comparative Analysis of the Impact of Hydrometeorological and Geological Disaster Events in Indonesia Seen from the Number of Victims and Damage (Study: Indonesian Disaster Event Data 2018)," *J. Disaster Management Dialogue*, vol. **10**, no. 1, pp. 12–21, 2019.

[5] F. Jul Fachri, M. Selintung, R. Karamma Civil Department, and A. Correspondence Fiqih Jul Fachri, "Analysis of River Hydrographs Using Hss In The Jeneberang Watershed Of Gowa Regency Analysis Of Hydrograph River Using Synthetic Unit Hydrograph (Hss) In The Jeneberang Watershed Gowa District," Makassar.

[6] A. Y. S. Siska Kania Oktapian, Suryana, "Flood disaster mitigation carried out by the community in Bojong Village, Majalaya District, Bandung regency," *Geoarea*, vol. **1**, no. 2, 2018.

[7] M. R. Mustamin, F. Maricar, and R. Karamma, "Hydrological Analysis In Selecting Flood Discharge Method In Watershed Of Kelara River," *INTEK J. Penelit.* , vol. **8**, no. 2, p. 141, 2021, doi: 10.31963/intek.v8i2.2874.

[8] A. Parandangi, R. T. Lopa, and B. Bakri, "Flood Management in Lake Tempe with Regulatory Ponds in Inflow," *J. Penelit. Enj.* , vol. **24**, no. 2, pp. 125–134, 2020, doi: 10.25042/jpe.112020.04.

The strategy of Tukad Badung pollution control using QUAL2Kw and AHP

Y Ciawi[12*], P M D Padilla[1], M I Yekti[1]

[1]Master of Civil Engineering Program Study, Faculty of Engineering, Udayana University, Jl.P.B, Sudirman, Denpasar-Bali, Indonesia
[2]Environmental engineering Department, Faculty of Engineering, Udayana University, Kampus Bukit Jimbaran, Bali, Indonesia

*Corresponding author: yenniciawi@unud.ac.id

Abstract. Tukad Badung is the major river flowing through Denpasar, the capital city of Bali Province, Indonesia. It is polluted by domestic dan industrial waste which is contributed by surrounding communities such as hotels, shopping area, animal husbandry, printing industries, tofu and tempeh industries industry, and housing area. This research aims to analyse the quality of the water and determine the best among options provided to overcome the pollution problem conducted by using two analytical programs, i.e.QUAL2Kw and Analytical Hierarchy Process (AHP). According to the results of QUAL2Kw analysis, several parameters of water quality of Badung River did not meet the Indonesian standards of water quality stated in Regulation Number 82 the Year 2001 about Water Quality Management and Water Pollution Control. The TSS reached 111.02 kg/day, the BOD reached 92.24 kg/day, and the COD reached 440.92 kg/day. Whereas the pollution load capacity is only 15.43 kg/day for TSS, 19.68 kg/day for BOD, and 7.45 kg/day for COD. This problem occurred was due to the home industries activities in the upstream part of the Badung River. The result of the Analytical Hierarchy Process (AHP) shows that "improving monitoring and supervision of waste disposal" has the highest composite weight value of 0.317. Among a series of solutions offered, this result suggests that the local government should take actions to impose the regulations on wastewater disposal to the river stream by surrounding communities especially the small and middle-size enterprises.

Keywords: *water quality, QUAL2Kw, water pollution control, analytical hierarchy process (AHP)*

1. Introduction

Tukad Badung is one of the major rivers that flow through Denpasar, the capital city of Bali. This river is polluted by solid waste and wastewater which are disposed of by surrounding communities. Increasing population growth on the river banks of Tukad Badung worsens this problem [1]. There are several groups of activities in the watershed area of Tukad Badung that contributed to the pollution of the river, i.e. hotels, commercial area, agriculture, animal husbandry, screen printing, tofu or tempeh industries, other small and medium enterprises, and housing areas [2][3].

The river pollution by organic matters is a very complex process. Therefore, a mathematical model is employed to set up water quality management programs to solve this problem [4] [5] [6] and to overcome the limitations of experimental and analytical approaches in making predictions of water quality along with river flows [7] [8]. On the other hand, the analytical hierarchy process (AHP), which was usually employed to solve a complex unstructured situation and determine the best priority to be carried out [9][10][11]. This method also has been implemented in calculating pollution load at Pudu River in Riau Province [12], in the budgeting of operation and maintenance of irrigation [13], and policy assessment of energy source [14].

Most researchers utilized QUAL2Kw or AHP for different approach separately, but this research try to employ both QUAL2Kw to analyze water quality problems and AHP to determine the best and the most urgent water pollution prevention strategies in Tukad Badung.

1.1. Water quality

Water quality is the nature of water and the content of substances, living things, energy, or other components in water [15]. River water quality can be expressed by physical, chemical, and biological parameters.

1.1.1. Physical Parameters, Temperature affects the metabolic activity of organisms, therefore the distribution of organisms both in marine and freshwater is limited by the temperature of the water [16].

1.1.2. Chemical and Biological Parameters, BOD Mays [17] defines BOD as the amount of oxygen used by the microbial population to degrade the organic matters content of the water, which means that the BOD value is directly proportional to the content of organic matters (biodegradable organics) in the water. COD A high COD level indicates a large number of organic and inorganic matters in the sample, which will reduce the dissolved oxygen (DO) level. Decreased DO can lead to anaerobic conditions, which can damage aquatic life. The COD is often used as an alternative to BOD because of its shorter analysis time [18][19]. Dissolved Oxygen/DO According to [20], the concentration of oxygen is mostly influenced by temperature, in which solubility is higher at a lower temperature. Oxygen is essential in water for the livelihood of aquaculture biota. Total coliform and E.coli Coliform act as an indicator organism for contamination of water by pathogens as these bacteria are also more tolerant against heat than other pathogens and easily isolated and cultured. Coliform are Gram-negative bacteria and usually found in the human colon [17]. Total Suspended Solid (TSS) TSS is the residue retained by the filter with a pore size of 2 μm [19]. The residue consists of clay, silt, metal oxides, sulphides, bacteria, fungi and algae. TSS can be removed by flocculation and filtering. According to [3], pH is the degree of acidity measured by the number of hydrogen ions using the formula pH = -log (H^+).

1.2. QUAL2Kw

QUAL2Kw simulates conventional (non-toxic) pollutant transfer and what occurs during the displacement. This model represents a river as a one-dimensional channel with a non-uniform shape, steady flow, and simulates the effect of the addition of pollutants from a point as well as from non-point sources [21][22][23][24][25][8].The QUAL2Kw model uses Microsoft Excel as an interface. This means that all input and output and model execution are carried out from within Excel [26].

1.3. Analytical Hierarchy Process (AHP)

AHP is a method for solving a complex unstructured situation into several components in a hierarchical arrangement, by giving a value to the importance of each relative variable (Table 1) and determining which variable has the highest priority affecting the results in that situation [9][10][11]. This method has been used by many researchers in many areas of application, e.g., an assessment of the suitability of land use [27], in the agri-food sector [28], selection of wastewater treatment process [29], energy source policy assessment [14], even in notebook selection [30].

Table 1 Scores for The Importance of Variable in Analytical Hierarchy Process (AHP) [31][32]

Importance Scale	Definition of Importance Scale
1	Equally Important Preferred
2	Equally to Moderately Important Preferred
3	Moderately Important Preferred
4	Moderately to Strongly Important Preferred
5	Strongly Important Preferred
6	Strongly to Very Strongly Important Preferred
7	Very Strongly Important Preferred
8	Very Strongly to Extremely Important Preferred
9	Extremely Important Preferred

1.4. *Bali Customary Law*
Bali is a unique place which has two regulations applicable in society, one is traditional customary law which bound the peoples according to their local wisdom and custom and the other is the official regulation under the national law [33]. In many cases, local traditional customary law has great impact on daily life [34][35], customary law also can be modernized according to the ongoing development, which does not only regulate traditional community life but also the modern society in Bali [36]. Imposing local customary sanctions in the area concerned is chosen as one of the alternative solution to overcome the problem of Badung river pollution in the running of the AHP program.

2. Methods

2.1. *Study area*
The study is carried out in the Denpasar area, i.e., Tukad Badung with a watershed area of + 35.80 km2 and the length of the main river is + 19.52 km. Denpasar is the capital city of Bali Province, Indonesia and situated in the southern part of Bali. Figure 1 is the map of the Tukad Badung watershed, which is divided into three segments, the upstream segment, located at the Bridge in Gajah Mada Road, the middle segment located in between the Bridge and Badung Motion Weir and the downstream segment is located at Badung Motion Weir-Estuary Dam.

2.2. *Data*
The data consist of primary and secondary data:
1. Map of the Tukad Badung River area, Jalan Gajah Mada Bridge, Badung Motion Weir, and Estuary Dam Temple, which is provided by Research and Development Agency (Bapenlitbang) Bali
2. Data on the discharge of the Tukad Badung River for Jalan Gajah Mada Bridge, Badung Motion Weir, and Estuary Dam from Regional Office of River Basin (BWS) Bali-Penida
3. Data on the water quality of the Tukad Badung river, Jalan Gajah Mada Bridge, Badung Motion Weir, and Estuary Dam provided by The Bali Provincial Environmental Office
4. Climatological data obtained from BWS Bali-Penida
5. Data related to local environmental conditions obtained from the Bali Provincial Environmental Office

Figure 1 Map and segmentation of Tukad Badung watershed and sampling points details [37]

2.3. *Water Quality Determination*

Determination of water quality was carried out using standard methods described in [38].

2.4. *Modelling Using QUAL2Kw*

Modelling activities include input of the 5 years data into the QUAL2Kw model cell, then in the form of MS-Excel based on the reach map and model calibration to align the model data and field data for each segment or reach (Figure 2). In the calibration of water quality, data are entered in the input of the WQ Data worksheet, diffuse sources (water quality data) for each reach. This requires trial and error on the coefficient value of the model. In the process of calibrating water quality data, the data that is tried out is the coefficient value of each parameter which is run into 1 unit or in other words, it cannot be calibrated individually, model must be calibrated with all parameters simultaneously with trial and error on the sheet "reach rates".

Determination of model coefficients: the model is simulated repeatedly to obtain a model that is almost or the same as the field conditions (actual conditions). The data is then entered into the QUAL2KW program and executed by pressing the "run" button. Results obtained automatically in graphs and tables format. To determine the coefficient, the model is run by trial and error repeatedly until the model results are close to the same conditions as the field.

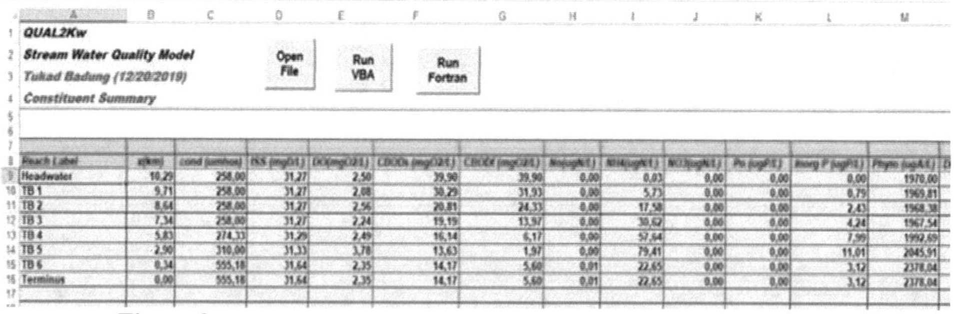

Reach Label	x(km)	cond (umhos)	ISS (mgD/L)	DO(mgO2/L)	CBODs (mgO2/L)	CBODf (mgO2/L)	No (ugN/L)	NH4(ugN/L)	NO3(ugN/L)	Po (ugP/L)	Inorg P (ugP/L)	Phyto (ugA/L)
Headwater	10,29	258,00	31,27	2,50	39,90	39,90	0,00	0,03	0,00	0,00	0,00	1970,00
TB 1	9,71	258,00	31,27	2,08	30,29	31,93	0,00	5,73	0,00	0,00	0,79	1969,81
TB 2	8,64	258,00	31,27	2,56	20,81	24,33	0,00	17,58	0,00	0,00	2,43	1968,38
TB 3	7,34	258,00	31,27	2,24	19,19	13,97	0,00	30,62	0,00	0,00	4,24	1967,54
TB 4	5,83	274,33	31,29	2,49	16,14	6,17	0,00	57,64	0,00	0,00	7,99	1992,69
TB 5	2,90	310,00	31,33	3,78	13,63	1,97	0,00	79,41	0,00	0,00	11,01	2045,91
TB 6	0,34	555,18	31,64	2,35	14,17	5,60	0,01	22,65	0,00	0,00	3,12	2378,04
Terminus	0,00	555,18	31,64	2,35	14,17	5,60	0,01	22,65	0,00	0,00	3,12	2378,04

Figure 2 Worksheet of Water Quality Output (TB=sampling point)

2.5. *Simulation using AHP*

In the AHP analysis, an alternative assessment is carried out by giving weight for each alternative as a solution to solve water pollution problems in Tukad Badung. The alternatives provided are as follows:
1. Imposing local customary sanctions in the area concerned.
2. Improve monitoring and supervision of wastewater disposal.
3. Imposing industrial waste tax and tightening waste disposal permits.
4. Imposing sanctions by the Law

3. Results and Discussion

3.1. *Simulation 1*

In simulation 1 (Table 2 and Table 3), the water quality data analysis in the upstream is adjusted to the quality standard of second-class water bodies, the pollutant comes from a non-point source. Domestic waste is assumed the same as present data. The goal is to obtain the value of the pollution load.

Table 2 Recapitulation of Water Quality Output Step 1

Reach Label	Distance (km)	TSS (mg/L)	DO (mg O_2/L)	BOD (mg O_2/L)	NH_4 (µg N/L)
Sampling point 1	9.71	31.39	2.34	38.10	0.042
Sampling point 2	8.64	37.64	2.40	24.35	0.029
Sampling point 3	7.34	39.10	2.52	21.59	0.120
Sampling point 4	5.83	36.71	2.20	19.50	0.190
Sampling point 5	2.90	35.12	2.41	15.27	0.210
Sampling point 6	0.34	32.10	2.32	14.52	0.200

Table 3 Recapitulation of Water Quality Output Step 2

Reach Label	Distance (km)	Phyto (µg/L)	Pathogen	COD (mg O_2/L)	pH
Sampling point 1	9.71	2,103	16	181	7.52
Sampling point 2	8.64	2,857	10	152	7.62
Sampling point 3	7.34	2,452	18	157	7.41
Sampling point 4	5.83	2,291	23	151	7.32
Sampling point 5	2.90	1,987	10	128	7.25
Sampling point 6	0.34	2,451	11	84	7.21

3.2. *Simulation 2*

In simulation 2 (Table 4-7), the parameters water quality at the upstream part is assumed as the standards of water for class two. The point source is assumed by trial and error until the water quality at the downstream part meets the water quality standard class two. Whereas the non-point source is assumed will not be higher than the existing condition. The aim of simulation two is to determine the maximum number of industrial plants allowed along the river to meet water quality class 2 at the downstream part of the river.

Table 4 Water Quality Output Part 1

Reach Label	Distance (km)	TSS (μg/L)	TSS standard	DO (mg O_2/L)	DO standard
Sampling point 1	9.71	34.5	50	2.34	4
Sampling point 2	8.64	34		2.4	
Sampling point 3	7.34	34.5		3.1	
Sampling point 4	5.83	32		6.2	
Sampling point 5	2.90	34.5		5.3	
Sampling point 6	0.34	22		5.3	

Table 5 Water Quality Output Part 2

Reach Label	Distance (km)	BOD (mg O_2/L)	BOD standard	NH_4 (μg N/L)	NH_4 standard
Sampling point 1	9.71	38.1	30	0.02	0.1
Sampling point 2	8.64	24.35		0.023	
Sampling point 3	7.34	21.59		0.025	
Sampling point 4	5.83	26.3		0.02	
Sampling point 5	2.90	28.1		0.02	
Sampling point 6	0.34	26.2		0.05	

Table 6 Water Quality Output Part 3

Reach Label	Distance (km)	Phyto (μg/L)	Pathogen	COD (mg O_2/L)	COD standard
Sampling point 1	9.71	2110	16	181	80
Sampling point 2	8.63	2110	19	152	
Sampling point 3	7.33	2110	18	157	
Sampling point 4	5.82	2110	23	151	
Sampling point 5	2.89	2110	22	128	
Sampling point 6	0.33	2120	18	84	

Table 7 Water Quality Output Part 4

Reach Label	Distance (km)	Ph	pH standard	Temperature
Sampling point 1	9.71	7.07	8.5	27.40
Sampling point 2	8.63	6.95		27.40
Sampling point 3	7.33	6.91		27.40
Sampling point 4	5.82	6.78		27.80
Sampling point 5	2.89	6.76		27.80
Sampling point 6	0.33	6.68		27.60

3.3. Simulation 3

Based on the trend of water quality in simulation 1 and simulation 2, the value of water quality parameters can be calculated using a mathematical approach with the following formula for the future trend as described in Table 8.

Table 8 Prediction Formulas for Water Quality in 2024

Parameter	Units	Formula
Temperature	C	$y = 0.45x^2-1.85x + 26.5$
pH	mg/L	$y = -0.985x^2 + 3.635x + 4.76$
DO	mg/L	$y = 1.76x^2 + 6.29x + 3.27$
BOD	mg/L	$y = -2x + 9.75$
COD	mg/L	$y = -12.76x + 50.73$
TSS	mg/L	$y = 53.49x-63.07$
NH_4	mg/L	$y = 0.29x + 3.22$

According to the above formula, DO value at year 6 can be predicted as follows:
y = 1.76x2 + 6.29x + 3.27
= 1.76 (6)2 + 6.29 (6) + 3.27
= -22.35
Where:
y = predicted quality
x = year of prediction (6 years)
In the next step, the calculation of the reduction of water quality from the year 2019 can be determined using the formula below. DO value in 2019 is 2.5 mg/L.

$$\frac{Prediction\ of\ DO\ Tukad\ Badung - DO\ Tukad\ Badung\ 2019}{DO\ Tukad\ Badung\ 2019} \text{ x } 100\ \%$$

$$= \frac{(-22.35)-2,5}{2,5} \text{x } 100\ \%$$

= -9.94

The other parameters of water quality are tabulated in Table 9 using the same formulas. When the water quality in 2024 is obtained, then the values are entered in the headwater worksheet. Then, simulation 3 is run to obtain a model. The percentage of loss in each parameter is as follows. It should be underlined that this decline has a negative character, if the percentage value of decline is negative, it means that the quality has increased in value, not the other way around. For example, the pH of -2.18 means that the pH has increased by 2.18 per cent. However, this applies normally if the percentage value shows a positive number, for example, the temperature is 0.16 means that the temperature increased by 0.16 per cent.

Table 9 Percentage Reduction of River Water Quality

Parameter	Units	Formula	% reduction in water quality
Temperature	0C	y = 0.45x2-1.85x + 26.5	0.16
pH	mg/L	y = -0.985x2 + 3.635x + 4.76	-2.18
DO	mg/L	y = 1.76x2 +6.29x + 3.27	-9.94
BOD	mg/L	y = -2x + 9.75	-1.06
COD	mg/L	y = -12.76x + 50.73	-1.69
TSS	mg/L	y = 53.49x-63.07	7.25
NH4	mg/L	y = 0.29x + 3.22	3.96

In simulation 3, data of water quality in the upstream part employing the data according to standards and in the downstream part, the water quality is expected to meet the water quality standards class 2. This simulation is run in trial and error mode with the polluting source of point source dan non-point source from domestic wastewater and hotel. Simulation 3 is aimed to determine the Tukad Badung water quality in the future (Table 10 and 11).

Table 10 Prediction of Tukad Badung water quality in the year 2024 part 1

Parameter	Units	Headwater	Sampling point 1	Sampling point 2	Sampling point 3
Temperature	^0C	27.34	27.44	27.44	27.44
pH	mg/L	7.34	6.92	6.80	6.76
DO	mg/L	2.25	2.11	2.16	2.79
BOD	mg/L	39.9	38.1	24.35	21.59
COD	mg/L	179.91	177.94	149.43	154.35
TSS	mg/L	33.54	37.00	36.47	37.00
NH_4	mg/L	0.031	0.021	0.024	0.026

Table 11 Prediction of Tukad Badung Water Quality in 2024 Part 2

Parameter	Units	Sampling point 4	Sampling point 5	Sampling point 6	Terminus
Temperature	^0C	27.84	27.84	27.64	27.74
pH	mg/L	6.63	6.61	6.53	7.22
DO	mg/L	5.58	4.77	4.77	4.59
BOD	mg/L	26.3	28.1	26.2	24.6
COD	mg/L	148.45	125.84	82.58	79.63
TSS	mg/L	34.32	37.00	23.60	24.13
NH_4	mg/L	0.021	0.021	0.052	0.073

3.4. Pollution Load and Pollution Capacity of Tukad Badung

Sources of water pollution in Tukad Badung come from industry, traditional markets, hospitals, and domestic waste. There are 393 small textile industries, 129 batik factories, 663 other fabric industries, 393 screen printing and printing industries, 132 tofu and tempeh industries, and 158 metal industries from Dauh Puri Kangin to Estuary Reservoir that dumped waste into Tukad Badung as much as 243,491.39 m3 per day [39][40]. The calculation of the pollution capacity of Tukad Badung uses the simulation output data in the source summary worksheet which is the calculation of the pollution load for each segment. The calculation of the carrying capacity of the pollution load uses simulation 1. Based on the output of simulation 1, the calculation of the capacity of the pollution load is obtained.

Pollution load analysis or source summary is the result of calculating the pollution load, discharge and water quality of each segment, where the results of running non-point source loads (Table 12 and 13). The conversion value here is the time conversion value which results in kg/day units which are worth 86.4. An example of a calculation is presented below while the other calculations are presented in tables.

Source summary = TSS x Inflow x 86.4 / 1000
= 31.27 x 4.3 x 86.4 / 1000
= 11,617 kg/day

Table 12 Pollution Load in Simulation 1

Reach Label	x(km)	*Flow Q, m3/s*	TSS (kg/day)	BOD (kg/day)	COD (kg/day)
Headwater	10.29	4.30	11.617	14.82	67.98
Sampling point 1	9.71	4.30	11.662	14.15	67.24
Sampling point 2	8.63	4.35	14.147	9.15	57.12
Sampling point 3	7.33	4.30	14.526	8.02	58.32
Sampling point 4	5.82	4.62	14.653	7.78	60.27
Sampling point 5	2.89	4.68	14.201	6.17	51.75
Sampling point 6	0.33	4.98	13.812	6.24	36.14
Terminus	0	6.01	16.398	6.95	42.06
Total			**111.017**	**73.310**	**440.924**

Table 13 Pollution Load in Simulation 2

Reach Label	x(km)	*Flow Q, m3/s*	TSS (kg/day)	BOD (kg/day)	COD (kg/day)
Headwater	10.29	4.30	11.62	14.82	66.84
Sampling point 1	9.71	4.30	12.82	14.15	66.11
Sampling point 2	8.63	4.35	12.78	9.15	56.16
Sampling point 3	7.33	4.30	12.82	8.02	57.34
Sampling point 4	5.82	4.62	12.77	10.50	59.26
Sampling point 5	2.89	4.68	13.95	11.36	50.88
Sampling point 6	0.33	4.98	9.47	11.27	35.53
Terminus	0	6.01	11.68	12.77	41.35
Total			**97.91**	**92.24**	**433.47**

Pollution load capacity can be calculated from the difference between simulation 1 and 2, which is presented in Table 14.

Table 14 Pollution Load Capacity of Tukad Badung

Reach Label	x(km)	*Flow Q m^3/s*	TSS (kg/day)	BOD (kg/day)	COD (kg/day)
Headwater	10.291	4.30	0.01	0.01	1.15
Sampling point 1	9.7145	4.30	1.16	0.41	1.14
Sampling point 2	8.636	4.35	1.37	0.38	0.97
Sampling point 3	7.336	4.30	1.71	0.15	0.99
Sampling point 4	5.8295	4.62	1.88	2.71	1.02
Sampling point 5	2.8985	4.68	0.25	5.19	0.87
Sampling point 6	0.338	4.98	4.35	5.03	0.61
Terminus	0	6.01	4.71	5.82	0.71
Total			**15.43**	**19.69**	**7.45**

3.5. Analytical Hierarchy Process (AHP)

In the AHP analysis, an alternative assessment is carried out by giving weight for each alternative as a solution to solve water pollution problems in Tukad Badung. This method also implemented in calculating pollution load at Pudu River in Riau Province [12] in the budgeting of operation and maintenance of irrigation [13] and policy assessment of energy source [41]. The alternatives provided are as follows:

1. Imposing local customary sanctions in the area concerned.
2. Improve monitoring and supervision of wastewater disposal.
3. Imposing industrial waste tax and tightening waste disposal permits.
4. Imposing sanctions by the Law

Table 15 Alternative Weight

Criteria	Weight	Imposing Local Customary Sanction	Improving Monitoring &Supervision	Imposing Waste Tax	Imposing sanction according to Law
Effective	0.312	0.124	0.309	0.429	0.510
Cost	0.099	0.345	0.150	0.048	0.053
Relevance	0.436	0.088	0.435	0.339	0.289
Time	0.153	0.443	0.106	0.185	0.147
CW		0.179	0.317	0.315	0.313

Table 15 shows in determining a solution to the pollution problem in Tukad Badung, alternative number 2: "Improving monitoring and supervision of wastewater disposal" has the highest Composite Weight, i.e., 0.317 which means that the authorities or related agencies should immediately take action to overcome the current pollution in Tukad Badung.

4. Conclusion

According to the Regulation of the Republic of Indonesia Number 82 the Year 2001 concerning Water Quality Management, some of the parameters of the water quality of Tukad Badung does not meet the Standards. The pollution load of Tukad Badung reaches 111.02 kg/day for TSS, 92.24 kg/day for BOD in the upstream, and 440.92 kg/day at the upstream, whereas the carrying capacity of pollution loads, is only 15.43 kg/day for TSS, 19.69 kg/day for BOD, and 7.45 kg/day for COD. This is due to the large number of cottage industries found in the upstream part of Tukad Badung. The river is no longer able to accommodate the pollution load, so it is necessary to carry out water quality management at Tukad Badung. The Analytical Hierarchy Process (AHP) result shows that pollution prevention with alternative number 2, i.e. "Improving monitoring and supervision of wastewater disposal" has the highest Composite Weight value of 0.317, which can be used as the basis for policymakers to take immediate action.

References

[1] As-syakur AR, Suarna IW, Adnyana IWS, Rusna IW, Laksmiwati IAA, Diara IW 2010 Study of Landuse Change in Badung Watershed *Jurnal Bumi Lestari* **10** 2 200-208 In Bahasa Indonesia

[2] Wijana, IMS, Dewi IGAKSP and As-syakur AR 2012 Study and Monitoring of Water Quality in Badung River Watershed Ecotrophic 7 1 54-58 ISSN 1907•5626

[3] Pertiwi IGAIM, Winaya NAP, Andayani KW, Kristinayanti WS 2019 Waste management system on Badung River area in Bali IOP Conf Ser Earth Environ Sci 351 012005

[4] Karim, MR dan Badruzzaman ABM 1999 M*odeling of Nutrient Transport and Dissolved Oxygen in The Water Column in River System Pollution* Pollut Res **3** 18 195-206

[5] Oliveira B, Bola J, Quinteiro P, Nadais H, Arroja L 2012 Application of QUAL2Kw model as a tool for water quality management: Cértima River as a case study *Environ Monit Assess* **184** 6197–6210 DOI 10.1007/s10661-011-2413-z

[6] Aher P, Jagarlapudi A and Gorantiwar S 2013 Prioritization of watersheds using multi-criteria evaluation through fuzzy analytical hierarchy process. *Agricultural Engineering International: The CIGR e-journal* **15** 1 11-18
[7] Kannel PR, Lee YS, Kanel SR and Pelletier GJ 2007 Application of Automated QUAL2Kw for Water Quality Modelling and Management in The Bagmati River *Ecological modelling* **202** 503–517 DOI: 10.1016/j.ecolmodel.2006.12.033
[8] Aliffia A and Karnaningroem 2019 Simulation of Pollution Load Capacity Using QUAL2Kw Model in Kali Surabaya River (Cangkir-Sepanjang Segment). *IOP Conf.Ser: Earth EnvironSci* **259** 012019
[9] Saaty TL and Vargas LG 2001 *Models, Method, Concept and Application of Analytic Hierarchy Process* Norwell Kluwer Academic Publishers
[10] Saaty TL 2008 Decision making with the Analytic Hierarchy Process *International Journal of Services Sciences* 1(1): 83-98 DOI: 10.1504/IJSSCI.2008.017590
[11] Thanki S, Govindan K and Thakkar J 2016 An investigation on lean-green implementation practices in Indian SMEs using analytical hierarchy process (AHP) approach *Journal of Cleaner Production* **135** 284-298, DOI: 10.1016/j.jclepro.2016.06.105
[12] Kusumawati I 2016 Evaluation of BOD and COD in Pudu River Subdistrict Mandau, District Bengkalis, Riau Province by Using QUAL2Kw *Journal of Env Engineering & Waste Management* **1** 2 107-118
[13] Sachro SS, Kurniawan V, Hadihardaja J, Hadihardaja IK 2018 Evaluating the budgeting approach of operation and maintenance of irrigation in Indonesia by using analytic hierarchy process *Agricultural Engineering International: CIGR Journal* **20** 2 1-8
[14] Zgür-Erol Ö and Kılkış B 2012 An energy source policy assessment using analytical hierarchy process Energy Conversion and Management 63 245-252 ISSN 0196-8904 https://doi.org/10.1016/j.enconman.2012 .01.040.
[15] Sahabuddin H, Harisuseno D and Yuliani E 2014 Analysis of Status of Water Quality and Pollution Load Capasity of River Wanggu Kendari City *Jurnal Teknik Pengairan* **5** 1 19–28
[16] Kordi, MGH and Andi BT 2009 *Management of Water Quality in Aquaculture* PT Rineka Cipta Jakarta
[17] Mays LW Editor 1996 *Water resources handbook* McGraw-Hill New York
[18] Ni'am MF, Othman F, Johan S and Fauzia Z 2007 Removal of COD and Turbidity to Improve Wastewater Quality Using Electrocoagulation Technique *Malaysian Journal of Analytical Sciences* **11** 1 198-205
[19] Aziz NIHA, Hanafiah MM, Halim NH and Fidri PAS 2020 Phytoremediation of TSS, NH_3-N and COD from Sewage Wastewater by *Lemna minor* L, *Salvinia minima, Ipomea aquatica and Centella asiatica* Appl Sci **10** 16 5397-5408 doi:10.3390/app10165397
[20] Salmin 2005 Dissolved Oxygen (DO) and Biological Oxygen Demand (BOD) as Indicator to Determine Water Quality *Oseana* **XXX** 3 21-26 ISSN 0216 1877
[21] Kori BB, Shashidhar T and Mise S 2013 Application of Automated QUAL2Kw for Water Quality Modeling in The River Karanja India *Global Journal of Bioscience and Biotechnology* **2** 2 193-203
[22] Marlina N, Kasam, and Juliani A 2015 Evaluation of Pollution Load Capacity Using Water Quality Model (Study Case Winongo River) *Jurnal Inovasi dan Kewirausahaan* **4** 2 78-86 Article in Bahasa Indonesia
[23] Lestari ADN, Sugiharto E and Siswanta D 2013 Application of QUAL2Kw Model to Determine the Strategy in Solving Gajahwong River Water Pollution Caused by Organic Matter *J Manusia dan Lingkungan* **20** 3 284-293 https://doi.org/10.22146/jml.18496
[24] Setiawan D, Dharma IGBS, Suyasa IWB 2017 Pollution Load Capacity of Badung River at Dauh Puri Village, Denpasar Town Using QUAL2Kw Model *Ecotrophic* **11** 2 116-124 https://doi.org/10.24843/EJES.2017.v11.i02.p10
[25] Setiawan AD, Widyastuti M and Hadi MP 2018 Water Quality Modeling for Pollutant Carrying

Capacity Assessment using QUAL2Kw in Bedog River *Indonesian Journal of Geography* **50** 1 49-56 DOI: http://dx.doi.org/10.22146/ijg.16429

[26] Hidayat YM 2016 The model of aquatic biota mortality as a function of detention time of the detergent wastewater and simple idea for controlling *Jurnal Sumber Daya Air* **12** 2 131-146 Doi: 10.32679/jsda.v12i2.60

[27] Rukanee D, Sangchan S and Choomjaihan P 2020 Assessment of the suitability of land use for agriculture by analytical hierarchy process: AHP in lower Prachinburi watershed, eastern Thailand *AgricEngInt CIGR Journal* **22** 3 19-26

[28] Nikkhah A, Ghnimi S, Firouzi S and Pelletier N 2018 Application of analytical hierarchy process to develop a weighting scheme for life cycle assessment of agri-food sector *CIGR, XIX World Congress, Abstracts* Presented at *The 19th World Congress of International Commission of Agricultural and Biosystems Engineering (CIGR)*

[29] Karimi AR, Mehrdadi N, Hashemian SJ, Bidhendi GRN, Moghaddam RT 2011 Selection of wastewater treatment process based on the analytical hierarchy process and fuzzy analytical hierarchy process methods *Int J Environ Sci Technol* 8: 267–280 https://doi.org/10.1007/BF03326215

[30] Abadi S, Huda M, Basiron B, Ihwani SS, Jasmi KA, Hehsan A, Safar J, Mohamed AK, Embong WHW, Mohamad AM, Noor SSM, Novita D, Maseleno A, Irviani R, Idris M, Muslihudin M. 2018. Implementation of fuzzy analytical hierarchy process on notebook selection. *International Journal of Engineering & Technology* **7** 227 238-243

[31] Taherdoost H 2018 Decision Making Using the Analytic Hierarchy Process (AHP); A Step by Step Approach *International Journal of Economics and Management Systems* 2 244-246

[32] Goepel, KD 2019 Comparison of Judgment Scales of the Analytical Hierarchy Process, A New Approach *International Journal of Information Technology & Decision Making* **17** 1-19 DOI: 10.1142/S0219622019500044

[33] Adharinalti. 2012 Existence of Customary Law in The Village Governance in Bali. Jurnal Rechtsvinding **1** 3 409-418 ISSN 2089-9009

[34] Aryawan BK 2006 *Implementation of Sanctions against Violations of Traditional Village Customary Law (Awig-awig) by Village Communities in Mengwi Village, Mengwi Subdistrict, Badung District, Bali Province* Thesis, Diponegoro University, Semarang, Indonesia

[35] Putri KAMP, Puspitasari NWF, Dewi NKK, Ekarini NW, Dewi IAPP, Mertadana DPK 2018 The Influence of Customary Law or Awig-awig on the Management of Village Fund In Village Banjar Subdistrict Banjar District of Buleleng, Bali Province Jurnal Ilmiah Akuntansi dan Humanika 8 1 ISSN: 2599-2651

[36] Wiryawan, IWG, Perbawa KSLP, Wiasta, IW 2015 Bali Customary Law in The Middle of Development Modernization and Global Culture Jurnal Bakti Saraswati 04 02 169-178 ISSN 2088-2149

[37] Rxerself 2022 Berkas:Bali topography plain.png. https://commons.wikimedia.org/wiki/User:RXerself. Accessed July 22nd 2022

[38] Rice EW, Baird RB and Eaton AD 2017 *Standard Methods for The Examination of Water and Wastewater* 23rd ed American Water Works Association ISBN: 9780875532875

[39] Eryani IGAP, Runa IW, Jayantari MW 2019 Water Potential Management and Arrangement of River Estuary Area for the Mitigation of the Climate Change in Bali *International Journal on Advanced Science Engineering Information Technology* 9 3 849-854

[40] Wijaya IKKA, Arthanaya IW and Suryani LP 2020 Utilization of Watershed Tukad Badung As Denpasar City Tourism (Environmental Law Perspective) *Proceedings of the First International Conference on Technology Management and Tourism*, ICTMT, 19 August, Kuala Lumpur, Malaysia http://dx.doi.org/10.4108/eai.19-8-2019.2294154

[41] Erol, Ö and Kılkıs B 2012 An energy source policy assessment using analytical hierarchy process *Energy Conversion and Management* 63 245-252 https://doi.org/10.1016/j.enconman.2012.01.040

Spatial Modelling for the Calculation of River Capacity: Case Study Downstream Area of Wanggu River Kendari

F Fadlin[1]*, M A Thaha[2], F Maricar and M P Hatta[2]

[1]Doctoral Student of Civil Engineering Department, Hasanuddin University, Makassar, Indonesia
[2]Civil Engineering Department, Hasanuddin University, Makassar, Indonesia

*Corresponding author: ferifadlin@gmail.com

Abstract. Spatial modelling of flood-prone areas will provide maximum results if it is supported by the accuracy of the data acquired, mainly related to elevation data or the area's topography. Spatial modelling generated from accurate topographic data can estimate the river's carrying capacity. This study built a spatial model using data from aerial, terrestrial, and hydrographic surveys. Aerial surveys were conducted using UAV corrected by terrestrial surveys, GCP, and ICP. Testing the accuracy of the spatial model is carried out by comparing the results of current field velocity with the results of 2D Hec-Ras numerical simulations using a variation of the manning coefficient. The combination of aerial, terrestrial, and hydrographic surveys produces a cross-sectional spatial model of the river, which is used in calculating the river's carrying capacity. The river's capacity is calculated using a 2D numerical simulation method using Hec-Ras software and verified by a mathematical approach based on the flood hydrograph curve. The results showed that the horizontal accuracy of the GCP was 2.8 cm and the vertical accuracy was 6.5 cm. The results of testing the vertical elevation accuracy of aerial photographs on terrestrial topographic data measured in the field (ICP) have a Mean Absolute Percentage Error (MAPE) value of 5.81%. According to the spatial model, the manning roughness value is 0.06-0.09. The river's capacity based on numerical simulations is 1.700.766 m3, and the results of the verification using a mathematical approach are 1.683.433 m^3 with a difference of 1.02%.

1. Introduction

One of the problems faced by the government of Kendari City in reducing the risk of flood disasters is the lack of spatial information about the condition of the area that has the potential to be affected by flooding, which can cause material and non-material losses. The unavailability of this information can exacerbate the losses incurred if this flood disaster occurs in the future. The next problem occurs in the managerial aspect, namely the stakeholders or agencies that have duties and functions in flood control. The problem is that the supporting data in controlling the destructive power of water or flooding are not integrated, so it is difficult to control. The impact of these problems is the determination of policies for the direction of the flood control program, including aspects of budgeting that become inappropriate due to inaccuracies in data and information. Policies that are constantly changing and the absence of a track record of data make the proposed activities in flood control unstructured and unintegrated.

Spatial modelling of flood-prone areas will provide maximum results if it is supported by the accuracy of the data acquired, mainly related to elevation data or the site's topography [1]. The extraction of altitude data to obtain a map of the regional situation can be earned using various methods or approaches, such as direct surveys or secondary data from satellite imagery with Digital Elevation Model (DEM) radar technology. These different methods have their respective advantages and disadvantages. DEM satellite images and topographic maps from Indonesian Geospatial Agency RBI maps are accessible data sources for a wide area. The data's weakness is the accuracy level. The RBI map provides data with a full scale currently available of 1:50.000 and DEM satellite imagery with a maximum spatial resolution of 0.27 arc second or about 8,3 meters. Field surveys can be carried out using measuring equipment (Total Station) with a relatively high level of accuracy if carried out with the correct method and high expertise from the surveyor. Still, it requires time and effort and also higher costs. The last method that can be used and is currently significantly developed in mapping the area is unmanned aircraft or Unmanned Aerial Vehicles (UAV) [2]. The use of UAV can provide information on the condition or situation of a site in great detail with a very high spatial resolution and can cover a large area in a shorter time [3]. The weakness of the UAV is that the acquired elevation data is Digital Surface Model (DSM) data or surface elevation features. To improve the accuracy of UAV acquisition data, DSM can be derived into Digital Terrain Model (DTM) using Bench Mark (BM) as the ground control point (GCP).

Research on spatial flood modelling has been carried out in various cities worldwide. One of the studies was conducted to prepare hydrodynamic models and flood inundation mapping using the Geographic Information System method [4]. The unit of analysis used in the research is the river using echosounder equipment for river bathymetry mapping. The resulting research output is an integrated terrain model between land topography data (DEM spatial resolution of 30 meters) and river bathymetry for flood inundation modelling. The combination of remote sensing methods and Geographic Information Systems (GIS) is also used to map flood risk downstream of the Togo River, West Africa [5]. The unit of analysis used is Watershed, with the main output being a flood risk map. In 2018 research on flood risk assessment using LiDAR technology was carried out using administrative areas as the unit of analysis [6]. The main output target of this study is to increase the accuracy of the primary data used in making flood inundation models. LIDAR technology is one of the topographic data acquisition technologies currently developing but requires very high costs in its utilization, so it isn't easy to use. Based on previous studies and literature reviews, the researchers intend to assess the risk of flooding using a photogrammetric approach using UAV technology data to compile a spatial model of the river. Flood tracking data and bathymetry information are collected in a spatial model of flood inundation simulation as the basis for compiling a flood risk map in Kendari City, where the initial study is to develop a spatial model for the downstream area for the calculation of river capacity.

2. Method

2.1. Location

This research was carried out in Kendari City in flood-affected areas in 2013/ 2017, namely downstream of the Wanggu river. The determination of the boundaries of the study area is carried out based on historical data on flood events at the research location. The research location can be seen in Figure 1.

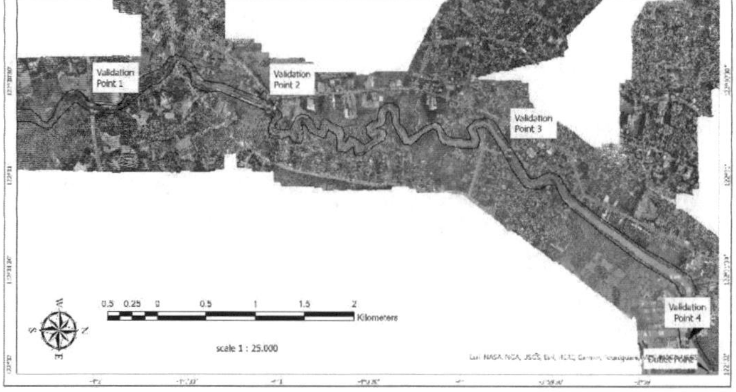

Figure 1. Research location

2.2. Terrestrial Survey
The Global Navigation Satellite System (GNSS) survey was conducted to obtain the coordinates of the Ground Control Point (GCP) for aerial photo correction, bathymetric and tidal measurement benchmarks. In general, 2 GNSS survey methods can be used, namely Real Time Kinematic (RTK) and Static methods. In principle, the measurement procedure uses a benchmark that has coordinates as a base and a location benchmark whose coordinates will be known as a rover. The command used in this study is the CORS (Continuously Operating Reference Stations) station and several stations of the nearest geodetic control point as an alternative.

2.3. Aerial Survey
The stages in mapping river landscapes using UAV technology are divided into 5: the initial survey of the location, pre-flight field work for GCP installation and UAV vehicle testing, UAV flight missions, quality check, data processing, and data extraction.

1. Initial Survey Map Site; This stage is the initial or preparation stage before data collection is carried out. The most critical aspects that must be carried out at this stage are identifying obstacles (obstacle identification), determining the position of the take-off and landing point, and designing the flight path. The trials included administrative practices, licensing arrangements, and the preparation of equipment and materials used in data collection.
2. Pre-flight fieldwork; The initial survey of the site map results is used to identify the Ground Control Point installation point (GCP) as the control point for developing orthomosaic photos. The GCP is installed on the flight path traversed by the UAV and avoids obstacles so that it is easily observed during the geometric correction of aerial photographs. Installation of GCP control points using the Global Navigation Satellite System (GNSS survey approach with the Real Time Kinematic GPS (RTK-GPS) method by taking the base at the nearest geodetic control network point from the research location (maximum distance of 2 km). The RTK-GPS method was chosen because it has a position accuracy level ranging from centimetres to decimeters [7]. At this stage, a UAV readiness check is also carried out, which includes flight altitude and camera settings.
3. Flight missions; The flight mission is carried out in several stages based on the design results (flight plan) on the Drone Deploy application. The UAV vehicle used in this mission is the DJI Mavic 2 Pro, with a maximum flight duration of 30 minutes at a safe distance of 500 meters.
4. Data Processing and Quality Check. Data processing and quality checks were carried out using ArcgisPro software. Data processing begins with flight path reconstruction, photo compilation, Ground Control Point correction, dense point cloud construction, and creating DEM and orthophoto.
5. Data Extraction. The data extracted from the results of data processing consists of raster data (aerial photos in tif and DEM format) and vector data in the form of points, lines, polygons, and objects. The data is then used as input in preparing the calculation of river capacity.

2.4. Hydrographic Survey
A hydrographic survey collects hydrographic information, including depth, currents, sediments, tides, and details of the situation and shoreline. The generalization begins with determining the boundaries of the area to be measured, preparing the route, preparing tools and vehicles (ships), and measuring tides. In general, the stages of this activity can be seen in Figure 2. In this study, the hydrographic survey focused on measuring the depth of the river and the tides, which also affect the water level in the river.

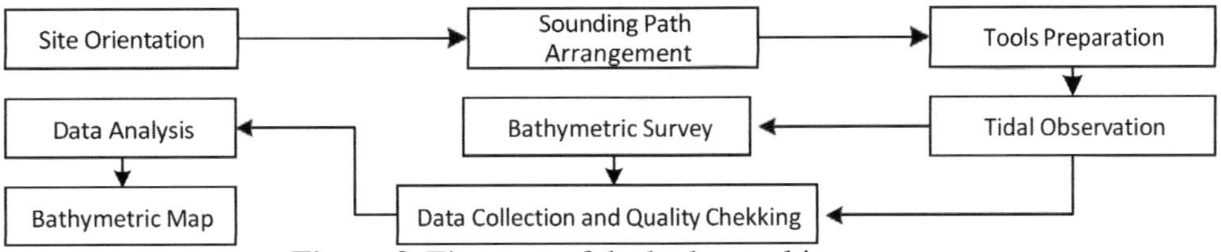

Figure 2. The steps of the hydrographic survey

3. Result and Discussion

The terrain model is a spatial model as the primary data used in calculating river capacity and flood simulations as a basis for determining flood hazards [8]. The terrain model in this study was compiled using data from aerial photo surveys, river bathymetry surveys, and terrestrial surveys/ topographic measurements directly in the field at several points. The topographic data generated from the processing of aerial photographs is then combined with bathymetric data and tested for accuracy using a sample of data from direct measurements in the field called the Independent Control Point (ICP). The graph of filed information, distribution of ICP points, and comparison of data between ICP and the result of aerial photos can be seen in Figure 3.

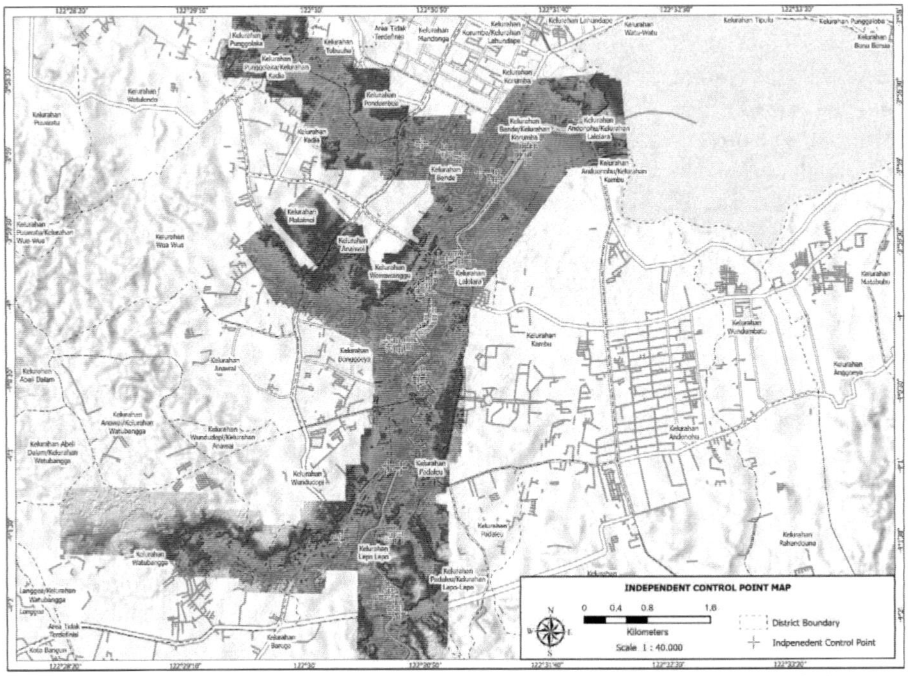

Figure 3. Terrain Model and The Distribution of Independent Control Point

Based on the Digital Surface Model (DSM) accuracy test of the downstream elevation, the average error data from aerial photography is 0,18 meters or 18 cm, and the mean absolute percentage error (MAPE) is 5.81 %. MAPE provides information on how significant the error value of the simulation data is to the measured data in the field [9]. The range of values used is the model's ability, or the prediction results are excellent if the MAPE value is less than 10 %, while the forecasting model is bad if the MAPE value is greater than 50%. Based on the data analysis, the Mean Absolute Percentage Error of terrain model elevation is 5.81% (Very Good). The comparison of field elevation data and terrain model can be seen in figure 4.

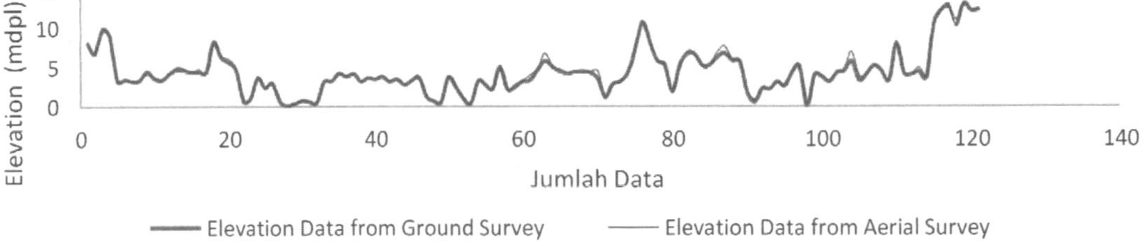

Figure 4. Comparison of Elevation Data from Aerial Photos and Field Data

The river cross-section spatial model data is then used to analyze and determine the volume of the downstream capacity of the Wanggu River. The capacity analysis is carried out by 2D Numerical simulation using Hec-Ras. Hec-Ras is one hydraulic model used widely in simulating flow patterns on the river [10]. Before carrying out the capacity analysis, the first step is to validate the simulation model to determine the appropriate computational grid size and the manning roughness coefficient according to the conditions in the field. The input data used is a cross-sectional spatial model of the river as a terrain model, upstream flow/discharge (Q25), and stage hydrograph using tidal data from secondary data. Data validation is carried out using the results of current velocity measurements at several measurement points. The current velocity measurement data is also used to determine the manning roughness coefficient at the bottom of the Wanggu River. The simulation scheme can be seen in Figure 1.

The type of riverbed at the research site is muddy soil. Based on the roughness table of the manning roughness coefficient data from the literature study results, the range of manning values for soil channels is 0.016 to 0.14. The selection of the value range was also based on the riverbed characteristics obtained from the bathymetric measurements using deeper sonar (Figure 5). Orange colour indicates the primary type of complex (hardest) channel in the form of subgrade, brown (medium), and black (smoothest). The green layer indicates the reflection of sonar waves with a weak signal indicating an organic layer above the surface of the riverbed. The simulation is carried out by inputting the discharge at the inflow and the tide at the outflow points. The results of the discharge measurement at the simulation input location can be seen in Figure 6. At the time of measuring the condition of the Kendari Bay waters towards high tide, the sea level elevation was at a position of 0.8 – 1.06 mdpl, and the measured discharge at the inflow point was 16,176 m3/s.

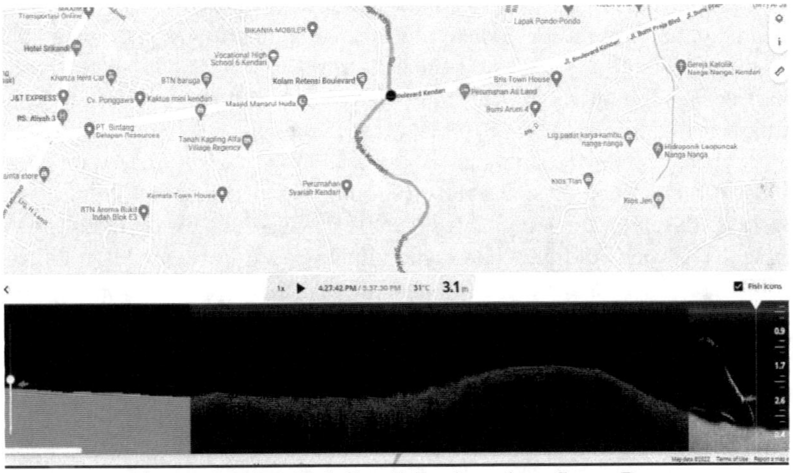

Figure 5. Riverbed Condition based on Sonar Data

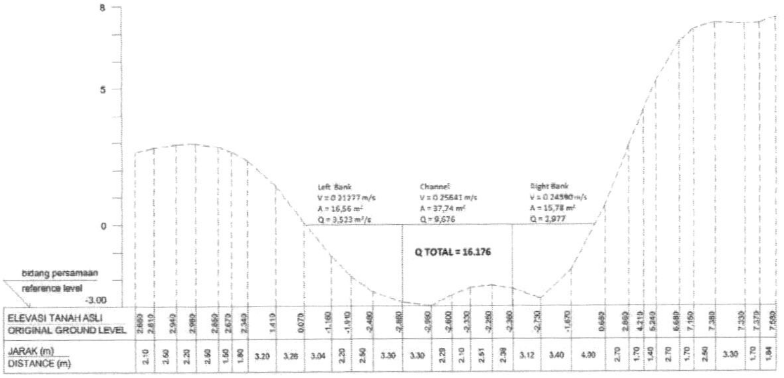

Figure 6. Cross Section at Upstream Boundary Condition Line

Validation is done by comparing the flow velocity of the measurement results in the field with the flow velocity of the simulation results at Point I and Point II. Model validation was also carried out using data from measurements in the area and evaluated using the Mean Absolute Percentage Error approach. In verifying the model, the manning roughness coefficient was also analyzed to determine the most suitable range of manning roughness for use in 2D flow simulations downstream of the Wanggu River, Kendari City. A comparison of flow velocity results from simulation and field measurements in several ranges of manning values (n = 0.01 – 0.14) can be seen in Figure 7.

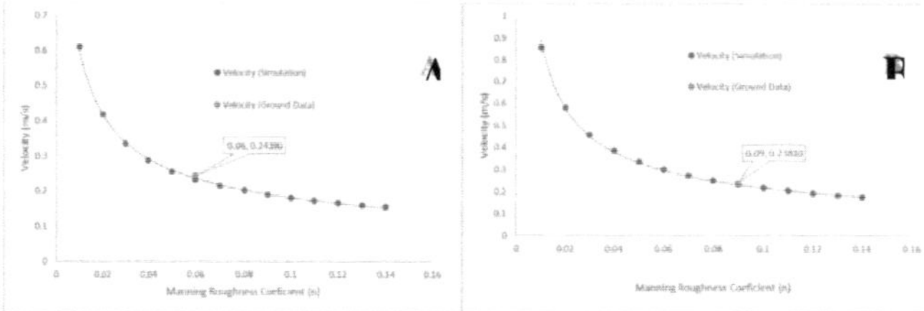

Figure 7. Comparison of Flow Velocity Simulation Results and Field Measurements in the Range n 0.01 – 0.14 Validation Points I (A) and II (B)

Based on the graph, it can be seen that at validation point I, the flow velocity of the field measurement results of 0.243 m/s has a comparable value to the simulation results using a manning coefficient of 0.06 with a Mean Absolute Percentage Error (MAPE) of 4.46%. Meanwhile, at validation point II, the flow velocity of the field measurement results has a value with the simulation results at 0.09 manning coefficient with a MAPE value of 2.56%. The data from the simulation shows that the spatial model of the downstream area of the Wanggu River resulting from aerial photography, topographic surveys, and bathymetry has a manning coefficient value range between 0.06 – 0.09. The spatial model that has been validated is then used in simulations to determine the downstream capacity of the Wanggu river. The flood discharge used is the flood discharge for 25 years in the Upper Wanggu Sub-watershed and tidal data in the downstream area. The determination of the carrying capacity is based on the simulation time, which shows an overflow in the river at 2 hours 45 minutes after the flow of the flood discharge at the 25-year return period. Based on the simulations, the river overflows when the upstream burst reaches 435,271 m^3/s, and the recorded volume accumulation in the downstream area is 1,700,766 m^3. Based on these data, it can be concluded that the maximum capacity of the downstream region of the Wanggu River is 1,700,766 m^3. In addition to the simulation method, the calculation of the river's carrying capacity is also carried out using a mathematical approach using a 25-year flood hydrograph curve graph used as input for simulation data, which can be seen in Figure 8.

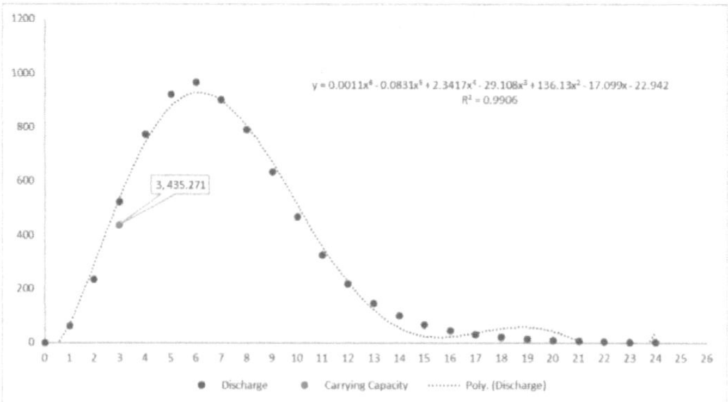

Figure 8. 25-year Return Periode Hydrograph

Based on the figure, it can be seen that the hydrograph curve approaches the polynomial equation pattern $y = 0.0011x^6 - 0.0831x^5 + 2.3417x^4 - 29.108x^3 + 136.13x^2 - 17.099x - 22.942$ f(x) with a coefficient of determination $R^2 = 99.06\%$ where the x-axis is time (t) and y-axis is discharged. From the simulation results, there was an overflow in the river when the discharge reached 435.271 (y = 435.271), precisely at the 2nd hour over 45 minutes (2.75 hours) which can be seen in the formula below. Calculate the volume of the river's capacity. It can be done by integrating the equation of the curve function concerning time on the X axis and then converting it to seconds and reducing the function at the maximum discharge during a flood event. The equations in calculating the volume of the downstream capacity of the Wanggu River are:

$$\int_{t_0}^{t_{overflow}} (f(x) - g(x) \, \partial x \quad (1)$$

Based on the calculations using the above equation, the downstream capacity of the Wanggu River is 1,683,433 m^3. Compared with the simulation data, there is a difference of 17,333 m^3 or 1.02%. Based on the results of the simulation and mathematical analysis, it can be concluded that the downstream capacity of the Wanggu River is in the range of 1,683,433 - 1,700,766 m^3.

4. Conclusion

The results showed that the horizontal accuracy of the GCP was 2.8 cm and the vertical accuracy was 6.5 cm. The results of testing the vertical elevation accuracy of aerial photographs on terrestrial topographic data measured in the field (ICP) have a Mean Absolute Percentage Error (MAPE) value of 5.81%. The manning roughness value of the spatial model is 0,06 -0,09. The river's capacity based on numerical simulations is 1.700.766 m^3, and the results of the verification using a mathematical approach are 1.683.433 m3 with a difference of 1.02%.

REFERENCES

[1] D. Molinari *et al.*, "Validation of flood risk models : current practice and possible improvements," *Int. J. Disaster Risk Reduct.*, 2018, doi: 10.1016/j.ijdrr.2018.10.022.

[2] M. La Salandra *et al.*, "Generating UAV high-resolution topographic data within a FOSS photogrammetric workflow using high-performance computing clusters," *Int. J. Appl. Earth Obs. Geoinf.*, **vol. 105**, no. June, p. 102600, 2021, doi: 10.1016/j.jag.2021.102600.

[3] Y. Watanabe and Y. Kawahara, "UAV Photogrammetry for Monitoring Changes in River Topography and Vegetation," *Procedia Eng.*, **vol. 154**, pp. 317–325, 2016, doi: 10.1016/j.proeng.2016.07.482.

[4] S. Saksena, V. Merwade, and P. J. Singhofen, "Flood inundation modelling and mapping by integrating surface and subsurface hydrology with river hydrodynamics," *J. Hydrol.*, **vol. 575**, no. June, pp. 1155–1177, 2019, doi: 10.1016/j.jhydrol.2019.06.024.

[5] J. Ntajal, B. L. Lamptey, B. Ibrahim, and B. K. Nyarko, "Flood Disaster Risk Mapping in the Lower Mono River Basin in Togo, West Africa," *Int. J. Disaster Risk Reduct.*, 2017, doi: 10.1016/j.ijdrr.2017.03.015.

[6] M. Rusnák, J. Sládek, A. Kidová, and M. Lehotský, "Template for high-resolution river landscape mapping using UAV," *Measurement*, 2017.

[7] S. Hemmelder, W. Marra, H. Markies, and S. M. De Jong, "Monitoring river morphology & bank erosion using UAV imagery – A case study of the river Buëch, Hautes-Alpes, France," *Int J Appl Earth Obs Geoinf.*, **vol. 73**, no. April, pp. 428–437, 2018, doi: 10.1016/j.jag.2018.07.016.

[8] C. Ben Khalfallah and S. Saidi, "Spatiotemporal floodplain mapping and prediction using HEC-RAS - GIS tools : Case of the Mejerda river, Tunisia," *J. African Earth Sci.*, **vol. 142**, pp. 44–51, 2018, doi: 10.1016/j.jafrearsci.2018.03.004.

[9] U. Khair, H. Fahmi, S. Al-Hakim, and R. Rahim, "Forecasting Error Calculation with Mean Absolute Deviation and Mean Absolute Percentage Error," *J. Phys. Conf. Ser.*, vol. 930, no. 1, 2017, doi: 10.1088/1742-6596/930/1/012002.

[10] R. Maatta, P. K. Joshi, and A. K. Jindal, "Solar power potential mapping in India using remote sensing inputs and environmental parameters," *Renew. Energy*, **vol. 71**, pp. 255–262, 2014, doi: 10.1016/j.renene.2014.05.037.

Experimental Study of the Effect of Morphological Changes due to the Permeable Type Groyne Structuren

A E D Massora[1*], M S Pallu[2] dan M P Hatta[3]

[1]Water Engineering, Faculty of Engineering, Hasanuddin University, Indonesian
[2]Water Engineering, Faculty of Engineering, Hasanuddin University, Indonesian
[3]Water Engineering, Faculty of Engineering, Hasanuddin University, Indonesian

Email: alfin.ekomassora@gmailcom

Abstract. The shape of a narrow river with a steep slope and high intensity of rainfall can affect the speed of the flow so that it has an impact on changes in flow dynamics that have the potential to cause erosion of the riverbed. This condition affects morphological changes with the emergence of degradation and agradation phenomena in certain areas. To determine the effect of morphological changes, laboratory tests were carried out using a permeable type groyne structure as an obstacle to reduce the flow velocity. The purpose of this study was to determine the effect of changes in basic morphology by changing the ratio of the groyne distance, which is 60 cm and 90 cm on variations in discharge and channel slope. The data required include flow velocity, flow depth, temperature and scour volume. In addition, the scour contours are drawn using Surfer software. The results showed that there was an effect of turbulent flow patterns on the use of groynes, causing local scour around the groynes and it was revealed that the use of groynes with a distance ratio of 90 cm was quite efficient in reducing scour.

Keywords: Morphological Changes, Permeable Groins, Scouring

1. Introduction

The flow that occurs in a river is usually accompanied by the occurrence of sediment transport and scouring processes. This scour process will form naturally due to the influence of river morphology or because of structures that block river flow. Meanwhile, sediment transport occurs because the flow of water in the river has a large enough energy to carry a number of sedimentary materials. If the incoming sediment is smaller than the outgoing sediment, there will be a decrease in the riverbed (degradation), but otherwise it will cause an increase in the river bed (aggradation) [1]. Characteristics of sediments that are carried away by the flow of water can be distinguished as basic sediments with bed loads and suspended loads. The bottom load moves in the river flow by rolling, sliding and jumping over the surface of the riverbed. While the floating charge consists of fine grains whose size is smaller than 0.1 mm and is constantly floating in the water flow. Moreover, very fine grains, even though the water is no longer flowing, the grains still do not settle and the water is still cloudy and this kind of sediment is called wash load [2] One of the factors that affect the amount of local scour is the depth of flow. The depth of flow will greatly affect the flow velocity that occurs. The deeper the flow that occurs, the less it is, if the depth of the flow is reduced, the velocity will increase, so that the amount of scour caused by the influence of the depth of flow will be different [3]. The handling of changes in flow velocity so far has been carried out using groynes, but it has not been an effective solution under certain conditions, so that changes in flow velocity have not been significant and have even caused new phenomena with flow turbulence resistance. The use of groynes must pay attention to the flow characteristics, if used as a flow speed regulation [4].

2. Research methods

2.1 Preparation of tools and materials

This research was conducted at the Hydraulics Laboratory of the Faculty of Engineering, Hasanuddin University with the implementation time between April to July 2021. The flume used is rectangular, width (b) 30.5 cm, height (h) 35 cm, length (L) 900 cm and equipped with inlet and outlet reservoirs as shown in Figure 1.

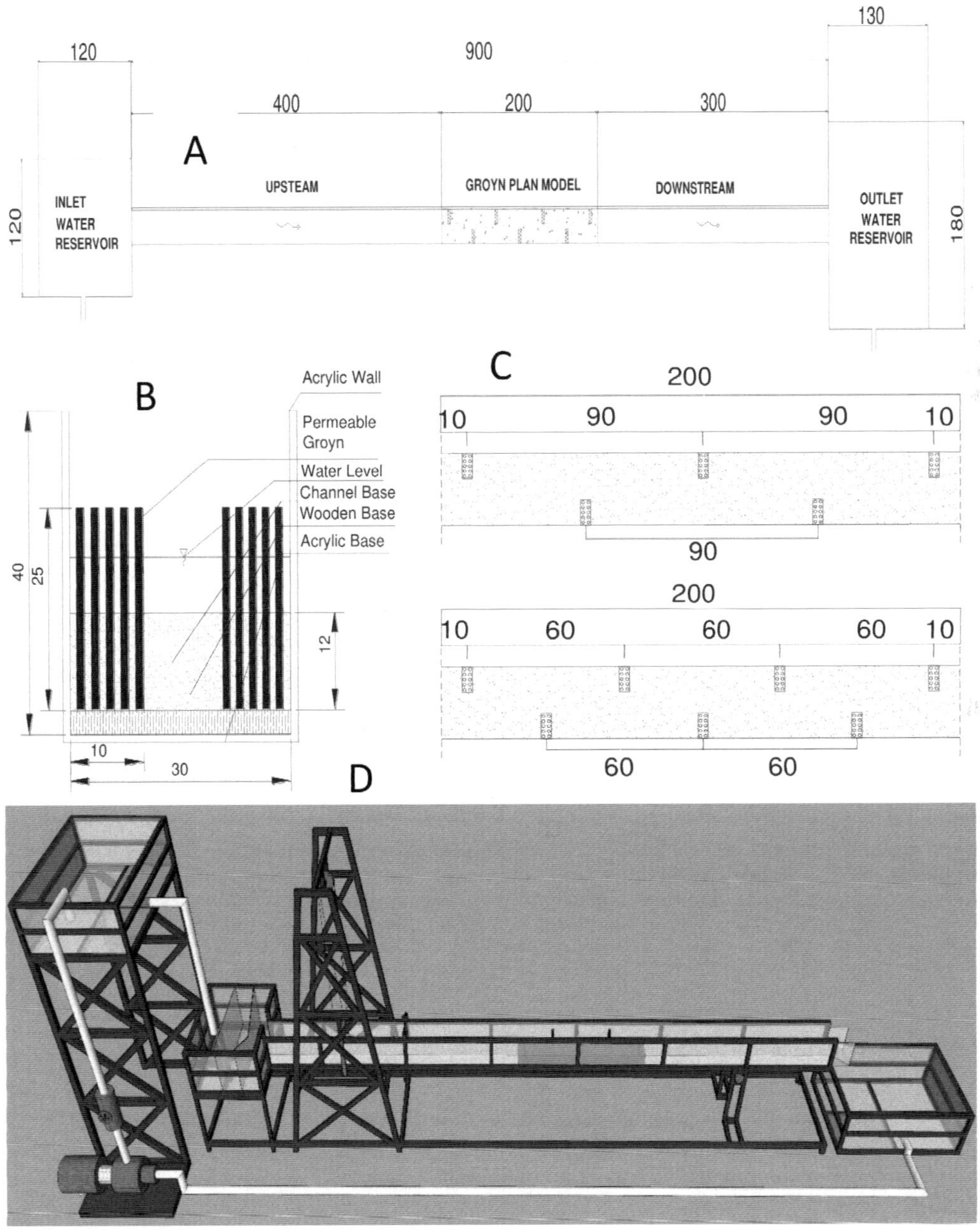

Figure 1. (a) Floor Plan of Flum Model, (b) Cross Section, (c) Installation of ribs at a distance of 60 cm; 90 cm distance and (d) 3D Flum Model

The test was carried out by laboratory experiments using flum as an open channel to observe the flow behavior on morphological changes due to the permeable groyn structure and analyze the relationship between independent variables as variables that can affect other variables. The permeable groyn material used is round wood with a diameter of 0.9 cm, a length of 20 cm and a density of 0.9 cm.

2.2 Method of collecting data

The method used in this research is experimental test. Flow testing is carried out directly to determine flow rate (Q), flow velocity (U), flow depth (h), scour depth and scour volume (vg).

2.3 Data analysis

Data obtained through flow testing is used to analyze flow characteristics and scour patterns. To determine the flow characteristics, the Froude number (Fr) and Reynods number (Re) approaches are used. In addition, scour contours are drawn using Surfer software.

3. Results and Discussion

3.1 Flow Characteristics

Based on the results of the study indicate that changes in flow velocity are influenced by the slope of the channel bottom and flow rate. Where the slope of the channel bottom is directly proportional to the flow velocity and inversely proportional to the Froude number (Fr). Flow conditions using a permeable type groyn structure indicate a change in flow pattern that causes flow turbulence. Groyn functions as a flow control and reduces flow velocity, from observations it can be seen that the groyn structure with installation distances of 0.6 m and 0.9 m can reduce the value of Froude number compared to without using Groyn. However, the use of a groyn structure can increase the value of turbulent flow so that flow resistance and back water occur around the groyn structure. Flow resistance and turbulence that have the potential to cause scour can affect changes in the morphology of the channel bottom.

Table 1. Flow Characteristics Based on Froude Number (Fr)

Channel Slope	Distance Discharge (m^3/s)	Groynes (cm)	Flow Velocity (m/s)	Water Depth (m)	Cross-Sectional Area (m^2)	Froude's Number (Fr)	Flow Characteristics
0,00056	0,0103	Without Krib	0,4266	0,078	0,0237	0,4909	Sub Critical
		0,6	0,3884	0,085	0,0258	0,4255	Sub Critical
		0,9	0,3909	0,089	0,0270	0,4193	Sub Critical
	0,0121	Without Krib	0,4315	0,084	0,0256	0,4965	Sub Critical
		0,6	0,3951	0,093	0,0283	0,4139	Sub Critical
		0,9	0,3962	0,092	0,0280	0,4175	Sub Critical
0,00189	0,0103	Without Krib	0,4734	0,073	0,0223	0,5447	Critical
		0,6	0,4308	0,082	0,0249	0,4804	Sub Critical
		0,9	0,4349	0,081	0,0246	0,4881	Sub Critical
	0,0121	Without Krib	0,4793	0,078	0,0237	0,5514	Critical
		0,6	0,4375	0,088	0,0269	0,4702	Sub Critical
		0,9	0,4414	0,090	0,0273	0,4710	Sub Critical
0,00267	0,0103	Without Krib	0,5008	0,070	0,0214	0,5762	Critical
		0,6	0,4566	0,079	0,0241	0,5177	Critical
		0,9	0,4610	0,078	0,0240	0,5241	Critical
	0,0121	Without Krib	0,5069	0,074	0,0225	0,5832	Critical
		0,6	0,4658	0,086	0,0262	0,5063	Critical
		0,9	0,4683	0,085	0,0258	0,5131	Critical

Table 2. Flow Characteristics Based on reynolds Number (Re)

Channel Slope	Distance Discharge (m^3/s)	Groynes (cm)	Flow Velocity (m/s)	Water Depth (m)	Cross-Sectional Area (m^2)	Reynold's Number (Re)	Flow Characteristics
0,00056	0,0103	Without Krib	0,4266	0,078	0,0237	24551,87	Turbulent
		0,6	0,3884	0,085	0,0258	24809,03	Turbulent
		0,9	0,3909	0,089	0,0270	25921,50	Turbulent
	0,0121	Without Krib	0,4315	0,084	0,0256	26341,34	Turbulent
		0,6	0,3951	0,093	0,0283	26674,59	Turbulent
		0,9	0,3962	0,092	0,0280	26884,34	Turbulent
0,00189	0,0103	Without Krib	0,4734	0,073	0,0223	26315,23	Turbulent
		0,6	0,4308	0,082	0,0249	26865,27	Turbulent
		0,9	0,4349	0,081	0,0246	27210,53	Turbulent
	0,0121	Without Krib	0,4793	0,078	0,0237	28139,85	Turbulent
		0,6	0,4375	0,088	0,0269	28262,32	Turbulent
		0,9	0,4414	0,090	0,0273	29610,75	Turbulent
0,00267	0,0103	Without Krib	0,5008	0,070	0,0214	27057,06	Turbulent
		0,6	0,4566	0,079	0,0241	27829,94	Turbulent
		0,9	0,4610	0,078	0,0240	28105,94	Turbulent
	0,0121	Without Krib	0,5069	0,074	0,0225	28775,82	Turbulent
		0,6	0,4658	0,086	0,0262	29693,11	Turbulent
		0,9	0,4683	0,085	0,0258	30275,59	Turbulent

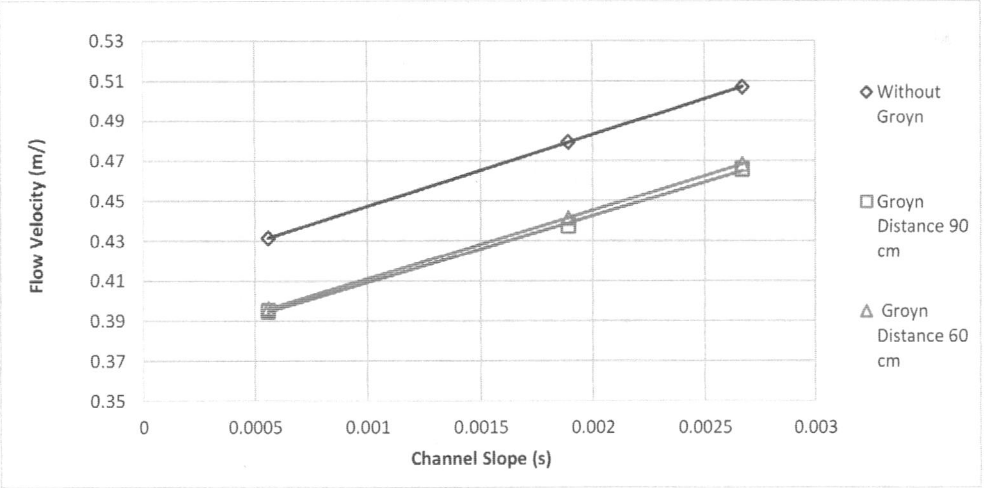

Figure 2. Graph relationship of Channel Slope (s) with Flow Velocity at Discharge 0.0103 m^3/s

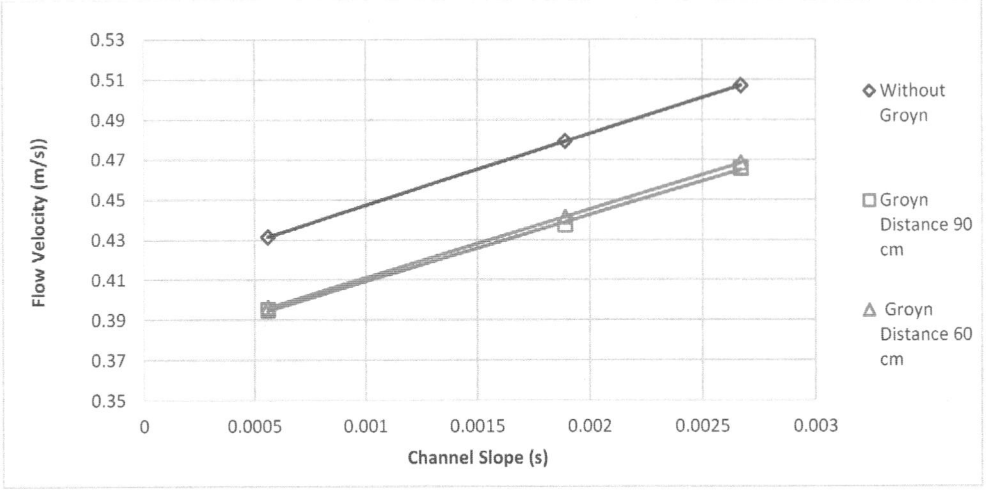

Figure 3. Graph relationship of Channel Slope (s) with Flow Velocity at Discharge 0.0121 m^3/s

Figures 2 and 3 show the effect of the channel bottom slope (s) on the flow velocity (U). Where the greater the slope of the channel bottom, the flow velocity increases. The use of groynes in reducing flow velocity is very effective, as can be seen from the observation that the value of flow velocity with a permeable groyne structure is smaller than the flow velocity without using a groyne. As a flow control structure, groyns can be used in rivers or canals that have heavy currents. This is due to the existence of obstacles to restrain the flow velocity so that it allows a change in the flow pattern from critical to sub critical.

3.2 Scouring Volume

The volume of scour and sediment can be calculated based on the change in the cross-sectional area of the channel and the change in weight of the initial sediment minus the weight of the final sediment (after flow). The following are the results of observations of scour volume using groynes and without using gryons on variations in slope and flow discharge.

Table 3. Recapitulation of scour volume

Channel Slope	Distance Discharge (m^3/s	Groynes (m)	Flow Velocity (m/s)	Water Depth (m)	Soil Volume (m^3)	Scour Volume (m^3)	Scour Precentage (%)
0,00056	0,0103	Without Krib	0,4266	0,078	0,0472	0,00407	8,62
		0,6	0,3884	0,085	0,0359	0,00744	20,71
		0,9	0,3909	0,089	0,0358	0,00706	19,71
	0,0121	Without Krib	0,4315	0,084	0,0477	0,00482	10,09
		0,6	0,3951	0,093	0,0353	0,00916	25,97
		0,9	0,3962	0,092	0,0369	0,00739	20,02
0,00189	0,0103	Without Krib	0,4734	0,073	0,0474	0,00477	10,07
		0,6	0,4308	0,082	0,0364	0,00832	22,89
		0,9	0,4349	0,081	0,0364	0,00827	22,72
	0,0121	Without Krib	0,4793	0,078	0,0508	0,00555	10,93
		0,6	0,4375	0,088	0,0361	0,01048	29,03

		0,9	0,4414	0,090	0,0360	0,00874	24,27
	0,0103	Without Krib	0,5008	0,070	0,0474	0,00523	11,05
		0,6	0,4566	0,079	0,0358	0,00920	25,67
		0,9	0,4610	0,078	0,0369	0,00919	24,88
0,00267	0,0121	Without Krib	0,5069	0,074	0,0503	0,00595	11,81
		0,6	0,4658	0,086	0,0359	0,01166	32,49
		0,9	0,4683	0,085	0,0362	0,00952	26,30

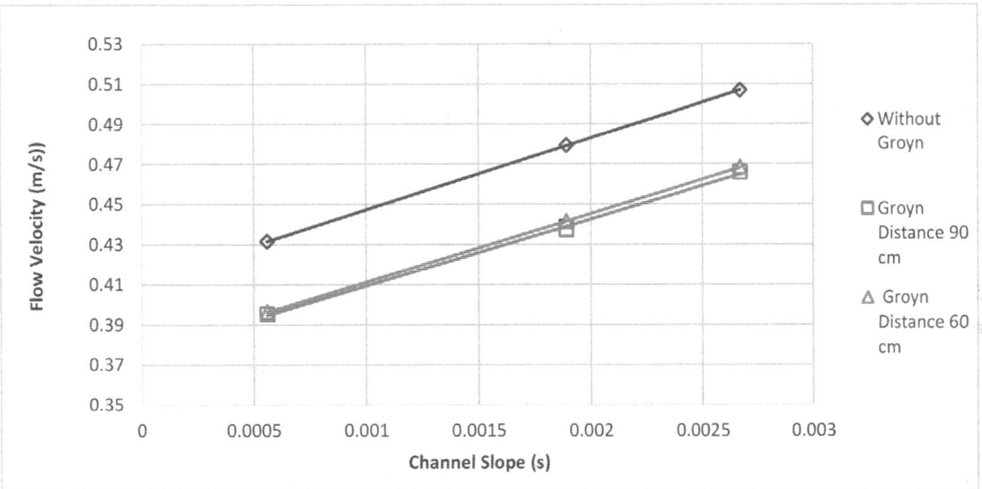

Figure 4. Relationship of Channel Slope (s) with Scouring Volume at Discharge of 0.0103 (m^3/s)

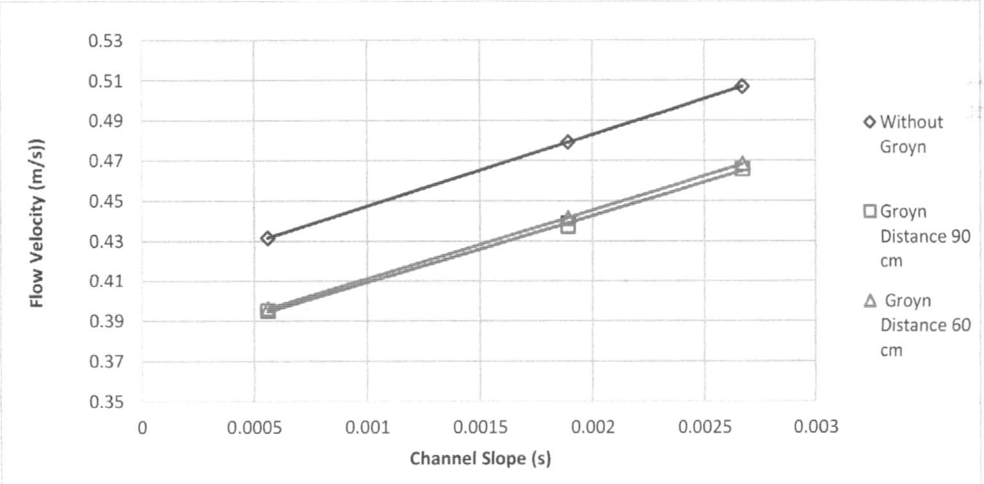

Figure 5. Relationship of Channel Slope (s) with Scouring Volume at Discharge of 0.0121 (m^3/s)

Based on Figures 2 and 3, it shows the effect of channel bed slope (s) on the scour volume (vg) using a groyne and without using a groyne. The results of the analysis stated that the use of groyns with an installation distance of 0.6 m experienced an increase in scour compared to the installation distance of 0.9 m. This is due to the pattern of flow resistance and flow turbulence at the groyn installation distance, thereby increasing the value of the basic shear stress. This condition affects the basic morphology changes because the scour rate is very high compared to without the use of groyns.

Therefore, a special study is needed to determine the type of groyne building that can reduce flow and reduce the level of scour so that morphological changes do not affect channels and rivers.

3.3 Scouring Pattern

The formation of the scour contour pattern can be described using the Surfer software as shown in the image below.

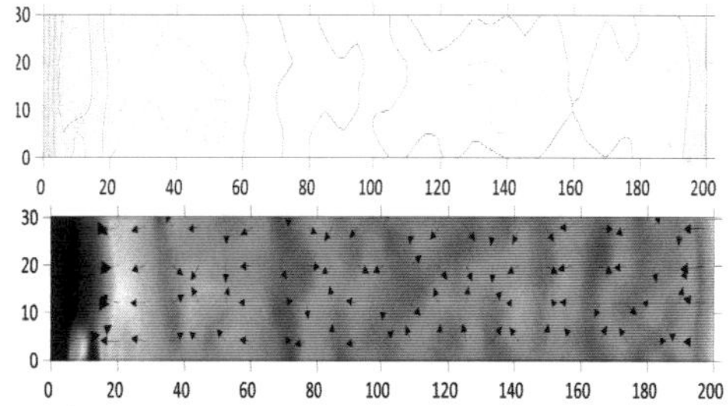

Figure 6. Scour Contour without Groyne on S = 0,00056 and Q = 0,0103 m^3/s

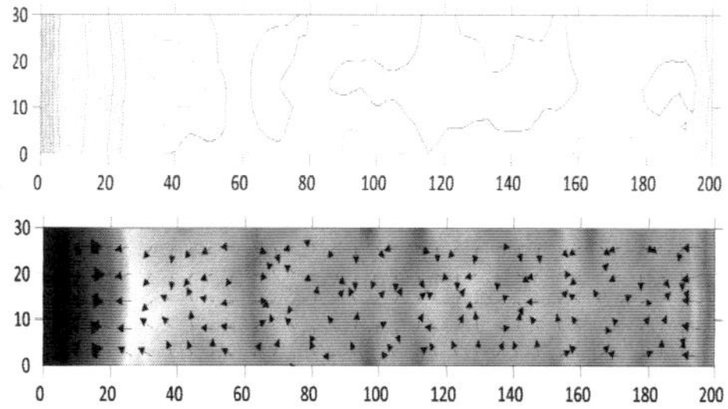

Figure 7. Scour Contour without Groyne on S = 0,00189 and Q = 0,0103 m^3/s

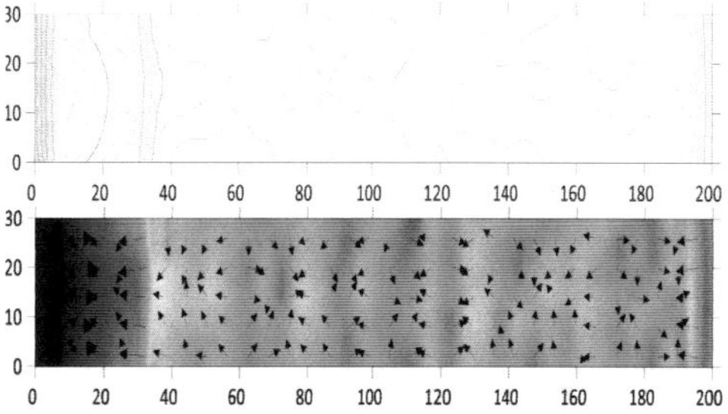

Figure 8. Scour Contour without Groyne on S = 0,00267 and Q = 0,0103 m^3/s

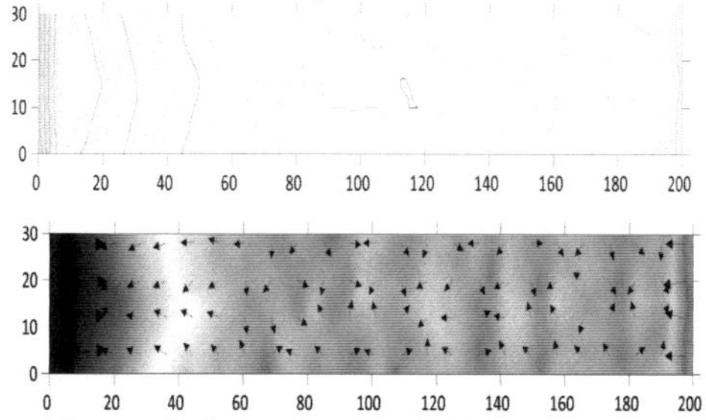

Figure 9. Scour Contour without Groyne on S = 0,00056 and Q = 0,0121 m^3/s

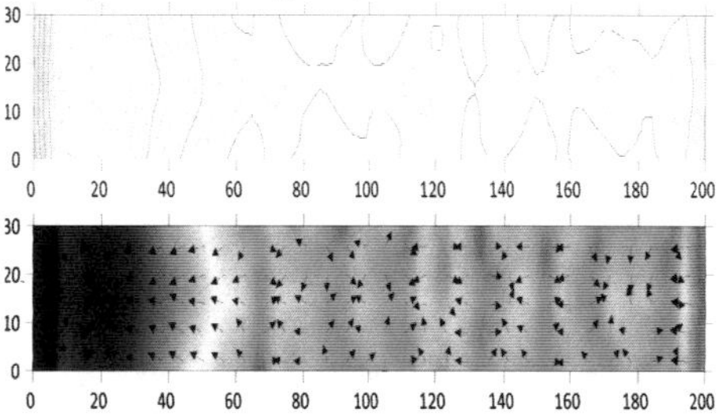

Figure 10. Scour Contour without Groyne on S = 0,00189 and Q = 0,0121 m^3/s

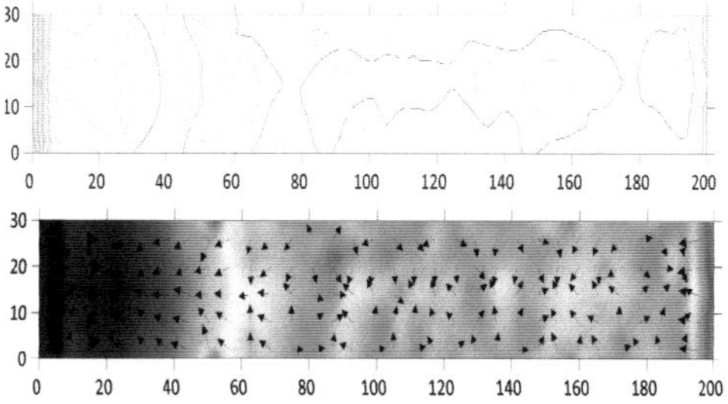

Figure 11. Scour Contour without Groyne on S = 0,00267 and Q = 0,0121 m^3/s

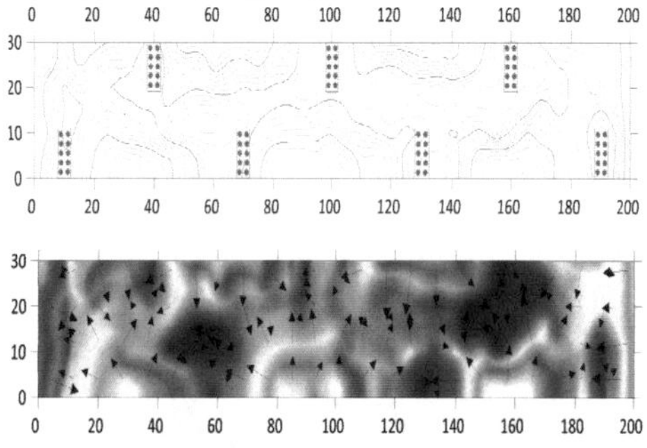

Figure 12. Scouring Contour with Groyne Distance 60 cm, S = 0,00056 ,Q = 0,0103 m^3/dtk

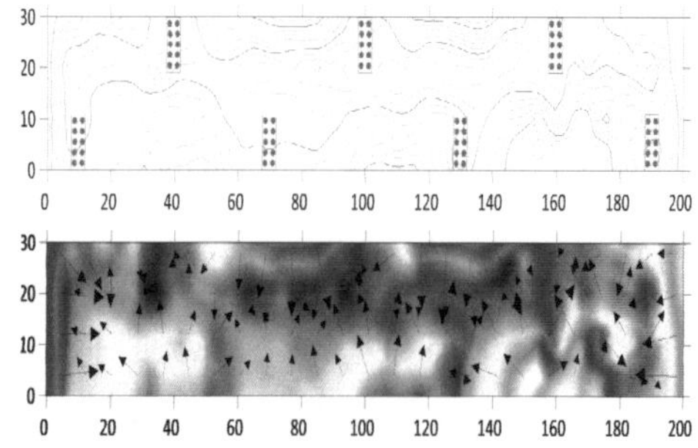

Figure 13. Scouring Contour with Groyne Distance 60 cm, S = 0,00189, Q = 0,0103 m^3/s

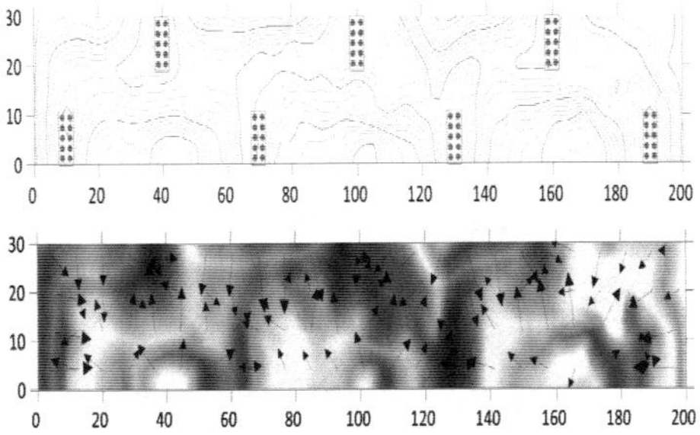

Figure 14. Scouring Contour with Groyne Distance 60 cm, S = 0,00267,Q = 0,0103 m^3/dtk

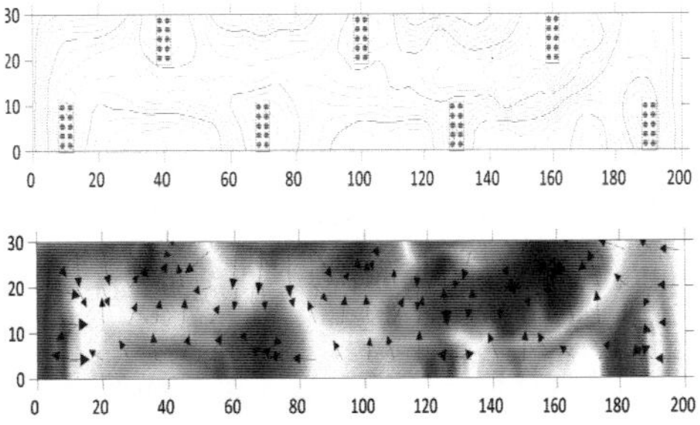

Figure 15. Scouring Contour with Groyne Distance 60 cm, S = 0,00056, Q = 0,0121 m^3/s

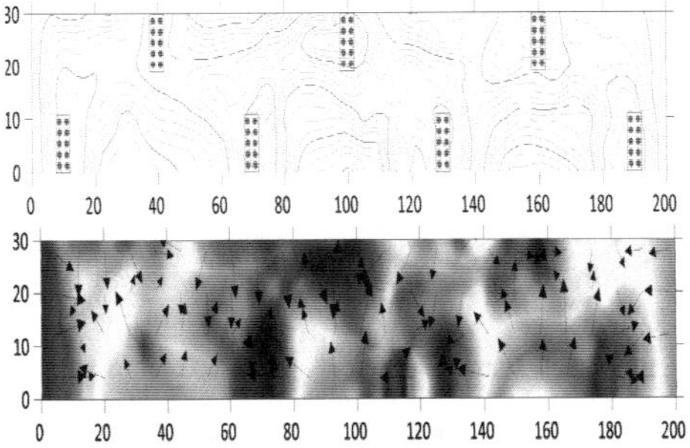

Figure 16. Scouring Contour with Groyne Distance 60 cm, S = 0,00189 ,Q = 0,0121 m^3/dtk

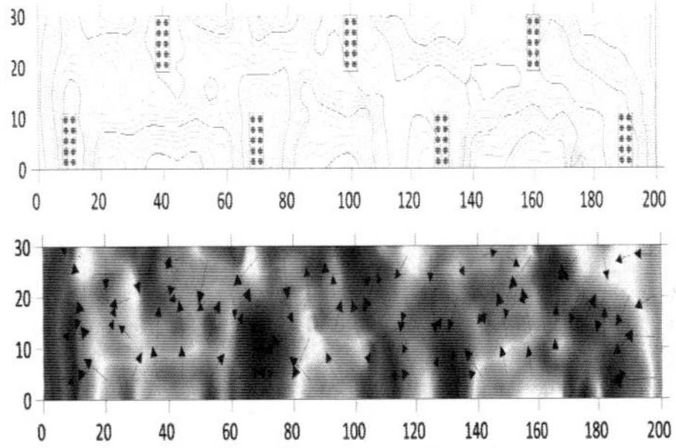

Figure 17. Scouring Contour with Groyne Distance 60 cm, S = 0,00267, Q = 0,0121 m^3/s

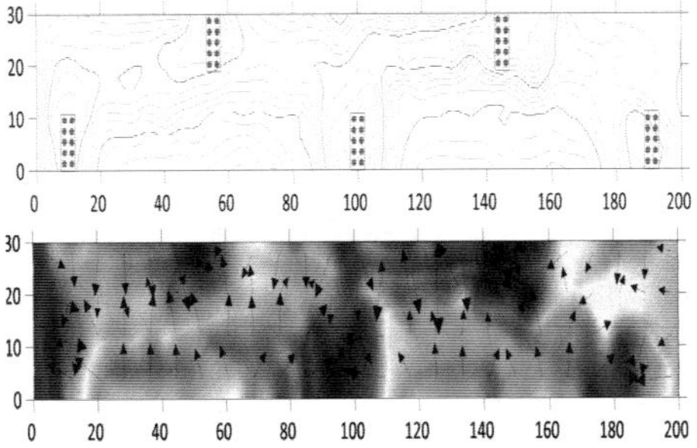

Figure 18. Scouring Contour with Groyne Distance 90 cm, S = 0,00056 ,Q = 0,0103 m^3/dtk

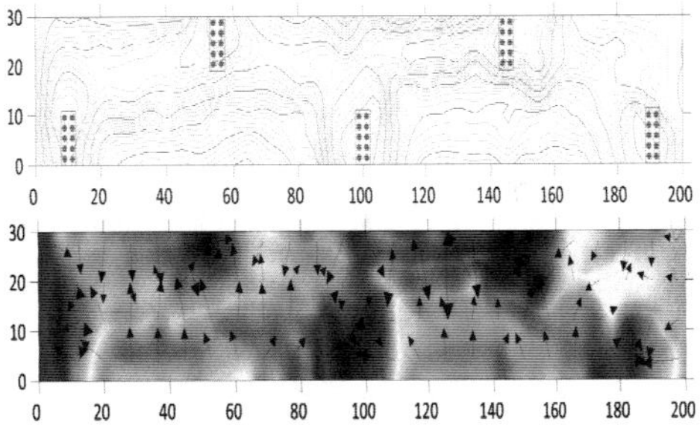

Figure 19. Scouring Contour with Groyne Distance 90 cm, S = 0,00189, Q = 0,0103 m^3/s

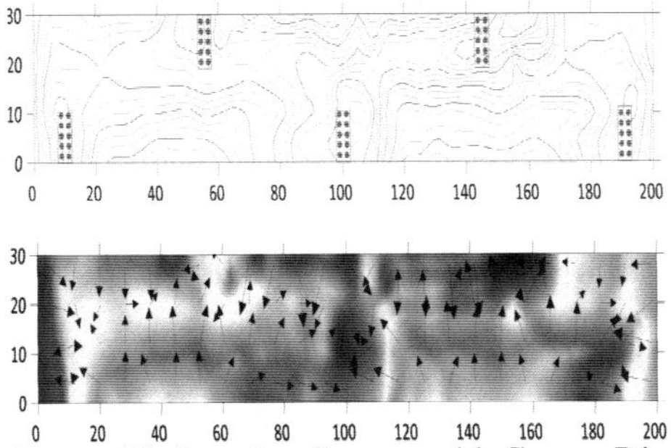

Figure 20. Scouring Contour with Groyne Distance S = 0,00267 ,Q = 0,0103 m^3/dtk

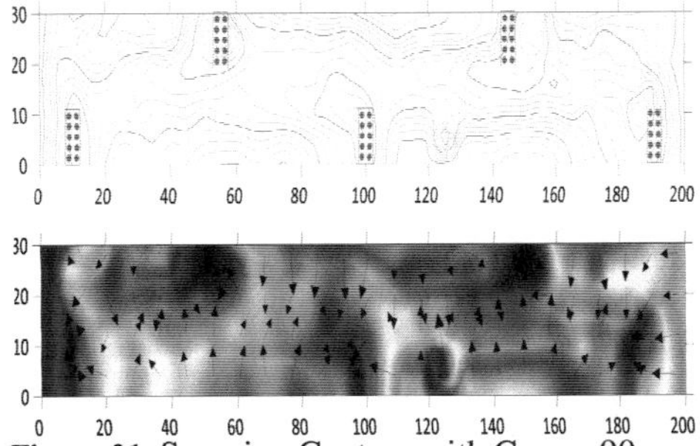

Figure 21. Scouring Contour with Groyne 90 cm, Distance 90 cm, S = 0,00056, Q = 0,0121 m^3/s

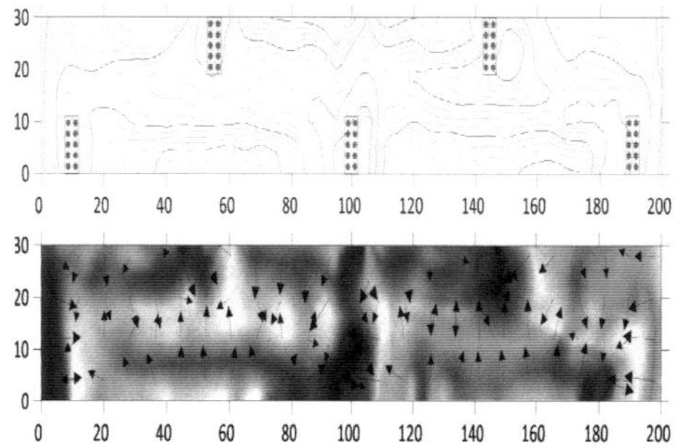

Figure 22. Scouring Contour with Groyne Distance S = 0,00189 ,Q = 0,0121 m^3/dtk

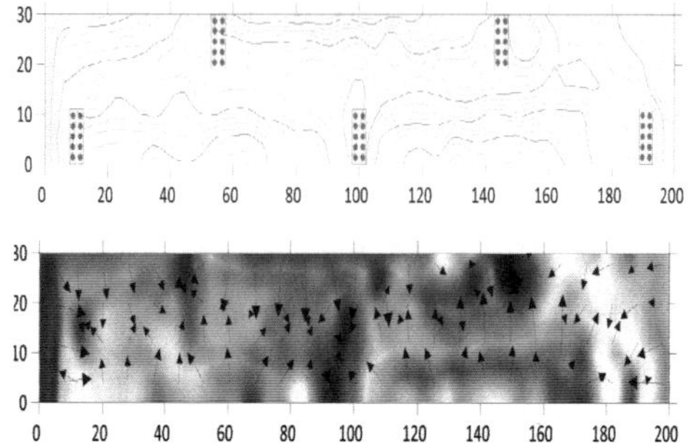

Figure 23. Scouring Contour with Groyne 90 cm, Distance 90 cm, S = 0,00267, Q = 0,0121 m^3/s

The result showed that there was an effect of morphological changes in the instalition of groynes at a distance of 60 cm and 90 cm, there was basic scour around the groynes. So that the scour problem still needs to be investigated by varing the distance and slope of the groynes.

4. Conclusion

The use of a permeable ground type causes the phenomenon of flow resistence and turbulence which has the potentialto cause basic scour around the ground for ground installation with a distance of 90 cm. Has a design that is more effective in reducing scour compared to using green at a distance od 90 cm.

References

[1] I. Eka S. Yulia, Agus Kironoto, "B3.Civil_Engineering_Vol_XV_1.pdf." Civil Engineering Department Faculty of Engineering Gadjah Mada University, Yogyakarta, hal. 1, 2005.

[2] dkk Ir. M, Gayo Yusuf, *Perbaikan dan Pengaturan Sungai*, I. Jakarta: PT Pradnya Paramita, 2008.

[3] F. Teknik *et al.*, "181 dari 272," *Pengaruh Tinggi Bukaan Pintu Air Terhadap Bilangan Froude Dengan Dasar Tanah Lempeng Pada Saluran Terbuka*, hal. 181–186, 2019.

[4] M. S. Pallu *et al.*, "Analisis pengaruh perubahan kecepatan aliran akibat krib permeabel pada saluran terbuka," *Anal. Pengaruh Perubahan Kecepatan Aliran Akibat Krib Permeabel Pada Saluran Terbuka*, hal. 202–210, 2021.

Engineering and Traffic Management Study for the Operation of the Makassar Elevated Toll Road

A I Kuku[1*], M I Ramli[2], S Rauf[3]

[1] Transportation Systems, Faculty of Engineering, Hasanuddin University
[2] Transportation Engineering, Faculty of Engineering, Hasanuddin University
[3] Transportation Engineering, Faculty of Engineering, Hasanuddin University

Abstrak. The development of the area, especially the city of Makassar, has a significant impact on traffic activities which causes traffic problems. For this reason, this study aims to see how much influence the operation of the Makassar elevated toll road has on the surrounding traffic, especially at the intersection of Jln. A.P. Pettarani - Jln. Boulevard and Jln A.P. Pettarani - Jln. Let. Jend. Hertasning. The data required in this study include surveys of traffic volume, vehicle speed, queue length, and geometric road data conducted on Tuesdays, Wednesdays, Saturdays, and Sundays, with survey periods starting at 07.00 – 09.00, 12.00 – 14.00, and 16.00 – 18.00. Engineering studies in the form of traffic simulations using vision software. From the results of traffic simulations on Wednesday, there were three intersections with the most significant queue length occurring at 16.00-18.00, namely the Boulevard short of 273.85 meters on Wednesday and 222.71 meters on Saturday. The application of a traffic design with an alternative change in traffic signal time at the A.P. Pettarani-Boulevard interchange was reduced by 26.21 meters and 23.75 meters from the existing conditions on Wednesday and Saturday. And for the alternative application of changing the direction of traffic movement at the A.P. Pettarani – Let. Jend. Hertasning, the decrease in the length of vehicle queues on Boulevard short was 21.31 meters and 17.56 from the existing conditions on Wednesday and Saturday.

Kata kunci : Rekayasa lalu lintas, Vissim, Tol layang makassar, Transportasi

1. Introduction

The development of the particular area of Makassar city significantly impacts traffic activities that cause traffic problems. The growth in the number of vehicles that is not proportional to the increase in road volume, which tends to be static, results in a slowdown in congestion on various road sections; therefore, the performance of an intersection is the main factor in determining the most appropriate handling to optimize the function of the corner. [1]. The rapid growth and development of a city without being followed by the procurement of an adequate transportation system is a form of significant demand for transportation needs compared to the provision of a transportation system, and vice versa, the rapid growth rate of the transportation system that does not match the size of a city's development is a form of provision that is so large from its demand [2]. There are several access roads to the center of Makassar, such as Jalan A.P. Pettarani is one of the links to the city of Makassar, which often occurs traffic problems, including congestion at several intersection points along the A.P. Pettarani road. The access road to the city center is currently inadequate despite traffic engineering, so an alternative route is needed, which can connect the city center with areas in the north and south of the city [3].

An intersection is part of a road network system that connects two or more roads at one point so that at a meeting, it can cause vehicle conflicts and cause the entire road network to become congested [4]. The Makassar passes several intersection points elevated toll road, especially the A.P. Pettarani road intersection – Boulevard road and the A.P. Pettarani road intersection – Alauddin road, which is a point for collecting data on traffic volume surveys, vehicle speed, queue length, and geometric road measurements because it is an intersection with many conflict points or existing traffic problems so that it makes the corner very much in need of handling in terms of engineering traffic and management of handling traffic problems. Currently, the elevated toll road will be operated in the city of Makassar, which is a link between the center of Makassar city and several surrounding areas as a transportation facility that can unravel traffic problems. Toll road implementation aims to improve the efficiency of distribution services to support economic improvement, especially in areas where the level of development is already high. [5].

The A.P. Pettarani road section is one of the main roads in Makassar city, so the activities on the road section are classified as congested; traffic management and engineering are needed to improve road section services and several intersection points as a whole. At specific points along the A.P. Pettarani road section, which may be disturbed due to access to exits and toll entrances, several traffic flows change direction from the network around the toll road to the toll road or vice versa was taken into account. Traffic engineering studies need to be carried out due to changes in the performance of the A.P. Pettarani road after the Makassar elevated toll road and studies related to traffic management in line with the increase in volume that will occur at a certain point after the opening of the A.P. Pettarani toll gate, to improve road performance due to the influence of existing traffic movements on the new road network system after the existence of the A.P. Pettarani toll road.

This study used *vissim* software as a tool to model traffic flow. The concept of simulation models is very often used in planning a transportation activity, especially those that are dynamic and very broad such as traffic conditions and situations that have a variety of characteristics and many parameters, so a simulation model approach is needed as a form of simplification of a complex problem. [6]. In *vissim* modeling requires several survey data parameters, namely recording vehicle traffic volume, vehicle speed data, vehicle queue length data, and geometric road data. *Vissim* is a microscopic simulation program used to model multi-modal traffic flow; this software can perform simulations for urban areas and highway traffic, including pedestrians, cyclists, and motor vehicles [7]. The construction of the Makassar elevated toll road cannot be separated from the influence of the existing road network system, namely, at several intersections along the A.P. Pettarani road, namely the A.P. Pettarani road intersection – Boulevard road and the A.P. Pettarani interchange – Let. Jend. Hertasning which became a traffic survey point for studies for traffic engineering and management after the introduction of the Makassar elevated toll road.

2. Research design

This research was conducted in Makassar City, and this type of research uses a descriptive approach to see the amount of traffic volume at a predetermined point and is simulated using PTV *Vissim*. To determine the alternatives of traffic engineering to be carried out.

3. Data collection methods

The survey method carried out in the study was surveying traffic to see the amount of traffic volume at predetermined points, namely the A.P. Pettarani road intersection-Boulevard road and the A.P. Pettarani road intersection- Let. Jend. Hertasning and simulated with PTV *Vissim*. For the selection of traffic engineering alternatives to be carried out.

The type of survey that will be carried out before being simulated with PTV *Vissim*, namely a geometric survey of intersections, is carried out to obtain information about land use conditions and dimensions at intersections. A survey of vehicle volumes where this survey is carried out to calculate the volume of vehicles according to the classification that has been determined in each intersection using surveyor power and distributed to several survey points in order to get results a good survey. A vehicle speed survey is carried out in conjunction with a vehicle volume survey. Finally, there is a queue length survey where this survey is carried out by looking at the queue length of vehicles that are waiting or queuing at intersections and then tracking the length of vehicle queues.

4. Data analysis

The data collected through traffic surveys are then analyzed using software such as *Ms. Excel* and PTV *Vissim*, which are used as simulator tools that will later produce interchange performance. For initial data analysis is to process data that has been obtained from direct surveys in the field, such as geometric interchange data, vehicle volume, speed data, and queue length data, then tidied up and recapitulated using Ms. Excel in the form of tables and graphs, then simulated using PTV *Vissim* software to get interchange simulation outputs that match the conditions in the field.

5. Traffic performance results

In the vehicle volume survey conducted, there are 18 types of vehicles categorized based on four categories of Vehicles, Light Vehicles (LV), Heavy Vehicles (HV), Motorcycles (Motorcycle (MC)), Motorized Vehicles (Unmotorized (UM)), where the categories of vehicles that are included in light vehicles (LV) are small city cars, big city cars, sedans, mpvs, SUVs, mini buses, pick ups, for the category of heavy vehicles (HV) namely small buses, medium buses, large buses, truck two axles, truck three axles, truck four axles, trailers, and motorcycles categorized as automatic motorcycles, duck motorcycles, and sports motorcycles.

Traffic volume at each survey point is at the intersection of A.P. Pettarani - Boulevard and A.P. Pettarani - Let. Jend. Hertasning, which is calculated throughout the period from 07.00 – 09.00, 12.00 – 14.00, and 16.00 – 18.00 WITA on Tuesday, Wednesday, Saturday, and Sunday, the highest traffic volume on weekdays is obtained on Wednesday with the highest vehicle volume and on holidays is on Saturday.

Based on the most significant traffic volume that occurs on weekdays occurred on Wednesday with 43325 vehicles dominated by two-wheeled vehicles with a percentage of 71.07% Mc, 28.30% Hv, and 0.63%, and the most significant traffic volume that occurred on holidays was on Saturdays, namely 39520 vehicles which were also dominated by two-wheeled vehicles with a percentage of 67.10% Mc, 32.26% Hv and 0.64%, so the conclusion is drawn in the analysis of later vehicle volumes on Wednesdays and Saturdays which will later be used based on the most significant volume of traffic on weekdays and holidays.

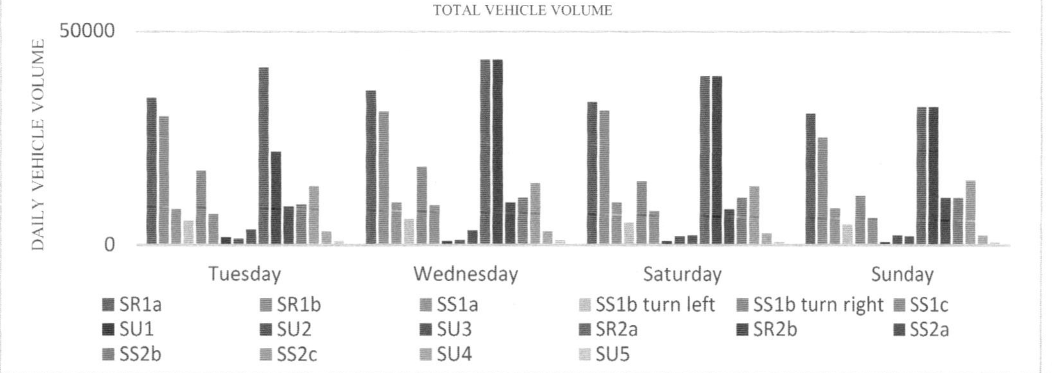

Figure 1. Total vehicle volume at each intersection

IOP Conf. Series: Earth and Environmental Science 1117 (2022) 012074 doi:10.1088/1755-1315/1117/1/012074

6. Traffic condition simulation results

Based on the calibration results for calibration results with *Geoffrey E. Havers* tests for all devices in each period, namely for the peak hour period of the morning, afternoon, and evening peak hours are qualified, where the *GEH* test value obtained is < 5, which means that the simulation model is acceptable or has been calibrated, but by looking at the peak hours of the traffic density level based on the length of the vehicle queue, Peak time occurs at 16.00 - 18.00. here are the calibration results with the GEH test for all shorts of intersection on Wednesday afternoon.

Table 1. Calibration results for Wednesday afternoon time with Geoffrey E. Havers test.

intersection arm	Volume (vehicle/hour)		Uji *GEH*	Conclusion
	Vissim model	Observation		
Urip 1	5490	5802	4.15	Accepted
Boulevard	4036	4348	4.82	Accepted
Alauddin 2	4719	5053	4.78	Accepted
Urip 2	4925	5204	3.92	Accepted
Urip 4	4891	5230	4.77	Accepted
Hertasning	2121	2147	0.56	Accepted
Alauddin 5	6168	6388	2.78	Accepted

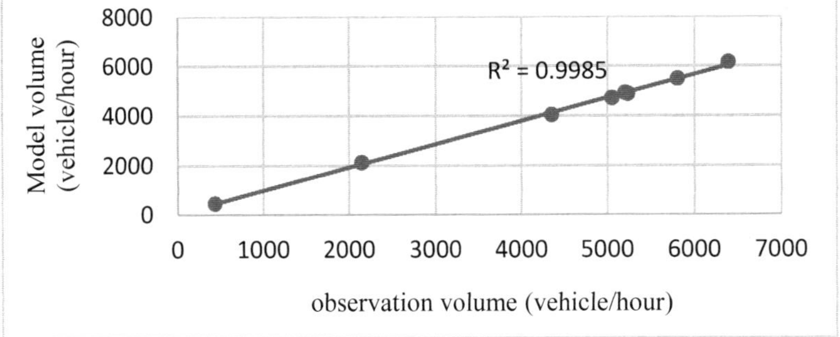

Figure 2. Graph of the relationship of the volume of the model with the results of observations

Here are the calibration results obtained with the Geoffrey E. Havers test for all interchange shorts on Saturday afternoon

Table 2. Calibration results for Saturday afternoon time with Geoffrey E. Havers test.

intersection arm	Volume (vehicle/hour)		Uji *GEH*	Conclusion
	Vissim model	Observation		
Urip 1	5254	5607	4.79	Accepted
Boulevard	3930	4226	4.64	Accepted
Alauddin 2	3937	3942	0.08	Accepted
Urip 2	4217	4535	4.81	Accepted
Urip 4	4287	4601	4.71	Accepted
Hertasning	2139	2164	0.54	Accepted
Alauddin 5	5077	5170	1.30	Accepted

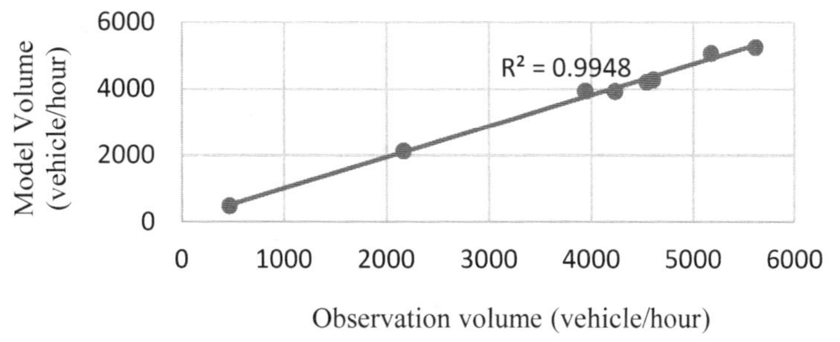

Figure 3. Graph of the relationship of the volume of the model with the results of observations

From the results of traffic simulations using *vissim* software that was calibrated on Wednesday, there are three intersection shorts with a maximum queue length that occurs at 16.00-18.00, namely the Urip Sumoharjo Road intersection with a queue length of 177.68 meters, the Boulevard Road intersection short of 245.90 meters, the Alauddin Road intersection short of 132.40 meters, and the traffic simulation on Saturday short of the Urip Sumoharjo Road intersection of 134.70 meters, the intersection of Jalan Boulevard is 200.20 meters. The short intersection of Jalan Alauddin is 109.60 meters.

Figure 4. The result of the calibration of the A.P. Pettarani - Boulevard interchange

Figure 5. The calibration results of the A.P. Pettarani – Let. Jend. Hertasning interchange

After calibrating traffic simulation data, the queue length is obtained at each intersection at the Boulevard - A.P. Pettarani road intersection and the A.P. Pettarani-Let. Jend. Hertasning, based on the results of traffic simulation data validation.

Table 3. Queue Length maximum vissim model vehicles and observation results on Wednesday

Period	Result (Maximum Queue Length (meters))					
	Model			Observation		
	Urip	Boulevard	Alauddin	Urip	Boulevard	Alauddin
08.00 - 09.00	100	91	82	90	84	76
13.00 - 14.00	111	150	86	97	136	76
17.00 - 18.00	178	274	155	159	246	132

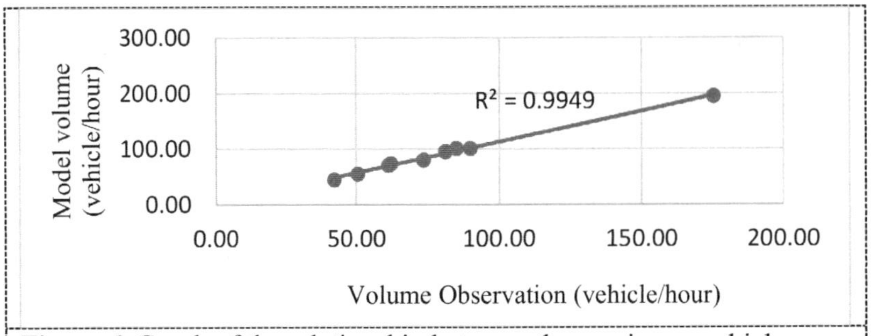

Figure 6. Graph of the relationship between the maximum vehicle queue length of observation results and vissim modeling on Wednesday

The average and maximum queue length on Wednesday was based on the chi-square test results with a Chi-square test confidence degree of 95% or a = 0.05, where the table value x2 in the Chi-square table was 9,488 while the chi-square table calculation result value for all the shorts qualified x^2 of the calculation result ≤ x^2 chi-square table result so that the model is declared valid.

Table 4. Chi-square Test Validation Results for Vehicle Rmaksimum Queue Length on Wednesda

Chi-Square Results		
Urip	Boulevard	Alauddin
1.05	0.53	0.45
1.83	1.47	1.1
1.88	2.85	3.27
4.76	4.85	4.82

Table 5. Maximum Queue Length of vissim model vehicles and observation results on Saturday

Period	Result (Maximum Queue Length (meters))					
	Model			Observation		
	Urip	Boulevard	Alauddin	Urip	Boulevard	Alauddin
08.00 - 09.00	85	77	70	76	71	64
13.00 - 14.00	94	88	80	84	79	71
17.00 - 18.00	152	223	127	135	200	110

Table 6. Chi-square Test Validation Results for Queue Length of Vehicles on Saturday

Chi-Square Results		
Urip	Boulevard	Alauddin
0.91	0.46	0.39
1.06	0.95	1.02
1.93	2.27	2.27
3.90	3.68	3.68

The average and maximum queue length on Saturdays is based on the chi-square test results with a Chi-square test confidence degree of 95% or a = 0.05, where the table x2 value on the Chi-square table is 9,488 while the chi-square table count result value for all the shorts meets the requirements x^2 of the calculation result ≤ x^2 chi-square table result so that the model is declared valid.

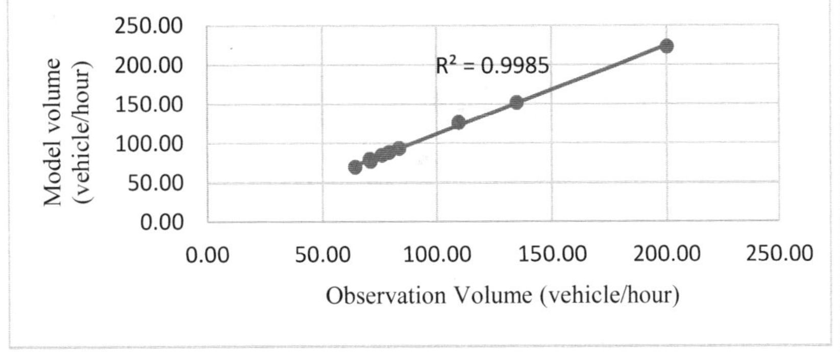

Figure 7. Graph of the relationship between the maximum vehicle queue length of the observation results and the vissim modeling on Saturday

Several intersection points with a significant level of vehicle volume can be seen using a calibrated traffic simulation so that traffic simulation results are obtained that are as similar as possible based on the results of the *vissim software* simulation..

Figure 8. The output of traffic simulation results from the A.P. Pettarani - Boulevard interchange.

Figure 9. The output of traffic simulation results from the A.P. Pettarani – Let. Jend. Hertasning interchange.

7. Traffic Engineering Alternatives

Hasil simulasi lalu lintas terdapat beberapa titik pendekat simpang yang menjadi pusat permasalahan lalu lintas pada simpang jalan A.P. Pettarani – jalan Boulevard dan jalan A.P. Pettarani – Let. Jend. Hertasning sehingga memerlukan rekayasa lalu lintas berupa alternatif rekayasa lalu lintas untuk mengurai permasalahan lalu lintas yang terjadi.

7.1 Alternatives 1

Traffic engineering at the A.P. Pettarani road intersection – Boulevard road against traffic movement for the first alternative is with a scenario of changing the time of the traffic cycle, where the green light and red light times increase per-phase.

Table 7. Alternative signal for traffic engineering A.P. Pettarani road interchange – Boulevard road

Alternative	intersection arm	Signal Time			Cycle Time (sec)
		green	red	yellow	
Existing	Urip Sumoharjo road short/phase I	40	78	2	120
	Boulevard road short/phase II	33	86	2	120

	Alauddin road shorthand/phase III	32	86	2	120
Alternative 1	Urip Sumoharjo road short/phase I	45	88	3	120
	Boulevard road short/phase II	40	89	3	120
	Alauddin road short/phase III	30	103	3	120

7.2 Alternative 2

Traffic engineering at the A.P. Pettarani – Let. Gen. Hertasning by closing the direction of movement directly from the direction of movement of traffic on Alauddin road to Let road. Gen. Hertasning so that traffic will take the direction of the next movement of traffic.

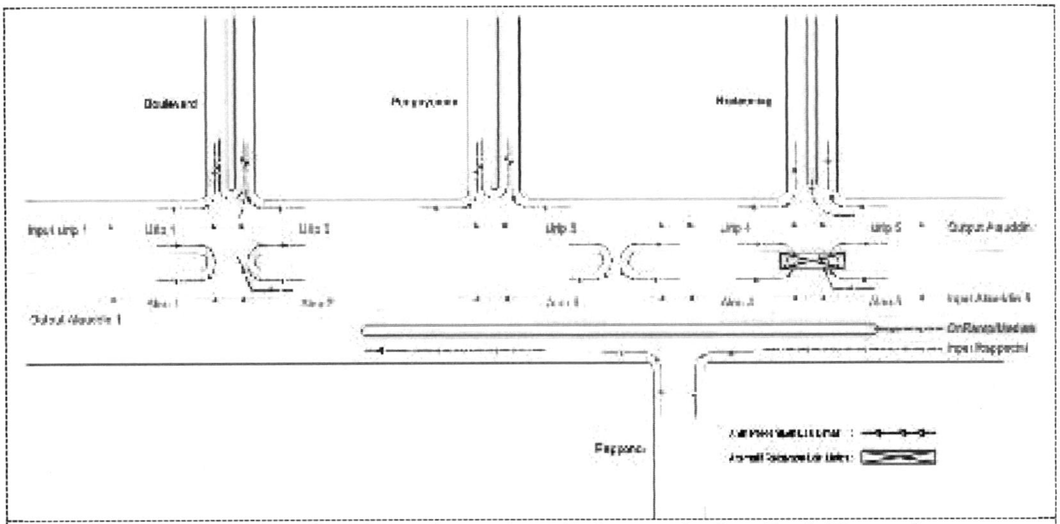

Figure 10. Traffic engineering alternative to the A.P. Pettarani – Let. Jend. Hertasning

7.3. Queue length output on each traffic engineering alternative

From traffic engineering, it was obtained that the output of the maximum vehicle queue length that occurred based on the results of traffic simulations, where the maximum queue length that occurred on Wednesday at the A. P. Pettarani road intersection – Boulevard road was on the boulevard road 247.64 meters for alternatives 1 and 250.17 meters for alternative 2. For the interchange of the A. P. Pettarani road – the Let. Gen. Hertasning, the maximum queue length occurred on the let's road. Gen. Hertasning is 61.38 meters for an alternative one, and for alternative 2, it is 54.34 meters.

Table 8. Traffic engineering output of A.P. Pettarani interchange - Boulevard on Wednesday

short of intersection	Peak Hours	Maximum Queue Length (meters)		
		Existing	Alternative 1	Alternative 2
Urip	08.00 - 09.00	99.85	87.52	88.48
	13.00 - 14.00	111.47	109.43	109.21
	17.00 - 18.00	177.68	171.37	170.73
Boulevard	08.00 - 09.00	90.66	86.14	85.79
	13.00 - 14.00	150.36	130.13	131.60
	17.00 - 18.00	273.85	247.64	250.17
Alauddin	08.00 - 09.00	81.90	98.59	98.93
	13.00 - 14.00	86.14	101.42	100.33
	17.00 - 18.00	154.89	175.58	177.46

Table 9. Traffic engineering output interchange A.P. Pettarani – Let. Gen. Hertasning on Wednesday

short of intersection	Peak Hours	Maximum Queue Length (meters)		
		Existing	Alternatif 1	Existing
Urip	08.00 - 09.00	22.54	25.09	27.64
	13.00 - 14.00	20.22	15.37	25.28
	17.00 - 18.00	19.92	27.36	24.43
Boulevard	08.00 - 09.00	48.78	48.78	38.74
	13.00 - 14.00	31.58	31.58	32.87
	17.00 - 18.00	51.28	61.38	54.34
Alauddin	08.00 - 09.00	33.60	33.60	29.40
	13.00 - 14.00	27.95	27.95	27.59
	17.00 - 18.00	35.26	37.09	33.25

From traffic engineering, the output of the maximum vehicle queue length occurred based on traffic simulation results, where the maximum queue length occurred on Saturday at the A. P. Pettarani road intersection – Boulevard road was on the 201.40 meters boulevard road for alternatives 1 and 205.15 meters for alternative 2. For the interchange of the A. P. Pettarani road – the let road. Gen. Hertasning, the maximum queue length occurred on the let's road. Gen. Hertasning is 41.25 meters for an alternative one, and for alternative 2, it is 42.61 meters.

Table 10. Traffic engineering output of A.P. Pettarani interchange – Boulevard on Saturday

short of intersection	Peak Hours	Maximum Queue Length (meters)		
		Existing	Alternatif 1	Alternatif 2
Urip	08.00 - 09.00	84.87	74.39	75.21
	13.00 - 14.00	93.58	94.34	91.68
	17.00 - 18.00	151.80	146.42	147.15
Boulevard	08.00 - 09.00	77.06	31.50	72.92
	13.00 - 14.00	88.36	76.47	77.34
	17.00 - 18.00	222.71	201.40	205.15
Alauddin	08.00 - 09.00	69.62	83.80	84.09
	13.00 - 14.00	79.85	94.01	93.00
	17.00 - 18.00	126.54	143.44	144.97

Table 11. Traffic engineering output interchange A.P. Pettarani – Let. Gen. Hertasning on Saturday

short of intersection	Peak Hours	Maximum Queue Length (meters)		
		Existing	Alternatif 1	Alternatif 2
Urip	08.00 - 09.00	13.91	19.10	23.72
	13.00 - 14.00	20.92	15.83	29.38
	17.00 - 18.00	15.17	19.17	33.94
Boulevard	08.00 - 09.00	41.25	41.25	42.61
	13.00 - 14.00	29.59	29.59	23.74
	17.00 - 18.00	28.65	37.92	28.95
Alauddin	08.00 - 09.00	31.23	31.23	28.53
	13.00 - 14.00	30.64	30.64	31.40
	17.00 - 18.00	34.62	34.62	24.69

8. Conclusion

The most significant traffic volume that occurs based on the survey results to see traffic problems that impact interchange performance for weekdays is on Wednesdays and for holidays on Saturdays, so that vehicle data for both days is used in traffic simulation inputs. From the results of traffic simulations, several intersections with the highest vehicle queue lengths were obtained, namely the Boulevard road short of 273.85 meters on Wednesday and for Saturdays as large as 222.71 meters which occurred in the traffic survey period of 16.00-18.00. Based on traffic simulations with such a large volume of traffic, traffic problems need to be solved through traffic engineering. For this reason, the traffic engineering taken is by experimenting with scenarios for changing traffic cycle times at the A.P. Pettarani – Boulevard intersection (alternative 1) and for the A.P. Pettarani – Let interchange. Gen. Hertasning is to shut down traffic movement to get into the Let. Jend. Hertasning from the vehicle input of Jalan Alauddin (Alternative 2). From the results of traffic engineering for the alternative one at the intersection with the highest queue length that occurred (existing) on Wednesday, namely at the Boulevard interchange, there was a reduction from 273.85 meters to 247.64 meters, and for Saturday, the length of the existing queue of 222.71 meters was reduced to 201.40 meters. while the traffic engineering results for alternative two at the intersection with the highest queue length that occurred (existing) on Wednesday, namely at the Boulevard interchange, there was a reduction from 273.85 meters to 250.17 meters, and for Saturday, the length of the existing queue was 222.71 meters reduced to 205.15 meters.

Daftar Pustaka

[1] Aly S.H., Hustim M., and Wahab A. A. (2018). Analysis of Vehicle Delays in Micro-Based Non-Signaled Interchange Simulations, in National Conference on Civil Engineering 12, September, pp. 18–19.

[2] Irawan M. Z., Princess N. H. (2015). Vissim Calibration for Microsimulation of Mixed Traffic Flow at Interchange (Case Study: Simpang Tugu, Yogyakarta)," J. Research. Transportation. Multimodal, vol. 13, no. 3, pp. 97–106.

[3] Ulfah F. D., Purwanti O. (2019). Performance Analysis of The Intersection of Jalan Laswi with Jalan Gatot Subroto, Bandung City Using PTV VISSIM 9.0. RekaRacana J. Tech. Civil, vol. 5, no. 3, p. 74-85, doi: 10.26760/rekaracana.v5i3.74.

[4] Lubis A. S., Muis Z. A., Nasution T. (2016). Pemodelan Hubungan Parameter Karakteristik Lalu Lintas pada Jalan Tol Belmera," Media Komun. Tek. Sipil, vol. 22, no. 2, p. 151, 2016, doi: 10.14710/mkts.v22i2.12878.

[5] Tappangrara, M., Tappangrara M. A., Buana C. (2013). Traffic Management Due to the Construction of Santika Gubeng Hotel," J. Tek. ITS, vol. 2, no. 2, pp. E78–E83, 2013, [Online]. Available: http://www.ejurnal.its.ac.id/index.php/teknik/article/view/3788.

[6] Ali H. N., Ramli M. I. (2005). Traffic Analysis on Jl. Abd. Dg. Sirua Due to The Construction Of Alternative Roads In Makassar City," J. Transp., vol. 5, no. 2, pp. 187–200, [Online]. Available: https://journal.unpar.ac.id/index.php/journaltransportasi/article/view/1793.

[7] Ulfah M., Aly S. H., Ramli M. I. (2017). Traffic Microsimulation at Simpang Tiga with Vissim Software (Case Study: Simpang Jl. A. P. Pettarani – Jl. Let. Jend. Hertasning Dan Simpang Jl. A. P. Pettarani – Jl. Rappocini Raya)," Pros. Simp. Stud Forum. Transp., no. November, pp. 4–5.

Performance Analysis of Cenderawasih Secondary Collector Road Section in Mimika Regency

O Batto[1], S Rauf[2], dan S H Aly[3]

[1]Transportation Engineering, Faculty of Engineering, Hasanuddin University
[2]Transportation Engineering, Faculty of Engineering, Hasanuddin University
[3]Transportation Engineering/ Environmental Engineering, Faculty of Engineering, Hasanuddin University

Email : oktabatto18@gmail.com

Abstract. This study aims to analyze the performance of the Cenderawasih Street section in the Mimika Regency. Traffic flow is observed by dividing the Cenderawasih Street into five posts. The traffic flow survey was obtained for three days, namely Tuesday, Wednesday, and Saturday, with observation periods at peak hours of 07.00-09.00, 12.00-14.00, and 16.00-18.00. The performance analysis of road sections includes traffic volume, travel time, average space speed, traffic density, road section capacity, degree of saturation, and level of service, referring to MKJI 1997. The highest traffic flow volume occurred at post 1 of Cenderawasih road ranging from 1428.20 pce / hour with a degree of saturation of 0.93. This shows that the Cenderawasih Road Section is at the service level with category E.

Keywords: Road Section Performance, Degree of saturation, Level of Service, Cenderawasih Street

1. Introduction

Transportation and the needs and activities of the community are things that are intertwined with each other, so it is very difficult to separate them. The impact that occurs in one sector will also increase in the other sector. The fundamental purpose of transportation is fundamental because it facilitates different locations. Thus, transportation has a role in the structure of space and territory that varies according to the level of development [1]. Traffic movement arises due to the process of meeting needs. Each land use or activity system has a certain type of activity that will arouse movement and attract movement in meeting needs. The location of activities is spread heterogeneously in the existing space, which eventually causes the need for movement used for the process of meeting needs [2].

The increase in community movement has led to an increase in the need for the availability of infrastructure and transportation facilities, the need for infrastructure and transportation facilities including increasing the length of roads, improving the quality of existing roads, increasing the number of vehicles and other facilities needed to support these activities, in determining the needs of infrastructure and transportation facilities require careful planning and actual data on traffic conditions in the area, considerable traffic movement causes the need for analysis related to road performance based on the value of the degree of road saturation following standards [3]. The increase in the volume of traffic flow will cause changes in the traffic behavior of a road section, especially on roads per city.

The increase in traffic volume is caused by the increase in the number of residents and the need for transportation facilities so that sufficient space is needed, such as road infrastructure, parking lots, and so on, the existence of transportation activities. There is a movement of traffic flow. [4]. The size and diversity, and level of development of urban areas show that the behavior of the driver and the vehicle population (age, power, condition of the vehicle, vehicle composition) are various characteristics these are included in the calculation procedure directly, through the size of the city with increasing the volume of traffic flow will cause changes in the traffic behavior of a road section, especially on urban roads. Meanwhile, the main characteristic that affects the performance of the road is when a section of the road is burdened by traffic [5].

The Cenderawasih Street section in Mimika Regency, Papua, is one of the longest road sections, namely approximatelly 13.6 km and a road width of 16 m with a 4/2 divided road type is a collector secondary road section. Several roads are directly related to the Cenderawasih Road section. One of the road sections is a one-way road whose exit point is on Cenderawasih Street, this section of the road is used as the main route connecting to main activities such as the center of government, in this case, the regent's office and related agencies and offices in other private sectors, the main route to Kuala Kencana and vice versa as the main access to the center of activities in Timika City. Therefore, at the node of Cenderawasih Street to Yos. Sudarso Street was often stacking at peak hours due to diverting one-way traffic flow from Budi Utomo Street to Cenderawasih Street.

Related to the development plan of the Papua region from district to province, synergize to create areas that balance each other. Suppose the performance of the road section is known. In that case, it is hoped that in the future, it can realize the ability of the road section to contribute to development which will certainly increase all community movements. The good thing is that a road network's performance greatly affects a city's development. When the road network has a good road network performance, there are many benefits obtained by the community. These benefits ultimately increase regional income and income, with the smooth movement and movement of goods and services, it will directly increase people's economic income, causing the movement of goods and services to be smooth so that the economic turnover process is even more efficient with the better road performance also facilitates people's activities in working, schooling and shopping so that in the end, a good road section performance succeeds in increasing community productivity and an increasingly productive society will improve welfare. [6]. The development of a region can be seen from the many improvements in each component of the region, such as the increasing number of cooperation established with the region and surrounding sub-regions, economic activity, and alternative sources of regional income [7].

The performance parameter of a road section consists of traffic volume, the number of vehicles based on a unit of time passing through a point at a given location, generally expressed in vehicles per hour, smp per hour, and vehicles per minute [8]. Travel Time is the time it takes for a vehicle to pass over a certain segment length, generally expressed in units of seconds or hours (seconds/3600). Traffic speed is a vehicle's movement rate calculated based on distance per unit of time. In [8], it uses travel speed as a key performance measure in road segments. It is easy to understand and measure and is an important input in the cost of using roads in economic analysis. There are two types of traffic flow speed analysis. Namely, time means speed, and Space means speed. Based on how it is retrieved and the type of data, the traffic speed uses Space mean speed (SMS), It is the average speed of a vehicle that meets a segment of a road or section of the road at a certain time interval, which defines the number of vehicles occupying a length of road or lane, generally expressed in vehicles per kilometer. As for this, the density in the field will be difficult to measure, so it is calculated by the relationship between the value of speed and current [8]. Capacity is defined as the maximum current passing at a freeway point, which can be maintained in the prevailing conditions. As for undivided freeways, capacity is a maximum two-way current (a combination of both directions), and a divided freeway capacity is a maximum current calculated from each lane [8]. The degree of saturation is the ratio of current to capacity, which is used as the main factor in determining the level of performance of intersections and road segments. The value of the ratio determines whether a road

IOP Conf. Series: Earth and Environmental Science 1117 (2022) 012075 doi:10.1088/1755-1315/1117/1/012075

segment has a capacity problem. The degree of saturation itself is calculated using current and capacity expressed in pce/hour. The degree of roughness is used in analyzing traffic behavior through speed. The ratio of volume to capacity is the main factor in determining the level of performance of intersections or road sections. Service levels generally measure the limiting influence caused by increasing traffic volume. [3].

2. Research Methods

2.1. Location and Research Design

The research was conducted on Cenderawasih Street, Timika, Mimika Regency, Papua. This type of research uses a quantitative descriptive approach method to determine the volume and travel time of the vehicle at a predetermined point or post, which then uses an explanation based on the MKJI MKJI so that the performance of the road section is known.

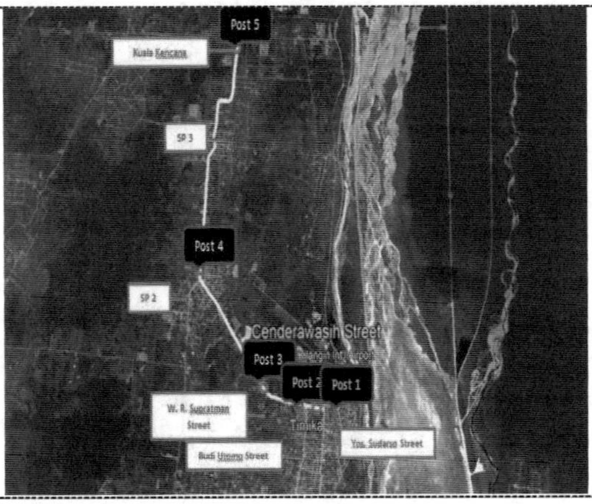

Figure 1. Cenderawasih Secondary Colector Road Survey Location

Figure 2. Post 1 (Yos Sudarso Street - Budi Utomo Street)

Figure 3. Post 2 (Budi Utomo Street - W.R. Supratman Street)

Figure 4. Post 3 (W.R. Supratman Street -SP 2)

Figure 5. Post 4 (SP 2-SP 3)

Figure 6. Post 5 (SP 3-Kuala Kencana)

2.2. Data Collection Methods

The survey method carried out in this study is a survey then at the five points or observation posts. The existing posts are given codes according to the intersection of road sections, namely Post 1 (Yos Sudarso Street - Budi Utomo Street), Post 2 (Budi Utomo Street - W.R. Supratman Street), Post 3 (W.R. Supratman Street - SP 2), Post 4 (SP 2 - SP 3), and Post 5 (SP 3 - Kuala Kencana) assuming that each post leading to the next intersection is the direction to Kuala Kencana and vice versa the 5th interchange leads to the 4th interchange is the direction to Timika..

The type of survey carried out to determine the performance of road sections is a survey of vehicle volume and vehicle travel time based on its type, as well as a geometric survey of road sections. The survey was conducted based on a preliminary survey to determine the peak traffic hours. Data collection was carried out for three days, namely Tuesday, Wednesday, and Saturday, with an observation period from 07.00-09.00, 12.00-14.00, and 16.00-18.00. Travel time data is obtained by marking the STA 0+100 road segment with a period mark on the STA 0+00 line and STA 0+100 at each observation post.

Using surveyor personnel to record data during the observation time, the traffic survey recording data is then recalculated based on the data type, namely vehicle volume and travel time. The data is recapitulated using a traffic counting tool and entered into the vehicle number form. At the same time, the travel time is calculated using a stopwatch tool and recorded per vehicle sample..

2.3. Data Analysis Methods

Based on the data collected through traffic surveys, then analysis was carried out using Ms. Excel's software. Ford preliminary data analysis, and data classification, are carried out, such as vehicle volume and vehicle travel time that have been distinguished based on the type of vehicle. The data is recapitulated and processed based on MKJI 1997 for traffic volume, the average speed of space, traffic density, the capacity of road sections, degree of saturation, and level of service. Then it is tidied up and made into a graph to find out the overall amount of road section performance according to field conditions.

3. Results and Discussion

3.1. Geometric Conditions of Cenderawasih Road Section

Table 1. Geometric Section of Cenderawasih Road

Direction	To Kuala Kencana (m)	To Timika (m)
Curb Width	1	1
Road Width	8	8

Median Width	0.5

3.2. Traffic Volume

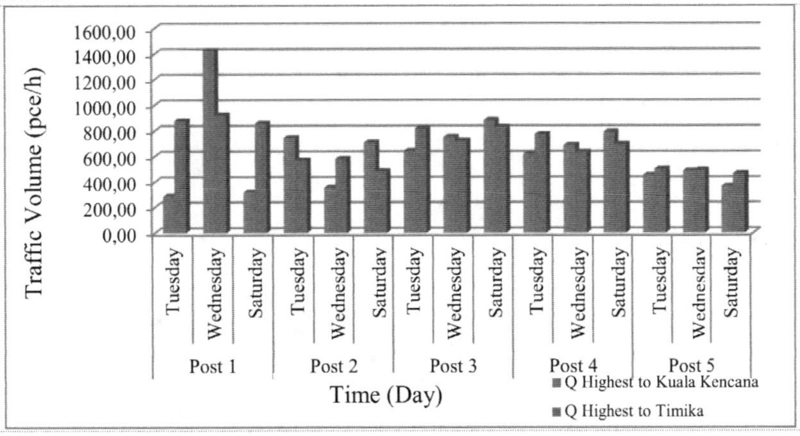

Figure 7. Maximum traffic volume Post 1-5

Based on the highest traffic volume calculated from post 1 to post 5, a volume comparison between posts was carried out, and it was found that the highest volume of traffic passing through the Cenderawasih Street section occurred at post one on Wednesday's observation, the direction of Yos Sudarso Street - Budi Utomo Street was 1428.20 pce / hour.

3.3. Vehicle Travel Time

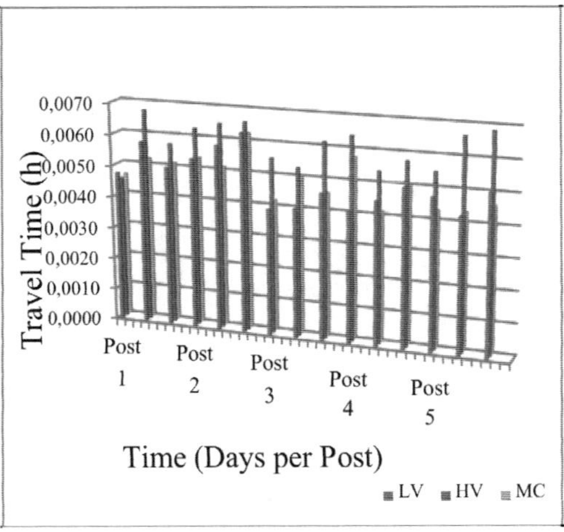

Figure 8. Maximum vehicle travel time towards Timika – Kuala Kencana

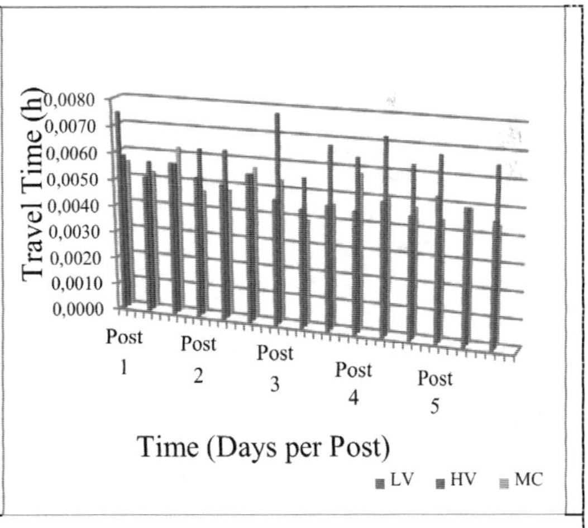

Figure 9. Maximum vehicle travel time towards Kuala Kencana - Timika

The travel time of a vehicle or the length of a vehicle to pass a road segment, based on its type at post one day Tuesday for the direction of Timika - Kuala Kencana (Yos Sudarso Street - Budi Utomo Street) travel time for light, heavy vehicles and motorcycles when passing through the road segment has a value with the same significance. However, overall for the comparison of the travel time of heavy vehicles at the same post, day, and direction, the travel time of heavy vehicles is the smallest, while for the travel time of heavy vehicles, the largest occurs at post three days Tuesday direction Kuala Kencana - Timika (W.R. Supratman Street - SP 2). At post one on Tuesday with the direction of Kuala Kencana - Timika (Budi Utomo Street - Yos Sudarso Street), it is known that the travel time of light vehicles is greater than other types of vehicles, for the shortest travel time there is a post of 3 days Tuesday in the direction of Timika - Kuala Kencana (W.R. Supratman Street - SP 2). For post one Saturday direction Kuala Kencana - Timika (Budi Utomo Street - Yos Sudarso Street), it is known that the travel time for motorcycle type vehicles is greater than other types of vehicles, while the shortest travel time occurs post three days Wednesday direction Kuala Kencana - Timika (SP 2 - W.R. Supratman Street).

3.4. Average Traffic Speed

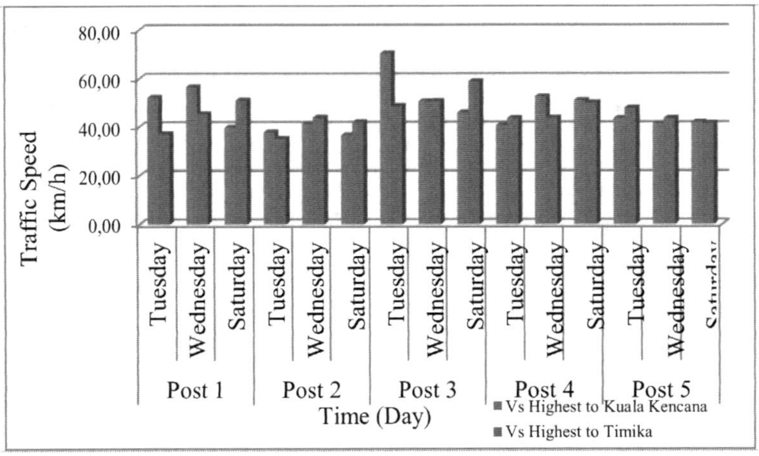

Figure 10. Graph of average speed of maximum space Posts 1-5

The average speed using Space Mean Speed (SMS) is a calculation of the average speed of all vehicles occupying a piece of road with a certain segment length over a substantial period of time. Based on the average speed of the maximum space of each post, it is known that the maximum speed occurred at the post 3 in the Timika - Kuala Kencana direction (W.R. Supratman Street- SP 3) in Tuesday's observation of 70.31 km/hour.

3.5. Traffic Density

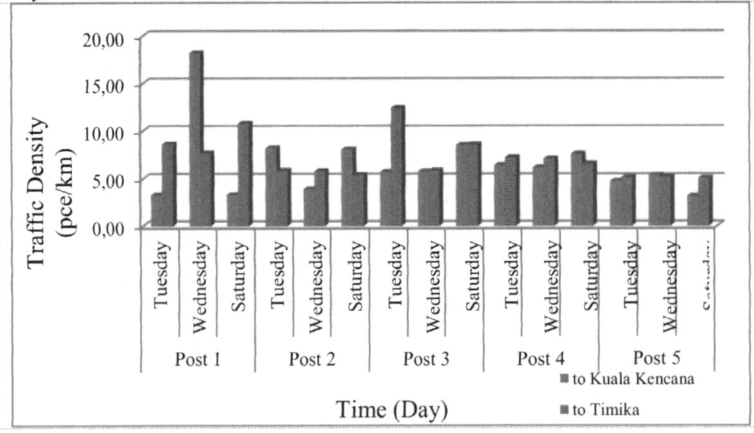

Figure 11. Maximum traffic density graph Post 1-5

Density is the number of vehicles occupying a section with a certain road length. In the comparison of the maximum density of post 1 to post 5, it was found that the highest density occurred at post 1 in Wednesday's observation towards Kuala Kencana - Timika (Yos Sudarso Street - Budi Utomo Street) of 18.22 pce / km.

3.6. *Road Section Capacity*

Based on a study referring to the MKJI 1997 for urban roads, with the geometric conditions of the Cenderawasih road section and Mimika Regency data in Figures with projections for the next five years, the actual road section capacity on the Cenderawasih road section is 1542,618 pce / hour smaller than 1650 pce / hour for the basic capacity of urban roads.

3.7. *Degree of Saturation and Level of Service*

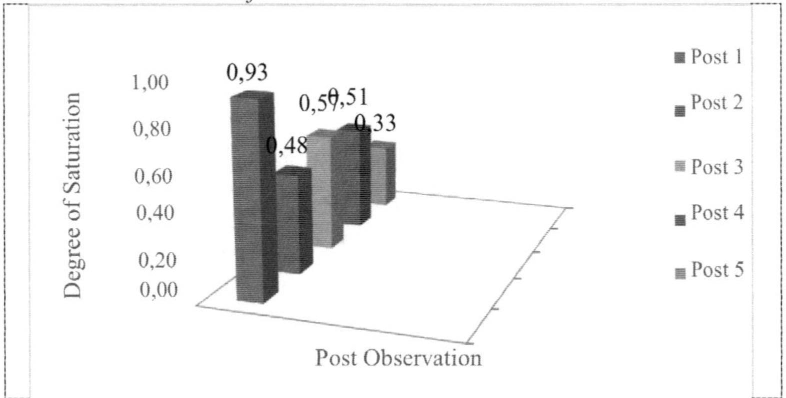

Figure 12. Max saturation degree graph Pos 1-5

Based on the calculation of the capacity of the road section in the projected next five years with existing conditions, the capacity of the Cenderawasih Road road section is 1542,618 smp / hour, slightly below the capacity of the urban road plan. As for the degree of saturation in each post that can be seen for post 1 of 0.93, post 2 of 0.48, post 3 of 0.57, post 4 of 0.51, and post 5 of 0.33, overall, the degree of maximum saturation of the five observations is known to occur in post 1 of 0.93.

4. Conclusions

The Cenderawasih Street section is divided into five observation segments as an observation post for the volume of traffic flow. The highest traffic volume was 1428.20 pce/hour, which occurred at post one, and the lowest traffic volume of 502.55 pce/hour occurred at post 5. The vehicle travel time is longer in the direction of Kuala Kencana - Timika at Pos 1 and Pos 3. The highest average speed occurred at post 3 of 70.31 km/hour, and the maximum density occurred at post 1 of 18.22 pce/km. The maximum degree of saturation value is 0.93. This shows that the Cenderaasih Street section is at the service level E.

Bibliography

[1] Rodrigue, J-P., Comtois, C., & Slack, Brian. 2006. *The Geografhy and Transport Systems*. Edisi 1. New York: Taylor & Francis Library
[2] Tamin, Ofyar Z. 2000. *Perencanaan dan Pemodelan Transportasi*. Institut Teknologi Bandung. 2(89-90)
[3] Rosmantyo, Windhy R. 2018. *Rancang bangun Sistim Informasi Perhitungan Kinerja Ruas Jalan Provinsi di Provinsi Jawa Timur*. Undergraduate Thesis. Surabaya: Repository Universitas 17 Agustus 1945
[4] Lalenoh, Rusdianto H. 2015. *Analisa Kapasitas Ruas Jalan Sam Ratulangi dengan Metode MKJI 1997 dan PKJI 2014*. Jurnal Sipil Statik. Vol 2 No 11 (737-746).
[5] Titirlobi, Angelina I. 2016. *Analisa Kinerja Ruas Jalan HasanuddinKota Manado*. Jurnal Sipil

Statik. Vol 4 No 7 (423-431)

[6] Koloway, Barry S. 2009. *Kinerja Ruas Jalan Perkotaan Jalan Prof. Dr. Satrio, DKI Jakarta.* Jurnal Perencanaan Wilayah dan Kota. Vol 2 No 3 (215-230)

[7] Riadi, A. 2015. *Analisis Perkembangan Wilayah dan Arahan Prioritas Penanganan Jaringan Jalan di Kabupaten Bogor*. Tugas Akhir. Bogor: Sekolah Pascasarjana-IPB

[8] Direktorat Jenderal Bina Marga. 1997. *Manual Kapasitas Jalan Indonesia (MKJI)*. Jakarta: Bina Karya

IOP Conf. Series: Earth and Environmental Science 1117 (2022) 012076 doi:10.1088/1755-1315/1117/1/012076

Selection Model of Use of the Elevated Toll Road Route in Makassar City

T S Roni[1], M I Ramli[1], S A Adisasmitha[1]

[1]Departement of Civil Engineering, Hasanuddin University, Gowa, South Celebes, Indonesia

Email : tulussetiady31@gmail.com

Abstract. This study aims to analyze the selection of the A.P.Pettarani Elevated Toll Road route. Data collection was carried out by distributing questionnaires and reviewing respondents to describe the travel characteristics of A.P.Pettarani Elevated Toll Road respondents using Ms. Excel to process the data. The results showed that in the selection of A.P.Pettarani elevated toll road to the airport, and the dominant respondents entered the toll road through the Bakti road where the traveler was aged 50-59 years old, with the last education being a bachelor's degree, working as an officer, with an income of IDR 4.5M-5.5M. The travel costs incurred are IDR20-30K, travel distance of 7-11 km, travel time of 10-20 minutes, and travel frequency of 15-20 times a month to travel for work/business. In the selection of travel routes using the A.P.Pettarani elevated toll road to the port, the dominant respondents entered the overpass toll road through the Rappocini road, where the traveler was aged 20-29 years old, with the last education being a bachelor's degree, working as an officer, with an income of IDR 4.5M-5.5M. The travel costs incurred are IDR20-30K, travel distance of 7-11 km, travel time of 20-30 minutes, and travel frequency of 15-20 times to travel for work/business.

Keywords: Transportation, Elevated toll road, routes selection

1. Introduction

Transportation is moving people and goods from one place to another using certain systems for a specific purpose. Transportation of people or goods is usually not the final destination; therefore, the demand for transportation services can be referred to as derived demand arising from the need for commodities or other services. Thus the demand for new transportation will exist if there are driving factors. The need for transportation services does not stand alone but is hidden behind other interests. (Molok, 1984) [1]. The role of transportation is not only to facilitate the flow of goods and human mobility but also to help achieve optimal allocation of economic resources (Nasution, 1996: 12)[2].

Traffic Congestion is a state of road condition if there is no balance between road capacity and the number of vehicles passing by. This symptom is characterized by low speed until the accumulation of cars occurs; the distance between one vehicle and another becomes tight, so the driver cannot run the car at the desired speed [3].

Like major cities in Indonesia, transportation problems are also profound, especially congestion problems faced by Makassar City. The rapid growth in the number of vehicles is considered not comparable to the provision of an adequate road network system to meet transportation needs, so currently, the congestion problem is becoming increasingly complex, especially on significant road sections in Makassar City, such as Jalan A.P. Pettarani.

The many congestion problems in Makassar City today make the government consider alternative solutions to minimize and solve the congestion problem. Thus, an idea was born by applying the concept of elevated toll roads as a solution in solving the problem of congestion in Makassar City, precisely on Jalan A.P. Pettarani [4].

The construction of the A.P. Pettarani Elevated Toll Road cannot be separated from the influence of the existing road network system, so a study is needed to predict traffic problems that may occur after the existence of the A.P. Pettarani Elevated Toll Road by looking at the route points and exits and entrances to the toll roads provided. The A.P. Pettarani road section is one of the main roads in Makassar City, so the activities on the road section are classified as congested; traffic management and traffic engineering are needed to be able to improve the overall level of road service at specific points the capacity of the road that may be disrupted due to the access of exits and toll entrances so that the traffic flows that move directly from the network around the toll road to the toll road or vice versa need to be taken into account. An engineering study of route selection needs to be carried out due to changes in road performance after the existence of the A.P. Pettarani Elevated Toll Road and studies related to traffic management in line with the increase in volume that will occur at a certain point after the opening of the A.P. Pettarani elevated toll gate, to improve road performance due to the influence of existing traffic movements on the Selection of Routes for Users of the A.P Petterani Elevated Toll Road.

2. The Literatur Review

2.1. Road definition

According to Law No. 38 of 2004 concerning Roads, the road is land transportation infrastructure that includes all parts of the road, including complementary buildings, and its equipment intended for traffic, which is at ground level, above ground level, below ground, and or water level, as well as above water level, except railway roads and cable roads. A public road is intended for general traffic; a particular route is built by an agency, business entity, individual, or community group for its benefit [5].

2.2. Toll roads

According to PP No. 15 of 2005 concerning toll roads, it is explained that the definition of a toll road is a public road that is part of the road network system and a national road whose use is required to pay tolls. Tolls are a certain amount of money paid for the use of toll roads. The amount of toll tariff is different for each class of a vehicle and these provisions have been determined based on a presidential decree. Meanwhile, toll road sections are parts or fragments of certain toll roads whose business can be carried out by certain business entities [6].

Based on the Indonesian Road Capacity Manual (MKJI 1997) it is explained about the definition of toll roads as roads for continuous traffic with full control of the entrance road, whether it is a divided or undivided road. The types of toll roads are two-lane two-way undivided (2/2 UD), four-lane two-way divided (4/2 D), and toll roads divided with more than four lanes [7].

2.3. The concept of effectiveness and customer satisfaction on toll roads

Poor toll road conditions, both in terms of services and facilities available, can cause inconvenience to users when crossing the toll road. This will lead to low user satisfaction due to the interest in toll road performance (because they feel that there is no difference between toll roads and public roads) which will further encourage users to leave the toll road. Thus, toll roads should be created and developed based on the perspective of their users (user-oriented). One of the key aspects of the concept of fulfilling consumer satisfaction is how consumers feel satisfactorily served. If an organization wants to satisfy its customers, the fundamental question is what can satisfy its customers and vice versa what things can disappoint its customers. Thus, meeting customer satisfaction depends on how the

IOP Conf. Series: Earth and Environmental Science 1117 (2022) 012076 doi:10.1088/1755-1315/1117/1/012076

organization balances the customer's expectations and experiences with the products and services received by that customer (Parasuraman et al. 1990)[8].

2.4. Route Selection Analysis Model
The existence of differences based on the purpose of movement results in the distribution of vehicles on each route, this is called a stochastic process (considering its role) in route selection. The method of analyzing the selection of routes used in the loading of the crossplay is very dependent on one part of the analysis. But on the contrary, if the stochastic element is removed, then the calculation of the load capacity (V/C) ratio is indispensable (Ofyar, 2000) [9].

2.5. Route selection process
Traffic flow on a road section in a network can be estimated as a result of the process of combining route selection information, description of network systems, and modeling route selection. The route selection procedure aims to model the behavior of movement actors in choosing the route that they think is the best route. In other words, in the process of route selection, the movement between two zones (which is obtained from the distribution of movement) for a certain mode (which is obtained from the stage of movement distribution) for a certain mode (which is obtained from the selection of modes) is charged to a certain route consisting of a certain network section (or public transport).

The main thing in the process of loading the route is to estimate the assumptions of road users regarding the best choice. Several factors influence the choice of the route when we travel. Some of them are travel time, cost distance (fuel and others), congestion and queues, the type of maneuver needed, the type of highway (toll road, arterial), scenery, completeness of traffic signs, and road markings, as well as habits.

One of the most frequently used approaches is to consider two main factors in the selection of a route, namely movement, and the time value of the cost of movement is considered proportional to the mileage. In some route selection models, it is possible to use different weights for the travel time factor and the mileage factor to describe the rider's perception of both factors. According to Ofyar (2000), There is strong evidence to suggest that travel time has a more dominant weight than mileage for movement within the city [10]. ofyar

3. The Study Methods

3.1. Types of research
This study uses a qualitative descriptive approach, and qualitative descriptive analysis to describe the results of the identification of variables related to the selection of the use of elevated toll road routes.

3.2. Location and Research Design
This research was conducted in Makassar City and this type of research uses a descriptive approach because it wants to explain and describe the current travel characteristics of elevated toll road users.

3.3. Population and Sample
The population is all users of the A.P. Pettarani elevated expressway. Whereas a study cannot take samples from all existing populations, this is due to limited time, effort, and costs. So in the study, it is necessary to determine the size of the sample needed by looking at the number of existing populations. Mathematically, sampling in this study uses the slovin formula, so that a large number of samples is obtained, namely 504 people from the total population.

3.4. Method of collecting data
The survey techniques and methods carried out in this study are distributing questionnaires and interviews directly to users of the A.P. Pettarani elevated toll road randomly. The variables selected as a hypothesis in this study are as follows:

3.4.1. Respondent characteristics The characteristics data collected are data on the gender of the respondent, the age of the respondent, the latest education, the occupation, the respondent's income, and vehicle ownership.

3.4.2. Trip Attribute For the collected trip characteristics, i.e. choice of elevated toll road route, travel cost, travel time, travel frequency, and travel destination.

3.5. Data analysis method

The data collected through the questionnaire is then assembled according to the classification of the data and the variables involved. Microsoft Excel software is used to organize data based on the items contained in the questionnaire. Analysis of the model of selecting the use of elevated toll road routes using graphs and charts to see the distribution of elevated toll road route selection based on respondents' characteristics and travel characteristics.

4. Result and Discussion

From the questionnaire, data on respondents' characteristics and decisions related to trip activities are generated.

4.1. Distribution of elevated toll route selection

Table 1 shows that the dominant respondents chose to use the elevated toll road with the A.P.Pettarani road route to the Airport.

Table 1. Distribution of elevated toll route selection

No	Category	Result	(%)
1	Elevated toll A.P. Pettarani road - Airport	384	76,2
2	Elevated toll A.P. Pettarani road - Port	120	23,8
	Number of respondents	**504**	**100**

4.2. Distribution of elevated toll route selection by passenger characteristics

4.2.1. Age distribution of respondents. From the age of the respondents, it is known that 20% of respondents aged 50-59 years old predominantly choose the elevated toll road route to the airport through the entrance of the service road. Meanwhile, the elevated toll road route to the port is dominated by respondents aged 20-29 years old (10%) through the entrance of the Rappocini road. The age distribution of respondents for the elevated toll to the airport route can be seen in Figure 1.

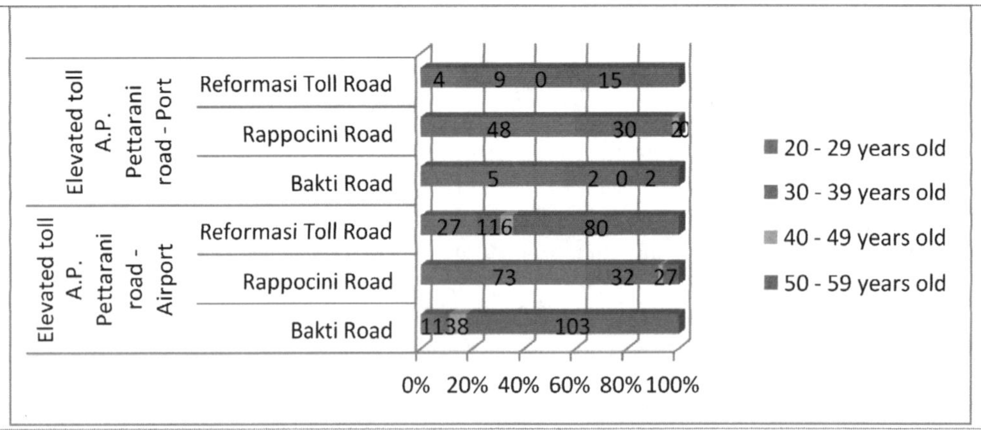

Figure 1. Elevated toll route selection by respondents age characteristics

4.2.2. The last educational distribution of respondents. Based on the last education of respondents, it was found that 16.3% of respondents with bachelor's degree education predominantly chose the elevated toll road route to the airport through the entrance of the reformed toll road. As for the elevated toll road route to the port, 10.9% of respondents with a bachelor's degree education use the elevated toll road through the entrance of the Rappocini road. The distribution of the last education of respondents can be seen in Figure 2.

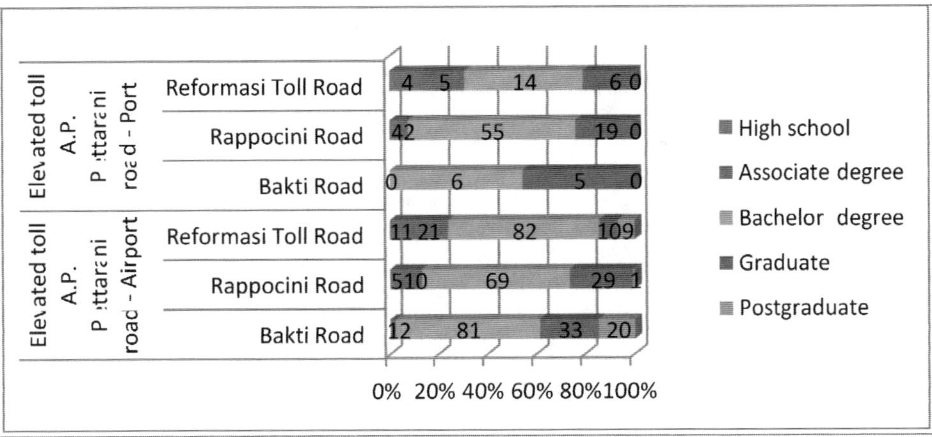

Figure 2. Elevated toll route selection by education

4.2.3. Distribution of occupation respondents. From the work of respondents, it is known that 15.1% of respondents chose to enter the A.P. Pettarani elevated expressway to the airport through the entrance of the reformed expressway. Meanwhile, 5.6% of respondents entered through the Rappocini road flyover toll gate to the port. The distribution of elevated toll road routes can be seen in Figure 3.

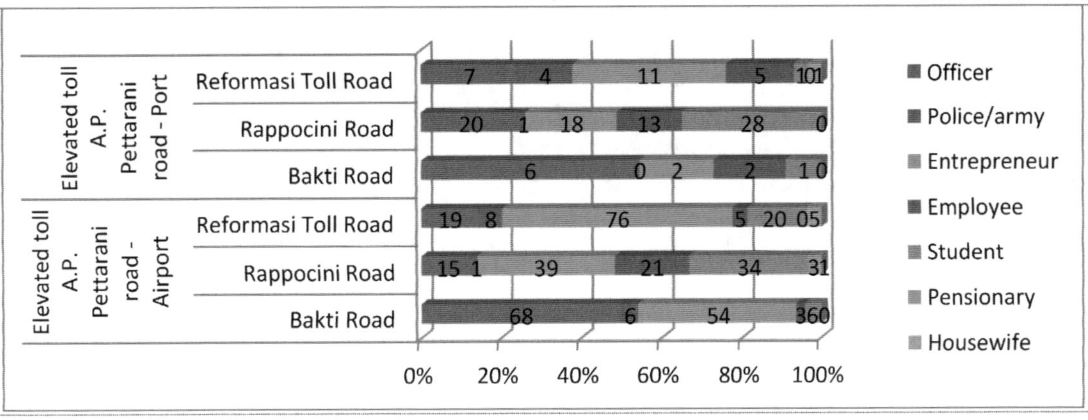

Figure 3. Elevated toll route selection by occupation

4.2.4. Distribution of the income level of airline respondents. From passenger income, it is known that respondents earning 4.5-5.5 million / month are respondents who chose the elevated toll road route to the airport entering from Bakti road (19%). Meanwhile, 7.5% of respondents earning 3.5-4.5 million / month choose the elevated toll road route to the port entering from Rappocini road. The distribution of elevated toll road routes can be seen in Figure 4.

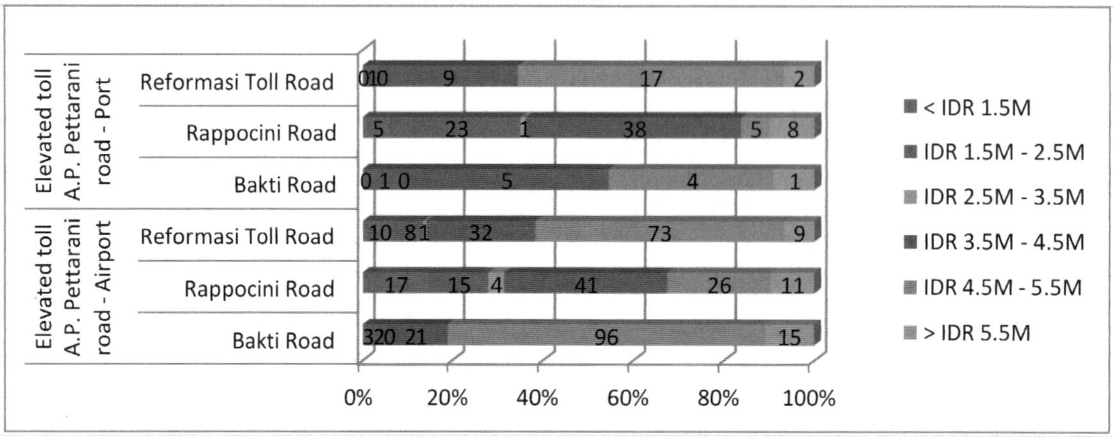

Figure 4. Elevated toll route selection by income

4.2.5. Distribution of respondents based on travel cost. Based on the respondent's travel costs, it was found that respondents who chose to use the A.P.Pettarani elevated toll road to the airport through the entrance of the reform toll road predominantly paid a travel fee of 20,000 to 30,000 rupiah. Meanwhile, respondents who chose to use the A.P.Pettarani elevated toll road to the port through the entrance of the Rappocini road predominantly paid a travel fee of 20,000 to 30,000 rupiah. The distribution of elevated toll road routes can be seen in Figure 5.

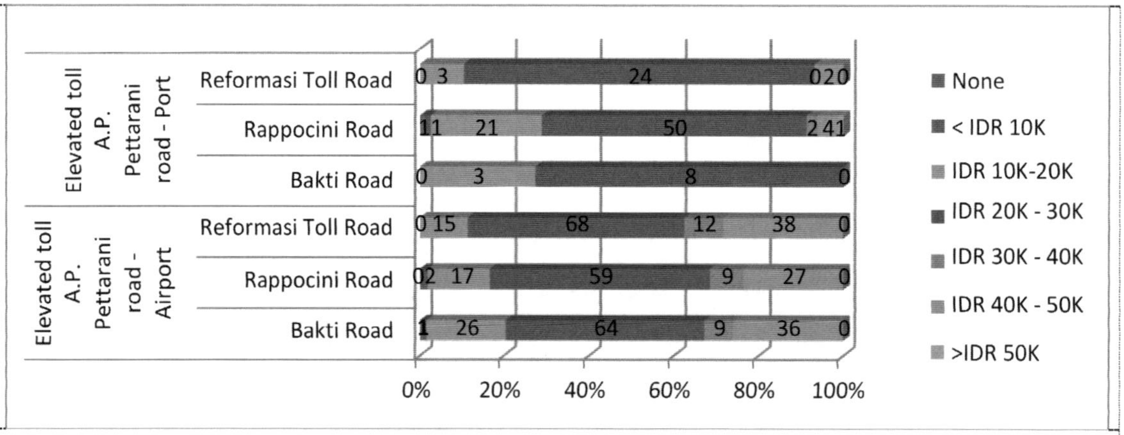

Figure 5. Elevated toll route selection by travel cost

4.2.6. Distribution of respondents based on travel distance. Based on the travel distance, it is known that as many as 26.5% of respondents who choose to use the A.P.Pettarani elevated toll road to the airport through the entrance of the Bakti road predominantly travel along 7-11 km to their destination. Meanwhile, among respondents who chose to use the A.P.Pettarani elevated toll road to the port 15% of respondents covered a distance of 7-11 km through the entrance of the Rappocini road. The distribution of elevated toll road routes can be seen in Figure 6.

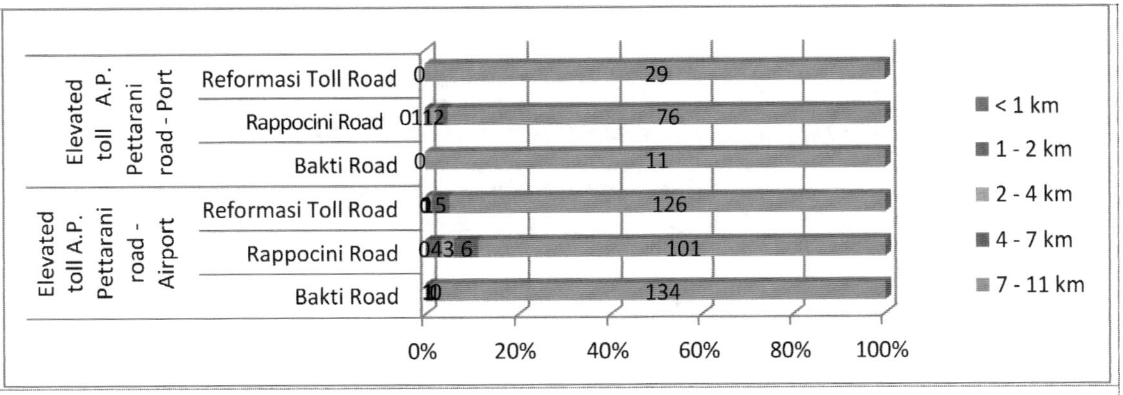

Figure 6. Elevated toll route selection by travel distance

4.2.7. Distribution of respondents based on travel time. Based on travel time, it is known that as many as 24.4% of respondents who choose to use the A.P.Pettarani elevated toll road to the airport through the entrance of the Bakti road predominantly travel for 10-20 minutes to their destination. Meanwhile, among respondents who chose to use the A.P.Pettarani elevated toll road to the port 13% of respondents traveled for 20-30 minutes through the entrance of the Rappocini road to their destination. The distribution of elevated toll road routes can be seen in Figure 7.

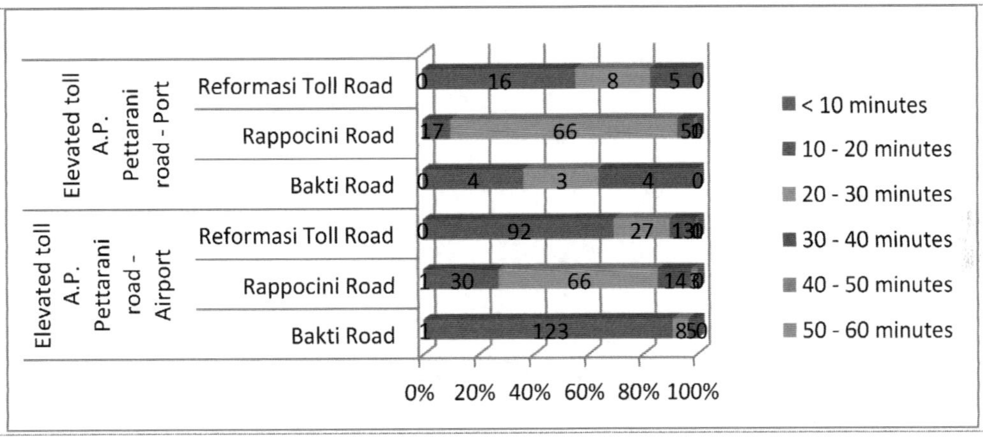

Figure 7. Elevated toll route selection by travel time

4.2.8. Distribution of respondents based on travel frequency. Based on the frequency of travel, it is known that as many as 22.4% of respondents who choose to use the A.P.Pettarani elevated toll road to the airport through the entrance of the Bakti road predominantly travel 15 to 20 times. Meanwhile, among respondents who chose to use the A.P.Pettarani elevated toll road to the port 10% of respondents traveled 15 to 20 times through the entrance of the Rappocini road to their destination. The distribution of elevated toll road routes can be seen in Figure 8.

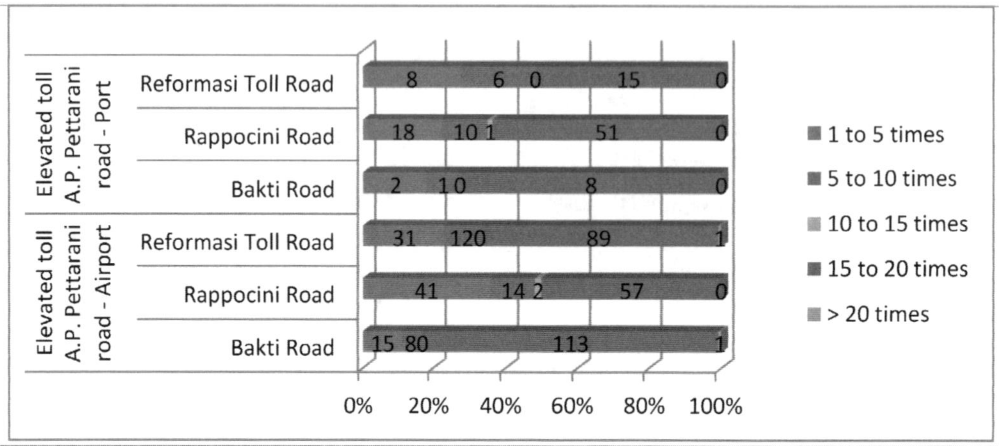

Figure 8. Elevated toll route selection by travel frequency

4.2.9. Distribution of respondents based on the travel destination. Based on the purpose of the trip, it is known that as many as 24.2% of respondents choose to use the A.P.Pettarani elevated toll road to the airport through the entrance of the Bakti road with the aim of work/business. Meanwhile, among respondents who chose to use the A.P.Pettarani elevated toll road to the port as many as 12% of respondents entered through the Rappocini road to their destination for work/business. The distribution of elevated toll road routes can be seen in Figure 9.

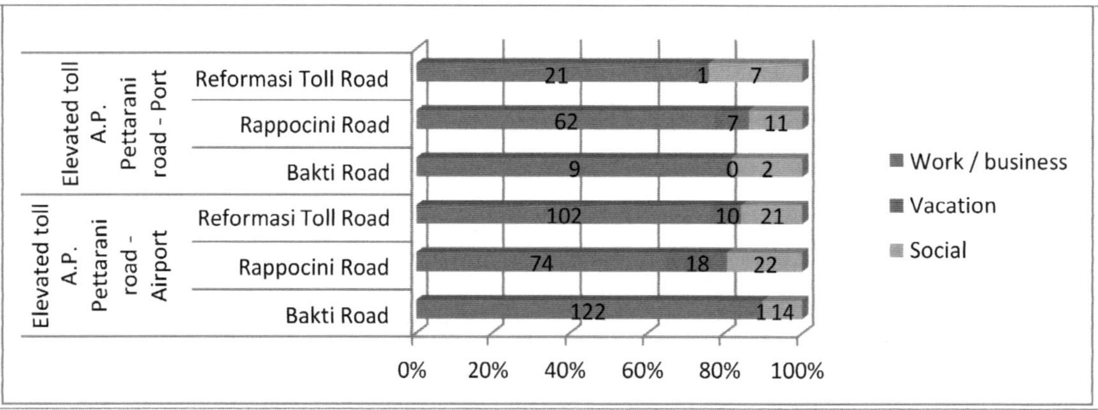

Figure 9. Elevated toll route selection by travel destination

5. Conclusions

Based on the results of research on the selection of travel routes using the A.P.Pettarani elevated toll road, it can be concluded that :

- Choosing a travel route using the A.P.Pettarani elevated toll road to the airport, dominant respondents entered the high toll road through the Bakti road where the traveler was aged 50-59 years, with the last education being a bachelor's degree, working as an officer, with an income of IDR 4.5M-5.5M. The travel costs incurred are 20,000-30,000 rupiah, with a distance of 7-11 km, travel time of 10-20 minutes, and travel frequency of 15 to 20 times a month to travel for work/business.

- Selection of travel routes using the A.P.Pettarani elevated toll road to the port, the dominant respondents entered the elevated toll road through the Rappocini road where the traveler was aged 20-29 years, with the last education being a bachelor's degree, working as an officer, with an income of IDR 4.5M-5.5M. The travel costs incurred are 20,000-30,000 rupiah, with a distance of 7-11 km, travel time of 20-30 minutes, and travel frequency of 15 to 20 times a month to travel for work/business.

References

[1] E.K.Morlok, " *Introduction to Transportation Engineering and Planning*," Jakarta : Erlangga, 1984.

[2] H.M.Nasution, " *Transportation Management*," Jakarta : Ghalia Indonesia, 1996.

[3] H.A.Hasanuddin, " *Identification of Congestion Points on A.P. Pettarani Road Sections Using GPS Technology,"* Makassar, 2018.

[4] A.Afandi, N. Pertiwi, and Rahmatan, " *Service Level Analysis of Jalan Andi Djema Makassar City,"* Makassar, 2017.

[5] Law Number 38 of 2004 concerning Roads, Indonesia, 2004.

[6] Government Regulation Number 15 of 2005 concerning Toll Roads, Indonesia 2005

[7] Indonesian Road Capacity Manual 1997.

[8] Parasuman, V. A.Zeithmal, L.L. Berry, *"Delivering Quality,"* 1990.

[9] O. Z. Tamin, Transportation Planning and Modeling, Intitut Teknologi Bandung, 2000.

[10] O. Z. Tamin, Planning and, Bandung: Bandung Institute of Technology (ITB), Transportation Modeling, Second Edition.

AUTHOR INDEX